POLYMER GLASSES

POLYMER GLASSES

edited by

Connie B. Roth

Emory University, Atlanta, Georgia

CRC Press
Taylor & Francis Group
Boca Raton London New York

CRC Press is an imprint of the
Taylor & Francis Group, an **informa** business

CRC Press
Taylor & Francis Group
6000 Broken Sound Parkway NW, Suite 300
Boca Raton, FL 33487-2742

First issued in paperback 2020

© 2017 by Taylor & Francis Group, LLC
CRC Press is an imprint of Taylor & Francis Group, an Informa business

No claim to original U.S. Government works

ISBN-13: 978-1-4987-1187-6 (hbk)
ISBN-13: 978-0-367-78243-6 (pbk)

Library of Congress Cataloging-in-Publication Data

Names: Roth, Connie B., 1974- author.
Title: Polymer glasses / Connie B. Roth.
Description: Boca Raton, FL : CRC Press, Taylor & Francis Group, [2016] |
Includes index.
Identifiers: LCCN 2016028864| ISBN 9781498711876 (hardback ; alk. paper) |
ISBN 1498711871 (hardback ; alk. paper)
Subjects: LCSH: Polymers--Properties. | Glass. | Polymeric composites. |
Viscoelasticity. | Glass transition temperature.
Classification: LCC QD381 .R65 2016 | DDC 547/.7--dc23
LC record available at https://lccn.loc.gov/2016028864

Visit the Taylor & Francis Web site at
http://www.taylorandfrancis.com

and the CRC Press Web site at
http://www.crcpress.com

To all those who simply find joy in learning science.

Contents

Foreword

Understanding the slow dynamics and mechanical response of bulk glass-forming liquids and suspensions (colloids, nanoparticles, molecules, metals, ceramics, polymers, biological matter) remains, as it has been for many decades, a grand scientific challenge of diverse technological relevance. At the heart of the difficulty is understanding strongly activated collective structural rearrangements on the nanometer scale governed by temperature-dependent barriers that are somehow related to Angstrom-scale interactions of the elementary constituents. The most fundamental question is the physical mechanism of vitrification—Is it driven by thermodynamics, structure, and/or purely kinetic considerations? Many diverse, often difficult to reconcile, theoretical proposals coexist in the literature, which range from microscopic to highly phenomenological in character. What is meant by enduring concepts such "cooperative motion" and "dynamic heterogeneity," and whether the latter is of zeroth-order importance for ensemble-averaged structural relaxation, viscosity, and diffusion, remains heavily debated. These issues apply even to the simplified models studied at relatively high temperatures using computer simulation, let alone in the chemically complex venue of deeply supercooled real materials. Conceptual convergence is frustrated by the inability of almost all theories to make quantitative and falsifiable predictions for real systems under the thermodynamic state conditions studied in the laboratory. The possibility remains that the leading order physics depends on material type and/or specific temperature regime probed.

The introduction of interfaces and surfaces modifies glassy dynamics in a spatially inhomogeneous manner with gradients of all dynamical properties induced by nonuniversal interfacial effects and geometric confinement. In engineering applications, glasses are generally used in the nonequilibrium solid state and are often subjected to strong mechanical forces of variable magnitude, symmetry, and temporal history. Additional difficult scientific questions then arise such as the nature of physical aging, deformation-accelerated relaxation and plastic flow, and materials failure. The out-of-equilibrium glass experiences competing driving forces that effectively move it up and down on the potential energy landscape. Boltzmann statistical mechanics no longer apply, presenting a huge challenge for formulating predictive descriptions.

The present book edited by Connie Roth is highly welcome since it provides an excellent snapshot of recent progress on the three broad topics sketched above in the context of arguably the most phenomenologically rich class of glass-forming materials, polymers. Consider first, the structural alpha relaxation in the equilibrated cold liquid. Chemically, one might say that polymers are relatively simple since they are typically constructed from nonpolar monomers. However, they bring much material-specific complexity associated with local conformational flexibility, backbone stiffness, nonspherical monomer shape, tacticity, and chain degree of polymerization. Understanding the most basic question of the mean alpha relaxation time over the typically measured 10–16 decades, and its Arrhenius to supra-Arrhenius evolution with cooling, is an especially large challenge for polymers due to the presence of coupled intra- and interchain degrees of freedom. The richness of polymer chemistry results in remarkably large variations of key dynamical quantities. For example, the glass transition temperature, T_g, can be tuned from roughly 150–500 Kelvin for long chains, and the dynamic fragility, which quantifies the rate of increase of the alpha time at kinetic vitrification, varies by nearly an order of magnitude from approximately 25 to over 200. These dynamic properties exhibit chemically specific changes (typically increasing) upon going from oligomers to long chains. This chain length sensitivity can be very large, and potentially provides a unique window on dynamical length-scale effects in the simplifying context of fixed chemistry and intermolecular forces. Although such chemical complexity might be viewed as undesirable from the point of view of comprehending the fundamental physics of glass-forming liquids, it is highly welcome from a materials science perspective, and I believe provides a powerful set of experimental constraints on the development of a predictive and broadly applicable fundamental theory of supercooled liquid dynamics.

A second theme of the book is confinement effects, particularly free-standing (vapor interfaces), supported (one solid, one vapor interface), and capped (two solid surfaces) thin films. The presence of interfaces can speed up or slow down relaxation and transport, and introduces dynamic anisotropy including steep spatial gradients of mobility and mechanical stiffness. The presence of such gradients renders understanding film-averaged properties particularly subtle given the need to average over heterogeneous motion in a manner consistent with the property of interest. Near an interface, polymer chains pack differently, experience a gradient of local density and backbone orientation, and can often strongly physically adsorb via polymer-surface cohesive attractions. These effects locally modify kinetic constraints and hence activation barriers, which are then somehow transmitted into the bulk of the film over surprisingly long distances. Remarkably, the elastic stiffness of the condensed phase boundaries that confine a supercooled liquid also matters, as documented in recent experimental and simulation studies. These fascinating substrate elasticity effects can be surprisingly large. They may be crucial in elucidating whether the alpha relaxation process is associated solely with compact rearranging domains of a few nanometers, or is intimately coupled to the spatially longer-range elasticity that emerges in a cold liquid and the high-frequency mechanical stiffness of confining (solid or liquid) boundaries. All these complexities are, in principle, present for small molecule systems. However, polymers bring qualitatively new scientific aspects, in addition to their practical advantage as excellent thin-film formers. For example, the monomers of a connected polymer chain experience a broad range of friction and local mobility in the spatially heterogeneous film, which modifies its macromolecular diffusivity, length-scale-dependent conformational dynamic modes, and viscoelastic response in a complicated and poorly understood manner.

A third theme of the book, is the below T_g nonequilibrium polymer glass. Even for nondeformed materials, properties are time-dependent due to physical aging, a challenging problem in the bulk and even more so in confined films with mobility gradients. The question of how aging and dynamic heterogeneity are coupled, and how the near-Arrhenius relaxation observed in quenched glasses evolves to supra-Arrhenius behavior at long enough times, is not well understood. Indeed, even the question of whether Arrhenius relaxation can be characteristic of the equilibrated state of some glassy polymers remains debated. When subjected to deformation, the most elementary and foundational question is how the nanometer-scale segmental relaxation process changes. At least four highly nonlinear, coupled processes come into play: external forces can directly reduce effective activation barriers, local structure can become more disordered (sometimes called "mechanical rejuvenation"), the distribution of relaxation times can be strongly distorted, and the physical aging rate becomes stress-dependent. Thus, understanding even the mean segmental relaxation time, its distribution due to dynamic heterogeneity, and the local plastic flow process as a function of stress, strain rate, temperature, aging protocol, and other control variables is a challenging problem in nonequilibrium physics. But again, these issues potentially arise in all glass-forming materials. What is particularly unique about polymer glasses is not only their widespread use as engineering thermoplastics, but the physical consequences of chain connectivity and entanglements on the nonlinear mechanical response. There are many phenomena with an important macromolecular component such as yielding, nonentropic large amplitude strain hardening, fracture, crazing, and the ductile-brittle transition. Addressing them requires confronting the thorny issue of the physical origin of stress, which in polymer glasses subjected to large deformation has both interchain (local forces, nonlocal entanglements) and intrachain origins. Ultimately, a synthesis of ideas from molten polymer rheology with the more solid-state concepts of local activated relaxation and plastic flow in glasses is required to make transformative progress. Addressing these formidable complexities in polymer glasses could be viewed as intractable for fundamental studies, or alternatively as fascinating scientific opportunities with high materials application relevance. Fortunately, polymer scientists adopt the latter perspective, with the present volume providing excellent examples of state-of-the-art efforts in these directions.

In conclusion, the present book will be of great value for both newcomers to the field and mature active researchers by serving as a coherent and timely introduction to some of the modern approaches, ideas, results, emerging understanding, and many open questions in this fascinating field of polymer glasses, supercooled liquids, and thin films.

Kenneth S. Schweizer
University of Illinois at Urbana-Champaign

Preface

Polymer glasses have become ubiquitous to our daily life, from the polycarbonate eyeglass lenses on the end of our nose to the large acrylic glass panes holding back the seawater in the Georgia Aquarium tanks. As polymers, they have the advantage of being lighter and easier to manufacture, while possessing the transparency and rigidity associated with glasses. Polymer glasses also have the additional advantage of being ductile, not brittle, after yielding, allowing the material to retain some functionality after failure, instead of disintegrating into a pile of shards. Your plastic water bottle may become dented, but it still holds its water. Given all these important practical uses, polymer glasses have been studied for decades. However, because of the complexities associated with understanding glasses at a fundamental molecular level, modeling properties of polymer glasses have frequently been limited to heuristic approaches.

The challenge with understanding glasses at the molecular level is that they are nonequilibrium materials whose properties depend on many-body interactions. Our traditional statistical mechanics approaches are for equilibrium systems and simple two-body interactions. Investigation of commonalities across different types of glass formers (polymers, small molecules, colloids, and granular materials) has enabled microscopic- and molecular-level frameworks to be developed for these complex systems. Despite their long-chain molecular nature, polymer glasses exhibit many of the same properties as other glass formers because the packing frustration that leads to kinetic arrest and rigidity during glass formation occurs at the segmental level. Thus, theoretical insight from how glass formers are modeled across different systems has led to treatments for polymer glasses with first principle-based approaches and molecular-level detail. These efforts have resulted in improved understanding and agreement between experiment and theoretical modeling that can increasingly be brought to bear on more complicated systems. It is these efforts that the present book aims to summarize and in so doing provide a foundation for research in this field.

The format of the present book has evolved from lively and stimulating sessions that occur yearly at the American Physical Society (APS) March Meeting. In recent years, sessions on polymer glasses have focused on geometrically confined systems such as thin films and on understanding thermo-mechanical deformations and failure mechanisms. Both of these are areas that have many relevant applications driving interest in the field, but are also motivated by fundamental importance. Studies of glass formers in confined geometries strive to access insight on the length scales associated with cooperative motion thought to control the inherent dynamic arrest occurring at the glass transition. The fundamental response of polymer glasses to deformations with different stress and strain stimuli probes the underlying potential energy landscape governing glassy mobility. In the same way that insight from other glass formers has informed our understanding about polymer glasses, studies of polymer glasses represent a rich, well-categorized system for testing theoretical ideas about glass formers in general.

The book is divided into three parts. The first part provides a summary of the fundamental characteristics of polymer glasses, including how they are measured and simulated. The second part covers polymer glasses in confined geometries, while the third part tackles polymer glasses under deformation. The various topics of polymer glasses are covered by experts in these areas: Sindee Simon on the measurement of structural recovery and physical aging (Chapter 2), Jörg Baschnagel on the approaches to computer modeling (Chapter 3), James Caruthers and Grigori Medvedev on the various thermo-mechanical characteristics exhibited by polymer glasses (Chapter 4) and how they are best modeled with various constitutive descriptions (Chapter 14), Connie Roth on the glass transition and physical aging in thin films (Chapter 5), Greg McKenna on the mechanical and viscoelastic properties of thin films (Chapter 6), Koji Fukao on dielectric relaxation spectroscopy studies (Chapter 7), Francis Starr and Jack Douglas on simulating polymers in thin films (Chapter 8), Didier Long on theoretical modeling of glassy thin films and nanocomposites (Chapter 9), Mark Ediger on measuring enhanced local mobility in deformed glasses (Chapter 10), Jörg Rottler on simulating local relaxations in polymer glasses under stress (Chapter 11), Shi-Qing Wang on the role of chain networks in yielding and failure behavior (Chapter 12), and Rob Hoy on modeling strain hardening (Chapter 13).

I am greatly indebted to all those who have contributed chapters to this book. As the reader will see, the authors went to great lengths to provide extensive summary and perspective of the various topics, explaining both the phenomena as well as providing an outlook for where the outstanding issues still remain. The book is much richer for all their efforts, and it is my hope that this book will provide a starting point for those scientists new to the field of polymer glasses.

Connie B. Roth
Emory University
Atlanta, Georgia

Editor

Connie B. Roth is currently an Associate Professor of Physics at Emory University, as well as the Director of Graduate Studies for the Physics Doctoral Program. She received her Ph.D. and M.Sc. in Physics from the University of Guelph, Canada. Her interest in polymers stems from her time working at Xerox Research Centre of Canada (XRCC) during summers while pursuing her B.Sc. in Physics at McMaster University in Canada. Following postdoctoral positions at Simon Frazier University, Vancouver, and Northwestern University, Chicago, she joined Emory's faculty in 2007. Her research lab studies the physical and mechanical properties of polymer glasses near interfaces, as well as the effects of stress, temperature, and miscibility. She has received a National Science Foundation (NSF) CAREER Award, American Chemical Society PRF Doctoral New Investigator grant, and was the 2009 recipient of the Division of Polymer Physics (DPOLY)/United Kingdom Polymer Physics Group (UKPPG) Polymer Lecture Exchange by the American Physical Society.

Contributors

Roman R. Baglay
Department of Physics
Emory University
Atlanta, Georgia

Jörg Baschnagel
Institut Charles Sadron
Université de Strasbourg
Strasbourg, France

Olivier Benzerara
Institut Charles Sadron
Université de Strasbourg
Strasbourg, France

James M. Caruthers
School of Chemical Engineering
Purdue University
West Lafayette, Indiana

Shiwang Cheng
Department of Polymer Science
University of Akron
Akron, Ohio

Alain Dequidt
Institut de Chimie
Université de Clermont-Ferrand
Clermont-Ferrand, France

Jack F. Douglas
Materials Science and Engineering
National Institute of Standards and
    Technology
Gaithersburg, Maryland

Mark D. Ediger
Department of Chemistry
University of Wisconsin–Madison
Madison, Wisconsin

Jean Farago
Institut Charles Sadron
Université de Strasbourg
Strasbourg, France

Koji Fukao
Department of Physics
Ritsumeikan University
Kusatsu, Shiga, Japan

Paul Z. Hanakata
Department of Physics
Wesleyan University
Middletown, Connecticut

Kelly Hebert
Department of Chemistry
University of Wisconsin–Madison
Madison, Wisconsin

Julian Helfferich
Physikalisches Institut
Albert-Ludwigs-Universität
Freiburg, Germany

Robert S. Hoy
Department of Physics
University of South Florida
Tampa, Florida

Ivan Kriuchevskyi
Institut Charles Sadron
Université de Strasbourg
Strasbourg, France

Didier R. Long
Laboratoire Polymères et Matériaux
    Avancés
CNRS/Solvay
Saint-Fons, France

Gregory B. McKenna
Department of Chemical Engineering
Texas Tech University
Lubbock, Texas

and

Laboratoire Sciences et Ingénierie de la
    Matière Molle
Paris, France

Grigori A. Medvedev
School of Chemical Engineering
Purdue University
West Lafayette, Indiana

Samy Merabia
Institut Lumière Matière
Université de Lyon
Lyon, France

Hendrik Meyer
Institut Charles Sadron
Université de Strasbourg
Strasbourg, France

Beatriz A. Pazmiño Betancourt
Department of Physics
Wesleyan University
Middletown, Connecticut

and

Materials Science and Engineering
National Institute of Standards and
    Technology
Gaithersburg, Maryland

Justin E. Pye
Department of Physics
Emory University
Atlanta, Georgia

Connie B. Roth
Department of Physics
Emory University
Atlanta, Georgia

Jörg Rottler
Department of Physics and Astronomy
The University of British Columbia
Vancouver, British Columbia, Canada

Céline Ruscher
Institut Charles Sadron
Université de Strasbourg
Strasbourg, France

Sindee L. Simon
Department of Chemical Engineering
Texas Tech University
Lubbock, Texas

Paul Sotta
Laboratoire Polymères et Matériaux
    Avancés
CNRS/Solvay
Saint-Fons, France

Francis W. Starr
Department of Physics
Wesleyan University
Middletown, Connecticut

Shi-Qing Wang
Department of Polymer Science
University of Akron
Akron, Ohio

Joachim P. Wittmer
Institut Charles Sadron
Université de Strasbourg
Strasbourg, France

Meiyu Zhai
Whitacre College of Engineering
Department of Chemical Engineering
Texas Tech University
Lubbock, Texas

WHAT MAKES POLYMER GLASSES UNIQUE?

Fundamentals of polymers and glasses

CONNIE B. ROTH AND ROMAN R. BAGLAY

1.1 POLYMERS AND GLASSES

Polymers are long-chain molecules, a feature that gives polymers many of their unique and important material properties such as rubber viscoelasticity. However, despite being long-chain molecules, in many cases, polymer glasses behave much the same as regular small-molecule glasses. They exhibit many of the same phenomena such as a non-Arrhenius temperature dependence of the α-relaxation and temporal evolution of the nonequilibrium glassy state, leading to the characteristic logarithmic time dependence of material properties typically referred to as physical aging. This universality with other glass formers is because the packing frustration associated with the glass transition occurs at the segmental level for polymers. In essence, the molecular dynamics of the chain segments are so limited by the tight packing of other neighboring segments that it does not really matter that some of the segments are covalently bonded together into long chains (with the exception of adding cross-links at a density higher than the entanglement density; see Figure 1.4). In this sense, the study of the polymer glass transition is informative to glasses in general. For polymers, the glass transition temperature (T_g) is independent of the chain length (molecular weight), above some minimum molecular weight (described in more detail below). The one major area where chain connectivity plays a significant role in glassy polymer behavior is in deformation where yielding leads to ductile flow instead of brittle fracture (see Chapters 12 and 13).

Given that chain connectivity plays a limited role in the glass transition, much can be learned from the study and comparison with other types of glass formers such as small molecules, colloids, and granular materials. However, despite these similarities, polymer glasses are frequently omitted from compendiums on glasses. This treatise focuses specifically on polymer glasses, summarizing and discussing various aspects currently driving research in the field. Many of the authors contributing to this book have expertise with other types of glass formers; thus, where possible, they have delineated what is common to other glass-forming systems or is unique to polymers. This chapter attempts to provide a brief overview of the phenomenology of glasses and polymers by summarizing some conceptual ideas used in their understanding and interpretation to create a common starting point for readers from different backgrounds. However, as both fields of polymers and glasses are vast, the reader is referred to various books for more background: for polymers [1–8], for glasses [9–13], and more specifically for polymer glasses [14–16].

The purpose of this book is to provide a summary of recent advances in understanding that have occurred in polymer glasses. Although the thermal, mechanical, and even geometrically confined behavior of polymer glasses have been studied for decades, significant advances in molecular-level understanding of these phenomena have occurred during the past 5–10 years that warrants a compendium. It is the hope that this book will enable new researchers, either from different fields or starting out their careers, to get up to speed by using it as a guide to the vast literature. For those researchers already in the field, perhaps this summary of collected works in a single location will aid the field in making connections between similar concepts and approaches across different experimental, computational, and theoretical methods.

1.1.1 What does it mean to be a glass?

Fundamentally, glasses are nonequilibrium materials formed because of packing frustration. As the material is cooled or the density increased, a point is reached eventually where the available thermal energy is insufficient to allow the system to explore all the different possible configurations on accessible time scales (i.e., the system becomes nonergodic). Thus, the system falls out of equilibrium and becomes trapped in some nonequilibrium state related to the conditions it experienced at the point of vitrification (glass formation). The most classic demonstration of this is the influence of cooling rate on the glassy state formed. Glasses that were formed by cooling quickly are less dense and more unstable than glasses that were formed by cooling slowly. It is this fact that makes the glass transition unusual, often being referred to as a kinetic transition and calling into question whether or not it is a "real" transition in the thermodynamic sense, despite exhibiting many of the same characteristics as a second-order phase transition. The debate of whether or not there exists a true thermodynamic transition hidden underneath the kinetic transition is one that has been raging for decades [10,17–21], having been dubbed "the deepest and most interesting unsolved problem" in condensed matter theory by Phillip Anderson in 1995 [22].

Glasses have the mechanical properties of a solid, while at the molecular level, the system is disordered, appearing to be structurally indistinguishable from a liquid. Where the molecules in a liquid move around continuously exploring all possible configurations (conformations in a chemistry sense), a glass is nearly static. How the material transitions on cooling from an easily flowing liquid to sluggish dynamics that are drastically slower and conceptually quite different is one of the great mysteries about glass formation. What causes the massive slowing down in dynamics with only modest increase in density, decrease in temperature?

This dramatic slowing down begins already well above the glass transition ($\sim T_g + 50$ K). Unlike at high temperatures, where molecules in the liquid state can easily slide past each other, molecular motions at temperatures not too far above T_g are referred to as having activated dynamics where the activation energy increases with decreasing temperature, thought to be caused by cooperative motion. In this supercooled regime, which assumes that crystallization was avoided, the motions of a given molecule are thought to become constrained by a progressively increasing number of its neighboring particles as T_g is approached on cooling. It is free to move locally about, but any large-scale translational motion is limited by the lack of available space. The molecule must wait until its neighbors shift collectively apart, freeing up space for it to move into. This is the notion behind cooperative motion. The neighboring molecules create a collective constraint that introduces a temperature-dependent free energy barrier that must be overcome before motion is allowed. The onset of this type of constrained dynamics is often associated with the so-called α–β splitting transition. α-relaxations, so called because they are the first to freeze-out on cooling, refer to this type of slow cooperative motion that requires collective movement of neighboring units before the molecule (or polymer segment) can "hop" to a new adjacent location. It is this collective motion that is frozen-out at the glass transition. β-relaxations are the local motions, rattling back and forth within the small region trapped by its neighbors. They can be observed as a distinct activated process with a constant activation energy, likely because the slow α-relaxation creates an almost static energy landscape for local motions.

One conceptually simple and illustrative picture that forms the foundation for the rather complicated mode-coupling theory (MCT) is the idea of caging. In this picture, molecules are treated as simple hard spheres with the onset for cooperative motion occurring when the density increases to the point where the central sphere in question becomes trapped within a cage formed by the surrounding spheres as illustrated in Figure 1.1. At short times, the central sphere rattles around trapped within its cage. Cooperative motion, or an α-relaxation, occurs when a shift in the surrounding spheres creates a break in the cage large enough for the central sphere to escape. The mathematical framework that leads to what is formally known as mode-coupling theory [23–25] is far from simple (see Chapter 3 for more details), and has received many additions and corrections over the years [25–29]. One of the fundamental limitations associated with MCT is that the predicted glass transition, kinetic arrest of α-relaxations, occurs at too high a temperature for comparison to real materials. As a consequence of the approximations made in the theory, the system simply locks into place and solidifies too easily. Yet, the basic idea of how caging of molecules can emerge on densification, leading to the onset of cooperative motion and α-relaxations, is a central feature of how the glass transition is thought of and treated in the literature.

In principle, any material or system that can become kinetically trapped exhibits some of the properties of glasses. Colloidal hard sphere glasses are an idealized example of this. Micron-sized hard plastic spheres are suspended in typically a density-matched fluid where Brownian motion of the particles creates the equivalence of molecular dynamics. The volume fraction ϕ of spheres in the system is increased, mimicking a decrease in temperature for a molecular glass, leading to transitions from liquid to supercooled fluid at $\phi = 0.494$, and to a glassy state at $\phi_g \approx 0.58$ [30]. To avoid crystallization of the spheres, polydisperse particles or binary-sphere mixtures are used (similar to strategies employed in computer simulations). One of the key advantages of colloidal glasses is that they can be imaged using confocal microscopy by using fluorescently tagged particles in an index-matched fluid to provide the coordinates of all the particles in real time, giving a visualization of what dynamics in supercooled liquids and glasses might look like if we could actually see molecules directly. In this sense, colloidal glasses have some of the same advantages of computer simulations where the position and dynamics of all the particles are known. Colloidal glasses also exhibit many of the same features of molecular glasses such as dynamic heterogeneity and a dramatic non-Arrhenius increase in viscosity with increasing volume fraction [30,31]. Perhaps unsurprising, mode-coupling theory actually describes hard sphere colloidal glasses fairly well [5,31]. However, as with all simplified systems, the analogy with molecular

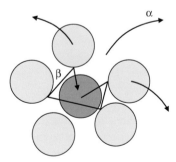

Figure 1.1 A simple illustration of cooperative motion (α-relaxation) in a supercooled liquid composed of hard spheres. Such simple systems are common in the glassy community since the dynamics are somewhat successfully treated mathematically with mode-coupling theory (MCT) and are experimentally accessible with colloidal suspensions. In the supercooled liquid regime (above the glass transition), the central particle in question is able to "hop" out of the "cage" formed by surrounding particles via an α-relaxation process, or "cage-breaking" event, where two or more surrounding particles shift to allow the central particle to escape. As the volume fraction increases (or equivalently the temperature decreases), this cooperative α-relaxation process eventually freezes out at T_c ($>T_g$), trapping all particles within their cages. The local "cage-rattling" relaxational motion of the particles trapped within their cage is often referred to as a (fast) β-relaxation, but note that this is distinctly different from the (slow) β-relaxation that occurs in molecular liquids at the α–β splitting transition shown in Figure 1.7 (see Reference 9).

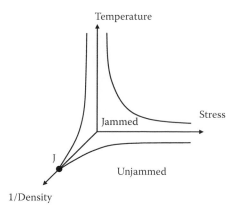

Figure 1.2 Jamming phase diagram depicting how the jammed state (gray area) can be accessed equivalently through temperature, stress, or density (inverse of specific volume). In general, granular systems operate in the zero-temperature plane since Brownian (thermally activated) motion ceases to drive particle motion for such large granular particles; in contrast, much smaller colloidal particles are continuously agitated by Brownian motion mimicking thermal motions in real molecular systems.

glasses is not perfect. Some of the limitations to making connections between colloidal glasses and molecular glasses are a result of experimental challenges. For example, with colloidal glasses, it is typically easier experimentally to initiate an aging experiment from a shear quench for a sample at constant volume fraction [30,32], while for molecular glasses, this is typically done by a temperature quench at constant pressure. The extent to which different glass formation pathways actually result in the same glassy state within the same system is still an outstanding issue [33–35] (see also Chapter 4). Nevertheless, the field of colloidal glasses continues to evolve, adding complexity in the form of particle shape [36–38], softness [39], interparticle interactions [30], and thermosensitive particles that can change size with temperature [34], which can be used to better understand how vitrification comes about in molecular glasses.

Granular materials also exhibit some properties similar to glasses. Rough particles like sand pack together in random configurations, easily becoming jammed into amorphous solids [13]. The main distinction for granular materials is that the particles are sufficiently large that thermal energy (k_BT) is insufficient to agitate them. This lack of thermal motion means that granular materials may be considered to be a glass in the limit of zero thermal energy, that is, in the limit $T \to 0$. An appealing framework that attempts to conceptually unify different types of glass formers is the jamming phase diagram proposed by Liu and Nagel [40,41]. Figure 1.2 illustrates the concept where temperature, density, and stress are proposed as three orthogonal axes that can each independently transform the system from a rigid solid to a flowing liquid. For ordinary molecular glasses and polymers, temperature is typically the primary control variable, while for colloidal and particulate systems, this is typically volume fraction ϕ (proportional to density). For a granular system, mechanical deformation (for example, shear stress) can lead to unjamming and yielding of a rigid, hard packed system. In some granular experiments, the application of physical agitation (shaking) has been imparted to mimic thermal motions [42]. Although the specific details of whether jamming (a zero-temperature phenomenon) can be strictly compared to the glass transition in materials are still under debate [41,43–46], the extent to which similar phenomena can arise in such disparate systems is informative. Even within polymers, understanding the response of the material to deformation and pressure isothermally can be used to illuminate the nature of the glassy state (see Chapter 4).

1.1.2 WHAT DOES IT MEAN FOR A POLYMER TO BE A GLASS?

Let's focus now specifically on polymers and discuss in more detail what molecular motions undergo dynamic arrest at T_g. For polymers, the packing frustration that leads to dynamic arrest is that of chain segments and monomers. Chain segments pack so tightly together that the motion is limited by the close presence of neighboring monomers that may or may not be on the same chain. Imagine being trapped in a room with so many people around you that you can no longer move. If you happen to be holding hands with your neighbors forming a long chain, that does not change the fact that you are still trapped by the close proximity of your immediate neighbors surrounding you. This is why chain connectivity does not play a dominant role in the glass transition and likely why the glass transition in polymers exhibits most of the same characteristics as other small-molecule glass formers. However, keep in mind that the absolute value of T_g exhibited by different polymers will be influenced by their local chemical structure related to torsional barriers and chain stiffness, similar to the way that the size and flexibility of small-molecule glass formers influence their specific T_g values.

Where the big difference for polymers comes in is when this packing frustration is released above T_g and the monomers are able to easily slide past each other. Then the resulting flow is dictated by chain connectivity: Rouse dynamics (bead–spring motions) for unentangled chains with constrained tube dynamics (primarily reptation) for entangled chains. Unlike small molecules that immediately transition to liquid flow above T_g, entangled polymers transition instead to a rubbery melt whose flow time scale is dictated by tube dynamics (primarily reptation, but also other relaxation processes such as contour length fluctuations [1]) that scales strongly with molecular weight M (chain length): viscosity $\eta \sim M^{3.4}$. Given that entanglement effects set in at already relatively modest molecular weights, $M_c \approx 30$ kg/mol for polystyrene (~300 monomers) [5,47], most polymers easily fall into this category. Thus, for polymers, the glass transition is typically decoupled from flow time scales. The extreme case being a cross-linked rubber (for example, car tire) where the material is above its T_g, but will never undergo flow.

The value of the glass transition temperature for polymers does not depend on molecular weight after an initial saturation value of a couple of hundred monomers. Figure 1.3a plots data of T_g for polystyrene as a function of number average molecular weight M_n where it is clear that the T_g reaches the limiting value of ~100°C at $M_n \approx 20$ kg/mol (~200 monomers) [2,48,49]. Such T_g data is found to scale linearly with $1/M_n$, where M_n is specifically the *number* average and not *weight* average molecular weight, indicating that it is the number density of chain ends that is the relevant parameter. This saturation behavior for T_g as a function of chain length has traditionally been treated as a simple free volume or density argument because the local density around chain ends is slightly less than that around monomers within the chain [2,50]. However, more

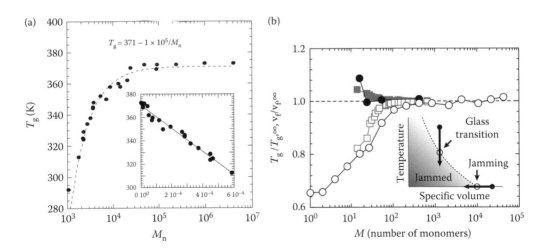

Figure 1.3 (a) Molecular weight dependence of the glass transition temperature T_g for polystyrene (data from References 47 and 48), showing that T_g saturates at a limiting value of ~100°C beyond $M_n \approx 20$ kg/mol (~200 monomers). The inset shows the linear dependence with $1/M_n$, where M_n is the *number* average molecular weight. (Figure reproduced with permission from P. C. Hiemenz and T. P. Lodge, *Polymer Chemistry*, 2nd edn., CRC Press, Copyright 2007, Taylor & Francis, Boca Raton, FL.) (b) Chain length dependence of the specific packing volume v_f (effective T_g) for granular ball-chains also demonstrating saturation at ~200 monomers (open circles represent linear chains). (From L.-N. Zou et al., *Science*, 326, 408–410, 2009. Reprinted with permission of AAAS.)

recent work has demonstrated that the saturation of T_g with chain length is more accurately correlated with differences in local chain flexibility near chain ends and saturates when the chain dynamics asymptotically display Gaussian behavior [51,52]. The T_g values of polymers for these higher molecular weights then depend on chemical structure, influenced by factors such as chain flexibility and side group bulkiness. Interestingly, granular experiments on simple ball-chains (commonly used as window shade pulls) find a similar relationship between specific packing volume $v_f = 1/\rho_f$ and chain length M, where this effective T_g is also found to saturate at ~200 monomers, as shown in Figure 1.3b [53].

The differences in relaxation times for polymers are frequently illustrated on a master curve of log(modulus) versus log(time), or equivalently temperature. Figure 1.4 illustrates schematically the shape of this curve for different cases. At short times or low temperatures, the material behaves as a glassy solid with a modulus ~10^9 Pa. As the temperature is increased, the material undergoes the glass transition. For small molecules or low molecular weight unentangled polymers, the modulus drops to zero as the material becomes a liquid and flows. However, for entangled polymers, the modulus drops only to the rubbery plateau modulus ~10^5–10^6 Pa. The extent of this plateau depends strongly on the molecular weight of the chains, with flow only occurring for times longer than the chain diffusion time scale $\tau_d \sim M^{3.4}$, typically dominated by reptation [1]. As T_g (~100 s) is independent of molecular weight for such entangled polymers, while flow (reptation) can easily reach out to hours, days, even weeks at modest temperatures above T_g depending on the molecular weight, flow is effectively decoupled from the glass transition.

The x-axis in Figure 1.4 can span 10^{10}–10^{12} orders of magnitude in time scales [3,7]. Thus, this master curve is not a plot that can be measured in a single experiment. Such master curves are constructed by time–temperature superposition, from multiple measurements each spanning 10^3 – 10^4 orders of magnitude in time (or frequency) collected at different temperatures and then superimposed by shifting the data horizontally to create a continuous curve. The shift factor $a_T = \eta(T)/\eta(T_{ref})$ follows a Williams, Landel, and Ferry (WLF) functional form:

$$\log(a_T) = \log\left[\frac{\eta(T)}{\eta(T_{ref})}\right] = -\frac{C_1(T - T_{ref})}{C_2 + (T - T_{ref})}, \tag{1.1}$$

where C_1 and C_2 for a given reference temperature T_{ref} (usually taken to be T_g of the polymer) are tabulated in polymer handbooks [54]. The WLF equation is mathematically equivalent to the Vogel–Fulcher–Tammann

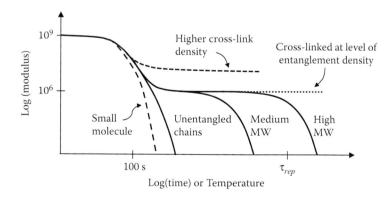

Figure 1.4 Modulus curves (on a log scale) as a function of temperature or logarithmic time for polymers of various molecular weights. Low molecular weight unentangled chains exhibit the same behavior as small molecules, transitioning from a glass (~10^9 Pa) directly to flow as the material goes through the glass transition (at ~100 s) with increasing temperature. Higher molecular weight entangled chains transition instead from a glass (~10^9 Pa) to a rubbery melt with the plateau modulus (~10^6 Pa) whose value depends on the entanglement density. The rubbery plateau modulus extends out as far as the reptation time ($\tau_{rep} \sim \tau_d \sim M^{3.4}$) before terminal flow finally occurs, which for very high molecular weights can easily reach hours, days, or even weeks at modest temperatures above T_g. In such a way, flow is effectively decoupled from the glass transition for high molecular weight entangled polymers, while permanently cross-linked polymers never exhibit flow.

(VFT) equation (see below), but more commonly used in polymer rheology studies [2,7]. Time–temperature superposition works as well as it does because effectively all the temperature dependence is treated by the same microscopic friction time $\tau_0 \approx \zeta b^2 / kT$, which forms the basis for local, Rouse, and reptation motions [1,2,6]:

$$\tau_p \approx \frac{N^2}{p^2}\left(\frac{\zeta b^2}{kT}\right), \quad \tau_{\text{Rouse}} \approx N^2\left(\frac{\zeta b^2}{kT}\right), \quad \tau_{\text{rep}} \approx \frac{N^3}{N_e}\left(\frac{\zeta b^2}{kT}\right), \tag{1.2}$$

where τ_p is the pth Rouse mode for $p = 1,2,3,\ldots$. Time–temperature superposition works remarkably well in overlapping data over many decades in time, allowing master curves spanning 10–14 decades in time to be constructed. However, strictly speaking, experiments that try to really test the limits of time–temperature superposition by measuring four plus decades in time do find discrepancies [55,56], meaning polymers are not perfectly thermorheologically simple in practice. For example, the scaling shift factor a_T can deviate substantially from WLF dependence near and below T_g [57–60] (see Chapter 4).

1.2 PROPERTIES AND MYSTERIES OF GLASSES

What makes the glass transition mysterious or hard to understand is the dramatic slowing down in dynamics, molecular relaxation times, that occurs over a rather narrow temperature range right above T_g, while there is comparatively little change in the structural variables of the material (such as density, structure factor, specific heat) over this same temperature range. This dramatic slowing down ultimately leads to dynamic arrest of the system and the formation of a glass, a solid material with an effective viscosity so large that it no longer flows on a time scale relevant with any time scale we can measure. For some materials, the viscosity or dominant relaxation time τ_α can change by more than 12 decades in time over only ~50 K (that can be as fast as a decade of time for every 3 K), while structural variables vary by only perhaps 10% or a factor of 2–3 over this same temperature interval [61,62]. From a structural point of view, we cannot currently take a picture of a position of all the atoms and molecules in a glass and a liquid, and then tell from all the particle positions based on established standard structural quantities (for example, the structure factor), which material is a liquid, with particles free to move around, or a glass, with the particles dynamically locked into place forming a solid. Numerous theories have been proposed over the years in an attempt to understand this complex behavior, to the extent that even a cursory summary would be a daunting task. David Weitz at Harvard has joked "There are more theories of the glass transition than there are theorists who propose them" [63]. However, it would be informative to at least discuss various approaches and conceptually introduce a number of important ideas used to describe and understand the glass transition.

1.2.1 GLASS TRANSITION AND PHYSICAL AGING

The glass transition temperature (T_g) has a number operational definitions that are used experimentally, and sometimes in an analogous fashion computationally, to measure T_g. The simplest to visualize intuitively is a plot of the temperature dependence of the specific volume v(T) or entropy as shown in Figure 1.5a (note that a plot of the temperature-dependent density $\rho(T) = 1/v(T)$ would simply be the reciprocal of this). At high temperature in the equilibrium liquid state, the v(T) curve is linear, representing the thermal expansion of the liquid. On cooling, assuming crystallization can be avoided, the liquid enters the supercooled regime where cooperative motion begins to set in. On cooling further, the available thermal energy is eventually insufficient to keep the system ergodic (able to explore all possible configurations) before the given cooling rate drops the temperature further and the system falls out of equilibrium at $T_g^{\,1}$ into a glassy state (glass 1). In the glassy state, the v(T) curve is again linear, reflecting the thermal expansion of the glass, typically roughly a factor of two less than that of the liquid. If the cooling rate is slower, then the system can travel further along the equilibrium liquid line before falling out of equilibrium at a lower temperature $T_g^{\,2}$ into a different glassy state (glass 2).

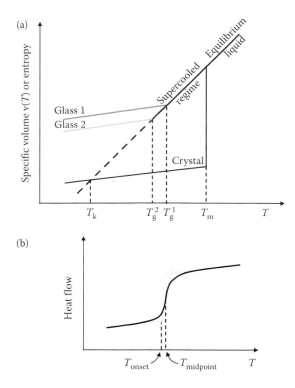

Figure 1.5 (a) Specific volume or entropy plotted as a function of temperature for a polymeric or small molecule glass-forming material. At high temperature, the system is in an equilibrium liquid state where the molecules easily slide past each other and sample all of coordinate space; consequently, the system is ergodic. Upon cooling, if the glass former is successfully able to avoid crystallization at T_m, the system enters into a supercooled liquid regime where the system is still ergodic, but the dynamics begin to slow as cooperative motion sets in. Upon cooling further, the system will fall out of equilibrium at T_g^1 into a glass (glass 1) whose thermal expansion is comparable to that of the crystal. If the system is cooled more slowly, it will reach a lower temperature T_g^2 before finally falling out of equilibrium into a glass (glass 2). Hypothetically, if the system could be cooled infinitely slowly, it is not clear if a thermodynamic glass transition would ever intervene such that the supercooled liquid could conceivably continue to much lower temperatures. If no glass transition intervenes, in principle, an entropy crisis could result at the Kauzmann temperature T_k where the entropy of the disordered liquid would become less than that of the ordered crystal. (b) DSC heat flow curve as a function of temperature, typically measured on heating, where convention usually defines T_g either by the onset T_{onset} (identified by the intersection of linear fits) or by the midpoint $T_{midpoint}$ of the transition.

It is clear from this definition that the glass transition is a kinetic transition (depending on the cooling rate), formally only defined on cooling. Many experimental techniques, such as for instance differential scanning calorimetry (DSC), typically measure the glass transition on heating, primarily for historical reasons because it was easier to control temperature on heating. Such a measurement on heating strictly measures what is called the fictive temperature T_f (see further discussion in Chapter 2), which is only experimentally equivalent to T_g (within error) if the heating rate and previous cooling rate are the same such that the same glassy line is used to evaluate both [64]. Figure 1.5b illustrates a typical plot of the glass transition response measured by DSC on heating where "T_g" is typically defined as either the onset (on heating) or the midpoint of the step change in the measured heat flow. A more accurate method (described in Chapter 2) is based on identifying the temperature at which equal areas occur between the measured heat flow curve and the linear extrapolation of the liquid and glassy regimes. Another definition of T_g, primarily used for small molecules, has been the extrapolated temperature at which the viscosity reaches $\eta(T) = 10^{12}$ Pa s, a somewhat arbitrary historical definition representing a viscosity so high that flow no longer occurs on a measurable time scale. Such a definition is less meaningful for entangled polymers where the viscosity and liquid flow are dominated by entanglements and thus decoupled from the glass transition.

Many polymers have irregular stereochemistries that prevent crystallization even at the slowest cooling rates. For small molecules, it can be relatively easy to avoid crystallization by cooling quickly past the point where crystallization usually occurs to enter the supercooled regime where molecular rearrangements become sufficiently sluggish that crystallization can be avoided kinetically. In this manner, it is possible to approach the glass transition temperature on cooling at very slow cooling rates. Such a possibility then begs the question that if a supercooled liquid can be cooled infinitely slowly, would it ever fall out of equilibrium? This seemingly innocuous question has important implications theoretically and has been one of the main driving issues motiving research on the glass transition for decades [17,19–21,65,66]. Back in 1948, Kauzmann postulated the conundrum [17,20,67], now known as Kauzmann's paradox or entropy crisis, that if the glass transition could be avoided by hypothetically cooling infinitely slowly, the material would remain in equilibrium presumably following the extrapolated liquid line down to lower and lower temperatures. If this occurred, then eventually the entropy of this disordered liquid state would become less than that of the ordered crystal; the Kauzmann temperature T_k refers to the point at which the entropy of this hypothetical liquid state crosses that of the crystal. The possibility of this counterintuitive occurrence spurred theorists to insist that some "true" thermodynamic glass transition to an "ideal glass" must intervene prior to T_k to prevent this entropy crisis from occurring [17,20,65]. Whether such a thermodynamic transition is necessary remains to be a primarily theoretical concern because for all practical purposes, an experimentally measured kinetic glass transition intervenes long before T_k. However, the philosophical opinion about whether the glass transition has an underlying thermodynamic or purely kinetic origin strongly influences how the glass transition is theoretically conceptualized to the extent that it is possible to broadly classify different theoretical approaches to modeling the glass transition as having either a thermodynamic or kinetic framework [19,65,66,68].

It is also worth noting that there is quite a bit of discussion about how various parameters (volume, density, relaxation times) extrapolate below T_g [57,58,60,69–73]. Do the equilibrium properties of the supercooled liquid regime continue to extrapolate smoothly below T_g if the glassy state is aged to equilibrium? Evidence appears to indicate that volume (density) and enthalpy continue to extrapolate linearly to temperatures far below where T_g typically occurs [69–71,74]. A couple of decades have been achieved [60,70,73], although this is still far above Kauzmann's hypothetical T_k. Some of the best evidence for this come from recent work on ultrastable glasses, thin glassy films formed slowly by physical vapor deposition at specific substrate temperatures where enhanced surface mobility enables molecular packing at far higher densities than is normally achieved on cooling [75]. In this manner, molecular packing indicative of glasses with far lower fictive temperatures can be achieved accessing stable glassy states far below the ordinary T_g typically measured [73,74,76].

Physical aging refers to the material evolution of the glassy state. Formally, the kinetic evolution of thermodynamic state variables such as volume and enthalpy are referred to as structural recovery, while concomitant changes in other material properties such as modulus, yield, or permeability are referred to as physical aging [77,78]. However, in practice, the term "physical aging" is often loosely used to refer to most structural evolution of the glassy state. Note that physical aging is distinct from chemical aging in that physical aging can always be completely reversed or eliminated by reheating the material back into the equilibrium liquid state; in contrast, chemical aging or degradation results in permanent chemical changes to the material.

For a glass that has been formed by a simple, "down jump" temperature quench from the equilibrium liquid state into a nonequilibrium glassy state (say along the path glass 1 in Figure 1.5a), the system is left trapped with a higher specific volume (reduced density) than it ideally would have in equilibrium [78]. Over time, the resulting glassy material will slowly densify on a logarithmic time scale until the equilibrium density is reached. The volumetric evolution of the material is typically characterized by the departure from equilibrium $\delta(t) = (v(t) - v_\infty)/v_\infty$, which usually follows the shape sketched in Figure 1.6a plotted as a function of the logarithm of the aging time [77–79]. The time scale needed to reach equilibrium increases drastically with decreasing temperature such that in practice, equilibrium can only feasibly be reached for temperatures a few tens of degrees below T_g [70,71], although recent works have used some rather creative means to investigate the equilibrium glassy state even further below the commonly measured T_g [60,73].

As a means of characterizing the stability of the glassy state, Struik [77] defined the physical aging rate

$$\beta = -\frac{1}{v_\infty}\frac{dv}{d\log t} = -\frac{d\delta}{d\log t} \tag{1.3}$$

representing the slope of the linear portion in Figure 1.6a, which is often the only region that can be measured unless one is close to T_g. A glass formed at a faster cooling rate leads to a less dense and less stable glass (a larger departure from equilibrium), and consequently experiences a larger aging rate. The evolution of the nonequilibrium glassy state is driven by the thermodynamic driving force to reach equilibrium, proportional to $\delta(t)$, but evolution of the material toward this equilibrium is strongly limited by kinetic factors. As one can see from Figure 1.5a, the distance from the glassy line to the equilibrium liquid line increases with decreasing temperature, such that the thermodynamic driving force for aging will grow with decreasing temperature starting from zero at T_g. However, as the temperature is decreased, the available thermal energy for motion also decreases, leading ultimately to a decrease in the aging rate. These competing factors typically result in a roughly parabolic profile for the temperature dependence of the physical aging rate as depicted in Figure 1.6b [72,77,80].

During a DSC measurement, physical aging manifests itself as an enthalpy overshoot at T_g, illustrated in Figure 1.6c, which can be characterized by the excess area under the heat flow curve [70,81] (see Chapter 2). In a manner analogous to the volumetric measurements, the time dependence of physical aging can be measured with DSC by studying the enthalpic departure from equilibrium $\delta_H(t) = H(t) - H_\infty$, which follows a similar relaxation curve as shown in Figure 1.6d, and an enthalpic aging rate can be equivalently defined

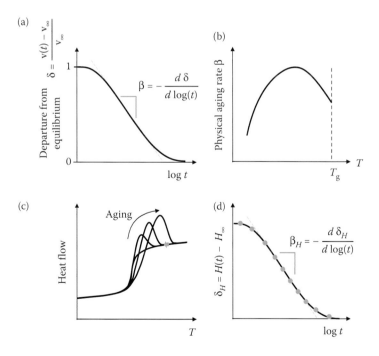

Figure 1.6 (a) Sketch of the volumetric departure from equilibrium δ as a function of logarithmic aging time. The physical aging rate β is defined as the slope of the linear region, which is often the only portion measurable because equilibrium can only reasonably be reached at temperatures a few degrees below T_g. (b) Characteristic temperature dependence of the physical aging rate β exhibits a roughly parabolic shape reflecting the competition between the enthalpic drive toward equilibrium and the available thermal energy to facilitate relaxation. (c) DSC heat flow curves collected on heating after different aging times at a given aging temperature. The amount of physical aging, enthalpic departure from equilibrium δ_H, is characterized from the excess area under the enthalpy overshoot peak (see Chapter 2). (d) Many individual measurements of δ_H are required to map out the aging curve, where an enthalpic physical aging rate β_H can be similarly defined.

as $\beta_H = -d\delta_H/d\log t$ [79]. It should be appreciated that such measurements are more time consuming than volumetric measurements because sampling the enthalpy overshoot with DSC at different aging times necessitates re-equilibrating the sample above T_g to obtain the reference baseline. Thus, each aging time point necessitates a different sample. It is interesting to note that enthalpic and volumetric aging behavior do not strictly match, with reports that the time to equilibrium can be perceived to be different depending on the type of measurement and cooling rate used to form the glass [79,82].

Finally, it should be made clear that such "down jump" aging measurements as described above provide only limited insight into the glassy state. More informative are so-called "up jump" and "memory" experiments that can better interrogate the thermal history of the sample [33,78] (see Chapters 2 and 4). Such experiments undeniably illustrate that physical aging dynamics are not simply slow because insufficient thermal energy is available for molecular motion, but the more complicated factor is the glassy material becomes structurally locked together such that steric constraints must slowly be released before motion and material evolution can occur. This impact of structural constraints is most evident when comparing the kinetic evolution of glasses that undergo an "up jump" and are driven to approach the equilibrium state from below (from a higher density) than a "down jump" approaching the equilibrium state from above (lower density); see discussion in Chapter 2.

1.2.2 NON-ARRHENIUS TEMPERATURE DEPENDENCE AND FRAGILITY

The supercooled regime, spanning the temperature range from T_g to approximately $T_g + 50$ K, is characterized by the dramatic slowing down of the dominant relaxation time τ_α and the viscosity η. As mentioned above, the cause of this large change in dynamics is one of the great mysteries of the glass transition. The data for these parameters are found to follow a non-Arrhenius (sometimes called super-Arrhenius) temperature dependence that is well fit by the empirical VFT equation:

$$\tau_\alpha = \tau_0 \exp\left(\frac{A}{T - T_0}\right) \sim \frac{\eta}{G_\infty}, \tag{1.4}$$

where τ_0, A, and T_0 are essentially material parameters to be fit. The viscosity η generally has the same temperature dependence because the high-frequency modulus G_∞ is fairly temperature independent for most glasses. The temperature T_0 at which the divergence occurs is referred to as the Vogel temperature and is typically located 30–50 K below T_g, far below where data are normally collected. Whether or not the relaxation time actually diverges below T_g, truly forming an amorphous solid ($\tau_\alpha \to \infty$), is still a matter for debate [57,83], with recent measurements on 20-million-year-old amber strongly suggesting that the VFT equation does not fit the data below T_g and that the relaxation time does not appear to diverge [60].

Arrhenius temperature dependence refers to a simple activation energy process:

$$\tau \propto \exp\left(\frac{E_a}{k_B T}\right), \tag{1.5}$$

where the activation energy barrier E_a is independent of temperature (k_B refers to Boltzmann's constant). Data are typically plotted on a so-called Arrhenius plot (log τ versus $1/T$), where the activation energy E_a can easily be determined from the slope of a linear fit. Following this convention, non-Arrhenius temperature-dependent data such as the α-relaxation times τ_α are also routinely plotted on such axes, where the non-Arrhenius behavior becomes easy to spot from the curvature of the data. Figure 1.7 sketches a typical Arrhenius plot for the α-relaxation time τ_α illustrating the strong non-Arrhenius curvature observed. T_g is often defined as the temperature at which τ_α reaches 100 s (log $\tau_\alpha = 2$). T_C refers to the dynamic crossover, ~$10^{-8} - 10^{-6}$ s, where the α- and β-relaxations split [84]. The more local β-relaxation follows a simple Arrhenius dependence. At sufficiently high temperature, T_A (~10^{-10} s), the behavior reverts to a simple liquid following an Arrhenius temperature dependence [28,85,86].

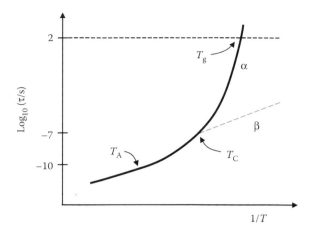

Figure 1.7 Typical Arrhenius plot (log τ versus $1/T$) of the α-relaxation time $\tau_\alpha(T)$ illustrating the VFT-like temperature dependence, the faster than Arrhenius (exponential) slowing down of the dynamics as the glass transition is approached from above. The dynamic T_g is often defined as the temperature when τ_α reaches 100 s (log $\tau = 2$). T_C represents the dynamic crossover at ~10^{-8} – 10^{-6} s, where the α- and β-processes split with the β-relaxation continuing on a simple Arrhenius trajectory. At very high temperature above T_A, the relaxation becomes Arrhenius again, characteristic of a simple liquid.

One can think of VFT behavior as an activation energy process with a temperature-dependent activation energy where the size of the energy barrier grows with decreasing temperature. This notion has been tied with the idea of cooperative motion and a growing length scale or growing number of units that must collectively be activated for motion to occur at lower temperature [87,88]. Conceptually, this was the basic idea put forward by Adam and Gibbs [89] where they introduced the notion of a cooperatively rearranging region (CRR) that referred to the number of units N_{CRR} that must collectively move together for motion to occur. Adam and Gibbs argued that the non-Arrhenius VFT behavior can be rationalized as an Arrhenius-activated process where the activation energy increases on cooling associated with the decrease in configurational entropy S_c and the onset of cooperative motion. This increase in activation energy was tied to the size of CRRs by effectively writing

$$\tau \propto \exp\left(\frac{zE_a^*}{k_B T}\right), \tag{1.6}$$

where E_a^* is the activation energy for a rearrangement of a single unit, such that a CRR made up of z units would have an activation energy of zE_a^* for a collective rearrangement (see Chapter 8 for a critical discussion of this idea).

Adam and Gibbs envisioned these CRRs as independent spherical regions of size $\xi_{CRR} \sim N_{CRR}^{1/3}$ ($N_{CRR} = z$). However, during the late 1990s, computer simulations of Lennard-Jones spheres in the supercooled regime were able to finally visualize collective rearrangements by particles and found that such cooperatively rearranging regions were string-like in shape [90,91]. Collective particle displacements were observed to follow a little train where a subsequent particle moved into the space vacated by the first. Similar behavior has also since been visualized in dense suspensions of spherical colloids [92–94]. It should be emphasized that such string-like cooperative motion, even when visualized in computer simulations of bead–spring polymers are not at all related to the connectivity of the particles. Unconnected spheres behave in a similar manner. Although the evidence of such string-like cooperative motion in computer simulations and dense colloidal suspensions are quite well established now [87,88,95], these studies are necessarily done at relatively high temperature, or equivalently low density. A 2008 paper argued on theoretical grounds that the string-like or fractal nature of the CRRs should become more compact and spherical as the temperature is decreased [96], with some experimental evidence in colloidal glasses now supporting this picture [97]. Recent works by Starr and Douglas [87,88,98] have incorporated such string-like collective motion into an activated free energy model similar to Equation 1.6, but with the collective activation energy zE_a^* now being proportional to the average string length $L(T)$ (see Chapter 8).

Some glasses, particularly network glasses such as SiO_2, do have an Arrhenius, or nearly Arrhenius, temperature dependence. To categorize such behavior between different types of glass formers, in the 1980s, C.A. Angell introduced the concept of fragility [61,99]

$$m = \frac{d\log\tau_\alpha}{d(T_g/T)}\bigg|_{T=T_g}, \tag{1.7}$$

which characterizes the slope of the log τ_α versus $1/T$ behavior at T_g. This fragility parameter m introduced the classification of *strong* (Arrhenius-like) and *fragile* (non-Arrhenius-like) glass formers, where most polymers fall in the fragile category [100–103]. Figure 1.8 graphs what is now called an Angell plot (log τ_α versus T_g/T), where the fragility m directly represents the slope of the data as $T_g/T \rightarrow 1$. Physically, the fragility of a glass represents the material's resistance to undergoing the glass transition when subjected to a small temperature perturbation, a measure of how quickly the α-relaxation time slows down with temperature as T_g is approached from above. Whereas strong liquids with an Arrhenius temperature dependence can be somewhat reasonably interpreted as simply a very viscous liquid (a large yet temperature-independent activation energy), fragile liquids are extremely sensitive to small temperature changes near T_g, suggesting their dynamics are more complicated [17]. (Note that the terms "fragile" or "strong" do not relate to the brittleness or toughness of the material.)

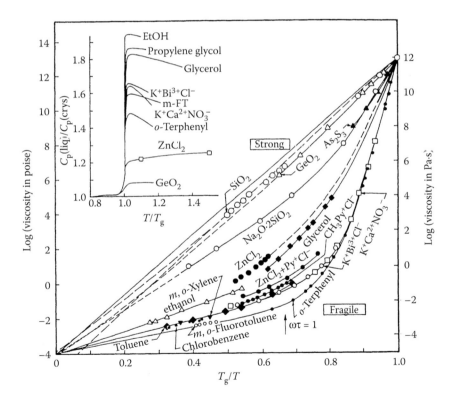

Figure 1.8 Angell plot (log η versus T_g/T) illustrating the concept of fragility m characterizing the slope of the log η or log τ_α versus $1/T$ behavior as T_g is approached for various glass-forming materials. *Strong* glass formers show an Arrhenius-like (linear) temperature dependence, while *fragile* glass formers, typical of most polymers, show a highly non-Arrhenius-like curvature indicative of the very fast slowing down of dynamical time scales with temperature. The inset shows the jump in heat capacity C_p at T_g, which is modest in comparison to the large change in dynamics occurring. (From C. A. Angell, *Science*, 267, 1924–1935, 1995. Reprinted with permission of AAAS.)

One can consider fragility and T_g to represent two different parameters characterizing the glass transition: T_g effectively representing the temperature when $\tau_\alpha = 100$ s, and the fragility m representing the local slope at that temperature. Historically, there have been several attempts to correlate fragility with T_g [100,103,104]; however, such comparisons have frequently indicated more variability than trend [103]. Recent work has found that fragility can be correlated with chemical structure in polymers, observing distinctions for polymers with stiff backbones versus bulky side groups [102,105]. Fragility has also received renewed interest in studies of thin films and confined systems with indications that fragility decreases upon confinement [106–112] (see Chapter 8).

1.2.3 DYNAMIC HETEROGENEITY

Another key signature of the supercooled regime is that the time evolution of relaxation functions, typically correlation functions of density such as the structure factor, have a final relaxation whose time dependence is slower than the typical single exponential decay most common in science [113,114]. These data are well fit to the Kohlrausch–Williams–Watts (KWW) function

$$\phi(t) = \phi_0 \exp\left[-(t/\tau)^{\beta_{KWW}}\right], \tag{1.8}$$

a stretched exponential equation where the exponent β_{KWW} is usually found to be less than one ($\beta_{KWW} \lesssim 0.5$) as T_g is approached. Mathematically, Equation 1.8 can be viewed as equivalent to a sum of single exponential decays:

$$\phi(t) \approx \sum_i \phi_i \exp\left[-t/\tau_i\right]$$

with a broad spectrum (distribution) of different relaxation times τ_i. Historically, there had been debate about whether the empirically observed macroscopic behavior of Equation 1.8 implied that microscopically the material relaxed homogeneously with all regions in the material relaxing in the same way according to Equation 1.8, or if the material was heterogeneous in nature with different regions each relaxing as a simple exponential decay with very different time scales τ_i [114]. We now know that this dynamical heterogeneity picture is correct [66,85], with experimental and computational evidence of dynamically fast and slow regions within the material [10,91,93,113]. Not only is the material spatially heterogeneous, but these dynamically distinct spatial clusters evolve in time whereby slow regions become fast regions, and vice versa. The β_{KWW} exponent (typically varying between 0.3 and 0.7) characterizes the breadth of the distribution of relaxation times, that is, the extent of the heterogeneity of the dynamics. Perhaps not surprisingly, more fragile liquids usually have a lower β_{KWW} exponent [115], providing some indication of the microscopic nature behind the non-Arrhenius temperature dependence.

The spatial and time-dependent heterogeneity of the dynamics forms one of the core aspects that make supercooled liquids and the glass transition phenomenon distinct from traditional homogeneous liquids [66]. Much of the properties and behaviors we associate with normal liquids are based on the assumption of homogeneous dynamics. For example, traditional correlations between transport coefficients such as translational and rotational diffusion coefficients or the so-called Stokes–Einstein relation between diffusion and viscosity, which are expected to scale in the same manner with temperature, are found to strongly deviate by 2–3 orders of magnitude on approaching T_g [113]. This translational–rotational decoupling of the dynamics is a manifestation of how broad distributions of spatially heterogeneous time scales are sampled by the two measures. Thus, violation of the Stokes–Einstein relation in supercooled liquids is an important consequence of dynamic heterogeneity and a key indication of the broad distribution of time scales (see also Chapter 9).

1.2.4 POTENTIAL ENERGY LANDSCAPE, AGING, AND REJUVENATION

Another important conceptual framework that has historically been used to understand the nature of the glass transition is energy landscapes. Potential energy landscapes (PELs) are representations of the potential energies of the system in configuration space where the glassy system is treated as evolving within this land-scape via activated dynamics, visualized as "hopping" over the various distributed energy barriers [116–119]. Formally, the potentially energy function can be calculated to determine the position of a computationally modeled system at any given point within this landscape. However, as this landscape is at its simplest a $3N$-dimensional surface for an N-body system, any visual representation of the PEL is typically done with a two-dimensional cartoon drawing, as shown in Figure 1.9. At high temperature, high above the PEL surface, the system is free to explore all possible configurations (the system is ergodic). As the system is cooled, its trajectory within the PEL eventually becomes hindered by the rugged energy barriers it must overcome. The various minima within the PEL are referred to as inherent states, with a local collection of such states sepa-rated by only small barriers referred to as metabasins [119]. Transitions between neighboring metabasins are referred to as α-relaxations, while the local transitions between inherent states within the same metabasin are referred to as β-relaxations. As the system is cooled further, now below T_g, eventually, it becomes trapped within a single metabasin. Physical aging, time-dependent structural relaxation, is then a further decrease of the system in the PEL within a single metabasin.

The PEL framework has been particularly informative for understanding how deformation alters the dynamics of glasses during plasticity and yielding [120–136]. The applied deformation, stress or strain, is typically treated as a modification to the thermally activated hopping events. Much of the work was initially informed by the idea of Eyring [137], where the height of the energy barrier in the direction of flow was treated as decreasing linearly with increasing stress, $E(\sigma) = E_0 - V_0\sigma$. The proportionality constant V_0 is referred to as the activation volume, thought of as the small region that needs to be "activated" for flow to occur [15,16]. One of the big issues driving this area has been to understand the extent to which deformation can act to "rejuve-nate" a glass (i.e., make it younger and in so doing, reset the aging clock) [35,120,124,125,131,135,138]. Work by Chen and Schweizer [122–125] in collaboration with experimental and computational efforts [127–132], as well as efforts by others [120,121,134–136], have found that initially, deformation *pre*-yield acts to "tilt" the landscape, hastening aging (driving the system deeper into the PEL), while later, deformation *post*-yield acts to raise the system within the PEL, giving the illusion of "rejuvenation." Although the mobility is temporarily increased by the application of stress pre-yield, the aging rate is found to return to the same value once the stress has been removed [129,131,136]. In contrast, deformations beyond yield alter the dynamics of the glass, narrowing the distribution of relaxation times [132,134,135] (see Chapters 10 and 11).

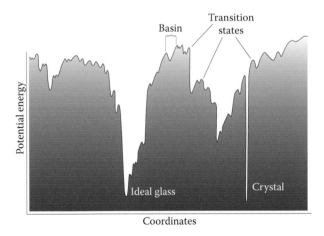

Figure 1.9 Cartoon illustration of a rugged potential energy landscape (PEL) characteristic of glass-forming materials. At high temperatures, the system is easily able to explore all possible configurations having enough thermal energy to overcome all energy barriers (the system is ergodic). As the temperature is lowered, the dynamics begin to slow down as the system can only explore the various minima (inherent states) and metaba-sins (local collection of minima) by activated dynamics, hopping over energy barriers. As the system is cooled further, it eventually becomes trapped within a single metabasin below T_g. (From P. G. Debenedetti and F. H. Stillinger, *Nature*, 410, 259–267, 2001. Reprinted with permission of Springer Nature.)

1.2.5 WHAT MAKES POLYMER GLASSES UNIQUE?

As described above, polymer glasses exhibit many of the same features that other nonequilibrium glasses possess, yet polymer glasses also manifest several differences unique to molecules with chain connectivity. Most obvious is that chain connectivity alters the nature of flow. During deformation, once the packing frustration associated with glass formation is overcome, which occurs with plastic yielding, glasses begin to "flow." For molecular glasses or short unentangled polymers, this is observed as sliding of the molecular units past each other, leading to fracture of the material. However, entangled polymers are able to undergo ductile yielding (plastic flow) whereby the material irreversibly deforms, but remains intact. For industrial applications, this has considerable advantages. The nature of how chain connectivity in polymers alters this behavior, along with the subsequent strain hardening, are areas that have seen a lot of advancement recently in our fundamental understanding [139–145] (see Chapters 12 and 13). Unlike the viscoelastic properties of polymer melt dynamics above T_g, which can be easily coarse-grained with bead–spring elasticity and a general friction factor, chemical structure plays a much more important role in the mechanical properties of glasses because packing and jamming of *inter*-chain segments often dominate or modify the effects of *intra*-chain connectivity. This likely explains one of the reasons why deformation of polymer glasses is so much harder to understand than the deformation of polymer melts, irrespective of the nonequilibrium nature of the material.

1.3 IMPORTANCE OF POLYMER GLASSES

The nature and properties of glasses are particularly important to polymers because polymers are very good glass formers. Their stereoirregularity and sluggish dynamics make polymers hard to crystallize. Even so-called crystalline polymers frequently have less than 60% crystalline content, making them formally semicrystalline with amorphous regions connecting the crystalline domains. In addition, many processing methods involve rapid cooling or solvent quench, leading to a higher prevalence for glass formation than crystallization. Thus, the nature and properties of polymer glasses are highly industrially relevant for many different applications.

The ease of processing polymers has led to their extensive use in industrial and technological applications. For these practical and industrial reasons, there has been a historical need to characterize and model property changes associated with complicated thermal histories and deformations of polymer glasses. To the extent that it has been possible, these efforts have been informed to varying degrees by conceptual or theoretical frameworks. However, the complexity of the system or behavior to be modeled has frequently led to heuristic approaches often in the form of constitutive equations (see Chapter 14). Until recently, the detailed scientific understanding at the microscopic level or the computational power to model the complexity accurately has been lacking. This is starting to change, and in recent years, microscopic- and molecular-level understanding has been brought to bear on increasingly more complicated systems with improved agreement between experiment and theoretical modeling. This book aims to provide a summary of the status of these efforts.

The main areas of focus in the study of polymer glasses have been their thermal and time-dependent behavior to confinement and deformation. Confinement refers to geometrically restricting the material to a small size such that its properties can become easily dominated by surface and interfacial interactions with neighboring components in the system. Deformation refers to the thermo-mechanical behavior exhibited by the material in response to externally applied stress or strains. The drive to study these areas is informed by both industrial needs and scientific relevance. For example, the theoretical expectation of a growing length scale associated with cooperative motion in glasses has prompted studies of glasses confined to small sample sizes, with the hope that restricting the material to a size comparable to this length scale would perturb the dynamics in such a way that would be informative. The ease of processing and industrial relevance of polymer thin films has influenced the prevalence of such studies to be done on this system. Similarly, the ease with which polymer glasses can be deformed and manipulated near their T_g has made them an excellent test bed for various scientific examinations of theoretical predictions. It is these types of features that this book explores.

The book is divided into three parts. Part 1 provides a summary and description of the status of experimental measures (Chapter 2), computational modeling approaches (Chapter 3), and the characteristic features exhibited by polymer glasses (Chapter 4). Part 2 addresses how the properties of polymer glasses change in thin films. Chapter 5 correlates changes in the glass transition and physical aging observed in thin polymer films, while Chapter 6 addresses the mechanical and viscoelastic properties. Studies using dielectric spectroscopy are summarized in Chapter 7. Theoretical considerations are covered with Chapter 8 focusing on molecular dynamics modeling of polymers in thin films and Chapter 9 considering efforts to model dynamics in polymer nanocomposites (another system with large surface-to-volume ratio). Part 3 addresses polymer glasses under deformation. How local segmental dynamics are accelerated by stress and strain, leading to yielding and ductile flow, is covered experimentally in Chapter 10 and theoretically in Chapter 11. Characteristics of larger deformations associated with yielding, ductile flow, and strain hardening are discussed in Chapter 12 from an experimental perspective and in Chapter 13 from a computational modeling approach. Chapter 14 ends the book with a detailed assessment of the successes and failures in modeling complicated polymer glass deformations. One of the main advantages in the scientific study of polymer glasses is that much of the experimental behavior has been historically characterized and cataloged, ripe for assessment of new theoretical ideas. It is the hope that this book will provide a springboard for stimulating new scientific advances into the nature of polymer glasses.

ACKNOWLEDGMENTS

The authors would like to thank Professors Eric Weeks and Jörg Baschnagel for useful and informative discussions. CBR gratefully acknowledges support from Emory University and the National Science Foundation, Division of Materials Research, Polymers Program in the form of a CAREER award (Grant No. DMR-1151646).

REFERENCES

1. M. Rubinstein and R. H. Colby, *Polymer Physics* (Oxford University Press, Oxford, UK), 2003.
2. P. C. Hiemenz and T. P. Lodge, *Polymer Chemistry* (CRC Press/Taylor & Francis Group, Boca Raton, FL), 2nd edn., 2007.
3. G. R. Strobl, *The Physics of Polymers: Concepts for Understanding Their Structures and Behavior* (Springer-Verlag, Berlin, Heidelberg), 3rd edn., 2007.
4. W. W. Graessley, *Polymeric Liquids & Networks: Structure and Properties* (Garland Science, Taylor & Francis Books, New York), 2004.
5. W. W. Graessley, *Polymeric Liquids & Networks: Dynamics and Rheology* (Garland Science, Taylor & Francis Books, New York), 2008.
6. M. Doi and S. F. Edwards, *The Theory of Polymer Dynamics* (Oxford University Press, Oxford, UK), 1986.
7. J. D. Ferry, *Viscoelastic Properties of Polymers* (John Wiley & Sons, New York), 3rd edn., 1980.
8. P. G. de Gennes, *Scaling Concepts in Polymer Physics* (Cornell University Press, Ithaca, NY), 1979.
9. E. Donth, *The Glass Transition: Relaxation Dynamics in Liquids and Disordered Materials* (Springer, Berlin, Heidelberg), 2001.
10. L. Berthier, G. Biroli, J. P. Bouchaud, L. Cipelletti, and W. van Saarloos, editors, *Dynamical Heterogeneities in Glasses, Colloids, and Granular Media* (Oxford University Press, Oxford, UK), 2011.
11. K. Binder and W. Kob, *Glassy Materials and Disordered Solids: An Introduction to Their Statistical Mechanics* (World Scientific Publishing, Singapore), 2011.
12. P. G. Wolynes and V. Lubchenko, editors, *Structural Glasses and Supercooled Liquids: Theory, Experiment, and Applications* (John Wiley & Sons, Hoboken, NJ), 2012.
13. A. J. Liu and S. R. Nagel, editors, *Jamming and Rheology: Constrained Dynamics on Microscopic and Macroscopic Scales* (Taylor & Francis, London), 2001.
14. R. N. Haward and R. J. Young, editors, *The Physics of Glassy Polymers* (Chapman & Hall, London), 2nd edn., 1997.

15. A. S. Argon, *The Physics of Deformation and Fracture of Polymers* (Cambridge University Press, Cambridge, UK), 2013.

16. I. M. Ward and J. Sweeney, *Mechanical Properties of Solid Polymers* (John Wiley & Sons, Ltd., West Sussex, UK), 3rd edn. 2013.

17. A. Cavagna, *Phys. Rep.*, 476, 51, 2009.

18. J. C. Dyre, *Rev. Mod. Phys.*, 78, 953, 2006.

19. L. Berthier and G. Biroli, *Rev. Mod. Phys.*, 83, 587, 2011.

20. M. D. Ediger, C. A. Angell, and S. R. Nagel, *J. Phys. Chem.*, 100, 13200, 1996.

21. C. A. Angell, K. L. Ngai, G. B. McKenna, P. F. McMillan, and S. W. Martin, *J. Appl. Phys.*, 88, 3113, 2000.

22. P. W. Anderson, *Science*, 267, 1615, 1995.

23. W. Gotze, *Complex Dynamics of Glass-Forming Liquids: A Mode-Coupling Theory* (Oxford University Press, Oxford, UK), 2009.

24. W. Gotze and L. Sjogren, *Rep. Prog. Phys.*, 55, 241, 1992.

25. S. P. Das, *Rev. Mod. Phys.*, 76, 785, 2004.

26. W. Gotze and L. Sjogren, *Transport Theory and Statistical Phys.*, 24, 801, 2006.

27. K. Schweizer, *J. Chem. Phys.*, 123, 244501, 2005.

28. S. Mirigian and K. S. Schweizer, *J. Phys. Chem. Lett.*, 4, 3648, 2013.

29. S. Mirigian and K. S. Schweizer, *J. Chem. Phys.*, 140, 194506, 2014.

30. G. L. Hunter and E. R. Weeks, *Rep. Prog. Phys.*, 75, 066501, 2012.

31. K. S. Schweizer, *Curr. Opin. Colloid Interface Sci.*, 12, 297, 2007.

32. R. E. Courtland and E. R. Weeks, *J. Phys. Condens. Matter*, 15, S359, 2003.

33. G. B. McKenna, *J. Non-Cryst. Solids*, 353, 3820, 2007.

34. X. Peng and G. B. McKenna, *Phys. Rev. E.*, 90, 050301, 2014.

35. G. B. McKenna, *J. Phys. Condens. Matter*, 15, S737, 2003.

36. V. N. Manoharan, M. T. Elsesser, and D. J. Pine, *Science*, 301, 483, 2003.

37. G. R. Yi, V. N. Manoharan, E. Michel, M. T. Elsesser, S. M. Yang, and D. J. Pine, *Adv. Mater.*, 16, 1204, 2004.

38. M. T. Elsesser, A. D. Hollingsworth, K. V. Edmond, and D. J. Pine, *Langmuir*, 27, 917, 2011.

39. J. Mattsson, H. M. Wyss, A. Fernandez-Nieves, K. Miyazaki, Z. Hu, D. R. Reichman, and D. A. Weitz, *Nature*, 462, 83, 2009.

40. A. J. Liu and S. R. Nagel, *Nature*, 396, 21, 1998.

41. A. J. Liu and S. R. Nagel, *Annu. Rev. Condens. Matter Phys.*, 1, 347, 2010.

42. H. M. Jaeger, S. R. Nagel, and R. P. Behringer, *Rev. Mod. Phys.*, 68, 1259, 1996.

43. M. Pica Ciamarra, M. Nicodemi, and A. Coniglio, *Soft Matter*, 6, 2871, 2010.

44. A. Ikeda, L. Berthier, and P. Sollich, *Phys. Rev. Lett.*, 109, 018301, 2012.

45. A. Ikeda, L. Berthier, and P. Sollich, *Soft Matter*, 9, 7669, 2013.

46. D. Bi, J. Zhang, B. Chakraborty, and R. P. Behringer, *Nature*, 480, 355, 2011.

47. H. J. Unidad, M. A. Goad, A. R. Bras, M. Zamponi, R. Faust, J. Allgaier, W. Pyckhout-Hintzen, A. Wischnewski, D. Richter, and L. J. Fetters, *Macromolecules*, 48, 6638, 2015.

48. T. G. Fox and P. J. Flory, *J. Polym. Sci.*, 14, 315, 1954.

49. P. G. Santangelo and C. M. Roland, *Macromolecules*, 31, 4581, 1998.

50. G. B. McKenna, Glass formation and glassy behavior, in *Comprehensive Polymer Science and Supplements, Vol. 2: Polymer Properties*, edited by C. Booth and C. Price (Pergamon Press, Oxford), 311–362, 1989.

51. Y. Ding, A. Kisliuk, and A. P. Sokolov, *Macromolecules*, 37, 161, 2004.

52. S. Mirigian and K. S. Schweizer, *Macromolecules*, 48, 1901, 2015.

53. L.-N. Zou, X. Cheng, M. L. Rivers, H. M. Jaeger, and S. R. Nagel, *Science*, 326, 408, 2009.

54. J. E. Mark, editor, *Physical Properties of Polymers Handbook* (AIP Press, New York), 1996.

55. D. J. Plazek, *J. Phys. Chem.*, 69, 3480, 1965.

56. K. L. Ngai, The glass transition and the glassy state, in *Physical Properties of Polymers*, edited by J. E. Mark, K. L. Ngai, W. W. Graessley, L. Mandelkern, E. Samulski, J. Koenig, and G. D. Wignall, 3rd edn. (Cambridge University Press, Cambridge, UK), 72–152, 2004.

57. G. B. McKenna, *Nat. Phys.*, 4, 673, 2008.

58. P. A. O'Connell and G. B. McKenna, *J. Chem. Phys.*, 110, 11054, 1999.

59. S. L. Simon, J. W. Sobieski, and D. J. Plazek, *Polymer*, 42, 2555, 2001.

60. J. Zhao, S. L. Simon, and G. B. McKenna, *Nat. Commun.*, 4, 1783, 2013.

61. C. A. Angell, *Science*, 267, 1924, 1995.

62. W. Kob, Supercooled liquids and glasses, in *Soft and Fragile Matter: Nonequilibrium Dynamics, Metastability and Flow*, edited by M. E. Cates and M. R. Evans (Scottish Universities Summer School in Physics & Institute of Physics Publishing, Bristol, Great Britain), 259–284, 2000.

63. Kenneth Chang, The nature of glass remains anything but clear, *The New York Times*, July 29, 2008.

64. P. Badrinarayanan, W. Zheng, Q. Li, and S. L. Simon, *J. Non-Cryst. Solids*, 353, 2603, 2007.

65. S. A. Kivelson and G. Tarjus, *Nat. Mater.*, 7, 831, 2008.

66. G. Biroli and J. P. Garrahan, *J. Chem. Phys.*, 138, 12A301, 2013.

67. W. Kauzmann, *Chem. Rev.*, 43, 219, 1948.

68. G. Tarjus, An overview of the theories of the glass transition, in *Dynamical Heterogeneities in Glasses, Colloids, and Granular Media*, edited by L. Berthier, G. Biroli, J. P. Bouchaud, L. Cipelletti, and W. van Saarloos (Oxford University Press, Oxford, UK), 39–67, 2011.

69. D. Huang, S. L. Simon, and G. B. McKenna, *J. Chem. Phys.*, 119, 3590, 2003.

70. Y. P. Koh and S. L. Simon, *Macromolecules*, 46, 5815, 2013.

71. Q. Li and S. L. Simon, *Polymer*, 47, 4781, 2006.

72. E. A. Baker, P. Rittigstein, J. M. Torkelson, and C. B. Roth, *J. Polym. Sci., Part B: Polym. Phys.*, 47, 2509, 2009.

73. K. L. Kearns, S. F. Swallen, M. D. Ediger, T. Wu, Y. Sun, and L. Yu, *J. Phys. Chem. B*, 112, 4934, 2008.

74. S. S. Dalal, Z. Fakhraai, and M. D. Ediger, *J. Phys. Chem. B*, 117, 15415, 2013.

75. S. F. Swallen, K. L. Kearns, M. K. Mapes, Y. S. Kim, R. J. McMahon, M. D. Ediger, T. Wu, L. Yu, and S. Satija, *Science*, 315, 353, 2007.

76. S. Singh, M. D. Ediger, and J. J. de Pablo, *Nat. Mater.*, 12, 139, 2013.

77. L. C. E. Struik, *Physical Aging in Amorphous Polymers and Other Materials* (Elsevier Scientific Publishing Company, Amsterdam), 1978.

78. G. B. McKenna, Physical aging in glasses and composites, in *Long-Term Durability of Polymeric Matrix Composites*, edited by K. V. Pochiraju, G. P. Tandon, and G. A. Schoeppner (Springer US, Boston, MA), 237–309, 2012.

79. J. M. Hutchinson, *Prog. Polym. Sci.*, 20, 703, 1995.

80. R. Greiner and F. R. Schwarzl, *Rheol. Acta*, 23, 378, 1984.

81. I. M. Hodge, *J. Non-Cryst. Solids*, 169, 211, 1994.

82. P. Badrinarayanan and S. L. Simon, *Polymer*, 48, 1464, 2007.

83. T. Hecksher, A. I. Nielsen, N. B. Olsen, and J. C. Dyre, *Nat. Phys.*, 4, 737, 2008.

84. V. N. Novikov and A. P. Sokolov, *Phys. Rev. E*, 67, 031507, 2003.

85. R. Richert, Supercooled liquid dynamics: Advances and challenges, in *Structural Glasses and Supercooled Liquids: Theory, Experiment, and Applications*, edited by P. G. Wolynes and V. Lubchenko (John Wiley & Sons, Hoboken, NJ), 1–30, 2012.

86. K. Chen, E. J. Saltzman, and K. S. Schweizer, *J. Phys. Condens. Matter*, 21, 503101, 2009.

87. F. W. Starr, J. F. Douglas, and S. Sastry, *J. Chem. Phys.*, 138, 12A541, 2013.

88. B. A. Pazmiño Betancourt, J. F. Douglas, and F. W. Starr, *J. Chem. Phys.*, 140, 204509, 2014.

89. G. Adam and J. H. Gibbs, *J. Chem. Phys.*, 43, 139, 1965.

90. C. Donati, J. Douglas, W. Kob, S. Plimpton, P. Poole, and S. Glotzer, *Phys. Rev. Lett.*, 80, 2338, 1998.

91. S. C. Glotzer, *J. Non-Cryst. Solids*, 274, 342, 2000.

92. A. H. Marcus, J. Schofield, and S. A. Rice, *Phys. Rev. E*, 60, 5725, 1999.

93. E. R. Weeks, J. C. Crocker, A. C. Levitt, A. Schofield, and D. A. Weitz, *Science*, 287, 627, 2000.

94. E. R. Weeks and D. A. Weitz, *Phys. Rev. Lett.*, 89, 095704, 2002.

95. A. Shavit, J. F. Douglas, and R. A. Riggleman, *J. Chem. Phys.*, 138, 12A528, 2013.

96. J. D. Stevenson, J. O. R. Schmalian, and P. G. Wolynes, *Nat. Phys.*, 2, 268, 2006.

97. K. H. Nagamanasa, S. Gokhale, A. K. Sood, and R. Ganapathy, *Nat. Phys.*, 11, 403, 2015.

98. B. A. Pazmiño Betancourt, P. Z. Hanakata, F. W. Starr, and J. F. Douglas, *Proc. Natl. Acad. Sci. USA*, 112, 2966, 2015.
99. C. A. Angell, *J. Non-Cryst. Solids*, 73, 1, 1985.
100. A. P. Sokolov, V. N. Novikov, and Y. Ding, *J. Phys. Condens. Matter*, 19, 205116, 2007.
101. L. Hong, V. N. Novikov, and A. P. Sokolov, *J. Non-Cryst. Solids*, 357, 351, 2011.
102. K. Kunal, C. G. Robertson, S. Pawlus, S. F. Hahn, and A. P. Sokolov, *Macromolecules*, 41, 7232, 2008.
103. Q. Qin and G. B. McKenna, *J. Non-Cryst. Solids*, 352, 2977, 2006.
104. L. Hong, P. D. Gujrati, V. N. Novikov, and A. P. Sokolov, *J. Chem. Phys.*, 131, 194511, 2009.
105. K. F. Freed, *Acc. Chem. Res.*, 44, 194, 2011.
106. P. Z. Hanakata, J. F. Douglas, and F. W. Starr, *J. Chem. Phys.*, 137, 244901, 2012.
107. P. Z. Hanakata, J. F. Douglas, and F. W. Starr, *Nat. Commun.*, 5, 4163, 2014.
108. P. Z. Hanakata, B. A. Pazmiño Betancourt, J. F. Douglas, and F. W. Starr, *J. Chem. Phys.*, 142, 234907, 2015.
109. C. M. Evans, H. Deng, W. F. Jager, and J. M. Torkelson, *Macromolecules*, 46, 6091, 2013.
110. T. Lan and J. M. Torkelson, *Macromolecules*, 49, 1331, 2016.
111. C. Zhang and R. D. Priestley, *Soft Matter*, 9, 7076, 2013.
112. C. Zhang, Y. Guo, K. B. Shepard, and R. D. Priestley, *J. Phys. Chem. Lett.*, 4, 431, 2013.
113. M. D. Ediger, *Annu. Rev. Phys. Chem.*, 51, 99, 2000.
114. R. Richert, *J. Phys. Condens. Matter*, 14, R703, 2002.
115. R. Böhmer, K. L. Ngai, C. A. Angell, and D. J. Plazek, *J. Chem. Phys.*, 99, 4201, 1993.
116. F. H. Stillinger, *Science*, 267, 1935, 1995.
117. F. H. Stillinger, *Phys. Rev. B*, 41, 2409, 1990.
118. P. G. Debenedetti and F. H. Stillinger, *Nature*, 410, 259, 2001.
119. A. Heuer, *J. Phys. Condens. Matter*, 20, 373101, 2008.
120. D. J. Lacks and M. J. Osborne, *Phys. Rev. Lett.*, 93, 255501, 2004.
121. Y. G. Chung and D. J. Lacks, *Macromolecules*, 45, 4416, 2012.
122. K. Chen and K. S. Schweizer, *Europhys. Lett.*, 79, 26006, 2007.
123. K. Chen and K. S. Schweizer, *Macromolecules*, 41, 5908, 2008.
124. K. Chen and K. S. Schweizer, *Macromolecules*, 44, 3988, 2011.
125. K. Chen and K. S. Schweizer, *Phys. Rev. E*, 82, 041804, 2010.
126. K. Chen, E. J. Saltzman, and K. S. Schweizer, *Annu. Rev. Condens. Mater. Phys.*, 1, 277, 2010.
127. R. A. Riggleman, H. N. Lee, M. D. Ediger, and J. J. de Pablo, *Phys. Rev. Lett.*, 99, 215501, 2007.
128. R. A. Riggleman, K. S. Schweizer, and J. J. de Pablo, *Macromolecules*, 41, 4969, 2008.
129. H. N. Lee, K. Paeng, S. F. Swallen, and M. D. Ediger, *Science*, 323, 231, 2009.
130. H.-N. Lee, K. Paeng, S. F. Swallen, M. D. Ediger, R. A. Stamm, G. A. Medvedev, and J. M. Caruthers, *J. Polym. Sci., Part B: Polym. Phys.*, 47, 1713, 2009.
131. H. N. Lee and M. D. Ediger, *Macromolecules*, 43, 5863, 2010.
132. R. A. Riggleman, H.-N. Lee, M. D. Ediger, and J. J. de Pablo, *Soft Matter*, 6, 287, 2010.
133. K. Hebert, B. Bending, J. Ricci, and M. D. Ediger, *Macromolecules*, 48, 6736, 2015.
134. M. Warren and J. Rottler, *Phys. Rev. Lett.*, 104, 205501, 2010.
135. M. Warren and J. Rottler, *J. Chem. Phys.*, 133, 164513, 2010.
136. A. Y. H. Liu and J. Rottler, *Soft Matter*, 6, 4858, 2010.
137. H. Eyring, *J. Chem. Phys.*, 4, 283, 1936.
138. Y. G. Chung and D. J. Lacks, *J. Chem. Phys.*, 136, 124907, 2012.
139. S.-Q. Wang, S. Cheng, P. Lin, and X. Li, *J. Chem. Phys.*, 141, 094905, 2014.
140. S.-Q. Wang, *Soft Matter*, 11, 1454, 2015.
141. S. Cheng and S.-Q. Wang, *Macromolecules*, 47, 3661, 2014.
142. G. D. Zartman, S. Cheng, X. Li, F. Lin, M. L. Becker, and S.-Q. Wang, *Macromolecules*, 45, 6719, 2012.
143. R. S. Hoy, *J. Polym. Sci., Part B: Polym. Phys.*, 49, 979, 2011.
144. R. S. Hoy and M. O. Robbins, *Phys. Rev. Lett.*, 99, 117801, 2007.
145. R. S. Hoy and C. S. O'Hern, *Phys. Rev. E*, 82, 041803, 2010.

Structural recovery and physical aging of polymeric glasses

SINDEE L. SIMON AND GREGORY B. McKENNA

2.1 INTRODUCTION

Amorphous polymeric materials are often used in the so-called glassy state, as is the case for polystyrene, polycarbonate, and poly(methyl methacrylate), as well as for epoxy and polycyanurate resins in composite systems. For such materials, the structural recovery and physical aging behavior is reasonably well understood and can be treated in the context of glass-forming materials as described below. On the other hand, for semicrystalline polymers, such as nylon or PEEK, both glassy amorphous domains and reinforcing crystalline domains are present. In addition, the amorphous phase in the neighborhood of the crystallites may be modified, leading to "composite" material behavior that can be quite complex. Furthermore, even "simple" composites exhibit complexity in their physical aging responses that is not fully understood. In the following sections, we first describe the important aspects of the glassy behavior as related to physical aging in amorphous materials, and then we discuss the physical aging process and its measurement in amorphous, glassy polymers. We then move on to reinforced amorphous materials and semicrystalline polymers. Finally, because materials are frequently used at stresses or deformations above the linear viscoelastic limit, we briefly discuss the aging behavior of materials in the nonlinear regime up to the point of yield. We end with brief conclusions.

2.2 GLASS FORMATION AND STRUCTURAL RECOVERY

One important aspect of glassy materials in terms of their long-term behavior is the fact that the glass is a nonequilibrium material. Furthermore, of special import in the case of engineering applications of glassy polymers is that they are used at a high fraction of their glass transition temperatures. Hence, this nonequilibrium state of the material can have important consequences on the behavior. In this section, we describe the phenomenon of structural recovery, which is the change of the thermodynamic properties of the glass as a function of the thermal history. More specifically, structural recovery is termed volume recovery and enthalpy recovery, respectively, when discussing the evolution of those particular properties. The changes in thermodynamic properties that take place during structural recovery are related to the changes in, for example, mechanical properties, such as the viscoelastic response, that are involved in the physical aging process.

Figure 2.1 shows a schematic diagram of the glass transition phenomenon as it relates to the volume or enthalpy of the glass-forming material [1–4]. The diagram should be considered by starting at high temperature and following the volume or enthalpy along the solid curve as one cools the sample. If the material is crystallizable, at point T_m, one observes a first-order phase transition accompanied by abrupt changes in material density and enthalpy. If crystallization is avoided (in a crystallizable material), the sample becomes supercooled and continues on the solid line with a liquid-like volume and enthalpy; for a sample that cannot crystallize (as is the case for many polymers), the liquid line is, in fact, the equilibrium line. For such a sample, as one continues down the cooling curve, at some point, the rate of cooling becomes similar to the rate of molecular movement and the material does not "relax" into equilibrium and the volume begins to deviate from the liquid volume. This point is related to the glass transition temperature T_g, an important characterizing parameter for amorphous materials, which can be defined in several ways [5–7], but in volume or enthalpy space, the value of the transition is generally taken to be defined as the temperature at which the extrapolated glass and liquid equilibrium lines cross. As one continues cooling along the solid line, the material is in the glassy state. If one now stops cooling at an aging temperature, T_a, and anneals the sample, the fact that it is not in equilibrium leads to a spontaneous evolution of volume and enthalpy toward their equilibrium values, known as structural recovery or structural relaxation [1–4,8–10]. Owing to the relaxation that can occur in the glassy state, T_g is correctly measured only on cooling from the equilibrium state [5–7].

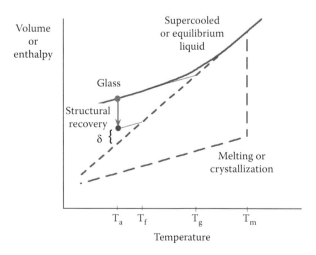

Figure 2.1 Schematic of the volume or enthalpy versus temperature for glass-forming materials.

The glass transition and structural recovery are manifestations of the same underlying kinetics, and these kinetics show both a strong temperature dependence and nonlinear behavior. The temperature dependence of the kinetics results in T_g depending strongly on the cooling rate, whereas nonlinearity arises from the fact that the rate of relaxation depends on the structure of the glass. In the equilibrium state, it is well known that the relaxation time depends only on two nonconjugate state variables, for example, temperature and pressure, but in the glassy state, it also depends on the structure of the glass, and perhaps even the full thermal history, as will be discussed later. Two quantities are generally used to describe the structure of the glass—the departure from equilibrium δ and the fictive temperature T_f. Both quantities are shown in Figure 2.1 for the glass structure midway between the glass and equilibrium lines demarcated by a dark gray point. The volumetric departure from equilibrium is the difference between a point in the glassy state and the extrapolated equilibrium line normalized by the equilibrium volume for the volume departure from equilibrium, that is, $\delta(t) = (v(t) - v_\infty)/v_\infty$, where v is the volume of the glass of interest and v_∞ is the equilibrium volume at the aging temperature (or the volume on the supercooled liquid line for a crystallizable material). The fictive temperature, on the other hand, is the temperature at which a given glass if heated or cooled along the glassy line would sit on the extrapolated equilibrium line, and it is thought of as the temperature corresponding to the liquid structure "frozen" into the glassy state.

In the 1960s, A. J. Kovacs [1] used volume dilatometry to catalog three signatures of structural recovery: the intrinsic isotherm, asymmetry of approach, and memory effect experiments. We describe these next as they provide insights into the nonlinearity and nonexponentiality of the structural recovery process and the impact of temperature history or memory that are key to modeling the phenomenon.

2.2.1 INTRINSIC ISOTHERMS

In the intrinsic isotherm experiment, the material is equilibrated at T_o, near or above the nominal glass transition temperature, and then the sample is "jumped" rapidly to a lower aging temperature in the glassy state where the volume recovery response is monitored as a function of time after the temperature jump. A family of intrinsic isotherms is created by performing the recovery experiment at different aging temperatures. Figure 2.2 shows schematically the volume–temperature surface corresponding to the intrinsic isotherm experiment. Figure 2.3 shows a set of intrinsic isotherms for a polystyrene [11] as a plot of the volume departure from equilibrium δ versus the logarithm of the aging time t after the quench. The total changes in volume are small, less than 1%. Note also that the curves at temperatures near polystyrene's nominal glass transition temperature of 100°C take little time to evolve into equilibrium, whereas even 10 K below

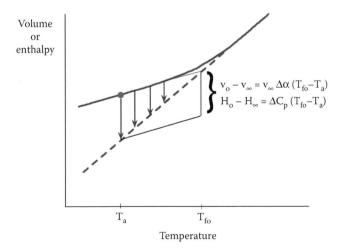

$$v_o - v_\infty = v_\infty \, \Delta\alpha \, (T_{fo} - T_a)$$
$$H_o - H_\infty = \Delta C_p \, (T_{fo} - T_a)$$

Figure 2.2 Schematic of the volume or enthalpy temperature paths used in the intrinsic isotherm experiment. Four different aging temperatures are shown. The total changes in volume and enthalpy are given for the lowest T_a, which is marked as such on the x-axis.

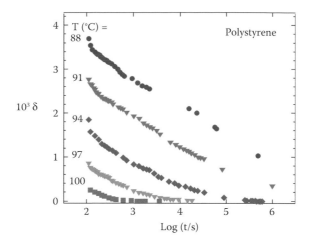

Figure 2.3 Intrinsic isotherms for a polystyrene. (After S. L. Simon, J. W. Sobieski, and D. J. Plazek, *Polymer*, 42 (6), 2555–2567, 2001.)

T_g, the time required to reach equilibrium is more than 1 week. Thus, one anticipates that the times required to achieve equilibrium far below the glass transition temperature will grow exponentially in spite of the fact that it appears that the equilibrium Williams–Landel–Ferry (WLF) temperature dependence [12] turns over to a more gentle temperature dependence below T_g [11,13–19].

In the ideal case, when the quench rate is fast enough and T_o is low enough, the initial fictive temperature T_{fo} for all of the isotherms is the same and equal to the initial temperature T_o. However, for finite quenches, T_{fo} can be less than T_o and equal to the glass transition temperature associated with the cooling rate used. For example, for the experiments shown in Figure 2.3, although the dilatometer was quickly transferred from an oil bath maintained at 104°C to a second bath at the aging temperature (similar to how Kovacs performed his seminal experiments), the temperature profile in the sample and its mean cooling rate depends on the aging temperature due to the relatively large size of the dilatometric samples, ranging from approximately 2 to 20 K/min through the glass-forming region for the highest and lowest aging temperatures, respectively.

Moreover, owing to the breadth of the transition, the fictive temperature decreases from the point of vitrification where $T_f = T$ to $T_f = T_{fo}$, as can be ascertained from Figure 2.2. This issue is important for understanding how the magnitude of the volume or enthalpy change is expected to change with aging temperature. Based on simple geometric arguments, the total changes in the volume and the enthalpy during structural recovery are given by

$$v_o - v_\infty = v_\infty \Delta\alpha(T_{fo} - T_a) \tag{2.1}$$

$$H_o - H_\infty = \Delta C_p(T_{fo} - T_a) \tag{2.2}$$

where subscript o and ∞ indicate the values initially and at equilibrium. In these equations, the step changes in the thermal expansion coefficient and heat capacity at T_g ($\Delta\alpha = \alpha_l - \alpha_g$ and $\Delta C_p = C_{pl} - C_{pg}$, where subscripts l and g indicate the properties in equilibrium and glassy states, respectively) are assumed to be constant; if they are not, the step change in α or C_p must be integrated over fictive temperature from T_{fo} to T_a. Appropriate application of Equations 2.1 and 2.2 in the glass transition region requires understanding that T_{fo} is not a constant in this region for finite cooling rates [20,21]. In addition, T_{fo} may depend on the temperature history for large samples where thermal gradients are present. Unfortunately, a number of authors have suggested that the extrapolated liquid line is not reached at equilibrium [22–29], based, in part, on incorrect assumptions regarding the value of T_{fo} and without accounting for uncertainty in the values for the step change in heat capacity at T_g. Importantly, all of the works suggesting that the equilibrium line is not reached deal with enthalpy measurements, which in contrast to volume measurements, are not absolute. For example, in an early work by Kovacs [30], it was confirmed that the volume reached at equilibrium was on the extrapolated liquid line.

2.2.2 ASYMMETRY OF APPROACH

The asymmetry of approach experiment is one of the great interest in the structural recovery catalog of Kovacs because it clearly demonstrates the nonlinear nature of the structural recovery process. The experiment consists of performing a down-jump and an up-jump of the same magnitude, as shown in Figure 2.4. If the responses were linear, one would find that the structural recovery at the aging temperature would be mirror images for the two jumps. Figure 2.5 shows clearly for a polycarbonate glass [6,31,32] that the responses are asymmetric with the down-jump experiments exhibiting much smaller departures from equilibrium at the shortest measured times and arriving into equilibrium faster than in the case of the up-jump experiments. The current understanding of the nonlinearity is that it arises from a dependence of the relaxation time on the departure from equilibrium with the positive departures (down-jump) having more

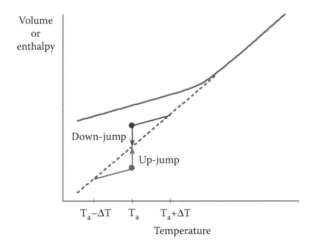

Figure 2.4 Schematic of the volume and enthalpy temperature paths used in the asymmetry of approach experiment.

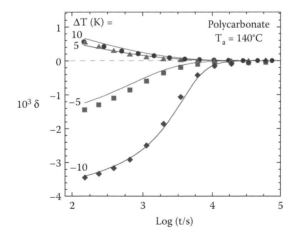

Figure 2.5 Asymmetry of approach results for a polycarbonate glass. (Data from C. R. Schultheisz and G. B. McKenna, *Proceedings North American Thermal Analysis Society, 25th Annual Meeting*, 366–373, 1997.)

rapid relaxations than do the negative departures (up-jump) [1–4,6]. The concept is frequently referred to as time–structure superposition, implying that the molecular mobility shifts with changing structure in the same fashion as one would consider time–temperature superposition as being due to changing mobility due to change in temperature. In both time–temperature and time–structure superposition, the shape of the response function is assumed to not change such that all relaxation processes are affected the same amount by either temperature or structure.

2.2.3 MEMORY EFFECT

The third important signature of structural recovery cataloged by Kovacs is the memory effect. In this case, the response is measured after experiments in which a sample is subjected to two sequential temperature steps. The first temperature step is a down-jump from equilibrium density (or enthalpy) in which the sample is allowed to undergo partial recovery toward equilibrium. The second temperature step is an up-jump to a point such that the sample is approximately on the extrapolated liquid line and the departure from equilibrium is (near to) zero. This is shown schematically in Figure 2.6. If the recovery response could be described as a single mechanism (single relaxation time process), the fact of being in equilibrium would lead to no volume (or enthalpy) change after the second step in temperature. However, as shown in Figure 2.7 for several different memory experiments for polystyrene [33], the behavior is very rich with the departure from equilibrium showing a nonmonotonic response, first increasing away from equilibrium to a maximum and then approaching equilibrium at a rate that is similar to the direct down-jump to T_a. This memory effect is a typical "viscoelastic" type of response and is accounted for in the current models of structural recovery using Boltzmann superposition and a relaxation function having nonexponential behavior, either through a distribution of relaxation times or through a stretched Kohlrausch [34]–Williams and Watts [35] (KWW) exponential function.

2.2.4 SIGNATURES OF STRUCTURAL RECOVERY IN ENTHALPY SPACE

The three signatures of structural recovery, the intrinsic isotherms, asymmetry of approach, and memory effect, were shown in Figures 2.3, 2.5, and 2.7 for volume measurements. Similar signatures are expected for enthalpy but the enthalpy experiments, which are typically accomplished using differential scanning calorimetry (DSC), are much more time consuming to perform. The reason for this is that volume is an absolute quantity and its value can be measured in situ and continuously as a function of time for any given thermal history. On the other hand, enthalpy is a relative quantity and the enthalpy changes associated with volume

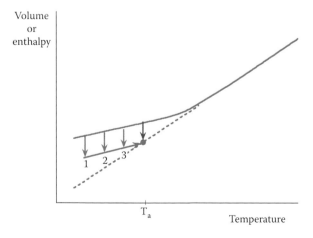

Figure 2.6 Schematic of the volume and enthalpy temperature paths used in the memory experiment. The arrows above points 1, 2, and 3 indicate three two-step histories, whereas the arrow at T_a is a conventional down-jump.

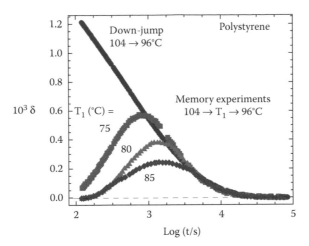

Figure 2.7 Memory response to three two-step temperature histories for a polystyrene and the associated down-jump. (After P. Bernazzani and S. L. Simon, *J. Non-Cryst. Solids*, 307–310C, 470–480, 2002.)

recovery are quite small. Thus, although adiabatic calorimeters are sensitive enough to measure the heat flow (i.e., the time derivative of the enthalpy) directly during structural recovery over a limited range of aging temperatures near T_g, DSC can only be used to measure the change in enthalpy of a given aged sample after aging. The procedure generally used is to compare two heating scans, one for the aged glass and one for an unaged glass, with the difference in areas under the curves being related to the enthalpy released during aging, as schematically shown in Figure 2.8 and first demonstrated by Petrie [36]. Since heating to above T_g erases the previous thermal history, a new experiment must be performed for every aging time in contrast to the volumetric experiments where volume is obtained in situ as a function of time. In addition, the enthalpy change is obtained by subtracting two heating scans and integrating the difference—small differences in the baseline will lead to a lack of superposition of the liquid and glass heat capacity (or heat flow) and will lead to errors. Thus, enthalpy recovery data have considerably more error in them than volume recovery data. Typical intrinsic isotherms are shown for a polystyrene [11,21,37] in Figure 2.9 and such data have been obtained for a number of polymer and nonpolymeric glass formers. On the other hand, enthalpic up-jump and memory experiments have not been performed often because of the long aging times generally required to either equilibrate at low temperature prior to an up-jump or at T_1 for the memory effect. Oguni successfully performed enthalpic up-jumps using adiabatic calorimetry (where measurements can be performed in situ)

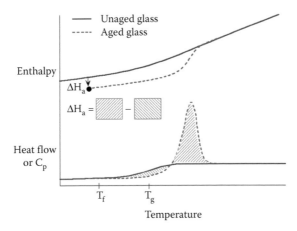

Figure 2.8 Schematic of how the change in enthalpy is determined from two DSC temperature scans, that is, for an aged and unaged glass.

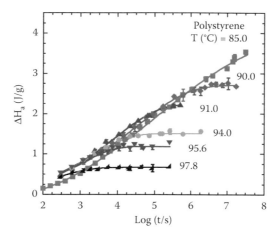

Figure 2.9 Enthalpic intrinsic isotherms for a polystyrene from DSC; lines are only a guide to the eye. (Data after Y. P. Koh and S. L. Simon, *Macromolecules*, 46 (14), 5815–5821, 2013.)

for low-molecular-weight glasses and small temperature jumps [38,39], and Boucher et al. [29] performed up-jumps for several polymers with DSC, although they did not perform the true asymmetry of approach experiment as they did not ensure that the up-jumps started from equilibrium. As for the enthalpic memory effect experiment, only two groups, to the best of our knowledge, have conducted such two-step temperature history experiments prior to our recent work: Adachi and Kotaka for polystyrene using adiabatic calorimetry [40] and Johari and coworkers for poly(methyl methacrylate) and several other glasses using conventional DSC [41].

In our own laboratory, we have recently exploited a rapid scanning nanocalorimeter to measure the enthalpy recovery of single polystyrene thin films [42–44], including all three signatures of structural recovery. The advantages of nanocalorimetry include high cooling rates and fast response times: the high cooling rates allow the investigation of enthalpy recovery at high aging temperatures, which lead to significantly shorter times required to reach equilibrium, whereas the fast response times allow the aging effects to be discerned at times as short as 0.01 s, allowing the full enthalpy recovery to be observed after a down-jump—the initial plateau at short aging times, the linear region where enthalpy changes approximately linearly with the logarithm of time, and the final plateau at long times when the material has achieved equilibrium. The initial plateau, which occurs when the aging time is shorter than the time scales associated with structural relaxation, is particularly interesting as it is related to the nonlinearity of the relaxation process [45]. However, although the induction or stabilization period had been described by others [1,46,47], its usefulness was deemed unclear [48] because the induction time could be obtained over only a limited range of temperatures deep in the glassy state. Exploiting the high cooling rates and short aging times of nanocalorimetry, however, has allowed the measurement of the induction time over a wide range of aging temperatures [44], as shown in Figure 2.10 for relaxation of polystyrene at 50.5 and 100.5°C. Here, the entire relaxation response can be observed at the higher aging temperature with the fictive temperature remaining constant for the initial half decade in time (at 117°C due to the 1000 K/s cooling rate), then decreasing approximately linearly with logarithmic time, and finally leveling off at equilibrium with $T_f = T_a = 100.5°C$. In comparison, for the aging temperature of 50.5°C, the induction time is nearly 30 times longer, and equilibrium is not reached within the time scale of the measurements. The inset shows that the induction time increases with decreasing aging temperature with an apparent activation energy of 108 kJ/mol.

In addition to the entire enthalpic response to a down-jump, the asymmetry of approach and memory effect measurements become much more feasible with nanocalorimetry, and we have successfully performed both of these experiments in enthalpy space [43]. As an example, the enthalpic memory effect is

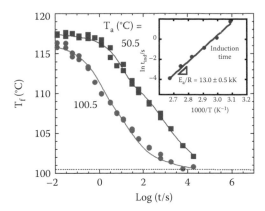

Figure 2.10 Enthalpic intrinsic isotherms at 50.5°C and 100.5°C for a polystyrene from Flash DSC. The inset shows an Arrhenius plot of the dependence of the induction time t_{ind} as a function of reciprocal temperature. (After Y. P. Koh, S. Y. Gao, and S. L. Simon, *Polymer*, 96, 182–187, 2016.)

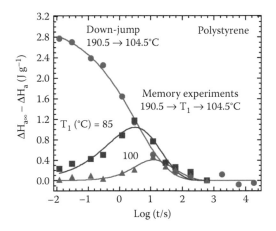

Figure 2.11 Enthalpic memory effect, plotted as the enthalpy departure from equilibrium for a polystyrene. (After E. Lopez and S. L. Simon, *Macromolecules*, 49, 2365–2374, 2016.)

shown in Figure 2.11 for a polystyrene at an aging temperature of 104.5°C, some 4.5 K above the nominal T_g. To give the reader an idea of the time gained by performing the experiments at a higher aging temperature, we can compare the time that the sample was held at $T_1 = 85$°C for the volumetric memory effect shown in Figure 2.7 where the final aging temperature T_a was 96°C relative to the times at T_1 for the enthalpic data for $T_a = 104.5$°C. We should recall that the sample is held at T_1 until the fictive temperature equals the final aging temperature, that is, such that the sample would be in equilibrium after jumping up to T_a. When initially aging at $T_1 = 85$°C, the time to reach $T_f = 96$°C was 30 h, whereas the time to reach $T_f = 104.5$°C was only 13 min (0.22 h), nearly 1000 times shorter. Similarly, the peak in the memory effect is shifted two or three decades to shorter times for $T_a = 104.5$°C, as is the time required to reach equilibrium. Thus, nanocalorimetry offers considerable promise for obtaining not only all of the signatures of enthalpy recovery in reasonable experimental times but also for performing other types of enthalpy recovery experiments.

2.2.5 τ_{eff}-PARADOX

An important aspect of the signatures of structural recovery was demonstrated by Kovacs [1] in comparing the down-jump response to the up-jump response, in addition to the asymmetry of approach behavior, is the so-called τ_{eff}-paradox. Here, Kovacs defined an effective retardation (relaxation) time for the structural recovery as

$$\tau_{eff}^{-1} = \left(\frac{-1}{\delta}\right)\left(\frac{d\delta}{dt}\right) \tag{2.3}$$

where δ is the departure from equilibrium as defined above and t is the time after the up- or down-jump in temperature. We remark that for a single exponential relaxation process, τ_{eff} is a constant, but this is not the case for relaxation in the glassy state. In addition to τ_{eff} evolving during structural recovery, as shown in Figure 2.12, Kovacs also reported that, for positive values of δ, the data for τ_{eff}^{-1} merge to a single value for each temperature of the isotherm, independent of the initial temperature of the jump. On the other hand, the data for the up-jump condition, that is, for negative values of δ, show an apparent dependence of the final value of τ_{eff}^{-1} on the initial temperature. Since the relaxation time in equilibrium, that is, when $\delta = 0$, must be independent of the path to equilibrium, Kovacs referred to this as a paradox [1]. While there has been some controversy [49] over the existence of the paradox, an extensive statistical analysis of the data by McKenna et al. [50] supports the contention that for the specific set of data for up-jumps to 40°C, the data are real. In addition, data from Kolla and Simon [51] for an epoxy sample were not only consistent with McKenna's reanalysis [50] of Kovacs' data [1], but showed the existence of the expansion gap for smaller temperature jumps and to smaller δ values due to the higher resolution of their volume measurements, coupled with taking the data linearly in time (thereby reducing the error in τ_{eff}^{-1} since that value depends on the derivative $d\delta/dt$). Why this is important stems not so much from the paradox, which at extremely long times must and does go away [51,52], but on the observed "expansion gap" and the way that the data in the up-jump experiment approach equilibrium. In fact, these and other experiments indicate that the relationship between the relaxation time and the glassy structure is nonlinear [33,53], and there is some evidence that it may depend on thermal history [53–55]. The phenomenological models of structural recovery, described in the next section, cannot describe this behavior unless a strong nonlinear dependence of the relaxation time on structure is incorporated [56], and newer models need to be developed, with those from the Caruthers group being important contributions and described in some detail in Chapters 4 and 14 of this book.

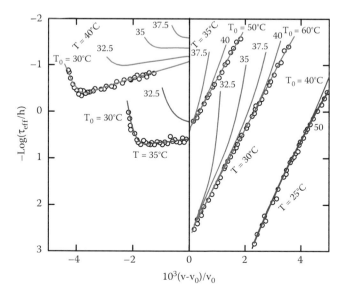

Figure 2.12 τ_{eff}^{-1} figure for poly(vinyl acetate) glass former showing the expansion gap for up-jump conditions. (Data from A. J. Kovacs, *Fortschr. Hochpolym.-Forsch.*, 3, 394–507, 1963.)

2.2.6 MODELS OF STRUCTURAL RECOVERY

The two major models of structural recovery that are widely used in the field are the Tool [57]–Narayanaswamy [58]–Moynihan [59] (TNM) and the Kovacs, Aklonis, Hutchinson, Ramos [60] (KAHR) models. These empirical models are phenomenologically equivalent although there are some differences in the way that the underlying physics are perceived. Both deal only with the kinetics associated with the glass transition and the structural recovery, as do recent modifications [56] of the models. A more general model based on an extension of the rational mechanics framework of Coleman and Noll [61–63] will also be briefly discussed.

The KAHR and TNM models have many similarities and can be made equivalent from a phenomenological perspective. The ingredients of both models arise from the essentials mandated by the observations cataloged above and include nonlinearity, that is, the molecular mobility or relaxation times depend on the volume or enthalpy of the glass at a given temperature and a given structure, nonexponentiality, or a distribution of relaxation times, and memory or a dependence on previous thermal history. In the case of the KAHR model, the nonlinearity is incorporated into the model by allowing the relaxation times to depend on the departure from equilibrium, whereas in the case of the TNM model, they depend on the fictive temperature. The nonexponential decay is accounted for using a sum of exponentials for the response function in the KAHR model and a stretched exponential Kohlrausch–Williams, Watts or KWW [34,35] function in the TNM model. Finally, in both models, the thermal history or memory is accounted for using a Boltzmann [64] superposition for the glassy structure as a function of temperature history. When we combine these ingredients, we have for the KAHR model for volume recovery for the case of constant $\Delta\alpha$:

$$\delta(z) = -\Delta\alpha \int_0^z R(z - z') \frac{dT}{dz'} dz' \tag{2.4}$$

where $\delta(z)$ is the departure from equilibrium defined above, and the KAHR response function is a sum of exponentials:

$$R(t) = \sum_{i=1}^n g_i e^{t/\lambda_i} \tag{2.5}$$

where g_i are weighting factors that add to 1.0, and λ_i are the relaxation times. The time argument z is the reduced time and represents the fact that the mobility of the glassy material depends on both temperature, glassy structure, as represented by δ in the KAHR model, and pressure (although in most experiments, pressure is constant at 1 bar). The reduced time is, thus, generally written as

$$z = \int_0^t \frac{d\xi}{a_T a_\delta} \tag{2.6}$$

where the shift factors a_T and a_δ correspond to the factor by which the relaxation times scale due to changing temperature or departure from equilibrium (structure). An additional term a_P would be required if pressure were also a variable. In the case of the TNM model, the general ideas are the same as the KAHR model, though the TNM model is generally formulated in differential rather than integral form. Furthermore, the TNM model relates the glassy structure to the fictive temperature T_f rather than to the departure from equilibrium. Hence, there are some subtle differences in the meaning of model parameters, but the phenomenology can be made equivalent.

The most important aspect of the physics of the structural recovery models has been the temperature dependence of the relaxation times and their structure dependence. In the case of the original KAHR model, these were taken as follows for volume recovery:

$$\ln\frac{\lambda(T,\delta)}{\lambda(T_{ref},\delta=0)} = \ln(a_T a_\delta) = -\theta(T-T_{ref}) - (1-x)\theta\delta/\Delta\alpha \tag{2.7}$$

Just as in linear thermoviscoelasticity [65], the reference temperature T_{ref} is arbitrary and the equilibrium value of the departure from equilibrium ($\delta=0$) is taken as a convenient reference point as well. The temperature dependence of the relaxation time is accounted for by θ and the nonlinearity or dependence on δ is accounted for by x. Here, an exponential temperature dependence of the equilibrium relaxation times is assumed, whereas in the TNM model, an Arrhenius temperature dependence is assumed:

$$\ln\frac{\lambda(T,T_f)}{\lambda(T_{ref},T_f=T)} = \ln(a_T a_{T_f}) = \frac{x\Delta h}{RT} + \frac{(1-x)\Delta h}{RT_f} - \frac{\Delta h}{RT_{ref}} \tag{2.8}$$

The normalized activation energy $\Delta h/R$ is related to the KAHR parameter $\theta \approx \Delta h/RT_g^2$. Neither model accounts for the WLF temperature dependence observed for the relaxation times at equilibrium density above T_g, as has been discussed by Hodge [66] in his extensive review. However, the use of the exponential or Arrhenius temperature dependence in the models appears to be adequate in most cases due to the limited range of temperatures (usually near T_g) where structural recovery is studied and modeled and due to the fact that the temperature dependence of many materials does seem to turn over toward more gentle Arrhenius-temperature dependence just below T_g [11,13–19]. A recent modification used successfully by Grassia and Simon [42,56] incorporates both the equilibrium WLF dependence above T_g and a more gentle temperature dependence of the equilibrium relaxation times below T_g, as well as using TV^γ scaling [67–72] in order to incorporate the effect of pressure into the model without the use of a separate parameter to account for pressure.

One important aspect of the KAHR and TNM models arises from the simple structure of the equations and the fact that they do capture the qualitative features of the kinetics associated with T_g and structural recovery, including the cooling rate and pressure dependence of T_g and the three signatures of structural recovery (intrinsic isotherms, asymmetry of approach, and memory effect). Thus, they provide a useful framework for understanding and predicting behavior in the glassy state, especially in the sense of the physical aging response described subsequently. However, because the models do still exhibit some difficulties in quantitative prediction with model parameters showing a dependence on thermal history, there remains considerable effort to improve upon them. One of the most important is the model that has come out of the works of Caruthers and his students at Purdue University [73–75], along with collaborations and extensions by Adolf and Chambers at Sandia National Laboratories [76–82]. The model developed uses the ideas of rational mechanics [83] and includes both the structural recovery aspects of the behavior and the nonlinear mechanics of glassy polymers. The model is much more complex than the TNM and KAHR models, and though it is powerful, it does not seem to do a better job of describing the structural recovery of the materials than do these models. Therefore, we leave the reader to examine the relevant literature with the idea that the approach is very useful, and probably more relevant to nonlinear viscoelastic and yielding behavior than to the structural recovery behavior of polymers.

2.2.7 BEST PRACTICES FOR MEASURING VOLUME AND ENTHALPY RECOVERY

Before leaving the topic of volume and enthalpy recovery to discuss physical aging, a few comments on the best practices for making volume and enthalpy recovery measurements are in order. We do not address different methods of measuring volume and enthalpy recovery, as these have been discussed in a prior work [6].

Here, we simply focus on best practices when performing volume measurements with mercury-confinement dilatometry and enthalpy measurements with differential scanning calorimetry. In particular, we focus on the following issues, many of which are also important in physical aging measurements:

- Temperature control and stability
- Temperature accuracy
- Sample size and thermal gradients
- Experiment reproducibility
- Analysis issues

The first and foremost issue is that of temperature control. Not only is the time required to reach equilibrium very sensitive to temperature, the total change in volume and enthalpy is also a strong function of temperature, as shown in Equations 2.1 and 2.2. Ideally, temperature control should be better than 0.01 K over the course of the structural recovery. For dilatometric experiments, Kovacs' temperature control was ±0.015 K over 100 h [50], Simon reported ±0.005 K over 3 days [33,53], and McKenna and coworkers [84] reported ±0.01 K over the course of week-long experiments for their larger torsional dilatometer with better control over shorter periods. Differential scanning calorimeters with controlled stable coolers should have similar degrees of temperature control over the course of several days, and thus, aging can be performed directly using the DSC as an oven. It is best, however, to use a custom-built cooling system for DSCs, which maintain the temperature of the cold sink above the dew point since when subambient cooling systems are used, drift of the temperature will occur with the buildup of condensate or frost on the cold sink. For this reason, it is essential to perform an unaged heating scan after every aged scan in order to ensure that the DSC calibration is maintained and to limit aging in the DSC to several days when a subambient cooling system is in use. The commercial Flash DSC nanocalorimeter seems to have more severe limitations in this regard and can only be used for aging runs of less than 24 h.

Enthalpy experiments for longer than a few days are generally performed outside the DSC in a temperature-controlled environment. Standard practice involves making a number of samples, weighing and identifying each, placing them in an oven or in sealed tubes in a constant-temperature bath, and taking them out periodically for measurement. If an oven is used, it is important to use a convection oven, to place samples in the center of the oven on a heavy metal plate (or in a heavy metal box) with a large heat capacity in order to minimize fluctuations when the oven door is opened periodically to remove sample. Vacuum ovens, although they are stable in use, are poor choices because they are known to have spatial differences in temperature and, in addition, heating rates are low, leading to long stabilization times after opening the door. Ideally, the oven temperature should be monitored throughout long-term experiments so that long-term temperature fluctuations are known. In addition, a minimum of three samples should be measured for every aging time. Finally, to reduce systematic errors involving temperature fluctuations, samples should not be introduced at the same time; for example, some samples should be put in the temperature chamber at time 0, others at week 1, and others at week 4.

In addition to temperature stability, the actual temperature should be known to at least ±0.1 K, especially if the comparison of different properties is of interest. Temperature calibration of temperature baths is straightforward, whereas isothermal calibration of scanning calorimeters requires more effort. Most laboratories calibrate their DSCs (monthly or less frequently) at a specific heating rate using, for example, melting standards and then checking the calibration regularly (for example, daily). For runs made at different rates, a correction is applied based on the melting of the standard at the condition of interest. In order to obtain the correction for the isothermal situation, the melting point is checked at various heating rates and the value is extrapolated to the zero rate (isothermal) condition. In previous work, we have found that using the 0.1 K/min rate as the isothermal condition is adequate [11]. If no correction is made, the DSC aging temperature may be as much as 1 K higher than the set point, depending on the DSC and the rate used in the general calibration; the result will be a shorter time required to reach equilibrium than expected, as well as a lower total change in enthalpy.

Temperature gradients within the sample are also an issue in both volume and enthalpy recovery measurements. Volume measurements generally require rather large samples (on the order of grams) with minimization of the thermal gradients achieved by ensuring one dimension is small. For example, for our dilatometric work [7,11], we use a cylindrical sample with an axial hole drilled through the center to minimize thermal lag; still, cooling and heating rates were kept low (<0.2 K/min) to ensure a thermal lag of less than 0.1 K [7]. The result is that temperature jumps are not, in fact, jumps, in the case of dilatometric down-jumps because relaxation will occur and the fictive temperature will change during the jump—this is an important point to take into account when analyzing and modeling the data [20,21]. For example, it has been erroneously suggested that Kovacs volumetric recovery data do not reach the extrapolated equilibrium line based on the change in volume observed and the temperature from which the jumps are made; however, this conclusion is only reached if it is assumed that T_{fo} in Equation 2.1 is equivalent to the temperature T_o from which the jump is made—in a typical dilatometric experiment, this is never the case for $T_o > T_g$—and in fact, as already mentioned, Kovacs [30] ensured that measurements reached the extrapolated equilibrium line based on the absolute value of the measurements.

Temperature gradients and thermal lag in DSC are also important issues, and considerable research has been devoted to studying their effects. Enthalpy overshoot peaks on heating are broadened and pushed to higher temperatures due to thermal lag effects [85–90] which arise in spite of the small calorimetric sample sizes (~10 mg for conventional DSC and ~100 ng for Flash DSC) because of the relatively high heating rates used. In addition, because the Flash DSC sample is not encapsulated in a metallic pan (which transfers heat around the conventional DSC sample such that both the top and bottom of the sample are generally considered to be at the same temperature), a gradient can exist through the thickness of the Flash DSC sample even at isothermal conditions. Thermal gradients in calorimetric samples can be minimized using thin samples and slow heating rates, for example, samples of less than 0.5 mm and heating rates of 2 or 5 K/min for conventional DSC or samples of less than 5 µm for Flash DSC. Thermal contact resistance between the sample and the furnace can also be minimized using flat, thin samples with a small amount of a high thermal conductivity grease applied between the sample pan and furnace for conventional DSC, or a small amount of oil applied between the sample and sensor in Flash DSC. Since thermal lag effects influence the shape of the enthalpy overshoots, the parameters obtained in fitting DSC heating scans to the models of structural recovery depend on whether the thermal gradients are incorporated into the calculations [86,87]. On the other hand, the values of the fictive temperature and the enthalpy change calculated from the enthalpy overshoot are unaffected by thermal lag if care is taken to ensure good contact at the DSC oven surface such that the bottom of the sample is at the oven temperature.

Volume recovery measurements are performed in situ, and volume is an absolute quantity. The largest source of error in volume recovery experiments generally arises from temperature fluctuations. In addition to fluctuations in the temperature bath, fluctuations in room temperature can result in fluctuations in the data if the mercury level in the dilatometric capillary is above the level of the bath (and thus, exposed to room temperature) and/or if electrical circuits used in the measurement of the capillary height are not shielded from temperature changes. On the other hand, enthalpy recovery measurements are generally performed ex situ, with the change in enthalpy that occurred during isothermal recovery determined from the enthalpy overshoot measured during a heating scan after aging. As already mentioned, the enthalpy change is generally obtained by subtracting the heating scan from that for an unaged run and integrating the difference—hence, small differences in the baseline will lead to a lack of superposition of the liquid and glass heat capacity (or heat flow) and will lead to errors. In conventional DSC, the errors are generally on the order of ±0.1 J/g and they are larger when using an oven for long aging times. Hence, it is advised to use three replicates in enthalpy recovery measurements for each aging time; standard errors using three replicates were reported to be ±0.05 and ±0.15 J/g for DSC- and oven-aged samples, respectively [21]. In passing, it is further noted that assuming that the nonreversing heat flow from temperature-modulated DSC (TMDSC) is equivalent to the enthalpy change on aging is incorrect due to erroneous assumptions made in the TMDSC analysis [91].

Finally, errors can be made in the analysis of data, particularly in enthalpy recovery experiments. A common error made by those unfamiliar with the kinetics of the glass transition is to define the glass transition temperature T_g as the onset or midpoint in the step change in the heat capacity (or heat flow) for a calorimetric scan measured on heating. This is incorrect: T_g is defined as the point of vitrification and should only be measured on cooling. On the other hand, the fictive temperature must be reported for heating scans.

A second common problem involves the drawing of liquid and glass lines for the determination of the change in enthalpy on aging and the fictive temperature. As already mentioned, the former is obtained by subtracting the heating scan obtained on aging from that of an unaged sample and integrating the difference, as shown in Figure 2.8. The fictive temperature, on the other hand, is obtained either from integrating the heat flow curve to obtain the enthalpy and then finding the intersection of the extrapolated glass and liquid lines or, more preferably, using the method of Moynihan [92]:

$$\int_{T_f}^{T \gg T_g} (C_{pl} - C_{pg})dT = \int_{T \ll T_g}^{T \gg T_g} (C_p - C_{pg})dT \tag{2.9}$$

where C_{pl} and C_{pg} are the liquid and glass heat capacity (or heat flow) lines, respectively, which may be a function of temperature and must be extrapolated as needed, and C_p is the actual sample heat capacity (or heat flow). Moynihan's method is preferred to integrating the heat flow curve and then extrapolating liquid and glass regions to find T_f at their intersection for two reasons: (1) the liquid and glass enthalpy lines are often curved, making long extrapolations difficult, and (2) the difference between the slopes of the liquid and glass regions in enthalpy space is difficult to discern because the difference between C_{pl} and C_{pg} is small compared to the absolute value of C_p. Hence, subtracting off the glass enthalpy, as is done in Equation 2.9, makes the analysis easier. It is also noted that sagacious drawing of the liquid and glass heat capacity lines is critical to obtaining good data, especially when large overshoots are present. In our laboratory, we superpose a series of aging scans in the glass and liquid regions, away from the overshoot region, then draw glass and liquid lines based on the superposed curves and use these same lines for the separate analysis of each scan. In the event that glass and liquid lines do not superpose from successive scans, extreme care should be taken, since this may very well mean that the sample has changed during aging (for example, by losing mass, degradation, or chemical reaction) or that the calorimeter, itself, has changed (for example, through frost buildup on the cold sink, a change in the gas flow rate, or contamination of the oven).

2.3 PHYSICAL AGING OF POLYMERS AND COMPOSITES

Associated with the changing thermodynamic state variables of volume and enthalpy in the glass are changes in mechanical properties. These changes are referred to as physical aging and, at least for the linear viscoelastic properties, the behavior is fairly well understood [93–96]. Kovacs et al. first reported the evolution of the dynamic mechanical properties accompanying structural recovery in 1963 [97]. Subsequently, Struik published his thesis in 1978 in which sequential creep measurements were used to follow the effects of physical aging on the viscoelastic creep compliance [98]. Figure 2.13 is a schematic of the temperature and stress history and resulting strain response for the periodic creep experiment after a down-jump in temperature. The experiment consists of cooling or quenching the sample from above T_g to an aging temperature T_a and then applying a load at approximately equal logarithmic increments of aging time. Hence, there are two time scales in the experiment: the aging or elapsed time that begins at the moment of the quench and the creep time that begins at the moment of loading. Creep times are generally performed for less than 10% of the aging time in order to ensure that no appreciable aging occurs during the creep test. Hence, the creep response measured is a "snap shot" of the viscoelastic response at a particular aging time and is referred to as a momentary creep curve. The time between creep tests generally doubles or triples, allowing sufficient time for creep recovery such that the strain change due to creep recovery from one loading/unloading cycle is negligible compared to the subsequent loading. The time-dependent viscoelastic response to an up-jump is measured in a similar fashion. Small stresses are used to ensure that the viscoelastic response is linear. The evolution of the modulus during physical aging is similarly monitored by applying small strains and measuring stress relaxation. For both creep and stress relaxation after temperature down-jumps, the momentary responses shift to longer times as aging progresses as a consequence of volume recovery and the concomitant increase in relaxation times.

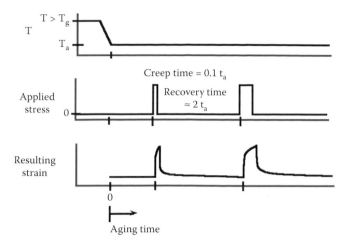

Figure 2.13 Schematic of the sequential creep test to measure the viscoelastic response as a function of aging time in the glassy state.

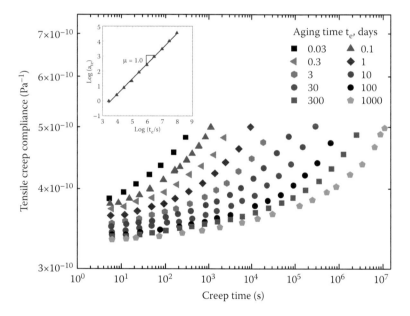

Figure 2.14 Physical aging data, in creep, for a poly(vinyl chloride) glass at 40°C (T$_g$ + 10°C). (Reprinted from *Physical Aging in Amorphous Polymers and Other Materials*, Elsevier, Amsterdam, L. C. E. Struik, Copyright 1978, with permission from Elsevier; inset after G. B. McKenna, *J. Rheol.*, 56, 113–158, 2012.)

A typical result from Struik's work [98] is shown in Figure 2.14 for a poly(vinyl chloride) (PVC) glass approximately 40°C below its glass transition temperature. The figure shows the evolution of the creep response as aging time increases from 0.03 to 1000 days. In addition, the inset [96] shows the shift factors that superimpose the curves at different aging times. Interestingly, this line gives a power law of unity over the full time range of experimentation. Two comments are in order about the behavior. First, because the sample was far enough below the glass transition temperature that aging did not cease in the time scale of the experiments, the long time asymptote for the shift factors is not achieved. Second, because the shortest aging times cannot be measured, the apparent behavior does not show a short aging time plateau or induction time, which it must. The power law result is simply a consequence of the material relaxation time being an exponential in the volume departure from equilibrium [99–101], which in the intermediate regime of structural recovery is approximately linear in the logarithm of the aging time. Then, $a_{te} \propto t_e^{\mu}$, where t_e is the elapsed or aging time, μ is the power law index or shift rate, and a_{te} is the scaled relaxation time or aging time shift factor. The shift rate is defined from the experiments as $\mu = d\log(a_{te})/d\log(t_e)$, and in the case of the inset in Figure 2.14, $\mu = 1.0$; typically, in the aging regime, $\mu \leq 1.0$ since volume or enthalpy changes are approximately linear with the logarithm of time and relaxation times grow longer proportionally to the time of aging. If one could cover the entire aging regime from short times to long times, the shift rate versus $\log t_e$ would be sigmoidal in form [102], similar to what is observed with enthalpy or volume when the entire response is observed (see, for example, the down-jump in Figure 2.10 obtained using nanocalorimetry).

Struik was also one of the first researchers to investigate the relationship between the structural recovery and mechanical/viscoelastic responses of the glassy material [103]. By comparing both volume recovery and creep measurements for a variety of single- and double-temperature jump histories, he claimed a unique relationship between the aging time shift factor and the instantaneous volume. For example, he showed that for a memory experiment similar to those depicted in Figures 2.7 and 2.10, the shift factors required to superimpose the creep curves were nonmonotonic similar to the volume and enthalpic departures from equilibrium. Hence, volume determines, to a first approximation, the relaxation times for the material. However, reality is more complicated than this. McKenna and coworkers [52,84,104–106] using a unique torsional dilatometer found that the mechanical relaxation times and the volumetric relaxation times were decoupled for an epoxy. In fact, a number of studies over the last 40 years have compared volume, enthalpy, and/or mechanical properties with differing results depending on which properties were compared and how [4,7,11,40,107–118]. A coherent picture from Simon's laboratory [15] indicates that properties diverge from one another a few degrees below the nominal glass transition unless small (nearly linear) temperature jumps are made. Furthermore, there is evidence that the structural relaxation time depends on the size of the temperature jump or the path [53–55]. Interestingly, the rational thermodynamics framework built by Caruthers and his students [73–76] and elaborated in collaboration with Adolf and Chambers at Sandia National Labs [77–82] also indicates that such a path dependence is not unreasonable. We again refer the reader to Chapters 4 and 14 in this book.

2.3.1 EFFECTIVE TIME THEORY

In his pioneering work, Struik [98] referred to an "effective time" concept that has since become known as an "effective time" theory. If used correctly, effective time theory can be a powerful tool for estimating the long time behavior of polymeric (or other glassy) materials that are undergoing aging in nonisothermal histories [98]. It is based on the early ideas of Hopkins [119], Leaderman [120], Ferry and coworkers [12,65,121], Tobolsky and Catsiff [122], and others [123,124]. In the general case, one considers the creep strain as a function of arbitrary load and thermal histories. Taking the Boltzmann integral representation of the creep as a function of the stress, the time-dependent creep strain $\varepsilon(t)$ is given by

$$\varepsilon(t) = \int_0^t D(t-t')\frac{d\sigma}{dt'}dt' \tag{2.10}$$

where $D(t)$ is the time-dependent creep compliance response function. For a general thermal history, one generally writes this in terms of a reduced time z and the relevant equation follows:

$$\varepsilon(z) = \int_0^z D(z-z')\frac{d\sigma(s)}{ds}ds = \int_0^z D(z-z')\frac{d\sigma}{dz'}dz' \tag{2.11}$$

where the reduced time argument is given by $z - z' = \int_s^t (d\xi)/a_T(\xi)$, the shift factor $a_T(\xi)$ is the temperature shift factor, and ξ is a dummy variable. When structural recovery occurs, there is also a shift factor for the changing structure δ (or T_f) a_δ or a_{Tf}. Then, Equation 2.11 can be used with the reduced time argument given as

$$z - z' = \int_s^t \frac{d\xi}{a_T(\xi)a_\delta} \text{ or as } z - z' = \int_s^t \frac{d\xi}{a_T(\xi)a_{Tf}}$$

Formally, these equations are those that describe the time, temperature, and structure dependence of the linear viscoelastic response (in terms of the creep compliance $D(t)$), and the physical justification is similar to that developed for time–temperature superposition in thermoviscoelasticity, similar to the justification in the KAHR and TNM models described above. However, a difficulty arises in the implementation of these equations because, in general, the volume and enthalpy recovery data are not available but the physical aging data are. Therefore, the effective time theory was developed to try to circumvent the problem. Hence, rather than explicitly writing the shift factors in terms of volume or enthalpy departures from equilibrium, it has become common practice to write the reduced time argument in terms of the aging time shift factor a_{te} as follows:

$$z - z' = \int_0^t \frac{d\xi}{a_T a_{te}} \tag{2.12}$$

where the structure shift factor a_δ is replaced by the aging time shift factor a_{te}. Physically, it is easy to understand how the relaxation time depends on the volume departure from equilibrium and, consequently, how the relaxation time depends on the aging time through the volume dependence on the aging time; and as already mentioned, based on Struik's experiments, to a first approximation this is valid. It is worth remarking that, in general, this approach is also problematic because the explicit use of the aging time shift factor leads to difficulty for complex thermal histories due to the fact that the general expressions for the aging time shift factor have been in terms of the aging time itself. This is not generally allowable in thermoviscoelasticity and leads to issues with time-frame invariance of the constitutive equations that result. Nonetheless, the method has been used with some success in simple (down-jump) conditions for long-term creep estimates.

At the same time, there is recent work by Guo et al. [125] and Guo and Bradshaw [100] that considers the effective time theory in terms of a construction based on the KAHR type of model in which the long-term creep in complex thermal and aging histories but at a constant stress is considered. We follow that development and show the success of these ideas. A comment of note is that the approach of Guo and Bradshaw [100,125] does not require a direct measurement of the volume departure from equilibrium; hence, while it is based on the ideas developed above, it builds a simpler experimental structure that requires only mechanical measurements.

The creep response is taken to be represented by a stretched exponential function as Struik proposed [98]:

$$D(t) = D_o e^{(t/\lambda)^\beta} \tag{2.13}$$

$$D(t, t_e) = D_o e^{(a_{te} t/\lambda)^\beta} \tag{2.14}$$

where D_o is the zero time or infinite frequency compliance, t is time, λ is a characteristic material time, and β is the KWW stretching exponent. We remark that other forms of creep function can also be used [54,126], and that Equations 2.13 and 2.14, although adequate for describing the short-time glassy response, are not valid in the long-time limit because they are missing both the limiting rubbery compliance and the t/η viscous flow term. The form of the compliance incorporating these mechanisms should be [65]

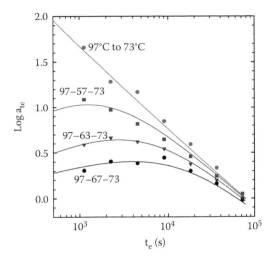

Figure 2.15 Time-aging time shift factors in two-step histories obtained from creep measurements showing memory effect in the mechanical response for a poly(phenylene sulfide) (PPS). Points are data and lines are KAHR-a_{te} model predictions [100,125]. (Data courtesy of Y. Guo and R. D. Bradshaw, *Polymer*, 50, 4048–4055, 2009.)

$$D(t) = D_o + (D_e - D_o)(1 - e^{-(t/\lambda)^\beta}) + t/\eta \qquad (2.15)$$

$$D(t, t_e) = D_o + (D_e - D_o)(1 - e^{-(a_{te}t/\lambda)^\beta}) + t/\eta \qquad (2.16)$$

However, regardless of the use of Equation 2.14 or 2.16, for any aging time and temperature, the material characteristic time would be treated in a fashion similar to Equation 2.6, but now with the shift factor being an aging time shift factor rather than that related to the volume departure, that is, we write a_{te} rather than a_δ. Again, the insight from this KAHR-a_{te} model is that the shift factors a_{te} and a_δ are approximately related as $\log a_{te} \propto \log a_\delta \propto \delta$. Though there has been some success from using this approach, as seen in Figures 2.15 and 2.16 for two-step (memory-type) and for long-time creep histories, respectively, the number of parameters necessary to modify the KAHR model to build the empirical relationship between a_{te} and δ (which is only an approximation) may be problematic. Further work and direct measurement of the volume and enthalpy along with the mechanical (physical aging) experiments should be undertaken on the same samples. We do expect that the problems found for the KAHR and TNM models will propagate through to the predictions of the physical aging response. For composite materials, semicrystalline polymers and for the nonlinear response behavior, the problem becomes more complicated and, as briefly summarized below, further models are needed.

2.3.2 AGING IN COMPOSITE MATERIALS

The above discussions of structural recovery and physical aging also apply to materials in which a glassy polymer matrix is reinforced by either particles or fibers. In principle, if one knows the behavior of the polymer matrix, then one should be able to determine the aging response of the composite using the same approaches as described above. As shown in Figure 2.17, it does seem that time–aging time superposition can hold for composite materials, in this case, for a polyphenylsulfone reinforced with an IM7 carbon fiber [127]. Furthermore, in the case of this study by Hastie and Morris [127], they were able to predict the long-term isothermal aging from the "momentary" creep curves using the effective time theory discussed above for this simple thermal history, although they did not attempt to carry out work in multiple-step temperature histories. However,

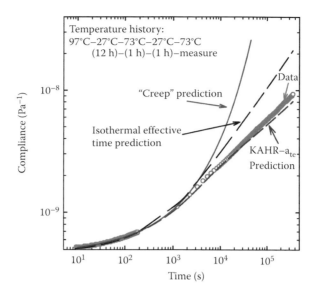

Figure 2.16 Comparison of data, real-time ("creep"), isothermal effective time, and KAHR-a_{te} model predictions for the creep response during the last temperature/aging step in a continuous creep, multiple-temperature step aging experiment for a PPS. (Data courtesy of Y. Guo and R. D. Bradshaw, *Polymer*, 50, 4048–4055, 2009.)

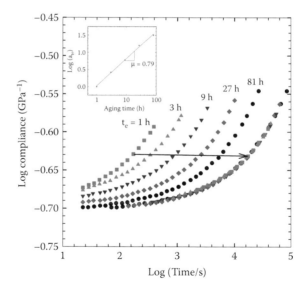

Figure 2.17 Creep compliance curves (shear direction) for a carbon fiber–reinforced thermoplastic polymer at different aging times. Inset shows aging time shift factors versus aging time. Master curve is offset for clarity. Arrow shows that small vertical shifting was necessary. (Data from R. L. Hastie, Jr. and D. H. Morris, The effects of physical aging on the creep response of a thermoplastic composite, in *High Temperature and Environmental Effects on Polymeric Composites, ASTM STP 1174*, edited by C. E. Harris and T. S. Gates, American Society of Testing and Materials, Philadelphia, PA, 63–185, 1993.)

such results do not seem to be the case for all composite materials. For example, Figure 2.18 shows the isothermal creep response for a polyimide reinforced with the same IM7 type of fiber [128], and now the response is not very well described by effective time theory in spite of the fact that the neat resin can be. Also, as shown by the shift rates as a function of temperature for the composite in Figure 2.19, the geometry of deformation may result in different apparent aging rates [128], although we note that the Hastie and Morris work was not

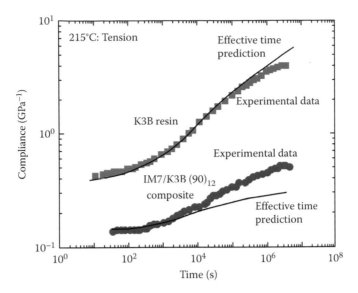

Figure 2.18 Creep compliance curves (tension) for a carbon fiber–reinforced thermoplastic polymer and the neat polymer in an isothermal aging experiment showing that effective time theory provides a reasonable representation of the resin response, but not for the composite. (Data from T. S. Gates, D. R. Veazie, and L. C. Brinson, *J. Comp. Matls.*, 31, 2478–2505, 1997.)

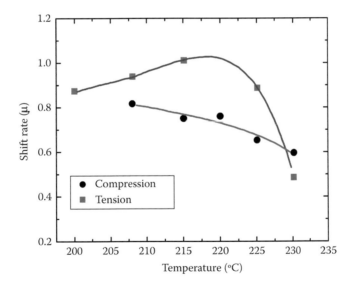

Figure 2.19 Aging time shift factors as a function of temperature of aging for a carbon fiber–reinforced polymer resin showing that aging in tension and compression can be different. (Data from T. S. Gates, D. R. Veazie, and L. C. Brinson, *J. Comp. Matls.*, 31, 2478–2505, 1997.)

conclusive on this point. The implication of such findings, if they hold, is that the polymer resin is somehow interacting with the fibers, either in the initial cure chemistry or through residual stresses, in a way that makes the composite system become effectively three phases (polymer matrix, reinforcing element, interface or interphase region). Regardless, the prediction of the long-term behavior of reinforced polymers remains an area that requires further investigation to fully account quantitatively for the aging response even in the nominally linear viscoelastic regime of the composite itself.

2.3.3 Aging in semicrystalline polymers

Aging in semicrystalline polymers is complicated by the fact that the system is like a nanocomposite material with very small crystalline lamellae that reinforce either a glassy polymer matrix or a rubbery polymer matrix [10,17,129–140]. In addition, it is known that in the case of many polymers, there is an additional "phase" called the rigid amorphous phase [141,142] that is associated with the transition region between the amorphous and crystalline phases. As a result, there is a three-phase structure, and this can lead to difficulties in describing the aging behavior of semicrystalline polymers because each phase has its own characteristics. Frequently, one considers the crystalline phase as somewhat inert; the amorphous phase as an ordinary glass-forming material, which can show physical aging when $T < T_g$ while being in equilibrium at higher temperatures; and the rigid amorphous fraction (RAF) or phase as a material that shows glassy-like physical aging at temperatures below T_g of the RAF, but well above the nominal macroscopic glass transition temperature of the amorphous material. As a result, the aging of semicrystalline polymers is quite complex. The simplest sort of behavior is found when the system has been well stabilized (secondary crystallization has taken place and the sample is no longer evolving at temperatures of ~$T_g + 20°C$). Then, normal physical aging below T_g occurs and the material can be treated as described above. Figure 2.20 shows typical aging of a semicrystalline syndiotactic polystyrene material below the glass transition temperature. This same material, though, above the glass transition temperature still ages as shown in Figure 2.21. However, now the aging is primarily through vertical shifts, possibly because there was some minor recrystallization taking place. It is also worth noting that for this material, the individual aging curves at intermediate temperatures (from T_g to $T_g + 10°C$) did not superimpose, and hence, the effective time theory would not apply to this material above T_g. In some instances, there is also aging of the rigid amorphous phase, which has a glassy type of response, but that differs from the general amorphous material because its T_g is higher. In principle, this aging can be treated by effective time theory, but there is a need to know the volume fraction of the rigid amorphous material. There is still need for quantitative assessment of such behavior both for semicrystalline polymers and for nanocomposites in which there is a "bound" surface layer on the particle surface. Some success has been reported for the latter case by Papon et al. [143]. The reader is also referred to the work of Struik [129,130,144,145] where the complexity of aging in semicrystalline polymers is well described in a series of papers on polyethylene.

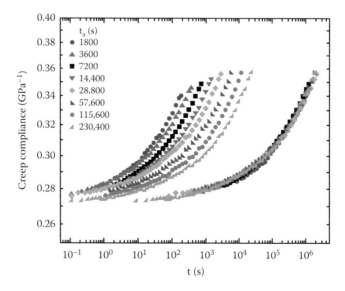

Figure 2.20 Creep compliance versus time at different aging times for a semicrystalline syndiotactic polystyrene material aged at 70°C (approximately $T_g - 30°C$). Note that master curve formed by time-aging time superposition is offset for clarity. (Data from J. Beckmann et al., *Polym. Eng. Sci.*, 37, 1459–1468, 1997.)

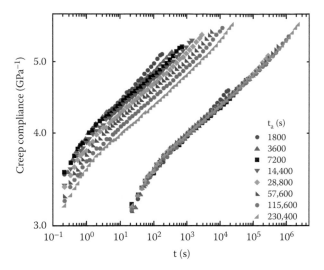

Figure 2.21 Creep compliance versus time at different aging times for a semicrystalline syndiotactic polystyrene material aged at 120°C (approximately T_g + 20°C). Note that master curve formed by time-aging time superposition via vertical shifting is offset for clarity. No horizontal shifts were used. (Data from J. Beckmann et al., *Polym. Eng. Sci.*, 37, 1459–1468, 1997.)

2.4 INFLUENCE OF STRUCTURAL RECOVERY ON NONLINEAR VISCOELASTIC PROPERTIES

The influence of mechanical deformations on mobility in glasses has been debated for quite a number of years with a number of studies showing that nonlinear deformation can accelerate the relaxation process [77,98,146–153]. For example, Struik [98] observed that the shift rate required for the time-aging time superposition of creep data decreased as the magnitude of the stress increased. Figure 2.22 shows a plot of the shift rate for a poly(vinyl chloride) as a function of applied stress for different temperatures from the work of Struik [98].

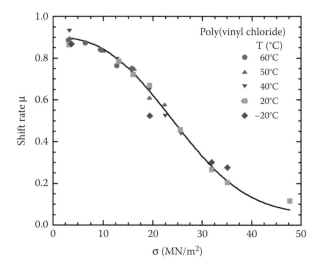

Figure 2.22 Shift rate versus applied stress for creep-aging experiments in a poly(vinyl chloride) glass. Quenched from 90°C to temperatures shown. (Reprinted from *Physical Aging in Amorphous Polymers and Other Materials*, Elsevier, Amsterdam, L. C. E. Struik, Copyright 1978, with permission from Elsevier.)

As seen in the figure, there is a significant reduction in the apparent aging due to the large change in mobility of the polymer at high stresses. Similar results are found in stress relaxation experiments where the shift rate decreases as the strain increases [98].

McKenna and Kovacs [154] were the first to use the term "rejuvenation" to describe the shift of creep curves toward shorter times, and the term was further popularized by Ricco and Smith [155]. McKenna later refuted this interpretation [93,96,102] citing the fact that for nonlinear deformations applied to measure momentary stress relaxation, mechanical relaxation and volume recovery were decoupled. In addition, for post-yield deformations, aging increased as strain increased and appeared to tend toward a different equilibrium state for both nonlinear uniaxial tensile and compressive deformations [156,157]. The latter results, which have been referred to as implosion [93], may be due to changes in the thermodynamic state of the glass such that the equilibrium state may no longer resemble the initial undeformed state. A related idea is that of a stress-induced tilting of the potential energy landscape [158–160], which can either make hopping between barriers easier at small strains or change barriers and the equilibrium state altogether at larger strains. Chen and Schweizer [161] have also suggested that in addition to the stress-induced tilting of the energy landscape, deformation may increase the amplitude of density fluctuations, resulting in mobility changing with strain and strain rate. Govaert and coworkers [162,163] have further invoked a deformation-dependent viscosity although this mechanism would not be expected to be operative for the cross-linked samples, and they have also suggested that the α- and β-processes are shifted differently by deformation.

A full description of the relationships between nonlinear response and the underlying glassy structure or energy landscape is beyond the scope of this chapter. It is worth noting, though, that beyond the briefly mentioned studies above, multiple other authors have addressed the nonlinear viscoelastic response of glasses through constitutive modeling, computer simulation and experiment. Some of the relevant references [73–82,93,96,98,102–106,147–163] have been provided for the reader. In addition, more recent developments are outlined in Chapters 4 and 14 in this book.

2.5 IMPACT OF STRUCTURAL RECOVERY AND PHYSICAL AGING

The volume changes incurred during structural recovery are typically less than 1% (as shown in Figure 2.3), and these are accompanied by decreases in the compliance and increases in modulus on the order of 25% (as shown in Figure 2.14). These changes are not insignificant due to their dramatic influence on failure and lifetime expectancies. Failure modes for many polymers change from ductile to brittle failure with aging [138,164–171], and decreases in the strain at break [138,164–170,172–179] and in the fracture energy [176,179–182] are observed. Lifetimes change, becoming shorter for polystyrene [174], but longer for poly(methyl methacrylate) [183] and poly(vinyl chloride) [184]. Although the effects of structural recovery and physical aging on failure are not universal, they obviously will impact the design, processing, and performance of glassy polymers, composites, and even semicrystalline materials used below T_g of the amorphous phase, including structural and long-term aerospace and automotive applications, xerographic applications, and glass-to-metal hermetic seals used in radioactive waste storage [2,93,185].

2.6 CONCLUSIONS

Structural recovery and physical aging in amorphous materials are well understood from phenomenological points of view. The three signatures of volume and enthalpy recovery, the intrinsic isotherms, asymmetry of approach, and memory effect, first cataloged by Kovacs [1], provide insight into the "essential" ingredients needed to model structural recovery, that is, nonlinearity, nonexponentiality, and memory or Boltzmann superposition. The structural recovery models can be extended to describe the physical aging response. The models are limited, however, in their abilities to predict volume and enthalpy recovery and physical aging effects outside of the range of conditions modeled, and although various approaches have been attempted, a clear understanding of where the shortcomings lie is elusive. The physical aging of composites and semicrystalline polymers is complicated by the aging of interphase material (either between polymer and filler or between the amorphous and crystalline phases). Similarly, the influence of structural recovery on nonlinear viscoelastic properties has not been thoroughly investigated.

ACKNOWLEDGMENT

The authors gratefully acknowledge financial support from the National Science Foundation NSF DMR-1006972 (SLS) and NSF DMR-1207070 (GBM), as well as support from the ESPCI ParisTech in Paris, France (GBM).

REFERENCES

1. A. J. Kovacs, Transition vitreuse dans les polymères amorphes. Etude phénoménologique, *Fortschr. Hochpolym.-Forsch.*, 3, 394–507, 1963.
2. G. B. McKenna, Glass formation and glassy behavior, in *Comprehensive Polymer Science, Vol. 2: Polymer Properties*, edited by C. Booth and C. Price (Pergamon, Oxford), 311–362, 1989.
3. C. A. Angell, K. L. Ngai, G. B. McKenna, P. F. McMillan, and S. W. Martin, Relaxation in glassforming liquids and amorphous solids, *J. Appl. Phys.*, 88, 3113–3157, 2000.
4. G. W. Scherer, *Relaxation in Glass and Composites* (Krieger Publishing Co., Malabar, FL), 1992.
5. D. J. Plazek and K. L. Ngai, The glass temperature, in *Physical Properties of Polymers Handbook*, edited by J. E. Mark (American Institute of Physics, Woodbury, NY), 139–159, 1996.
6. G. B. McKenna and S. L. Simon, The glass transition: Its measurement and underlying physics, in *Handbook of Thermal Analysis and Calorimetry, Applications to Polymers and Plastics*, edited by S. Z. D. Cheng Vol. 3 (Elsevier Science, Amsterdam), 49–109, 2002.
7. P. Badrinarayanan, W. Zheng, Q. X. Li, and S. L. Simon, The glass transition temperature versus the fictive temperature, *J. Non-Cryst. Solids*, 353, 2603–2612, 2007.
8. E. J. Donth, *The Glass Transition: Relaxation Dynamics in Liquids and Disordered Materials* (Springer-Verlag, New York), 2001.
9. S. V. Nemilov, *Thermodynamic and Kinetic Aspects of the Vitreous State* (CRC Press, Boca Raton, FL), 1995.
10. G. B. McKenna, Chapter 5: Dynamics of materials at the nanoscale: Small molecule liquids and polymer films, in *Polymer Physics: From Suspensions to Nanocomposites and Beyond*, edited by L. A. Utracki and A. M. Jamieson (John Wiley & Sons, Hoboken, NJ), 177–209, 2010.
11. S. L. Simon, J. W. Sobieski, and D. J. Plazek, Volume and enthalpy recovery of a polystyrene, *Polymer*, 42 (6), 2555–2567, 2001.
12. M. L. Williams, R. F. Landel, and J. D. Ferry, The temperature dependence of relaxation mechanism in amorphous polymers and other glass-forming liquids, *J. Am. Chem. Soc.*, 77, 3701–3707, 1955.
13. P. A. O'Connell and G. B. McKenna, Arrhenius like temperature dependence of the segmental relaxation below T_g, *J. Chem. Phys.*, 110, 11054–11060, 1999.
14. X. Shi, A. Mandanici, and G. B. McKenna, Shear stress relaxation and physical aging study on simple glass-forming materials, *J. Chem. Phys.*, 123, 174507, 2005.
15. P. Badrinarayanan and S. L. Simon, Origin of the divergence of the timescales for volume and enthalpy recovery, *Polymer*, 48, 1464–1470, 2007.
16. T. Hecksher, A. I. Nielsen, N. B. Olsen, and J. C. Dyre, Little evidence for dynamic divergences in ultra-viscous molecular liquids, *Nat. Phys.*, 4, 737–741, 2008.
17. G. B. McKenna, Chapter 7: Physical aging in glasses and composites, in *Long-Term Durability of Polymeric Matrix Composites*, edited by K. V. Pochiraju, G. P. Tandon, and G. A. Schoeppner (Springer, New York), 237–309, 2011.
18. J. Zhao, S. L. Simon, and G. B. McKenna, Using 20 million year old amber to test the super-Arrhenius behavior of glass-forming systems, *Nat. Commun.*, 4:1783, 1–6, 2013.

19. J. Zhao and G. B. McKenna, Accumulating evidence for non-diverging time-scales in glass-forming fluids, *J. Non-Cryst. Solids*, 407, 3–13, 2015.

20. J. M. Hutchinson and P. Kumar, Enthalpy relaxation in polyvinyl acetate, *Thermochim. Acta*, 391 (1–2), 197–217, 2002.

21. Q. X. Li and S. L. Simon, Enthalpy recovery of polymeric glasses: Is the theoretical limiting liquid line reached?, *Polymer*, 47 (13), 4781–4788, 2006.

22. J. M. G. Cowie and R. Ferguson, Physical aging of poly(methyl methacrylate) from enthalpy relaxation measurements, *Polymer*, 34 (10), 2135–2141, 1993.

23. J. M. G. Cowie, S. Harris, and I. J. McEwen, Physical ageing in poly(vinyl acetate): 1. Enthalpy relaxation, *J. Polym. Sci., Part B: Polym. Phys.*, 35 (7), 1107–1116, 1997.

24. N. R. Cameron, J. M. G. Cowie, R. Ferguson, and I. J. McEwen, Enthalpy relaxation of styrene-maleic anhydride (SMA) copolymers Part 1: Single component systems, *Polymer*, 41 (19), 7255–7262, 2000.

25. A. Brunacci, J. M. G. Cowie, R. Ferguson, J. L. Gomez Ribelles, and A. Vidaurre Garayo, Structural relaxation in polystyrene and some polystyrene derivatives, *Macromolecules*, 29 (24), 7976–7988, 1996.

26. J. L. Gomez Ribelles and M. Monleon Paradas, Structural relaxation of glass-forming polymers based on an equation for configurational entropy: 1. DSC experiments on polycarbonate, *Macromolecules*, 28 (17), 5867–5877, 1995.

27. L. Andreozzi, M. Faetti, M. Giordano, D. Palazzuoli, and F. Zulli, Enthalpy relaxation in polymers: A comparison among different multiparameter approaches extending the TNM/AGV model, *Macromolecules*, 36 (19), 7379–7387, 2003.

28. L. Andreozzi, M. Faetti, M. Giordano, and D. Palazzuoli, Enthalpy recovery in low molecular weight PMMA, *J. Non-Cryst. Solids*, 332 (1–3), 229–241, 2003.

29. V. M. Boucher, D. Cangialosi, A. Alegria, and J. Colmenero, Enthalpy recovery of glass polymers: Dramatic deviations from the extrapolated liquid like behavior, *Macromolecules*, 44 (20), 8333–8342, 2011.

30. A. J. Kovacs, La contraction isotherme du volume des polymers amorphes, *J. Polym. Sci.*, 30, 131–147, 1958.

31. C. R. Schultheisz and G. B. McKenna, Volume recovery, physical aging and the τ-effective paradox in glassy polycarbonate following temperature jumps, *Proceedings North American Thermal Analysis Society, 25th Annual Meeting*, McLean, Virginia, 366–373, 1997.

32. P. A. O'Connell, C. R. Schultheisz, and G. B. McKenna, The physics of glassy polycarbonate: Superposability and volume recovery, in *The Physics of Glassy Polymers*, edited by A. Hill and M. Tant (ACS Books, Washington, D.C.), 199–217, 1998.

33. P. Bernazzani and S. L. Simon, Volume recovery of polystyrene: Evolution of the characteristic relaxation time, *J. Non-Cryst. Solids*, 307–310C, 470–480, 2002.

34. R. Kohlrausch, Theorie des elektrischen rückstandes in der leidener flasche, *Annalen der Physik und Chemie von J. C. Poggendorff*, 91, 179–214, 1854.

35. G. Williams and D. C. Watts, Non-symmetrical dielectric relaxation behaviour arising from a simple empirical decay function, *Trans. Faraday Soc.*, 66, 80–85, 1970.

36. S. E. B. Petrie, Thermal behavior of annealed organic glasses, *J. Polym. Sci. A-2 Polym. Phys.*, 10 (7), 1255–1272, 1972.

37. Y. P. Koh and S. L. Simon, Enthalpy recovery of polystyrene: Does a long-term aging plateau exist? *Macromolecules*, 46 (14), 5815–5821, 2013.

38. H. Fujimori and M. Oguni, Non-exponentiality of the enthalpy relaxation under constant-temperature conditions in the time domain in supercooled liquids and glasses, *J. Non-Cryst. Solids*, 172–174, 601, 1994.

39. H. Fujimori, H. Fujita, and M. Oguni, Nonlinearity of the enthalpic response function to a temperature jump and its interpretation based on fragility in liquid glasses, *Bull. Chem. Soc. Jpn.*, 68, 447, 1995.

40. K. Adachi and T. Kotaka, Volume and enthalpy relaxation in polystyrene, *Polym. J.*, 14, 959–970, 1982.

41. K. Hofer, J. Perez, and G. P. Johari, Detecting enthalpy "cross-over" in vitrified solids by differential scanning calorimetry, *Philos. Mag. Lett.*, 64, 37–43, 1991.

42. Y. P. Koh, L. Grassia, and S. L. Simon, Structural recovery of a single polystyrene thin film using nanocalorimetry to extend the aging time and temperature range, *Thermochim. Acta*, 603, 135–141, 2015.

43. E. Lopez and S. L. Simon, Signatures of structural recovery in polystyrene by nanocalorimetry, *Macromolecules*, 49, 2365–2374, 2016.

44. Y. P. Koh, S. Y. Gao, and S. L. Simon, Structural recovery of a single polystyrene thin film using Flash DSC at low aging temperatures, *Polymer*, 96, 182–187, 2016.

45. J. Málek, Rate-determining factors for structural relaxation in non-crystalline materials I. Stabilization period of isothermal volume relaxation, *Thermochimica Acta*, 313, 181–190, 1998.

46. R. Greiner and F. R. Schwarzl, Thermal contraction and volume relaxation of amorphous polymers, *Rheol. Acta*, 23, 378–395, 1984.

47. J. Málek, Dilatometric study of structural relaxation in arsenic sulfide glass, *Thermochim. Acta*, 311, 183–198, 1998.

48. J. Málek and J. Shanelová, The effect of non-linearity contribution on the volume and enthalpy relaxation in amorphous materials, *J. Non-Cryst. Solids*, 307–310, 463–469, 2002.

49. L. C. E. Struik, Volume-recovery theory: 1. Kovacs' τ_{eff} paradox, *Polymer*, 38, 4677–4685, 1997.

50. G. B. McKenna, M. G. Vangel, A. L. Rukhin, S. D. Leigh, B. Lotz, and C. Straupe, The τ-effective paradox revisited: An extended analysis of Kovacs' volume recovery data on poly(vinyl acetate), *Polymer*, 40, 5183–5205, 1999.

51. S. Kolla and S. L. Simon, The τ-effective paradox: New measurements towards a resolution, *Polymer*, 46, 733–739, 2005.

52. G. B. McKenna, Y. Leterrier, and C. R. Schultheisz, The evolution of material properties during physical aging, *Polym. Eng. Sci.*, 35, 403–410, 1995.

53. S. L. Simon and P. Bernazzani, Structural relaxation in the glass: Evidence for a path dependence of the relaxation time, *J. Non-Cryst. Solids*, 352 (42–49), 4763–4768, 2006.

54. M. Alcoutlabi, F. Briatico-Vangosa, and G. B. McKenna. Effect of chemical activity jumps on the viscoelastic behavior of an epoxy resin: The physical aging response in carbon dioxide pressure jumps, *J. Polym. Sci., Part B: Polym. Phys.*, 40 (18), 2050–2064, 2002.

55. Y. Zheng and G. B. McKenna, Structural recovery in a model epoxy: Comparison of responses after temperature and relative humidity jumps, *Macromolecules*, 36 (7), 2387–2396, 2003.

56. L. Grassia and S. L. Simon, Modeling volume relaxation of amorphous polymers: Modification of the equation for the relaxation time in the KAHR model, *Polymer*, 53 (16), 3613–3620, 2012.

57. A. Q. Tool, Relation between inelastic deformability and thermal expansion of glass in its annealing range, *J. Am. Ceram. Soc.*, 29, 240–253, 1946; A. Q. Tool, Viscosity and the extraordinary heat effects in glass, *J. Res. Natl. Bureau Standards (USA)*, 37, 73–90, 1946.

58. O. S. Narayanaswamy, A model of structural relaxation in glass, *J. Am. Ceram. Soc.*, 54, 491–498, 1971.

59. C. T. Moynihan, P. B. Macedo, C. J. Montrose, P. K. Gupta, M. A. DeBolt, J. F. Dill et al. Structural relaxation in vitreous materials, *Ann. N.Y. Acad. Sci.*, 279, 15–35, 1976.

60. A. J. Kovacs, J. J. Aklonis, J. M. Hutchinson, and A. R. Ramos, Isobaric volume and enthalpy recovery of glasses. II. A transparent multiparameter model, *J. Polym. Sci. Polym. Phys. Ed.*, 17, 1097–1162, 1979.

61. B. D. Coleman, Thermodynamics of materials with memory, *Arch. Ration. Mech. Anal.*, 17, 1–46, 1964.

62. B. D. Coleman, On thermodynamics, strain impulses, and viscoelasticity, *Arch. Ration. Mech. Anal.*, 17, 230–254, 1964.

63. W. Noll, A mathematical theory of the mechanical behavior of continuous media, *Arch. Ration. Mech. Anal.*, 2, 197–226, 1958.

64. L. Boltzmann, Zur theorie der elastischen nachwirkung, *Sitzungsber. Akad. Wiss. Wien. Mathem. Naturwiss. Kl.*, 70, 2. Abt. 275–300, 1874.

65. J. D. Ferry, *Viscoelastic Properties of Polymers* (John Wiley, New York), 3rd ed., 1980.

66. I. M. Hodge, Enthalpy relaxation and recovery in amorphous materials, *J. Non-Cryst. Solids*, 169, 211–266, 1984.

67. G. Tarjus, D. Kivelson, S. Mossa and C. Alba-Simionesco, Disentangling density and temperature effects in the viscous slowing down of glassforming liquids, *J. Chem. Phys.*, 120 (13), 6135–6141, 2004.

68. R. Casalini and C. M. Roland, Themodynamical scaling of the glass transition dynamics, *Phys. Rev. E*, 69 (6), 062501, 2004.

69. R. Casalini and C. M. Roland, Scaling of the segmental relaxation times of polymers and its relation to the thermal expansivity, *Colloid Polym. Sci.*, 283 (1), 107–110, 2004.

70. R. Casalini and C. M. Roland, An equation for the description of volume and temperature dependences of the dynamics of supercooled liquids and polymer melts, *J. Non-Cryst. Solids*, 353 (41–43), 3936–3939, 2007.

71. C. Dreyfus, A. Aouadi, J. Gapinski, M. Matos-Lopes, W. Steffen, A. Patkowski, and R. M. Pick, Temperature and pressure study of Brillouin transverse modes in the organic glass-forming liquid orthoterphenyl, *Phys. Rev. E*, 68 (1), 011204, 2003.

72. C. Dreyfus, A. Le Grand, J. Gapinski, W. Steffen, and A. Patkowski, Scaling the alpha-relaxation time of supercooled fragile organic liquids, *Eur. Phys. J. B*, 42 (3), 309–319, 2004.

73. S. R. Lustig, A Continuum Thermodynamics Theory for Transport in Polymer/Fluid Systems, Ph.D. thesis, Purdue University, Lafayette, IN, 1989.

74. D. M. Colucci, The Effect of Temperature and Deformation on the Relaxation Behavior in the Glass Transition Region, Ph.D. thesis, Purdue University, Lafayette, IN, 1995.

75. D. S. McWilliams, Study of the Effect of Thermal History on the Structural Relaxation and Thermoviscoelasticity of Amorphous Polymers, Ph.D. thesis, Purdue University, Lafayette, IN, 1996.

76. S. R. Lustig, R. M. Shay, and J. M. Caruthers, Thermodynamic constitutive equations for materials with memory on a material time ccale, *J. Rheol.*, 40, 69–106, 1996.

77. J. M. Caruthers, D. B. Adolf, R. S. Chambers, and P. Shrikhande, A thermodynamically consistent, nonlinear viscoelastic approach for modeling glassy polymers, *Polymer*, 45, 4577–4597, 2004.

78. D. B. Adolf, R. S. Chambers, and J. M. Caruthers, Extensive validation of a thermodynamically consistent, nonlinear viscoelastic model for glassy polymers, *Polymer*, 45, 4599–4621, 2004.

79. D. B. Adolf, and R. S. Chambers, Application of a nonlinear viscoelastic model to glassy, particulate-filled polymers, *J. Polym. Sci., Part B: Polym. Phys.*, 43, 3135–3150, 2005.

80. D. B. Adolf and R. S. Chambers, A thermodynamically consistent, nonlinear viscoelastic approach for modeling thermosets during cure, *J. Rheol.*, 51, 23–50, 2007.

81. D. B. Adolf, R. S. Chambers, J. Flemming, J. Budzien, and J. McCoy, Potential energy clock model: Justification and challenging predictions, *J. Rheol.*, 51, 517–540, 2007.

82. D. B. Adolf, R. S. Chambers, and M. A. Neidigk, A simplified potential energy clock model for glassy polymers, *Polymer*, 50, 4257–4269, 2009.

83. C. Truesdell, *Rational Thermodynamics*, (Springer-Verlag, New York), 2nd ed., 1984.

84. M. M. Santore, R. S. Duran, and G. B. McKenna, Volume recovery in epoxy glasses subjected to torsional deformations—The question of rejuvenation, *Polymer*, 32 (13), 2377–2381, 1991.

85. I. M. Hodge and R. Heslin, Effects of thermal history on enthalpy relaxation in glasses. 7. Thermal time constants, *J. Non-Cryst. Solids*, 356, 1479–1487, 2010.

86. P. Badrinarayanan, S. L. Simon, R. J. Lyng, and J. M. O'Reilly, Effect of structure on enthalpy relaxation of polycarbonate: Experiments and modeling, *Polymer*, 49, 3554–3560, 2008.

87. S. L. Simon, Enthalpy recovery of polyetherimide: Experiment and model calculations incorporating thermal lag effects, *Macromolecules*, 30 (14), 4056–4063, 1997.

88. J. M. O'Reilly and I. M. Hodge, Effects of heating rate on enthalpy recovery in polystyrene, *J. Non-Cryst. Solids*, 131–133, 451–456, 1991.

89. J. E. K. Schawe and C. Schick, Influence of the heat conductivity of the sample on DSC curves and its correction, *Thermochim. Acta*, 187, 335–349, 1991.

90. I. M. Hodge, Effects of annealing and orior history on enthalpy relaxation in glassy polymers. 6. Adam-Gibbs formulation of nonlinearity, *Macromolecules*, 20, 2897–2908, 1987.

91. S. L. Simon and G. B. McKenna, Quantitative analysis of the errors in TMDSC in the glass transition region, *Thermochim. Acta*, 348 (1–2), 77–89, 2000.

92. C. T. Moynihan, A. J. Easteal, M. A. DeBolt, and J. Tucker, Dependence of the fictive temperature of glass on cooling rate, *J. Am. Ceram. Soc.*, 59, 12–16, 1976.

93. G. B. McKenna, On the physics required for the prediction of long term performance of polymers and their composites, *J. Res. NIST*, 99, 169–189, 1994.

94. S. L. Simon, Physical aging, in *Encyclopedia of Polymer Science* (John Wiley, New York), 2001.

95. J. M. Hutchinson, Physical aging in polymers, *Progr. Polym. Sci.*, 20, 703–760, 1995.

96. G. B. McKenna, Deformation and flow of matter: Interrogating the physics of materials using rheological methods, *J. Rheol.*, 56, 113–158, 2012.

97. A. J. Kovacs, R. A. Stratton, and J. D. Ferry, Dynamic mechanical properties of polyvinyl acetate in shear in the glass transition temperature range, *J. Phys. Chem.*, 67, 152–161, 1963.

98. L. C. E. Struik, *Physical Aging in Amorphous Polymers and Other Materials* (Elsevier, Amsterdam), 1978.

99. C. R. Schultheisz, G. B. McKenna, Y. Leterrier, and E. Stefanis, A comparison of structure a_δ and aging time ate shift factors from simultaneous volume and mechanical measurements, *Proc. Society of Experimental Mechanics*, Grand Rapids, Michigan, 329–335, 1995.

100. Y. Guo and R. D. Bradshaw, Long-term creep of polyphenylene sulfide (PPS) subjected to complex thermal histories: The effects of nonisothermal physical aging, *Polymer*, 50, 4048–4055, 2009.

101. Y. Zheng, R. D. Priestley, and G. B. McKenna, Physical aging of an epoxy subsequent to relative humidity jumps through the glass concentration, *J. Polym. Sci., Part B: Polym. Phys.*, 42, 2107–2121, 2004.

102. G. B. McKenna, Mechanical rejuvenation in polymer glasses: Fact or fallacy? *J. Phys. Condens. Matter*, 15, S737–S763, 2003.

103. L. C. E. Struik, Dependence of relaxation times of glassy-polymers on their specific volume, *Polymer*, 29 (8), 1347–1353, 1988.

104. G. B. McKenna and C. R. Schultheisz, Nonlinear viscoelastic analysis of the torque, axial normal force and volume change measured simultaneously in the national institute of standards and technology torsional dilatometer, *J. Rheol.* 46 (4), 901–925, 2002.

105. G. B. McKenna, C. R. Schultheisz, and Y. Leterrier, Volume recovery and physical aging: Dilatometric evidence for different kinetics, *Deformation, Yield and Fracture in Polymers, Proceedings of 9th International Conference*, Cambridge, UK, 31/1, 1994.

106. G. B. McKenna, Dilatometric evidence for the apparent decoupling of glassy structure from the mechanical-stress field, *J. Non-Cryst. Solids*, 172, 756–764, 1994.

107. A. Weitz and B. Wunderlich, Thermal analysis and dilatometry of glasses formed under elevated pressure, *J. Polym. Sci. Polym. Phys.*, 12, 2473, 1974.

108. H. Sasabe and C. T. Moynihan, Structural relaxation in poly(vinyl acetate), *J. Polym. Sci. Polym. Phys. Ed.*, 16, 1447, 1978.

109. R.-J. Roe and G. M. Millman, Physical aging in polystyrene: Comparison of the changes in creep behavior with the enthalpy relaxation, *Polym. Eng. Sci.*, 23 (6), 318, 1983.

110. J. M. G. Cowie, S. Elliott, R. Ferguson, and R. Simha, Physical ageing studies on poly(vinyl acetate): Enthalpy relaxation and its relation to volume recovery, *Polym. Commun.*, 28, 298, 1987.

111. E. F. Oleinik, Glassy polymers as matrices for advanced composites, *Polym. J.*, 19 (1), 105, 1987.

112. C. A. Bero and D. J. Plazek, Volume dependent rate processes in an epoxy resin, *J. Polym. Sci., Part B: Polym. Phys.* 29, 39–47, 1991.

113. J. Perez, J. Y. Cavaille, R. D. Calleja, J. L. G. Ribelles, M. M. Pradas, and A. R. Greus, Physical ageing of amorphous polymers. Theoretical analysis and experiments on poly(methyl methacrylate), *Makromol. Chem.*, 192, 2141–2161, 1991.

114. J. Mijovic and T. Ho, Proposed correlation between enthalpic and viscoelastic measurements of structural relaxation in glassy polymers, *Polymer*, 34, 3865, 1993.

115. I. Echeverria, P.-C. Su, S. L. Simon, and D. J. Plazek, Physical aging of a polyetherimide: Creep and DSC measurements, *J. Polym. Sci., Part B: Polym. Phys.*, 33, 2457–2468, 1995.

116. S. L. Simon, D. J. Plazek, J. W. Sobieski, and E. T. McGregor, Physical aging of a polyetherimide: Volume recovery and its comparison to creep and enthalpy measurements, *J. Polym. Sci. Polym. Phys.*, 35, 929–936, 1997.

117. I. Echeverria, P. L. Kolek, D. J. Plazek, and S. L. Simon, Enthalpy recovery, creep and creep-recovery measurements during physical aging of amorphous selenium, *J. Non-Cryst. Solids*, 324 (3), 242–255, 2003.

118. J. Zhao and G. B. McKenna, Temperature divergence of the dynamics of a poly(vinyl acetate) glass: Dielectric vs. mechanical behaviors, *J. Chem. Phys.*, 136 (15), 154901, 2012.

119. I. L. Hopkins, Stress relaxation or creep of linear viscoelastic substances under varying temperature, *J. Polym. Sci.*, 28, 631–633, 1958.

120. H. Leaderman, *Elastic and Creep Properties of Filamentous Materials* (Textile Foundation, Inc., Washington, D.C.), 1943.

121. R. S. Marvin, E. R. Fitzgerald, and J. D. Ferry, Measurements of mechanical properties of polyisobutylene at audiofrequencies by a twin transducer, *J. Appl. Phys.*, 21, 197–203, 1950.

122. A. Tobolsky and E. Catsiff, Elastoviscous properties of polyisobutylene (and other amorphous polymers) from stress-relaxation studies. IX. A summary of results, *J. Polym. Sci.*, 19, 111–121, 1956.

123. T. Alfrey, Jr., *Mechanical Behavior of High Polymers* (Interscience, New York), 1948.

124. L. W. Morland and E. H. Lee, Stress analysis for linear viscoelastic materials with temperature variation, *Trans. Soc. Rheol.*, 4, 233–263, 1960.

125. Y. Guo, N. Wang, R. D. Bradshaw, and L. C. Brinson, Modeling mechanical aging shift factors in glassy polymers during nonisothermal physical aging. I. Experiments and KAHR-a$_{t_e}$ model prediction, *J. Polym. Sci., Part B: Polym. Phys.*, 47, 340–352, 2009.

126. S. Kollengodu-Subramanian and G. B. McKenna, A dielectric study of poly(vinyl acetate) using a pulse-probe technique, *J. Therm. Anal. Calorim.*, 102, 477–484, 2010.

127. R. L. Hastie, Jr. and D. H. Morris, The effects of physical aging on the creep response of a thermoplastic composite, in *High Temperature and Environmental Effects on Polymeric Composites, ASTM STP 1174*, edited by C. E. Harris and T. S. Gates (American Society of Testing and Materials, Philadelphia, PA), 63–185, 1993.

128. T. S. Gates, D. R. Veazie, and L. C. Brinson, Creep and physical aging in a polymeric composite: Comparison of tension and compression, *J. Comp. Matls.*, 31, 2478–2505, 1997.

129. L. C. E. Struik, The mechanical and physical aging of semicrystalline polymers 1, *Polymer*, 28, 1521–1533, 1987.

130. L. C. E. Struik, The mechanical and physical aging of semicrystalline polymers 2, *Polymer*, 28, 1534–1542, 1987.

131. S. M. Aharoni, Increased glass transition temperature in motionally constrained semicrystalline polymers, *Polym. Adv. Tech.*, 9, 169–201, 1998.

132. J. Beckmann, G. B. McKenna, B. G. Landes, D. H. Bank, and R. A. Bubeck, Physical aging kinetics of syndiotactic polystyrene as determined from creep behavior, *Polym. Eng. Sci.*, 37, 1459–1468, 1997.

133. M. L. Cerrada and G. B. McKenna, Creep behavior in amorphous and semi-crystalline PEN, in *Time Dependent and Nonlinear Effects in Polymers and Composites*, edited by R. A. Schapery and C. T. Sun (American Society of Testing and Materials, West Conshohocken, PA, Special Technical Publication, STP 1357), 47–69, 2000.

134. B. E. Read, D. E. Tomlin, and G. D. Dean, Physical ageing and short-term creep in amorphous and semicrystalline polymers, *Polymer*, 31, 1204–1215, 1990.

135. J. Zhao, J. Wang, C. Li, and Q. Fan, Study of the amorphous phase in semicrystalline poly(ethylene terephthalate) via physical aging, *Macromolecules*, 35, 3097–3103, 2002.

136. D. P. N. Vlasveld, H. E. N. Bersee, and S. J. Picken, Creep and physical aging behaviour of PA6 nanocomposites, *Polymer*, 46, 12539–12545, 2005.

137. G. Vigier and J. Tatibouet, Physical ageing of amorphous and semicrystalline poly(ethylene terephthalate), *Polymer*, 34, 4257–4266, 1993.

138. M. R. Tant and G. L. Wilkes, Physical aging studies of semicrystalline poly(ethylene terephthalate, *J. Appl. Polym. Sci.*, 26, 2813–2825, 1981.

139. K. Banik and G. Mennig, Influence of the injection molding process on the creep behavior of semicrystalline PBT during aging below its glass transition temperature, *Mech. Time-Dependent Mater.*, 9, 247–257, 2006.

140. I. Spinu and G. B. McKenna, Physical aging of Nylon 66, *Polym. Eng. Sci.*, 34, 1808–1814, 1994.

141. J. Menczel and B. Wunderlich, Heat capacity hysteresis of semicrystalline macromolecular glasses, *J. Polym. Sci. Polym. Lett.*, 19, 261–264, 1981.

142. P. Huo and P. Cebe, Effects of thermal history on the rigid amorphous phase in poly(phenylene sulfide), *Coll. Polym. Sci.*, 270, 840–852, 1992.

143. A. Papon, H. Montes, M. Hanafi, F. Lequeux, L. Guy, and K. Saalwächter, Glass-transition temperature gradient in nanocomposites: Evidence from nuclear magnetic resonance and differential scanning calorimetry, *Phys. Rev. Lett.*, 108, 065702-1–065702-5, 2012.

144. L. C. E. Struik, Mechanical behavior and physical aging of semicrystalline polymers: 3. Prediction of long-term creep from short-time tests, *Polymer*, 30, 799–814, 1989.

145. L. C. E. Struik, Mechanical behavior and physical aging of semicrystalline polymers: 4, *Polymer*, 30, 815–830, 1989.

146. R. A. Schapery, On characterization of nonlinear viscoelastic materials, *Polym. Eng. Sci.*, 9, 295–310, 1969.

147. F. A. Myers, F. C. Cama, and S. S. Sternstein, Mechanically enhanced aging of glassy polymers, *Ann. N.Y. Acad. Sci.*, 279, 94–99, 1976.

148. R. Pixa, C. Goett, and D. Froelich, Influence of deformation on the physical ageing of polycarbonate. 1. Mechanical properties near ambient temperature, *Polym. Bull.*, 14, 53–60, 1985.

149. A. F. Yee, R. J. Bankert, K. L. Ngai, and R. W. Rendell, Strain and temperature accelerated relaxation in polycarbonate, *J. Polym. Sci., Part B: Polym. Phys.*, 26, 2463–2483, 1988.

150. B. Haidar and T. L. Smith, Physical ageing of stretch specimens of a polycarbonate film and its temperature dependence, *Polymer*, 31, 1904–1908, 1990.

151. J. Bartos, J. Muller, and J. H. Wendorff, Physical ageing of isotropic and anisotropic polycarbonate, *Polymer*, 31, 1678–1684, 1990.

152. W. G. Knauss and W. Zhu, Nonlinearly viscoelastic behavior of polycarbonate. I. Response under pure shear, *Mech. Time-Dependent Mater.*, 6, 231–269, 2002.

153. H. Kawakami, R. Otsuki, and Y. Nanzai, Structural relaxation and evolution of yield stress in epoxy glass aged under shear strain, *Polym. Eng. Sci.*, 45, 20–24, 2005.

154. G. B. McKenna and A. J. Kovacs, Physical ageing of poly(methyl methacrylate) in the nonlinear range: Torque and normal force measurements, *Polym. Eng. Sci.*, 24, 1131–1141, 1984.

155. T. Ricco and T. L. Smith, Rejuvenation and physical aging of a polycarbonate film subjected to finite tensile strains, *Polymer*, 26, 1979–1984, 1985.

156. D. M. Colucci, P. A. O'Connell, and G. B. McKenna, Stress relaxation experiments in polycarbonate: A comparison of volume changes for two commercial grades, *Polym. Eng. Sci.*, 37, 1469–1474, 1997.

157. G. B. McKenna, M. M. Santore, A. Lee, and R. S. Duran, Aging in glasses subjected to large stresses and deformations, *J. Non-Cryst. Solids*, 131, 497–504, 1991.

158. Y. C. G. Chung and D. J. Lacks, Atomic mobility in a polymer glass after shear and thermal cycles, *J. Phys. Chem. B*, 116, 14201–14205, 2011.

159. R. A. Riggleman, K. S. Schweizer, and J. J. de Pablo, Nonlinear creep in a polymer glass, *Macromolecules*, 41, 4969–4977, 2008.

160. Y. G. Chung and D. J. Lacks, How deformation enhances mobility in a polymer glass, *Macromolecules*, 45 (10), 4416–4421, 2012.

161. K. Chen and K. S. Schweizer, Theory of aging, rejuvenation, and the nonequilibrium steady state in deformed polymer glasses, *Phys. Rev. E*, 82, 041804, 2010.

162. D. J. A. Senden, S. Krop, J. A. W. van Dommelen, and L. E. Govaert, Rate- and temperature-dependent strain hardening of polycarbonate, *J. Polym. Sci., Part B: Polym. Phys.*, 50, 1680–1693, 2012.

163. L. C. A. van Breemen, T. A. P. Engels, E. T. J. Klompen, D. J. A. Senden, and L. E. Govaert, Rate- and temperature-dependent strain softening in solid polymers, *J. Polym. Sci., Part B: Polym. Phys.*, 50, 1757–1771, 2012.

164. S. E. B. Petrie, The effect of excess thermodynamic properties versus structure formation on the physical properties of glassy polymers, *J. Macromol. Sci. Phys. Ed.*, 12, 225–247, 1976.

165. A. Aref-Azar, F. Biddlestone, J. N. Hay, and R. N. Haward, The effect of physical aging on the properties of poly(ethylene-terephthalate), *Polymer*, 24, 1245–1251, 1983.

166. J. H. Golden, B. L. Hammant, and E. A. Hazell, Effect of thermal pretreatment on strength of polycarbonate, *J. Appl. Polym. Sci.*, 11, 1571–1579, 1967.

167. D. G. Legrand, Crazing, yielding, and fracture of polymers. I. Ductile brittle transition in polycarbonate, *J. Appl. Polym. Sci.*, 13, 2129–2147, 1969.

168. J. R. Flick and S. E. B. Petrie, Studies in physical and theoretical chemistry in structures and properties of amorphous materials, in *Proc. Second Symp. Macromol.*, Cleveland, Ohio, edited by A. G. Walton, 10, 145–163, 1978.

169. K. Strabala, S. Meagher, C. Landais, L. Delbreilh, M. J. M. Saiter, J. Turner, A. Ingram, and R. Golovchak, Anisotropic loss of toughness with physical aging of work toughened polycarbonate, *Polym. Eng. Sci.*, 54 (4), 794–804, 2014.

170. A. B. Brennan and F. Feller III, Physical aging behavior of a poly(arylene etherimide), *J. Rheol.*, 39 (2), 453–470, 1995.

171. A. C.-M. Yang, R. C. Wang, and J. H. Lin, Ductile-brittle transition induced by aging in poly(phenylene oxide) thin films, *Polymer*, 37 (25), 5751–5754, 1996.

172. Z. H. Ophir, J. A. Emerson, and G. L. Wilkes, Sub T_g annealing studies of rubber modified and unmodified epoxy systems, *J. Appl. Phys.*, 49 (10), 5032–5038, 1978.

173. C. G'Sell and G. B. McKenna, Influence of physical aging on the yield response of model DGEBA poly(propylene oxide) epoxy glasses, *Polymer*, 33 (10), 2103–2113, 1992.

174. J. C. Arnold, The influence of physical aging on the creep-rupture behavior of polystyrene, *J. Polym. Sci. Polym. Phys. Ed.*, B31, 1451–1458, 1993.

175. G. M. Gusler and G. B. McKenna, The craze initiation response of a polystyrene and a styrene-acrylonitrile copolymer during physical aging, *Polym. Eng. Sci.*, 37 (9), 1442–1448, 1997.

176. C-H. Liu and J. A. Nairn, Using the essential work of fracture method for studying physical aging in thin, ductile, polymeric films, *Polym. Eng. Sci.*, 38 (1), 186–193, 1998.

177. J. M. Hutchinson, S. Smith, B. Horne, and G. M. Gourlay, Physical aging of polycarbonate: Enthalpy relaxation, creep response, and yielding behavior, *Macromolecules*, 32, 5057–5061, 1999.

178. C. G. Robertson, J. E. Monat, and G. L. Wilkes, Physical aging of an amorphous polyimide: Enthalpy relaxation and mechanical property changes, *J. Polym. Sci., Part B: Polym. Phys.*, 37, 1931–1946, 1999.

179. W. Liu, J. Shen, F. Lu, and M. Xu, Effect of physical aging on fracture behavior of polyphenylquinoxaline films, *J. Appl. Polym. Sci.*, 78, 1275–1279, 2000.

180. R. M. Mininni, R. S. Moore, J. R. Flick, and S. E. B. Petrie, Effect of excess volume on molecular mobility and on mode of failure of glassy poly(ethylene terephthalate), *J. Macromol. Sci. B Phys.*, B008 (1–2), 343–359, 1973.

181. G. B. McKenna, J. M. Crissman, and A. Lee, Relationships between failure and other time dependent processes in polymeric materials, *Polym. Prep. V*, 29 (2), 128–129, 1988.

182. J. A. Manson, R. W. Hertzberg, S. L. Kim, and W. C. Wu, Toughness and brittleness of plastics, in *Advances in Chemistry Series No. 154*, edited by R. D. Deanin and A. M. Crugnola (American Chemical Society, Washington, D.C., 146), 1976.

183. J. M. Crissman and G. B. McKenna, Physical and chemical aging in PMMA and their effects on creep and creep-rupture behavior, *J. Polym. Sci., Part B: Polym. Phys.*, 28, 1463–1473, 1990.

184. H. A. Visser, T. C. Bor, M. Wolters, J. G. F. Wismans, and L. E. Govaert, Lifetime assessment of load-bearing polymer glasses: The influence of physical aging, *Macromol. Mater. Eng.*, 295, 1066–1081, 2010.

185. I. A. Hodge, Physical aging in polymer glasses, *Science*, 267, 1945–1947, 1995.

3

Glass transition and relaxation behavior of supercooled polymer melts
An introduction to modeling approaches by molecular dynamics simulations and to comparisons with mode-coupling theory

JÖRG BASCHNAGEL, IVAN KRIUCHEVSKYI, JULIAN HELFFERICH,
CÉLINE RUSCHER, HENDRIK MEYER, OLIVIER BENZERARA,
JEAN FARAGO, AND JOACHIM P. WITTMER

3.1 INTRODUCTION

3.1.1 POLYMER MELTS: ASPECTS OF THEIR STRUCTURE AND DYNAMICS

Polymer melts are liquids composed of macromolecular chains [1,2]. A macromolecule can have various microstructures (homopolymer, copolymer, etc.) and architectures (linear, ring, star, etc.). In this chapter, we will only be concerned with the simplest structure, that of a linear homopolymer, where N monomeric repeat units of the same type are connected to form a chain. Moreover, we will assume that the polymers in the melt are monodisperse (all have the same N), that their chain conformations are flexible, and that the nonbonded interactions are short-range and spatially isotropic. This is a standard chain model pertinent to many synthetic polymers [2]. Experimentally, the chain length N can be large, a typical range being $10^3 \lesssim N \lesssim 10^5$. Since the average size of a polymer grows with N, these large chain lengths also imply the chain size to be large. A possible measure of the chain size is the (average) end-to-end distance (R) of a polymer [1,2]. For $10^3 \lesssim N \lesssim 10^5$, the end-to-end distance is typically in the range of $R \sim 100 - 1000\,\text{Å}$. The size of a chain thus exceeds that of a monomer $(\sim 1\,\text{Å})$ by several orders of magnitude.

These widely distributed length scales manifest themselves in the structural and dynamic properties of a polymer melt. In the melt, the monomers spatially arrange in densely packed nearest-neighbor shells. On a local scale, this leads to an amorphous short-range order and on a macroscopic scale to a small compressibility. Both features are reflected in the static structure factor $S(q)$ that measures the collective density fluctuations in the melt (here, q denotes the modulus of the wave vector $\boldsymbol{q}$) [3]. $S(q)$ displays an "amorphous halo," a first sharp diffraction peak, at $q \approx 1-1.5\,\text{Å}^{-1}$ [4]. This q values corresponds to a small distance of $\lambda = 2\pi/q \sim 0.5$ nm and results from local packing of the monomers. With decreasing q, the structure factor decreases toward a low plateau value related to compressibility of the melt, $S(q \to 0) = k_\text{B} T \rho \kappa_T \lesssim 0.3$ [5], with k_B being the Boltzmann constant, T the absolute temperature, ρ the monomer (number) density, and κ_T the isothermal compressibility. This q dependence of $S(q)$ is characteristic of the liquid state and qualitatively agrees with that of nonpolymeric liquids [3].

For polymers, however, $S(q)$ has both intramolecular and intermolecular contributions [4,6,7]. The intramolecular contributions represent polymer-specific properties of the chain conformations. The conformation of a long polymer in a (three-dimensional) melt has an "open" structure, which allows other chains to penetrate into the volume occupied by the polymer [2,8]. On average, a polymer segment with s monomers $(1 \ll s \le N)$ has $\sqrt{s}$ contacts with other segments [2,9–11]. The number of these contacts becomes huge when $s \to N$ and N is large. This strong interpenetration of the chains has important consequences. For instance, as s increases, intrachain excluded volume interactions, which swell the polymer in dilute solution [1,2,8], are progressively screened by neighboring chains [9–11]. To first (and good) approximation, a polymer in a melt thus has an "ideal," random-walk-like, conformation. In particular, this implies that the end-to-end distance increases with chain length as $R = b\sqrt{N}$, where b is the effective bond length $(b = \sqrt{C_\infty \langle l^2 \rangle}$, where C_∞ is the Flory characteristic ratio for $N \to \infty$, l is the bond length, and $\langle...\rangle$ denotes the thermal average) [1,2]. Corrections to ideal chain behavior, due to residual excluded volume interactions in the melt, have been worked out and provide, for instance, extrapolation formulas allowing to determine b for long, but finite, chain lengths [9–11].

The strong interpenetration of the chains in the melt has a further consequence. Since the chains cannot cross each other, they form a temporary network of entanglements, if N is much larger than the entanglement length N_e. For $N > N_e$, entanglements strongly affect the relaxation of the polymers [1,2,8]. The classical approach to the modeling of polymer dynamics therefore takes two stages: If $N < N_e$, entanglements are supposed to be negligible so that the Rouse model applies [1,2,8]. This model adopts a coarse-grained description: A tagged polymer is represented by a Gaussian bead–spring model (BSM) with ideal chain statistics. The effect of the other surrounding chains is to create a structureless heat bath exerting a thermal noise and the same local friction on every "bead" (coarse-grained monomer) of the polymer. A key prediction of the Rouse model is that the polymer relaxation time (τ_N) considerably exceeds the bead relaxation time (τ_1), even for modest N, since $\tau_N = \tau_1 N^2$. This explains why melts of short chains are already much more viscous than nonpolymeric liquids. Viscosity and slow relaxation become strongly enhanced for $N \gg N_e$. The temporary network of entanglements confines the polymer motion to an effective tube, imposing the slithering of the chain along the tube axis ("reptation") as the dominant relaxation mechanism. This greatly slows down the polymer dynamics ($\tau_N = \tau_1 N^3 / N_e$) and also leads to pronounced viscoelastic effects [1,2,8]. Again, corrections to both the Rouse and reptation models have been discussed. For the Rouse model, they may be attributed to residual excluded volume interactions (see, for example, [10,12] and references therein) and hydrodynamic interactions caused by the viscoelasticity of the melt [13–15]. For the reptation model, the corrections are commonly described in terms of contour length fluctuations and (collective) constraint-release effects [1,2,16] (see also [17] for a critical discussion of the Rouse, reptation, and other single-chain models for describing nonentangled and entangled polymer melts as well as a detailed comparison with multichain molecular dynamics [MD] simulations).

The preceding overview about polymer melts may be summarized as follows. The collective structure of the melt, measured by $S(q)$, does not change when N increases; it remains liquid-like. By contrast, the dynamics is strongly affected. With increasing N, the monomer relaxation time τ_1 and the chain relaxation time τ_N become more and more separated. This opens a time window $\tau_1 < t < \tau_N$ where slow relaxation of the polymers occurs, before viscous flow sets in for $t > \tau_N$. Therefore, triggered by the increase of chain length, viscoelasticity emerges in polymer liquids, already at high temperature.

3.1.2 Polymer crystallization or glass formation

When cooled to low temperatures, the polymer liquid transforms into a solid that can be either amorphous (glassy) or semicrystalline [18]. Semicrystalline polymers have both amorphous and crystalline regions. In many cases, the crystalline regions are formed of lamellar sheets in which the polymers fold back and forth so that chain sections align parallel to each other. The sheets twist and branch as they grow outward from a nucleus, forming spherulitic structures at large scales [18]. The hierarchy of these morphological features range from the lamellar ordering of the chains (~10 nm) to the macroscopic packing of the spherulites (100 μm and larger). This reflects the complexity of the underlying crystallization process, which is not fully understood [19–21].

The ability to form crystals hinges on the microstructure of the polymer. Only chains with regular configurations (for example, isotatic or syndiotactic orientations of the sidegroups [2]) or chains without sidegroups (for example, polyethylene) can align parallel to each other so as to pack into crystalline lamellae. However, even in these favorable cases, full crystallization is hard to achieve (see, for example, [19]). This intrinsic difficulty of crystal formation implies that polymer melts are in general good glass formers [22–24]. Either they can be readily supercooled* or, due to irregular chain configurations, a crystalline phase does not exist at all. There are several examples for the latter case, for instance, homopolymers with an atactic orientation of (bulky) sidegroups (atactic polystyrene, etc.) or random copolymers, such as *cis–trans* polybutadiene, in

---

* A remark on the term "supercooled state" appears to be appropriate. Within the framework of first-order phase transitions, a supercooled state is defined as the region between the binodal and the spinodal lines in the phase diagram. There the system is in "metastable equilibrium": it is in a long-lived state protected by a free-energy barrier against transformation into the ordered phase. However, there are systems—like atatic polymer melts or polydisperse colloidal suspensions—that do not crystallize. These examples suggest that metastability relative to a crystalline phase is not always necessary for the emergence of glassy behavior. In the parlance of glass physics, a liquid is said to be "supercooled" if the temperature is so low that the slow structural relaxation processes, which ultimately lead to the glass transition, can be observed. See References 25 and 26 for further discussion of the latter point and also Reference 27 in particular for systems where a competition between crystallization and glass formation exists.

which monomers, having the same chemical composition, but different microstructures (*cis/trans* configuration of butadiene), are randomly concatenated.

3.1.3 GLASS-FORMING LIQUIDS: SOME KEY CHARACTERISTICS

These polymeric glass formers exhibit several properties that are also found in other nonpolymeric (intermediate and fragile [28,29]) glass-forming liquids and are thus typical of the glass transition in general [23,24,27,29–31]. For instance, as the liquid is cooled toward low temperatures, it becomes kinetically arrested below a characteristic temperature T_g and forms an amorphous solid for $T \ll T_g$.* Often, T_g is determined by heat capacity measurements (cf. Chapters 2 and 4) and is therefore also referred to as "calorimetric glass transition temperature." In the course of the vitrification process, standard static correlation functions, such as $S(q)$, change very little. The structure of the liquid and the glass is essentially identical[†]: A glass just appears to be a "frozen liquid," a liquid that has stopped to flow. This suggests to focus on the dynamic behavior and its T dependence as the glass transition is approached from the high-temperature side.

Indeed, on cooling toward T_g, characteristic dynamic features, sometimes collectively referred to as "glassy dynamics," emerge. Certainly, the most prominent feature is the massive increase of all structural relaxation times, over about 14 orders of magnitude, from $\sim 10^{-12}$ s at high T to $\sim 10^2$ s at T_g. In practice, the structural relaxation time can be determined, for example, by dielectric spectroscopy—measuring the relaxation of local electric dipoles—or dynamic neutron scattering—measuring the relaxation of dynamic density fluctuations for different q values, for instance, on the local scale corresponding to the first maximum of $S(q)$. This structural relaxation time, probing local relaxation processes, is called "α relaxation time" (τ_α). The strong increase of τ_α with decreasing temperature is often represented in an Arrhenius diagram plotting $\log(\tau_\alpha)$ versus $1/T$. In this representation, the Arrhenius law, $\tau_\alpha(T) \propto \exp(E_a/k_B T)$, yields a straight line whose slope is given by the activation energy E_a. However, fragile or intermediate glass-forming liquids, including polymer melts, show an upward curvature in the diagram, implying an effective activation energy $E_a(T)$ that increases when T is lowered toward T_g. Since this "super-Arrhenius" behavior of τ_α is the precursor to the glass transition, uncovering its molecular origin is one of the challenging problems in glass physics (see [24,27,29–36] and Chapters 8 and 9).

Along with the super-Arrhenius behavior of τ_α, another feature of glassy dynamics arises: the stretching of the structural relaxation over large time windows. The stretching is revealed in the t dependence of ensemble-averaged dynamic correlation functions, $\phi_A(t) = \langle A^\star(t)A(0)\rangle$, of structure-sensitive variables A, such as coherent or incoherent density fluctuations (here, $A^\star$ denotes the complex conjugate). These correlation functions relax in two steps where the first step occurs on a (weakly T dependent) microscopic time scale τ_{mic} and the second on the (strongly T dependent) α time scale τ_α. The steps are thus separated by an intermediate time window ($\tau_{mic} < t < \tau_\alpha$) that increases in size on cooling toward T_g. In the intermediate time window, the correlation function remains close to a plateau value f_A, that is, $|\phi_A(t) - f_A| \ll 1$ for $\tau_{mic} \ll t \ll \tau_\alpha$. This intermittence of structural relaxation implies that viscoelastic behavior emerges for $t < \tau_\alpha$ when a liquid is cooled toward T_g. The onset of viscous flow is postponed to times larger than the terminal relaxation time, which is τ_α in simple liquids. For polymer melts, however, τ_α is not the terminal relaxation time, since it is a local relaxation time on the order of τ_1 and viscous flow of the melt only occurs for $t > \tau_N$. Supercooled polymer melts therefore have two physically distinct sources of viscoelasticity, one related to cooling and

---

* The question of whether the state below T_g is a true amorphous solid is a matter of debate. A true solid should have infinite relaxation time so that it does not flow and is mechanically stable for all times. By contrast, below the nominal T_g glasses show slow relaxation processes toward equilibrium (called "physical aging," cf. Chapters 2 and 4) and perhaps even have an equilibrium structural relaxation time that diverges only at zero temperature (cf. Section 3.3.2.4). However, for $T \ll T_g$, this relaxation time is so long that the glass can be considered effectively as a solid with well-defined mechanical properties (for example, elastic constants).

† The conclusion that the structure of the liquid and glassy state is essentially identical is based on the observation of the static structure factor that measures spatial correlations between two particles ($S(q)$ is a "two-point correlation function"). This does not exclude the possibility of more subtle (higher-order) static correlations that would not be visible in $S(q)$. The search for such static correlations is an active topic of current research in glass physics [30].

the attendant slow monomer relaxation for $t \ll \tau_\alpha \sim \tau_1$ and the other related to slow chain relaxation for $\tau_1 \ll t \ll \tau_N$.

The two-step relaxation related to glassy dynamics already develops for moderate supercooling in the time regime of nanoseconds (and beyond) [24,27,29,32]. This time regime is well accessible to molecular modeling approaches by computer simulations. Computer simulation have therefore played an important role in exploring the onset of glassy dynamics [4,7,25,29,37]. This is the main topic of this chapter.

There is another field where computer simulations have played an important role. A relatively recent addition to the phenomenology of glassy dynamics is the observation of dynamic heterogeneities in supercooled liquids [27,29–31,38,39]. What does the term "dynamic heterogeneities" mean? Within the time interval preceding the α relaxation, some particles remain localized near their initial positions, whereas others move over (large) distances comparable to their size. These populations of "mobile" and "immobile" particles cluster in space and the size of the clusters grows with decreasing temperature. This heterogeneous dynamics, both in space and in time, is a main difference to the homogeneous behavior of the high-temperature liquid [27,29–31,38,39]. Dynamic heterogeneities will only be touched upon in this chapter; they are discussed in more detail in Chapters 8 and 9.

3.1.4 OUTLINE OF THE CHAPTER

In various chapters of this book, including the present one, molecular dynamics simulations are applied to study glass-forming polymer melts. Section 3.2 therefore gives an introduction to this simulation technique and also presents some polymer models commonly employed in the simulations.

In Section 3.3, our discussion revolves around two aspects: (i) the determination of T_g in the simulation and the dependence of T_g on pressure, chain length, or cooling rate (Section 3.3.1) and (ii) the description of the dynamics of supercooled polymer melts in thermal equilibrium (Section 3.3.2). Our analysis in Section 3.3.2 will often refer to the mode-coupling theory (MCT) for the structural glass transition. We give an introduction to this theory, discuss applications to polymer systems, and assess the strengths and weaknesses of MCT.

The final Section 3.4 revisits some of the points examined before, completes our discussion by mentioning further lines of current research, and also hints at possible future research directions.

3.2 COMPUTER SIMULATIONS OF GLASS-FORMING POLYMERS

Computer simulations in condensed matter physics aim at calculating (macroscopic) observables of a many-body system from the interactions between the constituent particles [40–42]. Conceptually, the simulations are therefore closely connected with statistical mechanics. In statistical mechanics, observables are identified with ensemble averages. These ensemble averages are estimated in the simulation by "time" averages over a (finite) sequence of microscopic configurations, which have to be generated with the equilibrium probability density corresponding to the considered thermodynamic ensemble. One simulation technique to achieve this goal is the molecular dynamics method. MD simulations numerically integrate Newton's equations of motion of classical many-body systems. Thereby, they allow to determine thermodynamic, structural, and dynamic properties of a system. This possibility to access static and transport features and the (continued) development of optimized codes for parallel computers, such as LAMMPS [43,44], GROMACS [45,46], or NAMD [47,48] (all public domain), have turned MD simulations into an important tool of modern research. Classical molecular dynamics is nowadays applied to a wide variety of problems [49], including soft matter systems [50] and glasses [29,39,51].

An in-depth discussion of the MD method is certainly beyond the scope of this chapter. Comprehensive, pedagogical descriptions may be found in excellent textbooks [41,42] and advances of the field have been the subject of topical reviews (see, for example, [49,52]). Here we only want to touch on a few aspects of the method, which should help to understand other parts of this chapter and of this book (cf. Chapters 8, 11, and 13).

3.2.1 MOLECULAR DYNAMICS SIMULATIONS

3.2.1.1 INTRODUCTION: EQUATIONS OF MOTION FOR ATOMIC SYSTEMS

Let's consider a classical system of identical particles (same mass m) in the microcanonical ensemble. This ensemble is defined as the set of all microscopic configurations (x) that are compatible with the macroscopic constraints of fixed particle number (N), volume (V), and total energy (E, sum of kinetic and potential energy). At any time t, the positions (r) and velocities (v) of all particles fully specify the microscopic configuration of the system: $x(t) = (r_1(t),...,r_N(t);v_1(t),...,v_N(t))$. To obtain $x(t)$, we have to integrate Newton's equations of motion for all particles:

$$\frac{d^2 r_i(t)}{dt^2} = \frac{1}{m} F_i(t) \quad (i=1,...,N), \tag{3.1}$$

where F_i denotes the total force on particle i. If we assume that no external fields (like gravity, etc.) act on the particle, F_i comprises two contributions: one due to the interaction potential $U_N(r_1,...,r_N)$ between the particles and one due to the interaction of the particle with the boundary of the system. To minimize surface effects, simulations typically employ periodic boundary conditions, that is, the system is periodically replicated in all spatial directions (cf. Figure 3.1). A particle at the boundary therefore sees copies of the other particles from inside the system. This implies that U also determines the boundary contribution and the total force is given by $F_i = -\partial U_N(r_1,...,r_N)/\partial r_i$.

Molecular dynamics simulations numerically integrate Equation 3.1 in an iterative way. Let's assume that we know the positions and velocities for all particles at time t. The positions $r_i(t)$ allow us to determine the forces $F_i(t)$. Then, a Taylor expansion gives an estimate of the new positions and velocities at a small time increment Δt later:

$$r_i(t + \Delta t) \approx r_i(t) + v_i(t)\Delta t + \frac{F_i(t)}{2m}(\Delta t)^2, \tag{3.2}$$

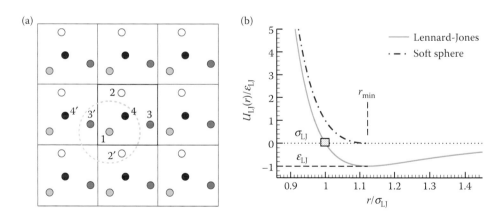

Figure 3.1 (a) Illustration of the periodic boundary conditions (PBCs). The central (gray) simulation box is replicated in all spatial directions. This replication creates identical copies ("image particles") of all particles in the central box (particles are indicated by spheres). The identity of a particle and all of its images implies that they move in the same way. For instance, if particle 3 leaves the central box to the right, all copies leave their boxes to the right. Thereby, particle 3' enters the central box. This concerted motion keeps the particle density constant. PBCs also influence which particles interact with one another. Let the interaction range of particle 1 be indicated by the dashed circle. While particle 2 is outside of this range, its image 2' is not. Owing to PBCs, particle 1 interacts with the nearest image particles—that is, particles 2', 3', and 4—and so do all particles ("minimum image convention" [40–42]). (b) Lennard-Jones potential (Equation 3.6) versus distance r (solid line). $U_{LJ}(r)$ is scaled by the energy minimum ε_{LJ} (indicated by a horizontal dashed line) and r is scaled by the particle diameter σ_{LJ} (indicated by a square). The vertical dashed line shows the position of the energy minimum, $r_{min} = 2^{1/6}\sigma_{LJ}$. The dash-dotted line depicts the (purely repulsive) soft-sphere potential derived from the LJ potential by truncating Equation 3.6 at r_{min} and shifting it to 0.

$$\boldsymbol{v}_i(t+\Delta t) \approx \boldsymbol{v}_i(t) + \frac{\boldsymbol{F}_i(t)}{m}\Delta t. \tag{3.3}$$

The time increment Δt is referred to as the "time step" in MD simulations. From $\boldsymbol{r}_i(t+\Delta t)$, we can obtain the new forces $\boldsymbol{F}_i(t+\Delta t)$, which completes the input for Equations 3.2 and 3.3 to calculate the positions and velocities at time $t + 2\Delta t$. Iteration of this procedure therefore furnishes a discretized trajectory of the system, $\boldsymbol{x}(t_k = k\Delta t)$, with $k = 0,1,2,\ldots,N_{max}$, starting from the initial configuration ($k = 0$) up to the final configuration for the maximum number N_{max} of time steps simulated.

While the preceding description illustrates the principle of the MD simulation, Equations 3.2 and 3.3 are not used in practice due to the following drawback: They are not invariant with respect to time reversal, whereas Equation 3.1 is. Time-reversal symmetry implies that the structure of the equations of motion is preserved, if we move backward in time. Assume that we know the positions and velocities at time $t + \Delta t$ and we want to calculate $\boldsymbol{r}_i(t)$ and $\boldsymbol{v}_i(t)$ by inserting the reverse time step $-\Delta t$ in Equations 3.2 and 3.3. This would alter the equations, thereby violating time-reversal symmetry.

To remedy this problem, assume again that we know the positions and velocities at time $t + \Delta t$. From the positions, we calculate the new forces $\boldsymbol{F}_i(t+\Delta t)$. Then, we can apply the Taylor expansion of the velocities in reverse:

$$\boldsymbol{v}_i(t) \approx \boldsymbol{v}_i(t+\Delta t) - \frac{\boldsymbol{F}_i(t+\Delta t)}{m}\Delta t \quad \Rightarrow \quad \boldsymbol{v}_i(t+\Delta t) \approx \boldsymbol{v}_i(t) + \frac{\boldsymbol{F}_i(t+\Delta t)}{m}\Delta t.$$

Adding this result to Equation 3.3 gives

$$\boldsymbol{v}_i(t+\Delta t) \approx \boldsymbol{v}_i(t) + \frac{\boldsymbol{F}_i(t) + \boldsymbol{F}_i(t+\Delta t)}{2m}\Delta t. \tag{3.4}$$

Now, Equation 3.4 is time reversible because it is symmetric with respect to t and $t + \Delta t$, contrary to Equation 3.3. Moreover, Equation 3.4 also makes the positions time reversible. To see this, let's apply the reverse time step $-\Delta t$ from $t + \Delta t$ to t in Equation 3.2:

$$\boldsymbol{r}_i(t) \approx \boldsymbol{r}_i(t+\Delta t) - \boldsymbol{v}_i(t+\Delta t)\Delta t + \frac{\boldsymbol{F}_i(t+\Delta t)}{2m}(\Delta t)^2.$$

Inserting Equation 3.4, we get

$$\boldsymbol{r}_i(t) \approx \boldsymbol{r}_i(t+\Delta t) - \boldsymbol{v}_i(t)\Delta t - \frac{\boldsymbol{F}_i(t)}{2m}(\Delta t)^2,$$

which can be rearranged to give back Equation 3.2. The resulting algorithm

$$\boldsymbol{r}_i(t+\Delta t) \approx \boldsymbol{r}_i(t) + \boldsymbol{v}_i(t)\Delta t + \frac{\boldsymbol{F}_i(t)}{2m}(\Delta t)^2,$$
$$\boldsymbol{v}_i(t+\Delta t) \approx \boldsymbol{v}_i(t) + \frac{\boldsymbol{F}_i(t) + \boldsymbol{F}_i(t+\Delta t)}{2m}\Delta t \tag{3.5}$$

is called a "velocity-Verlet algorithm" [41,42]. It is the most commonly used algorithm in MD simulations.

The preceding description of the MD method shows that the time step Δt is a crucial operational parameter. Certainly, we would like to take Δt as large as possible, since this allows to extend the longest simulation time, $t_{max} = N_{max}\Delta t$, at fixed computational effort N_{max}. However, the accuracy of the simulation may then suffer because the Taylor expansion in Equation 3.5 is only appropriate for small Δt. So, how should Δt be chosen?

The answer depends on the system, more precisely, on the interaction potential. To see this, let's assume that the particles interact pairwisely by a Lennard-Jones (LJ) potential

$$U_{LJ}(r) = 4\varepsilon_{LJ}\left[\left(\frac{\sigma_{LJ}}{r}\right)^{12} - \left(\frac{\sigma_{LJ}}{r}\right)^{6}\right] = \varepsilon_{LJ}U_{LJ}^{*}(r^{*} = r/\sigma_{LJ}),$$ (3.6)

where ε_{LJ} is the depth of the potential minimum and σ_{LJ} the particle diameter (cf. Figure 3.1). The constants ε_{LJ} and σ_{LJ} set, respectively, a scale for the energy and the length. So, we can introduce nondimensional ("reduced") quantities: $U_{LJ}^{*} = U_{LJ}/\varepsilon_{LJ}$ and distance $r^{*} = r/\sigma_{LJ}$ [41,42]. This has an immediate consequence for the force on particle i ($U_{N}(\mathbf{r}_{1},\ldots,\mathbf{r}_{N}) = \sum_{j>i} U_{LJ}(r_{ij})$ with $r_{ij} = |\mathbf{r}_{ij}|$ and $\mathbf{r}_{ij} = \mathbf{r}_{i} - \mathbf{r}_{j}$) [53]:

$$\mathbf{F}_{i} = -\frac{\partial U_{N}(\mathbf{r}_{1},\ldots,\mathbf{r}_{N})}{\partial \mathbf{r}_{i}} = -\sum_{j=1; j\neq i}^{N} \frac{\partial U_{LJ}(r_{ij})}{\partial r_{ij}} \frac{\mathbf{r}_{ij}}{r_{ij}}$$

$$= \frac{\varepsilon_{LJ}}{\sigma_{LJ}}\left[-\sum_{j=1; j\neq i}^{N} \frac{\partial U_{LJ}^{*}(r_{ij}^{*})}{\partial r_{ij}^{*}} \frac{\mathbf{r}_{ij}^{*}}{r_{ij}^{*}}\right] = \frac{\varepsilon_{LJ}}{\sigma_{LJ}} \mathbf{F}_{i}^{*} \quad (i = 1,\ldots,N).$$ (3.7)

We see that the scale of the force is set by $\varepsilon_{LJ}/\sigma_{LJ}$, a result that could have been obtained directly by a dimensional analysis (ε_{LJ} is the energy scale, σ_{LJ} the length scale; so $\varepsilon_{LJ}/\sigma_{LJ}$ is the scale of the force). Applying this analysis to the velocity, we get $\mathbf{v} = \sqrt{\varepsilon_{LJ}/m}\mathbf{v}^{*}$. With these results, we can write Equation 3.5 in nondimensional form:

$$\mathbf{r}_{i}^{*}(t+\Delta t) \approx \mathbf{r}_{i}^{*}(t) + \mathbf{v}_{i}^{*}(t)\sqrt{\frac{\varepsilon_{LJ}}{m\sigma_{LJ}^{2}}}\Delta t + \frac{\mathbf{F}_{i}^{*}(t)}{2}\frac{\varepsilon_{LJ}}{m\sigma_{LJ}^{2}}(\Delta t)^{2},$$

$$\mathbf{v}_{i}^{*}(t+\Delta t) \approx \mathbf{v}_{i}^{*}(t) + \frac{\mathbf{F}_{i}^{*}(t) + \mathbf{F}_{i}^{*}(t+\Delta t)}{2}\sqrt{\frac{\varepsilon_{LJ}}{m\sigma_{LJ}^{2}}}\Delta t.$$ (3.8)

This shows that the LJ potential also leads to a scale for the time in the simulation, the "Lennard-Jones time":

$$\tau_{LJ} = \sqrt{\frac{m\sigma_{LJ}^{2}}{\varepsilon_{LJ}}}.$$ (3.9)

Now, we have the desired criterion for the choice of the time step: The Taylor expansion (3.8) will be a good approximation to the classical trajectory, if $\Delta t/\tau_{LJ} \ll 1$. This completes our outline of the MD method.

This outline hints at a number of important practical questions: How does one realize the inequality $\Delta t \ll \tau_{LJ}$ in actual simulations? What is the longest simulation time t_{max} with current computer power? Is there a correlation between this time and the size of the system, that is, the total number of particles that can be simulated? We will address these questions in the following section.

3.2.1.2 SOME COMPUTATIONAL ASPECTS AND THEIR CONSEQUENCES FOR SIMULATED LENGTH AND TIME SCALES

3.2.1.2.1 Time step: Typical values and conversion to SI units for LJ systems

The question of how to choose the time step is at the heart of the MD method and has been extensively studied since its invention [41,42]. This accumulated experience suggests that an appropriate range is $10^{-3}\tau_{LJ} \leq \Delta t \leq 10^{-2}\tau_{LJ}$, not only in the microcanonical ensemble, but also for simulations in other ensembles, where special extensions of the classical equations of motion allow to impose, for example, constant temperature T (canonical NVT ensemble) or constant temperature and constant pressure P (NPT ensemble) [41,42]. Since these external constraints can be realized in experiment (contrary to NVE conditions), simulations in ensembles other than the microcanonical are often preferred. Examples can be found in this chapter and other chapters (cf. Chapters 8, 11, and 13).

The Lennard-Jones potential (3.6) was originally introduced as an interaction model for noble gas atoms, but has become a standard potential in statistical–mechanical theories [3,54] and molecular simulations [41,42]. The constants ε_{LJ} and σ_{LJ} are tabulated for a large collection of fluids [54], for example, for argon ($\varepsilon_{LJ}/k_B = 119.8$ K, $\sigma_{LJ} = 3.405 \times 10^{-10}$ m, $m = 6.63 \times 10^{-26}$ kg). On inserting these values into Equation 3.9, one finds that $\tau_{LJ} = 2.156 \times 10^{-12}$ s, implying a time step in the range of 10^{-15} s $\lesssim \Delta t \lesssim 10^{-14}$ s.

3.2.1.2.2 Longest simulation time and system size

For a given Δt, the longest simulation time t_{max} ($=N_{max}\Delta t$) is determined by the maximum number of time steps N_{max} that one can (reasonably) simulate with available computational resources. Certainly, N_{max} has to decrease, if the calculations per Δt consume more CPU time; and this is the case when the size of the system (i.e., N) increases.

To see that, let's return to the force calculation. Equation 3.7 shows that the force per particle involves a sum with $N - 1$ terms. This summation must be repeated N times to find the force on all particles, which is required by Equation 3.5 to advance their positions and velocities in a time step. The force calculation per time step is thus of order $O(N^2)$. Clearly, optimization procedures must address this (most time consuming) part of the MD simulation. The optimization starts from the observation that the LJ potential is small at large distances. It therefore appears legitimate to introduce a cutoff distance r_{cut} beyond which interactions are ignored. Usually, the true potential (3.6) is then replaced by a truncated and shifted potential $U_{LJ}^{ts}(r)$ [41,42], that is,

$$U_{LJ}^{ts}(r) = \begin{cases} U_{LJ}(r) - U_{LJ}(r_{cut}) & \text{for } r \leq r_{cut}, \\ 0 & \text{else,} \end{cases} \tag{3.10}$$

where the shift ensures the continuity of the potential at r_{cut} (but not of its derivatives [55–57]). In practice, r_{cut} is typically chosen in the range $r_{min} \leq r_{cut} \leq 2.5\sigma_{LJ}$, with $r_{min} = 2^{1/6}\sigma_{LJ}$ being the position of the minimum of the LJ potential [41,42]. If $r_{cut} = r_{min}$, the potential is purely repulsive ("soft-sphere potential," see Figure 3.1), whereas a major part of the attractive interaction is included for $r_{cut} = 2.5\sigma_{LJ}$ ($U_{LJ}(r_{cut}) \simeq -0.016\varepsilon_{LJ}$).* The key advantage of this procedure is that every particle only interacts with a finite number (not $\sim N$) of neighbors. Combined with special techniques (neighbor and cell lists [41,42]), the force calculation then becomes an operation of $O(N)$. This is the best scaling with N one can achieve. Under these most favorable conditions, the CPU time per Δt increases linearly with system size, since the forces must be evaluated once every time step. Therefore, with given computational resources, we cannot concurrently maximize the longest simulation time t_{max} and the particle number N. There must be a trade-off between both.

Simulations of glass-forming liquids aim at attaining long times in order to monitor the increase of the structural relaxation time with decreasing temperature to as low T as possible. At present, the longest simulation times are in the range of $10^5\tau_{LJ} \lesssim t_{max} \lesssim 10^6\tau_{LJ}$ (i.e., up to a few microseconds) for systems with $10^3 \lesssim N \lesssim 2 \times 10^4$ particles [62]. These are "practical" limits, typically found in current simulations, not the peak performance furnished by modern computers. Certainly, longer times ($t_{max} \sim 10^7\tau_{LJ}$) and larger systems ($N \sim 10^5$) [63–66] are already feasible and will be studied more frequently in the future, in particular due

---

* For condensed phases, the choice $r_{cut} = 2.5\sigma_{LJ}$ (or $r_{cut} \approx 2r_{min}$) is appropriate for model studies because it encompasses the first and second nearest-neighbor shells, which play a dominant role in determining the structure in (nonassociated) liquids [3]. Neglecting interactions with particles beyond r_{cut} affects, however, the thermodynamic properties of the system. For instance, the location of the critical point for the liquid–gas transition is reduced for a finite cutoff [42,58] and the melting temperature of the LJ crystal varies in a nontrivial way with r_{cut} [59]. Also the viscous slowing down of glass-forming LJ liquids is strongly influenced by the size of the cutoff (when comparing $r_{cut} = r_{min}$ with $r_{cut} = 2.5\sigma_{LJ}$) [60,61].

to the perspective to identify the (putative) true correlation length that grows on cooling toward T_g and is responsible for the slowdown of the dynamics [30,39,62–64,66].

3.2.1.2.3 Periodic boundary conditions and wave vectors

In the search for such a correlation length, simulations often explore the properties of glass-forming liquids as a function of the wave vector q [63,64,67]. The possible wave vectors for this analysis are determined by the periodic boundary conditions. To see this, let $r = (r_x, r_y, r_z)$ be the position of a particle. In x direction, this particle has a periodic image at position $r_x + L$, where L is the linear dimension of the (cubic) simulation box. Since the particle and its image are identical, this implies for the density fluctuation $\exp[i q \cdot r]$ that

$$\exp[iq_x r_x] = \exp[iq_x(r_x + L)] \quad \Rightarrow \quad q_x = \frac{2\pi}{L} n_x \quad \text{with} \quad n_x = 0, \pm 1, \pm 2, \dots. \tag{3.11}$$

Analogous equations also hold for the y and z coordinates of q. Therefore, the modulus of the smallest (finite) wave vector is given by $q_{min} = 2\pi/L$. Using the relation between the particle density ρ and L, that is, $\rho = N/L^3$, we find that q_{min} scales with the particle number as $q_{min} \sim 1/N^{1/3}$. An increase of N by a factor of 10 only allows to decrease q_{min} by a factor of about 2. Extracting a length scale that grows on cooling therefore represents a huge challenge for simulations: Large systems and long times are needed!

3.2.2 MODELS EMPLOYED IN POLYMER GLASS SIMULATIONS

The previous section introduced the MD method via the example of an atomic LJ system. However, the general considerations—for instance, about the correlation between the time step and interaction potential or the trade-off between the longest simulation time and the computational cost per time step—hold for all systems, also for polymers. This section extends the discussion in the latter direction, with a focus on glass-forming homopolymer melts. Polymers have both intramolecular and intermolecular interactions, which must be represented in a molecular model. In polymer glass simulations, two types of models have been employed—atomistic models and generic models [4,7,68–70]. Here, we introduce these models and consider their simulation in view of the discussion of the previous section. Applications are found in later sections and also in other parts of this book (cf. Chapters 8, 11, and 13).

3.2.2.1 ATOMISTIC MODELS

An atomistic model aims at capturing one specific polymer in chemical detail. The model is defined by its (classical) "force field," that is, the potential energies for valence terms (bond lengths, valence angles, torsional angles, etc.) and nonbonded interactions (short-range repulsion, van der Waals attraction, Coulomb interactions, etc.). The form of these potentials is postulated and the corresponding parameters (for example, equilibrium bond length, force constants) are determined from quantum-chemical calculations and experiments [68,71].

Throughout the past decades, there has been extensive work to develop accurate force fields for both all-atom and united-atom models (see [53,68] and references therein). An all-atom model treats every atom present in nature as a separate interaction site, while a united-atom model groups a small number of real atoms into one site (cf. Figure 3.2) [68,72]. Typical united atoms are CH, CH_2, and CH_3. Appealing back to Section 3.2.1.2, we can understand why such united-atom models may be advantageous for MD simulations. The reduction of the number of atoms per monomer implies less force calculations, thereby decreasing the computational cost per time step. Longer times and/or larger systems can thus be simulated than for all-atom models. However, this computational advantage carries a price. Lumping atoms together into one site is a coarse-graining step that tends to smooth the potential energy landscape (PEL) and to decrease the barriers that the coarse-grained site encounters when moving [73]. United-atom models therefore (often) exhibit enhanced dynamics compared to experiment and need careful fine-tuning of torsional barriers to (approximately) compensate for this effect [74]. This hints at a general point, valid for both united-atom and all-atom models: The force field, regardless of its source, should be thoroughly verified through comparison of

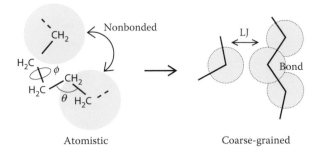

Atomistic   Coarse-grained

Figure 3.2 Schematic representation of different levels one may utilize to model polymers. The atomistic level models a specific polymer (here polyethylene) based on a chemically realistic force field. A force field is the total potential energy resulting from the interactions of all atoms ("explicit atom model") or from the interactions of spherical sites comprising several atoms (for example, CH_2; "united-atom model"). In the left figure, two united atoms are indicated by the shaded circles. Typically, a force field contains contributions from bonded and non-bonded interactions. Bonded interactions comprise potentials for the bond length (nearest neighbor), the bond angle θ (second nearest neighbor), and the torsional angle ϕ (third nearest neighbor). Between neighbors (atoms or united atoms) that are further apart along the backbone of the chain, nonbonded interactions are taken into account. For uncharged polymers, they are often modeled by a Lennard-Jones potential. Computationally, less demanding than atomistic models are simulations at the coarse-grained level. Here, a monomer is associated with a spherical site and the realistic potentials are replaced by simpler ones. This simplification, if carried out systematically, can lead to coarse-grained models for a specific polymer—recent approaches have been reviewed in Baschnagel et al. [71], Müller-Plathe [72], Peter and Kremer [73], and Müller-Plathe [298]. Otherwise, it leads to generic models, such as the bead–spring models described in Section 3.2.2.2.

thermodynamic, structural, and dynamic data between simulation and experiment. The importance of this validation is highlighted in the literature [68,73,75,76].

All-atom and united-atom models were utilized in the chemically realistic modeling of glass-forming polymers [68,77]. Examples include polyisoprene [78,79], atactic polystyrene (PS) [80–82], 1,4-polybutadiene (PBD) [68,83–86], atactic poly(propylene oxide) [87,88], poly(vinyl methyl ether) [89], or polyisobutylene (PIB) [90,91]. Let's consider the recent work on PIB to get a feeling for the length and time scales typically accessible on modern computers. References 90 and 91 employed the COMPASS force field (implemented in the commercial software package Materials Studio [92]), a general all-atom force field optimized for condensed-phase applications [75]. The bonded interactions of PIB are fairly "stiff," leading to short vibration times and so to a small time step. References 90 and 91 choose $\Delta t = 10^{-15}$ s—a usual value for atomistic simulations of polymers. With $\Delta t = 10^{-15}$ s, the longest runs correspond to $t_{max} = 100$ ns, implying a simulation over $N_{max} = 10^8$ time steps. This simulation was carried out for a system with 16,840 atoms, distributed over 20 PIB chains with a length of $N = 70$ monomers each (the PIB monomer, $CH_2–C(CH_3)_2$, has 12 atoms; each end monomer carries an additional H atom). Since the entanglement length of PIB is $N_e \approx 125$ monomers [93], the chains are nonentangled. They have an (average) end-to-end distance of $R \approx 43$ Å, which is comparable to the linear dimension of the (cubic and periodic) simulation box ($L \approx 54$ Å).* This box size implies that the smallest accessible wave vector has the modulus $q_{min} = 2\pi/L \approx 0.12$ Å^{-1} (cf. Equation 3.11). By limiting the analysis to $q \gtrsim 0.6$ Å^{-1}, good agreement between simulation and neutron scattering experiments was reported for various (local) structural and dynamic properties [90], thus supporting the validity of the employed force field.

---

* If the size of the simulation box becomes comparable to the chain dimension, correlations of a chain with its periodic images cannot be ruled out. Instead of having $\sqrt{N}$ contacts with other chains, as is typical of polymer melts [2], the environment of a chain will be (strongly) influenced by the images of the chain itself. Such finite system size effects may affect polymer conformation [10] and dynamics [15,94] at large scales, but do not appear to seriously affect local properties. A case in point is provided by an extreme situation: Consider a single long polymer with an unperturbed ideal conformation. Let the polymer be folded back into the simulation box by the periodic boundary conditions and assume the box size to be chosen such that the resulting monomer density agrees with the experimental value. Then, there are no interchain contacts; all nonbonded interactions come from the polymer itself. In this respect, the system resembles rather a collapsed chain than a polymer melt. Nevertheless, the (local) structure and dynamics of such a single-chain system can agree well with neutron scattering results on macroscopic samples (for wave vectors around the first peak of the collective structure factor, which typically is $q \sim 1$ Å^{-1}) [86].

3.2.2.2 GENERIC MODELS: BEAD–SPRING MODELS

Atomistic simulations of carefully designed models are the best way to explore specific materials, in particular, if quantitative agreement with experiment is sought. On the other hand, as pointed out in Section 3.1.3, many liquids—including polymer melts—form glasses and their glass transition is accompanied by several general features (for example, emergence of viscoelasticity, strong slowing down of the relaxation dynamics accompanied by only weak changes in the static structure factor). To explore such general features, a common tradition in statistical mechanics is to turn to simple models that qualitatively reproduce the phenomenology of real systems, but are easier to analyze. This approach has been strongly pursued in the computational studies of nonpolymeric glass formers [27,29,31,37,39]. Although atomistic simulations, for example, for amorphous silica [95,96], exist, the most extensive body of work is concerned with atomic liquids, a prime example being binary mixtures of LJ or soft-sphere particles [29,51].

In the same spirit, many simulations have also addressed the polymer glass transition and the properties of polymer glasses by means of generic, instead of atomistic, models (see [4,7,68–70,97] for reviews and Chapters 8, 11, and 13). A generic model only retains the most basic features of a polymer. For (electrically uncharged) linear polymers, these features are presumed to be the connectivity between the monomers, their excluded-volume interactions, and, possibly also, monomer–monomer attractions and/or stiffness along the chain backbone. Various such generic models have been studied by simulations [98,99]. An archetypal and widely employed example is the flexible bead–spring model (see Figure 3.2).

3.2.2.2.1 Flexible bead–spring model: Force fields and simulation aspects

In the flexible bead–spring model, a monomer is represented by a spherical particle ("bead"). For homopolymers, all N monomers of a chain are identical; they have the same mass m and diameter σ_{LJ}. The nonbonded interaction between the monomers is generally taken as a truncated and shifted LJ potential (cf. Equation 3.10)

$$U_{LJ}^{ts}(r) = \begin{cases} 4\varepsilon_{LJ}\left[\left(\sigma_{LJ}/r\right)^{12} - \left(\sigma_{LJ}/r\right)^{6}\right] - 4\varepsilon_{LJ}\left[\left(\sigma_{LJ}/r_{cut}\right)^{12} - \left(\sigma_{LJ}/r_{cut}\right)^{6}\right] & \text{for } r < r_{cut}, \\ 0 & \text{else}, \end{cases}$$

(3.12)

where r_{cut} denotes the cutoff distance. Typical values for r_{cut}, employed in polymer glass simulations [57,100–116], are $r_{min} \leq r_{cut} \leq 2.5\sigma_{LJ}$, with $r_{min} = 2^{1/6}\sigma_{LJ}$ being the position of the minimum of the LJ potential.* In addition to nonbonded interactions, nearest neighbors along the chain are connected by a bond potential. One possibility to introduce this chain connectivity is to confine the bond length (l) by a harmonic potential

$$U_{bond}(l) = \frac{1}{2}k(l - l_0)^2,$$

(3.13)

where k is the force constant and l_0 the equilibrium bond length. Typical values for these parameters are $1000\varepsilon_{LJ}/\sigma_{LJ}^2 \leq k \leq 2000\varepsilon_{LJ}/\sigma_{LJ}^2$ and $0.9\sigma_{LJ} \leq l_0 \leq 1\sigma_{LJ}$ (cf. [101,105–108,112–116] and Chapters 8 and 11). Besides Equation 3.13, another often-employed way to introduce chain connectivity (cf. [100,102–104,109–111] and Chapters 11 and 13) has its origin in an influential model proposed by Kremer and Grest [117,118] building on earlier studies by Ceperley et al. [119] and by Bishop et al. [120]. Kremer and

---

* Computational expedience suggests to work with a short-range potential because the number of neighbors n, with which a particle interacts, scales with the cutoff distance as $n \propto r_{cut}^3$ [41,42]. In this respect, purely repulsive interactions, where $r_{cut} = 2^{1/6}\sigma_{LJ}$ (like in the Kremer–Grest model [117]), are particularly advantageous; they were applied in some polymer glass simulations, for example, in Yamamoto and Onuki [100] and Durand [101]. Many simulations, however, have employed larger cutoffs. Often made choices are $r_{cut} = 1.5\sigma_{LJ}$ [102,103], $r_{cut} = 2 \times 2^{1/6}\sigma_{LJ}$ [7,104], $r_{cut} = 2.3\sigma_{LJ}$ [105–108], or (the classical value of) $r_{cut} = 2.5\sigma_{LJ}$ [57,109–116]. Larger cutoffs include part of the attractive LJ interactions, which is necessary, for example, for simulations of free interfaces in polymer films [108,113,114] or cavitation phenomena in deformed polymer glasses [111].

Grest represented nearest-neighbor interactions along the chain by a FENE (finitely extensible nonlinear elastic) potential

$$U_{\text{FENE}}(l) = \begin{cases} -\dfrac{1}{2} k R_0^2 \ln\left[1 - (l/R_0)^2\right] & \text{for } l < R_0, \\ \infty & \text{else,} \end{cases} \qquad (3.14)$$

where $k = 30\varepsilon_{\text{LJ}}/\sigma_{\text{LJ}}^2$ and $R_0 = 1.5\sigma_{\text{LJ}}$. Equation 3.14 diverges logarithmically if $l \to R_0$ ("finite extensibility") and vanishes parabolically close to the origin ("elastic behavior"). Thus, the FENE potential alone cannot prevent adjacent monomers from overlapping. Such overlaps are avoided by imposing Equation 3.12 also for nearest neighbors along the chain. The superposition of the FENE and the LJ potentials results in a steep effective bond potential with a minimum at $l \approx 0.96\sigma_{\text{LJ}}$ [99]. This value has motivated the choice of the interval for the equilibrium bond length l_0 in Equation 3.13.

The parameters of the harmonic bond potential or of the FENE springs strongly suppress deviations of l from its equilibrium value (for example, for Equation 3.13, we get that $\sqrt{\langle l^2 \rangle - \langle l \rangle^2} \sim \sqrt{k_{\text{B}} T/k} \sim 10^{-2}\,\sigma_{\text{LJ}}$ for $1000\varepsilon_{\text{LJ}}/\sigma_{\text{LJ}}^2 \le k \le 2000\varepsilon_{\text{LJ}}/\sigma_{\text{LJ}}^2$ and a typical thermal energy of $k_{\text{B}} T = 1\varepsilon_{\text{LJ}}$). Since the equilibrium bond length is smaller than the monomer diameter, the LJ potential prevents the chains from crossing each other. To see this, let's estimate the lowest LJ energy for two chains to cross. This energy corresponds to a configuration where two bonds of two chains are perpendicular and overlap in their middle. The distance between two monomers is then $r \approx 0.68\sigma_{\text{LJ}}$, leading to an energy cost of $4U_{\text{LJ}}(r \approx 0.68\sigma_{\text{LJ}}) \approx 1400\varepsilon_{\text{LJ}}$ for the four monomers involved in the two bonds. The energy cost is huge, making such intersections effectively impossible. This noncrossability imposes topological constraints [1], which ultimately lead to reptation-like dynamics in the limit of large chain length, an extensively studied topic by computer simulations [17,94,117,121–123].

On the other hand, the bond potential introduces the period of a bond oscillation τ_{bond} as a new time scale, not present in atomic liquids. Using Equations 3.13 and 3.9, we find

$$\tau_{\text{bond}} = 2\pi \sqrt{\frac{m}{k}} = \frac{2\pi}{\sqrt{|k|}}\,\tau_{\text{LJ}}, \qquad (3.15)$$

where $|k|$ denotes the dimensionless numerical value of k. Equation 3.15 implies that τ_{bond} is smaller than τ_{LJ}, by a factor of about 5 to 7 for $1000\varepsilon_{\text{LJ}}/\sigma_{\text{LJ}}^2 \le k \le 2000\varepsilon_{\text{LJ}}/\sigma_{\text{LJ}}^2$. This illustrates the remark made in the previous section on atomistic modeling: Stiff potentials—typically resulting from the valence terms in the force field—entail small characteristic times and should accordingly require a reduction of the time step Δt in the MD simulation (cf. Section 3.2.2.1). Practice, however, suggests that for bead–spring models, this problem is not as severe as for atomistic models, except in extreme cases, such as isolated chains [124], where the bond potential dominates the total force on a monomer. Typical time steps used in recent polymer glass simulations are still in the range $0.001\tau_{\text{LJ}} \le \Delta t \le 0.01\tau_{\text{LJ}}$ [57,101,103,105,106,111,113–116].

3.2.2.2.2 Lennard-Jones units and approximate mapping to SI units

In the discussion following Equation 3.6, it was pointed out that the parameters of the LJ potential define characteristic scales that allow to introduce nondimensional "reduced" quantities (denoted by the superscript "$\star$"). For instance, $r^\star = r/\sigma_{\text{LJ}}$, $t^\star = t/\tau_{\text{LJ}}$ with $\tau_{\text{LJ}} = (m\sigma_{\text{LJ}}^2/\varepsilon_{\text{LJ}})^{1/2}$ (Equation 3.9), or $T^\star = T/(\varepsilon_{\text{LJ}}/k_{\text{B}})$. It is common practice to report the simulation results in "LJ units" [41,42]. That is, one poses $\varepsilon_{\text{LJ}} = 1$, $\sigma_{\text{LJ}} = 1$, $m = 1$, and $k_{\text{B}} = 1$. Then, $r^\star = r$, $t^\star = t$, and so on, and we can drop the "$\star$" to simplify the notation. This notation and LJ units will be used in the following sections.

The mapping of LJ units to SI units has been discussed in several works [68,117,125–127]. Virnau et al. studied the phase separation kinetics of a mixture of hexadecane ($C_{16}H_{34}$) and carbon dioxide (CO_2) [125,126]. By matching the critical point of the liquid-gas transition in hexadecane with that of bead–spring chains (consisting of five monomers each), they obtain $\sigma_{\text{LJ}} \simeq 4.5 \times 10^{-10}$ m and $\varepsilon_{\text{LJ}} \simeq 5.8 \times 10^{-21}$ J ($\simeq 420$ K). Assuming $\sigma_{\text{LJ}} = 4 \times 10^{-10}$ m, Paul and Smith converted τ_{LJ} to seconds by comparing the late-time diffusive dynamics of chemically realistic models for nonentangled melts of polyethylene and polybutadiene with that of a bead–spring model [68]. The result is $\tau_{LJ} \simeq 2.1 \times 10^{-13}$ s. However, these values for σ_{LJ}, ε_{LJ}, and τ_{LJ} depend on the real polymer for which the conversion is carried out, and may vary a lot. This caveat is discussed in Peter and Kremer [73], Kremer and Grest [117] and Kröger [127]. In their seminal work on reptation dynamics [117], Kremer and Grest reported that an LJ bead corresponds to the range of 1/2–5 monomers for real polymers. This variation reflects differences in size and flexibility of the specific monomer, and entail fluctuations in the results for σ_{LJ}, ε_{LJ}, and τ_{LJ}. Typically, they fall in the range: 5×10^{-10} m $\lesssim \sigma_{LJ} \lesssim 13 \times 10^{-10}$ m, 300 K $\lesssim \varepsilon_{LJ}/k_B \lesssim 500$ K, and 2×10^{-12} s $\lesssim \tau_{LJ} \lesssim 2 \times 10^{-10}$ s [117].

Owing to these uncertainties, we will make the following choices when giving an approximate conversion from LJ to physical units in the remainder of this chapter: $\sigma_{LJ} = 5 \times 10^{-10}$ m, $\varepsilon_{LJ}/k_B = 450$ K, and $\tau_{LJ} = 2 \times 10^{-12}$ s. Owing to Equation 3.9, this implies a monomer mass of $m \simeq 60$ g/mol. With that we also obtain reference values for other quantities, for example, a mass density of 1 corresponds to 0.8 g/cm³ and a pressure of 1 corresponds to 497 bar. Since a pressure of 1 bar is very small in LJ units ($P \simeq 0.002$), vanishing reduced pressure ($P = 0$) is a good proxy for ambient pressure conditions.

3.2.2.2.3 Simulation of bead–spring models on current computers: An example

As done for atomistic models (cf. Section 3.2.2.1), it is interesting to conclude this section on generic models by an example for the system sizes and time scales that can be simulated with reasonable effort on current computers. As a case in point, let's use our own work on a bead–spring model defined by Equations 3.12 and 3.13 (with $r_{cut} = 2.3\sigma_{LJ}$) [105]. The simulations were carried out for systems with 12,288 monomers and different chain lengths between $N = 4$ and $N = 64$ (entanglement length $N_e \approx 70$ [128]). Even for $N = 64$, the linear dimension of the simulation box ($L \approx 23\sigma_{LJ} \approx 115$Å) is at least a factor of 2 larger than the end-to-end distance of the chains. This box size implies that the smallest accessible wave vector has the modulus $q_{min} = 2\pi/L \approx 0.27\,\sigma_{LJ}^{-1} \approx 0.05$ Å⁻¹ (cf. Equation 3.11). For these systems, current processors (consisting of multiple cores) need about 1 μs per monomer and time step. With $\Delta t = 0.005\tau_{LJ}$ [105], the integration of the equations of motion over the time of one τ_{LJ} takes $12,288 \times 200 \times 10^{-6}$ s ≈ 2.5 s. So, in a month (=30 days), one can simulate about $10^6 \tau_{LJ}$ or up to about 2 μs.

3.2.2.2.4 From flexible to semiflexible bead–spring models

In the flexible bead–spring model, backfolding of successive bonds along the chain backbone is suppressed by the repulsive part of the LJ potential, if the distance between the first and the last monomer of the bonds becomes smaller than σ_{LJ}. This renders the polymer chain very flexible.* Together with the fact that the bond and LJ potentials introduce two incompatible length scales, l_0 and r_{min}, the model can be readily supercooled. Note however that the incompatibility of l_0 and r_{min} does not exclude the mere possibility of a crystalline state. By fully extending the chains and aligning them parallel to each other, a crystalline state, satisfying both bond and LJ interactions, can be set up (by hand, see Figure 3.3). However, the flexible model has no energetic driving force to form such extended chain structures (such a force would be provided, for example, by a bond-angle potential favoring large angles [21,129,130] or by exposure to a nucleating surface [115,131], leading then to polymer crystallization). Therefore, bulk polymer liquids of the flexible bead–spring model can be easily supercooled and eventually form a polymer glass at low temperature (cf. Figure 3.3).

Nevertheless, flexible real polymers—and the corresponding atomistic models—are locally more rigid than flexible bead–spring models, due to polymer-specific valence terms (bond angles, torsional angles). Such chain stiffness effects can be included in the bead–spring model, for instance, by associating a potential $U_{ang}(\theta)$ with the angle θ between successive bonds in the chain. (Second nearest-neighbor monomers along the chain thus interact by $U_{ang}(\theta)$ and $U_{LJ}(r)$.) Following References 127 and 132–135, this bending potential may be taken as (cf. Chapter 13)

$$U_{ang}(\theta) = \varepsilon_\theta(1 + \cos\theta). \qquad (3.16)$$

---

* As a measure of chain flexibility, we can take the Flory characteristic ratio $C_\infty = \lim_{N \to \infty} R^2/(Nl_0^2)$ [2]. For flexible generic models, typical values are $1.5 \lesssim C_\infty \lesssim 1.9$ [9], whereas one finds $5 \lesssim C_\infty \lesssim 10$ for flexible synthetic polymers [2].

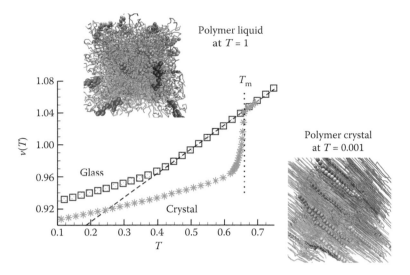

Figure 3.3 Temperature dependence of the specific volume v ($=1/\rho$, ρ being the monomer number density) for the melting of a polymer crystal (stars) and the vitrification of a polymer liquid upon cooling (squares). All data are given in LJ units. The results refer to a system with 1536 chains of length $N = 16$ and were obtained from constant-pressure MD simulations ($P = 0$) of a flexible bead–spring model defined by Equations 3.12 and 3.13 (with $r_{\mathrm{cut}} = 2.3$). Depending on T, this yields a linear dimension of the simulation box in the range 28–30. The upper inset depicts the snapshot of the polymer liquid at high temperature (at $T = 1$). For clarity purposes, most chains are represented by thin lines and only a few are shown by beads. The amorphous structure of melt is clearly visible. This disordered structure is maintained when cooling the melt continuously from $T = 1$ to $T = 0.001$ following the schedule (3.22) with rate $\Gamma = 5 \times 10^{-6}$. The dashed line indicates a linear fit at high temperature: $v(T) = 0.844 + 0.302T$. For $T < 0.4$ the data (squares) bend away from the extrapolated liquid line and form a glass. However, it is also possible to set up a polymer crystal for the flexible bead–spring model. There are several ways to achieve this. One way is to construct a crystal "by hand" by juxtaposing extended chains in a crystalline structure [144]. Another way is to add temporarily the bond-angle potential (3.16) with a very high energy parameter ($\varepsilon_\theta = 32$) to the flexible model. At high T, this creates a liquid crystal, which can be cooled into a crystalline solid with the same cooling schedule as for the glass. At $T = 0.001$, the bond-angle potential is then switched off (i.e., by setting $\varepsilon_\theta = 0$) and the system is reequilibrated at $P = 0$. The snapshot on the right shows the resulting structure of an extended chain crystal, typical of short chains, which cannot fold back onto one another [129,130]. When heated according to a continuous heating schedule with rate $\Gamma = 2 \times 10^{-5}$, the crystal melts at $T_{\mathrm{m}} \approx 0.66$.

We see that $U_{\mathrm{ang}}(\theta)$ decreases from a maximum value $2\varepsilon_\theta$, if successive bonds fold back onto each other ($\theta = 0$), to 0, if the bonds are colinear ($\theta = 180°$). Equation 3.16 corresponds to a discrete representation of the local energy controlling the curvature in the wormlike chain model for semiflexible polymers [135]. Semiflexible polymers can undergo a transition to a nematically ordered state [136] (see [68] for a review of simulation results). In order to avoid nematic ordering, the energy parameter ε_θ must not be too large. References 133 and 134 suggest that $\varepsilon_\theta = 2\varepsilon_{\mathrm{LJ}}$ is a good choice to preserve the amorphous state of the melt, also when cooling the melt through the glass transition into the glassy state [137].

Equation 3.16 is not the only extension of the bead–spring model to semiflexible chains in polymer glass simulations. For instance, References 138 and 139 considered, in addition to bond and LJ interactions, a bending potential that is harmonic in $\cos\theta$, and a dihedral potential to account for rotational-isomeric states [2] and torsional barriers. A variant of this model was extensively utilized in a systematic study of the influence of intramolecular barriers on the glassy dynamics of nonentangled polymer melts [140–142] (see also [4] for a topical review). In addition to the FENE potential (Equation 3.14), the force field of the model comprises two further valence terms: a bending potential

$$U_{\mathrm{ang}}(\theta) = \frac{K_{\mathrm{B}}}{2}\left(\cos\theta - \cos\theta_0\right)^2 \quad (\theta_0 = 109.5°) \tag{3.17}$$

and a torsional potential ($a_0 = 3$, $a_1 = -5.9$, $a_2 = 2.06$, $a_3 = 10.95$)

$$U_{\text{tor}}(\phi_{i,i+1}; \theta_i, \theta_{i+1}) = K_T \sin^3 \theta_i \sin^3 \theta_{i+1} \sum_{n=0}^{3} a_n \cos^n \phi_{i,i+1}. \qquad (3.18)$$

The torsional angle $\phi_{i,i+1}$ is defined for four monomers $i - 1, i, i + 1$, and $i + 2$. The bond angle terms in front of the sum in Equation 3.18 serve to avoid numerical instabilities when two consecutive bonds are aligned [138,140–142]. Variation of the energy parameters K_B and K_T allows to change chain stiffness. Nonbonded interactions are modeled by a (purely repulsive) soft-sphere (ss) potential

$$U_{\text{ss}}(r) = 4\varepsilon_{\text{LJ}} \left[(\sigma_{\text{LJ}}/r)^{12} - C_0 + C_2 (r/\sigma_{\text{LJ}})^2 \right] \quad (r_{\text{cut}} = c\sigma_{\text{LJ}} \quad \text{with} \quad c = 1.15). \qquad (3.19)$$

The constants $C_0 = 7c^{-12}$ and $C_2 = 6c^{-14}$ guarantee the continuity of the potential and force at r_{cut}.

3.3 GLASS TRANSITION AND STRUCTURAL RELAXATION OF SUPERCOOLED POLYMER MELTS

This section presents results on the glass transition and its precursor, the slow structural relaxation, of supercooled polymer melts. Our report will be mainly concerned with MD studies of two generic models: a flexible bead–spring model, defined by Equation 3.13 and Equation 3.12 with $r_{\text{cut}} = 2.3$, and a semiflexible bead–spring model where, in addition to Equations 3.12 and 3.13, Equation 3.16 with $\varepsilon_\theta = 2$ is also applied. In the following, we refer to these models as flexible BSM and semiflexible BSM, respectively. Work on other bead–spring models [4,140–142] and chemical realistic models [4,68] will also be discussed as we go foward.

3.3.1 GLASS TRANSITION TEMPERATURE: DEPENDENCE ON PRESSURE, CHAIN LENGTH, AND COOLING RATE

A classical experimental method to determine T_g is dilatometry (see [22,24,143] and Chapters 2 and 4). Dilatometry measures the temperature dependence of the specific volume $v(T)$ upon cooling. The glass transition temperature is located in the T interval where $v(T)$ changes slope, having a large thermal expansion coefficient, $\alpha = (\partial \ln v/\partial T)_P$, in the liquid state and small α in the glassy state. Operationally, T_g can be defined as the intersection of straight-line extrapolations from the glassy and liquid branches of the volume–temperature curve [22,24].

The same method was also employed in many simulation studies of polymeric and nonpolymeric glass formers [82,101,137,144–148]. Figure 3.4 shows an example from constant-pressure MD simulations of the flexible and semiflexible BSM for different chain lengths [137] and pressures [128]. If we assume that the expansion coefficients of the glass (α_g) and the liquid (α_l) are constant over some temperature interval near T_g, the logarithm of v is a linear function of T. Figure 3.4 supports this assumption for the BSM and furthermore reveals two features:

1. In the glass, the expansion coefficient varies little with N and P, but does so more strongly in the liquid phase, in particular with pressure. We see that α_l decreases with increasing P, as also seen in experiment [22]. For poly(vinyl acetate), typical experimental values of α_l/α_g range from $\alpha_l/\alpha_g = 2.55$ at 1 bar to 2.42 at 800 bar [22]. For a comparable range of pressures, the BSM gives similar values, for example, $\alpha_l/\alpha_g = 2.52$ at $P = 0.01$ (≈ 5 bar) and 2.27 at $P = 1.5$ (≈ 750 bar) for the flexible model [128].
2. Figure 3.4 shows that the volume–temperature trace bends from the liquid to the glassy state over a temperature interval of about $\Delta T = 0.05$. With $\varepsilon_{\text{LJ}}/k_B = 450$ K, this corresponds to about 23 degrees. Thus, the vitrification zone is on the order of "about some multiple of 10 degrees" [18], as typically found in

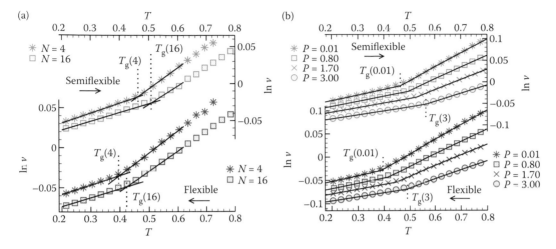

Figure 3.4 Logarithm of the specific volume v versus T for the flexible BSM (left ordinate) and the semiflexible BSM (right ordinate). (a) Results for two chain lengths, $N = 4$ and $N = 16$, which were obtained with the cooling rate $\Gamma = 2 \times 10^{-5}$ and at the pressure $P = 1$. (b) Dependence on pressure for one chain length, $N = 10$, and $\Gamma = 2 \times 10^{-5}$. In both panels, the vertical dotted lines indicate the determination of T_g by the intersection of linear extrapolations (solid lines) from the glass and liquid sides. (a: Adapted from B. Schnell et al., *Eur. Phys. J. E*, 34, 97, 2011.) (b: Adapted from B. Schnell, Numerical Studies of the Glass Transition and the Glassy State of Dense Amorphous Polymers: Mechanical Properties and Cavitation Phenomenon [in French], Ph.D. thesis, Université de Strasbourg, 2006, available from http://scd-theses.u-strasbg.fr/1131/01/SCNNELL2006.pdf.)

experiment, too. The position of the bend, determined by the intersection of linear extrapolations from the glass and liquid branches of ln v, defines T_g. Figure 3.4 shows that T_g increases with chain stiffness and, for given chain stiffness, with pressure and chain length. These features of the BSM are again in qualitative agreement with experiment (see [4,22,143,149,150] and Chapter 4).

3.3.1.1 DEPENDENCE OF THE GLASS TRANSITION TEMPERATURE ON PRESSURE AND CHAIN LENGTH

The dependence of T_g on pressure and chain length is analyzed in more detail in Figure 3.5. Experimentally, one finds that the increase of T_g with P can often be well parameterized by an empirical relation proposed by Andersson and Andersson [151]:

$$T_g(P) = k_1 \left(1 + \frac{k_2}{k_3} P \right)^{1/k_2},$$

(3.20)

where k_1 (in units of K), k_2 (dimensionless), and k_3 (in units of Pa) are material constants. As seen from Figure 3.5a, this relation also provides a good description of the BSM data. From the fit, we get $k_2 = 1$ (flexible model) and $k_2 = 3$ (semiflexible model). These values are within the experimental range of 1 to 10 (see [143] and references therein), where $k_2 \approx 1$ is often found for flexible polymers, like polyisoprene [152] or polydimethylsiloxane (PDMS) [153], while very stiff polymers, such as poly(p-phenylene) [154], can have large values ($k_2 \approx 10$). Moreover, for small pressures, experiments show that the change of T_g is linear in P, with a pressure coefficient given by $\lim_{P \to 0} dT_g/dP = k_1/k_3$. From Figure 3.5a, we obtain $k_1/k_3 \approx 305$ K/GPa (flexible BSM) and $k_1/k_3 \approx 403$ K/GPa (semiflexible BSM). Again, these values are in reasonable agreement with experimental results (cf. Table 1 in Roland et al. [143]).

Figure 3.5b depicts the N dependence of T_g for the flexible and semiflexible BSM at $P = 1$. As in experiments for many linear polymers (see [22,149,150] and references therein) and other simulations [82,101,155], we find that T_g increases with chain length and saturates at a stiffness-dependent value for large N. Fox and

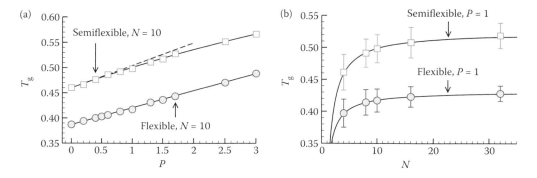

Figure 3.5 Dependence of T_g on pressure (a) and chain length (b) for the flexible and semiflexible BSM. The data of (a) are obtained for the chain length $N = 10$ and $\Gamma = 2 \times 10^{-5}$, those of (b) for $P = 1$ and cooling rate $\Gamma = 2 \times 10^{-5}$. The solid lines in (a) are fits to Equation 3.20 with $k_1 = 0.387$, $k_2 = 1$, $k_3 = 11.5$ for the flexible BSM and $k_1 = 0.46$, $k_2 = 3$, $k_3 = 10.34$ for the semiflexible BSM. For the semiflexible BSM the dashed line shows the linear approximation to Equation 3.20, valid for small P. The solid lines in (b) are fits to Equation 3.21 with $T_g^\infty = 0.432$, $K = 0.145$ for the flexible BSM and $T_g^\infty = 0.525$, $K = 0.264$ for the semiflexible BSM. (a: Adapted from B. Schnell, Numerical Studies of the Glass Transition and the Glassy State of Dense Amorphous Polymers: Mechanical Properties and Cavitation Phenomenon [in French], Ph.D. thesis, Université de Strasbourg, 2006, available from http://scd-theses.u-strasbg.fr/1131/01/SCNNELL2006.pdf.) (b: Adapted from B. Schnell et al., *Eur. Phys. J. E*, 34, 97, 2011.)

Flory were the first to explain this N dependence in terms of free-volume concepts [156,157]. For sufficiently long chains, they suggest

$$T_g(N) = T_g^\infty - \frac{K}{N},\tag{3.21}$$

with T_g^∞ being the glass transition temperature at infinite N and K a constant. Although the Fox–Flory interpretation has been criticized (see, for example, [22]) and other approaches were discussed [149,150] or can be justified theoretically [158,159], Equation 3.21 allows for a reasonable fit of the BSM data (Figure 3.5b) and also provides a simple means to compare simulated and experimental T_g values.

Figure 3.6 shows such a comparison to experimental data for PS [149,160], poly(α-methylstyrene) (PMS) [161], polyvinylchloride (PVC) [162], and PDMS [149]. The Fox–Flory equation suggests that plotting $T_g(N)/T_g^\infty$ versus N_g/N with $N_g = K/T_g^\infty$ gives a master curve. Figure 3.6 supports this assumption, provided T_g^∞ and N_g are determined from fits of Equation 3.21 to the low molecular weight data for PS, PMS, PVC, and PDMS [137]. For higher molecular weight, the experimental T_g values bend upward, with the exception of PDMS, which stays on the extrapolated line of the low molecular weight data. These deviations from the small-N behavior are polymer specific and could be related to structural details, such as backbone rigidity or bulkiness of the sidegroups, as suggested in Agapov and Sokolov [150].

3.3.1.2 DEPENDENCE OF THE GLASS TRANSITION TEMPERATURE ON COOLING RATE

The glass transition has a kinetic dimension: T_g depends on the rate with which the liquid is cooled into the glass (see [23,27,163] and Chapters 1, 2, and 4). This dependence can be rationalized as follows. Let's assume that, starting from some initial temperature $T_i \gg T_g$, we continuously cool the liquid with the rate Γ:

$$T(t) = T_i - \Gamma t \quad \Leftrightarrow \quad t(T) = \frac{T_i - T}{\Gamma}.\tag{3.22}$$

As T decreases, the time scale $t(T)$ associated with the cooling protocol increases, but not as quickly as the molecular relaxation time $\tau_\alpha(T)$ (the "α relaxation time") of the liquid. In the supercooled liquid, $\tau_\alpha(T)$

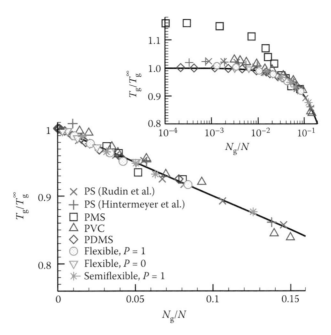

Figure 3.6 Scaling plot of T_g/T_g^∞ versus N_g/N with $N_g = K/T_g^\infty$, as suggested by Equation 3.21. The simulation results for the flexible and semiflexible BSMs (at pressures $P = 0$ and $P = 1$, cooling rate $\Gamma = 2 \times 10^{-5}$) are compared to experimental data of low-molecular-weight polystyrene (PS) [149,160], poly(α-methylstyrene) (PMS) [161], polyvinylchloride (PVC) [162], and polydimethylsiloxane (PDMS) [149]. The simulation results for N_g range from 0.2 to 0.5, the experimental results are larger ($0.6 \le N_g \le 2.2$) [137]. Inset: Same data as in the main figure on a logarithmic scale for N_g/N and including also high molecular weight data from experiment. Deviations from the low-molecular-weight extrapolation occur for all real polymers except for PDMS. (Figure adapted from B. Schnell et al., *Eur. Phys. J. E, 34*, 97, 2011.)

increases in a super-Arrhenius fashion [29], which can be fitted by the Vogel–Fulcher–Tammann (VFT) equation (see also Section 3.3.2.4 where the VFT equation is further discussed):

$$\tau_\alpha(T) = \tau_\infty \exp\left(\frac{B}{T - T_0}\right). \tag{3.23}$$

Here, τ_∞ is the asymptotic relaxation time at high temperature (formally, $\lim_{T\to\infty}\tau_\alpha(T) = \tau_\infty$), B is a material-characteristic temperature scale, and T_0 is the "Vogel–Fulcher temperature" at which the relaxation time appears to diverge [24,27,29]. As T traverses the region near T_g, $\tau_\alpha(T)$ becomes comparable to $t(T)$. For $T < T_g$, $\tau_\alpha(T)$ quickly exceeds $t(T)$ by many orders of magnitude so that the liquid freezes in a (nonequilibrium) amorphous solid on the time scales accessible to experiment or simulation. Operationally, we can therefore define T_g by the criterion $\tau_\alpha(T_g) = t(T_g)$. With Equations 3.22 and 3.23, we then get (up to corrections of order $\ln[1 + 1/\ln \Gamma]$)

$$T_g(\Gamma) = T_g^0 - \frac{B}{\ln(A\Gamma)}, \tag{3.24}$$

where A ($=\tau_\infty/[T_i - T_0]$) is a characteristic cooling rate and T_g^0 ($=T_0$) is the glass transition temperature for infinitely slow cooling. With this line of reasoning, we expect that T_g has a weak logarithmic dependence on cooling rate, as a consequence of the super-Arrhenius increase of the α relaxation time. In applications of Equation 3.24 to experimental [164] or simulation data [82,144,145] T_g^0, A and B are treated as adjustable parameters.

Figure 3.7 compares Equation 3.24 to $T_g(\Gamma)$ for an atomistic model of PS [82] and to $T_g^\infty(\Gamma)$ for the flexible BSM (T_g^∞ is the glass transition temperature extrapolated via Equation 3.21 to infinite N). In accord with

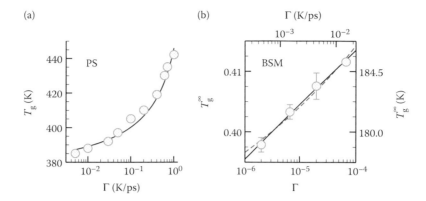

Figure 3.7 (a) Glass transition temperature (T_g) versus cooling rate (Γ) for an atomistic model of atactic polystyrene (PS) with $N = 80$ monomers per chain. The solid line shows a fit to Equation 3.24 with $T_g^0 = 371$ K, $B = 110$ K, and $A = 0.23$ ps/K. Data are taken from [82] and were provided by courtesy of A. Lyulin. (b) Cooling rate dependence of T_g^∞, the fit result from Equation 3.21, for the flexible BSM (at pressure $P = 0$). The results are given in LJ units (left ordinate and lower abscissa) and in real units (right ordinate and upper abscissa) with the conversions $\varepsilon_{LJ}/k_B = 450$ K and $\tau_{LJ} = 2 \times 10^{-12}$ s (cf. Section 3.2.2.2). The data may be fitted in two ways: by Equation 3.24 with $T_g^0 = 0.348$, $B = 0.832$, and $A = 0.033$ (dashed line) or by Equation 3.25 with $T_g^\infty(\Gamma_1) = 0.398$ and $B = 0.009$ for a reference cooling rate of $\Gamma_1 = 2 \times 10^{-6}$ (solid line).

many experimental results [164], we see that Equation 3.24 can describe the simulation data, but the fit is better for PS than for the BSM. In the range of cooling rates studied for the BSM, the change of T_g is rather weak and may also be approximated by a linear extrapolation in terms of $\log \Gamma$:

$$T_g^\infty(\Gamma) = T_g^\infty(\Gamma_1) + B \log(\Gamma/\Gamma_1), \quad B = \left. \frac{dT_g^\infty}{d \log \Gamma} \right|_{\Gamma_1}, \tag{3.25}$$

where Γ_1 is some reference rate from which the extrapolation is started. Figure 3.7 shows that $B \approx 4$ K, which is a reasonable value when compared to experiment [22].

Figure 3.7 also exemplifies the typical cooling rates employed in current simulations. They are in the range $10^9 - 10^{12}$ K/s, and thus orders of magnitude faster than in experiment. Typical experimental values are $10^{-3} - 10^0$ K/s, for materials with high glass-forming tendency, such as SiO_2, specific metallic alloys, or atactic PS [163,164] (see also Chapters 1 and 4). Despite this huge difference, the discussion of this section shows that the variations of T_g with chain length, pressure, or cooling rate qualitatively agree between experiment and simulation. This may appear paradoxical, but is (likely) related to the fact that both laboratory and computer studies probe the emergence of nonequilibrium phenomena when the monomer relaxation time becomes comparable to the "experimental" time scale set by the respective cooling process. Similarly, as T is lowered below T_g, the laboratory and numerical systems lose the ability to relax and freeze in a glassy solid. Like the experimental systems (cf. Chapters 1, 2, and 4), the simulated glasses show physical aging and a mechanical response characteristic of polymers [70,97,102]. Therefore, simulations may be used to obtain insight into phenomena pertaining to the glassy state (cf. Chapters 11 and 13).

3.3.2 STRUCTURAL RELAXATION OF SUPERCOOLED POLYMER MELTS IN THERMAL EQUILIBRIUM

Finite cooling rate effects and the attendant nonequilibrium behavior can be avoided for $T > T_g$. At high-enough temperature, the relaxation time of the polymer melt is smaller than the time scales accessible in simulations; the cooling process may then be carried out quasistatically, achieving thermal equilibrium at all T studied. With currently available computational resources, this allows to follow the initial increase of the α

relaxation time upon cooling, from tens of picoseconds up to the μs range. In the corresponding temperature regime of the "slightly" supercooled liquid, many dynamic features emerge that are not found in the normal liquid state. As mentioned in Section 3.1.3, they involve, for instance, a two-step decay of dynamic correlation functions, stretching of the structural (α) relaxation, a super-Arrhenius temperature dependence of structural relaxation times, and dynamic heterogeneities [27,29,31,32,51,165,166].

Two-step decay of dynamic correlation functions and stretching of the α relaxation are key predictions of the so-called mode-coupling theory for glassy dynamics [32,167–169]. In the past 30 years (the present year is 2016), MCT has strongly influenced the analysis of supercooled liquids, both in experiments [32,167,168] and simulations [4,7,25,37,170]. In the following, we therefore give an introduction to MCT and put its predictions into perspective with simulation results for polymer melts.

3.3.2.1 BRIEF SURVEY OF IDEALIZED MODE-COUPLING THEORY

Let's consider a single-component liquid consisting of N (atomic) particles in a volume V with density $\rho = N/V$. The liquid is a dense system with a low compressibility. Particles are in close contact with each other and structurally arranged in nearest-neighbor shells. The nearest neighbors form a "cage" around a tagged particle. However, this cage cannot be static. If it were, the particle would be permanently trapped and could only vibrate around its equilibrium position as in a solid. In a liquid, the particle diffuses over long distances, requiring a disruption of the cage and thus motion of the neighbors. The neighbors, however, are "caged" themselves (by the tagged particle and other particles) and can only move, if their neighbors give way, and so on. This cooperative motion between a tagged particle and its neighbors is called "cage effect" in liquid state theory [32,171]. We can reasonably expect that the relevance of the effect is enhanced at low temperature. As T decreases, the density of the liquid increases and the particles pack more tightly, rendering the thermal disruption of the cages difficult. This suggests that the cage effect could lead to glassy dynamics for supercooled liquids. It is this idea that is pursued by mode-coupling theory.

3.3.2.1.1 Idealized MCT: Background

The cage effect is a collective phenomenon related to temporal fluctuations of the spatial distribution of the particles. The simplest functions reflecting this many-particle dynamics are collective density fluctuations at wave vector $\boldsymbol{q}$:

$$\delta\rho(\boldsymbol{q},t) = \int \delta\rho(\boldsymbol{r},t)\exp(\mathrm{i}\boldsymbol{q}\cdot\boldsymbol{r})\mathrm{d}^3\boldsymbol{r}$$
$$= \sum_{i=1}^{N} \exp[\mathrm{i}\boldsymbol{q}\cdot\boldsymbol{r}_i(t)] \quad (\text{for } q \neq 0), \tag{3.26}$$

where $\boldsymbol{r}_i(t)$ is the position of particle i at time t and

$$\delta\rho(\boldsymbol{r},t) = \rho(\boldsymbol{r},t) - \rho, \quad \rho(\boldsymbol{r},t) = \sum_{i=1}^{N} \delta[\boldsymbol{r} - \boldsymbol{r}_i(t)], \quad \rho = \langle\rho(\boldsymbol{r},t)\rangle.$$

Here, $\langle\ldots\rangle$ denotes the ensemble average. The dynamic correlation function of $\delta\rho(\boldsymbol{q},t)$ defines the (experimentally relevant) coherent intermediate scattering function $\phi_q(t)$:

$$\phi_q(t) = \frac{F(q,t)}{S(q)}, \quad F(q,t) = \frac{1}{N}\langle\delta\rho^*(\boldsymbol{q},t)\delta\rho(\boldsymbol{q},0)\rangle, \quad F(q,0) = S(q), \tag{3.27}$$

where $q = |\boldsymbol{q}|$ is the modulus of the wave vector and $S(q)$ the static structure of the liquid. Qualitatively, we may interpret $\phi_q(t)$ as the ensemble-averaged overlap between the initial configuration of the liquid and its configuration at time t. Both configurations are fully specified by the respective set of positions ($\{\boldsymbol{r}_i\}$) and the overlap is probed on the length scale $1/q$. Liquid state theory now suggests that the cage effect is dominated by density fluctuations for wave vectors corresponding to the inverse interparticle distance, that is, $q \sim 2\pi/\sigma_{LJ}$ ($\sim 1\,\text{Å}^{-1}$) [171]. Therefore, MCT argues that the understanding of the structural relaxation in supercooled liquids requires a theory of $\phi_q(t)$ for wave numbers extending from macroscopic to microscopic scales.

The starting point of this theory is an exact equation of motion for $\phi_q(t)$ derived by the projection operator formalism of Zwanzig and Mori [171]. For classical particles of mass m obeying Newtonian dynamics, this equation reads [32,169]

$$\frac{\partial^2 \phi_q(t)}{\partial t^2} + \Omega_q^2 \phi_q(t) + \int_0^t dt' M(q,t-t')\frac{\partial \phi_q(t')}{\partial t'} = 0, \tag{3.28}$$

where the frequency Ω_q is given by

$$\Omega_q^2 = \frac{q^2 v^2}{S(q)} \quad \text{with} \quad v^2 = k_B T/m \tag{3.29}$$

and the initial conditions are $\phi_q(t=0)=1$ and $\partial_t \phi_q(t=0)=0$. Equation 3.28 is akin to the equation of motion of a damped harmonic oscillator: $\ddot{x} + \gamma \dot{x} + \omega_0^2 x = 0$. Ω_q corresponds to the angular frequency ω_0; it sets the time scale for the initial decay of $\phi_q(t)$ because $\phi_q(t) = 1 - \Omega_q^2 t^2/2$ for $t \to 0$. The integral term in Equation 3.28 corresponds to the viscous damping ($\gamma\dot{x}$) of the oscillator ($\gamma\dot{x}$ is obtained when choosing $M(q,t)=\gamma\delta(t)$). The damping coefficient $M(q,t-t')$ depends on the full history of the motion from time t' up to time t and is therefore called "memory kernel." The memory kernel is a complicated object. The Zwanzig–Mori theory provides a rigorous derivation of Equation 3.28 from the full Newtonian dynamics and yields an expression of $M(q,t)$ in terms of the "fluctuating force" $F^Q(q,0)$ (see below for an explicit expression within MCT) [32,169,172]:

$$M(q,t) = \frac{1}{mNk_B T}\langle F^{Q*}(q,0)e^{iQ\mathcal{L}Qt}F^Q(q,0)\rangle.$$

This expression links dissipation (i.e., M) and fluctuations (i.e., F^Q). However, the time evolution of the memory kernel is unusual; it is not given by the full Newtonian dynamics, but is determined by the "reduced Liouvillian" $Q\mathcal{L}Q$.

What does this mean? Consider some observable $A(x(t))$ depending on the microscopic configuration (x) of the system at time t. The time evolution of $A(t)$ is formally given by $A(t) = e^{i\mathcal{L}t}A(0)$, where $\mathcal{L}$ is the Liouvillian [171]. The essence of the Zwanzig–Mori formalism is to split this time evolution into two parts: (i) a "relevant" part due to variables that are supposed to dominate the dynamics of A and (ii) the rest. With the chosen relevant variables, a projection operator $\mathcal{P}$ can be defined and the "rest" is formally given by the complementary operator $Q = 1 - \mathcal{P}$. Since $\mathcal{P}$ is a projection operator, it satisfies $\mathcal{P}^2 = \mathcal{P}$ and so $\mathcal{P}Q = \mathcal{P} - \mathcal{P}^2 = 0$, that is, $\mathcal{P}$ and Q project onto orthogonal spaces. Writing then $\mathcal{L} = \mathcal{L}(\mathcal{P}+Q) = \mathcal{L}\mathcal{P} + \mathcal{L}Q$, the splitting of the full dynamics ($\mathcal{L}$) into a relevant part ($\mathcal{L}\mathcal{P}$) and a remaining part ($\mathcal{L}Q$) is achieved. The memory kernel $M(q,t)$ solely evolves with the time evolution operator of the remaining part, so in a "space" from which all contributions stemming from the dynamics of the relevant variables have been removed. As a result, the fluctuations of the degrees of freedom comprised in the remaining part act on the relevant variables only as a decoupled "noise"—the archetypal example of such a situation is that of a Brownian particle whose velocity is the sole relevant variable [171].

The Zwanzig–Mori approach therefore provides a formalism to extract the dynamics of the chosen relevant variables. However, the approach does not give a hint as to which variables are really relevant, nor does it suggest a tractable approximation for the memory kernel $M(q,t)$, which remains a complicated, fairly abstract, quantity. The (so-called) idealized MCT is an attempt to "fill this gap," based upon two suggestions (see [32,169] for mathematical details and also [172] for a lucid presentation of the derivation):

1. MCT chooses two relevant ("slow") variables, $\delta\rho(\boldsymbol{q},t)$ and the longitudinal current $j(\boldsymbol{q},t) = \boldsymbol{j}(\boldsymbol{q},t) \cdot (\boldsymbol{q}/q)$. (Owing to the continuity equation, $\partial_t \rho(\boldsymbol{r},t) = -\nabla \cdot \boldsymbol{j}(\boldsymbol{r},t)$, this implies that density fluctuations and their time derivatives are considered to be slow.) This leads to the following expression for the random force: $F^Q(\boldsymbol{q},0) = m\mathcal{QL}j(\boldsymbol{q},0) = m\mathcal{L}j(\boldsymbol{q},0) - [qk_{\mathrm{B}}T/S(q)]\delta\rho(\boldsymbol{q},0)$. So, the random force consists of two terms, a term due to density fluctuations and a term related to the longitudinal part of the force resulting from the fluctuations of the microscopic stress tensor ($\sigma_{\alpha\beta}$) in the system (since $\partial_t j(\boldsymbol{q},t) = \mathrm{i}\mathcal{L}j(\boldsymbol{q},t) = (\mathrm{i}/qm)\sum_{\alpha\beta} q_\alpha q_\beta \sigma_{\alpha\beta}(\boldsymbol{q},t)$ with $\alpha, \beta = x, y, z$).

2. MCT recognizes that Equation 3.28 expresses a one-to-one correspondence between the correlation functions of density fluctuations and of the random forces. If density fluctuations relax slow, so does the memory kernel. Density fluctuations may relax slowly because particles are transiently localized, possibly through mutual strong interactions between neighboring particles in the supercooled liquid. A first step to model this (cage) effect is to consider pairs of particles or—in terms of collective variables—pairs of densities. So the simplest, physically reasonable, hypothesis is to assume that $F^Q(\boldsymbol{q},t)$ has an overlap with the product of two density fluctuations, schematically with $\delta\rho(\boldsymbol{k},t)\delta\rho(\boldsymbol{p},t)$. MCT suggests to extract this contribution by a second Zwanzig–Mori step projecting $F^Q(\boldsymbol{q},t)$ onto $\delta\rho(\boldsymbol{k},t)\delta\rho(\boldsymbol{p},t)$. This splits the memory kernel into two parts: $M(q,t) = M^{\mathrm{reg}}(q,t) + M^{\mathrm{MC}}(q,t)$. After the second Zwanzig–Mori projection step, the "regular" part M^{reg} aggregates the remaining contributions to the fluctuating force after the removal of the fluctuations of density, longitudinal current, and density pairs. M^{reg} is assumed to condition the dynamics beyond the (deterministic) short-time regime (ruled by Ω_q^2), to lead to a fast decay of density fluctuations, and to vary only weakly with T upon cooling. In this sense, it is regular because no slow relaxation is expected to result from it. As a simple model, MCT poses $M^{\mathrm{reg}}(q,t) = \nu_q \delta(t)$ with the constant $\nu_q \geq 0$ (see however [173] in which the regular kernel was not dropped). The second part, M^{MC} models the cage effect via a coupling to pairs of density fluctuations. The memory kernel being a force–force correlation function involves the product of four density fluctuations evolving in time with the reduced Liouvillian $\mathcal{QLQ}$:

$$\langle \delta\rho^*(\boldsymbol{k},0)\delta\rho^*(\boldsymbol{p},0)e^{\mathrm{i}\mathcal{QLQ}t}\delta\rho(\boldsymbol{k},0)\delta\rho(\boldsymbol{p},0)\rangle \approx \langle \delta\rho^*(\boldsymbol{k},0)e^{\mathrm{i}\mathcal{L}t}\delta\rho(\boldsymbol{k},0)\rangle\langle \delta\rho^*(\boldsymbol{p},0)e^{\mathrm{i}\mathcal{L}t}\delta\rho(\boldsymbol{p},0)\rangle. \quad (3.30)$$

The right-hand side (rhs) of this equation is the core (and uncontrolled) approximation of the idealized MCT: The dynamic four-point correlation function of the densities is factorized into a product of two-point correlation functions and the reduced dynamics is replaced by the full one. In this way, Equation 3.28 for $\phi_q(t)$ is closed because the rhs of Equation 3.30 can be expressed in terms of intermediate scattering functions. With $M(q,t) \approx \nu_q \delta(t) + M^{\mathrm{MC}}(q,t) = \nu_q \delta(t) + \Omega_q^2 m_q^{\mathrm{MC}}(t)$, the final result of the idealized MCT reads

$$\frac{\partial^2 \phi_q(t)}{\partial t^2} + \nu_q \frac{\partial \phi_q(t)}{\partial t} + \Omega_q^2 \phi_q(t) + \Omega_q^2 \int_0^t \mathrm{d}t' m_q^{\mathrm{MC}}(t-t') \frac{\partial \phi_q(t')}{\partial t'} = 0, \quad (3.31)$$

with

$$m_q^{\mathrm{MC}}(t) \approx \mathcal{F}_q[\phi_k(t), \phi_p(t)] = \frac{1}{V}\sum_{k+p=q} V(\boldsymbol{q};\boldsymbol{k},\boldsymbol{p})\phi_k(t)\phi_p(t),$$

$$V(\boldsymbol{q};\boldsymbol{k},\boldsymbol{p}) = \frac{1}{2}\rho S(q)S(k)S(p)\left\{\boldsymbol{q}\cdot[\boldsymbol{k}c(k) + \boldsymbol{p}c(p)]/q^2 + \rho c_3(q,k,p)\right\}^2 \geq 0. \quad (3.32)$$

Here, $c(q)$ is the direct correlation function and $c_3(q,k,p)$ the triple-correlation function [3,32]. The direct correlation function is related to the static structure factor by the Ornstein–Zernike equation $\rho c(q) = 1 - 1/S(q)$; qualitatively, $c(q)$ can be interpreted as the effective interaction potential between two particles in the system [3]. The triple-correlation function is defined by the correlations of three static density fluctuations: $\langle \delta\rho^*(\boldsymbol{q})$ $\delta\rho(\boldsymbol{k})\delta\rho(\boldsymbol{p})\rangle = N\delta_{q,k+p}S(q)S(k)S(p)[1 + \rho^2 c_3(q,k,p)]$. Neglecting $\rho^2 c_3(q,k,p)$ with respect to 1 is known as convolution approximation [3]. This approximation appears to be justified for simple glass formers [174,175], but not for structurally more complicated ones, such as *ortho*-terphenyl [176] or silica [174] (for tests on polymer melts, see [142,177]).

Let's comment on Equations 3.31 and 3.32 by a few remarks:

1. For large times, the dynamics is dominated by the memory integral in Equation 3.31 and the first two terms, $\partial_t^2 \phi_q(t) + v_q \partial_t \phi_q(t)$, can be neglected [32]. Then, Ω_q^2 drops out and $\phi_q(t)$ no longer depends on the underlying microscopic dynamics. Therefore, the relaxation of $\phi_q(t)$ at long times should be the same for simulations with Newtonian, Brownian, or Monte Carlo dynamics. This prediction was verified for simple glass formers [178–181].

2. Equations 3.31 and 3.32 couple the dynamics of $\phi_q(t)$ (of the "mode q") to that of all products $\phi_k(t)\phi_p(t)$, obeying $\boldsymbol{k} + \boldsymbol{p} = \boldsymbol{q}$ (whence the name "mode-coupling theory"). The coupling coefficients $V(\boldsymbol{q};\boldsymbol{k},\boldsymbol{p})$ are determined by the equilibrium structure of the system, that is, by $S(q)$, $c(q)$, and $c_3(q,k,p)$. These structural quantities depend on external control parameters, such as temperature or density, but change only weakly on approach to the glass transition. Nevertheless, Equations 3.31 and 3.32 yield glassy dynamics because the nonlinear coupling between the density fluctuations greatly amplifies the weak structural changes.

3. Analysis of Equations 3.31 and 3.32 for a hard-sphere system* shows that the glassy dynamics is predominantly caused by the coupling of density fluctuations for intermediate q values, involving the first maximum of $S(q)$ and beyond [32,182]. This q interval is sensitive to the intermediate-range order in the liquid (nearest-neighbor shells). On the other hand, contributions to the memory kernel from small wave numbers ($q \to 0$) are negligible; hydrodynamic fluctuations are irrelevant for the MCT explanation of glassy dynamics, at least in hard-sphere-like systems (see page 253 of [32]).

4. Equations 3.31 and 3.32 can also be derived for polymer melts in the framework of a site formalism [183] associating an interaction site with every monomer of a chain [184,185]. Polymer-specific effects, such as chain stiffness [141,142] or chain length [105], enter the relaxation of $\phi_q(t)$ only via the equilibrium properties encapsulated in ρ, $S(q)$, and $c(q)$ (and in principle also in $c_3(q,k,p)$ [142,177,185]). Since this extension of MCT to the polymer melts preserves the general structure of the theory, all universal results concerning the MCT liquid–glass transition, originally developed for simple liquids (monatomic liquids or binary mixtures), are also valid for macromolecular systems.[†]

3.3.2.1.2 Some predictions of the idealized MCT

What does it mean that the MCT predicts "universal results?" These results do not depend on the precise functional form of the structural input (i.e., of $S(q)$, etc.), but are mathematical consequences of the fact that

---

* MCT proposes the hard-sphere system as an archetypal example of a glass-forming liquid. The reason is as follows: Different glass formers have an amorphous structure being characterized by static density fluctuations ($S(q)$, $c(q)$, etc.) that are qualitatively similar to the hard-sphere system. These fluctuations determine the memory kernel and so the glassy dynamics at long times, according to MCT. Hence, the hard-sphere system should be a simple, but relevant, representative for the class of glass-forming liquids.

† Within MCT, the slowing down of the relaxation results from the cage effect modeled by the memory kernel M^{MC}. If the direct correlation function $c(q)$ was zero, M^{MC} would vanish, and no dynamic anomalies would be predicted by the theory. In polymer physics, $c(q)$ can be identified with the excluded volume parameter v, if the spatial dependence is neglected [186] (more precisely, $c(q) \rightsquigarrow -v$, compare, for example, Equation 3.17 in [185] with Equation 3.19 in [12]). The excluded volume in polymer melts is known to be approximately screened so that the chains have nearly ideal conformations [10]. Roughly speaking, this implies $v = 0$ [1]. Nevertheless, long polymers have anomalous dynamics because they cannot cross one another (topological interactions), leading to chain entanglements [1,2]. Thus, the MCT based on the cage effect cannot describe the dynamics due to entanglements. Other concepts, such as slip-link or reptation models [17,122], are needed.

the mode-coupling functional $\mathcal{F}_q(f_k, f_p) = (1/V)\sum_{k+p=q} V(q;k,p) f_k f_p$ and its derivatives (with respect to f) are not negative [32,167–169]. Here, f_q denotes the long-time limit of the density fluctuations, $f_q = \lim_{t\to\infty}\phi_q(t)$, which obeys $0 \le f_q < 1$. The analysis of Equations 3.31 and 3.32 now reveals that f_q qualitatively changes behavior at a critical temperature T_c:

$$\lim_{t\to\infty} \phi_q(t) = \begin{cases} 0 & \text{for } T > T_c, \\ f_q(T) > 0 & \text{for } T \le T_c. \end{cases} \tag{3.33}$$

What does this mean? For $T > T_c$, density fluctuations fully relax and the system eventually loses the memory of its initial state. This is a characteristic feature of the liquid state. By contrast, density fluctuations cannot fully decay in an amorphous solid: a particle can only oscillate around its equilibrium position without being able to leave the nearest-neighbor cage. Then, a finite fraction of $\phi_q(t)$, $0 < f_q < 1$, survives in the long-time limit. Therefore, T_c is identified with the (ideal) glass transition in MCT. The value of $f_q(T \le T_c)$ increases (toward 1) with decreasing temperature; it can be interpreted as a measure for the "solidity" of the amorphous solid on the length scale $1/q$, like the Debye–Waller factor does in a crystal. MCT calls f_q the "Debye–Waller factor" or also the "nonergodicity parameter" because the glass phase below T_c is nonergodic, in the sense of Reference 187.

In addition to T_c, the analysis of Equations 3.31 and 3.32 yields two further numbers: (i) the "separation parameter" σ, which is a function of the reduced distance to the critical point, $\varepsilon = (T_c - T)/T_c$, and, to leading order in ε, is given by

$$\sigma = C\varepsilon \quad (C = \text{system-specific constant}), \tag{3.34}$$

and (ii) the "exponent parameter" λ, which determines and correlates the two key exponents, a and b, of the idealized theory (Γ is the gamma function):

$$\lambda = \frac{\Gamma(1-a)^2}{\Gamma(1-2a)} = \frac{\Gamma(1+b)^2}{\Gamma(1+2b)} \quad (1/2 \le \lambda < 1). \tag{3.35}$$

MCT refers to a as the "critical exponent" and to b as the "von Schweidler exponent." Since $1/2 \le \lambda < 1$, Equation 3.35 gives $0 < a < 0.3953$ and $0 < b \le 1$ [32,169].

On cooling toward T_c, the separation parameter and the exponents a and b determine the dynamics. Provided T is close to T_c, the following predictions can be obtained from Equations 3.31 and 3.32. The theory yields two time scales: the β relaxation time t_σ:

$$t_\sigma = \frac{t_0}{|\sigma|^{1/2a}} \quad (\text{for } T \to T_c^\pm) \tag{3.36}$$

and the α relaxation time t'_σ:

$$t'_\sigma = \frac{t_0}{(-\sigma)^\gamma} \quad (\text{only for } T \to T_c^+),$$
$$\gamma = \frac{1}{2a} + \frac{1}{2b} \quad (\gamma > 1.765). \tag{3.37}$$

Here, t_0 denotes a system-specific microscopic time. Equations 3.36 and 3.37 indicate that t'_σ diverges more strongly than t_σ on cooling toward T_c. Owing to this separation of the time scales, MCT predicts a two-step relaxation scenario. The first step describes the relaxation for times on the scale t_σ (i.e., for $t_0 \ll t \ll t'_\sigma$) for which $\phi_q(t)$ remains close to $f_q^c = f_q^c(T_c)$, the value of the nonergodicity parameter at T_c. This time regime, where $|\phi_q(t) - f_q^c| \ll 1$, is called β process (or β relaxation regime) by MCT. The second step corresponds to the α process (or α relaxation regime), where $\phi_q(t)$ decays from f_q^c to zero for times on the scale t_σ'. These definitions imply that the β and α processes overlap for $t_\sigma \ll t \ll t_\sigma'$, provided $T > T_c$. For $T < T_c$, the α process stops ($t_\sigma' = \infty$ for all $T \le T_c$); only the early β process remains, accounting for the relaxation of $\phi_q(t)$ toward the nonergodicity parameter $f_q(T) > f_q^c$ (note that t_σ decreases as T is lowered below T_c).

For both the α and β processes, MCT makes detailed predictions [32,169], many of which were tested in experiments and simulations (for reviews on these tests see, for example, [4,7,32,37,167,168,170]). Here, we only mention a few results. For the time regime of the early α process, that is, for the time regime roughly extending from $t \sim t_\sigma$ to $t \sim t_\sigma'$, MCT suggests that the relaxation of $\phi_q(t)$ can be expanded in terms of $(t/t_\sigma')^b$ [188]:

$$\phi_q(t) = f_q^c - h_q^{(1)} \left(\frac{t}{t_\sigma'} \right)^b + h_q^{(2)} \left(\frac{t}{t_\sigma'} \right)^{2b} \quad \text{(for } T \to T_c^+\text{)}. \tag{3.38}$$

Here, not only f_q^c and b are independent of temperature, but also the amplitudes $h_q^{(1)}, h_q^{(2)}$; the T dependence solely results from the time scale t_σ'.* This implies a time–temperature superposition principle (TTSP): The early α process is described by a T independent master curve, which is a function of the scaled time t/t_σ'. MCT predicts that this TTSP holds not only for the early α process (Equation 3.38), but for the whole α process, that is, for all $t \gtrsim t_\sigma$. Model calculations furthermore reveal that the MCT α relaxation is stretched. As for experimental or simulation data, this stretched relaxation can be fitted well by a Kohlrausch–Williams–Watts (KWW) function:

$$\phi_q(t) \simeq f_q^K \exp\left[-(t/\tau_q^K)^{\beta_q^K} \right] \quad (t \ge t_\sigma), \tag{3.39}$$

where f_q^K is an amplitude, τ_q^K the KWW relaxation time, and $\beta_q^K \le 1$ the Kohlrausch (stretching) exponent. Although the KWW function is a well-suited fit function, it is in general not a solution of the MCT α process, except in the special limit of large q. In this limit, it was proved [189] that there is a time interval $t/t_\sigma' \ll t_q^K/t_\sigma' \le 1$ in which the α process obeys

$$\lim_{q \to \infty} \phi_q(t) = f_q^c \exp\left[-\Gamma_q(t/t_\sigma')^b \right], \quad \Gamma_q \propto q. \tag{3.40}$$

This implies

$$\lim_{q \to \infty} f_q^K = f_q^c, \quad \lim_{q \to \infty} \beta_q^K = b, \quad \lim_{q \to \infty} \tau_q^K \propto q^{-1/b} t_\sigma'. \tag{3.41}$$

Equation 3.40 can be interpreted in terms of the theory of stochastic processes, as a consequence of Lévy's generalization of the central limit theorem [189,190].

3.3.2.2 GLASSY AND POLYMER DYNAMICS IN NONENTANGLED MELTS: COMPARISON BETWEEN THEORY AND SIMULATIONS OF A FLEXIBLE BEAD–SPRING MODEL

As temperature decreases toward T_c, MCT predicts intermittence of all structural relaxation processes. This intermittence manifests itself, not only in $\phi_q(t)$, but in all quantities that couple to density fluctuations. Examples involve correlation functions related to the single-particle dynamics, like the mean-square displacement (MSD) $g_0(t)$ or the incoherent intermediate scattering function $\phi_q^s(t)$, and also other collective correlation functions, such as the shear relaxation modulus $G(t)$ [32,169].

---

* This sets constraints on the data analysis. Still, the fit procedure is fairly intricate and so prone to criticism; for a discussion of these technical aspects see, for example, [7,29,105].

3.3.2.2.1 Tagged particle dynamics and stress relaxation: Background

Let's consider a polymer melt consisting of n chains with N monomers each. We denote the total number of monomers in the melt by $M = nN$. Single-particle density fluctuations at wave vector q are given by (cf. Equation 3.26)

$$\delta\rho_i(q,t) = \int \delta\rho_i(r,t)\exp(iq\cdot r)\mathrm{d}^3r = \exp[iq\cdot r_i(t)],$$

where $r_i(t)$ is the position of monomer i at time t and

$$\delta\rho_i(r,t) = \rho_i(r,t) - \langle\rho_i(r,t)\rangle = \rho_i(r,t) - \frac{1}{V}$$
$$= \rho_i(r,t) \quad \text{(thermodynamic limit)}.$$

In analogy to Equation 3.27, the dynamic correlation function of $\rho_i(q,t)$ defines the (experimentally relevant) incoherent intermediate scattering function $\phi_q^s(t)$:

$$\phi_q^s(t) = \frac{1}{M}\sum_{i=1}^{M}\langle\rho_i^*(q,t)\rho_i(q,0)\rangle = \frac{1}{M}\sum_{i=1}^{M}\langle\exp\{-iq\cdot[r_i(t)-r_i(0)]\}\rangle, \tag{3.42}$$

which measures the decorrelation of the positions of an individual monomer with time on the length scale $1/q$. Since the polymer melt is an isotropic system and also (expected to be) invariant with respect to inversion of the position coordinates ($r_i \to -r_i$, $i = 1,...,M$), the expansion of Equation 3.42 for small q reads $\phi_q^s(t) = 1 - q^2 g_0(t)/6 + O(q^4)$. Therefore, in the limit of vanishing wave vectors, $\phi_q^s(t)$ is related to the mean-square displacement, $g_0(t)$, of a monomer

$$g_0(t) = \frac{1}{M}\sum_{i=1}^{M}\langle[r_i(t)-r_i(0)]^2\rangle = \lim_{q\to0}\frac{6[1-\phi_q^s(t)]}{q^2}. \tag{3.43}$$

Let's further assume that the intramolecular and intermolecular interactions between the monomers are pair potentials so that the total potential energy of the melt may be written as $U_M(r_1,...,r_M) = \sum_{j>i}U(r_{ij})$. Here, $r_{ij} = |r_{ij}|$, with $r_{ij} = r_i - r_j$. The components of the vector r_{ij} shall be denoted by $r_{\alpha,ij}$, $\alpha = x,y,z$. Then, the microscopic expression of the stress tensor $\sigma_{\alpha\beta}$ is given by the Kirkwood formula [32,171]

$$\sigma_{\alpha\beta} = \frac{1}{V}\left[\sum_{i=1}^{M}mv_{\alpha,i}v_{\beta,i} - \sum_{j>i}\left(\frac{\partial U}{\partial r_{ij}}\right)\frac{r_{\alpha,ij}r_{\beta,ij}}{r_{ij}}\right], \tag{3.44}$$

where v_i is the velocity of the ith monomer. In Equation 3.44, the first term on the rhs is the contribution from the kinetic energy to the stress, the second the virial contribution. The off-diagonal elements of $\sigma_{\alpha\beta}$ characterize the viscoelastic properties of the melt. Assume that an infinitesimal step strain of magnitude γ is applied at $t = 0$ in the x direction. Instantaneously, this creates a shear stress $\langle\sigma_{xy}(t = 0)\rangle$. However, this stress cannot persist in the polymer liquid. As time increases, the stress $\langle\sigma_{xy}(t)\rangle$ must decrease toward 0 because the melt flows. The ratio $G(t) = \langle\sigma_{xy}(t)\rangle/\gamma$ is independent of γ (in the linear response regime) and defines a material function, the shear relaxation modulus $G(t)$ [1,2]. In the (polymer) liquid [191], $G(t)$ may be expressed in terms of the autocorrelation function of the shear stress [15,171]:

$$G(t) = \frac{V}{k_B T}\langle\sigma_{xy}(t)\sigma_{xy}(0)\rangle. \tag{3.45}$$

For all mentioned quantities—$\phi_q^s(t)$, $g_0(t)$, and $G(t)$—MCT makes specific predictions in the limit $T \to T_c^+$. For the early α relaxation, they take the same form as Equation 3.38 [32,169,188,192]:

$$\phi_q^s(t) = f_q^{sc} - h_q^{s(1)}\left(\frac{t}{t'_\sigma}\right)^b + h_q^{s(2)}\left(\frac{t}{t'_\sigma}\right)^{2b},$$

$$g_0(t) = 6r_{sc}^2 + 6h_{msd}^{(1)}\left(\frac{t}{t'_\sigma}\right)^b - 6h_{msd}^{(2)}\left(\frac{t}{t'_\sigma}\right)^{2b}, \qquad (3.46)$$

$$G(t) = G_T^c - h_T^{(1)}\left(\frac{t}{t'_\sigma}\right)^b + h_T^{(2)}\left(\frac{t}{t'_\sigma}\right)^{2b}.$$

Here, the von Schweidler exponent b and the α relaxation time $t'_\sigma(T)$ are the same as in Equation 3.38. Only the nonergodicity parameters (f_q^{sc}, r_{sc}, G_T^c) and the amplitudes ($h_q^{s(1)}$, $h_{msd}^{(1)}$, ...) depend on the quantity considered.

The MSD and the shear relaxation modulus are also key quantities discussed in polymer physics. If the chains are nonentangled, the Rouse model [1,2] is expected to give a good description of the dynamics in polymer melts [12,13,17]. For the MSD, the model predicts [1,2]

$$g_0(t) = \begin{cases} b^2\sqrt{Wt} & \text{for } \tau_1 < t < \tau_N, \\ 6Dt & \text{for } t > \tau_N, \end{cases} \qquad (3.47)$$

where

$$\tau_1 = \frac{4}{\pi^3 W} \quad \text{and} \quad \tau_N = \tau_1 N^2. \qquad (3.48)$$

Here, b denotes the effective bond length, W the monomer relaxation rate, τ_1 the monomer relaxation time, τ_N the Rouse time, and D the diffusion coefficient of the polymer. Equation 3.47 can be rationalized by a scaling argument [2]: A polymer chain in the melt is a self-similar object, implying that a segment with s monomers ($1 \le s \le N$) just looks the same as the whole chain. The segment therefore has the relaxation time $\tau_s = \tau_1 s^2$ (cf. Equation 3.48). For $\tau_1 < t < \tau_N$, the time t probes all segmental relaxation times so that $s(t) = \sqrt{t/\tau_1}$. The segment length $s(t)$ corresponds to the internal distance $R(s) = b\sqrt{s}$ because a polymer in the melt obeys (approximately [10]) ideal chain statistics. In the time interval $\tau_1 < t < \tau_N$, a monomer will therefore move in space over a typical distance of order $R(s)$ and so we get $g_0(t) \sim R^2(s) \sim b^2\sqrt{Wt}$. Chain connectivity leads to subdiffusive monomer motion, as long as the monomer explores distances smaller than the chain dimension $R(N)$. For larger distances and so for times $t > \tau_N$, the monomer moves with the chain as a whole and its motion becomes diffusive ($g_0(t) \sim Dt$).

The Rouse model also provides an expression for the shear relaxation modulus, in terms of the superposition of the relaxation of all Rouse modes [1,2,193]. An excellent approximation to this exact form is furnished by the following formula (see Equation 8.48 of [2]):

$$G(t) = \frac{k_B T \rho}{\sqrt{t/\tau_1}}\exp\left(-\frac{t}{\tau_N}\right) \quad \text{for } t > \tau_1. \qquad (3.49)$$

This result may be interpreted as follows [1,2]: At the monomer relaxation time τ_1, the level of the shear relaxation modulus is proportional to the thermal energy $k_B T$ and the total monomer density ρ. This implies that all monomers in the melt bear the stress at time τ_1. As time increases, the chains relax and the density of stress-bearing units decreases to $\rho/s(t)$ so that $G(t) \sim k_B T\rho/s(t) = k_B T\rho/\sqrt{t/\tau_1}$ for $\tau_1 < t < \tau_N$. At τ_N, the density of stress-bearing units is the density of chains, ρ/N. For $t > \tau_N$, the melt then behaves as a Maxwell fluid with initial shear modulus $k_B T \rho/N$ and relaxation time τ_N.

3.3.2.2.2 Simulation results for the flexible BSM

Figure 3.8 depicts $G(t)$ (Figure 3.8a), $g_0(t)$ (Figure 3.8b), and $\phi_q^s(t)$ (Figure 3.8c). The data are obtained from constant-pressure ($P = 0$) MD simulations for the flexible BSM with chains of length $N = 64$ ($<N_e \approx 70$ [128]). Figure 3.8a and b show the evolution of the structural relaxation as T is lowered toward T_c ($=0.416$ [105]), whereas Figure 3.8c focuses on the length-scale dependence of the single-monomer relaxation at $T = 0.44$, a low temperature near T_c. Since the Newtonian dynamics underlying the MD simulation is invariant with respect to time inversion ($t \to -t$), the initial time dependence of all quantities is proportional to t^2. This is illustrated for the monomer MSD: Initially, a monomer behaves as if it was a free particle, solely subjected

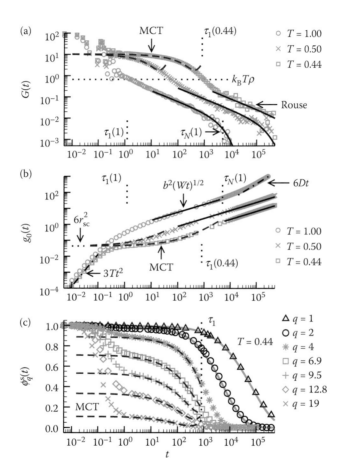

Figure 3.8 Structural relaxation of a polymer melt with $N = 64$ for the flexible BSM: (a) shear relaxation modulus $G(t)$ (symbols) for $T = 1$, 0.5, 0.44, (b) monomer MSD $g_0(t)$ (symbols) for $T = 1$, 0.5, 0.44, (c) incoherent scattering function $\phi_q^s(t)$ (symbols) for different moduli q of the wave vector at $T = 0.44$. For $N = 64$ the critical temperature of MCT is $T_c = 0.416$ [105]. In all panels, the dashed lines present the fit results to the late-β/early-α process of MCT, Equation 3.46 (for details on the fit, see [105]). The (black) solid lines in (a) and (b) show the Rouse model predictions, Equations 3.47 and 3.49. Two parameters enter these equations, b and W. b is determined first from a fit to the internal distances $R(s)$ [299] and then W can be obtained by fitting the MSD to Equation 3.47. In (c), the (light) solid lines for $q = 1$ and $q = 2$ show the Gaussian approximation (Equation 3.52). In all panels, the vertical dotted lines indicate the monomer relaxation time $\tau_1(T)$ at $T = 0.44$ and $T = 1$. For $T = 1$, the Rouse relaxation time $\tau_N(T)$ is also shown as a vertical dotted line. In (b), the dash-dotted lines depict, respectively, the ballistic behavior at short times ($g_0(t) = 3v^2 t^2$ with the thermal velocity $v^2 = k_B T/m$) and the diffusive behavior at long times ($g_0(t) = 6Dt$, where D is fitted to the data). (b and c: Adapted from S. Frey et al., *Eur. Phys. E*, 38, 11, 2015.)

to the inertia force, $m\ddot{\mathbf{r}} = \mathbf{0}$. Integration gives $g_0(t) = 3(k_B T/m)t^2$ because the (constant) velocity is given by the thermal velocity $(k_B T/m)^{1/2}$. This free ballistic motion cannot last for long; intrachain and interchain interactions slow down the dynamics in a complicated way.* Therefore, the monomer MSD gradually flattens, as time increases toward $t = 1$. Similarly, for $t < 1$, we see that the shear relaxation modulus features damped oscillations, which may be attributed to the interplay of inertia with the (harmonic) bonding and other forces [123,193]. In the polymer liquid phase at $T = 1$, the monomer relaxation time is estimated to be $\tau_1 \approx 1$ from Equation 3.48 (see Figure 3.8). For longer times, both $g_0(t)$ and $G(t)$ are well described by the Rouse model, Equations 3.47 and 3.49 (solid lines in Figure 3.8a and b). This finding agrees with expectations from polymer theory [1,2]: For long times, microscopic details (of the chain conformations and in time) should average out, making a coarse-grained description, such as the Rouse model, pertinent.

The good agreement between simulation and polymer theory hints at the significance of flexible chain models for polymer physics. Owing to their flexibility, the number of Kuhn monomers (n_K) is close to the actual chain length (N) so that the limit of long chains, where universal features determine the polymer behavior, is reached earlier than for more complicated models. This argument may be made more precise by estimating n_K (see Equation 2.15 of [2]):

$$n_K = \frac{R_{max}^2}{C_\infty n l_0^2},$$

(3.50)

where n is the number of bonds of the underlying chain and R_{max} its maximum length. For flexible models, we have $n = N - 1$ and $R_{max} = n l_0$, leading to $n_K = (N - 1)/C_\infty$. Using $1.5 \lesssim C_\infty \lesssim 1.9$ [9], we find that a value of $n_K = 10^3$, which for real polymers corresponds to high molecular masses of a few hundred thousand g/mol [2], requires chain lengths of about $1500 \lesssim N \lesssim 2000$. Such chain lengths are well accessible in current simulations of polymer liquids at high T. Simulations of flexible models have therefore been extensively used to test polymer theories and to help improve on them. Examples involve the study of the molecular rheology of entangled polymer chains [17,122,123], of the conformation and dynamics of ring polymers [194–197], of chain conformations of linear and living homopolymers [10], or of the influence of hydrodynamic interactions in the melt on the chain's center-of-mass dynamics [13,15,198].

Let's return to Figure 3.8 and to the evolution of the structural relaxation upon cooling. The solid lines in Figure 3.8a and b show that the Rouse predictions, Equations 3.47 and 3.49, also provide a good description of the simulation data at low temperature, but the onset of the description is continuously shifted to longer times as T decreases toward T_c. Temperature enters the Rouse theory through $b(T)$ and $\tau_1(T)$ (in addition to $k_B T \rho$). For the flexible BSM, chain conformations remain, in very good approximation, unaltered on cooling [105]. By contrast, the monomer relaxation time is strongly affected by temperature: τ_1 increases by about three orders of magnitude in the interval from $T = 1$ to $T = 0.44$. This opens an intermediate time window between the microscopic regime ($t < 1$) and the Rouse regime ($t > \tau_1$) where a plateau progressively emerges with decreasing T. Qualitatively, this agrees with the expected signature of early α process, if the plateau is identified with the nonergodicity parameter of the considered correlation function. This suggests to perform an MCT analysis. The dashed lines in Figure 3.8 present the fit results to Equation 3.46, where the T dependence only enters through the α relaxation time t'_α (Equation 3.37) that is common to all dynamic correlation functions. For T near T_c, we see that Equation 3.46 can indeed describe the data well in an intermediate time window, not only for the collective shear stress relaxation, but also for the single-monomer relaxation where finer details, such as the q dependence of the incoherent scattering function (Figure 3.8c), can also be reproduced. In all cases, the von Schweidler exponent b is the same and T independent. We find $b = 0.582 < 1$ [105]; this implies stretched relaxation for $G(t)$ and $\phi_q^s(t)$, and subdiffusive behavior for the monomer MSD.

How can we interpret this intermediate regime of the MCT β/α process? Let's consider the MSD as an example. As T decreases toward T_c, the intermediate regime expands in time. There, $g_0(t)$ increases only slowly with t, remaining close to $6r_{sc}^2$, the value of which is very small. From Figure 3.8b, we read off: $g_0(t) \approx 6r_{sc}^2 \approx 0.06$.

---

* Appealing back to Section 3.3.2.1, the description of the dynamics beyond the early time t^2 regime requires a modeling of the regular kernel M^{reg} and of precursors of the cage effect, which even for simple liquids is very sophisticated and often employs heuristic approaches [171,173].

So, $r_{sc} \approx 0.1$, that is, r_{sc} is about 10% of the monomer diameter. This result is reminiscent of the Lindemann criterion of melting [199–201]. Lindemann proposed that the melting of a crystal is associated with a vibrational instability of the crystalline lattice, which occurs if the root-mean-square displacement of the atoms reaches a critical value of about 10% of the interparticle distance. Finding this critical value here is an evidence for incipient glass formation. In the cold melt near T_c, liquid-like transport, involving large displacements, is suspended. Monomers are transiently localized in their nearest-neighbor cage and begin to exhibit solid-like dynamic features: They "oscillate" around their "equilibrium" positions with a typical fluctuation (displacement) of order $\sqrt{6}r_{sc}$ (r_{sc} is therefore sometimes referred to as the "Lindemann localization length"). For $T > T_c$, this solid-like dynamics cannot persist; eventually, the cages will decay. The late-β/early-α process, Equation 3.46, describes the onset of this "melting" of the temporarily stable amorphous solid.

Not only the MSD but also other correlation functions display similar solid-like features. The nonergodicity parameter of $G(t)$ signals the onset of temporary shear elasticity [63,191,202]. Perhaps this is related to the enhanced elasticity that has long been known to exist in cold polymer melts [203] and was more recently attributed to transient segmental localization near T_c [204,205]. Furthermore, the nonergodicity parameters (f_q^{sc}) of $\phi_q^s(t)$ reflect the solid-like fluctuations of the monomers around their "equilibrium" positions. To see this, let's assume that these solid-like fluctuations (of size r) are Gaussian distributed with variance $6r_{sc}^2$:

$$F_{sc}^{G}(r) = \left(\frac{3}{2\pi(6r_{sc}^2)} \right)^{3/2} \exp\left(-\frac{3r^2}{2(6r_{sc}^2)} \right),$$

Fourier transformation gives a Gaussian approximation for f_q^{sc} [192]:

$$f_q^{sc,G} = \exp(-q^2 r_{sc}^2). \tag{3.51}$$

Qualitatively, this explains why f_q^{sc} decreases with increasing q. At a quantitative level, Equation 3.51 compares well with the fit results for f_q^{sc}, if $q \lesssim 8$ (see [105] for details). From Equation 3.43, we would also expect that the Gaussian approximation

$$\phi_q^{s,G}(t) = \exp\left[-\frac{1}{6}q^2 g_0(t) \right] \tag{3.52}$$

can provide a good description of the time dependence of $\phi_q^s(t)$ in the small-q limit. The (gray) solid lines in Figure 3.8c show that this expectation is indeed borne out, if $q \lesssim 2$.

The fit to the MCT prediction (3.37) also furnishes the α relaxation time $t'_\sigma(T)$. Inserting the fit result for the von Schweidler exponent b into Equation 3.35 and using Equations 3.34 and 3.37, the critical temperature T_c can be determined from $t'_\sigma(T)$. The result of such an analysis for different chain lengths ($4 \leq N \leq 64$) is shown in Figure 3.9 [105]. As for T_g, the critical temperature increases with N and the increase can be fitted by the Fox–Flory equation:

$$T_c(N) = T_c^\infty - \frac{K}{N}, \tag{3.53}$$

where T_c^∞ is the critical temperature for infinite chain length and K a constant.

How can we interpret this result? In MCT, the critical temperature is a state variable characterizing the equilibrium state of the system. For a one-component, one-phase system—as the studied polymer melt—the equilibrium state is specified by two intensive variables, say, density and pressure. The equation of state then gives $T = T(\rho, P)$. Here, the simulations are done at constant pressure ($P = 0$). Along this isobar, T_c is a function of density only: $T_c = T_c(\rho_c)$ with ρ_c being the density at the critical point of MCT. Quite generally,

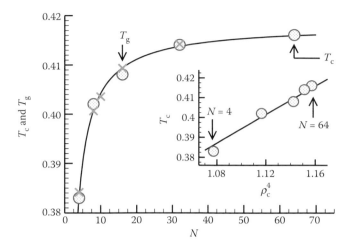

Figure 3.9 T_c versus N for the flexible BSM at pressure $P = 0$ (filled circles). The solid line shows a fit to Equation 3.53 with $T_c^\infty = 0.418$ and $K = 0.139$. The crosses present the results for T_g at pressure $P = 1$ from Figure 3.5b; the data are scaled by 0.968 ($= l_c^\infty / T_g^\infty$ with $T_g^\infty = 0.432$). Inset: T_c versus ρ_c^4 (filled circles), where ρ_c denotes the monomer density at T_c. The linear regression (solid line) gives $\Gamma_c^{-4} = 0.358$, that is, a coupling parameter of $\Gamma_c = 1.29$. (Figure adapted from S. Frey et al., *Eur. Phys. E*, 38, 11, 2015.)

the density defines an average distance (r_c) between the particles, $r_c = \rho_c^{-1/3}$. Typically, r_c is (slightly) smaller than the particle diameter (because $\rho_c \gtrsim 1$ [105]), implying that the behavior of the system should be dominated by the repulsive part of the interaction potential between the particles. This suggests to map the liquid onto an effective soft-sphere system [143,206–208]. A soft-sphere system has a purely repulsive inverse-power potential of the form $U(r) = 4\varepsilon_{LJ}(\sigma_{LJ}/r)^{3\gamma}$ with γ being a material-specific constant. Using $r = \rho^{-1/3}$, we get $U(r)/k_BT \propto \Gamma^\gamma$, where $\Gamma = \rho T^{-1/\gamma}$ combines density and temperature effects and is the only relevant control parameter of the soft-sphere system. For LJ potentials, a natural choice is $\gamma = 4$ (cf. Equation 3.6). Specified to the critical point, this suggests $\Gamma_c = \rho_c T_c^{-1/4}$. So, T_c is proportional to ρ_c^4, provided Γ_c is constant [206]. The inset of Figure 3.9 tests this idea for the flexible BSM by plotting T_c versus ρ_c^4. We see that the data are indeed compatible with a linear relationship (a fit yields $\Gamma_c = 1.29^*$). Since the simulations furthermore show that the monomer density for all temperatures increases with N as $\rho_\infty - \rho(N) \sim 1/N$ (ρ_∞ being the density extrapolated to infinite chain length), this explains the N dependence of T_c in terms of a density effect [105].

A similar increase of T_c with the density is reported for atomic LJ-type liquids [60,206,210]. LJ liquids are a representative of a newly defined class of simple liquids—the so-called "Roskilde liquid"—which is characterized by a strong correlation between the fluctuations of the potential energy and those of the virial contribution to the pressure in the canonical ensemble [207,211]. A Roskilde liquid has invariant structure and dynamics along specific curves, called "isomorphs," in the ρ–T plane. It is plausible that this striking scaling behavior of the Roskilde liquid is also pertinent for the flexible BSM, due to the dominant role played by the intermolecular LJ forces in the model. Indeed, recent constant-volume simulations of a flexible BSM similar to one discussed here support this idea [209]. It would be interesting to explore further this connection between the Roskilde liquid and the BSM, in particular, if chain stiffness effects are additionally taken into account.

---

* The value $\Gamma_c = 1.29$ is within the error bars of the result $\Gamma_c = 1.27 \pm 0.02$ obtained in simulations at different pressure for the flexible BSM with $N = 10$ ($P = 0$ [108], $P = 0.5,1,2$ [300]). Nevertheless, a best fit to the data of Figure 3.9 gives $T_c \sim \rho_c^{4.5}$, that is, an exponent $\gamma > 4$ [105]. Such larger exponents are also found for glass-forming LJ liquids [207,208] and for another flexible BSM model [209]. They can be explained by the contribution of the attractive term of the LJ potential ($\sim -r^{-6}$), which makes the effective repulsion steeper. However, the reported values are larger than $\gamma = 4.5$, typically $5 \lesssim \gamma \lesssim 7$ [61,207–209].

3.3.2.3 INFLUENCE OF POLYMER-SPECIFIC INTRAMOLECULAR STRUCTURE ON THE GLASSY DYNAMICS

Atomistic polymer models have intramolecular degrees of freedom—for instance, bond angles and torsional angles—which are absent in the flexible BSM (cf. Section 3.2.2). Inclusion of these valence terms in the potential energy renders these models much stiffer than the flexible BSM and also introduces intramolecular (torsional) barriers.

What is the influence of the intramolecular degrees of freedom on the local structure and dynamics of a supercooled polymer melt? An example is shown in Figure 3.10. The figure compares simulation results for two models of 1,4-polybutadiene melt at $T = 273$ K ($T_c = 214$ K [83]) [212,213]: a carefully validated united-atom model, which reproduces the experimentally found structure and dynamics of the melt (denoted by "CRC" for chemically realistic chain in Figure 3.10), and the same model with the torsional potential switched off (denoted by "FRC" for freely rotating chain in Figure 3.10). Apparently, the torsional potential has no discernible influence on the structure of the melt; the collective static structure factor $S(q)$ (Equation 3.27) and the interchain structure factor, the form factor $F(q)$ [2], are the same for the CRC and FRC models. However, their dynamics is very different. The inset in Figure 3.10 shows the monomer MSD. An intermediate plateau regime is only visible for the CRC model, but not for the FRC model. The suppression of the torsional potential thus leads to a considerable acceleration of the local dynamics of the monomers. The FRC model appears to be much farther away from T_c than the CRC model, although the ensemble-averaged local packing of both models is the same. This challenges the structure–dynamics correlation proposed by MCT.

Does this mean that MCT cannot be applied to real polymers? Not necessarily. In the past years, the asymptotic predictions for the β and α processes (Equations 3.33 through 3.41) were employed in the analysis of various chemically realistic models, including 1,4-polybutadiene [83,85], poly(vinyl methyl ether) [89], and

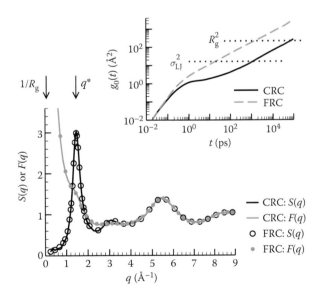

Figure 3.10 Simulation results for a melt of *cis–trans* 1,4-polybutadiene consisting of 40 chains with 30 repeat units each [212]. Data are taken from Reference 7 and were provided by courtesy of W. Paul. Main panel: Collective static structure factor $S(q)$ and single-chain structure factor (form factor) $F(q)$ as a function of the modulus of the wave vector q at $T = 273$ K ($T_c \approx 214$ K [83]). Results for two united-atom models are shown: a chemically realistic chain (CRC) model and the same model but without torsional potential (freely rotating chain, FRC). The vertical arrows indicate the q values associated with the radius of gyration R_g and with the first maximum of $S(q)$ ("amorphous halo"). The maximum occurs at $q^* \simeq 1.47\,\text{Å}^{-1}$. In real space, this corresponds to an intermonomer distance of $\approx 4.3\,\text{Å}$, which roughly agrees with the average Lennard-Jones diameter of the model ($\sigma_{LJ} \approx 3.8\,\text{Å}$). Inset: Monomer MSD $g_0(t)$ versus time for the CRC and FRC models at $T = 273$ K. The horizontal dotted lines indicate σ_{LJ}^2 and the radius of gyration $R_g^2 = 218\,\text{Å}^2$ (which is the same for both models [213]). (Figure adapted from J. Baschnagel and F. Varnik, *J. Phys. Condens. Matter*, 17, R851, 2005.)

polyisobutylene [91] (see [4] for a recent review). These studies confirm the applicability of the asymptotic predictions, thereby lending credence to the relevance of the cage effect also for real polymers. However, the studies also hint at an important (quantitative) difference between the flexible BSM and atomistic models. The relaxation dynamics of chemically realistic models is strongly stretched, implying substantially smaller values for the von Schweidler exponent (b) and larger values for the exponent parameter (λ) (Equation 3.35) than for the flexible BSM [4]. For the flexible BSM, one finds $b \simeq 0.58$ and $\lambda \simeq 0.74$ (cf. Table 3.1)— that is, values typical of hard-sphere systems [32,167,188,192] and atomic or molecular glass-forming liquids [4,32,167,170]—whereas atomistic models rather yield $b \approx 0.3$ and $\lambda \approx 0.9$ [4,83,85,89,91]. This finding has a meaning within MCT. The theory expects large λ values near 1, if different arrest mechanisms simultaneously operate [32,214]. Such competing arrest mechanisms were found in simulations and experiments. Examples involve packing constraints and bond formation in short-ranged attractive colloids [215] or packing constraints and spatial confinement in statically [216] or strong dynamically asymmetric binary mixtures [217,218]. For chemically realistic polymer models, the different mechanisms could correspond to intermolecular packing—as in simple liquids or in the flexible BSM—and intramolecular barriers for the relaxation of the chain conformations [4,85,89,91].

The interpretation of coexisting arrest mechanisms in polymer melts is supported by a combined simulation/MCT study of the bead–spring model defined by Equations 3.17 through 3.19 [4,140–142]. This study examines the glassy dynamics in melts of tenmers (at constant density $\rho = 1$) when chain stiffness is varied by increasing the amplitudes of the bond-angle and torsional potentials. We will refer to this model as "BSM with variable stiffness." Chain stiffness is quantified by the characteristic ratio $C_N = R^2/(Nl_0^2)$ with $N = 10$ and ranges from $C_{10} = 1.3$ for the fully flexible model to $C_{10} = 4.2$ for a model with large amplitudes for the bond-angle and torsional potentials. References 4 and 140–142 show that MCT provides a pertinent framework for the analysis of the dynamics on cooling toward T_c. The results for T_c and λ systematically increase with chain stiffness (cf. Table 3.2). For the stiffest chains ($C_{10} = 4.2$), the analysis yields the large value $\lambda = 0.86$, a value typical of chemically realistic models and experiments [4].

3.3.2.4 TEMPERATURE DEPENDENCE OF THE α RELAXATION TIME

It is interesting to compare the flexible BSM and the BSM with variable stiffness in more detail. Here, we will do so for the temperature dependence of the α relaxation time. This will also allow us to hint at a major weakness of the idealized MCT.

For the flexible BSM, we define the α relaxation time (τ_α) as the time taken by a monomer to move over its size, that is, by $g_0(\tau_\alpha) = 1$, whereas for the BSM with variable stiffness, Reference 4 defines τ_α from the incoherent scattering function $\phi_q^s(t)$ at the peak position ($q^\star$) of $S(q)$ via the criterion $\phi_{q^\star}^s(\tau_\alpha) = 0.2$.* Figure 3.11 shows an Arrhenius plot (i.e., log τ_α versus $1/T$) for both models. In either case, the temperature axis is scaled by T_c (from Tables 3.1 and 3.2), showing that current simulations can explore supercooled melts down to $T \approx 0.95T_c$ (i.e., up to $T_c/T \approx 1.05$). For this T regime, we discuss the temperature dependence of τ_α and put it into perspective with other work. Five points will be addressed in the following:

1. For the flexible BSM, plotting τ_α against T_c/T allows to collapse the data for different N onto a master curve (Figure 3.11a). Since we know from Figure 3.9 that $T_c \propto \rho_c^{\gamma=4}$, this scaling is akin to the

---

* These different definitions for $\tau_\alpha(T)$ give the same T dependence, if the time–temperature superposition principle (TTSP) holds. Then, all relaxation times are proportional to one common time scale underlying the α process. The TTSP is a main prediction of MCT (cf. Section 3.3.2.1). However, simulations, for example, for the flexible BSM [7] and also for other glass formers [179,182,219–222], find deviations from the TTSP, which emerge already for T near (but above) T_c, where, otherwise, MCT can describe many dynamic properties. In particular, the simulations show that the temperature dependence of τ_α is weaker for $q < q^\star$ than for $q \gtrsim q^\star$. This difference is likely also related to the violation of the Stokes–Einstein relation [224,225,227], implying that the product $D\eta/T$ of the self-diffusion coefficient D—corresponding to the limit $q \to 0$ of the single-particle dynamics (see Equation 3.43)—and of the viscosity η—expected to be proportional to τ_α at $q^\star$ [226]—is not independent of T. The violation of the Stokes–Einstein relation has been interpreted as a signature of increasingly heterogeneous dynamics in the liquid on cooling toward the glass transition, see, for example, [27,29,38,39, 225,227,228] and also Chapter 9 for further discussion.

Table 3.1 Survey of MCT and VFT parameters for the flexible BSM

N	4	8	16	32	64
T_c	0.383	0.402	0.408	0.414	0.416
$b = 0.582$,	$\gamma = 2.464$,	$\lambda = 0.735$,	$T_c/T_g = 1.197$,	$B = 3.184T_c$,	$T_0 = 0.740T_c$

N is the chain length and T_c the critical temperature of MCT. The exponent parameter λ, the von Schweidler exponent b (Equation 3.35), and the exponent γ (Equation 3.37) are found to be independent of N [105]. B and T_0 are, respectively, the "activation temperature" and the Vogel–Fulcher temperature of the VFT equation (Equation 3.23). They are found to be proportional to T_c. The VFT equation may be employed to extrapolate the simulation data for τ_α toward T_g by using the common laboratory definition of T_g as $\tau_\alpha(T_g) = 100$ s and $\tau_{LJ} = 2 \times 10^{-12}$ s (cf. Section 3.2.2.2). This allows to estimate the ratio T_c/T_g (see Figure 3.11 and the corresponding discussion).

Table 3.2 Survey of MCT and VFT parameters for the BSM with variable stiffness [4,140–142]

$C_{10}(T = 1.3)$	b	γ	λ	T_c	T_c/T_g	B	T_0
1.30	0.54	2.60	0.76	0.48	1.30	2.320	0.29
1.50	0.53	2.63	0.77	0.54	1.25	2.134	0.36
1.80	0.52	2.67	0.77	0.62	1.24	2.213	0.43
2.35	0.50	2.74	0.78	0.75	1.19	2.451	0.55
2.82	0.43	3.06	0.83	0.92	1.16	2.773	0.70
4.00	0.40	3.24	0.84	1.02	1.16	3.018	0.78
4.20	0.37	3.43	0.86	1.23	1.16	3.216	0.96

The values given here are taken from Tables 1 and 3 of Reference 4. $C_{10}(T = 1.3)$ is the characteristic ratio for the simulated chain length $N = 10$, obtained from constant-volume MD simulations at $T = 1.3$. The exponent parameter λ, the von Schweidler exponent b (Equation 3.35), and the exponent γ (Equation 3.37) are found to depend on chain stiffness. B and T_0 are, respectively, the "activation temperature" and the Vogel–Fulcher temperature of the VFT equation (Equation 3.23); they also depend on chain stiffness. The VFT equation was employed to extrapolate the simulation data for τ_α toward T_g by using the definition of T_g as $\tau_\alpha(T_g)/\tau_\infty = 10^{13}$, where τ_∞ is the prefactor in Equation 3.23 [4]. This allows to estimate the ratio T_c/T_g (see Figure 3.11 and also Reference 4 for further discussion).

(experimental) finding that the α relaxation times for various molecular and polymeric glass formers can be expressed as a unique function of ρ^γ/T, where γ is a material-specific constant (not the MCT exponent) [143,208,209]. Figure 3.11a therefore shows that change of chain length in constant-pressure simulations of the flexible BSM only shifts the density, but does not influence the steepness with which τ_α grows. The steepness of this growth is, however, affected by chain stiffness. Figure 3.11b presents the results of constant-volume simulations for the melt of tenmers from the BSM with variable stiffness [4]. The stiffer the chain (i.e., the larger C_{10}), the steeper the increase of τ_α on cooling. In the constant-volume simulations of References 4 and 140–142, the increase of T_c is related to a change of intramolecular factors, not of density. This may be the reason why the superpositioning of τ_α does not work when scaling temperature with T_c.

2. At high-enough temperatures, a traditional way of fitting relaxation times (or transport coefficients) is to employ an Arrhenius formula [229]:

$$\tau_\alpha(T) = \tau_\infty \exp\left(\frac{E_a}{k_B T}\right) \quad (\text{for} \quad T > T_A), \tag{3.54}$$

where τ_∞ is a microscopic time scale and E_a an activation energy. Numerous studies of glass-forming liquids [24,61,230,231] have found Equation 3.54 to apply for T larger than some "onset temperature" T_A, which is interpreted as the threshold below which cooperative molecular motion, leading to glassy

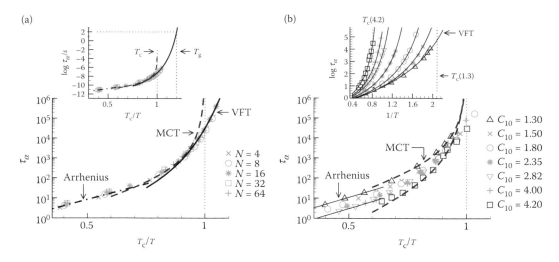

Figure 3.11 (a) Arrhenius plot of the α relaxation time (τ_α) for different chains lengths. The data (symbols) are obtained from constant-pressure MD simulations ($P = 0$) of the flexible BSM. τ_α is defined by the time it takes a monomer to move over its own size, that is, $g_0(\tau_\alpha) = 1$. Temperature is scaled by $T_c(N)$ (Figure 3.9). Results from three fits are included: (i) Arrhenius law (Equation 3.54) with $\tau_\infty = 0.233$ and $E_a = 7.181 k_B T_c$ (dash-dotted line), (ii) MCT critical power law (Equation 3.55) with $\tau_\infty = 4.1$ and $\gamma = 2.464$ (dashed line), (iii) VFT equation (Equation 3.23) with $\tau_\infty = 0.15$, $B = 3.184 T_c$, $T_0 = 0.74 T_c$ (solid line). The inset shows the same data after conversion of the LJ time to real units ($\tau_{LJ} = 2 \times 10^{-12}$ s). The ordinate is expanded up to 10^3 s. Defining T_g by the criterion $\tau_0(T_g) = 100$ s and using the VFT equation to extrapolate to 100 s, we get $T_c/T_g = 1.197$. The relaxation time at T_c is $\tau_\alpha(T_c) \approx 6 \times 10^{-8}$ s. (b) Arrhenius plot of τ_α for different chain stiffness. The data (symbol) are scanned from Figure 7 of Reference 4 and refer to constant-volume MD simulations ($\rho = 1$) of the bead–spring model defined by Equation 3.14 and Equations 3.17 through 3.19. Chain stiffness is quantified by the characteristic ratio $C_N(T = 1.3) = R^2/(N l_0^2)$, where $N = 10$ (Table 1 of Reference 4). τ_α is defined by the time at which $\phi_q^s(\tau_\alpha) = 0.2$ (for $q \approx 7$, where $S(q)$ peaks). Temperature is scaled by T_c (cf. Table 3 of Reference 4). Results from two fits are included: (i) Arrhenius law with $\tau_\infty = 0.198$ ($C_{10} = 1.3$) and $\tau_\infty = 0.058$ ($C_{10} = 2.82$) as well as $E_a = 8.022 k_B T_c$ for both stiffnesses (thin solid lines) and (ii) MCT critical power law with $\tau_\infty = 12.18$, $\gamma = 2.6$ for $C_{10} = 1.3$ and $\tau_\infty = 0.5$, $\gamma = 3.43$ for $C_{10} = 4.2$ (dashed lines). The values for γ are taken from Table 3 of Reference 4; τ_∞ was fitted. The inset reproduces Figure 3.7 of Reference 4: τ_α (symbols) and VFT equation (lines). Chain stiffness increases from right to left. The vertical dotted lines indicate $T_c(C_{10} = 1.3) = 0.48$ and $T_c(C_{10} = 4.2) = 1.23$.

dynamics, emerges (see [116,231,232] and Chapter 8). Figure 3.11 shows that τ_α for the bead–spring models can also be fitted by an Arrhenius law at high T. The fits yield an activation energy $E_a \approx (7–8)$ $k_B T_c$ for both models and furthermore allow to read off $T_A \approx 1.5 T_c$ (i.e., $T_A/T_c \approx 0.65$), roughly estimated here as the temperature when Equation 3.54 noticeably departs from the simulation data (for different approaches to estimate T_A, see [116,231] and Chapter 8).

While Equation 3.54 is generally taken as an appropriate representation of τ_α at high T (this habit is not uncontested, see the interesting exchange of views in [233,234]), it lacks a commonly accepted theoretical explanation [60,61]. Indeed, the standard explanation invokes the idea that particles are trapped in a potential well and must overcome a barrier by thermal activation in order to move. However, where should the barrier come from at high T? It is unlikely that dynamic fluctuations, allowing to separate the particles into "slow" ones producing the barrier and "fast" ones hopping over it, are the cause. Such "dynamic heterogeneities" emerge in the supercooled liquid and grow on cooling toward T_g (cf. Chapters 8 and 9), but are not observed for $T > T_A$ [166]. Textbooks on liquid state theory [3,171] rather interpret the dynamics in the high-T liquid as a compound effect aggregating contributions from the rattling motion of the particles in their neighbor cages, from statistical averaging over cage restructuring, and from the building up of flow patterns. On the other hand, it could be that looking for dynamic fluctuations is not the right way to uncover the origin of the barrier. There is ample experimental evidence that the activation energy E_a correlates with the heat of vaporization of the liquid (see Chapter 8). This suggests that the barrier is related to the free energy cost of removing a particle from its local environment and has both enthalpic and entropic contributions. This point of view has been developed in the framework of the string model for the cooperative dynamics in glass-forming liquids [116,231]. Recent advances are described in Chapter 8.

3. For $T < T_A$, deviations from the high-T Arrhenius behavior occur. Figure 3.11 shows that the initial rise of τ_α above Equation 3.54 can be fitted by the MCT prediction (3.37):

$$\tau_\alpha(T) = \frac{\tau_\infty}{[(T - T_c)/T_c]^\gamma} \quad \text{(for } T_c < T < T_A\text{)},$$ (3.55)

using the values for γ and T_c from Tables 3.1 and 3.2. The exponent γ quantifies the steepness of the temperature dependence. From Figure 3.11, we see that γ systematically increases with chain stiffness, from $2.4 \lesssim \gamma \lesssim 2.6$ for the flexible models to $\gamma = 3.43$ for $C_{10} = 4.2$, the largest stiffness studied in Colmenero [4]. The former values are typical of atomic and molecular glass formers, while the γ value for $C_{10} = 4.2$ is in the range commonly found in simulations of chemically realistic models and in experiments [4]. This illustrates the impact of polymer-specific intramolecular stiffness effects on the glassy slowing down of the α relaxation, as discussed at the end of Section 3.3.2.3.

According to Equation 3.55, τ_α diverges at T_c. However, this kinetic arrest is not observed in simulations or experiments [7,25,27,29,32,37,167–170]. Rather than diverging, the α relaxation time continuously increases, as T crosses T_c and is lowered further. Figure 3.11 illustrates this point for the bead–spring models. Therefore, the idealized MCT predicts the glass transition to occur at T_c, but empirically $T_c > T_g$. This weakness of the idealized theory has been known for a long time [26,168], but nevertheless MCT has remained an important theoretical approach in contemporary glass physics [29,172,235–237]. At first sight, this state of affairs may seem paradoxical; we give some explanation for that in Section 3.3.2.5.

Here, we rather want to mention an interesting analogy between the simulations and experiments. An extensive survey [238] of experimental data, including molecular, ionic, and polymer systems, suggests that the α relaxation time at T_c has a nearly universal value:

$$\tau_\alpha(T_c) \approx 10^{-7 \pm 1} \text{s}.$$ (3.56)

The simulation results for the bead–spring models support this near universality. From Figure 3.11, we see that $10^4 < \tau_\alpha(T_c) < 10^5$. Using the approximate conversion, $\tau_{LJ} = 2 \times 10^{-12}$ s (cf. Section 3.2.2.2), this range for $\tau_\alpha(T_c)$ agrees with Equation 3.56.

4. To fit the super-Arrhenius behavior for $T < T_A$, the most frequently applied equation [23,24,230] is the Vogel–Fulcher–Tammann equation (cf. Equation 3.23):

$$\tau_\alpha(T) = \tau_\infty \exp\left(\frac{B}{T - T_0}\right),$$

which predicts the relaxation time to diverge at the Vogel–Fulcher temperature T_0. Figure 3.11 shows the applicability of the VFT equation to the BSM data. From a numerical point of view, the quality of the fit is very good for $T < T_A$. The fit gives a Vogel–Fulcher temperature that is much lower than the T range where equilibrium data are available (Table 3.1 and 3.2). The VFT equation therefore allows to extrapolate τ_α from the T interval accessible to equilibrium simulations to lower temperatures. Using the common laboratory definition of T_g as $\tau_\alpha(T_g = 100$ s) and $\tau_{LJ} = 2 \times 10^{-12}$ s (cf. Section 3.2.2.2), the extrapolation gives $T_c/T_g \approx 1.2$ for the flexible BSM (Figure 3.11a). For the BSM with variable stiffness, similar extrapolations, taken from Reference 4, yield values in the range $1.16 \leq T_c/T_g \leq 1.3$, depending on chain stiffness (Table 3.2). The results from these extrapolations are in good agreement with experimental data (see Table 1 of [238]).

Nevertheless, the extrapolations should be taken with a grain of salt. In contrast to experiment, the application of the VFT equation is limited to fairly high temperatures in the simulation. This may raise concerns because it is known that the VFT parameters depend on the T window used for the fit. A case in point is provided by dielectric relaxation experiments on salol [239]. In this study, the available 14 decades of data for τ_α are divided into three decade segments and VFT fits are carried out for each segment. This analysis shows that T_0 increases as the segment is displaced to smaller relaxation times. In the range currently covered by simulation ($\tau_\alpha \lesssim 10^{-6}$ s), the extrapolated T_0 is found to be near T_g, that is, about 80 degrees larger than the T_0 obtained when the segment probes relaxation times close to the glass transition. Physically, the finding $T_0 \approx T_g$ is not acceptable; τ_α does not diverge at T_g. Similar results are also reported from other critical analyses of the VFT equation [240,241] (see also [23] for a comparative discussion of the results from [239] and [240,241] or [24,230] for comprehensive overviews). Based on these experimental studies, we should therefore expect that the extrapolations in Figure 3.11 give a too large value for T_0, and along with that also for T_g. Consequently, the results for T_c/T_g are likely too small and the good agreement with experiment might be fortuitous.

Indeed, the extrapolations need not necessarily employ the VFT equation. Other functional forms for the α relaxation time could be used (and can be justified by theoretical arguments, see [34–36,231] and [232] for a topical review). For instance, the string model for the dynamics of glass-forming liquids predicts a functional form for τ_α that is numerically almost indistinguishable from the VFT equation at temperatures where simulation data are available, but returns to an Arrhenius temperature dependence at low T. The extrapolation based on string model therefore predicts a much lower T_g than that of the VFT equation (see [231] and Chapter 8 for details).

5. If the T_0 obtained from simulations is problematic, how reliable is then the extrapolation in experiments? For experimental systems, one may expect the results to be compelling. Instead of only 2–4 decades as in simulations, equilibrium relaxation times over 14 orders of magnitude are available in experiments. Fitting the VFT equation to τ_α, say, between 1 μs to 10^2 s should yield definite values for T_0 and the other VFT parameters. However, this expectation is not unambiguously confirmed. Earlier experimental work, reviewed in Angell et al. [23] and Ngai [230], and also recent studies for a large range of glass-forming liquids [34,242,243], including polymers [244], challenge the divergence of τ_α at a finite T_0 predicted by the VFT equation.* Other functional forms, avoiding the singularity at a finite temperature, have been proposed and applied to experimental data [34,36,233,234]. For instance, an excellent description of the super-Arrhenius increase of τ_α—for 58 experimental systems and 5 simulated systems (all systems are nonpolymeric)—is provided by the following parabolic form [34]:

$$\log\left(\frac{\tau_\alpha(T)}{\tau_0}\right) = J^2 \left(\frac{1}{T} - \frac{1}{T_0}\right)^2 \quad (T_0 > T > T_\times), \tag{3.57}$$

for which no singular behavior occurs at any finite temperature. As the VFT equation, Equation 3.57 has three adjustable parameters: the onset temperature T_0 (not to be confused with the Vogel–Fulcher temperature) above which liquid transport has a weaker T dependence, the corresponding relaxation time $\tau_0 = \tau_\alpha(T_0)$, and the temperature scale J. The validity range of Equation 3.57 is restricted from above by T_0 and from below by $T_\times$, where τ_α is expected to cross over to an Arrhenius temperature dependence [34] (likely $T_\times \lesssim T_g$ [243]).

How can one understand Equation 3.57? Originally, Equation 3.57 was derived from studies of (a class of) so-called kinetically constrained models (KCMs) [34,245,246]. KCMs adopt a point of view completely different from MCT: Since static (two-point) correlations, encoded in $S(q)$, vary only weakly

---

* Nevertheless, the singularity of the α relaxation time at a finite temperature is attractive from a theoretical perspective. It could be indicative of an underlying—though (likely) experimentally unreachable—thermodynamic liquid–glass phase transition. Such a thermodynamic transition is expected from the random first-order transition (RFOT) theory, a modern mean-field approach to the glass transition. Excellent recent reviews are available; we refer the interested reader to References 27, 29, and 33.

on cooling toward T_g, KCMs presume that they are unimportant for the slow relaxation and can be neglected altogether. Complex dynamics is rather assumed to result from purely kinetic constraints. These constraints may allow an excitation to occur in some local region; the excitation then facilitates the dynamics in neighboring regions, which in turn facilitate the motion of their neighbors, and so on. In this picture, glassy dynamics is a consequence of localized excitations and their hierarchical correlation over larger length scales via dynamic facilitation [30,246].

Testing this mechanism in molecular simulations is predicated upon some prescription to identify the localized excitations. As a diagnostic for them, Reference 245 suggests to analyze single-particle trajectories. At supercooled conditions, where the plateau in dynamic correlation functions emerges, the trajectories exhibit features that are unusual for diffusive, random-walk-like, motion [29]: Long periods, where the particle is trapped and only fluctuates about a temporary equilibrium position, are interrupted by short periods of time (much shorter than τ_α) where the particle moves substantially, over a distance comparable to its diameter. Such "jump-like" motion was detected in many supercooled liquids [245,247], including the flexible BSM [248–251], and also in glasses (see [70,252,253] and Chapter 11). If a jump is not immediately reversed by a backward jump, but leads to a persistent ("irreversible") change of positions, it may be taken as a localized excitation in the sense of KCMs [245].

By an analysis of five different simple glass formers, the so-defined excitations are found to have the following properties [245]: (i) For T below the onset temperature T_0, the concentration (c_a) of jumps of size a follows a Boltzmann distribution, $c_a \propto \exp[-J_a(1/T - 1/T_0)]$, where $k_B J_a$ is the energy to create an excitation of size a. (ii) J_a depends logarithmically on a. More precisely, it is found that $J_a - J_\sigma \propto J_\sigma \ln(a/\sigma)$, with σ being the particle diameter. (iii) c_a defines the mean distance (ℓ_a) separating excitations by $\ell_a/a = (c_a a^d)^{-1/d_f}$, where the fractal dimension d_f is found to be close to the spatial dimension d. The length ℓ_a grows on cooling; it is suggested to be the principal length governing the slowing down of the relaxation on approach to T_g. To this end, it is argued that relaxation requires an excitation of size a to connect to neighboring excitations of the same size at distance ℓ_a. Since a and J_a are related, the lengths a and ℓ_a determine the activation energy, $k_B(J_{\ell_a} - J_a)$, for the relaxation to occur. So, we have for $a = \sigma$: $J_{\ell_\sigma} - J_\sigma \propto J_\sigma \ln(\ell_\sigma/\sigma) = -(J_\sigma/d_f)\ln(c_\sigma \sigma^d) \sim J_\sigma^2(1/T - 1/T_0)$. Together with the Arrhenius law, $\tau_\alpha(T)/\tau_0 = \exp[(J_{\ell_\sigma} - J_\sigma)(1/T - 1/T_0)]$, this gives Equation 3.57 with $J \propto J_\sigma$.

3.3.2.5 DISCUSSION: CURRENT STATUS OF MODE-COUPLING THEORY AND IDEAS ABOUT ITS EXTENSION

The previous sections discussed applications of MCT to polymer simulation data, revealing successes of the theory (cf. Figure 3.8) and also clear limitations (cf. Figure 3.11). This is not particular to the studied polymer systems; similar successes and limitations are commonly found in simulations and experiments of glass-forming liquids [29,32,37,167,168]. So, before concluding this chapter, it appears appropriate to revisit the strengths and weaknesses of the theory and to assess its current status. Our discussion is not exhaustive, but is meant to highlight some points that (we think) are important (for a lucid and more in-depth discussion; see, for example, [29]).

Strengths

- MCT provides an understanding for the emergence of the two-step relaxation generally observed in supercooled liquids. This relaxation is predicted to have remarkable properties, for instance, the time–temperature superposition property of the α process or the factorization property of Equation 3.46 implying a common time dependence for all correlation functions (coupling to density fluctuations). These properties, as well as other detailed predictions, some of which are summarized in Section 3.3.2.1, hold to good approximation in practical applications, provided T is near and above the (extrapolated) T_c [4,7,25,32,37,167,168,170].
- For the structural glass transition, MCT relates the glassy dynamics to one freezing mechanism, the cage effect. However, the theory is more comprehensive [32]. For instance, it also predicts complex relaxation patterns for systems with competing arrest mechanisms [32,214]. As argued in Section

3.3.2.3, chemically realistic polymer models (could) provide an example featuring such competing mechanisms (packing constraints due to the cage effect and polymer-specific intrachain barriers). But this is not the only example. Competing arrest mechanisms have been observed—and compared favorably to MCT—in systems of very different nature, including short-ranged attractive colloids [215,254] (packing constraints and bond formation) or binary mixtures [217,218] and polymer blends [255–257] with strong dynamic asymmetry (packing constraints and matrix-induced spatial confinement).

- MCT establishes a link between the equilibrium structure of a glass former, encapsulated in $S(q)$ and related quantities, and its dynamics. Provided the required static input can be determined with sufficiently high precision, this opens the possibility to predict the full dynamics (not only the asymptotic behavior near T_c as discussed in Section 3.3.2.1) and to test these predictions against the measured dynamics. Simulations have been instrumental for such ab initio tests. In addition to bead–spring polymer models [141,142,185], examples of tested systems involve network-forming strong liquids [174,258], binary mixtures [60,173,219,220], a model for *ortho*-terphenyl [176,259], and polydisperse hard-sphere(-like) systems [179,182]. These tests reveal that the idealized MCT fares well, at least a qualitative level, for many quantities (not all, see below), if $T > T_c$. This is a particular achievement that makes MCT practically useful: For any new system, the same formalism can be applied to identify the pertinent structural input and to obtain from this at least a qualitative insight into the relaxation behavior of the system. Therefore, the idealized theory continues to be developed (despite well-recognized limitations, see below). Recent examples involve the extension of MCT to inhomogeneous situations, including a discussion of multipoint (three-point or four-point) correlation functions and the discovery of a diverging correlation length associated with the MCT singularity [29,260], the extension to spatially confined glass-forming systems (films [261], fluids in disordered porous media [262,263], partially pinned fluid systems [264]) or the extension to nonequilibrium situations (colloidal suspensions under shear [265–267], residual stresses in colloidal glasses [268]).

Weaknesses

- MCT predicts the divergence of the α relaxation time at T_c. But this divergence is not observed in experiments or simulations (cf. Figure 3.11), since $T_c > T_g$. On the practical side, this limits the applicability of the idealized theory to the T interval above the fitted T_c—a relatively narrow interval [29,60] compared to the increase of τ_α by 14 orders of magnitude when cooling the liquid to its T_g. On the theoretical side, this raises questions about the physical significance of T_c. Should T_c be interpreted as an "avoided critical point" [232] marking the crossover from liquid-like (cage-effect dominated) transport above T_c to thermally activated (potential-energy-barrier dominated) transport below T_c [27,269–271]? Or is T_c just a consequence of the factorization approximation for the memory kernel (Equation 3.30) and would perhaps even vanish, if the factorization approximation was avoided [236]?
- MCT calculations for the critical temperature, based on structural input from simulations, are not very accurate. For simple or molecular glass formers [60,174,176,182,272] and silica [174], MCT usually overestimates T_c, sometimes up to a factor of about 2, compared to the empirical T_c determined from the fits to simulation data (as discussed in Section 3.3.2.2).* This implies that the calculated values for the exponent parameter (λ), the associated dynamic exponents (a, b, γ), and the amplitudes ($h_q^{(1)}$, $h_q^{(2)}$, etc.)— all of which are functionals of $S(q)$ at T_c—cannot also be fully accurate. Yet, in practice, the exponents and amplitudes turn out to be fairly close to their fitted counterparts so that simulated and predicted dynamics for many (two-point) correlation functions agree well after temperature rescaling to account for the difference in T_c. Still, the need to resort to a temperature rescaling is disturbing; it indicates that the ideal theory must be corrected and enhanced [235,237].

---

* By contrast, for bead–spring polymer models, the current site-formalism-based version of MCT typically underestimates T_c, sometimes also by a factor of about 2 [141,142,185]. However, one may argue that the level of accuracy of the polymer MCT is lower than for simple liquids—the polymer MCT does not describe correctly all qualitative influences of chain stiffness [141,142] and chain length [105] on the glassy dynamics. Certainly, the theory needs to be enhanced.

- The idealized theory predicts that one time scale underlies the α process, $t'_\sigma(T)$ (Equation 3.37). All other relaxation times and transport coefficients are proportional to $t'_\sigma(T)$. In particular, this implies that the Stokes–Einstein relation holds: The product $D\eta/T$ is independent of T, with D being the (single-particle) diffusion coefficient and η the (collective) shear viscosity. However, simulations show that diffusive and viscous time scales progressively decouple on cooling toward T_c [7,182,226,227].

- On the time scale τ_α, simulations [101,107,182,220–222] and experiments [222,273] find that the distribution of particle displacements is strongly non-Gaussian. This effect is not correctly captured by the idealized MCT [182,220,221,226,274]. One way to highlight the non-Gaussian character is to consider the probability distribution $P(\ln r;t)$ for the logarithm of the particle displacement in time t [220,221]. This distribution is defined by

$$P(\ln r;t) = 4\pi r^3 G_s(r,t). \tag{3.58}$$

Here, $G_s(r,t)$ is the self-part of the van Hove correlation function [3,171], which measures the probability to find a particle at time t at distance r from its initial position. If $G_s(r,t)$ is Gaussian, we have $P(\ln r;t) = 4\pi(3r^2/2\pi g_0(t))^{3/2} \exp[-3r^2/2g_0(t)]$, with $g_0(t)$ being the mean-square displacement. In this case, the shape of P is independent of t (P is a scaling function of $r/\sqrt{g_0(t)}$) and exhibits a single peak with height equal to $\sqrt{54/\pi}e^{-3/2} \simeq 0.925$. In supercooled liquids, there are clear deviations from this Gaussian behavior for $t \sim \tau_\alpha$ [101,107,182,220–222,273]. When T is lowered toward T_c, the distribution $P(\ln r;t)$ first broadens, exhibiting small ($r \gtrsim \sqrt{6}r_{sc} \approx \sqrt{6} \cdot 0.1 = 0.25$) and large ($r \gtrsim 1$) displacements. On cooling further, the broad distribution develops a pronounced shoulder at large displacements or even a double-peak structure. This is indicative of large disparities in particle mobility. Two populations of particles coexist on the α time scale, "slow" ones, which have not moved much farther than 10% of their diameter, and "fast" ones, which have left their cage and covered a distance of about their diameter or more, perhaps by a jump-like motion. This bimodal character of $P(\ln r;t)$ is not predicted by the idealized MCT [182,220], but reflects [222] the growing heterogeneity of the dynamics when approaching T_g (see [30,31,38,166] and Chapters 8 and 9 for a discussion on dynamic heterogeneities).

The above strengths and weaknesses reveal the problematic status of MCT. On the one hand, the theory offers a first principles approach, correlating the slow dynamics with the underlying structure, and uncovers many complex relaxation features of supercooled liquids. On the other hand, MCT suffers from several shortcomings and needs to be extended. How could such an extension be done? Several attempts in this direction have been pursued in the past [35,36,165,236,237,226,275–279]. We briefly outline some of the ideas here.

Already in the mid-1980s, extensions of MCT were developed [275,276] (see also [277,278] for reviews). These extended versions of the theory considered an enlarged set of "relevant variables" (cf. Section 3.3.2.1). In addition to density fluctuations as in idealized MCT, the coupling between density and current modes was also included. This approach appeared very promising. The density–current coupling was shown to avoid the ideal glass transition and to restore ergodicity, that is, the full relaxation of dynamic correlation functions, for $T \leq T_c$ [275,276]. Later, the extended MCT equations were solved to leading order in the β regime [280] and the solutions were compared to experimental data [167]. However, the status of these tests is unclear, due to both experimental and theoretical problems [167,281]. Moreover, in recent years, further arguments have been advanced that challenge the validity of the extended MCT [282]. One argument comes from simulations: Simulations can be carried out with Newtonian dynamics (which conserves density and momentum) or with Brownian dynamics (BD) or Monte Carlo (MC) dynamics (which only conserve density). In BD or MC simulations, the density–current coupling should thus be suppressed. Nevertheless, BD [220,221] or MC simulations [181] show the same deviations as the Newtonian case on cooling toward T_c. Similarly, the density–current coupling should not be important in colloidal suspensions of interacting Brownian particles, which should thus exhibit the ideal MCT scenario, but this has become increasingly doubtful too (see the interesting comment [283] and reply [284] on [223]).

Recent attempts to extend MCT therefore avoid the coupling to currents and build a theory entirely based on density modes [236,237,279]. How can this work? According to the discussion in Section 3.3.2.1 the memory kernel $M(q,t)$ in the exact equation of motion for $\phi_q(t)$ represents a force–force correlation function, the dominant contribution to which is the product of four density fluctuations. Instead of factorizing this four-point correlation function as in Equation 3.30, the main idea is now to apply the projection operator formalism again to derive an equation of motion for the four-point correlation function. This equation of motion takes a form analogous to Equation 3.31 and introduces a new memory kernel. It is then argued that this new memory kernel contains, to leading order, the product of six density fluctuations, that is, that it is a six-point correlation function. Either the factorization approximation is now applied to this six-point correlation function [279] or the scheme is continued to include higher-order correlations [237]. In this way, the factorization approximation can be delayed, allowing to explore corrections to the idealized MCT. The most recent implementation of this scheme treats static correlations as in idealized MCT (i.e., only $S(q)$ enters), but shifts the dynamic factorization approximation to the level of the eight-point correlation function [237]. This allowed for an ab initio comparison with the simulation data for $\phi_q(t)$ from the polydisperse hard-sphere-like system of Reference 182, using the simulated $S(q)$ as the sole static input. Interestingly, this approach gives semiquantitative agreement with the simulation from low to moderate supercooling, in particular, in the regime where the idealized theory predicts kinetic arrest and therefore fails [237]. The inclusion of higher-order (dynamic) density correlations can thus capture aspects of thermal activation that the idealized MCT misses. Therefore, this is a promising approach, although there is the hazard that the infinite hierarchy of all higher-order correlations might be required to faithfully describe the super-Arrhenius behavior of the dynamics. Whether this would be tractable is unknown at the time of writing (2016).

Another approach to extend MCT and to account for activated processes has been developed by Schweizer and collaborators throughout the last 10 years [35,36,165]. The theory does not attempt to address the collective dynamics of density fluctuations, but rather focuses on the single-particle dynamics. The starting point is a nonlinear Langevin equation (NLE) for the scalar displacement, $r(t)$, of a tagged particle at time t. In the overdamped limit (no inertia, i.e., $\ddot{r}(t) = 0$), the NLE reads

$$\zeta_s \frac{\partial r(t)}{\partial t} = -\frac{\delta F_{dyn}[r(t)]}{\delta r(t)} + \delta f(t), \quad r(t=0) = 0, \tag{3.59}$$

where $\delta f(t)$ is a Gaussian white noise, that is, a random force with zero mean and second moment related to the friction coefficient ζ_s by $\langle \delta f(t)\delta f(0) \rangle = 2k_B T \zeta_s \delta(t)$. The "dynamic free energy" $F_{dyn}[r(t)]$ depends nonlinearly on r and contains competing contributions that favor delocalization (fluid state) and localization (solid state). The form of $F_{dyn}[r(t)]$ is motivated by MCT: In the absence of noise, the minimum of the free energy yields an approximation for the expression of the Lindemann localization length r_{sc} known from idealized MCT (compare Equation 3.9 in [35] and Equation 3.23 in [192]). As in MCT, the NLE theory therefore implements caging constraints on the single-particle dynamics via $S(q)$. Upon cooling, the free energy changes shape from a monotonically decaying function of r, with attendant fluid-like (Brownian) particle trajectories, to an asymmetric, Morse-like, potential featuring a minimum at r_{sc} ($\approx 0.1\sigma$ with σ being the particle diameter) and an energy barrier at a larger position r_B ($\approx 0.35\sigma$). Physically, particles become then localized near r_{sc} for some period of time until rare, thermal noise–driven, fluctuations allow them to escape over the barrier by an activated process and to cross over to diffusive transport. The NLE theory therefore avoids the mode-coupling singularity at T_c; instead, it finds a super-Arrhenius increase of the α relaxation time and also a strongly non-Gaussian character of the particles trajectories [165,274]. These salient features have led—by a combination of mathematical analysis, physical arguments, and insights from experiments—to extensions of the theory, for example, to polymer melts, for the glassy dynamics on cooling toward T_g [5] and also for nonequilibrium phenomena of polymer glasses (physical aging, nonlinear mechanical response) [285].

Later work, however, pointed out that the NLE theory underestimates the super-Arrhenius increase of τ_α with progressive supercooling [210]. This drawback of the theory has recently been attributed to the fact that the theory implements a local cage picture that does not capture longer-range collective effects that presumably become progressively more important in the deeply supercooled regime near T_g [35,36]. For deep supercooling, a particle displacement must occur in a very viscous, temporarily almost rigid, environment that is supposed to elastically react on the deformation created by the particle displacement. The actual activation barrier for τ_α is thus suggested to consist of two (additive) contributions: a contribution from the (local) NLE theory and a contribution due to the (collective) elastic distortion of the medium surrounding the particle. This "elastically collective nonlinear Langevin equation (ECNLE) theory" has been developed for hard-sphere [35] and thermally driven glass formers [36], compared to experimental data for τ_α and other theories [36], and recently also extended to supercooled polymer melts [159].

3.4 CONCLUSIONS AND OUTLOOK

This chapter has two main parts, a methodological part on MD simulations and polymer models (Section 3.2) and an applicative part presenting results on the glass transition and the structural relaxation of supercooled polymer melts (Section 3.3). In this final section, we revisit some of the points examined before, complete our discussion by mentioning further lines of current research, and also hint at possible future research directions:

- Fifteen years ago (in 2000), simulations studied systems with $N \sim 10^3$ particles and times up to $t_{max} \approx 10^5 \tau_{LJ}$ ($\approx$10 ns). These were "practical" limits, commonly found in these numerical studies, not the peak performance of the computers at that time. Today, owing to the increase of the computational power, the practical limits are much larger: typically, $N \sim 10^4$ and $t_{max} \lesssim 10^6 \tau_{LJ}$ (cf. Section 3.2.1.2). Assuming that this increase continues over the next 15 years, one may anticipate the practical limits to be $N \sim 10^5$ and $t_{max} \lesssim 10^7 \tau_{LJ} \approx 10^{-5}$ s in 2030. Since the extrapolated mode-coupling T_c appears to correspond to a nearly universal α relaxation time of $\tau_\alpha(T_c) \approx 10^{-7\pm1}$ s (cf. Equation 3.56), simulations should be capable of exploring the equilibrium dynamics over the first few decades in time for $T < T_c$.

- The ability to extend the simulations to temperatures well below T_c could be an important step because it has been suggested that a change of transport mechanism occurs at T_c [27,269–271]. Evidence for this viewpoint comes from simulation studies of the total potential energy $U_N(\mathbf{r}_1,\dots\mathbf{r}_N)$—generally called the "potential energy landscape"—for simple glass-forming liquids with N particles (see [286–288] for reviews). Important topographic features of the PEL are minima and saddle points. They can be distinguished by the eigenvalues of the Hessian matrix (the matrix of the second derivatives of U_N): A minimum has only positive eigenvalues, whereas a saddle also exhibits negative eigenvalues corresponding to downward curvatures of the PEL and so to unstable directions. Analysis of the number (n_s) of negative eigenvalues in the simulations reveals that n_s decreases when cooling the liquid toward T_c and appears to vanish at T_c [27,270,271,287,288]. This progressive disappearance of the unstable directions suggests that T_c marks the crossover temperature for a change in transport mechanism: Above T_c, alternative mechanisms (unstable directions and thermal activation) control the slowing down of the dynamics, whereas the relaxation is governed by thermally activated processes below T_c. Simulation studies for $T < T_c$ may therefore be instrumental to test and to help develop further recent theoretical approaches (see, for example, [33,35,36,237,245] and Chapters 8 and 9) addressing the deeply supercooled regime below T_c. In these tests, it can be beneficial that the simulations allow to compare many observables for the same system because "a single or a small number of experimental observations are not enough—yet—to distinguish between leading theories," as pointed out in a recent perspective on the glass transition (page 3 of [30]).

- Simulation studies for $T < T_c$ may also be instrumental for another, important, line of recent research. A long-standing question in the field is whether the glass transition is connected with the growth of an underlying correlation length [24,27,29–31,37,39]. Although we have not discussed this aspect here, we want to briefly mention two main ideas. On the one hand, extensive work in the past two decades has found that the dynamics on the α time scale is spatially heterogeneous (see [30,31,38,39] and Chapters 8 and 9): Particles temporarily cluster into "slow" and "fast" domains (for an illustration, see [245]).

When T is lowered toward T_g, these spatiotemporal fluctuations become more pronounced and can be characterized by growing dynamical correlation length scales [39]. On the other hand, there could also be a static correlation length scale that grows in parallel. To evidence the existence of such static length scales, it was suggested to determine so-called point-to-set correlation functions [27,30]. What does this mean? Standard static correlation functions, such as $S(q)$, measure the correlation between two particles, that is, two points in space. They may thus be called "point-to-point" (or two-point) correlation functions. On approach to T_g, these correlation functions change only weakly with temperature. On the other hand, higher-order correlation functions, measuring the spatial correlation between a particle and a set of other particles, may behave differently, possibly evidencing the growth of a static multi-point ("point-to-set") correlation length underlying the collective processes that eventually lead to glass formation. The technical procedure to turn this idea into a measurement guideline for simulations is to spatially confine the glass former by specifically chosen amorphous boundary conditions [27,30,289]. More precisely, the point-to-set correlations are determined by first pinning a set of particles (chosen from the equilibrated glass former) and then measuring how the probability of finding another (not pinned) particle at some position depends on the pinning. The spatial range of this influence allows to extract point-to-set correlation length scales. Currently, however, it is unclear how (and if at all) these static length scales are related to the dynamic ones from the studies of dynamic heterogeneities (see [62] and references therein). It is possible that the answer to this question depends on the level of supercooling. Extending equilibrium simulations to the regime $T < T_c$ could therefore be of particular importance to clarify the picture.

- The simulations on point-to-set correlations explore atomic glass formers in spatial confinement. Here, it is argued that the specific design of the boundary conditions—by pinning a set of particles chosen from the glass-forming liquid that was equilibrated before—is such that the bulk behavior is not disturbed, thus opening the way to probe genuine bulk correlations [289]. On the other hand, there has been much work on the impact of spatial confinement on the glass transition [7,290], in particular for polymer films (see, for example, [7,290–293] for reviews and also Chapters 5 through 9), where substrate effects are thought to be important (or even dominant). Albeit difficult (due to the design of the boundary conditions), it might be interesting to attempt to extend the work on polymer films in the direction of the studies on point-to-set correlations, also because theoretical concepts to understand the statistical mechanics of such partly pinned fluid systems become available [294,295].

- Simulations offer the possibility to manipulate the studied models in order to identify the importance of several factors contributing concurrently to the slowing down of the relaxation. A case in point is provided by the bead–spring models with variable stiffness [4,140–142] discussed in Section 3.3.2.3 and 3.3.2.4. Increasing gradually the amplitudes of the valence terms in the potential energy, the influence of growing intramolecular barriers on the glassy dynamics could be explored. The results of References 4 and 140–142 reveal that substantial stiffness must be added to the flexible BSM before the model approximates well the behavior of real (flexible) polymers. This supports earlier results on an atomistic model for polybutadiene for which the dihedral potential was switched off (cf. Figure 3.10) [212]. Systematic work on the interplay of polymer-specific intramolecular features and packing constraints should be continued, for instance, for polymer films also. For polymer films, an interesting result, again for the atomistic PBD model, has recently been obtained [296]. The equilibrium dynamics of PBD, embedded in a nanoscopic film between two crystalline graphite layers (film thickness: 10 –20 nm), yielded no discernable deviation for τ_α from the bulk behavior, except for the 1–2 nm next to the graphite substrate [296]. This finding is attributed to the strong influence of the dihedral barriers on the dynamics, which is supposed to dominate over packing constraints. This influence is missing in flexible bead–spring models without intramolecular barriers and could explain why these models find deviations from bulk behavior in thin films (see [7,297] and Chapter 8). A systematic variation of the strength of these barriers, as done in Colmenero [4] and Bernabei [140–142] for the bulk, would help to clarify the issue.

- Owing to the coexistence of two arrest mechanisms—intramolecular constraints and intermolecular packing—polymers could be different from low-molecular-weight glass formers, where packing constraints appear to dominate [4]. Existing extensions of MCT to polymers are inspired by the theory for simple liquids and attempt to account for polymer-specific features (chain stiffness, chain length) by including intrachain structure factors [184]. This is not fully satisfactory in several respects: (i) T_c is underestimated by MCT calculations [105,141,142,185], not overestimated as for simple liquids. Apparently, the theory misses slow modes contributing to the glassy dynamics. (ii) MCT calculations predict T_c to increase with chain stiffness, but the increase levels off for large stiffness, whereas empirical fits show that T_c continues to increase [141,142]. (iii) MCT calculations predict T_c to decrease with increasing chain length. The opposite trend is found from empirical fits [105]. How to improve on these drawbacks is still a largely open question.

ACKNOWLEDGMENTS

Several results reported in this chapter were obtained in fruitful collaboration with S. Frey, M. Fuchs, and B. Schnell. It is a pleasure to thank all of them. We gratefully acknowledge financial support from the IRTG "Soft Matter Science," the DFH/UFA, and the DFG (HE7429/1–1). We also thank the computing clusters in Strasbourg (ENIAC and University of Strasbourg) for generous grants of computer time.

REFERENCES

1. M. Doi and S. F. Edwards, *The Theory of Polymer Dynamics* (Oxford University Press, Oxford), 1986.
2. M. Rubinstein and R. H. Colby, *Polymer Physics* (Oxford University Press, Oxford), 2003.
3. J. P. Hansen and I. R. McDonald, *Theory of Simple Liquids* (Academic Press, London), 1986.
4. J. Colmenero, *J. Phys. Condens. Matter* 27, 103101, 2015.
5. K. S. Schweizer and E. J. Saltzman, *J. Chem. Phys.*, 121, 1984, 2004.
6. K. S. Schweizer and J. G. Curro, *Adv. Polym. Sci.*, 116, 319, 1994.
7. J. Baschnagel and F. Varnik, *J. Phys. Condens. Matter*, 17, R851, 2005.
8. P.-G. de Gennes, *Scaling Concepts in Polymer Physics* (Cornell University Press, Ithaca), 1996.
9. J. P. Wittmer, P. Beckrich, H. Meyer, A. Cavallo, A. Johner, and J. Baschnagel, *Phys. Rev. E*, 76, 011803, 2007.
10. J. P. Wittmer, A. Cavallo, H. Xu, J. E. Zabel, P. Polińska, N. Schulmann et al., *J. Stat. Phys.*, 145, 1017, 2011.
11. A. N. Semenov and I. A. Nyrkova, Statistical description of chain molecules, in *Polymer Science: A Comprehensive Reference*, Vol. 1, edited by K. Matyjaszewski and M. Moller (Elsevier, Amsterdam), 3–29, 2012.
12. J. Farago, A. N. Semenov, H. Meyer, J. P. Wittmer, A. Johner, and J. Baschnagel, *Phys. Rev. E*, 85, 051806, 2012.
13. J. Farago, H. Meyer, J. Baschnagel, and A. N. Semenov, *Phys. Rev. E*, 85, 051807, 2012.
14. J. Farago, H. Meyer, and A. N. Semenov, *J. Phys. Condens. Matter*, 24, 284105, 2012.
15. A. N. Semenov, J. Farago, and H. Meyer, *J. Chem. Phys.*, 136, 244905, 2012.
16. T. C. B. McLeish, *Adv. Phys.*, 51, 1379, 2002.
17. A. E. Likhtman, Viscoelasticity and molecular rheology, in *Polymer Science: A Comprehensive Reference*, Vol. 1, edited by K. Matyjaszewski and M. Moller (Elsevier, Amsterdam), 133–179, 2012.
18. G. Strobl, *The Physics of Polymers: Concepts for Understanding Their Structures and Behavior* (Springer, Berlin–Heidelberg), 2007.
19. M. Muthukumar, *Adv. Chem. Phys.*, 128, 1, 2004.
20. G. Strobl, *Rev. Mod. Phys.*, 81, 1287, 2009.
21. C. Luo and J.-U. Sommer, *Phys. Rev. Lett.*, 112, 195702, 2014.
22. G. B. McKenna, Glass formation and glassy behavior, in *Comprehensive Polymer Science*, Vol. 2, edited by C. Booth and C. Price (Pergamon, New York), 311–362, 1986.
23. C. A. Angell, K. L. Ngai, G. B. McKenna, P. F. McMillan, and S. W. Martin, *J. Appl. Phys.*, 88, 3113, 2000.

24. E. Donth, *The Glass Transition* (Springer, Berlin–Heidelberg), 2001.

25. W. Kob, *J. Phys. Condens. Matter*, 11, R85, 1999.

26. W. Götze, *Condens. Matter Phys.*, 1, 873, 1998.

27. A. Cavagna, *Phys. Rep.*, 476, 51, 2009.

28. R. Böhmer, K. L. Ngai, C. A. Angell, and D. J. Plazek, *J. Chem. Phys.*, 99, 4201, 1993.

29. L. Berthier and G. Biroli, *Rev. Mod. Phys.*, 83, 587, 2011.

30. G. Biroli and J. P. Garrahan, *J. Chem. Phys.*, 138, 12A301, 2013.

31. M. D. Ediger and P. Harrowell, *J. Chem. Phys.*, 137, 080901, 2012.

32. W. Götze, *Complex Dynamics of Glass-Forming Liquids: A Mode-Coupling Theory* (Oxford University Press, Oxford), 2009.

33. V. Lubchenko, *Adv. Phys.*, 64, 283, 2015.

34. Y. S. Elmatad, D. Chandler, and J. P. Garrahan, *J. Phys. Chem. B*, 113, 5563, 2009.

35. S. Mirigian and K. S. Schweizer, *J. Chem. Phys.*, 140, 194506, 2014.

36. S. Mirigian and K. S. Schweizer, *J. Chem. Phys.*, 140, 194507, 2014.

37. K. Binder and W. Kob, *Glassy Materials and Disordered Solids* (World Scientific, Singapore), 2011.

38. S. C. Glotzer, *J. Non-Cryst. Solids*, 274, 342, 2000.

39. L. Berthier, G. Biroli, J.-P. Bouchaud, L. Cipelletti, and W. van Saarloos, *Dynamical Heterogeneities in Glasses, Colloids and Granular Media* (Oxford University Press, Oxford), 2011.

40. D. P. Landau and K. Binder, *A Guide to Monte Carlo Simulations in Statistical Physics* (Cambridge University Press, Cambridge), 2013.

41. M. P. Allen and D. J. Tildesley, *Computer Simulation of Liquids* (Clarendon Press, Oxford), 1987.

42. D. Frenkel and B. Smit, *Understanding Molecular Simulation* (Academic Press, London), 2nd edn., 2002.

43. S. C. Plimpton, *Comput. Phys.*, 117, 1, 1995.

44. LAMMPS (Large-scale Atomic/Molecular Massively Parallel Simulator), http://lammps.sandia.gov.

45. D. Van der Spoel, E. Lindahl, B. Hess, G. Groenhof, A. E. Mark, and H. J. C. Berendsen, *J. Comput. Chem.*, 26, 1701, 2005.

46. GROMACS (Groningen MAchine for Chemical Simulation), http://www.gromacs.org/.

47. J. C. Phillips, R. Braun, W. Wang, J. Gumbart, E. Tajkhorshid, E. V. C. Chipot, R. D. Skeel, L. Kale, and K. Schulten, *J. Comput. Chem.*, 26, 1781, 2005.

48. NAMD (NAnoscale Molecular Dynamics), http://www.ks.uiuc.edu/Research/namd/.

49. K. Binder, J. Horbach, W. Kob, W. Paul, and F. Varnik, *J. Phys. Condens. Matter*, 16, S429, 2004.

50. N. Attig, K. Binder, H. Grubmüller, and K. Kremer, *Computational Soft Matter: From Synthetic Polymers to Proteins* (NIC Series, Jülich, 2004), available from http://www.fzjuelich.de/nic-series.

51. H. C. Andersen, *PNAS*, 102, 6686, 2005.

52. G. Sutmann, Molecular dynamics—Extending the scale from microscopic to mesoscopic, in *Multiscale Simulation Methods in Molecular Sciences*, Vol. 42, edited by J. Grotendorst, N. Attig, S. Blügel, and D. Marx (NIC Series, Jülich), 1–49, 2009, available from http://www.fz-juelich.de/nic-series.

53. E. B. Tadmor and R. E. Miller, *Modeling Materials: Continuum, Atomistic and Multiscale Techniques* (Cambridge University Press, Cambridge), 2014.

54. H. T. Davis, *Statistical Mechanics of Phases, Interfaces and Thin Films* (Wiley-VCH, New York), 1995.

55. S. Toxvaerd and J. C. Dyre, *J. Chem. Phys.*, 134, 081102, 2011.

56. H. Xu, J. P. Wittmer, P. Polińska, and J. Baschnagel, *Phys. Rev. E*, 86, 046705, 2012.

57. C. Batistakis and A. V. Lyulin, *Comput. Phys. Commun.*, 185, 1223, 2014.

58. S. Toxvaerd, *Condens. Matter Phys.*, 18, 13002, 2015.

59. E. A. Mastny and J. J. de Pablo, *J. Chem. Phys.*, 127, 104504, 2007.

60. L. Berthier and G. Tarjus, *Phys. Rev. E*, 82, 031502, 2010.

61. L. Berthier and G. Tarjus, *Phys. Rev. Lett.*, 103, 170601, 2009.

62. L. Berthier, G. Biroli, D. Coslovich, W. Kob, and C. Toninelli, *Phys. Rev. E*, 86, 031502, 2012.

63. E. Flenner and G. Szamel, *Phys. Rev. Lett.*, 114, 025501, 2015.

64. E. Flenner and G. Szamel, *J. Phys. Chem. B*, 119, 9188, 2015.

65. F. Puosi, J. Rottler, and J.-L. Barrat, *Phys. Rev. E*, 89, 042303, 2014.

66. A. Lemaître, *Phys. Rev. Lett.*, 113, 245702, 2014.

67. C. Bennemann, C. Donati, J. Baschnagel, and S. C. Glotzer, *Nature*, 399, 246, 1999.

68. W. Paul and G. D. Smith, *Rep. Prog. Phys.*, 67, 1117, 2004.

69. K. Binder, J. Baschnagel, and W. Paul, *Prog. Polym. Sci.*, 28, 115, 2003.

70. J. Rottler, *J. Phys. Condens. Matter*, 21, 463101, 2009.

71. J. Baschnagel, K. Binder, P. Doruker, A. A. Gusev, O. Hahn, K. Kremer et al., *Adv. Polym. Sci.*, 152, 41, 2000.

72. F. Müller-Plathe, *ChemPhysChem*, 3, 754, 2002.

73. C. Peter and K. Kremer, *Soft Matter*, 5, 4357, 2009.

74. G. D. Smith, W. Paul, M. Monkenbusch, D. Richter, X. H. Qiu, and M. D. Ediger, *Macromolecules*, 32, 8857, 1999.

75. H. Sun, *J. Phys. Chem. B*, 102, 7338, 1998.

76. D. N. Theodorou, *Chem. Eng. Sci.*, 67, 5697, 2007.

77. J. H. R. Clarke, *Curr. Opin. Solid State Matter*, 3, 596, 1998.

78. J. Colmenero, F. Alvarez, and A. Arbe, *Phys. Rev. E*, 65, 041804, 2002.

79. J. Colmenero, A. Arbe, F. Alvarez, M. Monkenbusch, D. Richter, B. Farago, and B. Frick, *J. Phys. Condens. Matter*, 15, S1127, 2003.

80. A. V. Lyulin and M. A. J. Michels, *Macromolecules*, 35, 1463, 2002.

81. A. V. Lyulin and M. A. J. Michels, *Macromolecules*, 35, 9595, 2002.

82. A. V. Lyulin, N. K. Balabaev, and M. A. J. Michels, *Macromolecules*, 36, 8574, 2003.

83. W. Paul, D. Bedrov, and G. D. Smith, *Phys. Rev. E*, 74, 021501, 2006.

84. G. D. Smith and D. Bedrov, *J. Polym. Sci., Part B: Polym. Phys.*, 45, 627, 2007.

85. J. Colmenero, A. Narros, F. Alvarez, A. Arbe, and A. J. Moreno, *J. Phys. Condens. Matter*, 19, 205127, 2007.

86. A. Narros, A. Arbe, F. Alvarez, J. Colmenero, and D. Richter, *J. Chem. Phys.*, 128, 224905, 2008.

87. M. Vogel, *Macromolecules*, 41, 2949, 2008.

88. A. Bormuth, P. Henritzi, and M. Vogel, *Macromolecules*, 43, 8985, 2010.

89. S. Capponi, A. Arbe, F. Alvarez, J. Colmenero, B. Frick, and J. P. Embs, *J. Chem. Phys.*, 131, 204901, 2009.

90. Y. Khairy, F. Alvarez, A. Arbe, and J. Colmenero, *Macromolecules*, 47, 447, 2013.

91. Y. Khairy, F. Alvarez, A. Arbe, and J. Colmenero, *Phys. Rev. E*, 88, 042302, 2013.

92. http://www.accelrys.com/products/materials-studio/index.html

93. L. J. Fetters, D. J. Lohse, D. Richter, T. A. Witten, and A. Zirkel, *Macromolecules*, 27, 4639, 1994.

94. A. E. Likhtman, *J. Non-Newtonian Fluid Mech.*, 158, 158, 2009.

95. K. Binder, J. Horbach, H. Knoth, and P. Pfleiderer, *J. Phys. Condens. Matter*, 19, 205102, 2007.

96. K. Vollmayr, J. A. Roman, and J. Horbach, *Phys. Rev. E*, 81, 061203, 2010.

97. J.-L. Barrat, J. Baschnagel, and A. Lyulin, *Soft Matter*, 6, 3420, 2010.

98. K. Binder, Introduction: General aspects of computer simulation techniques and their applications in polymer science, in *Monte Carlo and Molecular Dynamics Simulations in Polymer Science*, edited by K. Binder (Oxford University Press, New York) 3–46, 1995.

99. J. Baschnagel, J. P. Wittmer, and H. Meyer, Monte Carlo simulation of polymers: Coarse-Greained models, in *Computational Soft Matter: From Synthetic Polymers to Proteins*, Vol. 23, edited by N. Attig, K. Binder, H. Grubmüller, and K. Kremer (NIC Series, Jülich) 83–140, 2004, available from http://www.fz-juelich.de/nic-series.

100. R. Yamamoto and A. Onuki, *J. Chem. Phys.*, 117, 2359, 2002.

101. M. Durand, H. Meyer, O. Benzerara, J. Baschnagel, and O. Vitrac, *J. Chem. Phys.*, 132, 194902, 2010.

102. R. S. Hoy, *J. Polym. Sci. B*, 49, 979, 2011.

103. A. Smessaert and J. Rottler, *Phys. Rev. E*, 88, 022314, 2013.

104. C. Bennemann, W. Paul, K. Binder, and B. Dünweg, *Phys. Rev. E*, 57, 843, 1998.

105. S. Frey, F. Weyßer, H. Meyer, J. Farago, M. Fuchs, and J. Baschnagel, *Eur. Phys. E*, 38, 11, 2015.

106. J. Farago, A. Semenov, S. Frey, and J. Baschnagel, *Eur. Phys. E*, 37, 46, 2014.

107. S. Peter, H. Meyer, and J. Baschnagel, *Eur. Phys. J. E*, 28, 147, 2009.

108. S. Peter, H. Meyer, and J. Baschnagel, *J. Polym. Sci. B*, 44, 2951, 2006.
109. F. Puosi and D. Leporini, *J. Chem. Phys.*, 136, 211101, 2012.
110. F. Puosi and D. Leporini, *J. Phys. Chem. B*, 115, 14046, 2011.
111. A. Makke, M. Perez, J. Rottler, O. Lame, and J.-L. Barrat, *Macromol. Theory Simul.*, 20, 826, 2011.
112. G. J. Papakonstantopoulos, R. A. Riggleman, J.-L. Barrat, and J. J. de Pablo, *Phys. Rev. E*, 77, 041502, 2008.
113. A. Shavit and R. A. Riggleman, *J. Phys. Chem.*, 118, 9096, 2014.
114. A. Shavit and R. A. Riggleman, *Macromolecules*, 46, 5044, 2013.
115. P. Z. Hanakata, J. F. Douglas, and F. W. Starr, *Nat. Commun.*, 5, 4163, 2014.
116. B. A. Pazmiño Betancourt, P. Z. Hanakata, F. W. Starr, and J. F. Douglas, *PNAS*, 112, 2966, 2015.
117. K. Kremer and G. S. Grest, *J. Chem. Phys.*, 92, 5057, 1990.
118. G. S. Grest and K. Kremer, *Phys. Rev. A*, 33, 3628, 1986.
119. D. Ceperley, M. H. Kalos, and J. L. Lebowitz, *Phys. Rev. Lett.*, 41, 1978, 1978.
120. M. Bishop, M. H. Kalos, and H. L. Frisch, *J. Chem. Phys.*, 70, 1299, 1979.
121. M. Pütz, K. Kremer, and G. S. Grest, *Europhys. Lett.*, 49, 735, 2000, see also the Comment and Reply in *Europhys. Lett.*, 52, 719, 2000, *Europhys. Lett.*, 52, 721, 2000.
122. T. Kreer, J. Baschnagel, M. Muller, and K. Binder, *Macromolecules*, 34, 1105, 2001.
123. A. E. Likhtman, S. K. Sukumaran, and J. Ramirez, *Macromolecules*, 40, 6748, 2007.
124. N. Schulmann, H. Xu, H. Meyer, P. Polińska, J. Baschnagel, and J. P. Wittmer, *Eur. Phys. J. E*, 35, 93, 2012.
125. P. Virnau, M. Müller, L. Gonzalez MacDowell, and K. Binder, *New J. Phys.*, 6, 7, 2004.
126. P. Virnau, M. Müller, L. G. MacDowell, and K. Binder, *J. Chem. Phys.*, 121, 2169, 2004.
127. M. Kröger, *Phys. Rep.*, 390, 453, 2004.
128. B. Schnell, Numerical Studies of the Glass Transition and the Glassy State of Dense Amorphous Polymers: Mechanical Properties and Cavitation Phenomenon (in French), Ph.D. thesis, Université de Strasbourg, 2006, available from http://scd-theses.u-strasbg.fr/1131/01/SCNNELL2006.pdf.
129. T. Vettorel and H. Meyer, *J. Chem. Theory Comput.*, 2, 616, 2006.
130. T. Vettorel, H. Meyer, J. Baschnagel, and M. Fuchs, *Phys. Rev. E*, 75, 041801, 2007.
131. M. E. Mackura and D. S. Simmons, *J. Polym. Sci. B*, 52, 134, 2014.
132. R. Auhl, R. Everaers, G. S. Grest, K. Kremer, and S. J. Plimpton, *J. Chem. Phys.*, 119, 12718, 2003.
133. R. Faller, F. Müller-Plathe, and A. Heuer, *Macromolecules*, 33, 6602, 2000.
134. R. Faller, A. Kolb, and F. Müller-Plathe, *Phys. Chem. Chem. Phys.*, 1, 2071, 1999.
135. K. G. Honnell, J. G. Curro, and K. S. Schweizer, *Macromolecules*, 23, 3496, 1990.
136. A. R. Khokhlov and A. N. Semenov, *Macromolecules*, 19, 373, 1986.
137. B. Schnell, H. Meyer, C. Fond, J. Wittmer, and J. Baschnagel, *Eur. Phys. J. E*, 34, 97, 2011.
138. M. Bulacu and E. van der Giessen, *J. Chem. Phys.*, 123, 114901, 2005.
139. M. Bulacu and E. van der Giessen, *Phys. Rev. E*, 76, 011807, 2006.
140. M. Bernabei, A. J. Moreno, and J. Colmenero, *Phys. Rev. Lett.*, 101, 255701, 2008.
141. M. Bernabei, A. J. Moreno, and J. Colmenero, *J. Chem. Phys.*, 131, 204502, 2009.
142. M. Bernabei, A. J. Moreno, E. Zaccarelli, F. Sciortino, and J. Colmenero, *J. Chem. Phys.*, 134, 024523, 2011.
143. C. M. Roland, S. Hensel-Bielowka, M. Paluch, and R. Casalini, *Rep. Prog. Phys.*, 68, 1405, 2005.
144. J. Buchholz, W. Paul, F. Varnik, and K. Binder, *J. Chem. Phys.*, 117, 7364, 2002.
145. K. Vollmayr, W. Kob, and K. Binder, *J. Chem. Phys.*, 105, 4714, 1996.
146. K. Vollmayr, W. Kob, and K. Binder, *J. Chem. Phys.*, 54, 15808, 1996.
147. A. V. Lyulin, B. Vorselaars, M. A. Mazo, N. K. Balabaev, and M. A. J. Michels, *Europhys. Lett.*, 71, 618, 2005.
148. J. P. Wittmer, H. Xu, P. Polińska, F. Weysser, and J. Baschnagel, *J. Chem. Phys.*, 138, 12A533, 2013.
149. J. Hintermeyer, A. Herrmann, R. Kahlau, C. Goiceanu, and E. A. Rössler, *Macromolecules*, 41, 9335, 2008.
150. A. L. Agapov and A. P. Sokolov, *Macromolecules*, 42, 2877, 2009.

151. S. P. Andersson and O. Andersson, *Macromolecules*, 31, 2999, 1998.
152. E. N. Dalal and P. J. Philipps, *Macromolecules*, 16, 890, 1983.
153. J. E. Roots, K. T. Ma, J. S. Higgins, and V. Arrighi, *Phys. Chem. Chem. Phys.* 1, 137, 1999.
154. A. Gitsas, G. Floudas, and G. Wegner, *Phys. Rev. E*, 69, 041802, 2004.
155. B. Lobe and J. Baschnagel, *J. Chem. Phys.*, 101, 1616, 1994.
156. T. G. Fox and P. J. Flory, *J. Polym. Sci.*, 14, 315, 1954.
157. T. G. Fox and S. Loshaek, *J. Polym. Sci.*, 15, 371, 1955.
158. J. Dudowicz, K. F. Freed, and J. F. Douglas, *Adv. Chem. Phys.*, 137, 125, 2008.
159. S. Mirigian and K. S. Schweizer, *Macromolecules*, 48, 1901, 2015.
160. A. Rudin and D. Burgin, *Polymer*, 16, 291, 1975.
161. J. M. G. Cowie and P. M. Toporowski, *Eur. Polym. J.*, 4, 621, 1968.
162. G. Pezzin, F. Zilio-Grandi, and P. Sanmartin, *Eur. Polym. J.*, 6, 1053, 1970.
163. R. Zallen, *The Physics of Amorphous Solids* (Wiley, New York), 1983.
164. R. Brüning and K. Samwer, *Phys. Rev. B*, 46, 11318, 1992.
165. K. S. Schweizer, *Curr. Opin. Colloid Interface Sci.*, 12, 297, 2007.
166. L. Berthier, G. Biroli, J.-P. Bouchaud, and R. L. Jack, Overview of different characterizations of dynamic heterogeneity, in *Dynamical Heterogeneities in Glasses, Colloids and Granular Media*, edited by L. Berthier, G. Biroli, J.-P. Bouchaud, L. Cipelletti, and W. van Saarloos (Oxford University Press, Oxford), 69–109, 2011.
167. W. Götze, *J. Phys. Condens. Matter*, 11, A1, 1999.
168. W. Götze and L. Sjögren, *Rep. Prog. Phys.*, 55, 241, 1992.
169. W. Götze, Aspects of structural glass transitions, in *Proceedings of the Les Houches Summer School of Theoretical Physics, Les Houches 1989, Session LI*, edited by J. P. Hansen, D. Levesque, and J. Zinn-Justin (North-Holland, Amsterdam), 287–503, 1991.
170. W. Kob, Supercooled liquids, the glass transition, and computer simulations, in *Slow Relaxations and Nonequilibrium Dynamics in Condensed Matter*, edited by J.-L. Barrat, M. Feigelmann, J. Kurchan, and J. Dalibard (EDP Sciences/Springer, Les Ulis/Berlin), 201–269, 2003.
171. U. Balucani and M. Zoppi, *Dynamics of the Liquid State* (Oxford University Press, Oxford), 2003.
172. D. Reichman and P. Charbonneau, *J. Stat. Mech. Theor. Exp.*, P05013, 2005.
173. W. Kob, M. Nauroth, and F. Sciortino, *J. Non-Cryst. Solids*, 307–310, 181, 2002.
174. F. Sciortino and W. Kob, *Phys. Rev. Lett.*, 86, 648, 2001.
175. J.-L. Barrat, W. Götze, and A. Latz, *J. Phys. Condens. Matter*, 1, 7163, 1989.
176. A. Rinaldi, F. Sciortino, and P. Tartaglia, *Phys. Rev. E*, 63, 061210, 2001.
177. M. Aichele, S.-H. Chong, J. Baschnagel, and M. Fuchs, *Phys. Rev. E*, 69, 061801, 2004.
178. T. Gleim, W. Kob, and K. Binder, *Phys. Rev. Lett.*, 81, 4404, 1998.
179. T. Voigtmann, A. M. Puertas, and M. Fuchs, *Phys. Rev. E*, 70, 061506, 2004.
180. S.-H. Chong and F. Sciortino, *Europhys. Lett.*, 64, 197, 2003.
181. L. Berthier and W. Kob, *J. Phys. Condens. Matter*, 19, 205130, 2007.
182. F. Weysser, A. M. Puertas, M. Fuchs, and T. Voigtmann, *Phys. Rev. E*, 82, 011504, 2010.
183. K. S. Schweizer and J. G. Curro, *Adv. Chem. Phys.*, 98, 1, 1997.
184. S.-H. Chong and M. Fuchs, *Phys. Rev. Lett.*, 88, 185702, 2002.
185. S.-H. Chong, M. Aichele, H. Meyer, M. Fuchs, and J. Baschnagel, *Phys. Rev. E*, 76, 051806, 2007.
186. A. N. Semenov and S. P. Obukhov, *J. Phys. Condens. Matter*, 17, S1747, 2005.
187. R. G. Palmer, *Adv. Phys.*, 6, 669, 1982.
188. T. Franosch, M. Fuchs, W. Götze, M. R. Mayr, and A. P. Singh, *Phys. Rev. E*, 55, 7153, 1997.
189. M. Fuchs, *J. Non-Cryst. Solids*, 172–174, 241, 1994.
190. W. Paul and J. Baschnagel, *Stochastic Processes: From Physics to Finance* (Springer, Berlin–Heidelberg), 2013.
191. J. P. Wittmer, H. Xu, and J. Baschnagel, *Phys. Rev. E*, 91, 022107, 2015.
192. M. Fuchs, W. Götze, and M. R. Mayr, *Phys. Rev. E*, 58, 3384, 1998.
193. M. Vladkov and J.-L. Barrat, *Macromol. Theory Simul.*, 15, 252, 2006.

194. J. D. Halverson, J. Smrek, K. Kremer, and A. Y. Grosberg, *Rep. Prog. Phys.*, 77, 022601, 2014.

195. J. D. Halverson, W. B. Lee, G. S. Grest, A. Y. Grosberg, and K. Kremer, *J. Chem. Phys.*, 134, 204904, 2011.

196. J. D. Halverson, W. B. Lee, G. S. Grest, A. Y. Grosberg, and K. Kremer, *J. Chem. Phys.*, 134, 204905, 2011.

197. S. Obukhov, A. Johner, J. Baschnagel, H. Meyer, and J. P. Wittmer, *Europhys. Lett.*, 105, 48005, 2014.

198. A. N. Semenov and H. Meyer, *Soft Matter*, 9, 4249, 2013.

199. R. W. Cahn, *Nature*, 413, 582, 2001.

200. F. H. Stillinger, *Science*, 267, 1935, 1995.

201. Y. Zhou, M. Karplus, K. D. Ball, and R. S. Berry, *J. Chem. Phys.*, 116, 2323, 2002.

202. C. L. Klix, F. Ebert, F. Weysser, M. Fuchs, G. Maret, and P. Keim, *Phys. Rev. Lett.*, 109, 178301, 2012.

203. R. F. Boyer, $T_{//}$ and related liquid-state transitions and relaxations, in *Encyclopedia of Polymer Science and Engineering*, Vol. 17, edited by J. Kroschwitz (Wiley, New York), 23–47, 1989.

204. A. Kisliuk, R. T. Mathers, and A. P. Sokolov, *J. Polym. Sci. Polym. Phys.*, 38, 2785, 2000.

205. K. Chen and K. S. Schweizer, *J. Chem. Phys.*, 126, 014904, 2007.

206. T. Voigtmann, *Phys. Rev. Lett.*, 101, 095701, 2008.

207. N. Gnan, T. B. Schrøder, U. R. Pedersen, N. P. Bailey, and J. C. Dyre, *J. Chem. Phys.*, 131, 234504, 2009.

208. D. Coslovich and C. M. Roland, *J. Phys. Chem. B*, 112, 1329, 2008.

209. A. A. Veldhorst, J. C. Dyre, and T. B. Schrøder, *J. Chem. Phys.*, 141, 054904, 2014.

210. L. Berthier and G. Tarjus, *Eur. Phys. J. E*, 34, 96, 2011.

211. N. P. Bailey, L. Bøhling, A. A. Veldhorst, T. B. Schrøder, and J. C. Dyre, *J. Chem. Phys.*, 139, 184506, 2013.

212. S. Krushev and W. Paul, *Phys. Rev. E*, 67, 021806, 2003.

213. S. Krushev, W. Paul, and G. D. Smith, *Macromolecules*, 35, 4198, 2002.

214. M. Sperl, *Phys. Rev. E*, 68, 031405, 2003.

215. K. N. Pham, A. M. Puertas, J. Bergenholtz, S. U. Egelhaaf, P. N. Moussaïd, P. N. Pusey, A. B. Schofield, M. E. Cates, M. Fuchs, and W. C. K. Poon, *Science*, 269, 104, 2002.

216. T. Voigtmann, *Europhys. Lett.*, 96, 36006, 2011.

217. A. J. Moreno and J. Colmenero, *Phys. Rev. E*, 74, 021409, 2006.

218. A. J. Moreno and J. Colmenero, *J. Chem. Phys.*, 124, 184906, 2006.

219. G. Foffi, W. Götze, F. Sciortino, P. Tartaglia, and T. Voigtmann, *Phys. Rev. E*, 69, 011505, 2004.

220. E. Flenner and G. Szamel, *Phys. Rev. E*, 72, 031508, 2005.

221. E. Flenner and G. Szamel, *Phys. Rev. E*, 72, 011205, 2005.

222. P. Chaudhuri, L. Berthier, and W. Kob, *Phys. Rev. Lett.*, 99, 060604, 2007.

223. G. Brambilla, D. El Masri, M. Pierno, L. Berthier, L. Cipelletti, G. Petekidis, and A. B. Schofield, *Phys. Rev. Lett.*, 102, 085703, 2009.

224. S. Merabia and D. Long, *Eur. Phys. J. E*, 9, 195, 2002.

225. A. P. Sokolov and K. S. Schweizer, *Phys. Rev. Lett.*, 102, 248301, 2009, see also the Comment and Reply, *Phys. Rev. Lett.*, 103, 159801, 159802, 2009.

226. S.-H. Chong, *Phys. Rev. E*, 78, 041501, 2008.

227. S. K. Kumar, G. Szamel, and J. F. Douglas, *J. Chem. Phys.*, 124, 214501, 2006.

228. M. D. Ediger, *Annu. Rev. Phys. Chem.*, 51, 99, 2000.

229. J. Frenkel, *Kinetic Theory of Liquids* (Dover, New York), 1955.

230. K. L. Ngai, *Relaxation and Diffusion in Complex Systems* (Springer, New York), 2011.

231. B. A. Pazmiño Betancourt, J. F. Douglas, and F. W. Starr, *J. Chem. Phys.*, 140, 204509, 2014.

232. G. Tarjus, An overview of the theories of the glass transition, in *Dynamical Heterogeneities in Glasses, Colloids and Granular Media*, edited by L. Berthier, G. Biroli, J.-P. Bouchaud, L. Cipelletti, and W. van Saarloos (Oxford University Press, Oxford), 39–67, 2011.

233. H. Z. Cummins, *Phys. Rev. E*, 54, 5870, 1996.

234. D. Kivelson, G. Tarjus, X.-L. Xiao, and S. A. Kivelson, *Phys. Rev. E*, 54, 5873, 1996.

235. A. Andreanov, G. Biroli, and J.-P. Bouchaud, *Europhys. Lett.*, 88, 16001, 2009.

236. P. Meyer, K. Miyazaki, and D. Reichman, *Phys. Rev. Lett.*, 97, 095702, 2006.

237. L. M. C. Janssen and D. Reichman, *Phys. Rev. Lett.*, 115, 205701, 2015.

238. V. N. Novikov and A. P. Sokolov, *Phys. Rev. E*, 67, 031507, 2003.

239. P. K. Dixon, *Phys. Rev. B*, 42, 8179, 1990.

240. F. Stickel, E. W. Fischer, and R. Richert, *J. Chem. Phys.*, 102, 6251, 1995.

241. F. Stickel, E. W. Fischer, and R. Richert, *J. Chem. Phys.*, 104, 2043, 1996.

242. T. Hecksher, A. I. Nielsen, N. B. Olsen, and J. C. Dyre, *Nat. Phys.*, 4, 737, 2008.

243. J. Zhao, S. L. Simon, and G. B. McKenna, *Nat. Commun.*, 4, 1783, 2013.

244. G. B. McKenna, *Nat. Phys.*, 4, 673, 2008.

245. A. S. Keys, L. O. Hedges, J. P. Garrahan, S. C. Glotzer, and D. Chandler, *Phys. Rev. X*, 1, 021013, 2011.

246. D. Chandler and J. P. Garrahan, *Annu. Rev. Phys. Chem.*, 191, 191, 2010.

247. M. P. Ciamarra, R. Pastore, and A. Coniglio, *Soft Matter*, 12, 358, 2016.

248. J. Helfferich, F. Ziebert, S. Frey, H. Meyer, J. Farago, A. Blumen, and J. Baschnagel, *Phys. Rev. E*, 89, 042603, 2014.

249. J. Helfferich, F. Ziebert, S. Frey, H. Meyer, J. Farago, A. Blumen, and J. Baschnagel, *Phys. Rev. E*, 89, 042604, 2014.

250. J. Helfferich, *Eur. Phys. J. E*, 37, 73, 2014.

251. J. Helfferich, K. Vollmayr-Lee, F. Ziebert, H. Meyer, and J. Baschnagel, *Europhys. Lett.*, 109, 36004, 2015.

252. K. Vollmayr-Lee, *J. Chem. Phys.*, 121, 4781, 2004.

253. K. Vollmayr-Lee, R. Bjorkquist, and L. M. Chambers, *Phys. Rev. Lett.*, 110, 017801, 2013.

254. K. Dawson, G. Foffi, M. Fuchs, W. Götze, F. Sciortino, M. Sperl, P. Tartaglia, T. Voigtmann, and E. Zaccarelli, *Phys. Rev. E*, 63, 011401, 2001.

255. A. J. Moreno and J. Colmenero, *Phys. Rev. Lett.*, 100, 126001, 2008.

256. A. J. Moreno and J. Colmenero, *J. Phys. Condens. Matter*, 19, 466112, 2007.

257. J. Colmenero and A. Arbe, *Soft Matter*, 3, 1474, 2007.

258. T. Voigtmann and J. Horbach, *Europhys. Lett.*, 74, 459, 2006.

259. S.-H. Chong and F. Sciortino, *Phys. Rev. E*, 69, 051202, 2004.

260. G. Biroli, J.-P. Bouchaud, K. Miyazaki, and D. R. Reichman, *Phys. Rev. Lett.*, 97, 195701, 2006.

261. S. Lang, R. Schilling, V. Krakoviack, and T. Franosch, *Phys. Rev. E*, 86, 021502, 2012.

262. V. Krakoviack, *Phys. Rev. E*, 75, 031503, 2007.

263. V. Krakoviack, *Phys. Rev. E*, 79, 061501, 2009.

264. V. Krakoviack, *Phys. Rev. E*, 84, 050501(R), 2011.

265. M. Fuchs, *Adv. Polym. Sci.*, 36, 55, 2010.

266. J. M. Brader, M. E. Cates, and M. Fuchs, *Phys. Rev. E*, 86, 021403, 2012.

267. T. Voigtmann, *Curr. Opin. Colloid Interface Sci.*, 19, 549, 2014.

268. S. Fritschi, M. Fuchs, and T. Voigtmann, *Soft Matter*, 10, 4822, 2014.

269. T. Franosch and W. Götze, *J. Phys. Chem. B*, 103, 4011, 1999.

270. L. Angelani, G. Ruocco, M. Sampoli, and F. Sciortino, *J. Chem. Phys.*, 119, 2120, 2003.

271. L. Angelani, C. De Michele, G. Ruocco, and F. Sciortino, *J. Chem. Phys.*, 121, 7533, 2004.

272. M. Nauroth and W. Kob, *Phys. Rev. E*, 55, 657, 1997.

273. P. Chaudhuri, Y. Gao, L. Berthier, M. Kilfoil, and W. Kob, *J. Phys. Condens. Matter*, 20, 244126, 2008.

274. E. J. Saltzman and K. S. Schweizer, *Phys. Rev. E*, 77, 051504, 2008.

275. S. P. Das and G. F. Mazenko, *Phys. Rev. A*, 34, 2265, 1986.

276. W. Götze and L. Sjögren, *Z. Phys. B*, 65, 415, 1987.

277. W. Götze and L. Sjögren, *Transport Theory Stat. Phys.*, 24, 801, 1995.

278. S. P. Das, *Rev. Mod. Phys.*, 76, 826, 2004.

279. G. Szamel, *Phys. Rev. Lett.*, 90, 228301, 2003.

280. M. Fuchs, W. Götze, S. Hildebrand, and A. Latz, *J. Phys. Condens. Matter*, 4, 7709, 1992.

281. W. Götze and T. Voigtmann, *Phys. Rev. E*, 61, 4133, 2000.

282. M. E. Cates and S. Ramaswamy, *Phys. Rev. Lett.*, 96, 135701, 2006.

283. J. Reinhardt, F. Weysser, and M. Fuchs, *Phys. Rev. Lett.*, 105, 199604, 2010.

284. G. Brambilla, D. El Masri, M. Pierno, L. Berthier, and L. Cipelletti, *Phys. Rev. Lett.*, 105, 199605, 2010.

285. K. Chen, E. J. Saltzman, and K. S. Schweizer, *J. Phys. Condens. Matter*, 21, 50301, 2009.

286. P. G. Debenedetti and F. H. Stillinger, *Nature*, 410, 259, 2001.

287. F. Sciortino, *J. Stat. Mech.*, 5, P05015, 2005.

288. A. Heuer, *J. Phys. Condens. Matter*, 20, 373101, 2008.

289. L. Berthier and W. Kob, *Phys. Rev. E*, 85, 011102, 2012.

290. M. Alcoutlabi and G. B. McKenna, *J. Phys. Condens. Matter*, 17, R461, 2005.

291. M. D. Ediger and J. A. Forrest, *Macromolecules*, 41, 471, 2014.

292. S. Napolitano, S. Capponi, and B. Vanroy, *Eur. Phys. J. E*, 36, 61, 2013.

293. M. Solar, K. Binder, and W. Paul, Dielectric relaxation of a polybutadiene melt at a crystalline graphite surface: Atomistic molecular dynamics simulations, in *Dynamics in Geometrical Confinement*, edited by F. Kremer (Springer, Berlin, Heidelberg, New York), 1–15, 2014.

294. V. Krakoviack, *Phys. Rev. E*, 82, 061501, 2010.

295. V. Krakoviack, *J. Chem. Phys.*, 141, 104504, 2014.

296. M. Solar, E. U. Mapesa, F. Kremer, K. Binder, and W. Paul, *Europhys. Lett.*, 104, 66004, 2013.

297. S. Peter, S. Napolitano, H. Meyer, M. Wübbenhorst, and J. Baschnagel, *Macromolecules*, 41, 7729, 2008.

298. F. Müller-Plathe, *Soft Mater.*, 1, 1, 2003.

299. S. Frey, Viscoelastic Properties of Glass-Forming Polymer Melts, Ph.D. thesis, Université de Strasbourg, Strasbourg, 2012, available from http://www.sudoc.fr/165862653.

300. C. Bennemann, W. Paul, J. Baschnagel, and K. Binder, *J. Phys. Condens. Matter*, 11, 2179, 1999.

Thermo-mechanical signatures of polymeric glasses

JAMES M. CARUTHERS AND GRIGORI A. MEDVEDEV

INTRODUCTION

Polymeric glasses are some of the most important advanced engineering materials, because of their unique combination of thermo-mechanical properties and their ability to be economically processed into complex shapes. Because of their engineering significance, the physical behavior of polymeric glasses has been extensively studied over the last 100 years. Although much is known, a fundamental understanding of the origin of the thermo-physical behavior of polymeric glasses is far from complete. In fact, the glassy state, of which polymeric glasses are one of the most important examples, is considered one of the outstanding problems in condensed matter physics [1,2]. Even if the objective is only a phenomenological description rather than fundamental understanding, currently there is no constitutive model that can capture the diversity of thermo-physical behaviors exhibited by polymeric glasses. However, there are a number of constitutive models that can describe at least limited portions of the observed thermo-mechanical response of polymeric glasses (see Chapter 14 for a more detailed discussion of various constitutive models that have been proposed to describe polymeric glasses).

In this chapter, we will critically review the diversity of thermo-mechanical and associated physicochemical behaviors of polymeric glasses, where the objective is to identify *signatures* of the thermo-physical response that are universal, or nearly universal, features of polymeric glasses. Since polymeric glasses will exhibit all of these signature behaviors, any fundamental model of the glassy state needs to address all phenomena, for example, a fundamental model of the glassy state cannot just address the linear viscoelastic (LVE) response but remain silent concerning enthalpy relaxation, since both processes are occurring simultaneously. With respect to the development of phenomenological constitutive models, the literature is replete with examples of models that describe one limited type of experiment, but are qualitatively incorrect or completely silent on related thermo-physical responses, for example, a constitutive model that has been developed to describe the nonlinear stress–strain behavior at a constant strain rate, but fails to describe nonlinear creep. The objective is not to denigrate the considerable constitutive modeling efforts to date; rather, it is to identify the signature behaviors exhibited by glassy polymers in order to provide a clear target for future constitutive model development and to show the diversity of behaviors that must be captured by a full fundamental understanding of the glassy state.

By *signature* we mean the qualitative behavior of the data exhibited by most, if not all, polymeric glasses. For example, consider the volume relaxation following a temperature jump from an equilibrium state in the glass transition region, where the response after a down-jump is qualitatively different from the response after an up-jump (see Section 4.3). This asymmetry of approach is a *signature* of volume relaxation in the glassy state that has been observed in various polymeric and low-molecular-weight glasses, where any constitutive model of polymeric glasses needs to address this *signature* behavior or at least explicitly acknowledge that the model is silent on this particular feature of the data. The qualitative nature of the signatures is most important, since the ability to capture the qualitative behavior provides the most discriminating evaluation of a constitutive model of the glassy state. For example, polymeric glasses exhibit a stress–strain curve with significant post-yield softening; where a signature of the glassy state is that the magnitude of the post-yield softening increases with physical aging (see Section 4.6.3). Consequently, an important step in constitutive model development is to determine if the model predicts post-yield softening that increases with aging, where the qualitative nature of this feature provides serious restrictions on the structure of the model prior to any quantitative fitting of the data. We believe that a comprehensive review of the *signatures* of the thermo-mechanical and related behavior of polymeric glasses will provide a solid foundation for the future development of improved fundamental models and constitutive descriptions of polymeric glasses.

In this chapter the primary focus is on the thermo-mechanical behavior of polymeric glasses. By thermo-mechanical behavior we mean all of the physical behavior that is needed for a thermodynamically complete description of the polymer. Various thermodynamically consistent constitutive frameworks have been proposed [3–5] and it is not the purpose of this chapter to attempt to evaluate the merits of the various proposals. However, all the various approaches require at least (i) a tensor-based multiaxial description, (ii) material objectivity (i.e., independence of the frame of reference) for finite deformations, and (iii) the inclusion of thermal contributions needed to construct a free energy. Thus, the objective of this communication is to completely identify the signatures of polymeric glasses for the range of phenomena that are addressed by a thermodynamically consistent model. There are also molecular signatures associated with the glass (for example, the rate of molecular reorientation as detected by dielectric relaxation) that may inform the development of a constitutive model. In this communication we will briefly review some of these molecular signatures of polymeric glasses, but a complete review of the molecular signatures of the glass is beyond the scope of this chapter.

A number of reviews of some aspects of the thermo-mechanical properties are given in Table 4.1, where this chapter will draw extensively on this excellent collection of reviews. In contrast to these reviews, in this chapter, we will (i) attempt to clearly identify the *signatures* of each phenomena as well as provide references to various experimental supporting studies and (ii) perform this analysis for the full range of thermo-mechanical and related phenomena. Thus, we anticipate that this chapter will provide a clear and concise summary of essential behaviors exhibited by polymeric glasses.

The rest of the chapter will be organized as follows:

1. Pressure–volume–temperature (PVT) behavior
2. Linear viscoelastic behavior
3. Nonlinear volume relaxation
4. Enthalpy relaxation
5. Coupled volume and mechanical relaxation
6. Nonlinear stress–strain and stress relaxation
7. Nonlinear creep
8. More complex thermo-deformational histories
9. Combined mechanical and enthalpy relaxation experiments
10. Molecular measurements of relaxation behavior
11. Discussion

Table 4.1 Existing reviews of thermo-mechanical behavior of polymeric glasses

Topic	References
PVT behavior; volume relaxation; Tg; and physical aging	McKenna [6]; Angell et al. [7]; McKenna [8]
Enthalpy relaxation	Hodge [9]
Linear viscoelasticity	Ferry [10]
Nonlinear stress–strain	Bowden [11]; Haward [12]; Crist [13]; Boyce and Haward [14]; Ward and Sweeney [15]; Argon [16]
Nonlinear creep	Turner [17]; Findlay et al. [18]
Multiaxial deformation	Bowden [11]

In Sections 4.6 through 4.10, specific thermo-mechanical phenomena will be discussed, including identification of the *signatures* of that phenomenon along with typical experimental data for a typical polymer system and references to other polymer systems that also exhibit this signature behavior. In Section 4.11, we will summarize some of the experiments that probe the underlying relaxation process that may inform the development of constitutive models; however, there will be no attempt at completeness for the molecular signatures of the glassy state. In the final section, there will be a summary of the thermo-mechanical signatures and thoughts on future challenges.

4.1 PRESSURE–VOLUME–TEMPERATURE BEHAVIOR

A glass is formed when a liquid is cooled and crystallization is suppressed. A key signature of the glass is the change in the temperature dependence of the specific volume over a relatively narrow temperature region as shown in Figure 4.1a for an epoxy resin. If the specific volume versus temperature response is fit with two straight lines, the intersection of the two lines is traditionally identified as the isobaric glass transition temperature Tg. However, unlike a first-order thermodynamic transition such as melting, the transition from the liquid to glassy state does not occur at a unique temperature as shown in Figure 4.1b, where Tg is a function of the cooling rate. As shown in Figure 4.2, Tg increases linearly with the log(cooling rate) for four glass-forming polymers, where for each order of magnitude increase in the cooling rate, Tg increases by 2.5°C for PC, 2.9°C for PS, 3.2°C for PVC, and 3.3°C for PMMA [20]. Tg is a linear function of the logarithm of the cooling rate for the range of cooling rates achievable in dilatometry experiments, which is limited due to large thermal

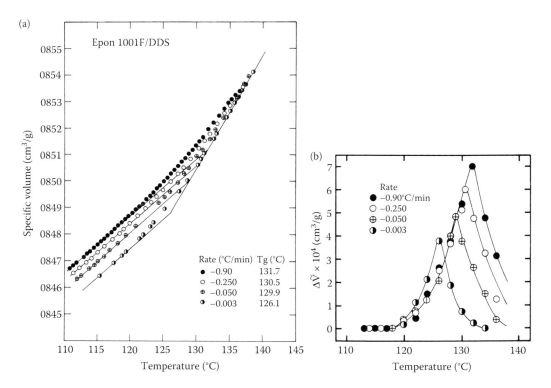

Figure 4.1 (a) Isobaric specific volume versus temperature dependence for Epon 1001F epoxy resin cured with diaminodiphenyl sulfone (DDS) at four different cooling rates of 0.90, 0.25, 0.050, and 0.003°C/min. (b) Deviation of specific volume with the linear glassy/liquid volume–temperature lines. (C. A. Bero and D. J. Plazek, *J. Polym. Sci. Polym. Phys.*, 1991, 29, 39. Copyright Wiley-VCH Verlag GmbH & Co. KGaA. Reproduced with permission.)

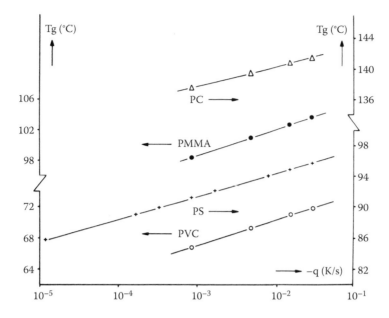

Figure 4.2 Cooling rate dependence of Tg for PC, PMMA, PS, and PVC. (With kind permission from Springer Science+Business Media: *Rheologica Acta*, 23, 1984, 378, R. Greiner and F. R. Schwarzl.)

mass of the dilatometer. It is expected that this linearity will probably not continue at higher cooling rates in a manner similar to the nonlinear cooling rate dependence observed in calorimetric experiments using Flash differential scanning calorimetry (DSC) (see Section 4.4). In addition, the breadth of the transition increases with the cooling rate as shown in Figure 4.1b, where the transition occurs over approximately 15°C for a 0.003°C/min cooling rate versus more than 25°C when the cooling rate is 0.90°C/min. The change in the temperature dependence of the specific volume in the Tg region as well as the cooling rate dependence of Tg has also been observed for small-molecule glasses [21].

Pressure–volume–temperature (PVT) behavior has been traditionally studied in pressure dilatometers using two distinct procedures: (i) isobarically, where volume is continuously monitored at a constant rate of cooling/heating while the pressure is held constant or (ii) isothermally, where at a given temperature, a series of discrete pressure changes are applied and the volume is recorded after each pressure change after allowing for the volume relaxation to cease. Experimentally, it is more difficult to achieve a constant cooling rate due to a large thermal mass of a pressure dilatometer; thus, the isothermal protocol is typically employed. However, one must be mindful that even for an equilibrium material above Tg, the two procedures may result in slightly different PVT surfaces (especially at the highest pressures) as shown in Figure 4.3a for PMMA, which is surely an experimental artifact. In the glassy state, the difference between the isothermal and isobaric data becomes significant and is a clear signature of how the current state of the glass is a function of the path by which the material enters the glassy state. Specifically, for PMMA shown in Figure 4.3, the "isothermal" volume response at 20°C at 200 MPa is for a material that entered the glassy state at 0.1 MPa; in contrast, the "isobaric" volume data at 20°C at 200 MPa is for a material that entered the glassy state at 200 MPa. Thus, the difference in the glassy PVT behavior for the "isothermal" and "isobaric" experimental protocols is a consequence of the nonequilibrium nature of the glassy state and its inherent dependence upon formation history. The general features of the PVT response shown in Figure 4.3 have been observed for a number of glassy polymers [22].

Pressure affects Tg as shown in Figure 4.3a, where Tg at each pressure can be defined as the intercept of the asymptotic equilibrium and glass volume–temperature curves. Tg versus pressure is nearly linear for up to 200 MPa as shown in Figure 4.3b for PMMA, but for a larger pressure range, the curvature is observed as shown for PS. The pressure dependence of Tg shown in Figure 4.3 is for the isobaric data, which determines the pressure dependence of Tg as the material enters the glass from the equilibrium rubber/liquid state.

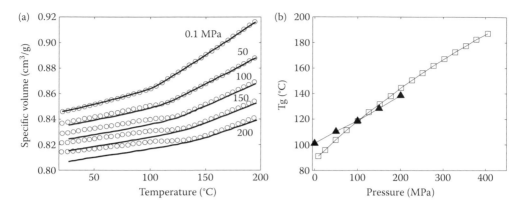

Figure 4.3 (a) PVT behavior of PMMA; circles—data obtained isothermally via discrete changes in pressure at each temperature, solid line—PVT data obtained isobarically via cooling at 0.5°C/min at indicated pressures. (Data from M. Schmidt and F. H. J. Maurer, *Macromolecules*, 33, 3879, 2000.) (b) Pressure dependence of Tg: for PMMA—triangles, that is, from the isobars in (a), for PS—squares. (Data from H.-J. Oels and G. Rehage, *Macromolecules*, 10, 1036, 1977.)

There is an alternative definition of the "Tg versus pressure line" as the intercept between the equilibrium PVT surface and the extrapolated PVT surface determined using the "isothermal" protocol with its associated formation pressure. McKinney and Goldstein [25] preformed an extensive characterization of the PVT behavior of poly(vinyl acetate), that is, PVAc. The resulting PVT surfaces are shown schematically in Figure 4.4, where there are two different glass surfaces for two different formation pressures: Glass 1 for a small formation pressure P_f (0.1 MPa in McKinney and Goldstein [25]) and Glass 2 for a much larger P_f (80 MPa in McKinney and Goldstein [25]). There would be infinity of glassy PVT surfaces between the Glass 1 and Glass 2 surfaces corresponding to different glass formation pressures. The pressure dependence of Tg shown in Figure 4.3b is for the line between Tg1 and Tg2 in Figure 4.4 and, thus, is Tg for the transition from equilibrium liquid to the glass. The PVT surfaces illustrated schematically in Figure 4.4 assume that there is no relaxation when the glassy material is near Tg for that glass. However, the "isothermal" protocol used to measure the PVT behavior does not always avoid relaxation in the Tg region, where the curvature of the constant pressure lines in Figure 4.3a is a result of relaxation of the glass as material moves into/out of the glass in the Tg region during the temperature–pressure history used in the "isothermal" PVT experiment. Similar types of PVT surfaces have been measured for several polymers including PVC [26], PMMA [23], and a styrene–acrylonitrile copolymer [27], where earlier work is reviewed in Rodgers [28].

The glass transition has typically been determined as the material is isobarically cooled into the glassy state. However, the glassy state can also be entered via an isochoric (i.e., constant volume) thermal history

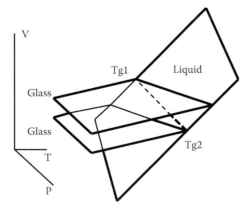

Figure 4.4 Schematics of PVT surfaces for polymer cooled into the glass at different formation pressures. Dashed line—Tg(P_f).

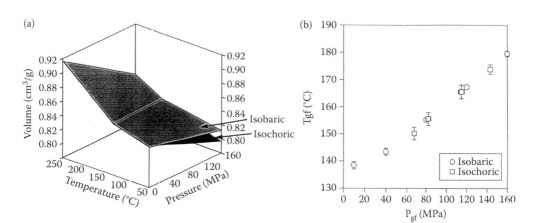

Figure 4.5 (a) Isochorically and isobarically formed glassy surfaces for PC. (b) Pressure dependence of Tg. (D. M. Colucci et al., *J. Polym. Sci. Polym. Phys.*, 1997, 35, 1561. Copyright Wiley-VCH Verlag GmbH & Co. KGaA. Reproduced with permission.)

as shown in Figure 4.5 for poly(carbonate). In these experiments [29], the volume was kept constant as the temperature was decreased, where there was a change in the slope of the pressure versus temperature data that was identified as Tg. As shown in Figure 4.5b, the isochoric and isobaric Tg are the same; however, there is a difference in the slope of the PVT surfaces of the isobarically formed glass versus that of the isochorically formed glass. Consequently, the state of the material as it enters the glass is not truly at equilibrium, because for an equilibrium process, an isochoric versus isobaric path would not matter; thus, $Tg(P_f)$ is an important, but not the only, descriptor of the glassy state.

SIGNATURES 1

Signatures of the PVT behavior of the glass-to-rubber transition include:

S1.1 There is a smooth change with temperature in the coefficient of thermal expansion in the glass-to-liquid transition. The extrapolated intersection of the glass and liquid volume–temperature lines is identified with Tg, which depends upon the glass formation conditions.

S1.2 Tg increases linearly with log(cooling rate) for the range of cooling rates accessible by dilatometry. The breadth of the glass-to-liquid transition increases as the cooling rate increases.

S1.3 Tg increases nearly linearly with the glass formation pressure.

S1.4 The glassy PVT surface depends upon the formation pressure. Both the coefficient of thermal expansion and the bulk modulus of the glass increase with formation pressure.

S1.5 The pressure dependence of Tg is the same for an isobarically and isochorically formed glass, although the resulting glassy PVT surface of the isochorically formed glass has a larger coefficient of thermal expansion and isothermal compressibility.

4.2 LINEAR VISCOELASTIC BEHAVIOR

The three-dimensional linear viscoelastic (LVE) behavior of an isotropic material is given by

$$\boldsymbol{\sigma}(t) = \int_{-\infty}^{t} 2G(t-\xi)\frac{d}{d\xi}\left[\boldsymbol{\varepsilon}(\xi) - \frac{1}{3}\boldsymbol{I}tr\boldsymbol{\varepsilon}(\xi)\right]d\xi$$
$$+ \int_{-\infty}^{t} K(t-\xi)\boldsymbol{I}\frac{dtr\boldsymbol{\varepsilon}(\xi)}{d\xi}d\xi + \int_{-\infty}^{t} A(t-\xi)\boldsymbol{I}\frac{dT(\xi)}{d\xi}d\xi$$

(4.1)

where $\boldsymbol{\sigma}(t)$ is the stress response to an arbitrary strain $\boldsymbol{\varepsilon}(t)$ and temperature $T(t)$ history. $\boldsymbol{\sigma}(t)$ and $\boldsymbol{\varepsilon}(t)$ are respectively the infinitesimal stress and strain tensors, $\boldsymbol{I}$ is the identity tensor, and tr indicates the trace of its tensor argument. $G(t)$ is the shear modulus; $K(t)$ is the bulk modulus; and, $A(t)$ is the thermal stress, which is the isotropic stress that develops when a material is heated while the volume is held constant. Analogous to Equation 4.1, the compliance form of linear viscoelasticity is given by

$$
\begin{aligned}
\boldsymbol{\varepsilon}(t) = & \int_{-\infty}^{t} 2J(t-\xi)\frac{d}{d\xi}\left[\boldsymbol{\sigma}(\xi)-\frac{1}{3}\boldsymbol{I}tr\boldsymbol{\sigma}(\xi)\right]d\xi \\
& + \int_{-\infty}^{t} B(t-\xi)\boldsymbol{I}\frac{dtr\boldsymbol{\sigma}(\xi)}{d\xi}d\xi + \frac{1}{3}\int_{-\infty}^{t}\alpha(t-\xi)\boldsymbol{I}\frac{dT(\xi)}{d\xi}d\xi
\end{aligned}
\tag{4.2}
$$

where $J(t)$ is the shear compliance, $B(t)$ is the bulk compliance, and $\alpha(t)$ is the coefficient of thermal expansion. For an LVE material, additional material functions can be determined from the three materials functions in Equation 4.1 (or equivalently from those in Equation 4.2), for example, the viscoelastic tensile modulus $E(t)$ and Poisson's ratio $\nu(t)$ can be determined for a material initially at equilibrium from $G(t)$ and $K(t)$, using well-known interconversion formulas [30]. A note of caution: the standard interconversion formulas assume that the material is initially in an equilibrium state, which is often a poor assumption for glassy materials where the time to reach equilibrium can be exceedingly long. The objective of this section is to describe the various linear viscoelastic material functions exhibited by glass-forming polymers.

The LVE description given in Equations 4.1 and 4.2 makes three implicit assumptions: (i) the perturbations are small enough that the response is linear, (ii) the small changes in temperature and deformation do not affect the $G(t)$, $K(t)$, and so on, material properties, and (iii) that the thermo-deformational process starts from an equilibrium state. It is well understood that strain and stress must be sufficiently small for an LVE description, but it is often not appreciated that the temperature perturbations must also be small. This is not an issue for the isothermal deformation of an equilibrium material, where the thermal history terms in Equations 4.1 and 4.2 remain zero. However, when a material is cooled into the glass, the temperature perturbation that accumulates prior to the deformation is large, where the thermal history term is present and may subsequently relax. If the material is deep enough in the glass so that the thermal history term does not relax during the course of deformation experiment, it may be possible to "subtract off" the thermal contributions to the strain/stress and just employ effective isothermal versions of Equations 4.1 or 4.2. However, to assume that somehow the shear and bulk processes are relaxing, but those related to the thermal history are frozen may not be physically reasonable.

The $G(t)$, $K(t)$, and other LVE material functions are defined at a specific temperature, where the only independent variable is time. However, it is common practice in the glass research literature to define an effective linear viscoelastic property, for example, $J(t; T, t_{ae})$—the linear creep compliance at a given temperature and sub-Tg aging time t_{ae}. This is formally incorrect, since a material property cannot depend upon the details of a specific thermal history; moreover, it hides the fact that material response may include contributions from the thermal history convolution integral in Equation 4.2 in addition to those from $J(t)$ and $B(t)$. This also creates confusion since the material function $G(t)$ is experimentally obtained as $\sigma_{12}(t)/\gamma_o$ for a small step shear strain deformation γ_o provided the experiment is carried out on an equilibrated material. On the other hand, if the material has not been fully equilibrated then the quantity $\sigma_{12}(t)/\gamma_o$ is not the same as $G(t)$, that is, the material function that is present in the constitutive Equation 4.1. Unfortunately, in the literature what is really $\sigma_{12}(t)/\gamma_o$ is often called $G(t)$ without ascertaining first that the contribution from the glass formation history is not part of the measured $\sigma_{12}(t)$ response.

4.2.1 TIME–TEMPERATURE AND TIME-AGING TIME SUPERPOSITION

The linear viscoelastic response of amorphous polymers is traditionally analyzed via time–temperature superposition [10], where there typically is acceptable superposition of isotherms at least for the range of

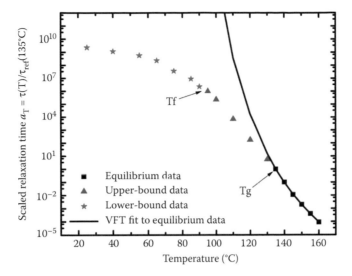

Figure 4.6 Temperature dependence of the $loga_T$ shift function for 20-million-year-old amber. Squares—equilibrium data; triangles—upper-bound data; stars—lower-bound data; and line—VTF fit to equilibrium data. (Reprinted by permission from Springer Nature. *Nat. Commun.*, J. Zhao, S. L. Simon, and G. B. McKenna, Copyright 2013.)

times/frequencies that are experimentally accessible. Above the conventional Tg (i.e., determined via cooling at a rate of 1°/min), the temperature dependence of the $loga_T$ shift function is described by Vogel–Tammann–Fulcher (VTF) [31–33] or equivalently the Williams–Landel–Ferry (WLF) [34] equations. Below this conventional Tg, there is a transition to approximately Arrhenian behavior as shown in Figure 4.6. It should be emphasized that the $loga_T$ response being discussed here is for the equilibrium material both above and below the conventional Tg. Obviously, equilibrating the material below the conventional Tg requires exceedingly long times so that only temperatures of several degrees below conventional Tg are normally accessible. McKenna and coworkers [35] have recently measured the relaxation response of amber that has been in the glassy state for times up to 20 million years (which resulted in an equilibrated material at Tg-43.6°C), where the equilibrium $loga_T$ temperature dependence was most accurately described by an empirical quadratic expression proposed by Elmatad et al. [36] versus the traditional VTF/WLF temperature dependence. The quadratic temperature dependence of $loga_T$ is of potential significance if a fundamental basis for its origin can be established, because it does not have a singularity at a finite temperature like the VTF/WLF expression. The VTF/WLF temperature dependence above Tg has been observed for a wide variety of glass-forming polymers [1] and the Arrhenian-like temperature dependence of $loga_T$ for equilibrated material below conventional Tg has also been observed for various compounds, including small-molecule glass formers selenium [37], glycerol, m-toluidine, and sucrose benzoate [38,39], and for polymeric glass formers PC [40], PS [41], and PVAc [42].

The linear viscoelastic response of a glassy material below Tg that has not been equilibrated often accommodates superposition of the different isotherms (at least for the limited range of time/frequency that is experimentally available). The underlying assumption is that at temperatures sufficiently below Tg, the rate of physical aging during the deformation is negligible so that the isotherms are for a constant age material. The resulting $loga_T$ temperature shift factor typically follows an Arrhenian dependence (that is different from the Arrhenian dependence for the equilibrium material described in the previous paragraph). The linear viscoelastic compliance response becomes stiffer with sub-Tg aging as shown in Figure 4.7 for PS, where the $J(t; T, t_{ae})$ compliance curves on a log time plot can be shifted to effect superposition in a manner reminiscent of time-temperature superposition.

The time/aging time shift factor $loga_{t_e}$ for PC is shown in Figure 4.8 for a series of sub-Tg annealing temperatures. As the annealing temperature decreases from 135°C (Tg-6°C) to 124°C (Tg-17°C), it takes progressively longer aging time to reach the saturated, that is, equilibrium, value of $loga_{t_e}$. Although not often

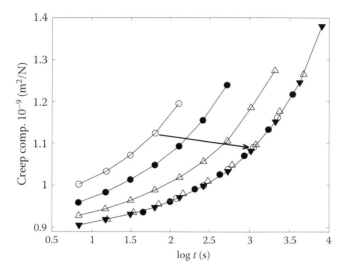

Figure 4.7 Time-aging time superposition of torsional creep compliance isotherms for PS quenched from 100°C to 50°C. Aging times are 0.17 h—open circles, 0.53 h—filled circles, 1.95 h—open triangles, and 5.4 h—filled triangles. The master curve is obtained by shifting to the 5.4 h data. Note that shifting is both vertical and horizontal as indicated by the arrow. (Adapted from L. C. E. Struik, *Polym. Eng. Sci.*, 17, 165, 1977.)

reported, a vertical shift is also needed to affect superposition of the compliance isotherms. Similar time/aging time superposition has also been observed in analyzing linear viscoelastic stress relaxation for PC [44] and an epoxy polymer [45]; and, also for dynamic linear viscoelastic data for PC, PMMA [46], and poly(arylene-etherimide) [47]. However, in case of the dynamic mechanical analysis, the shifting with $loga_{te}$ generally failed to effect simultaneous superposition of the storage and loss moduli.

Time–temperature superposition implicitly assumes that there is no change in the shape of the relaxation spectrum with temperature, that is, thermorheological simplicity, although a $\rho T/\rho_o T_o$ vertical shift is typically employed to account for the purely elastic temperature dependence of the polymer chains [10]. McCrum et al. [48] pointed out that the $\rho T/\rho_o T_o$ vertical shift assumes that the temperature dependencies of relaxed and unrelaxed compliances/moduli are the same, where they proposed a procedure to account for the more general case. McCrum and Pogany [49] were able to obtain reasonable superposition both of rubbery and glassy linear

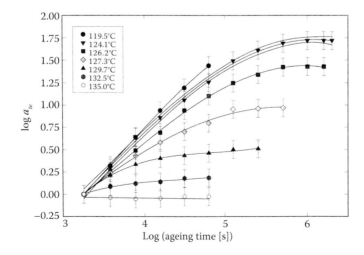

Figure 4.8 Temperature and aging time dependence of *logate* shift function for PC (Tg = 141°C). (Reprinted with permission from P. A. O'Connell and G. B. McKenna, *J. Chem. Phys.*, 110, 11054. Copyright 1999, American Institute of Physics.)

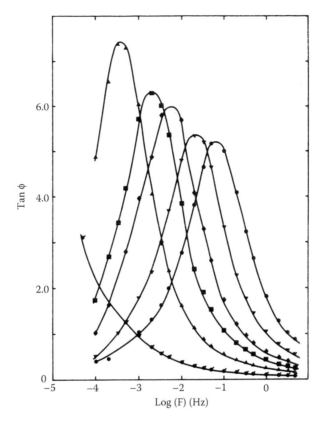

Figure 4.9 tan δ isotherms for atactic PS with a Tg of 96°C at 86.6°C, 91.4°C, 94.4°C, 95.9°C, 98.5°C, and 100.8°C (from right to left). (J. Y. Cavaille et al., *J. Polym. Sci. Polym. Phys.*, 1987, 25, 1235. Copyright Wiley-VCH Verlag GmbH & Co. KGaA. Reproduced with permission.)

viscoelastic compliance isotherms for an epoxy resin, using the McCrum–Morris procedure. However, more careful analysis of the linear viscoelastic behavior in the glass transition region indicates that thermorheological simplicity may be only an approximation. The dynamic modulus was determined by Cavaille et al. [50] for atactic polystyrene in the glass transition region, where G' and G'' isotherms appeared to superpose at least on the log(modulus) versus log(frequency) plot; however, when using the more sensitive measure of *tan*δ, there is clear lack of superposition as shown in Figure 4.9. A similar lack of superposition of the glassy *tan*δ isotherms was observed for an epoxy resin [51]. Unfortunately, the thermal history was not specified in Cavaille et al. [50] and Mikolajczak et al. [51] and there is a concern that samples were not fully equilibrated prior to the test, which in turns leaves the possibility that the material was aging during the course of the experiment, since all the experiments were very near Tg and the lowest frequency used was 10^{-4} Hz. The appearance of thermorheological complexity when approaching Tg region has also been observed in creep compliance measurements by Plazek for PS [52,53]. More recently, lack of superposition of the glassy *tan*δ isotherms near Tg was observed by Guo et al. [54] and Tao and Simon [55]. In summary, the principle of time–temperature superposition with the implicit assumption that the shape of the relaxation spectrum is invariant can be used as a zero-order approximation when analyzing time-dependent modulus/compliance isotherms and the storage/loss modulus/compliance data; however, more careful examination clearly shows that glassy materials are really thermorheologically complex.

4.2.2 Volumetric response

Most studies of linear viscoelastic behavior have been in uniaxial extension/compression or in shear; however, there are a few studies of the linear viscoelastic bulk modulus. McKinney and Belcher [56] measured the dynamic bulk compliance for poly(vinyl acetate) over a range of temperatures, pressures, and frequencies, where the data were analyzed via reduced variables method (i.e., time–temperature superposition with the addition of pressure). Volume relaxation following pressure jumps in PS was studied by Goldbach and Rehage [57] and also by Tribone et al. [58], although in the latter case most of the data is for large temperature jumps that are in the nonlinear region. Meng and Simon [59] have measured the transient bulk compliance of PS, where the linear viscoelastic compliance isotherms could be shifted to form a master curve as shown in Figure 4.10. The bulk response changes by only a factor of 3 as the material relaxes from the glass to the equilibrium liquid versus the compliance/modulus response in shear or in uniaxial deformation, where the change can be three or four orders-of-magnitude. In addition, the bulk relaxation response occurs over approximately five logarithmic decades as compared to the shear/tensile response that occurs over 15 or more logarithmic decades. Nevertheless, as shown in Figure 4.11, the $\log a_T$ shift factor needed to construct the bulk master curve is the same as those determined from shear compliance studies [60] using the same PS.

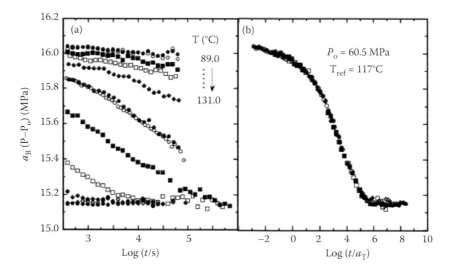

Figure 4.10 Dynamic bulk modulus isotherms (a) and associated master curve (b) for PS. Isotherms are at a nominal pressure of 60.5 MPa. (Y. Meng and S. L. Simon, *J. Polym. Sci. Polym. Phys.*, 2007, 45, 3375. Copyright Wiley-VCH Verlag GmbH & Co. KGaA. Reproduced with permission.)

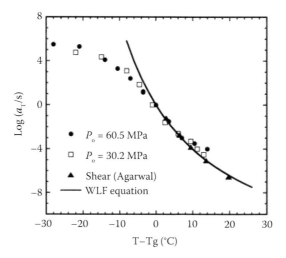

Figure 4.11 Shift factors used in constructing shear and bulk master curves shown in Figure 4.10 for PS. (Y. Meng and S. L. Simon, *J. Polym. Sci. Polym. Phys.*, 2007, 45, 3375. Copyright Wiley-VCH Verlag GmbH & Co. KGaA. Reproduced with permission.)

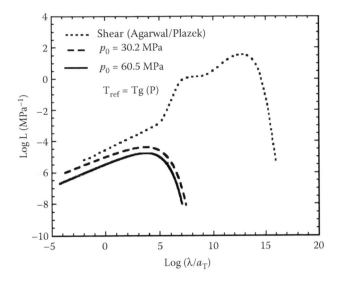

Figure 4.12 Shear and bulk retardation spectra computed from shear and bulk master curves for PS, where the bulk spectra are at the indicated hydrostatic pressures. (Y. Meng and S. L. Simon, *J. Polym. Sci. Polym. Phys.*, 2007, 45, 3375. Copyright Wiley-VCH Verlag GmbH & Co. KGaA. Reproduced with permission.)

The retardation spectra for both the bulk and shear compliance master curves are shown in Figure 4.12. The retardation spectrum at short times is very similar for the bulk and shear; however, at longer times, the shear spectrum is significantly broader than the bulk spectrum, where the latter has no contributions at longer times. It has been speculated that in the glass (i.e., at retardation times less than 10^5 in Figure 4.12), the relaxation processes for bulk and shear are quite similar; whereas, at long retardation times, the Rouse modes associated with segmental rearrangement become active for the shear deformation, but they make no significant contribution to the bulk viscoelastic response. The bulk and shear moduli have also been measured by Guo et al. for star PS [54] and by Tao and Simon for silica nanoparticle-filled PS [55] where the results are in qualitative agreement with those for linear PS shown in Figure 4.12. Deng and Knauss [61] and Sane and Knauss [62] studied viscoelastic bulk modulus for PVAc and PMMA and compared it to shear modulus for these materials, where they also observed that the shear relaxation had contributions of much longer relaxation times than the bulk relaxation response.

Recently, Hecksher et al. [63] measured the dynamic bulk and shear moduli for the small-molecule glass formers, where they found no difference between bulk and shear relaxation response. If confirmed, this result is important, because it addresses the question concerning the origin of the shear relaxation response at longer times; specifically, must the material be polymeric to exhibit long time relaxation in shear but not in bulk. However, there is a question if the dynamic experiment used in Hecksher et al. [63] is accurate at long times (or low frequencies). Specifically, Zondervan et al. [64] observed that a small-molecule glass former above Tg exhibited solid-like behavior at very low shear stresses, but these solid-like structures were easily destroyed at larger (but still very small) stresses. So, perhaps, there is a longer time component in shear relaxation of small-molecule glasses similar to polymeric glasses if the deformation is truly infinitesimal, but where the structure leading to the long time response is destroyed for small, but not infinitesimal, deformations.

The volumetric response in uniaxial extension is described using Poisson's ratio, ν. In the rubbery state, the material is nearly incompressible with $\nu = 0.5$; however, in the glassy state, ν is typically between 0.2 and 0.4. The temperature dependence of ν in the glassy state has been measured for a variety of polymers as reviewed in Tschoegl et al. [65], where it was observed to increase with temperature. Dynamic measurements of Poisson's ratio were conducted for PMMA [65] and epoxy polymers [66,67], where the frequency response $\nu(\omega)$ is qualitatively consistent with the time response $\nu(t)$. The linear viscoelastic lateral contraction ratio master curve for an epoxy resin is shown in Figure 4.13. The lateral contraction ratio in a linear viscoelastic creep experiment is analogous to Poisson's ratio in a stress relaxation experiment, where it increases from 0.4

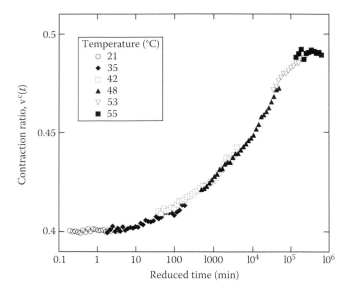

Figure 4.13 Lateral contraction ratio master curve for an epoxy resin at a reference temperature of 21°C (Tg-29°C). Isotherms are shifted with the same $loga_T$ used to construct the tensile compliance master curve. (With kind permission from Springer Science+Business Media: *Exp. Mech.*, 47, 2007, 237, D. J. O'Brien, N. R. Sottos, and S. R. White.)

to 0.49. Theocaris and Hadjijoseph [66] measured Poisson's ratio in both stress relaxation, that is, $\nu_R(t)$ and in creep, that is, $\nu_C(t)$, where they observed that (i) both responses started at 0.364 and went to 0.485 at long time and (ii) $\nu_C(t)$ reached the long time limit value before $\nu_R(t)$.

4.2.3 LINEAR VISCOELASTIC THERMAL EXPANSION RELAXATION

Glasses also exhibit linear viscoelastic behavior when subjected to a thermal perturbation as shown in Figure 4.14 for polystyrene that is at equilibrium prior to the step change in temperature. The isobaric volume relaxation following small temperature jumps for PS is essentially linear with respect to the magnitude

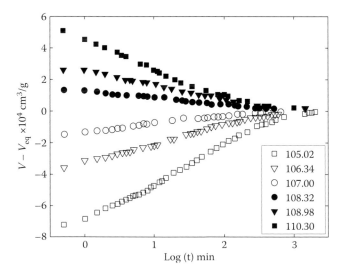

Figure 4.14 Isobaric volume relaxation of PS from indicated initial temperatures to a final temperature of 107.66°C. (Adapted from G. Goldbach and G. Rehage, *Rheologica Acta*, 6, 30, 1967.)

of the temperature jump if this magnitude, ΔT, is less than $\pm 1°C$, where significant nonlinearity appears when $\Delta T \geq \pm 1°C$. Examining Equation 4.2, for an isotropic, isobaric deformation of a material initially at equilibrium, only the thermal history integral is nonzero; consequently, for these conditions, the $\alpha(t)$ linear viscoelastic material property can be determined from the data as long as it is linear. The width of the $\alpha(t)$ relaxation response is relatively narrow, that is, it is only three to four logarithmic decades in width similar to the bulk modulus discussed in Section 4.2.2. Linear viscoelastic, volume relaxation data have also been measured for an epoxy resin (when $\Delta T \leq \pm 2.5°C$) and a relatively narrow thermal expansion retardation spectrum was also observed [19].

One must be careful in analyzing temperature jump experiments, because (as pointed out in the introduction to Section 4.2) interpretation of temperature jump data for a nonequilibrium glass is not straightforward; specifically, when temperature jumps are large, significant nonlinearity can occur, which is the subject of Section 4.3.

SIGNATURES 2

Signatures of the linear viscoelastic behavior of the polymeric glasses include:

S2.1   Time–temperature superposition of linear viscoelastic isotherms works well above Tg and reasonably well below Tg, at least for the range of times/frequencies that are usually employed experimentally, assuming that there are no sub-Tg relaxation processes.

S2.2   The temperature dependence of the $loga_T$ shift function for an *equilibrated* material above Tg is super-Arrhenian; however, for *equilibrated* material below conventional Tg, the $loga_T$ shift function is Arrhenian or possibly quadratic with respect to inverse temperature. Above Tg, the empirical VTF, WLF, and quadratic equations all describe the data equally well.

S2.3   Sub-Tg annealing causes the time-dependent linear viscoelastic compliance/modulus response to shift to longer times, i.e., the material becomes stiffer. The time-aging time shift function a_{te} depends upon both the sub-Tg annealing time and temperature, where $loga_{te}$ increases linearly with log(aging time) until the material reaches equilibrium.

S2.4   The shape of the linear viscoelastic material functions and associated relaxation spectra change slightly with temperature, i.e., glassy materials are thermorheologically complex. For the relatively limited time/frequency range that is experimentally convenient, thermorheological complexity maybe difficult to detect.

S2.5   The bulk compliance master curve can be constructed using time–temperature superposition, where the $loga_T$ shift function is the same as that observed for shear modulus isotherms.

S2.6   The bulk compliance retardation spectrum is similar to the short retardation time spectrum observed for shear deformation. However, for polymeric glasses, there are significant contributions to the shear retardation spectrum at long times; contributions that are absent in the long time bulk retardation spectrum.

S2.7   Poisson's ratio exhibits linear viscoelastic relaxation from its glassy value to the equilibrium value of nearly 0.5, where the width of the relaxation response is consistent with the bulk relaxation response. Poisson's ratio, which is associated with a stress relaxation experiment, relaxes slower than the analogous lateral contraction ratio associated with a creep deformation.

S2.8   The linear viscoelastic coefficient of thermal expansion for a fully equilibrated material can be determined from very small temperature jumps under isobaric conditions, where the width of the relaxation response has approximately the same width as the bulk modulus/compliance. The volumetric response rapidly becomes nonlinear as the magnitude of the temperature jump increases.

4.3 NONLINEAR VOLUME RELAXATION

In the classic study by Kovacs [68], the isobaric volume relaxation for a series of nearly step change thermal histories has been measured for poly(vinyl acetate), PVAc. In Figure 4.15, the intrinsic isotherms for a series of

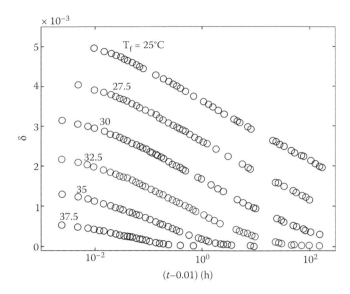

Figure 4.15 Intrinsic isotherms of PVAc volume relaxation following a temperature jump from 40°C to the indicated final temperature. (Adapted from A. J. Kovacs, *Fortschr. Hochpolym.-Forsch.*, 3, 394, 1963.)

temperature down-jumps from 40°C to temperatures from 37.5°C to 25°C are shown. Over a significant portion of the relaxation process $\delta = (V - V_\infty)/V_\infty$, that is, the normalized departure from the equilibrium volume V_∞, is a linear function of log(time). If the behavior is to be described via a first-order differential equation, the linearity of δ versus log(time) implies a very significant nonlinearity in the dependence of the relaxation time on δ; specifically, Kovacs showed that the intrinsic isotherms in Figure 4.15 could be described using an exponential integral [68]. Recently, a different approach that incorporates nanoscale fluctuations has been developed that is also able to quantitatively describe the intrinsic volume relaxation isotherms (see Chapter 14) [69,70].

The nonlinearity of the volume response is clearly shown when up- and down-temperature jumps of equal magnitude are compared as shown in Figure 4.16. Two features are apparent in the specific volume response:

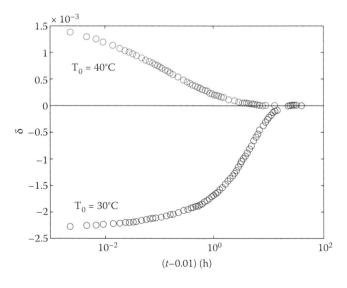

Figure 4.16 Asymmetry of approach for PVAc after a temperature down-jump from 40°C to 35°C (upper curve) and after a temperature up-jump from 30°C to 35°C (lower curve). (Adapted from A. J. Kovacs, *Fortschr. Hochpolym.-Forsch.*, 3, 394, 1963.)

first, the initial departure from equilibrium for the temperature up-jump is greater than for the down-jump; and second, the shape of approach to equilibrium is asymmetric. Specifically, for the temperature down-jump, the volume smoothly relaxes toward equilibrium at an increasingly slower rate; in contrast, the rate of volume relaxation after an up-jump is very slow for a long period of time and then rapidly accelerates before slowing down again as equilibrium is approached. If the volumetric response was linear, the up-jump and down-jump responses would be mirror images, like in the linear volume relaxation behavior shown in Figure 4.14 for $\Delta T \leq 1°C$ [57]. The difference in the initial departure seen in Figure 4.16 is a consequence of the rapid relaxation for a material that is initially at a high temperature when it is cooled at a finite rate into the glass, where no difference in the initial departures would be expected if the cooling was instantaneous.

Single temperature jumps for a material initially at equilibrium expose some features of the nonlinear volume relaxation, but multiple-step experiments provide a more challenging probe of relaxation in the glassy state. The volume relaxation following a temperature down-jump, annealing below Tg for various times and an up-jump back to the original temperature are shown in Figure 4.17 for the same PVAc material shown in Figures 4.15 and 4.16. Although the relaxation response looks relatively benign, the modeling of short time annealing results has been quite difficult (see Section 14.6 in Chapter 14). A visually more interesting experiment is the memory experiment shown in Figure 4.18. The thermal history is as follows: the material is (i) initially at equilibrium, (ii) quenched into the glass, (iii) allowed to anneal where the volume relaxes at a constant temperature, (iv) a temperature up-jump is applied such that the volume immediately following the up-jump is the equilibrium volume at the final temperature, and (v) the specific volume evolution is then measured. As shown in Figure 4.18b, the specific volume does not remain constant, but first increases and then decreases to the equilibrium value. The "memory effect" in volume relaxation is a signature of polymeric glasses, although it is not a particularly discriminating experiment because even a linear viscoelastic material with multiple relaxation times will exhibit memory-like behavior. This illustrates an important point when considering a critical study of the glassy state: it may be possible for a model to describe a single experiment or even several of experiments, but when one considers a larger set of experiments, even a relatively innocuous-looking result, for example, the short time annealing response in Figure 4.17, can become very discriminating with respect to potential candidate models.

The final signature of volume relaxation in glassy polymers is the tau-effective (or equivalently the expansion gap) paradox. Tau-effective is the normalized rate of relaxation defined by $\tau_{eff}^{-1} = -(1/\delta)d\delta/dt$. In Figure 4.19, τ_{eff} is plotted as a function of δ for a series of temperature up-jumps and down-jumps. Focusing

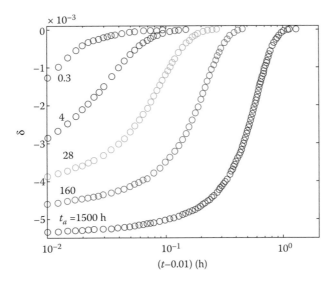

Figure 4.17 Annealing experiments for PVAc. Volume relaxation following a temperature down-jump from 40°C to 25°C, annealing at 25°C for the time indicated in the figure, and a temperature up-jump to the initial temperature of 40°C. (Adapted from A. J. Kovacs, *Fortschr. Hochpolym.-Forsch.*, 3, 394, 1963.)

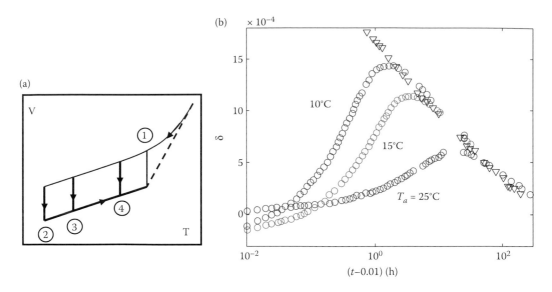

Figure 4.18 Memory experiment for PVAc. (a) Volume response for various thermal histories (thick solid lines): (1) temperature down-jump from 40°C to 30°C, volume relaxation at 30°C; (2) temperature down-jump from 40°C to 10°C, annealing for 160 h, temperature up-jump to 30°C, and volume relaxation at 30°C; (3) temperature down-jump from 40°C to 15°C, annealing for 140 h, temperature up-jump to 30°C, and volume relaxation at 30°C; and (4) temperature down-jump from 40°C to 25°C, annealing for 90 h, temperature up-jump to 30°C, and volume relaxation at 30°C. (b) Volume responses versus time: triangles—(1), red circles—(2), green circles—(3), and blue circles—(4). (Adapted from A. J. Kovacs, *Fortschr. Hochpolym.-Forsch.*, 3, 394, 1963.)

on the up-jumps to 40°C on the left side of Figure 4.19, as δ approaches zero, τ_{eff} is different by up to a factor of 4 depending upon the initial temperature, even though at the end of the relaxation, both the temperature and the specific volume are the same (at least within $\delta \sim 1.6 \times 10^{-4}$). This difference in τ_{eff} for different thermal histories, but at the same final temperature and essentially the same volume, is called the τ_{eff} expansion gap paradox. Specifically, this is an apparent paradox, because if the rate of relaxation depends upon only the current volume and temperature, then, τ_{eff} as $\delta \to 0$ should be the same irrespective of the history. The significant

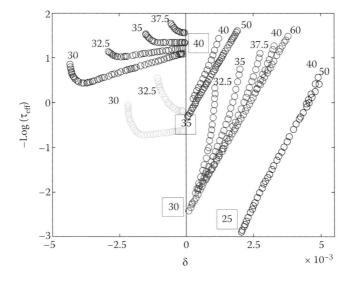

Figure 4.19 τ-effective expansion gap for PVAc in the glass transition region. Initial temperatures (in °C) for up- and down-jumps are indicated at the beginning of each curve. Final temperatures (in °C) are indicated in boxes. (Adapted from A. J. Kovacs, *Fortschr. Hochpolym.-Forsch.*, 3, 394, 1963.)

difference in τ_{eff} at very small δ for different histories is an important signature of volume relaxation in polymeric glasses. The τ_{eff} expansion gap has been extremely difficult to model [71], to the point that the validity of the Kovacs' data has been questioned [72]; however, McKenna subjected Kovacs' original data as well as some unpublished data from the Kovacs notebooks to rigorous statistical analysis and concluded that the data clearly support the τ_{eff} paradox down to $\delta = 1.6 \times 10^{-4}$ [73]. The τ_{eff} paradox was also observed by Kolla and Simon for an epoxy polymer [74].

The Kovacs data shown in Figures 4.15 through 4.19 are for PVAc [68] and it is the most complete and precise measurements of volume relaxation in polymeric glasses. However, there are other data sets that also support the general nature of the signatures of volume relaxation described above. In a separate study, Kovacs measured the intrinsic isotherms for a slightly different PVAc material as well as for PS, obtaining similar results to the ones described above [75]. Delin et al. [76] also measured the volume relaxation of PVAc, where they observed intrinsic isotherms similar to that in Figure 4.15 and a single up-jump experiment that shows similar nonlinearity and asymmetry as that shown in Figure 4.16. Goldbach and Rehage [57] performed an extensive set of up-jumps and down-jumps for uncross-linked and cross-linked PS, although the temperature jumps were over a smaller range than that in the Kovacs data. They observed (i) the expected linear response for small temperature jumps and (ii) the emergence of asymmetry and nonlinearity. Greiner and Schwarzl [77] investigated volume relaxation in a commercial PS, where they reported (i) an extensive set of temperature down-jump data with similar behavior as shown in Figure 4.15 and (ii) three up-jumps from an equilibrated material below Tg that were similar in shape to the up-jump shown in Figure 4.16. McKenna and coworkers [78] investigated volume relaxation in PC, where they clearly observed nonlinearity and asymmetry of approach and a small, but noticeable, expansion gap [79]. In a careful study, Bero and Plazek [19] measured the volume relaxation in the glass transition region for an epoxy thermoset system where they observed nonlinearity and asymmetry, although the effects were not as large as that observed by Kovacs [68]. The specific volume relaxation features reported in Figures 4.15 through 4.19 are universal signatures of the glassy state.

SIGNATURES 3

S3.1 *Intrinsic volume relaxation*: Isotherms from temperature down-jumps are linear with respect to log time over a significant portion of the relaxation response.

S3.2 *Nonlinearity*: The initial departure from equilibrium for temperature jumps of the same magnitude is larger for up-jump experiments than down-jumps. This is a consequence of fast relaxation during the finite time needed to impose the temperature change in the down-jump experiments.

S3.3 *Asymmetry*: The relaxation from a down-temperature jump is not the mirror image of that from an up-jump of equal magnitude. For a down-jump, the relaxation rate decreases monotonically with increasing time versus the up-jump where the relaxation rate is initially quite small, but then, there is an acceleration of the rate of relaxation prior to the decrease in relaxation rate as the material approaches equilibrium.

S3.4 *Memory*: When an equilibrium material is quenched in the glass, annealed below Tg, and rapidly reheated to a temperature such that the instantaneous specific volume upon the reheating equals the equilibrium specific volume at that temperature, the volumetric response is not constant but rather first increases and then relaxes back to equilibrium.

S3.5 *τ-Effective expansion gap*: The effective rates of volumetric relaxation for a series of temperature up-jumps to the same final temperature depend upon the initial temperature even when deviation from the final volume is very small.

4.4 ENTHALPY RELAXATION

4.4.1 LINEAR VISCOELASTIC RESPONSE

The frequency dependence of the enthalpic response to a small temperature variation has been probed using specific heat spectroscopy [80]. Specific heat spectroscopy measures the product of $\rho \kappa C_p$, where ρ

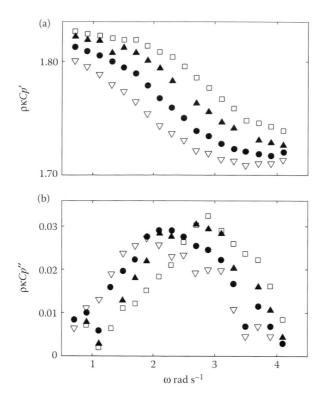

Figure 4.20 In-phase (a) and out-of-phase (b) dynamic heat capacity (units—J^2 cm^{-4}s^{-1}K^{-2} × 10^3) for PS with a Tg = 103°C. Isotherms are 110.1°C—open triangles, 113.3°C—circles, 116.2°C—filled triangles, and 119.2°C—squares. (Adapted from J. Korus et al., *Thermochimica Acta*, 304/305, 99, 1997.)

is the density, κ is the thermal conductivity, and C_p is the heat capacity. If it is assumed that ρ and κ are independent of frequency (although they can be a function of temperature), then, the frequency-dependent C_p can be extracted from the data. The dynamic heat capacity of PS is shown in Figure 4.20. Although the isotherms in Figure 4.20 appear similar, time–temperature superposition generally does not hold as evidenced by the fact that in case of the imaginary part of $\rho\kappa C_p$, that is, $\rho\kappa C_p''$, the peak on the 119.2°C curve (open squares) is higher than that on the 110.1°C curve (open triangles). The heat capacity response occurs over approximately four logarithmic decades, which is similar to the bulk modulus response for PS shown in Figure 4.10 that occurs over approximately five logarithmic decades. Thus, the linear viscoelastic heat capacity is quite similar to the viscoelastic bulk modulus; although there appears to be a small difference in the widths of the relaxation spectra, where additional study is needed to determine if this difference is real. Specific heat spectroscopy has been used to study a number of glass-forming polymers including styrene–butadiene rubber [82], poly(isobutylene) [82], PC [83], and poly(ethylene 1,4-cyclohexylenedimethylene terephthalate glycol) [83]. The pressure dependence of the dynamic heat capacity was determined for the small-molecule orthoterphenyl from 0.1 to 100 MPa, where temperature and pressure dependence of the mean relaxation time agrees with that observed via viscosity measurements [84]. To the best of our knowledge, the effect of sub-Tg aging on the linear viscoelastic heat capacity as determined by specific heat spectroscopy has not been measured.

4.4.2 NONLINEAR ENTHALPIC RELAXATION

The enthalpic response of polymer glasses can be probed by differential scanning calorimetry (DSC) using readily available commercial instruments; consequently, there are large numbers of DSC studies of relaxation in polymeric glasses which have been discussed in several reviews [9,85]. The temperature changes in these experiments are large; thus, the response is highly nonlinear. In this section, we will highlight the important features of enthalpy relaxation with no attempt to be complete due to the size of this literature.

Enthalpy relaxation is similar to volume relaxation, because both enthalpy and volume are first-order derivatives of the underlying free energy. The DSC does not measure enthalpy, but rather the change in heat ΔQ in an increment of time Δt as a sample is heated or cooled at a rate $\dot{T}$, where the instantaneous heat capacity C_p is defined as

$$\frac{\Delta Q}{\Delta t} = \frac{dQ}{dt} = \frac{dQ}{dT}\frac{dT}{dt} = C_p\dot{T} \tag{4.3}$$

$C_p(t)$ is a function of time that depends upon the thermal history used to form the glass as well as the instantaneous heating rate. Consequently, $C_p(t)$ as determined via DSC is non-isothermal property that is more difficult to interpret than the isothermal volume, stress relaxation, and creep properties discussed elsewhere in this chapter. Nevertheless, the effect of thermal history on $C_p(t)$ does provide an important view on the enthalpic relaxation processes that occur in glass-forming materials. Modulated DSC is an experimental method where a small sinusoidal temperature oscillation is applied on top of the larger temperature history and is a popular technique now that this capability is available on commercial DSC instruments. Even though the temperature oscillations are small, the effects are inherently nonlinear, because the underlying thermal history involves large temperature changes that induce nonlinear effects [85]; consequently, interpretation of modulated DSC is difficult and will not be considered further in this chapter.

Typical DSC traces upon heating are shown in Figure 4.21 for a poly(ether imide), PEI. For the unaged material, there is just an endothermic change in the heat flow upon heating through the glass transition; in contrast, for a sub-Tg- aged glass, a substantial heat flow peak is observed, where the difference in peak areas is the enthalpy difference between an aged and unaged glass. The heat flow peak increases with the sub-Tg annealing time as shown in Figure 4.21 and also with a decrease in cooling rate as shown in Figure 4.22. Note that the peak height and position also change with the heating rate (at fixed cooling rate, annealing time, and temperature).

In addition to the enthalpy response observed during constant rate heating and cooling, the characteristic features observed by Kovacs for volume relaxation such as nonlinearity, asymmetry of approach, and memory (see Section 4.3) have recently been observed for enthalpy for PS by Simon and coworkers (see Chapter 2).

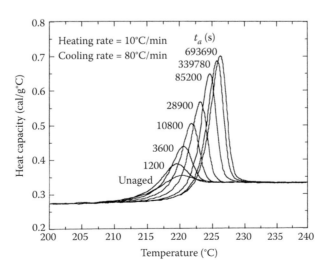

Figure 4.21 DSC heating scan for PEI cooled from 260°C at 80°C/min to 201.3°C (Tg-6°C) and annealed for indicated times. (I. Echeverria et al., *J. Polym. Sci. Polym. Phys.*, 1995, 33, 2457. Copyright Wiley-VCH Verlag GmbH & Co. KGaA. Reproduced with permission.)

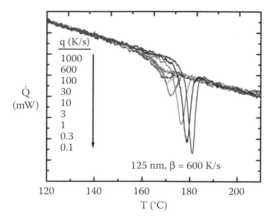

Figure 4.22 Flash DSC heat flow for a 125 nm thick PC film. Cooling (q) and heating (β) rates are given in the figure. (N. Shamim et al., *J. Polym. Sci. Polym. Phys.*, 2014, 52, 1462. Copyright Wiley-VCH Verlag GmbH & Co. KGaA. Reproduced with permission.)

4.4.3 ENTHALPY VERSUS VOLUME RELAXATION

As described above, dynamic specific heat spectroscopy indicates that the time-dependent heat capacity response is qualitatively similar to that of specific volume when subjected to the same thermo-deformational conditions. Of particular importance is the question of the connection between volume and enthalpy relaxation, because both processes are simultaneously occurring when a sample is subjected to any given temperature history. There have been several reports where the times to reach equilibrium for enthalpy relaxation, volume relaxation, and creep diverge below Tg (see Reference 88 for references and discussion); however, Badrinarayanan and Simon demonstrated that if the enthalpy and volume are measured during cooling, where the nonlinear effects are less pronounced, then, the time to reach equilibrium is the same within experimental uncertainty [89] for enthalpy and volume relaxation (Figure 4.23). In addition, as shown in

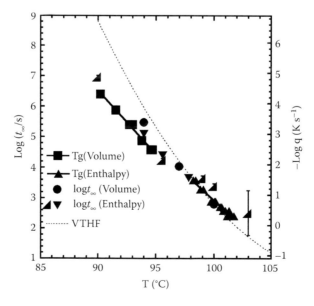

Figure 4.23 Time to reach equilibrium and the cooling rate dependence of Tg for enthalpy and volume relaxation for PS. (Reprinted from *Polymer*, 48, P. Badrinarayanan and S. L. Simon, 1464, Copyright 2007, with permission from Elsevier.)

Figure 4.23, the cooling rate dependence of Tg determined via enthalpy and volume is the same. However, one should remember that if the volumetric and enthalpic relaxation processes are nonlinear, then, there is no a priori physical reason that the two processes have to be exactly in sync [41] until the final approach to equilibrium.

SIGNATURES 4

S4.1   The width of the linear viscoelastic heat capacity response is similar to that observed for the bulk modulus.

S4.2   Upon heating from the glass, the specific heat capacity can exhibit a significant overshoot in the glass transition region as compared to the specific heat upon cooling that shows no maximum.

S4.3   With increased sub-Tg annealing and/or decreased cooling rate, (i) the magnitude of the heat capacity overshoot increases and (ii) the overshoot's peak temperature increases.

S4.4   Volume and enthalpy relaxation exhibit the same (i) time to reach equilibrium and (ii) cooling rate dependence of Tg.

4.5 COUPLED VOLUME AND MECHANICAL RELAXATION

All deformations are inherently three dimensional, for example, in uniaxial extension/compression, there is lateral deformation as given by Poisson's ratio (or equivalently the change in specific volume) in addition to the axial strain and stress. In most nonlinear deformations, the volume change induced by the deformation will overwhelm the volume relaxation that results from the formation of the glass; however, in an important experiment by McKenna and coworkers, the individual contributions of the thermally and deformation-ally induced volume changes were examined [90]. The experiment employed a specially designed torsional dilatometer [91], where a cylindrical specimen was subjected to a torsion deformation, while the length was held constant and the change in volume monitored by a mercury dilatometer. The torque, axial force, and specific volume were measured for an epoxy that had been quenched into the glass and then annealed for various lengths of time. Because the volume change resulting from a shear deformation is second order in the applied strain (vs. a uniaxial deformation where the volume change is first order in the axial strain), the volume change from the torsion is of the same order as that due to cooling the material into the glassy state. The thermo-deformational history employed was (i) the specimen was cooled into the glass, (ii) a twist was applied to the sample for a short period of time, (iii) the twist was reversed and the specimen was allowed to relax, (iv) sub-Tg annealing continued, and (v) a new cycle of twist, reversal of twist, and annealing was applied. Following the Struik protocol [92], the times for application of twist and twist reversal were a small fraction of the accumulated aging time. As shown in Figure 4.24, the volume response for a nearly unde-formed material (i.e., the blue triangles in Figure 4.24 corresponding to a small twist strain of 0.0025) exhibits the typical volume relaxation following a quench into the glass (see Section 4.3). At each annealing time, the volumetric response during the torsional deformation in Figure 4.24 shows two peaks corresponding to the twist and twist reversal, where after each twist, there is volume relaxation. What is significant is that after the torsional deformation, the specific volume rapidly returns to the underlying relaxation associated with just the thermal quench into the glass. This implies that the deformation contributions to the volumetric response are uncoupled from the thermal contribution to the volumetric response. The volumetric response is surprising considering that the shear-induced change in the specific volume is approximately of the same magnitude as the volume change due to quenching into the glass, where the twist strain of 3% is a significant fraction of yield [90].

The sub-Tg-aging time dependence of the torque response shown in Figure 4.25a can be shifted using time-aging time superposition to form a master curve; thus, the effect of physical aging on the shear response is in general what is expected for a glassy material (see Section 4.2.1). Note, however, that the applied strain of 3% used in Figure 4.25a is outside the linear range and as a result, the time-aging time shift factors needed to effect superposition are smaller than the time-aging time shift factors employed in case when the twist strains

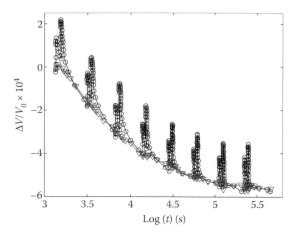

Figure 4.24 Specific volume relaxation at 33.5°C following a quench from 44.5°C for an epoxy glass DGEBA/D400 (Tg = 42.4°C) accompanied by back-and-forth twist strains of 0.25%—blue triangles and 3%—black circles; thin solid lines are a guide to the eye. V_0 is the specific volume immediately after the quench; $\Delta V = V - V_0$. (Reprinted from *Polymer*, 32, M. M. Santore, R. S. Duran, and G. B. McKenna, 2377, Copyright 1991, with permission from Elsevier.)

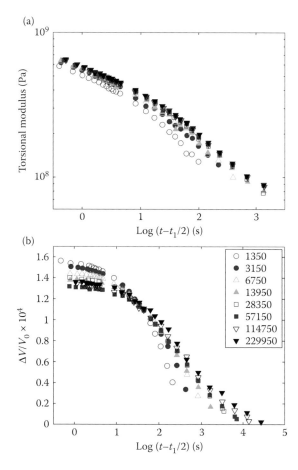

Figure 4.25 Torque (a) and specific volume relaxation (b) after twist of 3% of strain for torsional dilatometry experiment at 33.5°C following quench from 44.5°C, that is, the same experiment shown in Figure 4.24 as black circles. Aging times (in seconds) are given in the insert in (b). Time t_1 is 0.5 s. (Adapted from M. M. Santore, R. S. Duran, and G. B. McKenna, *Polymer*, 32, 2377, 1991.)

were less than 1%. Also, the torque curves in Figure 4.25a stopped evolving with physical aging after the aging times of approximately 10^4 s. The volume relaxation after the twist (Figure 4.25b) shows that unlike the torque response, the volume responses cannot be superposed by shifting along the time axis, where the curves corresponding to different annealing times cross. The volume relaxation response broadens with annealing and, unlike the torque response, continues to evolve as the annealing time increases. Despite these differences, the torque relaxation and the volumetric relaxation due to torque occur on a similar time scale of 10^4 s, which is strikingly different from the time scale of the volumetric relaxation due to temperature jump shown in Figure 4.24.

The torsional dilatometry data poses two key challenges:

1. Why does the volumetric response after application of the two twists always return to intrinsic volume relaxation as if no twist had been applied, if the rate of relaxation depends upon the volume?
2. If volume is the variable controlling mobility, then, the response along a constant ΔV line in Figure 4.24 must be the same for all situations whether for an undeformed material or for a material at the beginning, middle, or end of the relaxation as shown in Figure 4.25. However, this is not observed experimentally.

To the best of our knowledge, the torsional dilatometry data remains an unanswered challenge for the constitutive modeling of polymeric glasses.

SIGNATURES 5

S5.1 In torsional loading of a glass, the mechanically induced change in specific volume is uncoupled from the thermally induced change in specific volume, where the deformation-induced excess in specific volume relaxes orders-of-magnitude faster than the excess volume due to the thermal quench into the glass.

S5.2 The deformation-induced volume relaxation significantly broadens as the sub-Tg annealing progresses; in contrast, the associated torque relaxation has the same shape as the annealing time increases.

S5.3 The deformation-induced volume relaxation evolves as long as the underlying structural relaxation continues; in contrast, the torque response reaches its limiting response well before the structural relaxation ends.

4.6 NONLINEAR STRESS–STRAIN BEHAVIOR AND STRESS RELAXATION

Without question, the most studied thermo-mechanical behavior of polymeric glasses is the nonlinear stress–strain behavior, including the effect of deformation rate and thermal history on the nonlinear stress–strain curve. In this chapter, we are only concerned with the intrinsic material response, and thus will not discuss the effect of nonuniform kinematic fields on the mechanical response, for example, uniaxial extension experiments where necking was observed. Also, this chapter will not discuss ultimate failure, where the origin of the brittle-to-ductile transition is still a subject of intense study [15,16,93].

4.6.1 SINGLE-STEP CONSTANT STRAIN RATE

Polymeric glasses exhibit a generic nonlinear stress–strain curve as shown in Figure 4.26. Key features of the stress–strain curve include (i) an initial linear region followed by a slight curvature, that is, a nonlinear response, (ii) a yield point where the slope of the stress versus strain response is zero, that is, $d\sigma/d\varepsilon = 0$, (iii) a post-yield softening region where $d\sigma/d\varepsilon < 0$, (iv) a flow state where $d\sigma/d\varepsilon \approx 0$, and eventually (v) a hardening region where $d\sigma/d\varepsilon > 0$. Depending upon the temperature and deformation rate, the initial stress response may only appear to be linear, where the stress response will be curved if viscoelastic effects are present. The

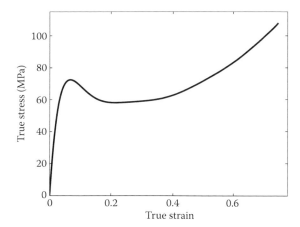

Figure 4.26 Illustration of the key features of the nonlinear stress–strain curve for a glassy polymer. PC at 23°C (Tg-124°C); uniaxial compression at strain rate of 10^{-3} s^{-1}. (Data from C. Dreistadt et al., *Mater. Des.*, 30, 3126, 2009.) The true stress is calculated assuming the deformation is isochoric.

nonlinear stress curve depends upon both the rate of deformation and the thermal history used to form the glass prior to deformation. The glassy stress–strain curve for a particular thermal history may not exhibit all of the features shown in Figure 4.26 [94], for example, if the material is rapidly quenched into the glass, there may be minimal post-yield softening, where the stress response directly transitions from yield into post-yield flow and/or hardening. The factors that affect the shape of the stress overshoot peak will be discussed in more detail in Section 4.6.3. The nonlinear stress–strain curve shown in Figure 4.26 is a ubiquitous feature of polymeric glasses having been observed for PMMA [95], PC [94], PS [96], poly(vinylchloride), PVC [97], poly(ethylene terephthalate), PET [98], poly(ethylene terephthalate)-glycol, PETG [98], poly(phenyleneoxide), PPO [99], and epoxy resins [100]. This is a partial list of the polymeric glasses that exhibit the nonlinear stress–strain behavior shown in Figure 4.26, where a more extensive compilation can be found in Bowden [11] and Crist [13].

Although the stress response shown in Figure 4.26 is generally reported as a function of strain, it is more appropriate to think of it as being a function of time, where time and strain are linearly related for a constant strain rate deformation (see Section 4.6.5, where the stress–strain curve for a multiple-step strain rate history is qualitatively different than that for a single-step constant strain rate history). For a constant strain rate deformation, the generic nonlinear stress–strain curve is observed in uniaxial extension [101], uniaxial compression [99], shear [100,102], and plane strain [98].

Yield has also been observed in the dilatationally dominated longitudinal deformation (as compared to most deformations that are deviatorically dominated) as shown in Figure 4.27, where the yield is identified with a change in the slope of the stress response and post-yield softening is not observed [103].

In uniaxial deformation, the axial stress is typically the only variable measured; however, the lateral strain (or equivalently the specific volume) also exhibits a time/strain- dependent response as shown in Figure 4.28. There is initially a volume increase followed by a decrease, which begins prior to the yield point. Qualitatively similar volumetric behavior was observed in uniaxial compression for PMMA, PS, PC, and poly(vinyl fluoride), PVF, where initially the volume decreased until yield and then remained constant [105]. The post-yield volumetric response is an important signature; specifically, even though the volume decreases or remains constant post-yield, the mobility appears to increase as indicated by the post-yield stress softening—a combination of responses that is difficult to resolve via simple free volume arguments.

4.6.2 Yield stress

The most commonly reported mechanical property for the constant strain rate deformation of glassy polymers is the yield stress, σ_y, and its dependence upon strain rate, temperature, and sub-Tg annealing. The temperature

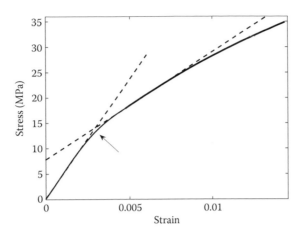

Figure 4.27 Stress–strain curve (black) for the longitudinal deformation of DGENG-4,4'MDA epoxy at a strain rate of 1.2×10^{-5} s^{-1} at 62°C (Tg-10°C). The arrow indicates yield. Linear extrapolation lines used to determine the yield point are dashed. (Adapted from J. W. Kim, G. A. Medvedev, and J. M. Caruthers, *Polymer*, 54, 2821, 2013.)

dependence of yield stress is shown in Figure 4.29a for PVC, where σ_Y decreases in a nearly linear manner with increasing temperature and goes to zero at Tg. The yield stress increases with the log(strain rate) as shown in Figure 4.29b. A typical strain rate dependence of σ_Y is nearly a linear function of log(strain rate) for up to four logarithmic decades until some curvature begins to appear at higher strain rates, which for most materials can only be accessed using the split Hopkinson bar experiments. However, for PMMA, the change in slope in the yield stress versus log(strain rate) occurs at strain rates accessible by standard test equipment [106]. The log(strain rate) dependence of yield stress has been modeled as two linear processes that have been associated with the α and β relaxation processes in polymeric glasses [107]. The temperature and log(strain rate) dependence of σ_Y have been measured for numerous polymeric glasses and extensively reviewed in Bowden [11].

The yield stress depends upon the sub-Tg annealing time as shown in Figure 4.30 for an epoxy polymer [100]. The yield stress increases linearly with the log(annealing time) until it levels off at a value associated with a fully equilibrated material. The leveling off of σ_Y with increase in sub-Tg annealing is possible to observe within experimentally accessible time only for the temperatures sufficiently close to Tg. Similar results were obtained for PC [108], where the yield stress saturated after annealing for less than 1 h at 145°C

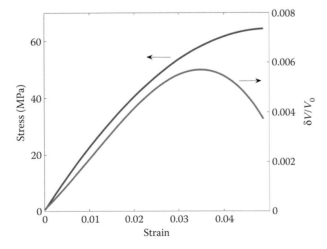

Figure 4.28 Specific volume response during uniaxial extension of PC at 25°C (i.e., Tg-122°C) at a constant strain rate of 7.3×10^{-5} s^{-1}. (Data from J. M. Powers and R. M. Caddell, *Polym. Eng. Sci.*, 12, 432, 1972.)

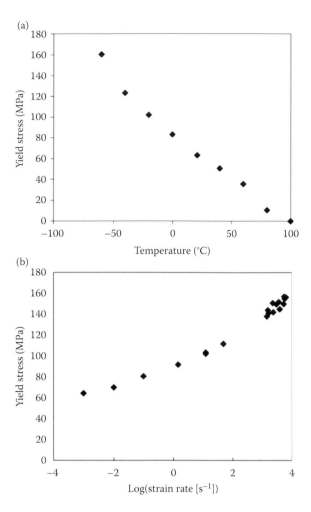

Figure 4.29 (a) Temperature dependence and (b) strain rate dependence of yield stress for PVC deformed in uniaxial compression. (a) Strain rate is 10^{-2} s^{-1}; (b) temperature is 20°C (Tg-60°C). Data at strain rates above 10^2 s^{-1} were obtained using a split Hopkinson pressure bar. (Reproduced from M. J. Kendall and C. R. Siviour, *P. Roy. Soc. A—Math. Phy.*, 470, 20140012, 2014, with permission from Royal Society Publishing.)

(Tg-2°C), but was still increasing linearly with log(annealing time) at 135°C (Tg-12°C) for the annealing times of up to 2.8×10^3 h. The annealing becomes less effective at temperatures significantly below Tg so that the slope of σ_Y versus log(annealing time) decreases from the maximum value reached at 130°C (Tg-17°C) by roughly a factor of 3 as the annealing temperature is decreased to 80°C (Tg-67°C) for PC [109]. Earlier works on the dependence of the yield stress on the sub-Tg annealing time are reviewed in Bowden [11].

The effect of hydrostatic pressure on yield is described via the pressure-modified von Mises criterion [110,111]

$$\tau_{oct} = \tau_{oct}^0 + \mu P \tag{4.4}$$

where the octahedral stress, σ_{oct}, and the hydrostatic pressure, P, are given in terms of the principal components of the stress tensor, σ_i, by

$$\tau_{oct} = 1/3[(\sigma_1 - \sigma_2)^2 + (\sigma_1 - \sigma_3)^2 + (\sigma_2 - \sigma_3)^2]^{1/2} \qquad P = -(\sigma_1 + \sigma_2 + \sigma_3)/3 \tag{4.5}$$

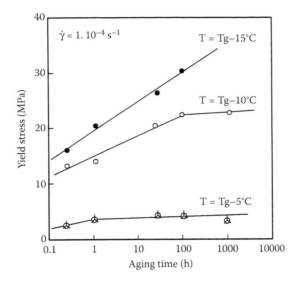

Figure 4.30 Effect of sub-Tg annealing time on σ_y in shear for DER332/Jeffamine epoxy (Tg = 42.4°C). (Reprinted from *J. Non-Cryst. Solids*, 172–174, M. Aboulfaraj et al., 615, Copyright 1994, with permission from Elsevier.)

τ_{oct}^0 is the octahedral yield stress at zero/atmospheric pressure. The yield surface described by Equation 4.4 in the principal stresses space is a cone as shown in Figure 4.31, where the cone apex is located at a negative pressure (i.e., the stress associated with an equal triaxial extension) of $P = -\tau_{oct}^0/\mu$. When $\sigma_3 = 0$, the yield stress locus is an ellipse in the $\sigma_1 - \sigma_2$ plane. Multiaxial yield has been studied using a thin-wall tube, where a combination of axial compression and internal pressure is applied. Using this method, the validity of the pressure-modified von Mises criterion in Equation 4.4 was established for PVC and PC [112]. The validity of the pressure-modified von Mises yield criterion has also been tested by plotting the uniaxial extension yield data obtained at elevated pressures as shown in Figure 4.32a for several glassy amorphous and semicrystalline polymers [113]. Thus, the linearity of the octahedral yield stress as a function of pressure as given in Equation 4.4 is established for PC in the pressure range from atmospheric pressure to a positive pressure of 800 MPa. Recently, the pressure range has been extended to negative pressures using a longitudinal deformation, also known as the poker chip test, where there is a very significant isotropic dilation [103]. As shown in Figure 4.32b, the linear pressure dependence of the modified von Mises criterion holds even for these large negative pressures.

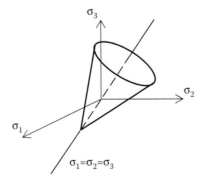

Figure 4.31 Schematic representation of yield envelope for the pressure-modified von Mises yield criterion.

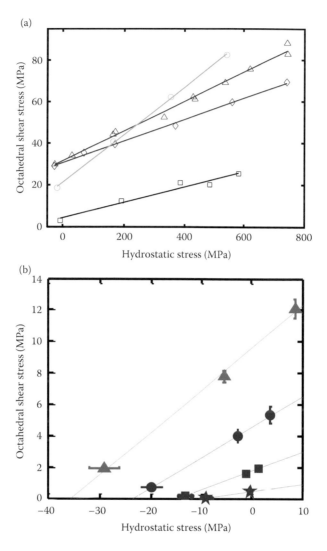

Figure 4.32 Evaluation of the pressure-modified von Mises yield criterion given in Equation 4.4 via the pressure dependence of the octahedral stress. (a) Yield in uniaxial extension at strain rate 1.7×10^{-4} s^{-1} and 23°C in the presence of hydrostatic pressure generated using castor oil as the pressure-transmitting fluid for semicrystalline poly(chlorotrifluoroethylene)—green circles, amorphous PC—red triangles, amorphous poly (ethylene terephthalate)—blue diamonds, and semicrystalline poly(tetrafluoroethylene)—black squares. Negative hydrostatic stress under 1atm pressure is due to dilation. (Adapted from A. W. Christiansen, E. Baer, and S. V. Radcliffe, *Philos. Mag.*, 24, 451, 1971, with permission from Taylor & Francis.) (b) Yield in longitudinal extension at strain rate 1.2×10^{-5} s^{-1} in for an epoxy resin DGENG-44'MDA (Tg = 72°C): stars—Tg-5°C; squares—Tg-10°C; circles—Tg-15°C; and triangles—Tg-20°C. (Reprinted from *Polymer*, 54, J. W. Kim, G. A. Medvedev, and J. M. Caruthers, 2821, Copyright 2013, with permission from Elsevier.)

4.6.3 POST-YIELD SOFTENING AND FLOW STRESS

Post-yield softening is an intrinsic feature of the nonlinear stress–strain response observed in a constant strain rate deformation. After the stress softening is complete, the stress becomes nearly constant, which is called either the flow stress or the lower yield. The flow stage ends when the stress begins to increase again, that is, exhibit strain hardening. As shown in Figure 4.33, the magnitude of the post-yield softening increases significantly with both strain rate [95] and the sub-Tg aging time [100]. With respect to strain rate, both the yield stress (i.e., the upper yield) and the flow stress (i.e., the lower yield) increase with strain rate, assuming

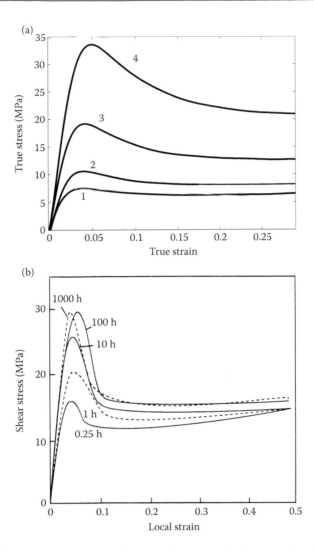

Figure 4.33 Effects of strain rate (a) and sub-Tg aging (b) on post-yield stress softening in a constant strain rate deformation. (a) PMMA at 110°C (Tg-5°C) in uniaxial compression with an annealing time of 1 h; strain rates are $1–3 \times 10^{-4}$ s^{-1}, $2–10^{-3}$ s^{-1}, $3–10^{-2}$ s^{-1}, and $4–10^{-1}$ s^{-1}. (Data from Ames, N.M., *Int. J. Plast.*, 25, 1495, 2009.) (b) DER 332 + Jeffamine epoxy resin at 32.2°C (Tg-10°C) at a shear strain rate of 5×10^{-4} s^{-1} with sub-Tg annealing times given in the figure. (Reprinted from *J. Non-Cryst. Solids*, 172–174, M. Aboulfaraj et al., 615, Copyright 1994, with permission from Elsevier.)

that the rest of the thermal history is constant. The increase in σ_Y with strain rate is larger than the increase in the flow stress with strain rate, resulting in an increase in the magnitude of the post-yield softening with strain rate as shown in Figure 4.34. The effect of the strain rate on the magnitude of post-yield strain softening is more pronounced at temperatures closer to Tg, as shown in Figure 4.33a for PMMA at Tg-5°C. Similarly, the effect of the strain rate on post-yield stress softening in PC is virtually unnoticeable at 25°C (Tg-122°C), but very large at 125°C (Tg-22°C) [101], where a similar trend was observed for PETG [98]. At higher strain rates (and at temperatures well below Tg, where the stress and hence the work of deformation is large), there may be adiabatic heating, which if present needs to be taken into account [12].

With respect to sub-Tg annealing, very little post-yield softening is observed for a sample quenched into the glassy state; in contrast, a very substantial overshoot peak emerges when the material is aged below Tg [13]. The effect of physical aging is most apparent when the material is aged at 10 to 30°C below Tg, although one anticipates that post-yield softening will emerge at lower temperatures if the sub-Tg annealing time is

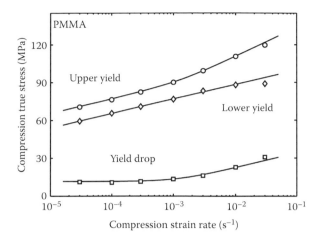

Figure 4.34 Effect of strain rate on the upper and lower yield for PMMA at 23°C (Tg-92°C). (L. C. A. van Breemen, *J. Polym. Sci. Polym. Phys.*, 2012, 50, 1757. Copyright Wiley-VCH Verlag GmbH & Co. KGaA. Reproduced with permission.)

long enough. Earlier works on the effect of sub-Tg aging on the post-yield softening for multiple materials and deformations are reviewed in Boyce and Haward [14], where the results are consistent with Figure 4.33b.

The post-yield flow stress increases with both a decrease in temperature [107] and an increase in strain rate as shown in Figures 4.33a and 4.34, but it is independent (within experimental uncertainty such as sample-to-sample variation, etc.) of the aging time as shown in Figure 4.33b and also in Senden et al. [108] and Hasan et al. [114]. It can be concluded that by the time the flow state is reached, the thermal history has been essentially erased.

4.6.4 STRAIN HARDENING

The final feature of the nonlinear stress–strain curve to be considered is the strain hardening that occurs if the specimen does not prematurely rupture. Post-yield hardening has been observed for a variety of polymeric glasses—PMMA, PC, and PVC in shear [102], PC, PS, PMMA, and PVC in uniaxial extension [101], and PMMA, PPO, PC, and PS in uniaxial compression [99], where additional examples and a review of earlier work are given in Haward [12] and Boyce and Haward [14]. Also, strain hardening was observed for PEGT and PET in plane strain compression [98], where the strain hardening was stronger than that observed in uniaxial compression. The strain-hardening modulus at least in uniaxial compression appears to increase with strain rate [115]. See Chapter 13 in this book for a more extensive discussion of strain hardening in polymeric glasses.

4.6.5 MULTISTEP STRAIN CONTROL EXPERIMENTS

Constant strain rate loading is the most common deformation history used to study the mechanical behavior of polymeric glasses; however, multistep experiments provide an additional perspective on the glassy state.

4.6.5.1 STRAIN RATE SWITCH EXPERIMENTS

Nanzai [116] systematically investigated the effect of changing strain rate during a uniaxial compressive deformation, where the strain rate during the second stage was either 100 times faster or 100 times slower than the strain rate during the first deformation stage. Of particular interest is the situation where the first stage proceeds to a strain large enough to produce both yield and post-yield softening. As shown in Figure 4.35, there is overshoot when the strain rate is increased by two orders-of-magnitude and a significant undershoot when the strain rate is decreased by two orders-of-magnitude. The data in Figure 4.35 were at 100°C (i.e., Tg-15°C), where qualitatively similar results were obtained at 80°C. The ability to predict this significant overshoot and undershoot is a challenge for constitutive models (see Chapter 14); thus, the strain rate switching experiments are an important thermo-mechanical signature of polymeric glasses.

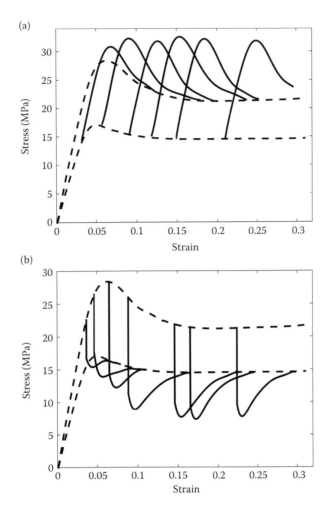

Figure 4.35 Two-stage loading experiments for PMMA in uniaxial compression at 100°C. Monotonic loading at a slow strain rate of 8.3×10^{-6} s^{-1}—bottom dashed line and at a fast strain rate of 8.3×10^{-4} s^{-1}—top dashed line. Switch from slow to fast strain rate (a) and from fast to slow strain rate (b) at various strains—solid lines. (Y. Nanzai, *Polym. Eng. Sci.*, 1990, 30, 96. Copyright Wiley-VCH Verlag GmbH & Co. KGaA. Reproduced with permission.)

4.6.5.2 LOADING–UNLOADING–RELOADING

The stress response of polycarbonate when subjected to cyclic loading–unloading–relaxation–reloading deformations [94] is shown in Figure 4.36, where the maximum strain just prior to unloading increases with each cycle. Key features of the response are:

1. The slope of the stress–strain response upon unloading is approximately the same as the slope of the pre-yield stress–strain response.
2. The unloading curve is smooth with no "second yield" whereas the latter is spuriously predicted by some constitutive models.
3. There is a significant permanent set that increases with the strain reached prior to unloading.
4. The initial tangent modulus upon reloading is essentially the same as in first loading.
5. Upon reloading, the stress approaches the underlying stress–strain curve for a single loading experiment (i.e., the "monotonic" curve in Figure 4.36) without any overshoot, indicating that memory of the unloading–reloading is completely erased.

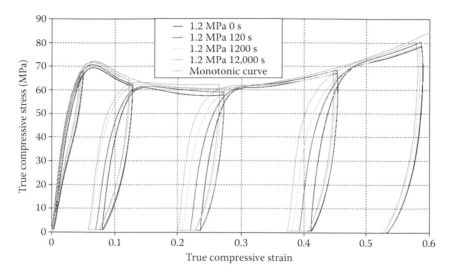

Figure 4.36 Cyclic compressive deformation of PC at 23°C (Tg-124°C). Loading strain rate—10⁻³s⁻¹, where loading continues up to true strains of 0.05, 0.13, 0.27, 0.45, and 0.59; unloading stress rate −2.3 MPa/s; and rest stress of 1.2 MPa, where the duration of the rest is given in the figure; loading resumes at the same strain rate of 10⁻³ s⁻¹. (Reprinted from *Mater. Des.*, 30, C. Dreistadt et al., 3126, Copyright 2009, with permission from Elsevier.)

Similar loading–unloading behavior has also been observed in uniaxial compression for PMMA at 23°C (Tg-92°C) with a strain rate of 10⁻³ s⁻¹ [114].

Now, consider the case when the specimen is (i) only partially unloaded, (ii) the stress is then held constant, that is, the specimen is allowed to creep, and (iii) the specimen is reloaded at a constant strain rate. As shown in Figure 4.37, the stress response upon reloading for the partially unloaded specimen is qualitatively different than the stress response when the material was completely unloaded prior to reloading; specifically, for the partially unloaded specimen, a significant stress overshoot is observed upon reloading at a constant

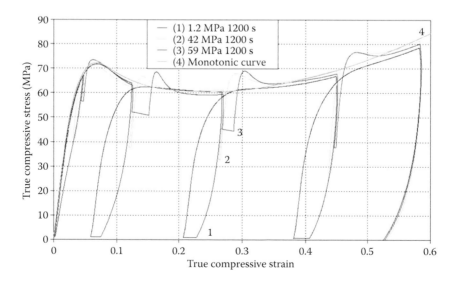

Figure 4.37 Cyclic compressive deformation of PC at 23°C (Tg-124°C). Loading at the strain rate of 10⁻³ s⁻¹ to true strains of 0.05, 0.13, 0.27, 0.45, and 0.59; unloading stress rate −2.3 MPa/s; rest stresses: 1.2, 42, and 59 MPa; and dwell time 1200 s. (Reprinted from *Mater. Des.*, 30, C. Dreistadt et al., 3126, Copyright 2009, with permission from Elsevier.)

strain rate. Examining the measured creep response during the constant stress stage of the deformation, the amount of creep depends on where the partial unloading occurs. Specifically, if the partial unloading occurs during the post-yield softening or the flow regions the creep strain increases; in contrast, if the stress is held constant for a material in the strain-hardening region, the creep strain decreases. Irrespective of whether the strain is increasing or decreasing during the constant nominal stress period, the magnitude of the stress over-shoot on reloading increases with the duration of the unloading period. The stress overshoot is followed by strain softening and eventual return to the stress–strain curve obtained in the single-step loading. The very distinctive features manifested during this complex deformation history that includes both strain and stress control provide qualitative signatures that will challenge any potential constitutive model.

4.6.5.3 EFFECT OF PRE-DEFORMATION

The effect of mechanical history on the stress–strain response for PC in shear is shown in Figure 4.38, where the first constant strain rate loading (i.e., curve *a*) exhibits a pronounced post-yield strain softening; however, after the specimen is returned to the original state of approximately zero strain (i.e., curve *b*), the second load-ing at the same strain rate (i.e., curve *c*) results in no strain softening. Similar results were obtained for the shear deformation of an epoxy polymer [100]. The effect of mechanical preconditioning in PS was studied by Govaert et al. [118], where tensile bar specimens were passed through a two-roll mill prior to uniaxial exten-sion tests. As a result of the deformation imposed by the two-roll mill, the specimen thickness decreased by 32% and the length increased by 36%. It was found that the appearance of the post-yield strain softening, that is, the height of the stress overshoot peak, depended on the length of time period between the preconditioning and the beginning of the uniaxial extension test, where the peak for a 20 min period is significantly higher than for 10 min. Similar results were obtained when axis-symmetrical bar specimens were subjected to large back-and-forth torsional deformations of 720° prior to the uniaxial extension experiments [119]. As com-pared to the untreated specimens, the mechanical preconditioning specimens displayed no stress overshoot; however, the effect was temporary, where the stress peak and the strain softening reappeared with aging. The results of these experiments are consistent with the observation made in the previous paragraph, where the unloading to zero stress resulted in the elimination of the stress overshoot upon reloading. This makes even more remarkable the finding of Dreistadt et al. [94] shown in Figure 4.37 that partial unloading (as opposed to unloading to zero stress) followed by creep instead enhances the stress overshoot upon reloading.

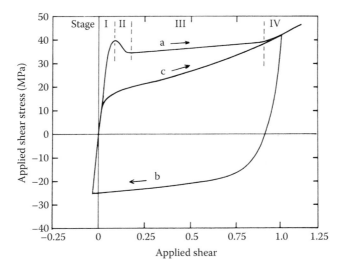

Figure 4.38 Effect of pre-deformation on the stress–strain response in shear deformation of PC at 23°C (Tg-124°C). The sequence of deformation steps is indicated by arrows: (a) first loading from zero strain, (b) unloading, and (c) second loading. (Reprinted from *Mat. Sci. Eng. A—Struct.*, 110, C. G'Sell et al., 223, Copyright 1989, with permission from Elsevier.)

4.6.6 Bauschinger effect

The Bauschinger effect is closely related to the pre-deformation effects described in the previous section. In its classic form, the Bauschinger effect is associated with a two-step deformation protocol, where the specimen is first deformed at a constant strain rate in uniaxial extension through yield and then deformed in the opposite direction back to zero strain so that stress becomes compressive after zero stress has been reached. The compressive portion of the stress–strain response obtained in the second step is compared to the stress–strain curve observed in a single-step compression of the same material. Both stress–strain responses exhibit compressive yield, but (and this is the essence of the Bauschinger effect) the value of the yield stress for the specimen that has been pre-deformed in extension is significantly lower. Another important observation is that the specimen that has first been subjected to the uniaxial extension does not exhibit any stress overshoot and hence post-yield softening. Implementing the Bauschinger protocol is technically difficult as the specimen most suited for uniaxial extension (for example, a slender rod) will buckle during the compressive part of the deformation history. To overcome this difficulty, Senden et al. [120] followed a slightly modified protocol, where (i) a rod-like specimen was used for the initial extension, (ii) the rod was then unloaded to zero stress, (iii) a short cylindrical plug was cut from the rod, and (iv) a compressive deformation was performed on this plug. It was assumed that no relaxation occurred during the time the specimen spent at room temperature, was cut, and then was remounted for the compression experiment. With this caveat, the results of the experiment are shown in Figure 4.39 for PC at room temperature, where the yield in compression is higher than that in extension [121]. Specifically, for a similar temperature and strain rate, the compressive yield stress is 50 MPa using the traditional uniaxial compression deformation history [120]; however, if the specimen was first deformed in uniaxial extension, the compressive yield stress is approximately 38 MPa as shown in Figure 4.39. The PC specimens used in the experiments shown in Figure 4.39 were mechanically preconditioned in torsion using the procedure described in the previous section (back-and-forth torsion with a twist angle of 990°) so that they do not exhibit stress overshoot with post-yield stress softening.

Anand and Ames [122] performed Bauschinger-like experiments on PMMA, but where the order of steps was reversed as compared to the standard Bauschinger protocol; specifically, a cylindrical sample was first subjected to uniaxial compression past the yield point and then deformed in extension. The resulting

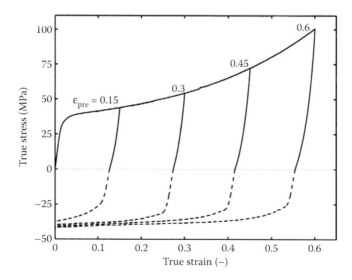

Figure 4.39 Stress–strain response of PC at 23°C (Tg-124°C) in uniaxial extension and compression at a strain rate of 10^{-4} s^{-1}. Specimens were first deformed in uniaxial extension to the strains indicated and then unloaded to zero stress. Then, the specimens for compression tests were cut out and the uniaxial compression experiments were carried out. (Reproduced from D. J. A. Senden, J. A. W. van Dommelen, and L. E. Govaert, *J. Polym. Sci. Polym. Phys.*, 2010, 48, 1483. Copyright Wiley-VCH Verlag GmbH & Co. KGaA. Reproduced with permission.) (True stress is calculated assuming isochoric deformation.)

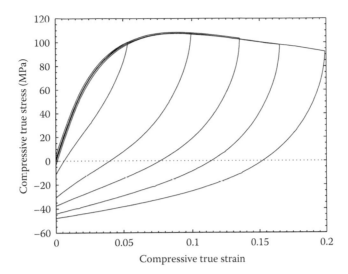

Figure 4.40 Stress–strain response of PMMA at 23°C (Tg-92°C) in uniaxial compression and extension at a strain rate of 3×10^{-4} s^{-1}. Specimens were first deformed in uniaxial compression and then unloaded. (Reprinted from *Int. J. Plast.*, 22, L. Anand and N. M. Ames, 1123, Copyright 2006, with permission from Elsevier.) (True stress is calculated assuming isochoric deformation.)

stress–strain curves are shown in Figure 4.40. The stress response in extension (i.e., the negative stress response in Figure 4.40) is very different from the stress–strain curve observed in uniaxial extension without pre-compression, where the extension yield behavior, if it is there at all, occurs over a significant range of strains with only a gradual change in the tangent modulus. Just like in the traditional Bauschinger experiment shown in Figure 4.39, the pre-compression in the first deformation step has a significant effect on the nonlinear mechanical response in the second step as compared to a material that did not experience the first step. Although there is no reason at this time to question either of the reported experimental results [121,122], these are technically difficult experiments. At this time, we provisionally accept the two different types of Bauschinger results shown in Figures 4.39 and 4.40; however, confirmation of these results would be useful—especially since some of the constitutive models for polymeric glasses predict behavior like that in Figure 4.39 and others like that in Figure 4.40 (see Chapter 14).

4.6.7 STRESS RELAXATION

The nonlinear stress relaxation response of a glassy epoxy resin is shown in Figure 4.41, where the material was cooled from above Tg to Tg-15°C and annealed for 30 min prior to applying a constant strain rate deformation in uniaxial extension [123]. After loading at a constant strain rate, the strain at various points along the stress–strain curve was held constant and the subsequent stress relaxation was monitored. The normalized stress relaxation response is shown in Figure 4.41b. When the loading strain rate is 1.5×10^{-4} s^{-1} as in Figure 4.41a, the normalized rate of stress relaxation increases with strain during the pre-yield part of the deformation and then becomes nearly constant at yield and during subsequent post-yield deformation. The increase in the rate of relaxation with strain is consistent with the concept that deformation increases the mobility in glassy polymers. However, when the strain rate is lowered by an order-of-magnitude to 1.2×10^{-5} s^{-1}, the trend is reversed as shown by the red curves in Figure 4.42b, where the relative rate of stress relaxation decreases as the deformation increases. A similar inversion in the rate of nonlinear stress relaxation with loading strain rate has been observed for the same material in uniaxial compression [123]. This inversion of the strain dependence of the rate of nonlinear stress relaxation is an unexpected signature of glassy polymers. Specifically, most constitutive models for polymer glasses assume that the mobility only increases with deformation, where the slowing down of the stress relaxation response with deformation at the slower strain rate is counterintuitive.

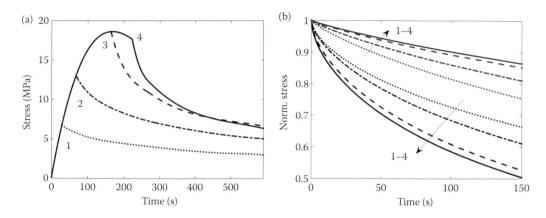

Figure 4.41 (a) Nonlinear stress relaxation following loading in uniaxial extension at a constant strain rate of 1.5×10^{-4} s^{-1} of an epoxy resin DGENG-44'MDA. Specimens annealed for 30 min at 57°C (Tg-15°C) prior to constant strain rate loading; stress at the beginning of stress relaxation: 1(dotted)—$\sigma_Y/3$, 2(dash-dotted)—$\sigma_Y2/3$, 3(dashed)—$\sigma_Y = 18.5$ MPa, and 4(solid)—17.7 MPa. (b) Normalized stress relaxation response versus time from the beginning of stress relaxation; black lines correspond to 1–4 responses in (a) in the order indicated by an arrow; red lines are for loading at a constant strain rate of 1.2×10^{-5} s^{-1} corresponding to stresses at the beginning of stress relaxation: 1(dotted)—$\sigma_Y/3$, 2(dash-dotted)—$\sigma_Y2/3$, 3(dashed)—$\sigma_Y = 9.1$ MPa, and 4(solid)—9.8 MPa in the order indicated by an arrow. (Adapted from J. W. Kim, G. A. Medvedev, and J. M. Caruthers, *Polymer*, 54, 3949, 2013.)

4.6.8 Stress memory

The stress memory is a multistep strain-controlled experiment, where the specimen is (i) loaded at a constant strain rate to a predetermined strain, (ii) unloaded to zero stress at a constant strain rate, and (iii) held constant at the strain reached in the previous step, where the evolution of stress during the third stage of deformation is called the stress memory response. A representative stress memory response for a glassy epoxy resin is shown in Figure 4.42a, where the stress during step (iii) passes through a maximum and then begins to slowly decrease to eventually reach zero. Stress memory responses for two different loading/unloading strain rates are shown in Figure 4.42b. The general features of the stress memory response are not surprising, where a linear viscoelastic material with more than one relaxation time subjected to the deformation history

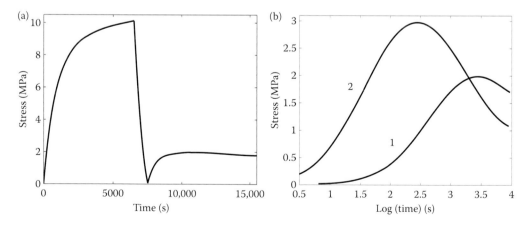

Figure 4.42 Stress memory experiment for a DGENG-44'MDA epoxy resin. The specimens were annealed for 30 min at 57°C (Tg-15°C) prior to the deformation. (a) Complete stress history where loading/unloading strain rate is 1.2×10^{-5} s^{-1}. (b) Stress memory response; zero time is from the moment the sample reaches zero stress on unloading. Curve 1—corresponds to (a) with a loading/unloading strain rate of 1.2×10^{-5} s^{-1}, curve 2—loading/unloading strain rate of 1.5×10^{-4} s^{-1}. (Adapted from J. W. Kim, G. A. Medvedev, and J. M. Caruthers, *Polymer*, 54, 5993, 2013.)

described above will exhibit an increasing and then decreasing stress memory response as shown in Figure 4.42b [124]. What is significant is that the peak in the stress memory overshoot exhibits a strong dependence on the loading/unloading strain rate, where there is an order-of-magnitude difference in the time of the stress overshoot for the two different loading/unloading strain rates shown in Figure 4.42b. This is significant from the modeling perspective for two reasons: First, a linear viscoelastic model predicts that the peak position is independent of the strain rate when unloading is in the post-yield strain range. Second, at the start of the stress memory response, the temperature and stress are the same for the two loading/unloading rate experiments and the strains can be made the same by controlling the strains when the unloading is begun. Consequently, nonlinear constitutive models where the mobility is a function of temperature, stress and/or strain will have difficulty describing the difference between stress memory responses 1 and 2 in Figure 4.42b, because the stress, strain, and temperature are the same at the start of the stress memory response.

SIGNATURES 6

S6.1 *Nonlinear stress–strain curve with yield and post-yield softening.* The stress response for a constant strain rate deformation in deviatorically dominated deformations, includes (i) a stress–strain response that is initially linear and then nonlinear, (ii) a yield point defined as $d\sigma/d\varepsilon = 0$, (iii) post-yield softening where $d\sigma/d\varepsilon < 0$, (iv) a flow state where $d\sigma/d\varepsilon \approx 0$, and finally (v) hardening where $d\sigma/d\varepsilon > 0$. For the dilatation-dominated longitudinal, deformation yield is associated with change in the slope of the tangent modulus with no post-yield softening.

S6.2 *Volume change during uniaxial deformation.* The specific volume increases/decreases slightly in, respectively, extension/compression and then remains largely unchanged during post-yield softening and flow.

S6.3 *Yield stress* in a constant strain rate deformation is a function of temperature, strain rate, and sub-Tg annealing time. Specific dependences include:
  a. σ_Y is a nearly linear decreasing function of temperature that reaches zero at Tg.
  b. σ_Y is an increasing function of log(strain rate) that can be linear, curved, or bilinear.
  c. σ_Y increases linearly with the log(annealing time) for a given sub-Tg annealing temperature until there is saturation for the fully equilibrated material.

S6.4 The yield surface is described by the pressure-modified von Mises criterion, which holds from large negative (observed in a longitudinal deformation) to large positive hydrostatic stresses.

S6.5 Magnitude of post-yield softening increases with strain rate.

S6.6 Magnitude of post-yield softening increases with sub-Tg aging.

S6.7 Post-yield flow stress increases linearly with the log(strain rate).

S6.8 Post-yield flow stress is independent of the thermal history (within experimental error).

S6.9 The strain-hardening modulus is weakly strain rate dependent.

S6.10 *Loading–unloading experiments.* The post-yield unloading stress–strain response is similar to the pre-yield loading stress–strain response. The unloading curve is smooth with no "second yield."

S6.11 *Loading–unloading–reloading experiments.* The initial modulus on reloading at all strains is similar to that of the original loading modulus. The reloading response returns to the underlying constant strain rate stress–strain curve.

S6.12 *Loading–partial unloading–reloading experiments.* Stress overshoot and strain softening is observed upon reloading if the specimen is only partially unloaded. The magnitude of the stress overshoot increases with (i) the partial unloading stress and (ii) the time spent in the partially unloaded state.

S6.13 *Mechanical preconditioning.* Large mechanical deformation prior to constant strain rate experiment diminishes, or eliminates entirely, the stress overshoot. The effect of mechanical preconditioning decreases with the time between application of the precondition and the constant strain rate deformation.

S6.14 *Bauschinger effect.* For a specimen pre-deformed in uniaxial extension and then subjected to uniaxial compression, the yield stress in the compressive part of the deformation is less than the compressive yield stress when there is no previous extension.

S6.15 The strain dependence of nonlinear stress relaxation response inverts with the magnitude of loading strain rate. At high strain rates, the rate of the normalized stress relaxation increases with deformation; in contrast, the rate of stress relaxation decreases with deformation at lower strain rates.

S6.16 *Stress memory experiment.* The location of the stress overshoot maximum on the time axis is very sensitive to the loading/unloading strain rate.

4.7 NONLINEAR CREEP

Stress-controlled experiments are the obvious counterpoint to the strain-controlled experiments discussed in the previous section. For linear viscoelastic materials, the compliance is formally related to the viscoelastic modulus, where there is no new information in a compliance experiment as compared to a stress relaxation experiment. However, one should remember that since the experimental data are obtained for a limited time range, there are quantities that can be more robustly extracted from the compliance experiments—for example, viscosity is more robustly determined from the long time creep compliance than from the long time stress relaxation response. In contrast to the linear case, for nonlinear deformations, there is no direct relationship between the time-dependent strain response in creep and the stress response in a strain-controlled experiment. Thus, nonlinear creep provides a complementary and equally important perspective on the behavior of glassy materials. Unfortunately and somewhat surprisingly, there are relatively few studies of the nonlinear creep response of glassy materials, where perhaps tellingly there is a chapter on creep in the first edition of *The Physics of Glassy Polymers* edited by Haward [17], but there is no chapter on creep in the 1997 second edition [125].

Nonlinear primary creep in uniaxial extension is shown in Figure 4.43 for PMMA, where the compliance curves at the smallest applied stresses superpose—at least for times less than $10^{3.5}$ s, which was the maximum creep time studied. In contrast, the compliance curves at higher applied stresses do not superpose. Similar results are observed for PVC in shear [127]. Primary creep has been studied for a variety of glassy polymers, where the earlier studies are reviewed in Turner [17] and Findley et al. [18]. The rate of creep is affected by both the applied stress as shown in Figure 4.43 and the thermal history (i.e., the temperature and sub-Tg aging time) where the rate of creep decreases as temperature decreases and as the aging time increases.

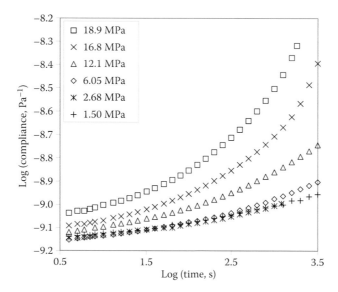

Figure 4.43 Primary creep of PMMA at 95°C (Tg-26°C) in uniaxial extension. The applied stresses are indicated in the figure. (Adapted from R. A. Martin, Purdue University, 2007.)

4.7.1 SINGLE-STEP CREEP

The complete nonlinear strain response to a step change in the axial stress is shown in Figure 4.44 along with the associated log(strain rate) versus strain curve, which more clearly delineates the various stages of creep. It should be emphasized that the deformation in Figure 4.44 is a homogeneous uniaxial compression so that the true material behavior is observed. We recently [129] proposed the following classification for the stages of nonlinear creep: In Stage I, or primary, creep, the time-dependent strain is concave downward just like in linear viscoelastic creep, although the response is nonlinear. In Stage II, or secondary, creep, the strain rate is nearly constant; however, it should be noted that there is a smooth transition between the various stages so that the assignment of secondary creep is somewhat arbitrary. In Stage III, or tertiary, creep, the strain rate accelerates, resulting in a concave upward strain versus time curve. Although not fully appreciated, careful examination of experimental data (see below) shows that following tertiary creep, the strain rate again becomes nearly constant in what we designate as Stage IV creep. Finally, in Stage V creep, the strain rate begins to slow down. Historically, Stages I and II were the primary focus of study, where some authors postulated (incorrectly) that the tertiary creep was an onset of failure rather than intrinsic material response [18]. All five stages of creep were observed by Lee et al. [130] in uniaxial extension of 25 μm thick PMMA films, where the axial and lateral strains were measured locally (thus, allowing extraction of the local strain history) even though the macroscopic sample did undergo necking. An important observation is that when flipped vertically, the log(strain rate) versus strain curve in Figure 4.44b qualitatively resembles the stress versus strain curve for constant strain rate loading (see Figure 4.26). From this perspective, Stage I corresponds to the initial stress buildup; the strain rate minimum corresponds to the yield point; Stage III is similar to the post-yield softening; Stage IV corresponds to the flow region; and, Stage V corresponds to the strain-hardening region. The correspondence between the yield point and the inflection point on the strain versus time creep curve was recognized by several authors, including Ender [131] and Bowden [11]. The observation that the inverted log(strain rate) versus strain curve for creep is a "mirror image" of the stress–strain curve for constant strain rate deformation for uniaxial compression of PMMA was made by Nanzai [132].

As discussed in Section 4.6, the features of the stress–strain curve such as yield stress, the post-yield stress softening, and the post-yield flow strongly depend on the strain rate and details of the thermal history, in particular, the sub-Tg aging time. Thus, in order to make the aforementioned correspondence between log(strain rate) versus strain curve for creep and the stress–strain curve for constant strain rate deformation more compelling, it is instructive to verify that similar dependencies on deformation rate and thermal history also hold for the Stages II, III, and IV of creep. Obviously, since the strain rate is changing during the creep experiment, a direct comparison to a strain rate-controlled experiment is not possible. However, if one accepts

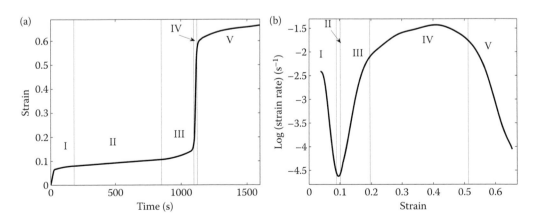

Figure 4.44 Stages of nonlinear uniaxial creep: (a) strain versus time; (b) log (strain rate) versus strain. The material is PC in compression at 23°C (Tg-124°C) with an applied stress of 64 MPa. (Data from Klompen, E.T.J., *Macromolecules*, 38, 7009, 2005.) (Reprinted from *Polymer*, 74, G. A. Medvedev and J. M. Caruthers, 235, Copyright 2015, with permission from Elsevier.)

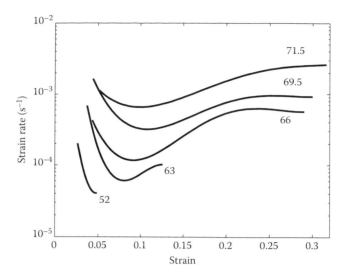

Figure 4.45 Strain rate versus strain data for creep of PMMA in uniaxial compression at 50°C (Tg-60°C). Applied axial stresses in MPa are indicated in the figure. (Data from O. A. Hasan and M. C. Boyce, *Polym. Eng. Sci.*, 35, 331, 1995; Reprinted from *Polymer*, 74, G. A. Medvedev and J. M. Caruthers, 235, Copyright 2015, with permission from Elsevier.)

the correspondence between the minimum point on the log(strain rate) versus strain curve and the yield point, it is expected that the former will depend on stress in a similar way to how the latter depends on the strain rate (i.e., Figure 4.33a). This is indeed the case as shown in Figure 4.45 for PMMA in uniaxial compression creep, where the value of the strain rate at the minimum increases with increasing stress. Flipping Figure 4.45 vertically, one observes an increase in the "yield stress" with the strain rate in qualitative agreement with Figure 4.33a. The data in Figure 4.45 also clearly indicate that after increasing for a period of time (i.e., the Stage III creep), the strain rate does become constant at larger strains (i.e., Stage IV creep). In a manner similar to how the flow stress in the strain control experiments increases with the applied strain rate (see Figure 4.33), the strain rate of the Stage IV creep increases with the applied stress. As shown in Figure 4.33b, the yield stress in constant strain rate experiment increases with sub-Tg aging time. Corresponding creep experiments in uniaxial extension were performed by Ender, where the aging time prior to application of the creep load was varied from 0.24 to 240 h. As seen in Figure 4.46, the minimum in the log (strain rate) versus strain curve (i.e., the "inverse yield" curve) decreases dramatically with aging time. Unfortunately, the data in Figure 4.46 do not extend to larger strains, where it would be possible to determine if the curves for different aging times merge as is the case for the stress versus strain curves in Figure 4.33b. Also, examination of Figures 4.45 and 4.46 suggests that the tertiary, that is, Stage III, creep can be eliminated altogether by the appropriate choice of the experimental conditions—namely for a sufficiently small stress and minimal sub-Tg aging.

4.7.2 CREEP RECOVERY

More complex creep experiments employ several steps, where a different stress is applied during each step. This includes the two-step creep-recovery experiment, where the applied stress during the second step is zero. For even a slightly cross-linked material, the creep recovery for a linear viscoelastic deformation is complete after removal of the load in a time that is of the same order as the time of the original creep deformation; thus, the specimen returns to its original dimensions that existed prior to deformation (neglecting any change in length associated with the long-term volume relaxation resulting from sub-Tg physical aging). The ability to recover fully is a good test for whether the creep deformation is within the linear range. Recovery from linear and nonlinear creep is illustrated in Figure 4.47 for an epoxy resin. For the lowest stresses, not only is the recovery complete but the characteristic time of recovery has the same time scale as the initial creep response.

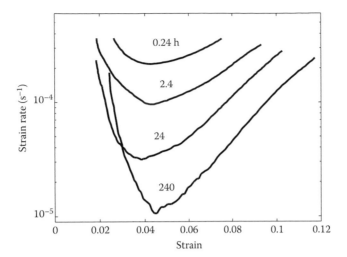

Figure 4.46 Uniaxial extension creep for PMMA at 80°C (Tg-35°C) for an applied stress of 27.5 MPa. The annealing temperature is 100°C with annealing times indicated in the figure. (Data from D. H. Ender, *J. Macromol. Sci. B*, 4, 635, 1970.) (Reprinted from *Polymer*, 74, G. A. Medvedev and J. M. Caruthers, 235, Copyright 2015, with permission from Elsevier.)

For nonlinear creep in the glassy state, the recovery often takes an exceedingly long time; however, heating the sample to a temperature above Tg results in complete recovery even from Stages III, IV, and V of creep.

In the recovery experiment, the time-dependent portion of the response that follows the instantaneous drop accompanying the removal of the load is important. As shown in Figure 4.48, a smooth recovery curve is observed, which is a discriminating signature because many constitutive models predict an abrupt transition with minimal additional relaxation after the instantaneous drop in strain upon removal of the load. The recovery response depends on all the conditions under which the creep and recovery experiment is conducted, including the thermal history prior to deformation, the stress during the initial creep step, and the duration of the creep [126,137]. An early study of the recovery from nonlinear creep attempted to find a scaling that would collapse the recovery curves obtained under all conditions into a master curve; however, this and other similar attempts were not successful [138].

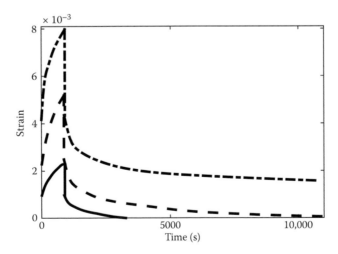

Figure 4.47 Creep and recovery for 1001F/DDS epoxy resin (Tg = 127°C) in uniaxial extension. The temperature is 118.3°C and the applied stresses are 1.17 MPa (solid), 2.62 MPa (dashed), and 3.79 MPa (dash-dotted). (Adapted from D. M. Colucci, Purdue University, 1992.)

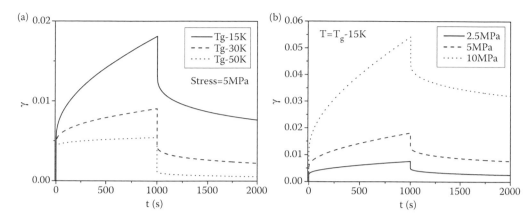

Figure 4.48 Creep and recovery for PMMA (Tg = 121°C) in uniaxial extension. The temperature and applied stress are indicated in the figures. (Reprinted with permission from K. Chen et al., *J. Chem. Phys.*, 129, 184904. Copyright 2008, American Institute of Physics.)

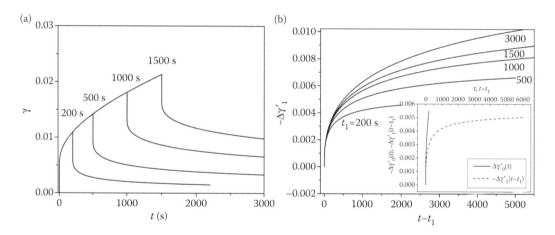

Figure 4.49 Creep and recovery for PMMA (Tg = 121°C) in uniaxial extension at 90°C with an applied stress of 5 MPa. The creep times prior to removal of the stress are indicated in the figures. (a) Strain versus time and (b) recovered strain versus time. (Reprinted with permission from K. Chen et al., *J. Chem. Phys.*, 129, 184904. Copyright 2008, American Institute of Physics.)

As expected, the rate of recovery from nonlinear creep decreases as the temperature is decreased as shown in Figure 4.48a, keeping constant the applied stress and time of the initial creep loading. Also, the initial rate of recovery increases with the applied creep stress as shown in Figure 4.48b, where the temperature, aging time prior to deformation, and the creep time are held constant. The effect of the creep time on the recovery response is shown in Figure 4.49. Both the rate of strain recovery and the amount of strain recovered increase with the time of the initial creep deformation, although the magnitude of the initial recovery response is independent of the creep time. The effect of the thermal history on nonlinear creep recovery is known to be important for primary creep, but where the effect of thermal history on creep recovery following Stages III, IV, and V of creep needs additional investigation.

SIGNATURES 7

S7.1 *Nonlinear creep*. At sufficiently large stresses, the creep response is nonlinear, where the compliance curves for different stresses only overlap in the linear viscoelastic limit.

S7.2 *Five stages of nonlinear creep.* For a sufficiently large applied stress, single-step creep exhibits five stages of creep, where the inverted vertically log(strain rate) versus strain dependence for the creep experiment is qualitatively similar to the stress versus strain curve for constant strain rate experiment. Specifically,

    a. Stage I is when the strain versus time response is concave downward.

    b. Stage II is the region around the minimum of the log(strain rate) versus strain curve and corresponds to yield point in a constant strain rate experiment.

    c. In Stage III creep, the strain versus time curve is concave upward where the strain rate accelerates, which corresponds to post-yield softening in a constant strain rate experiment.

    d   The strain rate is nearly constant in Stage IV creep, which corresponds to the post-yield flow region in a constant strain rate experiment.

    e. In Stage V creep, the strain rate decreases, which corresponds to strain hardening in a constant strain rate experiment.

The five stages of nonlinear creep have been observed in both uniaxial extension and compression.

S7.3 *Dependence on stress and thermal history.* The magnitude of the Stage III creep increases with stress similarly to how the magnitude of the post-yield softening increases with strain rate. The magnitude of the Stage III creep increases with the sub-Tg aging time similarly to how the magnitude of the post-yield softening increases with the sub-Tg aging time.

S7.4 *Recovery from nonlinear creep.* Removal of the load results in partial strain recovery, where at sub-Tg temperatures, the residual strain does not relax over experimentally accessible time. When heated above Tg, a specimen recovers completely even from later stages of nonlinear creep.

S7.5 *Rate of recovery from nonlinear creep.* Initial rate and amount of strain recovery increase with the stress and duration of creep.

4.8 MORE COMPLEX THERMO-DEFORMATIONAL HISTORIES

4.8.1 TICKLE EXPERIMENTS

Nonlinear viscoelastic relaxation of glassy polymers has been probed by performing a small deformation, that is, a "tickle," on top of a nonlinear deformation, where the intent is that the tickle measures the state of the structure as it evolves during the nonlinear deformation. Because the additive decomposition of the "linear" tickle response from the underlying nonlinear response cannot be formally justified, this linearized interpretation of the data is potentially problematic; nevertheless, the tickle data provide an interesting signature of the glassy state. A small oscillatory axial strain was imposed upon a PC specimen at 50°C (Tg-95°C) that was undergoing nonlinear stress relaxation [139], where the in-phase and out-of-phase components of the oscillatory stress response were reported as the dynamic moduli E' and E'' respectively as shown in Figure 4.50. The E' and E'' moduli 10 s after the beginning of the stress relaxation are shown in Figure 4.51, where the E' elastic modulus decreases with increasing strain up to a strain of about 4% after which $E'(t = 10 \text{ s})$ becomes constant. This is consistent with the observation that the pre-yield tangent modulus of the underlying stress–strain curve decreases with increasing strain; however, the reason that $E'(t = 10 \text{ s})$ becomes independent of strain at the larger axial strains is not clear. The loss tickle modulus E'' $(t = 10 \text{ s})$ increases with the axial strain, where it also becomes independent of the axial strain for strains larger than 4%.

In an analogous set of experiments, a small stress relaxation step strain was applied during the nonlinear stress relaxation of PC, where it was observed that rate of the incremental stress relaxation due to the step strain tickle (i) decreased during the course of the underlying nonlinear stress relaxation and (ii) increased with the nonlinear strain [140]. The data are consistent with the idea that nonlinear deformation increases the rate of relaxation, where the increased mobility relaxes during nonlinear stress relaxation. To the best of our knowledge, the effect of thermal history (i.e., aging time, temperature, and so on) on the tickle response has not been studied.

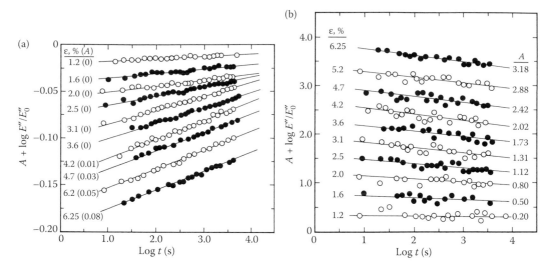

Figure 4.50 Time dependence of small oscillatory (0.1% strain) dynamic moduli at 10 Hz during nonlinear stress relaxation (a) E'/E'_0 and (b) E''/E''_0 for PC in uniaxial extension at indicated axial strains. The storage and loss moduli of the "undeformed" sample—E'_0 and E''_0, respectively, were determined at the static axial strain of 0.3%. The storage modulus E'_0 is of the order of the static Young's modulus $E_0 = 2$ GPa; vertical shift A indicated in the figures is used for visual clarity. (Reprinted from *Polymer*, 26, T. Ricco and T. L. Smith, 1985, Copyright 1979, with permission from Elsevier.)

Torsional oscillatory tickle experiments have also been performed on a sample subjected to large axial creep for both epoxy [141] and polyester [142] glasses. As shown in Figure 4.52, at the highest applied axial stress, the storage shear modulus decreased with the magnitude of axial creep, while the loss shear modulus increased. These observations are generally consistent with the tickle experiments discussed in the previous paragraph; however, one should note the decrease in the loss modulus at the intermediate axial stress in Figure 4.52c. After the axial load was removed and the axial strain has partially recovered, the recovery of the storage and loss shear moduli was essentially complete for the case of the highest load; however, in a later report, the loss modulus did not recover completely [142]. These observations need to be verified for other glass-forming polymers and for a more complete set of stresses and thermal histories.

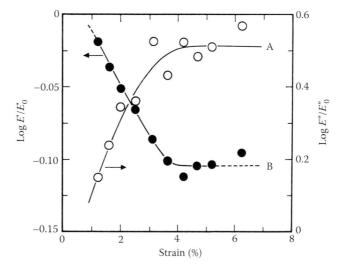

Figure 4.51 Strain dependence of E'/E'_0 and E''/E''_0 for dynamic data in Figure 4.50 at an elapsed time of 10 s. (Reprinted from *Polymer*, 26, T. Ricco and T. L. Smith, 1985, Copyright 1979, with permission from Elsevier.)

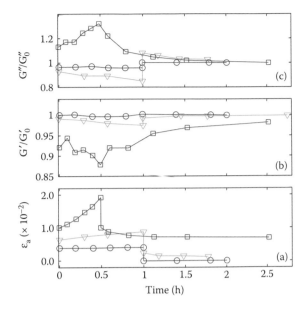

Figure 4.52 Evolution of axial strain (a), torsional storage modulus ($G_0' = 320$ MPa) (b), and torsional loss modulus ($G_0'' = 10.5$ MPa) (c) of an epoxy glass at 70°C for the indicated axial strains; frequency 0.5 Hz; torsional strain 0.1%. (Adapted from A. I. Isayev, D. Katz, and Y. Smooha, *Polym. Eng. Sci.*, 21, 566, 1981.)

4.8.2 THERMALLY STIMULATED CREEP

Thermally stimulated creep (TSC) is a nonisothermal experiment that is designed to isolate relaxation processes that become active over a specific temperature range [143,144]. The thermo-deformational history is shown in Figure 4.53; specifically,

1. A specimen is cooled at a constant rate from above Tg into the glassy state.
2. As the material is being cooled, a stress is applied over a predetermined range of temperatures, where the material undergoes creep and then partial recovery after the stress is removed.
3. The cooling continues after the removal of the stress, where the relaxation processes that were active during the time/temperature range when the stress was applied are now frozen in as evidenced by incomplete creep recovery.
4. Finally, the sample is heated at a constant rate, where those processes frozen during the cooling/ deformation history now relax as the temperature passes through the range at which the stress was previously applied.

The creep recovery response during heating is called thermally stimulated creep. An analogous experiment is when an electric field is applied during cooling rather than a creep stress in order to orientate polar molecules, where the subsequent depolarization is monitored upon heating—this is called the thermally stimulated depolarization current (TSDC). The derivative of the TSC response shows a peak that is typically analyzed as a single relaxation time process with activation energy. The TSC response in the Tg region is shown in Figure 4.54, where TSC peaks move to higher temperatures as the temperature at which the stress was applied increases. The TSC behavior shown in Figure 4.54 is exactly what would be expected if the material has a distribution of thermally activated processes, where the main objective of the TSC protocol is to determine an activation energy distribution. As discussed in Section 4.4 with respect to enthalpy relaxation, the analysis of nonisothermal experiments is more challenging than isothermal experiments, because of the convolution of the time/frequency nature of the relaxation process with the now time-dependent thermal effects. It is not clear that these nonisothermal experiments provide a more effective probe of the glassy state than the nonlinear, single and multiple step, stress relaxation and creep experiments described earlier in this

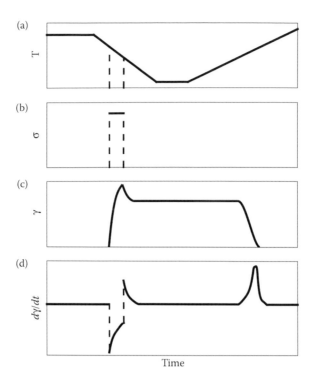

Figure 4.53 Schematic of time profiles during the thermally stimulated creep for (a) temperature, (b) stress, (c) creep strain, and (d) creep strain rate. (Adapted from J. J. del Val et al., *J. Appl. Phys.*, 59, 3829, 1986.)

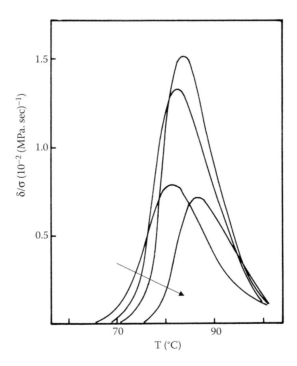

Figure 4.54 Thermally stimulated creep of PVC. For peaks with maximum from left to right (indicated by the arrow), the stress was applied for 74–69°C, 77.5–72.5°C, 82.5–77.5°C, and 85–80°C temperature intervals. (Reprinted with permission from J. J. del Val et al., *J. Appl. Phys.*, 59, 3829. Copyright 1986, American Institute of Physics.)

chapter. Nevertheless, it might be interesting to examine how the TSC response depends upon the magnitude of the applied stress and the thermal history.

4.8.3 Shrinkage strain/stress of drawn polymers

The nonlinear stress relaxation analog to TSC is to measure the shrinkage strain or stress after uniaxially drawing the polymer well beyond yield while it is in the glassy state. When the stress is released, the material is frozen in an extended configuration, where the polymer will relax once the temperature is raised into the Tg region. Andrews [145] observed relaxation in the overall length of atactic PS specimens with a Tg of 101°C that had been extended to a draw ratio between 1.2 and 5, where the rate of thermal shrinkage significantly increases as the temperature approaches Tg. Perhaps, a more informative experiment was performed by Pakula and Trznadel [146], where after uniaxially extending polycarbonate at 23°C (Tg-124°C) to a draw ratio of 1.8, the length of the specimen was held constant and the axial stress was monitored as the material was heated at constant rate of 3°C/min through Tg. The stress evolution during heating for PC1 is shown in Figure 4.55. During the initial heating, there is a small negative stress due to thermal expansion that is followed by a large increase in the axial stress, and finally, the axial stress begins to relax as the temperature is increased toward Tg. Since the polycarbonate was not cross-linked, the retractive force relaxed completely above Tg; alternatively, a cross-linked material would exhibit a retractive force associated with rubber elasticity. The effect of sub-Tg annealing prior to the temperature scan is also shown in Figure 4.55, where the maximum in the retractive stress shifts roughly to the temperature of the sub-Tg annealing and the height of the maximum decreases.

The time evolution of the retractive stress at a fixed temperature and thermal history was studied by Trznadel [147], where the results are shown in Figure 4.56. There is a significant induction period prior to the occurrence of the retractive force. The length of the induction time increases as (i) the annealing temperature is decreased and (ii) the length of the sub-Tg annealing time increases. The presence of such a significant induction time indicates that after the drawing process, the material does not have relaxation times shorter than the induction time, where decreasing the test temperature and increasing the annealing time shift the relaxation spectra to long times. Retractive stresses after cold drawing have been observed by Cheng and Wang [148] for PC, where they extensively investigated the effects of thermal history; in addition, more

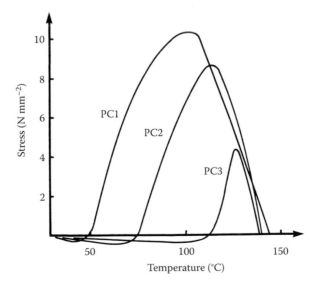

Figure 4.55 Evolution of shrinkage stress for polycarbonate (Tg = 147°C) with a draw ratio of 1.8 at room temperature that was subjected to heating at 3°C/min. Annealing histories: a specimen without annealing (PC1), annealed for 3 h at 75°C (PC2), and annealed for 3 h at 120°C (PC3) prior to reheating. (Reprinted from *Polymer*, 26, T. Pakula and M. Trznadel, 1011, Copyright 1985, with permission from Elsevier.)

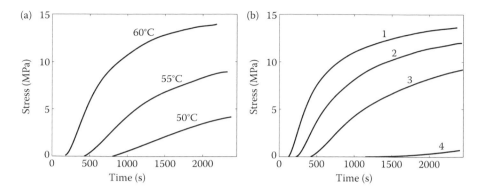

Figure 4.56 Thermal shrinkage stress versus time for PC (Tg = 147°C) extended to a draw ratio of 1.8 at room temperature. (a) Specimen was annealed at room temperature for 14 days and then tested at indicated temperature; (b) specimen was annealed at room temperature for 1–2 days, 2–5 days, 3–14 days, and 4–74 days and then tested at 55°C. (Reproduced from M. Trznadel, *J. Macromol. Sci. B*, 28, 285, 1989, with permission from Taylor & Francis.)

limited studies on drawn PMMA, poly(ethylene terephthalate), and poly(2,6-diemethyl-1,2-phenylene oxide) glasses indicate that they also exhibit similar retractive force behavior [147].

The thermo-mechanical behavior of drawn glassy polymers has been included in this chapter for completeness; however, one must realize that the retractive force measurements are for specimens that have undergone necking. Necking itself is an obvious mechanical instability that leads to a spatially nonuniform deformation field, where the experimentally measured stress–strain behavior during necking is not a true material property. However, the reported retractive force measurements were for pre-necked materials, where the test specimens were spatially uniform—at least at a macroscopic level. If the material in the necked region is microscopically uniform, then, the retractive force measurements are an important component of the overall characterization of the glassy state, albeit for a material that is on the large deformation side of a macroscopic mechanical instability. A word of caution: if the pre-necked material has microscopic heterogeneity, for example, micro-shear bands, then, it is not appropriate to include these retractive force measurements as inherent material behavior that should be part of the constitutive description of the glassy state.

4.8.4 TENSION–TORSION DEFORMATION

All deformations are inherently multiaxial even if the stimulus and response are only measured in one direction. The lateral strain (or equivalently the specific volume) response during uniaxial deformation was discussed previously in Section 4.6.1; a multiaxial criterion for yield was discussed in Section 4.6.2; and, the torsional dilatometry experiment was discussed in Section 4.5, where the shear, normal force, and volume relaxations were simultaneously measured. However, there are relatively few studies of the multiaxial viscoelastic behavior of glassy polymers, where there is a major stress/strain perturbation in more than one direction. Findley and Lai [149] studied the nonlinear primary creep in tension–torsion for PVC as well as the response in just simple tension and simple compression for various loading, unloading, and reloading deformation histories. No distinctive features emerge from the Findley nonlinear tension–torsion experiments, except that there is coupling between shear and tension relaxation.

Nonlinear viscoelastic relaxation in tension–torsion creep of PMMA was studied at a variety of temperatures and loading conditions by Lu and Knauss [150]. As shown in Figure 4.57a for a temperature well below Tg, axial deformation accelerates the rate of the shear creep in a tension–torsion experiment, where tension has a larger effect than compression. The situation changes in the immediate region of Tg as shown in Figure 4.57b, where uniaxial extension accelerates the shear creep more than uniaxial compression; however, the response in compression is slightly retarded as compared to a pure torsion deformation when the axial stress is zero. To the best of our knowledge, the effect of sub-Tg annealing on time-dependent relaxation in a multiaxial deformation has not been investigated.

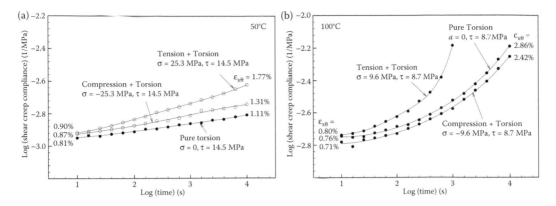

Figure 4.57 Tension–torsion creep of PMMA (Tg = 105°C). Shear compliance at 50°C (a) and at 100°C (b). (Reproduced from H. Lu and W. G. Knauss, *Mech. Time-Depend. Mat.*, 2, 307, 1998, with permission from Kluwer.)

In a variation on the traditional tension–torsion experiment, a cylindrical rod is subjected to a finite torsional deformation, where the length of the rod is held constant and consequently a normal force must be applied to the faces of the rod. The normal force was measured simultaneously with the torque for PMMA, PC, poly(ethyl methacrylate), and polysulfone [151,152]. In case of infinitesimal/linear deformations, the normal force is zero, because it is second order in the torsional shear strain. Thus, the normal force was measured for finite torsional deformation, but at strains that were, however, still below the yield strain. In a series of nonlinear stress relaxation experiments, it was shown that the torsional/shear stress and the normal stress relax on the same time scale [151], indicating that the same underlying physical mechanism governs both processes. Also, it was found that the ratio of the normal stress to the shear stress varies greatly from material to material, where at the same torsional strain, the ratio is nearly 2 times larger for PMMA than for PC. In an earlier study, McKenna and Kovacs [153] observed that time-aging time superposition failed for both PMMA and PC at the smallest torsional deformations, but worked at larger torsional strains of 7%. The simultaneous measurement of torque and normal force does provide a multiaxial probe into nonlinear mechanical relaxation in the glassy state; however, the interpretation of the torque and normal force for nonlinear deformations is quite difficult, because the torsional strain field is nonuniform being zero along the axis of the cylinder and a maximum at the outer radial surface of the cylinder.

SIGNATURES 8

S8.1 *Axial oscillatory "tickle" experiments on top of a nonlinear, uniaxial stress relaxation.* The magnitude of E' for the oscillatory component of the stress response decreases with the strain of the underlying nonlinear deformation, but then becomes constant; in contrast, the magnitude of E'' increases with applied strain before becoming constant. As the underlying relaxation progresses, both E' and E'' move toward their linear viscoelastic values.

S8.2 *Torsional oscillatory "tickle" experiments on top of nonlinear, uniaxial creep recovery.* In a nonlinear creep experiment, the torsional G' decreases with the applied stress, while G'' increases. When the axial load is removed, G' and G'' relax toward their values prior to the axial creep.

S8.3 *Thermally stimulated creep.* When a material is cooled into the glass and a stress pulse is applied during the cooling, the strain recovers upon reheating the material through the glass transition, where the shape and location of the recovery response depend upon the details of the thermal-stress history.

S8.4 *Shrinkage strain/stress relaxation of a drawn polymeric glass.* The polymer is drawn at a temperature below Tg, annealed, and then held at constant length.

    a. When the stress response is monitored as the material is heated, the stress exhibits a maximum at the annealing temperature.

    b. If temperature is maintained constant, the stress versus time dependence has a maximum; where there is an induction period before the stress increases, where the induction time increases with (i) the sub-Tg annealing time and (ii) lowering of the test temperature.

S8.5 *Shear compliance in nonlinear tension–torsion deformation.* When both a uniaxial stress and a shear stress are simultaneously applied

    a. The nonlinear shear compliance is greater and increases faster with both axial extension and compression at temperatures well below Tg.

    b. For temperatures near Tg, the nonlinear shear compliance is greater and increases faster with time for extension, but is smaller and increases slower with time for compression.

S8.6 *Normal force and torque in a torsional deformation.* When solid rod is subjected to a finite twist and the length is constrained to be constant, both the torque and normal force exhibit relaxation, where the rate of relaxation is similar for both the normal force and torque. The nonuniform torsional strain field makes interpretation of the macroscopic torque and normal force difficult.

4.9 COMBINED MECHANICAL AND ENTHALPY RELAXATION

When any material is subjected to a thermo-deformational perturbation, there is simultaneously a mechanical response and an enthalpic response, where the energy associated with these two responses is often comparable. Thus, a fundamental constitutive description of the glassy state will need to include both the mechanical and enthalpic response.

4.9.1 ENTHALPY RELAXATION AND YIELD

For polymeric glasses, both the enthalpy relaxation peak upon heating (Section 4.4) and the yield stress (Section 4.6) increase with sub-Tg annealing. The net difference between the DSC endotherms of annealed and quenched materials, ΔH, is shown in Figure 4.58a for PC. As shown in Figure 4.58b, ΔH becomes a linear function of $\ln t_a$ after an initial induction period, where t_a is the annealing time. As shown in Figure 4.58c, the annealing time dependence of the additional yield stress (i.e., as compared to the yield stress for a quenched sample) also shows a linear dependence on $\ln t_a$. In addition, as the annealing temperature approaches Tg, the evolution of both ΔH and $\Delta \sigma_Y$ with annealing time saturates [154,155], indicating that the equilibrium has been reached. A formal connection between the enthalpy relaxation prior to deformation and the changes in the yield behavior would be important; however, what Figure 4.58 shows is a correlation suggesting that both the additional yield stress and ΔH may have the same underlying cause.

4.9.2 DEFORMATION CALORIMETRY

Deformation calorimetry is an experiment where the heat flow is monitored during a mechanical deformation. These experiments are technically quite difficult and only few reports have appeared to date. Early deformation calorimetry experiments were reported by Adams and Farris [156,157], where the heat flow from the specimen was inferred from the changes in pressure of the surrounding gas; however, the accuracy of this technique may not have been sufficient. Oleinik and collaborators have carried out the most accurate deformation calorimetry measurements on glassy polymers, using two deformation calorimeters—one intended for large nonlinear deformations through yield and into the flow regime [158] and another of higher sensitivity designed to study smaller pre-yield deformations [159]. Analysis of the heat evolved during uniaxial compression has been summarized by Salamatina et al. [158] for a variety of glassy polymers including PS, PC, PMMA, amorphous poly(ethylene terephthalate), poly(phenylene oxide), and epoxy resins. A typical loading response for a constant strain rate deformation is shown in Figure 4.59, where the sample is cooled into the glassy state and then deformed isothermally. The work W is determined via integration of the stress–strain curve. The deformation heat Q_{def} is the difference between the response when the deformation calorimeter has a sample and when the calorimeter does not have a sample. The difference between W and Q_{def} is the internal energy U_{def} associated with the deformation.

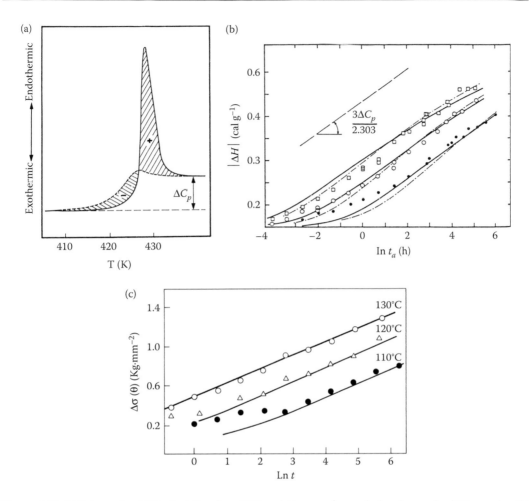

Figure 4.58 (a) Schematic of ΔH calculation from DSC endotherms of annealed and quenched PC samples; (b) annealing time, t_a, dependence of ΔH for PC at 130°C (open squares), 120°C (open circles), and 110°C (filled circles); and (c) annealing time dependence of the additional yield stress (as compared to the quenched sample). (Reprinted from *Polymer*, 23, C. Bauwens-Crowet and J.-C. Bauwens, 1599, Copyright 1982, with permission from Elsevier; *Polymer*, 27, C. Bauwens-Crowet and J.-C. Bauwens, 709, Copyright 1986, with permission from Elsevier.)

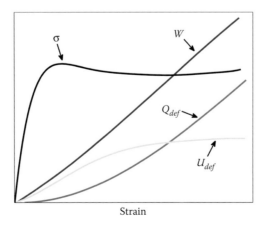

Figure 4.59 Stress–strain, heat flow, and other quantities during compressive constant strain rate loading (see text for details). (Adapted from O. B. Salamatina et al., *Thermochimica Acta*, 247, 1, 1994.)

Examining Figure 4.59 and similar figures in Salamatina et al. [158], there are a number of important observations:

1. The most striking feature of Figure 4.59 is that the Q_{def} versus strain response does not undergo any dramatic change in behavior at the yield point; in contrast, the stress–strain curve undergoes a very dramatic change at yield. The U_{def} versus strain curve exhibits an inflection point approximately at yield.
2. While the stress decreases during post-yield softening and then becomes constant during the flow stage, the internal energy U_{def} only reaches its steady state at significantly higher strains.
3. At the point where the internal energy U_{def} saturates (i.e., when the W and Q_{def} curves become parallel to each other), U_{def} constitutes 30%–50% of the mechanical work W. The percentage of W that goes into U_{def} is an order-of-magnitude greater than the corresponding percentage for metals (see references in Salamatina et al. [158]), where during plastic flow, the overwhelming majority of the work of deformation is transferred into heat.

At this point, there is no consensus regarding either the mechanism of dissipation underlying the heat response to deformation or a molecular/mesoscopic picture of how the internal energy is stored in the deformed glass. Because the heat generated is a significant fraction of the overall mechanical work, a constitutive model for glassy polymers should address both the stress and heat generated during deformation. Thus, the deformation calorimetry data provide very important signatures of the thermo-mechanical behavior of polymeric glasses—features that have not been fully considered to date in the development of constitutive models for the glassy state.

4.9.3 ENTHALPY RELAXATION FOLLOWING DEFORMATION

Deformation calorimetry as reported in the previous section is an extremely demanding experiment that requires special instrumentation. Fortunately, the information on the residual stored internal energy ΔU_{res} can be obtained by measuring heat capacity relaxation in a conventional DSC for glassy materials that have previously

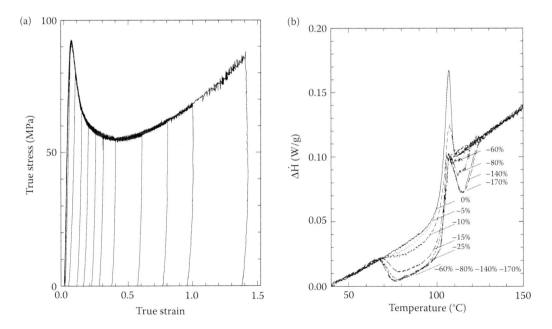

Figure 4.60 (a) Stress response of PS as a function of true (i.e., Hencky) strain, when loaded to various strains and then unloaded to zero stress at T_{def} = 23°C (Tg-77°C). Loading and unloading strain rate is 10^{-3} s^{-1}. (b) Heat capacity after the loading/unloading deformation determined by DSC upon heating at 10°C/min. The maximum compressive strain during loading is indicated in the figure. (Reprinted from *Polymer*, 34, O. A. Hasan and M. C. Boyce, 5085, Copyright 1993, with permission from Elsevier.)

been deformed. Examining the data in Figure 4.60a for PS, the specimen was (i) cooled from above Tg into the glass, (ii) deformed in uniaxial compression at a constant strain rate to a predetermined strain, (iii) unloaded at the same strain rate until the stress was zero, and (iv) then stored at room temperature to eliminate further relaxation. A small piece of the unloaded material was then cut from the mechanical test specimen and placed into a DSC apparatus, where the heat flow was measured upon heating through Tg at a constant rate. The heat capacity response upon heating depends upon both the thermal history used to form the glass and the uniaxial deformation history. Examining the data in Figure 4.60b, one observes a significant exotherm that develops at 70°C (i.e., Tg-30°C) and continues until slightly above Tg, where the magnitude of the exotherm increases with the maximum strain up through strains of 25% compression, which is near the minimum in the post-yield strain-softening response. If the unloading takes place in the strain-hardening region, a second exotherm appears in the DSC response, but now at temperatures greater than Tg, where the magnitude of the exotherm increases with the extent of the post-yield deformation. The appearance of two distinct exotherms indicates that there are different enthalpic processes operating in the pre-yield, yield, and post-yield softening region versus the strain-hardening region. Similar heat capacity relaxation curves are reported in Hasan and Boyce [160] for PC and PMMA and by other authors for PC [161], PS [162], and PMMA [163]. What is surprising is that the enthalpy relaxation exotherms due to deformation (i) begin so far below Tg and (ii) occur over such a wide range of temperatures. In contrast, the enthalpy relaxation endotherm of the undeformed material only begins near Tg and is relatively narrow, that is, the curve labeled 0% in Figure 4.60b. For the data in Figure 4.60b, all the deformations were carried out at the same temperature, where all the subsequent exotherms begin at the same temperature independent of the maximum strains. In contrast, when the deformation temperature is varied while keeping the maximum strain constant, the temperature when the exotherm begins, T_{begin}, changes significantly as shown in Figure 4.61; specifically, $T_{begin} \sim 45°C$ for $T_{def} = 23°C$, $T_{begin} \sim 70°C$ for $T_{def} = 65°C$, and $T_{begin} \sim 85°C$ for $T_{def} = 80°C$.

A related experiment is the measurement of the strain recovery upon heating after a material in the glassy state has been deformed past yield as is shown in Figure 4.61. The strain recovery rate versus temperature curves exhibit a broad, low-temperature peak that corresponds to the enthalpic exotherm. The strain recovery also exhibits a strong and relatively narrow peak in the Tg region, although no such peak is observed in the heat capacity relaxation. As seen in Figure 4.61, the height of the strain recovery rate peak located at Tg decreases as the deformation temperature T_{def} decreases. For an epoxy polymer (Tg = 145°C), the Tg region peak of the strain recovery rate versus temperature curve disappeared completely for $T_{def} = -85°C$ when the compressive uniaxial creep deformation was 12% [164]. The appearance of both a low- and high-temperature relaxation processes in the strain recovery of a glass deformed in uniaxial compression has also been observed in PMMA [165]. Only the low-temperature strain relaxation is observed for strains up to 20%, that

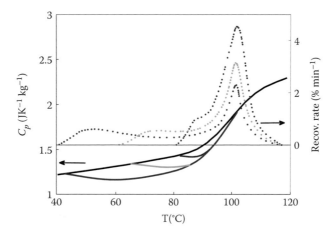

Figure 4.61 Heat capacity relaxation and the rate of thermal strain recovery upon heating at 20°C/min for a PS glass (Tg = 100°C) that was deformed to 13% compressive strain and unloaded at $T_{def} = 23°C$ (blue), 65°C (green), and 80°C (red). (Adapted from E. F. Oleinik, *Prog. Colloid Polym. Sci.*, 80, 140, 1989.)

is, for deformations in the post-yield softening and flow range of strains. At larger strains in the post-yield hardening range, a second high-temperature relaxation emerges that occurs in the Tg region, which is absent for deformations less than 20%. This strain recovery trends for PMMA are consistent with the ones observed in PS [164]. An interesting point with respect to the data in Figure 4.61 is that both the low-temperature and high-temperature processes are observed for PS which presumably does not have a significant sub-Tg relaxation process. Thus, the two relaxation processes observed via DSC and strain recovery must be part of the underlying thermo-mechanics of the glassy state, and not a result of a distinct sub-Tg β-relaxation.

SIGNATURES 9

S9.1 *Heat of deformation.* The heat generated during a uniaxial constant strain rate deformation does not exhibit any dramatic changes at the yield point or during post-yield softening.

S9.2 *Internal energy from deformation.* The fraction of the work of deformation stored as internal energy can be as large as 50%, which is an order-of-magnitude larger in polymeric glasses undergoing yield and flow than in crystalline metals undergoing plastic flow.

S9.3 *Internal energy evolution during deformation.* The internal energy generated in a constant strain rate deformation keeps increasing through yield, post-yield softening, and the beginning of the flow regime only reaching steady state at large strains.

S9.4 *Enthalpy relaxation of glass that is loaded/unloaded at a constant strain rate.* In contrast to the DSC heat flow upon heating of an undeformed polymer glass that has an endothermic peak near Tg, the heating DSC heat flow of a deformed and then unloaded polymer glass has two exothermic peaks—a broad peak below Tg and narrow peak above Tg. Specific features include:

    a. Broad exothermic peak below Tg: The beginning of the broad exothermic peak is located slightly above the temperature at which the deformation has been carried out, where its magnitude initially increases with the strain just prior to unloading, but then saturates during the post-yield flow region.

    b. Exothermic peak above Tg: The exothermic peak above Tg is only observed when the strain prior to unloading is in the post-yield hardening region, where the peak magnitude increases with strain.

S9.5 *Temperature-stimulated strain recovery.* The rate of the strain recovery versus temperature response is consistent with the enthalpy versus temperature behavior, where two peaks emerge—(i) a broad peak below Tg whose location depends on the temperature of the deformation and (ii) a narrow peak near Tg. The narrow peak only exists if (i) the temperature of the deformation is significantly lower than Tg and (ii) if the strain just prior to unloading is in the strain-hardening region.

4.10 MOLECULAR MEASUREMENTS OF RELAXATION BEHAVIOR

Thus far, the focus of this chapter has been on macroscopic properties, where a number of thermo-mechanical signatures of the glassy state have been identified. Since these signatures have molecular origins, it is appropriate to assess if the associated molecular relaxation can inform our understanding of the macroscopic thermo-mechanical response.

The measurement of molecular relaxation in the glassy state has been an active area of research for decades. The molecular motion in the glassy state has been studied using nuclear magnetic resonance (NMR) [166–172], light scattering [173–178], x-ray scattering [179], neutron scattering [180–182], dielectric relaxation [168,178,183–187], and rotational mobility of probe molecules [130,188,189]. These molecular relaxation studies provide valuable understanding about the glassy state, where a complete review of these findings is well beyond the scope of this chapter. Thus, we will only review a very limited portion of this extensive literature—selections that we believe most directly relate to the thermo-mechanical signatures discussed in the previous sections. In order to more directly connect to the previous sections, the following will not be organized around the experimental techniques, but rather around three features of the relaxation: (1) the effect of temperature and pressure on segmental mobility in the isotropic state, (2) the effect of active deformation on segmental mobility, and (3) dynamic heterogeneity.

4.10.1 Segmental mobility as a function of temperature and pressure

There is a vast literature on the segmental relaxation time and its dependence on temperature and pressure for equilibrium liquids. Significantly less data are available regarding the segmental mobility for a material in the glassy state. This is understandable as the segmental relaxation time (i.e., the α-process) is typically too long to be directly measured by most experimental techniques. However, it is sometimes possible to take advantage of coupling between the α-relaxation process and higher-frequency processes such as β-relaxation that may be more experimentally accessible, where the segmental relaxation time can be estimated using this coupling [186]. As expected, sub-Tg aging has a noticeable effect, for example, sub-Tg aging causes a decrease in the β-relaxation dielectric loss ε'' at a given frequency [186]. Another method used to evaluate the α-relaxation time below Tg is the dye molecules orientation probed by the second-harmonic generation optical methods [190]. The essential findings regarding the segmental relaxation time in both equilibrium and glassy state have been recently reviewed by Roland [135], where references to the original experimental studies can be found.

The techniques used to measure the segmental relaxation time include primarily dielectric spectroscopy and the NMR. The first question is the relationship between the segmental relaxation time obtained using these (and other) techniques and the shift factors resulting from the time–temperature superposition analysis of linear viscoelastic data. A compilation of both linear viscoelastic and molecular relaxation data for atactic polypropylene is shown in Figure 4.62 [185], where the shift factors have also been shifted vertically because none of the techniques provide absolute values of the relaxation time. Nevertheless, the main feature is unambiguous—the temperature dependencies of the relaxation time as viewed by the different experimental techniques are not the same in the glass transition region. One should note that this was not the main point that Roland et al.[185] were making; rather, their emphasis was on the fact that the terminal relaxation behavior (i.e., at temperatures much higher than Tg) determined by different techniques has the same temperature dependence. The difference in the Tg region then presents a challenge that has not yet been addressed. One possible explanation is a change in the shape of the relaxation spectrum as Tg is approached, where the reasoning is as follows: Since the various experimental techniques have differing sensitivities to the

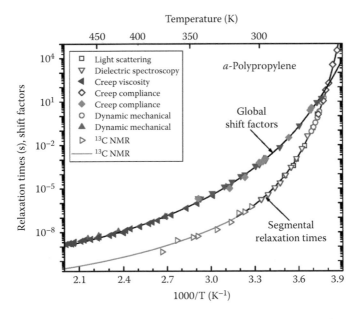

Figure 4.62 Comparison between mechanical and segmental relaxation times for atactic polypropylene in the region above Tg. (Reprinted with permission from C. M. Roland et al., *Macromolecules*, 34, 6159. Copyright 2001, American Chemical Society.)

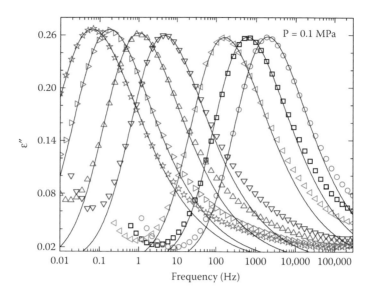

Figure 4.63 Dielectric loss peak in the α-relaxation region for poly(cyclohexyl methacrylate) (Tg = 75°C) at various temperatures from left to right: 87°C, 90°C, 96°C, 101°C, 118°C, 125°C, and 134°C. (Reprinted from *J. Non-Cryst. Solids*, 353, C. M. Roland and R. Casalini, 3996, Copyright 2007, with permission from Elsevier.)

underlying relaxation processes, the experimentally measured relaxation time is differentially weighted by those relaxation processes seen in a particular experimental method. As a result, a change in the width of the spectrum may affect different experiments differently. This leads to the question of whether this postulated change in the shape of the relaxation spectrum with temperature and pressure above Tg can be observed. Using dielectric spectroscopy, it was seen that the dielectric loss peaks at different temperatures above Tg cannot be superposed [135]; however, the mismatch is generally small as shown in Figure 4.63. Remarkably, when both temperature and pressure are varied in such a way that the position of the dielectric loss peak along the frequency axis (and hence the α-relaxation time) is kept constant, the shape of the entire curve is also constant and the curves corresponding to such temperature–pressure pairs lie on top of each other [135]. In the glassy state, the shape of the dielectric loss α-peak is difficult to ascertain as the relevant frequencies are too low. The β-peak, which is broad and asymmetric, decreases in height and shifts toward lower frequencies as temperature decreases; however, the effect of aging on the β-peak is weak [186]. In summary, the effect of temperature and pressure on segmental motion as directly measured via dielectric and other methods has been well established when the material is not in the glassy state, where there is a similar temperature dependence of the molecular and viscoelastic relaxation times well above Tg. However, there are differences between the molecular and mechanical relaxation times as Tg is approached [135]. There are very limited direct measurements of segmental mobility in the glassy state due to the very slow relaxation process, where there are perhaps indications that there is a weak aging time dependence [135], but this dielectric data is for the β versus the α-process.

4.10.2 EFFECT OF DEFORMATION ON MOBILITY

By definition, linear deformations do not change the relaxation time spectrum of a polymeric glass, which is the basis of linear viscoelasticity. On the other hand, it is generally believed that large deformations do affect the rate of relaxation, which is the generally accepted mechanism for yielding in polymeric glasses. Consequently, there have been several studies to assess the effect of large deformation on mobility using both mechanical and spectroscopic techniques.

A purely mechanical probe of the relaxation spectra are the "tickle" experiments, where a small magnitude probing deformation is applied on top of large nonlinear deformation as discussed previously in Section 4.8.1; however, the linear and nonlinear contributions to the overall response cannot be unambiguously decoupled [191]. Another example of a mechanical probing of the instantaneous relaxation time/spectra is stress relaxation following a constant strain rate deformation as discussed in Section 4.6.7, where the initial rate of stress relaxation is a reporter of the state of mobility in the material just prior to the moment when the stress relaxation has begun [123,192]. However, recent comparison of the characteristic relaxation time extracted from nonlinear stress relaxation with the characteristic relaxation time obtained by the photobleaching method [193] shows significant difference between the mechanical and molecular mobility measurements. In the purely mechanical experiments, it is clear that the macroscopic mobility increases with deformation; however, direct quantitative determination of the instantaneous relaxation spectrum is problematic. Specifically, since the overall relaxation behavior is nonlinear, it is not possibile to unambiguously decouple the instantaneous relaxation response due to the small mechanical perturbation from the underlying nonlinear mechanical response. Independent measurement of mobility using molecular probes may provide valuable information on how the relaxation spectrum changes with deformation.

Loo et al. [194] reported mobility enhancement due to deformation using deuterium NMR. They found that a uniaxial extension of a Nylon 6 specimen near the glass transition temperature of the amorphous phase caused a change in the ratio of specific peaks in the deuterium NMR spectrum, which indicated increased mobility in the amorphous regions of the material. Although the increase in mobility from the deuterium NMR was unambiguous, the results were qualitative, where the extent of mobility enhancement was difficult to quantify.

By far, the most detailed information on local mobility in the glassy state has been obtained using an optical fluorescent probe reorientation technique, known as "photobleaching," pioneered by Ediger and coworkers [130,188]. The photobleaching technique monitors rotation of dye molecules dispersed in a medium under the assumption that the relative rate of the rotation is a reporter of the local molecular mobility of the medium, that is, the polymeric glass [189]. The photobleaching experiment consists of the following steps: first, a high-intensity linearly polarized beam is used to photobleach the fraction of the dye molecules whose dipoles are at that instant parallel to the polarization of the beam; second, a low-intensity circularly polarized "reading" beam is applied, which induces fluorescence; and third, the parallel and perpendicular components (with respect to the polarization of the bleaching beam) of the fluorescent light are collected and analyzed. Because of the photobleaching, the intensity of the parallel component of the fluorescent light is initially lower than that of the perpendicular component. With time, the dye molecules rotate so that the effect of the bleaching is erased and the distribution of the dye molecules orientation becomes uniform. The experimentally measured decay of the dipole orientation autocorrelation function is a reporter of the ensemble-averaged molecular rotation of the dye molecule and hence of the "frozen" nature of the dye molecule in the surrounding glassy polymer environment. Note: the thin-film specimen used in the photobleaching experiments often exhibits necking, where the local deformation in the region of the photobleaching laser beams was measured via relative displacement of fiduciary marks; thus, the local strain and strain rate are not identical to macroscopic strain and strain rate, where the local strain and strain rate are used in the analysis of the photobleaching relaxation data.

Three assumptions are implicit in using the photobleaching method as a reporter of molecular mobility during deformation: (i) the concentration of the dye molecules is low so that all hindrance to their rotational motion is caused by the surrounding polymer matrix and not interactions with other dye molecules, and (ii) the plasticizing effect of the dye molecules on the surrounding polymer matrix is negligible. If these two assumptions are satisfied, the key question is: Does the dye rotational correlation time correspond to the same segmental relaxation time that is observed via dielectric and NMR experiments? Ediger and coworkers [189] demonstrated that the rotational correlation time of the dye molecules in the photobleaching experiment is proportional to the segmental relaxation time of the polymer. However, one should remember that the dye rotational dynamics as probed via the photobleaching experiment are not a "segmental measurement" such as dielectric relaxation or NMR, but rather a measurement of the rotation of a rather large molecule through a spatial region comprised of a significant number of polymer chain segments. Thus, there is a third assumption: (iii) the correlation between the dye's rotational dynamics and segmental mobility that was observed in the absence of deformation still holds when the polymer is deformed. With these three assumptions, the photobleaching technique determines the evolution of segmental mobility during deformation. A single photobleaching measurement typically takes 300 s for the autocorrelation function to decay sufficiently in order to be robustly fit by the KWW function, which gives both the value of the average correlation time τ_c^{KWW} and the width of the spectrum via the β^{KWW} parameter. This means that the rate of deformation has to be slow enough that the state of the material does not change significantly over 300 s. With this caveat and the three assumptions, the photobleaching technique is a powerful and versatile tool for measuring segmental mobility in polymer glasses for a wide range of thermal and deformational histories.

The decrease in the rotational correlation time τ_c^{KWW} and decrease in the width of the spectrum as characterized by β^{KWW} was observed in extension in both nonlinear creep [188] and constant strain rate deformations for PMMA [195]. Necking typically occurred in these experiments; however, it was possible to extract the local strain and strain rate history in the spot where the photobleaching experiment was taking place. A representative example of the results obtained via photobleaching technique is shown in Figure 4.64 for nonlinear creep of bulk-polymerized PMMA (57% syndiotactic) at Tg-14°C. Two values of applied stress were

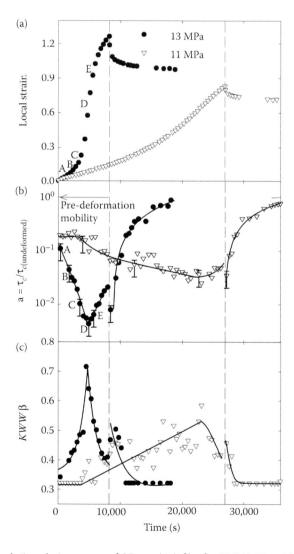

Figure 4.64 Mobility evolution during creep of 25 μm thick film for PMMA (Tg = 121°C) at Tg-14°C. Applied stresses are indicated in the figure. The stresses were removed at the times indicated by vertical dashed lines. (a) Local strain, (b) relative rotational correlation time, and (c) width of the spectrum as parameterized by the KWW parameter β. (H.-N. Lee et al., *J. Polym. Sci. Polym. Phys.*, 2009, 47, 1713. Copyright Wiley-VCH Verlag GmbH & Co. KGaA. Reproduced with permission.)

used, where both produce all five stages of creep. For the larger stress of 13 MPa, the increase in strain rate during the Stage III creep is much more pronounced. Examining Figure 4.64 [196], the rotational correlation time:

1. Instantaneously drops by less than an order-of-magnitude when the load is applied.
2. Decreases gradually during the Stage II (secondary) creep, which is seen clearly in case of the smaller applied stress of 11 MPa.
3. Reaches a minimum during the Stage III (tertiary) creep, where for the largest stress of 13 MPa, the rotational correlation time has decreased by up to three order-of-magnitudes as compared to that in the undeformed sample.
4. Increases again during Stages IV and V creep.
5. Instantaneously drops upon removal of the load, although this behavior is only seen if strain value is large [197].
6. Gradually increases back to its pre-deformation value.

The β^{KWW} parameter describes the width of the rotational correlation response and is correlated with the rotational correlation time so that a decrease in τ_c^{KWW} is accompanied by an increase in β^{KWW}. When the stress is 13 MPa, the maximum value is $\beta^{KWW} = 0.71$ as compared to the undeformed sample, where $\beta^{KWW} = 0.32$—this is a dramatic narrowing with deformation of the relaxation spectrum by approximately 3.5 logarithmic decades.

The photobleaching technique can be used to address the question of what physical quantities correlate with the segmental relaxation, which is a critical question in understanding the glassy state and has important consequences with respect to constitutive modeling. In Figure 4.65, the correlation between τ_c^{KWW} and the current, true stress value is shown for a creep deformation. It should be noted that due to changing cross-sectional area, the value of true stress changes during the course of a single creep experiment. In addition, the measurements obtained during recovery (i.e., at zero applied stress) are also shown in Figure 4.65. Several conclusions can be drawn: (i) stress is clearly not the variable that controls molecular mobility in glass, since during creep recovery, the stress is zero and yet the rotational correlation time changes by more than two orders-of-magnitude, and (ii) even during a single-step creep, the Eyring model only describes the behavior of the rotational correlation time at small stresses. Specifically, the main feature of the Eyring model is that the relaxation time (or more precisely the nonlinear viscosity to which the relaxation time is directly proportional) is approximately an exponential function of stress [198] as shown by the dashed line in Figure 4.65. The failure of the Eyring postulate to describe the relaxation time during deformation is critical, because a large number of constitutive models are constructed using the Erying postulate (see Chapter 14).

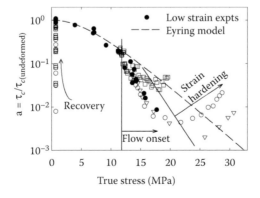

Figure 4.65 Mobility evolution during creep and recovery for PMMA (Tg = 121°C) film at Tg-19°C. The nominal applied stress is diamonds—13.5 MPa, squares—15.5 MPa, and triangles—16 MPa. (H.-N. Lee et al., *J. Polym. Sci. Polym. Phys.*, 2009, 47, 1713. Copyright Wiley-VCH Verlag GmbH & Co. KGaA. Reproduced with permission.)

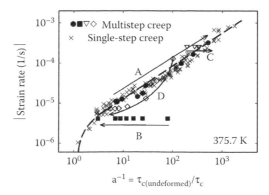

Figure 4.66 Mobility by photobleaching versus local strain rate in multi-step creep experiments for PMMA at Tg-19°C. A—first step— nominal stress 15 MPa, B—second—8 MPa, C—third—15 MPa, and D—fourth—0 MPa. (Reprinted with permission from H.-N. Lee et al., *Macromolecules*, 42, 4328. Copyright 2009, American Chemical Society.)

Another candidate variable for controlling mobility is the strain rate. In Figure 4.66, τ_c^{KWW} as determined from the photobleaching experiment is plotted as a function of the instantaneous strain rate, where the data are obtained from the single- and multistep creep experiments [196,197]. The correlation is reasonable if only single-step creep experiments are considered, where τ_c^{KWW} is a function of the strain rate. Photobleaching studies during constant strain rate deformation were also performed using the same PMMA material [195], where the strain rate dependence of τ_c^{KWW} was the same as seen in the creep experiments shown in Figure 4.66. Thus, the average rotational correlation time determined via the photobleaching experiments shows a strong correlation with the instantaneous strain rate for both single-step nonlinear creep and constant strain rate deformations. However, once the multistep creep experiments are considered, the apparent correlation between the rotational correlation time and the strain rate fails dramatically as shown in Figure 4.66. Specifically, the multistep creep experiments include partial unloading to a stress where the creep and recovery processes are nearly balanced, resulting in a nearly constant strain rate (i.e., step B in Figure 4.66). Even though the strain rate is small and constant, the rotational correlation time decreased by nearly an order-of-magnitude, clearly showing that the rotational correlation time is not a function of the strain rate. Thus, an important conclusion from the experiments of Ediger and coworkers that measured the rotational mobility of a probe dye during active deformations is that neither stress nor strain rate controls molecular mobility. Strain is also easily dismissed as the mobility-controlling variable by examining Figure 4.64, where the same value of strain reached during creep under 13 MPa and 11 MPa stress corresponds to vastly different values of the rotational correlation time.

In addition to the photobleaching technique, dielectric spectroscopy has also been employed to study the effect of deformation on mobility in polymeric glasses. Kalfus et al. [199] observed that the in-phase dielectric permittivity (normalized by the value without deformation) was unaffected by deformation versus the out-of-phase dielectric permittivity that was affected by a constant strain rate deformation as shown in Figure 4.67. Presumably the increase in ε'' indicates higher rate of segmental relaxation during deformation. This increase in the out-of-phase dielectric permittivity apparently reaches saturation at yield, where the maximum value increases with the strain rate.

4.10.3 Dynamic heterogeneity

"Dynamic heterogeneity" is used by the glass research community to designate the well-established observation that regions in glass separated by a few nanometers exhibit dramatically different mobility. Fluctuations in mobility at a nanometer scale are not unexpected, since in condensed matter, local thermodynamic quantities do fluctuate; hence, the mobility, which is presumably controlled by these quantities, will fluctuate as well. What makes the glass situation peculiar is that the spatial variations in mobility persist over

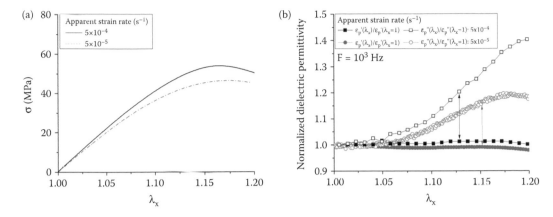

Figure 4.67 Effect of deformation on dielectric permittivity for PVC at 20°C. (a) Stress versus stretch ratio and (b) dielectric permittivities ε′ and ε″ at 1 kHz versus stretch ratio. (Reprinted with permission from J. Kalfus et al., *Macromolecules*, 45, 4839. Copyright 2012, American Chemical Society.)

astonishingly long times—at least as long (and perhaps even longer) than the characteristic time of the α-relaxation. In the case of polymeric glasses, this distribution of motional dynamics has been observed using multidimensional NMR in PVAc [200–202], polystyrene [203], and polycarbonate [204]. Using this and other techniques, the length scale of the heterogeneity has been estimated to be between 2 and 4 nm at the glass transition. The compilation of the observations of dynamic heterogeneity by various experimental techniques and of the studies using the molecular dynamics (MD) simulations as well as discussion of possible physical mechanisms responsible for the dynamic heterogeneity can be found in several extensive reviews [2,205–208]. Although the phenomenon of dynamic heterogeneity is not in doubt, its causes are a subject of intense debate.

While most researchers agree that dynamic heterogeneity is important, it is not obvious how it manifests itself in macroscopic, thermo-mechanical experiments. In fact, the overwhelming majority of the constitutive models to date employs the traditional continuum assumption, where variations at nanometer scale are not even acknowledged. Perhaps, the inability of the current set of constitutive models to describe the range of thermo-mechanical relaxation phenomena reported in this chapter is an indication that neglecting dynamic heterogeneity may be not such a good assumption (see discussion in Chapter 14). With this comment in mind, we will include the dynamic heterogeneity on the list of important thermo-mechanical features of polymeric glasses.

SIGNATURES 10

S10.1 The segmental relaxation time obtained by various molecular techniques exhibits different temperature dependence in the Tg range as compared to the relaxation time obtained from mechanical experiments.

S10.2 The molecular mobility as probed by probe dye rotation and dielectric spectroscopy is enhanced by active deformation; specifically,

 a. The decrease in the average relaxation time with deformation is accompanied by the narrowing of the relaxation spectrum.

 b. For both nonlinear single-step creep and constant strain rate deformations, the magnitude of the decrease in the average relaxation time with deformation is correlated with the instantaneous strain rate.

 c. In creep experiments, the mobility is highest during tertiary creep.

 d. For multistep creep deformations, the average relaxation time is not correlated with stress, strain rate, or strain.

S10.3 Dynamic heterogeneity at nanometer scale persists over times equal to or exceeding the α-relaxation time.

4.11 DISCUSSION

Polymeric glasses exhibit diverse, time-dependent, thermo-mechanical behavior as summarized in this chapter. The key point of the chapter is that there should be a unified perspective of the glassy state, where the diversity of phenomena observed experimentally has a common origin. For example, one would not expect a different set of molecular relaxation processes to control nonlinear creep than those that control yield in a constant strain rate deformation or nonlinear stress relaxation. In the same vein, one would expect that the same underlying molecular processes control both volume relaxation as measured in a dilatometer and the enthalpy relaxation as measured in a DSC. Thus, the research community should embrace the diversity of phenomena exhibited by materials in the glassy state and not focus on a limited set of experiments, where all phenomena must be engaged in a unified manner if understanding is to emerge.

There is no point in repeating here the signatures that are given at the end of each section. There are, however, several features worth keeping in mind when conducting experimental studies of polymeric glasses:

1. The thermal history, i.e., cooling rate, temperature, sub-Tg annealing time, etc., affects the subsequent thermo-mechanical properties.
2. Unlike the case of elastic materials, the thermo-mechanical behavior depends upon (i) whether the deformation is strain or stress controlled, (ii) what is the rate of deformation, and (iii) the deformation history, e.g., the rate of the previous loading history affects the stress memory response even though the strain, stress, and strain rate are the same (see Section 4.7.8).
3. With the exception of isotropic volume relaxation, all deformations are three dimensional, where all components of the stress and strain tensors are needed to fully describe the material response, for example, the volumetric response in a constant strain rate uniaxial deformation provides valuable insight about whether volume is the controlling variable for yield (see Figure 4.28).

There are perhaps other common features, but these three are among the most significant. Concerning the list of thermo-mechanical signatures identified in this chapter, this is not a final list. The signature list should be taken as a proposal that needs to be expanded, pruned, and refined by the research community.

A major point of this chapter is that if the objective is to develop a fundamental understanding of the thermo-mechanical behavior of polymeric glasses, one should not focus on a limited set of experiments—even if they are experimentally convenient. As an example, free volume models have been extensively used to describe relaxation in polymeric glasses, where they can effectively rationalize specific volume relaxation [209], the temperature dependence of $\log a_T$ in the region above Tg [34,210], and even the occurrence of yield in uniaxial extension [211]. However, the free volume model fails miserably when uniaxial compression is considered. Specifically, the specific volume (and hence the free volume) decreases during compression and thus according to the free volume model, a material in compression should harden and not exhibit yield, but this prediction is clearly contrary to what is observed experimentally. If compression yield had always been a major part of the data set, the free volume model would have been discarded by the polymer research community decades ago. There are other examples of where the consideration of a more extensive data set can provide valuable insight into proposed mechanisms for describing the physics of glasses.

Sometimes a single experiment is discriminating. For example, the reversal in the rate of stress relaxation with the loading strain rate (see Figure 4.41) calls into question the basic idea that the mobility always increases with the extent of deformation—the key idea behind a number of the nonlinear viscoplastic and viscoelastic constitutive models (see Chapter 14). In contrast, sometimes it is the full set of experiments that provide the key challenge to an idea. For example, consider the short time annealing experiments in the Kovacs volume relaxation data (see Figure 4.17), which by itself appears innocuous (if truth be told, almost boring). However, when considered in concert with the rest of the Kovacs volume relaxation data, it is impossible to predict using the traditional type of relaxation models no matter what is the temperature and volume dependence of the $\log a(T,V)$ shift factor, which calls into question the basic assumption of thermorheological simplicity that is an inherent component to a whole class of constitutive models (see Chapter 14). Researchers should always be looking for a single definitive experiment that provides unique insight, for example, the recent Ediger photobleaching experiments [130] where the KWW-β changed substantially with deformation, thus, clearly showing that the shape of the relaxation spectrum is not constant during deformation. However, in most cases, it will be the diverse set of experiments versus a single experiment that provides the basis for insight.

Examining the data reported in this chapter, one observes that most of the data sets are for different materials. This diversity of materials used in the polymeric glass mechanics research community makes it impossible to compare data across research groups as well as extend data sets taken in the past. This is a significant problem. Contrast this situation with that of the polymer rheology community that uses a series of well-characterized, monodisperse polymers when developing fundamental understanding. Moreover, many interesting thermo-mechanical studies use industrially produced materials that are not characterized with respect to molecular composition, the presence of additives, molecular weight distribution, and so on, where even if this information is available it is not always reported. This situation is understandable when considering (i) the commercial importance of these materials and thus research support from industry and (ii) that many of the earlier researchers in this field came from the mechanical arts versus the chemical sciences where a more rigorous chemical characterization of materials is expected. However, fundamental understanding of the thermo-mechanical behavior of polymeric glasses will require studies done with better-characterized materials— materials that can be robustly reproduced in other research laboratories. In order to address this challenge, our group has recently developed an amine-cured polymeric glass with an experimentally convenient Tg of 75°C that can be reproducibly made, where the starting materials are both readily available and can be easily evaluated for purity [103]. In our opinion, the glassy mechanics community needs to develop several other reference polymeric glasses that can be shared.

One area of study that is immensely important for developing understanding, but that has been sorely neglected, is the combined measurement of stress/strain and enthalpy during deformation. The deformation calorimetry results reported in Section 4.10.2, show that a significant fraction of the overall work of deformation goes into heat. Any thermodynamically consistent treatment of deformation cannot ignore the contribution of heat to the constitutive description. Yet, there is a dearth of combined stress and enthalpy studies of the large deformation behavior of polymeric glasses. The very informative deformation calorimetry experiments of Oleinik and coworkers [158] are probably not doable for most research groups; however, important information can be obtained via DSC studies *after* the material has been deformed (see Figure 4.60 and associated discussion) using commercial mechanical and DSC instruments. The current data is quite challenging, where to the best of our knowledge, there are no real explanations for even the most basic feature of the deformation-enthalpy data—for example, why does the heat capacity after deformation exhibit such a broad exotherm that starts well before Tg? What about similar deformation-induced enthalpy effects for nonlinear creep? As a function of sub-Tg aging time? For different loading/unloading deformations? And so on?

A final issue concerns the effect of chemical composition on the nonlinear thermo-mechanical behavior. Interestingly, for a few signature properties where diverse compositions have been studied (for example, temperature dependence of the segmental relaxation time above Tg), universal-like behaviors have emerged, suggesting that perhaps the fundamental mechanisms of glassy response maybe insensitive to chemical composition—this would have deep implications for modeling. One should note that this is the case in polymer rheology, where the flow behavior depends upon the molecular weight (or equivalently the chain length), topology (i.e., linear, star, branched, etc.), and chain stiffness versus specific molecular details. To conclude that there is a universality that applies to all the thermo-mechanical signatures of polymeric glasses listed in the previous sections would be a powerful statement indeed. Unfortunately, the data required for this type of generalization are lacking. As mentioned above, a single well-characterized system does not exist for which a sufficiently diverse suite of experiments has been obtained, where the establishment of universality requires several such systems. Contrast this with the situation in polymer rheology, where there are numerous systematic studies of the effect of molecular weight, degree/type of long-chain branching, chain stiffness, solvent strength, and so on on a wide range of nonlinear rheological properties of polymer melts and solutions. It has been these experimental studies that have provided the foundation for the fundamental models that describe the molecular origins of polymer rheology. There are also numerous systematic studies of the effect of molecular architecture on the linear viscoelastic behavior of polymers in the transition region above Tg [10]; however, there are few studies of how systematic changes in the molecular architecture affect the nonlinear thermo-mechanical behavior of polymeric glasses. These studies are sorely needed, because a fundamental model of polymeric glasses needs to be able to describe more than just how Tg changes with molecular structure, although even this is currently a significant challenge.

The concerns expressed in the last several paragraphs should not be taken as pejorative comments on what is a vigorous, strong research community. As shown in this chapter, there is a wealth of information that has already been obtained with respect to the thermo-mechanical behavior of polymeric glasses—experimental studies that already over-challenge all existing models of the glassy state and provide important insight. Rather, the hope is that this chapter will bring together in one place the diversity of thermo-mechanical phenomena that have already been observed for polymeric glasses and stimulate researchers to look for new points of intersection between the different types of data that can provide additional understanding and suggest new discriminating experiments. We believe this approach will stimulate the research community to work toward a more fundamental understanding of what is surely one of the most important challenges in condensed matter physics—why do polymeric glasses exhibit such interesting properties and how can this interesting physical behavior be described?

ACKNOWLEDGMENTS

This chapter was supported by National Science Foundation Grant Number 1363326-CMMI. This manuscript was started when JMC was on sabbatical at the Technische Universiteit Eindhoven in the Netherlands with Professor Leon Govaert, where both the stimulating conversations and financial support are gratefully acknowledged.

REFERENCES

1. C. A. Angell, *Science*, 267, 1924, 1995.
2. L. Berthier and G. Biroli, *Rev. Mod. Phys.*, 83, 587, 2011.
3. B. D. Coleman, *Arch. Ration. Mech. Anal.*, 17, 1, 1964.
4. D. Jou, J. Casas-Vazquez, and G. Lebon, *Extended Irreversible Thermodynamics* (Springer, Berlin), 4th edn., 2010.
5. M. Grmela and H. C. Ottinger, *Phys. Rev. E*, 56, 6620, 1997.
6. G. B. McKenna, in *Comprehensive Polymer Science. Polymer Properties*, edited by C. Booth and C. Price (Pergamon Press, Oxford), 311, 1989.
7. C. A. Angell, K. L. Ngai, G. B. McKenna, P. F. McMillan, and S. W. Martin, *J. Appl. Phys.*, 88, 3113, 2000.
8. G. B. McKenna, in *Long-Term Durability of Polymeric Matrix Composites*, edited by K. V. Pochiraju, G. Tandon, and G. A. Schoeppner (Springer, New York) , 237, 2012.
9. I. M. Hodge, *J. Non-Cryst. Solids*, 169, 211, 1994.
10. J. D. Ferry, *Viscoelastic Properties of Polymers* (John Wiley & Sons, New York), 3rd edn., 1980.
11. P. B. Bowden, in *The Physics of Glassy Polymers*, edited by R. N. Haward (Applied Science Publishers Ltd, London), 279, 1973.
12. R. N. Haward, in *The Physics of Glassy Polymers*, edited by R. N. Haward (Applied Science Publishers, Ltd, London), 340, 1973.
13. B. Crist, in *The Physics of Glassy Polymers*, edited by R. N. Haward and R. J. Young (Chapman & Hall, London), 155, 1997.
14. M. C. Boyce and R. N. Haward, in *The Physics of Glassy Polymers*, edited by R. N. Haward and R. J. Young (Chapman & Hall, London), 213, 1997.
15. I. M. Ward and J. Sweeney, *An Introduction to the Mechanical Properties of Solid Polymers* (John Wiley & Sons Ltd, Chichester), 2nd edn., 2004.
16. A. S. Argon, *The Physics of Deformation and Fracture of Polymers* (Cambridge University Press, Cambridge, UK), 2013.

17. S. Turner, in *The Physics of Glassy Polymers*, edited by R. N. Haward (Applied Science Publishers, Ltd, London), 223, 1973.

18. W. N. Findley, J. S. Lai, and K. Onaran, *Creep and Relaxation of Nonlinear Viscoelastic Materials* (Dover, New York), 1989.

19. C. A. Bero and D. J. Plazek, *J. Polym. Sci. Polym. Phys.*, 29, 39, 1991.

20. R. Greiner and F. R. Schwarzl, *Rheologica Acta*, 23, 378, 1984.

21. J. Malek, R. Svoboda, P. Pustkova, and P. Cicmanec, *J. Non-Cryst. Solids*, 355, 264, 2009.

22. J. Brandrup, E. H. Immergut, and E. A. Grulke, *Polymer Handbook* (Wiley, New York), 2336, 2003.

23. M. Schmidt and F. H. J. Maurer, *Macromolecules*, 33, 3879, 2000.

24. H.-J. Oels and G. Rehage, *Macromolecules*, 10, 1036, 1977.

25. J. E. McKinney and M. Goldstein, *J. Res. NBS A Phys. Chem.*, 78, 331, 1974.

26. M. Naoki, H. Mori, and A. Owada, *Macromolecules*, 14, 1567, 1981.

27. F. Briatico-Vangosa and M. Rink, *J. Polym. Sci. Polym. Phys.*, 43, 1904, 2005.

28. P. A. Rodgers, *J. Appl. Polym. Sci.*, 48, 1061, 1993.

29. D. M. Colucci, G. B. McKenna, J. J. Filliben, A. Lee, D. B. Curliss, K. B. Bowman, and J. D. Russell, *J. Polym. Sci. Polym. Phys.*, 35, 1561, 1997.

30. N. W. Tschoegl, *The Phenomenological Theory of Linear Viscoelastic Behavior: An Introduction* (Springer-Verlag, Berlin) 1989.

31. H. Vogel, *Phys. Z.*, 22, 645, 1921.

32. G. Tammann, *J. Soc. Glass Technol.*, 9, 166, 1925.

33. G. S. Fulcher, *J. Am. Ceram. Soc.*, 8, 339, 1925.

34. M. L. Williams, R. F. Landel, and J. D. Ferry, *J. Am. Chem. Soc.*, 77, 3701, 1955.

35. J. Zhao, S. L. Simon, and G. B. McKenna, *Nat. Commun.*, 4, 1783, 2013.

36. Y. C. Elmatad, D. Chandler, and J. P. Garrahan, *J. Phys. Chem. B*, 113, 5563, 2009.

37. I. Echeverria, P. L. Kolek, D. J. Plazek, and S. L. Simon, *J. Non-Cryst. Solids*, 324, 242, 2003.

38. X. Shi, A. Mandanici, and G. B. McKenna, *J. Chem. Phys.*, 123, 174507, 2005.

39. S. A. Hutcheson and G. B. McKenna, *J. Chem. Phys.*, 129, 074502, 2008.

40. P. A. O'Connell and G. B. McKenna, *J. Chem. Phys.*, 110, 11054, 1999.

41. S. L. Simon, J. W. Sobieski, and D. J. Plazek, *Polymer*, 42, 2555, 2001.

42. J. Zhao and G. B. McKenna, *J. Chem. Phys.*, 136, 154901, 2012.

43. L. C. E. Struik, *Polym. Eng. Sci.*, 17, 165, 1977.

44. P. A. O'Connell and G. B. McKenna, *Polym. Eng. Sci.*, 37, 1485, 1997.

45. A. Lee and G. B. McKenna, *Polymer*, 29, 1812, 1988.

46. L. Guerdoux, R. A. Duckett, and D. Froelich, *Polymer*, 25, 1392, 1984.

47. A. B. Brennan and F. Feller III, *J. Rheol.*, 39, 453, 1995.

48. N. G. McCrum, E. L. Morris, and P. Roy, *Soc. Lond. A Math.*, 281, 258, (1964.

49. N. G. McCrum and G. A. Pogany, *J. Macromol. Sci. Phys.*, 4, 109, 1970.

50. J. Y. Cavaille, C. Jourdan, J. Perez, L. Monnerie, and J. P. Johari, *J. Polym. Sci. Polym. Phys.*, 25, 1235, 1987.

51. G. Mikolajczak, J. Y. Cavaille, and J. Perez, *Polymer*, 28, 2023, 1987.

52. D. J. Plazek, *J. Phys. Chem.*, 69, 3480, 1965.

53. D. J. Plazek, *J. Polym. Sci. Polym. Phys.*, 6, 621, 1968.

54. J. Guo, L. Grassia, and S. L. Simon, *J. Polym. Sci. Polym. Phys.*, 50, 1233, 2012.

55. R. Tao and S. L. Simon, *J. Polym. Sci. Polym. Phys.*, 53, 621, 2015.

56. J. E. McKinney and H. V. Belcher, *J. Res. NBS A Phys. Chem.*, 67, 43, 1963.

57. G. Goldbach and G. Rehage, *Rheologica Acta*, 6, 30, 1967.

58. J. J. Tribone, J. M. O'Reilly, and J. Greener, *J. Polym. Sci. Polym. Phys.*, 27, 837, 1989.

59. Y. Meng and S. L. Simon, *J. Polym. Sci. Polym. Phys.*, 45, 3375, 2007.

60. P. K. Agarwal, Ph.D. Thesis, University of Pittsburgh, 1975.

61. D. H. Deng and W. G. Knauss, *Mech. Time-Depend. Mat.*, 1, 33, 1997.

62. S. B. Sane and W. G. Knauss, *Mech. Time-Depend. Mat.*, 5, 293, 2001.

63. T. Hecksher, N. B. Olsen, K. A. Nelson, J. C. Dyre, and T. Christensen, *J. Chem. Phys.*, 138, 12A543, 2013.

64. R. Zondervan, T. Xia, H. van der Meer, C. Storm, F. Kulzer, W. van Saarloos, and M. Orrit, *P. Natl. Acad. Sci. USA*, 105, 4993, 2008.

65. N. W. Tschoegl, W. G. Knauss, and I. Emri, *Mech. Time-Depend. Mat.*, 6, 3, 2002.

66. P. S. Theocaris and C. Hadjijoseph, *Colloid Polym. Sci.*, 202, 133, 1965.

67. D. J. O'Brien, N. R. Sottos, and S. R. White, *Exp. Mech.*, 47, 237, 2007.

68. A. J. Kovacs, *Fortschr. Hochpolym.-Forsch.*, 3, 394, 1963.

69. G. A. Medvedev, A. B. Starry, D. Ramkrishna, and J. M. Caruthers, *Macromolecules*, 45, 7237, 2012.

70. G. A. Medvedev and J. M. Caruthers, *Macromolecules*, 48, 788, 2015.

71. D. Ng and J. J. Aklonis, in *Relaxation in Complex Systems*, edited by K. L. Ngai and G. B. Wright (Naval Research Laboratory, Springfield, VA), 53, 1985.

72. L. C. E. Struik, *Polymer*, 38, 4677, 1997.

73. G. B. McKenna, M. G. Vangel, A. L. Rukhin, S. D. Leigh, B. Lotz, and C. Straupe, *Polymer*, 40, 5183, 1999.

74. S. Kolla and S. L. Simon, *Polymer*, 46, 733, 2005.

75. A. J. Kovacs, *J. Polym. Sci.*, 30, 131, 1958.

76. M. Delin, R. W. Rychwalski, J. Kubat, C. Klason, and J. M. Hutchinson, *Polym. Eng. Sci.*, 36, 2955, 1996.

77. R. Greiner and F. R. Schwarzl, *Colloid Polym. Sci.*, 267, 39, 1989.

78. C. R. Schultheisz and G. B. McKenna, in *North American Thermal Analysis Society (NATAS) 25th Annual Conference* McLean, VA, 366, 1997.

79. P. A. O'Connell, C. R. Schultheisz, and G. B. McKenna, in *Structure and Properties of Glassy Polymers*, edited by A. Hill and M. Trant (ACS Books, Oxford), 199, 1999.

80. N. O. Birge and S. R. Nagel, *Phys. Rev. Lett.*, 54, 2674, 1985.

81. J. Korus, M. Beiner, K. Busse, S. Kahle, R. Unger, and E. Donth, *Thermochimica Acta*, 304/305, 99, 1997.

82. H. Huth, M. Beiner, and E. Donth, *Phys. Rev. B*, 61, 15092, 2000.

83. A. Saiter, L. Delbreilh, H. Couderc, K. Arabeche, A. Schonhals, and J.-M. Saiter, *Phys. Rev. E*, 81, 041805, 2010.

84. H. Leyser, A. Schulte, W. Doster, and W. Petry, *Phys. Rev. E*, 51, 5899, 1995.

85. G. B. McKenna and S. L. Simon, in *Handbook of Thermal Analysis and Calorimetry*, edited by S. Z. D. Cheng (Elsevier, Amsterdam), 49, 2002.

86. I. Echeverria, P.-C. Su, S. L. Simon, and D. J. Plazek, *J. Polym. Sci. Polym. Phys.*, 33, 2457, 1995.

87. N. Shamim, Y. P. Koh, S. L. Simon, and G. B. McKenna, *J. Polym. Sci. Polym. Phys.*, 52, 1462, 2014.

88. P. Badrinarayanan and S. L. Simon, *Polymer*, 48, 1464, 2007.

89. P. Badrinarayanan, W. Zheng, Q. Li, and S. L. Simon, *J. Non-Cryst. Solids*, 353, 2603, 2007.

90. M. M. Santore, R. S. Duran, and G. B. McKenna, *Polymer*, 32, 2377, 1991.

91. R. S. Duran and G. B. McKenna, *J. Rheol.*, 34, 813, 1990.

92. L. C. E. Struik, *Physical Aging in Amorphous Polymers and Other Materials* (Elsevier, Amsterdam), 1978.

93. E. J. Kramer and L. L. Berger, in *Crazing in Polymers*, edited by H. H. Kausch (Springer, Berlin), 1, 1990.

94. C. Dreistadt, A.-S. Bonnet, P. Chevrier, and P. Lipinski, *Mater. Des.*, 30, 3126, 2009.

95. N. M. Ames, V. Srivastava, S. A. Chester, and L. Anand, *Int. J. Plast.*, 25, 1495, 2009.

96. H. G. H. van Melick, L. E. Govaert, and H. E. H. Mejer, *Polymer*, 44, 3579, 2003.

97. M. J. Kendall and C. R. Siviour, *P. Roy. Soc. A—Math. Phy.*, 470, 20140012, 2014.

98. R. B. Dupaix and M. C. Boyce, *Polymer*, 46, 4827, 2005.

99. M. Wendlandt, T. A. Tervoort, and U. W. Suter, *Polymer*, 46, 11786, 2005.

100. M. Aboulfaraj, C. G'Sell, D. Mangelinck, and G. B. McKenna, *J. Non-Cryst. Solids*, 172–174, 615, 1994.

101. C. G'Sell, J. M. Hiver, A. Dahoun, and A. Souahi, *J. Mater. Sci.*, 27, 5031, 1992.

102. C. G'Sell, S. Boni, and S. Shrivastava, *J. Mater. Sci.*, 18, 903, 1983.

103. J. W. Kim, G. A. Medvedev, and J. M. Caruthers, *Polymer*, 54, 2821, 2013.

104. J. M. Powers and R. M. Caddell, *Polym. Eng. Sci.*, 12, 432, 1972.

105. W. Whitney and R. D. Andrews, *J. Polym. Sci. Pol. Sym.*, 16, 2981, 1967.

106. J. A. Roetling, *Polymer*, 6, 311, 1965.
107. L. C. A. van Breemen, T. A. P. Engels, E. T. J. Klompen, D. J. A. Senden, and L. E. Govaert, *J. Polym. Sci. Polym. Phys.*, 50, 1757, 2012.
108. D. J. A. Senden, J. A. W. van Dommelen, and L. E. Govaert, *J. Polym. Sci. Polym. Phys.*, 50, 1589, 2012.
109. E. T. J. Klompen, T. A. P. Engels, L. E. Govaert, and H. E. H. Mejer, *Macromolecules*, 38, 6997, 2005.
110. I. M. Ward, *J. Polym. Sci. Pol. Sym.*, 32, 195, 1971.
111. P. B. Bowden and J. A. Jukes, *J. Mater. Sci.*, 7, 52, 1972.
112. R. Raghava, R. M. Caddell, and G. S. Y. Yeh, *J. Mater. Sci.*, 8, 225, 1973.
113. A. W. Christiansen, E. Baer, and S. V. Radcliffe, *Philos. Mag.*, 24, 451, 1971.
114. O. A. Hasan, M. C. Boyce, Z. S. Li, and S. Berko, *J. Polym. Sci. Polym. Phys.*, 31, 185, 1993.
115. L. E. Govaert, T. A. P. Engels, M. Wendlandt, T. A. Tervoort, and U. W. Suter, *J. Polym. Sci. Polym. Phys.*, 46, 2475, 2008.
116. Y. Nanzai, *Polym. Eng. Sci.*, 30, 96, 1990.
117. C. G'Sell, H. El Bari, J. Perez, J. Y. Cavaille, and G. P. Johari, *Mat. Sci. Eng. A—Struct.*, 110, 223, 1989.
118. L. E. Govaert, H. G. H. van Melick, and H. E. H. Mejer, *Polymer*, 42, 1271, 2001.
119. T. A. Tervoort and L. E. Govaert, *J. Rheol.*, 44, 1263, 2000.
120. D. J. A. Senden, J. A. W. van Dommelen, and L. E. Govaert, *J. Polym. Sci. Polym. Phys.*, 48, 1483, 2010.
121. S. Rabinowitz, I. M. Ward, and J. C. S. Parry, *J. Mater. Sci.*, 5, 29, 1970.
122. L. Anand and N. M. Ames, *Int. J. Plast.*, 22, 1123, 2006.
123. J. W. Kim, G. A. Medvedev, and J. M. Caruthers, *Polymer*, 54, 3949, 2013.
124. J. W. Kim, G. A. Medvedev, and J. M. Caruthers, *Polymer*, 54, 5993, 2013.
125. R. N. Haward and R. J. Young, *The Physics of Glassy Polymers* (Chapman & Hall, London), 1997.
126. R. A. Martin, MS Thesis, Purdue University, 2007.
127. P. J. Mallon and P. P. Benham, *Plast. Polym.*, 40, 22, 1972.
128. E. T. J. Klompen, T. A. P. Engels, L. C. A. van Breemen, P. J. G. Schreurs, L. E. Govaert, and H. E. H. Meijer, *Macromolecules*, 38, 7009, 2005.
129. G. A. Medvedev and J. M. Caruthers, *Polymer*, 74, 235, 2015.
130. H.-N. Lee, K. Paeng, S. F. Swallen, and M. D. Ediger, *Science*, 323, 231, 2009.
131. D. H. Ender, *J. Appl. Phys.*, 39, 4877, 1968.
132. Y. Nanzai, *JSME Int. J. A—Mech. M.*, 37, 149, 1994.
133. O. A. Hasan and M. C. Boyce, *Polym. Eng. Sci.*, 35, 331, 1995.
134. D. H. Ender, *J. Macromol. Sci. B*, 4, 635, 1970.
135. C. M. Roland, *Macromolecules*, 43, 7875, 2010.
136. D. M. Colucci, MS Thesis, Purdue University, 1992.
137. K. Chen, K. S. Schweizer, R. A. Stamm, E. Lee, and J. M. Caruthers, *J. Chem. Phys.*, 129, 184904, 2008.
138. S. Turner, *Polym. Eng. Sci.*, 6, 306, 1966.
139. T. Ricco and T. L. Smith, *Polymer*, 26, 1979, 1985.
140. A. F. Yee, R. J. Bankert, K. L. Ngai, and R. W. Rendell, *J. Polym. Sci. Polym. Phys.*, 26, 2463, 1988.
141. A. I. Isayev, D. Katz, and Y. Smooha, *Polym. Eng. Sci.*, 21, 566, 1981.
142. D. Katz and Y. Smooha, *J. Mater. Sci.*, 18, 1482, 1983.
143. J. J. del Val, A. Alegria, J. Colmenero, and C. Lacabanne, *J. Appl. Phys.*, 59, 3829, 1986.
144. S. Doulut, C. Bacharan, P. Demont, A. Bernes, and C. Lacabanne, *J. Non-Cryst. Solids*, 235–237, 645, 1998.
145. R. D. Andrews, *J. Appl. Phys.*, 26, 1061, 1955.
146. T. Pakula and M. Trznadel, *Polymer*, 26, 1011, 1985.
147. M. Trznadel, *J. Macromol. Sci. B*, 28, 285, 1989.
148. S. Cheng and S.-Q. Wang, *Macromolecules*, 47, 3661, 2014.
149. W. N. Findley and J. S. Y. Lai, *J. Rheol.*, 11, 361, 1967.
150. H. Lu and W. G. Knauss, *Mech. Time-Depend. Mat.*, 2, 307, 1998.
151. G. B. McKenna, *J. Rheol.*, 56, 113, 2012.
152. G. B. McKenna, *Int. J. Non-Linear Mech.*, 68, 37, 2015.

153. G. B. McKenna and A. J. Kovacs, *Polym. Eng. Sci.*, 24, 1138, 1984.

154. C. Bauwens-Crowet and J.-C. Bauwens, *Polymer*, 23, 1599, 1982.

155. C. Bauwens-Crowet and J.-C. Bauwens, *Polymer*, 27, 709, 1986.

156. G. W. Adams and R. J. Farris, *J. Polym. Sci. Polym. Phys.*, 26, 433, 1988.

157. G. W. Adams and R. J. Farris, *Polymer*, 30, 1824, 1989.

158. O. B. Salamatina, G. W. H. Hohne, S. N. Rudnev, and E. F. Oleinik, *Thermochimica Acta*, 247, 1, 1994.

159. S. V. Shenogin, G. W. H. Hohne, and E. F. Oleinik, *Thermochimica Acta*, 391, 13, 2002.

160. O. A. Hasan and M. C. Boyce, *Polymer*, 34, 5085, 1993.

161. J. B. Park and D. R. Uhlmann, *J. Appl. Phys.*, 44, 201, 1973.

162. V. A. Bershtein, V. M. Yegorov, L. G. Razgulyayeva, and V. A. Stepanov, *Vysokomol. Soedin. A*, 20, 2278, 1978.

163. Y. Nanzai, A. Miwa, and S. Z. Cui, *JSME Int. J. A—Mech. M.*, 42, 479, 1999.

164. E. F. Oleinik, *Prog. Colloid Polym. Sci.*, 80, 140, 1989.

165. M. S. Arzhakov and S. A. Arzhakov, *Int. J. Polym. Mater.*, 40, 133, 1998.

166. A.-C. Genix and F. Laupretre, *Macromolecules*, 39, 7313, 2006.

167. U. Tracht, A. Heuer, and H. W. Spiess, *J. Chem. Phys.*, 111, 3720, 1999.

168. S. A. Lusceac, C. Gainaru, M. Vogel, C. Koplin, P. Medick, and E. A. Rossler, *Macromolecules*, 38, 5625, 2005.

169. A. S. Maxwell, I. M. Ward, F. Laupretre, and L. Monnerie, *Polymer*, 39, 6835, 1998.

170. A. G. S. Hollander and K. O. Prins, *J. Non-Cryst. Solids*, 286, 12, 2001.

171. O. F. Pascui and D. Reichert, *Appl. Magn. Reson.*, 27, 419, 2004.

172. A. S. Kulik, H. W. Beckham, K. Schmidt-Rohr, D. Radloff, U. Pawelzik, C. Boeffel, and H. W. Spiess, *Macromolecules*, 27, 4746, 1994.

173. G. Fytas, *Macromolecules*, 22, 211, 1989.

174. V. N. Novikov, S. V. Adichtchev, N. V. Surovtsev, J. Wiedersich, A. Brodin, and E. A. Rossler, *J. Non-Cryst. Solids*, 353, 1491, 2007.

175. G. D. Patterson and P. J. Carroll, *J. Polym. Sci. Polym. Phys.*, 21, 1897, 1983.

176. L. Hong, B. Begen, A. Kisliuk, V. N. Novikov, and A. P. Sokolov, *Phys. Rev. B*, 81, 104207, 2010.

177. R. Bergman, L. Borjesson, L. M. Torell, and A. Fontana, *Phys. Rev. B*, 56, 11619, 1997.

178. G. Floudas and P. Stepanek, *Macromolecules*, 31, 6951, 1998.

179. T. Kanaya, R. Inoue, M. Saito, M. Seto, and Y. Yoda, *J. Chem. Phys.*, 140, 144906, 2014.

180. T. Kanaya, I. Tsukushi, K. Kaji, B. J. Gabrys, S. A. Bennington, and H. Furuya, *Phys. Rev. B*, 64, 144202, 2001.

181. E. Duval, L. Saviot, L. David, S. Etienne, and J. F. Jal, *Europhys. Lett.*, 63, 778, 2003.

182. A.-C. Genix, A. Arbe, J. Colmenero, J. Wuttke, and D. Richter, *Macromolecules*, 45, 2522, 2012.

183. A.-C. Genix and F. Laupretre, *Macromolecules*, 38, 2786, 2005.

184. C. M. Roland, M. J. Schroeder, J. J. Fontanella, and K. L. Ngai, *Macromolecules*, 37, 2630, 2004.

185. C. M. Roland, K. L. Ngai, P. G. Santangelo, X. H. Qiu, M. D. Ediger, and D. J. Plazek, *Macromolecules*, 34, 6159, 2001.

186. R. Casalini and C. M. Roland, *J. Non-Cryst. Solids*, 357, 282, 2011.

187. C. M. Roland and R. Casalini, *J. Non-Cryst. Solids*, 353, 3996, 2007.

188. H.-N. Lee, K. Paeng, S. F. Swallen, and M. D. Ediger, *J. Chem. Phys.*, 128, 134902, 2008.

189. T. Inoue, M. T. Cicerone, and M. D. Ediger, *Macromolecules*, 28, 3425, 1995.

190. A. Dhinojwala, G. K. Wong, and J. M. Torkelson, *J. Chem. Phys.*, 100, 6046, 1994.

191. G. B. McKenna, *J. Phys.: Condens. Matter*, 15, S737, 2003.

192. E. W. Lee, G. A. Medvedev, and J. M. Caruthers, *J. Polym. Sci. Polym. Phys.*, 48, 2399, 2010.

193. B. Bending and M. D. Ediger, *J Polym. Sci. Polym. Phys.*, 54, 1957, 2016.

194. L. S. Loo, R. E. Cohen, and K. K. Gleason, *Science*, 288, 116, 2000.

195. B. Bending, K. Christison, J. Ricci, and M. D. Ediger, *Macromolecules*, 47, 800, 2014.

196. H.-N. Lee, K. Paeng, S. F. Swallen, M. D. Ediger, R. A. Stamm, G. A. Medvedev, and J. M. Caruthers, *J. Polym. Sci. Polym. Phys.*, 47, 1713, 2009.

197. H.-N. Lee, R. A. Riggleman, J. J. de Pablo, and M. D. Ediger, *Macromolecules*, 42, 4328, 2009.

198. H. Eyring, *J. Chem. Phys.*, 4, 283, 1936.

199. J. Kalfus, A. Detwiler, and A. J. Lesser, *Macromolecules*, 45, 4839, 2012.

200. K. Schmidt-Rohr and H. W. Spiess, *Phys. Rev. Lett.*, 66, 3020, 1991.

201. U. Tracht, M. Wilhelm, A. Heuer, H. Feng, K. Schmidt-Rohr, and H. W. Spiess, *Phys. Rev. Lett.*, 81, 2727, 1998.

202. U. Tracht, M. Wilhelm, A. Heuer, and H. W. Spiess, *J. Magn. Reson.*, 140, 460, 1999.

203. J. Leisen, K. Schmidt-Rohr, and H. W. Spiess, *J. Non-Cryst. Solids*, 172, 737, 1994.

204. K. L. Li, A. A. Jones, P. T. Inglefield, and A. D. English, *Macromolecules*, 22, 4198, 1989.

205. L. Berthier, G. Biroli, J.-P. Bouchaud, L. Cipelletti, and W. van Saarloos, *Dynamical Heterogeneities in Glasses, Colloids, and Granular Media* (Oxford University Press, Oxford), 464, 2011.

206. M. D. Ediger, *Annu. Rev. Phys. Chem.*, 51, 99, 2000.

207 R. Richert, *J. Phys.: Condens. Matter*, 14, R703, 2002.

208. H. Sillescu, *J. Non-Cryst. Solids*, 243, 81, 1999.

209. A. J. Kovacs, J. J. Aklonis, J. M. Hutchinson, and A. R. Ramos, *J. Polym. Sci. Polym. Phys.*, 17, 1097, 1979.

210. A. K. Doolittle, *J. Appl. Phys.*, 22, 1471, 1951.

211. M. H. Litt, P. J. Koch, and A. V. Tobolsky, *J. Macromol. Sci. Phys.*, 1, 587, 1967.

POLYMER GLASSES IN CONFINEMENT

Correlating glass transition and physical aging in thin polymer films

CONNIE B. ROTH, JUSTIN E. PYE, AND ROMAN R. BAGLAY

5.1 INTRODUCTION

For over two decades, numerous studies have reported large shifts in the glass transition temperature T_g with confinement of the system to nanoscale dimensions [1–13]. These shifts in T_g can be several tens of kelvin in size and occur in either direction (up or down relative to bulk T_g) leading to enhanced or suppressed dynamics depending on the system. Although most heavily studied in polymers, similar phenomena are observed in small-molecule glass formers [1,8,14,15] and colloidal systems [16–21], indicating some universality to these effects. Other material properties related to T_g such as physical aging [22–29] and modulus [30–40] also tend to show changes with confinement at the nanoscale. Knowledge of these effects is important to our fundamental understanding of the glass transition, the development of technological applications such as microelectronic circuits [41–44] and gas separation membranes [29,45], and water dynamics in biological processes [46,47].

Such confinement effects of T_g have been frequently cited as "controversial" because of the wide variety of effects reported in the literature [48–53]. The difficulty is related to the fact that the glass transition is not yet understood well in bulk systems [54–59]. By its very nature, the glass transition is a nonequilibrium, many-body phenomenon: T_g represents the temperature upon cooling at which the cooperative dynamics in the system become too slow to remain in equilibrium given the available thermal energy before the temperature decreases further for the given cooling rate. This definition, although well defined experimentally, makes the glass transition a dynamic phenomenon depending on the cooling rate (speed) at which it is measured, which naturally leads to questions about the true existence of the transition from a thermodynamic sense [54–56]. Historically, most of our scientific foundation of statistical mechanics has been built on equilibrium systems with two-body interactions, while we have not yet fully developed the tools needed to understand the nonequilibrium behavior of complex, many-body cooperative phenomena.

One of the original motivations for the investigation of the glass transition in thin films or other confined geometries was the idea that reducing the sample size to values comparable to those of cooperative length scales associated with the glass transition should surely perturb those length scales and allow us to understand them better [1,7]. This idea has not proven to be so simple. Predominately, the phenomena and changes in material properties we associate with decreasing sample size appear to be the result of perturbing influences at the interfaces and boundaries of the material (for example, free surface or substrate interactions) [60]. As the surface-to-volume ratio of the material is increased, the perturbed dynamics and properties occurring at the boundary have a greater influence on the average properties displayed by the material. This is the fundamental idea behind polymer nanocomposites, reduce the filler size to nanometer length scales, resulting in large amounts of surface area between the polymer matrix and nanofiller surface [61], meaning there are considerable similarities between polymer properties in thin films and those in polymer nanocomposites [62–65]. As these boundary effects tend to dominate the observed phenomena, there is still considerable question and debate about whether or not there exists some "intrinsic" finite size effect, that is, change of the material properties simply from reducing the sample dimensions.

Polymers are further complicated by their wide range of length scales from segment or monomer size to that of the entire chain. Given that polymers can have unperturbed values for the radius of gyration R_g in bulk that are tens of nanometers in size, it is easy to investigate thin polymer films where the total film thickness is less than the unperturbed R_g making nonequilibrium chain conformations unavoidable. Yet, even when the overall polymer chain conformation may be distorted, it is usually possible to relax local conformations of chain segments leading to reproducible and reversible measurements of T_g [66,67]. Such annealing of local chain conformations appears to be sufficient for the local motions associated with T_g and physical aging that are unaffected by chain connectivity, and that show similar effects to small molecules and colloidal systems. However, distorted conformations of the overall chain can affect more macromolecular flow motions such as dewetting, diffusion, and viscosity [68–70].

This chapter attempts to provide a summary of our existing understanding of property changes associated with decreasing film thickness or sample size, the so-called "confinement effect." There have been many studies in this field, now spanning more than two decades; thus, to keep this attempt tractable, we focus here on the glass transition and physical aging, providing a summary of experimental results. Other chapters of this book will cover other material properties, as well as theoretical and computer simulation efforts. In particular, our focus here will be on what is known about the mechanism(s) that transport perturbed mobility from a boundary (i.e., free surface or other interface) into the material: the size of this length scale and its temperature dependence, and how it might relate to length scales associated with the glass transition. The goal will be to highlight what ideas have been asked and tested during the past two decades of intensive research efforts and what questions still remain open. Of the many experimental studies in this field over the years, there have certainly been some conflicting reports. In an effort to form a constructive framework for moving forward, the focus of this chapter will be to show that the majority of the studies do paint a coherent picture that we believe can form a viable foundation for developing a working understanding of these effects. Finally, we mention that a number of excellent reviews already exist for this vast field of research [5–10,51,71–73], and the reader is referred to them for a more complete picture.

5.2 LARGE SHIFTS IN THE AVERAGE GLASS TRANSITION TEMPERATURE WITH CONFINEMENT

Historically, this area of research started with the work of Jackson and McKenna investigating small-molecule glass formers, o-terphenyl (oTP) and benzyl alcohol, confined to controlled pore glasses with pore diameters of 4–73 nm [1]. They found the T_g of oTP confined to these nanoporous glasses, as measured by differential scanning calorimetry (DSC), to decrease by 18 K relative to the bulk value with decreasing pore size down to 4 nm; similar, but weaker effects were also observed for benzyl alcohol. This followed an earlier work [74] in which they had already observed decreases in the melting temperature T_m for a number of compounds, including oTP, upon confinement into these same nanoporous glasses. Over the years, such work on small molecules has continued (see, for instance, the review [8]), but the number of studies on thin polymer films has become more prevalent for primarily two reasons: (1) it is easier to control the confinement by simply changing the film thickness, and (2) polymer thin films are frequently used in industrial applications.

The first changes to T_g reported for thin polymer films with decreasing film thickness h were observed by Keddie, Jones, and Cory [2,3]. The $T_g(h)$ for supported polymer films were measured by ellipsometry by identifying T_g from the change in thermal expansion of the material. Polystyrene (PS) films supported on silicon showed large decreases in T_g of 20–25 K for films 10–15 nm thick [2], relative to the bulk value of T_g ($T_g^{bulk} \approx 100°C$ for PS) measured for thicker films. For poly(methyl methacrylate) (PMMA) films supported on silicon, $T_g(h)$ was found to increase slightly by a few degrees, while for PMMA supported on gold substrates, $T_g(h)$ decreased by ~5 K [3]. These initial studies made it clear that surface or boundary effects played a key role in the observed phenomenon. The data were fit to what has become known as the Keddie, Jones, and Cory functional form [2]:

$$T_g(h) = T_g^{bulk}\left[1 - \left(\frac{\alpha}{h}\right)^{\delta}\right],\tag{5.1}$$

a mostly empirical equation with two primary fitting parameters α and δ. To rationalize this equation, Keddie, Jones, and Cory proposed that the presence of the free surface might result in a "region of enhanced mobility," where the thickness ξ of this near-surface mobile layer would diverge at T_g^{bulk} following a typical power law behavior: $\xi(T) = \alpha[1 - (T/T_g^{bulk})]^{-1/\delta}$ [2]. A film thickness dependent $T_g(h)$ following Equation 5.1 would result when the thickness of this mobile layer grows to encompass the entire film: $\xi(T) = h$. Other functional forms, scaling as simply $1/h$, have been proposed with some theoretical basis [75–77]; however, they do not always produce the best fits to the experimental data.

Since these original measurements in 1994, this basic $T_g(h)$ behavior has been replicated by numerous groups and using various experimental techniques. Figure 5.1 shows $T_g(h)$ data for PS films supported on silica, collated in Roth and Dutcher [6], for a number of different experimental techniques collected by numerous groups [2,12,42,78–84]. The spread in the data can be explained by the observation that the transition tends to broaden with decreasing film thickness. The upper end-point of the transition appears to stay mostly fixed at T_g^{bulk}, while it is the lower end of the transition that shifts progressively to lower temperatures [81]. This broadening of the transition is believed to reflect the large gradient in dynamics that is established in thin films [81,85], from liquid- or near-liquid-like dynamics at the free surface to still basically bulk-like dynamics near the substrate [12]. Along with this broadening, the measured thermal expansion of the glassy state for

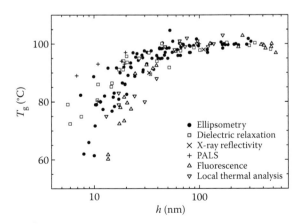

Figure 5.1 Film thickness dependence of the average glass transition temperature $T_g(h)$ measured for polystyrene (PS) films supported on a variety of substrates from a number of groups using different experimental techniques: ellipsometry [2,78–81], dielectric spectroscopy [82], x-ray reflectivity [83], positron annihilation spectroscopy (PALS) [84], fluorescence [12], and local thermal analysis [42]. (Reproduced from *J. Electroanal. Chem.*, 584, C. B. Roth and J. R. Dutcher, 13–22, Copyright 2005, with permission from Elsevier.)

thin films tends to become more liquid-like, while that in the liquid state tends to be unaffected by decreasing film thickness [81], consistent with this interpretation. This is also consistent with dielectric spectroscopy measurements that tend to find broadening of the transition with decreasing film thickness [86]. As different experimental techniques would sample this broad transition differently, the reported $T_g(h)$ values measured by the different techniques, as shown in Figure 5.1, would be expected to vary [85]. What is also interesting to note is that the $T_g(h)$ decrease shown in Figure 5.1 is only observed when measurements are made sufficiently slowly, typically at cooling rates of 0.5–3 K/min, whereas little to no $T_g(h)$ decrease is observed for cooling rates in excess of 130 K/min [87]. This rate dependence has also been observed in nanocalorimetry measurements [88–91] and in dielectric spectroscopy measurements [92,93], as discussed in Chapter 7. Interestingly, addition of plasticizer to the polymer also tends to reduce, or even eliminate, the thickness dependence to T_g changes [67,94–97].

In general, the length scale at which deviations in T_g from the bulk value are observed with decreasing film thickness occurs at quite large values of thicknesses ~100 nm, varying somewhat for different polymers [60,98,99]. Because this length scale of tens of nanometers can be comparable to polymer chain sizes, it has been suggested countless times that $T_g(h)$ changes might be due to changes in chain conformation (i.e., some chain confinement effect), and hence related to R_g. However, this notion has been disproven numerous times with studies demonstrating no observed molecular weight (MW) dependence to the shifts in $T_g(h)$ for supported films. The original Keddie, Jones, and Cory data for PS films were collected for three different molecular weights (120, 501, and 2900 kg/mol) showing no experimental difference in $T_g(h)$ [2], although perhaps there might have been the hint of a difference within the noise of the data, which is maybe why the notion has persisted. Over the years, several studies have rigorously tested for a possible MW dependence [80,98]. The same $T_g(h)$ decrease has been measured for PS, which deviates from bulk behavior below ~60 nm, for MWs as low as 2000 g/mol with $R_g \approx 2$ nm [100] and as high as 3×10^6 g/mol with $R_g \approx 50$ nm [98]. Thus, even very low MW chains that can be equilibrated to undistorted chain conformations demonstrate the same $T_g(h)$ decrease with confinement. These values also span the critical MW for entanglement, demonstrating that chain entanglements do not influence the $T_g(h)$ behavior [80,101]. However, a few studies have recently suggested that some MW dependence may be present for samples with very short chains [102,103] or measured at very slow cooling rates [104].

Possible changes in chain entanglements with decreasing film thickness have received renewed interest in recent years [105–107], with particular focus on how such changes might affect mechanical film properties such as modulus and yielding [37,108–110]. Somewhat related to this is whether chain connectivity in polymers plays any role in how dynamics can be perturbed near a boundary. Recent studies on polymers with different chain architectures suggest that entropic differences associated with restricted chain conformations of branched polymers near an interface affect material properties [34,40,111–115], with $T_g(h)$ even increasing with decreasing thickness for supported PS films of short, highly branched molecules [111]. These studies indicate there is another facet to these effects beyond simple linear chains that needs further exploration.

The main factors that do alter the $T_g(h)$ length scale for linear chains are monomer size and chain stiffness. In 2000, Zin et al. compared $T_g(h)$ for PS, poly(α-methyl styrene), and polysulfone, reporting that the decrease in $T_g(h)$ correlated with persistence length and not chain size [116]. However, more recent studies indicate that such correlations with $T_g(h)$ are less straightforward [33,117]. Torkelson et al. [98] have varied the monomer size studying PS, poly(4-methyl styrene), and poly(*tert*-butyl styrene) to compare confinement effects with the size of the cooperatively rearranging region (CRR), while Vogt et al. [118] studied a series of poly(n-alkyl methacrylate)s with varying CRR size. Yet, despite these careful studies, no direct correlation was found. More recently, efforts have turned to fragility, another important characteristic of glass-forming materials [99,119–122]. Fragility $m = (\partial \log \eta / \partial (T_g/T))|_{T=T_g}$ classifies glass-forming materials based on how quickly the dynamics slow down as T_g is approached [123], with most polymers being so-called fragile glass formers [124,125]. There are some indications that fragility also changes with confinement [92,119–121]; however, how that exactly tracks with $T_g(h)$ changes is not clear, but T_g and fragility are not strictly correlated in bulk systems either [125,126]. Nonetheless, studies of fragility in confined systems should provide us with useful insight into how the material's properties are changing. As many of the factors that affect $T_g(h)$

behavior are related to glass formation, there is reason to believe that studies of glassy properties in thin films can provide some insight into the glass transition and its length scales.

5.3 LOCAL T_g NEAR INTERFACES: GRADIENT IN DYNAMICS

Measurements of the average glass transition temperature $T_g(h)$ as a function of film thickness h for polymer films supported on silica (glass or silicon wafers that have a thin native oxide layer giving the same surface chemistry of hydroxyl groups) find that $T_g(h)$ can decrease [2,12,98], increase [60,127,128], or not change very much [3,44] with decreasing film thickness, depending on the particular polymer being measured. Figure 5.2a shows such $T_g(h)$ data for three representative polymers, PS, PMMA, and poly(2-vinyl pyridine) (P2VP), where differences in the bulk value of T_g for the different polymers have been accounted for by plotting $T_g(h) - T_g^{bulk}$ to highlight only the shift in T_g with decreasing thickness [60]. The behavior of PS and PMMA was discussed in the previous section, while here we can see that P2VP shows an increase in $T_g(h)$ as large as the decrease is for PS. Based on measurements of the local glass transition temperature near these interfaces [12,129], we now understand this behavior as a result of competing effects emanating from the free surface and substrate interfaces that average out to the measured $T_g(h)$ for the entire film [129]. The free surface tends to lead to an increase in local mobility resulting in a decrease in local T_g, while the substrate interface can reduce the local mobility, especially if hydrogen bonding between the polymer and substrate can occur, leading to a local increase in T_g. The strength of the interaction at each interface can vary depending on the particular polymer, leading to a variety of effects for the measured average $T_g(h)$. For example, it has been found that the strength of the free surface effect for PMMA is approximately a third that for PS [130,131].

Over the years, a great deal of information has been uncovered about the specific conditions at the boundaries of the free surface or substrate interface for different polymers. The most direct measurements have been done using a labeled fluorescence method to measure the local T_g of thin 10–15 nm layers placed at the free surface or substrate interface [12,129]. This work, pioneered by Torkelson and coworkers [12,132], was the first to demonstrate conclusively that the free surface was the source of the T_g reduction in thin films. A 14-nm thick free surface layer placed at the top of bulk PS showed a T_g decrease of −32 K relative to the bulk value, while an equivalent labeled layer placed within the film or at the substrate interface measured bulk T_g [12], as illustrated in Figure 5.2b. Subsequent measurements on a series of different polymers confirmed the notion that the average T_g of the film reflects the competition of boundary effects between enhanced mobility originating from the free surface and reduced mobility from the substrate interface [129]. For example, for PMMA a 12-nm thick layer at the free surface showed a local T_g decrease of −7 K, while a layer next to the silica substrate showed a local T_g increase of +10 K, explaining how competing boundary effects at the two interfaces can lead to an average $T_g(h)$ that only increases slightly with decreasing film thickness [129].

Figure 5.2b summarizes what is known about the conditions at the boundaries for the three polymers whose average $T_g(h)$ behavior is shown in Figure 5.2a. The magnitude of the T_g shift at both the free surface and the substrate interface varies with chemical structure of the polymer, giving rise to average $T_g(h)$ values that can shift up or down with confinement. It seems intuitively reasonable that the magnitude of the local T_g increase at the substrate interface would vary with chemical structure of the polymer. In most cases, inspection of the polymer's chemical structure easily allows identification of possible chemical groups that can lead to hydrogen bonding with the hydroxyl groups on the substrate's surface. PMMA contains polar ester groups in its side group and the pyridine ring in P2VP can also facilitate bonding with the substrate, while no such specific chemical groups are present in PS.

What is more puzzling is why the magnitude of the local T_g reduction at a free surface would also vary with chemical structure of the polymer. Several studies on free-standing films have confirmed that the magnitude of the enhanced mobility at the free surface of PMMA is approximately a third that of PS [44,130,131,133], consistent with the local T_g fluorescence measurements of layers at the free surface [12,129]. Surprisingly, a recent study on free-standing films by Paeng and Ediger [133] indicates that P2VP exhibits an increase in mobility at the free surface comparable to that of PS, which would imply that the local T_g increase of P2VP at the silica substrate [60] must be very strong indeed to achieve average $T_g(h)$ values that increase so

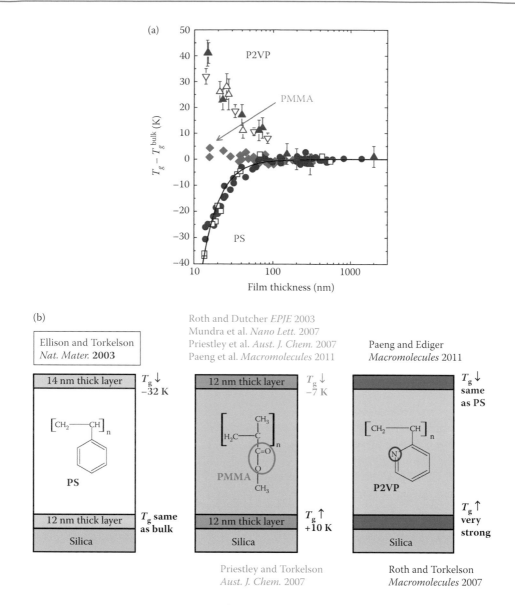

Figure 5.2 (a) Shifts in the average glass transition temperature, $T_g(h) - T_g^{\text{bulk}}$, as a function of film thickness for three characteristic polymers all supported on silica substrates: PS measured by fluorescence [12,98], poly(methyl methacrylate) (PMMA) by ellipsometry [3], and poly(2-vinyl pyridine) (P2VP) by fluorescence [60], x-ray reflectivity [127], and ellipsometry [128]. (Reprinted with permission from C. B. Roth et al., *Macromolecules*, 40, 2568–2574. Copyright 2007, American Chemical Society.) (b) Schematic summarizing local T_g values at the free surface and substrate interface for the three polymers from part (a). The local T_g of 12–14 nm thick layers have been explicitly measured by fluorescence for PS [12] and PMMA [129]. Several other studies have also inferred that the free surface effect is weaker for PMMA than PS by roughly a factor of three [44,130,131,133]. For P2VP, Paeng and Ediger [133] have measured the strength of the free surface effect for free-standing films and found P2VP to be the same as PS; thus, based on the large T_g increase observed for P2VP on silica [60], we can infer that the local T_g of P2VP near the silica substrate must be very strongly increased.

dramatically with confinement, as shown in Figure 5.2a. It is not clear why a seemingly neutral boundary like a free surface would have effects that depend on chemical structure. PMMA and PS have nearly identical values of surface tension near T_g, as well as comparable monomer size and persistence length [131]. If the ability of chain segments to locally slide around each other at the free surface is important, one possible factor that might explain the strength of the free surface effect is the value of the monomeric friction coefficient ζ, which is ~50% larger for PMMA than for PS at 100°C [131,134]. A recent study by White and Lipson using a thermodynamic model is able to predict a weaker $T_g(h)$ decrease for free-standing PMMA films compared to PS [135]. They attribute the effect to a stronger cohesive energy density for PMMA and that differences in the temperature dependence of the surface tension for the two polymers predict that the effective fraction of missing contacts for a PS segment located at the free surface relative to within the interior of the film is larger for PS than that for PMMA. Although the free surface is conceptually a well-defined idea and prevalent in experimental samples, theoretical treatment of the local increase in mobility at the free surface by estimating the number of missing interactions is nontrivial.

The importance of local perturbations at a boundary as the source of the T_g changes in confined systems is further supported by studies of capped films. Sharp and Forrest studied PS capped with Al or Au layers down to films as thin as 8 nm [11]. When the films were "properly" capped, removing the free surface, they exhibited T_g values to within ±1 K of the bulk value even for the thinnest films. Only films with a free surface exhibited T_g reductions with decreasing film thickness. A difficulty in capping the films arose because the evaporation of aluminum directly onto the polymer films led to the partial delamination of the Al capping layer and the creation of free surface. To avoid these concerns with the evaporation of the capping layer directly onto the polymer films, capped films were produced by a sandwich method in which two films of half thickness ($h/2$) were spin-coated separately and then sandwiched together to form so-called $2(h/2)$ films that were then annealed to create a well-capped, consolidated film. These $2(h/2)$ films did not show a T_g reduction when capped, but the T_g decrease could be recovered by subsequent removal of the capping layer and reintroduction of the free surface. These subtle differences associated with the evaporation of metallic electrodes atop polymer films that can eliminate or retain the presence of the free surface may explain why some dielectric spectroscopy studies show confinement effects, while others do not. In contrast, capped PMMA and P2VP films sandwiched between two silica substrates report very large T_g increases with confinement, starting at hundreds of nanometers in thickness, because of the strong attractive interactions between these polymers and the substrate interface [62]. These studies indicate that in the absence of interactions perturbing the dynamics at the boundary, pure confinement of the material to small dimensions does not appear to alter T_g, suggesting that there may be no intrinsic finite size effect. The work of Sharp and Forrest [11], along with that of Ellison and Torkelson [12], also demonstrates that dividing the sample into multiple layers, even with polymer chain conformations being highly distorted to be locally confined within these layers, does not appear to alter T_g either.

The local fluorescence measurements by Ellison and Torkelson showing that the local T_g of a 14-nm thick labeled layer at the free surface is reduced by −32 K also included measurements of buried layers, demonstrating that this local T_g is still perturbed relative to bulk quite deep into the material [12]. A 14-nm labeled layer located at a distance of ~15 nm from the free surface still had a local T_g reduction of −7 K. In fact, a depth of 30–40 nm from the free surface was required before the local T_g value returned to the bulk value for PS [12]. These measurements clearly demonstrate a long-range gradient in the local T_g of the material. This result is not to be confused with other types of measurements in the literature investigating liquid flow properties such as viscosity that often find liquid-like surface layers of only a few nanometers [136–138]. A region of reduced local T_g near the free surface may still be glassy, and therefore not liquid-like at the temperature being investigated, but it may also still not report bulk-like glassy dynamics and therefore be referred to as having enhanced mobility. This distinction has usually not been carefully made in the literature with both cases being rather loosely referred to as 'enhanced mobile surface layer.' In comparing different studies in the literature, it appears that a layer of a few nanometers located right at the free surface may always be liquid-like, while below that there appears to be a more extended, glassy region exhibiting a reduced local T_g whose gradient can extend for several tens of nanometers into the material before bulk glassy dynamics are recovered. The surface layer thickness of these two different types (or definitions) of enhanced mobile surface layers both have a temperature dependence, but are qualitatively opposite each other [139]. The thickness of liquid-like surface behavior naturally grows with increasing temperature as bulk T_g is approached from below [113,133,138], while the depth to which glassy dynamics report a reduced T_g grows with decreasing temperature below bulk T_g [28,140,141]. These differences are discussed in more detail below.

As explained, there is much evidence in the literature to support the conclusion that shifts in $T_g(h)$ with decreasing film thickness, so-called confinement effects, result from competing perturbations at interfaces that either increase or decrease the local mobility near the boundary, which is then propagated further into the material, resulting in a local gradient in dynamics. What mechanism controls or dictates the length scale of this gradient is not yet clear. For glassy dynamics such as physical aging (discussed below) or a gradient in local T_g, there is much evidence to support the idea that this length scale is related to cooperative motion and glass formation [28,139,141]. In contrast, for liquid flow properties, typically measured above T_g, this perturbation in dynamics is often found to be localized to within only a few nanometers of the free surface and depends significantly on the molecular weight, particularly if it is above or below entanglement [69,137,142,143]. In this chapter, we will focus our discussion on glassy dynamics, describing our understanding of gradients in local T_g and physical aging.

There have been numerous theoretical efforts over the years to describe the observed $T_g(h)$ behavior of thin films, a challenging task given that we are still lacking a fundamental understanding of the glass transition in bulk materials. One study that perhaps highlights some of the successes and challenges is by Lipson and Milner [144], who adapted the percolation model for glass formation developed by Long and Lequeux [145], to describe the local T_g gradient near a free surface and substrate interface. The percolation model describes heterogeneous dynamics in van der Waals liquids by dividing the material into fast and slow domains representing thermally induced density fluctuations, where the glass transition is controlled by the percolation of slow domains across the material [145]. Within this framework one could rationalize the $T_g(h)$ behavior of thin films by recognizing that for a free-standing film at the same temperature where bulk T_g occurs, slow domains within the film do not necessarily form percolating macroscopic slow aggregates, and conversely, boundaries with attractive interactions were more likely to facilitate percolation of slow domains. In the study by Lipson and Milner [144], they adapted the ideas of this percolation model to calculate the local $T_g(z)$ as a function of distance z from the boundary. For the attractive substrate interactions, the model did quite well. The difficulty was in comparing to experimental data because explicit data of $T_g(z)$ did not exist. Instead, Lipson and Milner needed to address the issue of how to average the local $T_g(z)$ values into thicker layers for comparison with the experimental data. For the free surface boundary, this model was unable to account for the large decreases in local T_g observed experimentally. This study, as well as others, highlights that the magnitude of the T_g reduction at the free surface is not well understood and not clear how to characterize theoretically [135,146–148].

More recently, studies have branched out to investigate different kinds of interfaces. It was originally assumed that an interface with a liquid would behave very similar to a free surface. However, surprisingly Wang and McKenna have demonstrated that a PS film floating on the surface of glycerol or an ionic liquid exhibits the same $T_g(h)$ as a PS film supported on silica [38,39]. Yet, PS spheres suspended in water demonstrate a large T_g decrease with decreasing sphere diameter consistent with the PS/water interface behaving like a free surface [13]. Studies have also investigated the perturbing effects of interfaces with another polymer [60,149–156]. Early studies showed that in some instances an underlying polymer–polymer interface can override the effects of the free surface [60], and there appears to be no difference between block copolymer interfaces and those between equivalent homopolymers [149]. Interfaces with a lower T_g polymer tend to lead to larger reductions in T_g as one might expect [150,155], but that is not true of all soft polymers, as polydimethyl siloxane (PDMS) appears to be an important exception [155,156]. Based on computer simulations, Simmons et al. [154] has recently proposed that the Debye–Waller factor (related to high-frequency shear modulus) may be the controlling factor, but this has yet to be correlated with experimental data for different systems.

If one avoids the free surface entirely and focuses on just a single polymer–polymer interface, the perturbations in local T_g can be quite dramatic and long ranged. Recently, Baglay and Roth [151] measured the profile in local $T_g(z)$ as a function of distance z through an interface between two semi-infinite layers of PS and poly(n-butyl methacrylate) (PnBMA) in contact. The two polymers differ in bulk T_g by 80 K; thus, the profile in $T_g(z)$, shown in Figure 5.3, transitions from $T_g^{bulk}{}_{PnBMA} = 20°C$ far from the interface in one direction to $T_g^{bulk}{}_{PS} = 100°C$ far from the interface on the opposite side. What is surprising about the data is that the local $T_g(z)$ is perturbed a great distance from the PS/PnBMA interface, and does not correlate with the local

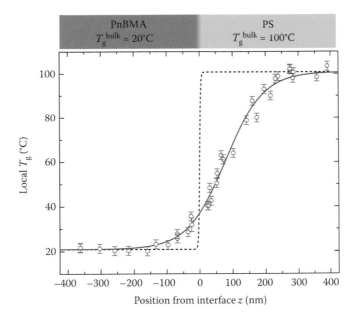

Figure 5.3 Profile of the local glass transition temperature $T_g(z)$ as a function of position z across a PS/PnBMA interface measured using fluorescence for 10–15 nm thick labeled layers placed at different distances from the interface by assembling samples from multiple high molecular weight layers and then annealing to form a consolidated material. The $T_g(z)$ profile representing the local cooperative dynamics transitions from $T_g^{bulk} = 20°C$ of poly(n-butyl methacrylate) (PnBMA) to $T_g^{bulk} = 100°C$ of PS over a much larger distance, and is asymmetric, relative to the composition profile (dashed curve) with 7 nm interfacial width. (Reprinted with permission from R. R. Baglay and C. B. Roth, *J. Chem. Phys.*, 143, 111101, 2015. Copyright 2015, American Institute of Physics.)

composition. The composition profile transitions sharply for this weakly immiscible blend with an interfacial width measured to be only 7 nm [150,151,157], yet it takes 350–400 nm for the local $T_g(z)$ value to transition from one bulk T_g value to another. Also striking is the asymmetric nature of the dynamical $T_g(z)$ perturbation. If we think of the PS/PnBMA interface as imparting a local dynamical perturbation, this perturbation propagates much further into the glassy PS side than the rubbery PnBMA side. The asymmetric behavior of the mobility profile across a rubbery/glassy interface was actually predicted by Tito et al. [153] using a limited mobility model that encodes how neighboring glassy regions can exchange local free volume or mobility. It is possible to qualitatively rationalize that the dynamical perturbation from an interface propagates further into hard glassy material, relative to soft rubbery material, by considering the recent theory by Mirigian and Schweizer [148,158,159] that has added a local elasticity term to allow for α-relaxation (cage breaking) events at lower temperatures. This suggests that the local compressibility of the material may influence how long ranged dynamical perturbations propagate in from an interface. Although the Mirigian and Schweizer [148,158,159] and Tito et al. [153] works both agree with the asymmetry being biased toward the glassy PS side, neither provide a good understanding why the length scale of the perturbation should propagate so far from the interface.

5.4 HIGH MOLECULAR WEIGHT FREE-STANDING FILMS: A SECOND MECHANISM

Because of the difficulty of trying to deconvolve competing effects from the free surface and substrate, much of the early work on characterizing $T_g(h)$ behavior was done on free-standing films. It was believed that this symmetric film geometry would simplify the analysis and facilitate an understanding of the free surface perturbation that could then be applied to understanding supported films. Instead, free-standing films turned out to have an unexpected richness of behavior in $T_g(h)$ that was not present in supported films. Contrary to

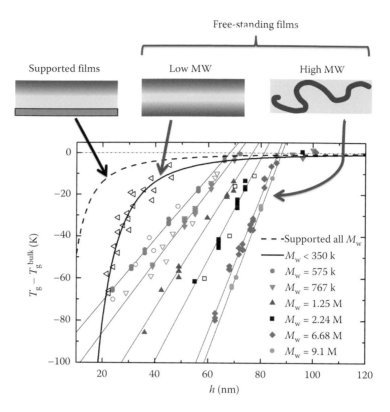

Figure 5.4 Compilation of $T_g(h)$ behavior for free-standing PS films with different molecular weights from data collected by Forrest, Dalnoki-Veress, Dutcher, Mattsson, and coworkers [4,5,79,140,160]. Low MW (M_w < 350 kg/mol) free-standing films (open triangles) show the same Keddie, Jones, and Cory $T_g(h)$ functional form as supported PS films (represented by dashed curve), but with a larger T_g reduction corresponding roughly to that for a supported film of half the thickness (i.e., $T_g(h)$ free-standing ≈ $T_g(h/2)$ supported) [140]. In contrast, high MW free-standing films (various solid symbols) show a linear $T_g(h)$ decrease with a M_w-dependent slope [160].

films supported on substrates, free-standing PS films exhibited a linear $T_g(h)$ reduction that was molecular weight dependent for high MW values, M_w > 350 kg/mol [4,79,140,160]. This behavior was extensively characterized by Forrest and Dalnoki-Veress [5].

Figure 5.4 summarizes the $T_g(h)$ behavior for PS films as it was classified in 2010 [5,6,140,160]. For low MW free-standing PS films with M_w < 350 kg/mol, there is no MW dependence to the T_g reduction and $T_g(h)$ follows the same functional form as that for supported PS films, the Keddie, Jones, and Cory equation previously given in Equation 5.1. The only difference is that the T_g reduction is larger, consistent with free-standing films having twice the amount of free surface as supported films. Forrest and Mattsson demonstrated that for a given free-standing film with film thickness h, the $T_g(h)$ reduction of a free-standing film was the same as that for a supported PS film with half the thickness, $h/2$ [140,161]. This observation agrees very nicely with our understanding of T_g reductions in supported polymer films described above; T_g reductions are caused by a gradient in dynamics emanating from the free surface such that free-standing films with twice the amount of free surface would be expected to have twice the magnitude in T_g reduction. It is higher MW free-standing films that behave qualitatively different. For free-standing PS films with M_w > 350 kg/mol, the $T_g(h)$ reduction is linear in h with a slope that is MW dependent [5,160]. In fact, the $T_g(h)$ data are found to scale exceedingly well with weight average molecular weight (M_w) such that all the reduced $T_g(h)$ data can be collapsed onto a single master curve [5,160]. Similar $T_g(h)$ reductions for high MW free-standing PS films have been observed using a number of different experimental techniques [162–166]. Qualitatively the same MW-dependent behavior is observed for free-standing PMMA films, except that the magnitude of the T_g reductions are approximately a third that of PS films [130,131].

The qualitative difference in $T_g(h)$ behavior between low and high MW free-standing films led to the suggestion that there may exist two different mechanisms that impart enhanced mobility from the free surface deeper into the film [131,145,161]. For the high MW, linear $T_g(h)$ reductions observed in free-standing PS films, the theoretical idea that has received the most attention was a "sliding" mode proposed by de Gennes [167,168]. The basic idea was that portions of the chain between two points located at the free surface could "slide" within its tube more easily because chain ends, which typically hinder the motion, did not have to poke into new material. This idea by de Gennes qualitatively agreed with the linear $T_g(h)$ reduction and suggested a way by which chain length dependence could occur. However, since then, this sliding mode idea has effectively been disproven. Lipson and Milner have taken this basic idea proposed by de Gennes and implemented it quantitatively, explicitly calculating the predicted MW-dependent $T_g(h)$ for this proposed sliding mode [169,170]. In this delayed glassification model, they demonstrate that the predicted $T_g(h)$ for different MW free-standing films merge for thinner films in complete contrast to the experimental data shown in Figure 5.4 where the $T_g(h)$ data for different MWs diverge with decreasing thickness. Kim and Torkelson experimentally tested de Gennes' sliding mode idea by constructing free-standing films out of multiple layers such that chains do not span across the thickness of the film [166]. Fluorescence was used to measure the local T_g within surface and middle layers within these free-standing films, finding that the same T_g reductions are observed whether chains span the thickness of the film from one free surface to the next, or not.

In 2011, Pye and Roth reported that both $T_g(h)$ behaviors could be observed in single high MW free-standing PS films [171]. Using ellipsometry to measure the thermal expansion of free-standing PS ($M_w = 934$ and 2257 kg/mol) films over an extended temperature range, they observed not only the MW-dependent $T_g(h)$ that had been previously reported for these films [4,79,140,160,162–166], but also a MW-independent $T_g(h)$ that corresponded well with the MW-independent $T_g(h)$ previously reported only for low MW ($M_w < 350$ kg/mol) free-standing films [140]. Figure 5.5 shows the two transitions observed and how they correspond to existing measurements. This observation in high MW free-standing films of a MW-independent $T_g(h)$ reduction following the Keddie, Jones, and Cory functional form, Equation 5.1, common to both supported films and low MW free-standing films, brings consistency to the literature suggesting that all films show $T_g(h)$ reductions compatible with a gradient in dynamics emanating from the free surface. With this perspective, additional measurements of high MW free-standing films from the literature could be viewed as being consistent with this picture. O'Connell and McKenna identified $T_g(h)$ for PS $M_w = 994$ kg/mol free-standing films [36,172] using a nanobubble inflation method [35], where the T_g reduction was in better agreement with the previously observed low MW free-standing film $T_g(h)$ behavior, as shown in Figure 5.5b. Paeng and Ediger also found their $T_g(h)$ data for free-standing PS films with molecular weights of $M_w = 8991, 1014, 168$ kg/mol to be MW independent and in line with the low MW free-standing film $T_g(h)$ behavior [138]. Physical aging measurements by Pye and Roth confirm that the film is glassy, undergoing physical aging below both transitions, and in particular also below the upper transition, but still above the lower transition [173].

As both $T_g(h)$ transitions were visible in the same film, Pye and Roth were able to compare the strength of the two transitions in Figure 5.5a by looking at the change in the thermal expansion coefficient occurring at the transitions [171,173]. On cooling from the liquid state, 80%–90% of the film was found to solidify at the upper transition, leaving only 10%–20% of the film mobile to lower temperatures. The upper transition shows no MW dependence and follows the same $T_g(h)$ functional form as the Keddie, Jones, and Cory equation 5.1 that has been previously associated with a gradient in dynamics emanating from the free surface. As this upper transition appears to encompass the free surface region, it is not clear where the remaining 10%–20% mobile fraction is located in the film. Thus, the picture emerging is that there is a primary confinement mechanism affecting all films that is responsible for propagating locally perturbed dynamics at the interfaces deeper into the material, as described for supported films above. This primary mechanism appears to control solidification of the majority of the film, or "matrix." Viscoelastic mechanical measurements on high MW free-standing films are consistent with the $T_g(h)$ of the film being the upper transition, which makes sense as these measurements would be dominated by the material response from this 80%–90% matrix. As shown in Figure 5.5b, the viscoelastic compliance measurements by O'Connell and McKenna using their nanobubble inflation method report $T_g(h)$ values consistent with the upper transition [36,172]. In addition, viscosity measurements using hole growth studies done on high MW free-standing PS films show only small shifts in the

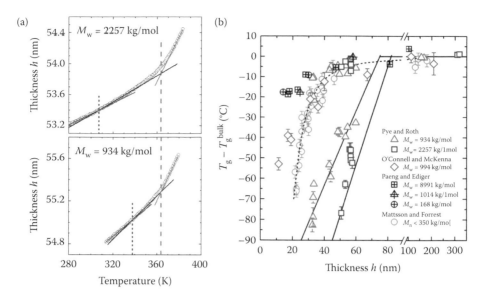

Figure 5.5 (a) Temperature-dependent thickness measurements by Pye and Roth [171] using transmission ellipsometry over an extended temperature range for high molecular weight free-standing PS films. Within single films, two transitions were observed: the weaker lower transition is M_w dependent in agreement with the literature data presented in Figure 5.4 [160]; the stronger upper transition was not previously observed in high MW free-standing films, but agrees with the MW-independent $T_g(h)$ previously observed in low MW free-standing films [140]. (b) Plot of $T_g(h) - T_g^{bulk}$ as a function of film thickness for free-standing PS films collated from a number of different studies. Both the lower and upper transitions measured by Pye and Roth [171] for free-standing films of $M_w = 934$ kg/mol (open triangles) and $M_w = 2257$ kg/mol (open squares), showing that the lower transition agrees with the M_w-dependent prediction by the Dalnoki-Veress scaling [5,160] (lines), while the upper transition agrees with the original MW-independent $T_g(h)$ data by Mattsson and Forrest [140] (open circles and dashed curve). $T_g(h)$ data (open diamonds) from O'Connell and McKenna [36,172] nanobubble inflation creep compliance mechanical measurements for high MW free-standing films of $M_w = 994$ kg/mol, and $T_g(h)$ data from Paeng and Ediger [138] dye reorientation measurements for $M_w = 8991$, 1014, and 168 kg/mol (crossed squares, triangles, and circles) are also shown, and found to agree reasonably well with the MW-independent $T_g(h)$ data for free-standing films (dashed curve). (J. E. Pye and C. B. Roth, *J. Polym. Sci., Part B: Polym. Phys.,* 2015, 53, 64–75. Copyright Wiley-VCH Verlag GmbH & Co. KGaA. Reproduced with permission.)

characteristic hole growth times $\tau(T)$ consistent with the $T_g(h)$ reductions of the upper transition that are MW independent [174,175]. For example, relative to bulk, the $\tau(T)$ values are shifted by -4.7 K for both 54 nm thick films of $M_w = 717$ kg/mol PS and 51 nm thick films of $M_w = 2240$ kg/mol [174,175], consistent with the upper $T_g(h)$ transition reduced by -5 K [171]. In contrast, the lower MW-dependent $T_g(h)$ transition is reduced by -28 K and -72 K relative to T_g^{bulk} for these two films, respectively [160].

The lower, weaker transition shows a linear $T_g(h)$ behavior whose slope is MW dependent. To date, it has only been observed in high MW free-standing films suggesting that this second mechanism is related to chain connectivity and perhaps the need for two free surfaces. In fact, Pye and Roth [171,173] have pointed out that interestingly the Dalnoki-Veress scaling [5,160], which perfectly collapses all the MW-dependent T_g reductions for high MW free-standing films, only collapses the data using the *weight* average molecular weight M_w, but does not collapse it if the *number* averaged molecular weight M_n is used, despite that all these measurements are done with narrow MW distribution samples. This observation further supports the notion that this second mechanism is related to chain connectivity, dominated by the largest chains in the system.

5.5 PHYSICAL AGING IN THIN POLYMER FILMS

Below the glass transition temperature, the material exists in a nonequilibrium glassy state having become kinetically trapped, "frozen in," at T_g during thermal cooling because the available thermal energy is insufficient to equilibrate the material. Held at a fixed temperature below T_g, this nonequilibrium state continues to evolve with time trying to reach equilibrium on a logarithmic time scale, a process generally referred to as physical aging [176–179]. Strictly speaking, evolution of thermodynamic state variables such as volume and enthalpy are referred to as structural recovery of the material, while other property changes such as mechanical behavior like modulus are classified as physical aging [180], but the term physical aging is commonly used to loosely and broadly refer to nearly all time-dependent evolution of the glassy state. In practice, equilibrium can only be reached on accessible time scales for perhaps temperatures 10–20 K below T_g [181,182].

These glassy dynamics can be measured to learn about the nonequilibrium nature of the material and the stability of the glassy state. In most cases, glass formation results in the material having a lower density than it would ideally have at equilibrium; thus, the most direct method of characterizing glassy dynamics is to measure the densification of the material with time. The overall volume decrease that occurs in the glassy state is exceedingly small, typically much less than 1%, but usually leads to much more substantial and adverse changes in various material properties. Glassy state dynamics are often characterized by measuring a physical aging rate, originally defined by Struik for volumetric measurements as the slope in the volume decrease with logarithmic time, $\beta = -d(V/V_\infty)/d(\log t)$ [176]. Similar aging rates are defined for other experimental measures such as enthalpy [183], and although these different measures of aging for a given material are related, they do not correlate exactly [177]. The temperature dependence of the physical aging rate β follows a roughly parabolic shape starting from zero aging above T_g to some maximum aging rate typically tens of degrees below T_g where the thermodynamic driving force to reach equilibrium is strong and sufficient thermal energy is still available to lead to the evolution of the material; at lower temperatures, the aging rate again drops off as the available thermal energy decreases [176,178].

Given that physical aging is inherently tied to the glass transition, studies have naturally been done on the physical aging of thin films that exhibit $T_g(h)$ changes [22–26,28,29,113,150,184]. Kawana and Jones found that very thin films exhibit reduced or no physical aging when held at temperatures below bulk T_g, but above the reduced $T_g(h)$ of the film [22]. With PMMA, Priestley et al. was able to observe physical aging for thin films above bulk T_g when the $T_g(h)$ of the film was elevated [23]. In most cases, the physical aging behavior of the film does not simply scale with the shifted average $T_g(h)$ of the film [25], but instead the local aging rate at a given depth within the film is found to correlate with the local T_g at that position [24,28,113]. For example, Priestley et al. used fluorescence of labeled rotor dyes that are sensitive to local changes in density and free volume to measure the local aging rate near a free surface and silica substrate interface within PMMA films [24]. The local aging rate was found to be reduced near the free surface, consistent with a local T_g reduction. Having to account for local gradients in the physical aging rate with depth $\beta(z)$ significantly complicates the analysis when experiments are only able to measure an average aging rate for the entire film [28,113,150].

Roth and coworkers have developed an efficient ellipsometry method to characterize the physical aging rate of polymer films. Baker et al. demonstrated that the film thickness could be used as an equivalent measure for the volumetric aging rate, $\beta = -d(h/h_0)/d(\log t)$ [185]. Characterizing the densification of the film by the decrease in thickness on a logarithmic time scale gave a temperature-dependent aging rate equivalent to that from volume dilatometry measurements [178]. Alternatively, the refractive index could also be used to define an aging rate, but as these gave identical measures to within experimental error, as well as being more laborious, the thickness measure was preferred [185].

Pye and Roth investigated the effect of decreasing film thickness on the temperature-dependent physical aging rate $\beta(T)$ for PS films supported on silicon [28]. Figure 5.6 compares the physical aging rate for ~2400 nm thick (bulk) films and ~30 nm thin films of PS supported on silicon. Figure 5.6a plots the normalized film thickness h/h_0 as a function of the logarithm of time, where the aging rate can be identified as the slope of the data, for 2300 and 29 nm thick PS films aged at 338 K, the peak in the temperature-dependent aging rate for bulk films. The physical aging rate for the 29 nm thick film is less than that of bulk, consistent with the observations of Kawana and Jones [22]. Figure 5.6b shows that the physical aging rate $\beta(T)$ is reduced at all temperatures for thin 30 nm films relative to bulk. This observation cannot be simply interpreted as thin films having a lower average glass transition temperature, $T_g(h = 30 \text{ nm})$ reduced by 5–10 K relative to bulk [6] (see Figure 5.1). If that were the case, then one would expect the 30 nm data in Figure 5.6b

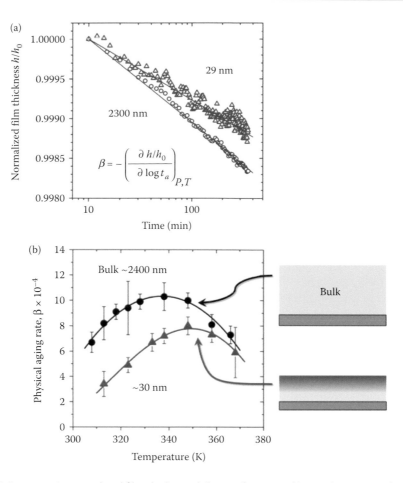

Figure 5.6 (a) Decrease in normalized film thickness h/h_0 as a function of logarithmic time after a temperature quench to 65°C from above T_g for PS films supported on silicon. The physical aging rate β, given by the slope of the data, is found to be less for thinner films (29 nm) than bulk (2300 nm). (Figure reproduced with permission from J. E. Pye et al., *Macromolecules*, 43, 8296, 2010.) (b) Temperature dependence of the physical aging rate $\beta(T)$ for ~2400 nm and ~30 nm thick films showing that physical aging is reduced at all temperatures in thin films. These data cannot be explained by a simple shift in the average $T_g(h)$ for thin films, but can be explained by a gradient in dynamics where the near free surface region does not undergo physical aging. (Reprinted with permission from J. E. Pye et al., *Macromolecules*, 43, 8296–8303, 2010. Copyright 2010, American Chemical Society.)

to be the same as the bulk data, but simply shifted to the left by 5–10 K to account for a shift in T_g. Instead, the 30 nm data are consistent with some fraction of the film not contributing to the aging signal. Presumably some region near the free surface that has a locally reduced T_g lower than the aging temperature is not aging because it is in equilibrium. To interpret this data correctly we must account for a depth-dependent aging rate $\beta(z)$ [28].

For thin PS films supported on silicon, we know quite a bit about the local depth dependent glass transition temperature, $T_g(z)$. Ellison and Torkelson showed that a 14-nm thick layer located at the free surface has a local T_g reduced by −32 K relative to bulk, while a layer located at the silica substrate interface reports a local T_g equivalent to bulk [12]. Some studies have estimated the local T_g right at the free surface to be ~300 K [161,186], while others have argued that the free surface may always be liquid-like [10,104,187]. Based on this information, we may estimate what the local depth-dependent aging rate $\beta(z)$ might be. Although the explicit functional form of $\beta(z)$ is not known, we do know that $\beta(z = 0)$ must go to zero at the free surface and return to the bulk aging rate deep into the material $\beta(z \to \infty) = \beta_{\text{bulk}}(T)$. Pye and Roth [28] modeled the depth-dependent aging rate $\beta(z)$ in two ways: (a) a simple two-layer model where some near free surface region of thickness $\Lambda(T)$ has zero aging rate, while the remainder of the film exhibits bulk-like aging dynamics; and (b) a gradient model where the local aging rate transitions exponentially between the two known limits at the free surface and deep into the material as $\beta(z,T) = \beta_{bulk}(T) \{1 - \exp[-z/\lambda(T)]\}$. Frieberg et al. [113] has argued that technically $\beta(z)$ may be nonmonotonic with depth at low aging temperatures because of the parabolic shape of $\beta_{bulk}(T)$, assuming the local gradient in $T_g(z)$ is monotonic with depth. However, they estimate that this nonmonotonic behavior in $\beta(z)$ would be located only within a narrow ~5 nm region near the free surface. Thus, even if the local aging rate $\beta(z)$ can be very different within regions only a couple of nanometers apart, the monotonic gradient approximation of $\beta(z,T)$ by Pye and Roth is likely a reasonable estimate. In practice, Pye and Roth [28] determined $A(T)$ for the two-layer model by identifying what equivalent layer thickness of the film exhibited bulk-like aging dynamics, and then subtracting this layer thickness from the total film thickness.

In the gradient model, the characteristic length scale $\lambda(T)$ plays the same role as $A(T)$, providing some measure of the depth to which the local aging rate is perturbed by the presence of the free surface before bulk-like aging dynamics are recovered deeper into the film. The temperature dependence of $A(T)$ and $\lambda(T)$ are plotted in Figure 5.7a, along with a similar measure of this length scale from low MW free-standing PS film data by Forrest and Mattsson [161]. To within experimental error, all measures of this length scale give roughly the same result, the length scale grows on cooling from a few nanometers near bulk T_g to a few tens of nanometers far below T_g^{bulk}. The data in Figure 5.7a would suggest that measurements of dynamics made at

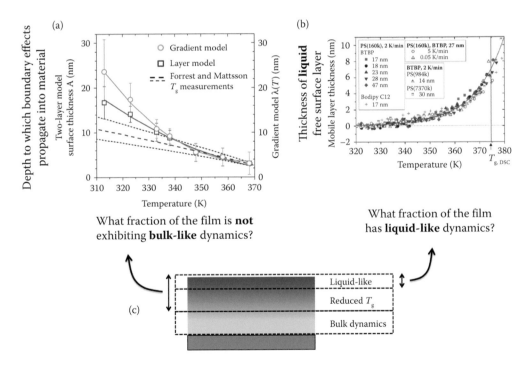

Figure 5.7 (a) Measure of the depth to which boundary effects are propagated into the material as a function of temperature based on analysis of the thin film $\beta(T)$ data from Figure 5.6b using a gradient model and two-layer model [28] (see text for details), along with a similar measure from Forrest and Mattsson [161] for $T_g(h)$ data from free-standing PS films. (Reprinted with permission from J. E. Pye et al., *Macromolecules* 43, 8296–8303, 2010. Copyright 2010, American Chemical Society.) (b) Measure of the thickness of a liquid-like free surface layer as a function of temperature based on analysis of dye reorientation measurements of free-standing PS films by Paeng and Ediger [138]. (Reprinted with permission from K. Paeng, S. F. Swallen, and M. D. Ediger, *J. Am. Chem. Soc.*, 133, 8444–8447, 2011. Copyright 2011, American Chemical Society.) (c) Schematic representation of the gradient in dynamics near a free surface where a temperature-dependent liquid-like layer exists right at the free surface, followed by a temperature-dependent glassy layer with locally reduced T_g underneath, and finally bulk dynamics far from the interface. Experimental measurements and analyses that identify *what fraction of the film has liquid-like dynamics* obtain temperature-dependent data like that in (b), while those that identify *what fraction of the film is not exhibiting bulk-like dynamics* obtain temperature-dependent data like that in (a).

elevated temperatures close to bulk T_g would be expected to only experience non-bulk-like dynamics within a few nanometers of the free surface. This may help to explain why various studies in the literature, mostly from surface viscosity measurements [136,137], report anomalous dynamics within only the first few nanometers of a free surface. Similarly, so-called "dynamic T_g" measurements [188], which are typically not found to show shifts in T_g relative to bulk [51], are usually done by measuring the temperature dependence of the mean α-relaxation time $\tau_\alpha(T)$ above T_g, for example using dielectric spectroscopy, and then extrapolating the data down to $\tau_\alpha(T) = 100$ s to identify a dynamic T_g value. According to the data shown in Figure 5.7a, measurements done above T_g would not be expected to exhibit dynamics different from bulk beyond a very narrow surface layer of perhaps a couple of nanometers or less.

Other studies have also measured the temperature dependence of a free surface layer, obtaining qualitatively opposite results [113,133,138,189]. Figure 5.7b graphs data of the "mobile layer thickness," assumed to be located at the free surface, as a function of temperature [138]. The data here are from Paeng and Ediger's work using dye-reorientation to measure the effective α-relaxation within thin free-standing PS films and then estimating the fraction of the film already liquid-like as a function of temperature. As shown in Figure 5.7b, the liquid-like surface layer thickness grows on increasing temperature from basically zero far from T_g to ~7 nm for PS films at T_g^{bulk}. Intuitively this makes sense because naturally the fraction of the film that is liquid-like must grow on increasing temperature [9]. How is it we get both the data in Figure 5.7a and 5.7b as the temperature dependence for some measure of the free surface mobile layer? On first consideration they appear to completely contradict each other. The answer to this lies in precisely how the free surface mobile layer was defined. In Figure 5.7a, the mobile surface layer was defined as *what fraction of the film is not exhibiting bulk-like dynamics*? While in Figure 5.7b, the mobile surface layer was defined as *what fraction of the film has liquid-like dynamics*? Interestingly, Frieberg et al. [113] collected the same data as Pye and Roth [28] measuring the temperature-dependent physical aging rate of thin PS films, which is what gave the data in Figure 5.7a, but Frieberg et al. interpreted their data differently by extracting a mobile surface layer that was defined as what fraction was exhibiting zero aging rate (i.e., was liquid-like) and in so doing arrived at a temperature dependence for their surface layer thickness that qualitatively matches the data shown in Figure 5.7b. This demonstrates that it is possible to take the same experimental data and depending on what criterion one uses to define the mobile surface layer, one can arrive at a temperature dependence for the mobile layer thickness that can either follow Figure 5.7a or 5.7b. Thus, we need to start being more precise in how we define the 'mobile surface layer' (very vague wording indeed) because clearly not all authors are defining it the same way. Lang and Simmons have also recently made a similar point based on how computer simulation data are interpreted [139].

What are the data shown in Figure 5.7a and 5.7b telling us about the temperature dependence of the gradient in dynamics near the free surface? The data from Figure 5.7b says that the liquid layer right at the free surface is only a few to several nanometers in thickness and gets smaller as the temperature decreases further below T_g^{bulk}. Although the experiments of Figure 5.7b do not explicitly demonstrate that this liquid-like mobile fraction exists at the free surface, other studies directly measuring the surface dynamics do [10,136,137]. However, at the same time, the data from Figure 5.7a tells us that the fraction of the film that is exhibiting bulk-like dynamics decreases with decreasing temperature. Even though less of the film is behaving like a liquid at lower temperatures, also less of the film is exhibiting bulk dynamics; thus, a larger fraction of the material must be behaving like a glass with a locally reduced T_g (i.e., not bulk-like). This conclusion is illustrated in Figure 5.7c where we can interpret the data from Figure 5.7a and 5.7b as implying that the gradient in dynamics near the free surface can be considered to be composed of a surface layer exhibiting liquid-like dynamics, followed by a near surface layer that is glassy but with a locally reduced T_g, and then finally followed by bulk-like dynamics deeper into the film. Both the liquid-like surface layer and the glassy layer with locally reduced T_g have a temperature dependence as described by the data shown in Figure 5.7b and 5.7a, respectively. Depending on the experiments being performed and how the data are being interpreted, vastly different conclusions can be drawn about the thickness of the 'mobile surface layer.' Measurements of flow properties such as viscosity or diffusion will observe liquid-like surface properties that only penetrate a few nanometers into the film, especially if the measurements are performed at elevated temperatures near T_g^{bulk}. Here, the data in Figure 5.7a and 5.7b both say that the film will have a liquid-like layer of a few nanometers with nearly bulk-like glassy dynamics underneath, consistent with experimental observations [137,138,190].

However, if local T_g or glassy physical aging dynamics are measured at temperatures far from T_g, the data in Figure 5.7a and 5.7b say that most of the film is glassy, but the local T_g is reduced for several tens of nanometers from the free surface. This indicates that the influence of the dynamical perturbation at the free surface altering glassy dynamics penetrates deeper into the film at lower temperatures. It is these types of observations of a growing length scale of the gradient in dynamics, along with MD simulations, that suggest that the mechanism responsible for propagating enhanced glassy dynamics from an interface deeper into the material is related to the glass transition and cooperative motion [28,139,141,151]. (See also Chapter 8 for further discussion of this point based on MD simulations.)

The physical aging studies described thus far in this section are for studies that observed deviations in physical aging behavior relative to bulk for film thicknesses less than ~100 nm, comparable to thicknesses where changes in T_g have been observed. However, a number of other studies over the years have reported faster physical aging (so-called accelerated aging) for thinner films with deviations from bulk behavior occurring already for film thicknesses of several microns [191–197]. A study by Gray et al. [198], who argued that changes in physical aging at micron length scales were too large to be related to T_g changes caused by boundary effects, systematically compared a number of possible factors and arrived at the conclusion that increased stress present during the thermal quench can be responsible for increasing the aging rate in thinner films. Pye and Roth were later able to conclusively demonstrate this by measuring the physical aging rate for free-standing PS films held by rigid sample holders [199]. For a free-standing polymer film supported on its perimeter by a rigid holder of a given material, an in-plane stress is imparted to the polymer film on cooling because of the thermal expansion mismatch between the polymer film and sample holder material. This in-plane stress is independent of thickness and can be explicitly calculated from the difference in thermal expansion between the polymer film and holder from [199]

$$\sigma = \frac{E_{PS}}{1 - v_{PS}}\left[\alpha_{PS} - \alpha_{holder}\right]\Delta T. \tag{5.2}$$

As the thermal expansion α_{PS} of polystyrene is larger than that of the sample holder α_{holder}, tension is applied to the film on cooling between the glass transition and the aging temperature, $\Delta T = T_g - T_{aging}$. Formally, as the thermal expansion, modulus E_{PS}, and Poisson's ratio v_{PS} of PS are strongly temperature dependent in this temperature range, Equation 5.2 must be evaluated as an integral [199]. Figure 5.8 shows data from Pye and Roth's study [199] demonstrating that when the stress is held constant the physical aging rate of free-standing PS films is independent of film thickness above 220 nm, in the range where no $T_g(h)$ reductions are observed. In addition, by changing the sample holder material, they were able to show that the physical aging rate of 500 nm thick free-standing PS films correlated with the stress imparted to the film on cooling following Equation 5.2. Although this has only been demonstrated for PS films, there is no reason to believe that similar results would not be obtained for other polymers. Why stress present on thermal cooling, corresponding to formation of the glassy state, leads to a less stable glass with faster physical aging is something that was further studied by Gray and Roth [200]. Following many of the ideas about how deformation (stress or strain) imparts mobility in glasses (as described in Part 3 of this book), they have hypothesized that stress present during glass formation (vitrification) may distort ("tilt") the potential energy landscape (PEL) such that for large enough stress, the system may become trapped within a different, higher-energy metabasin; thus forming a less stable glass exhibiting faster physical aging [200].

5.6 FUTURE OUTLOOK

As this chapter has tried to outline, there seems to be a coherent picture emerging from the literature about how changes in the glass transition and physical aging with decreasing sample size, confinement, can be understood. There appears to be a primary confinement mechanism [171,173] that is responsible for propagating dynamical perturbations at an interface deeper into the material resulting in a gradient in dynamics [12,28,151]. This mechanism has been observed in a number of glass forming systems [8,14,18] and is affected

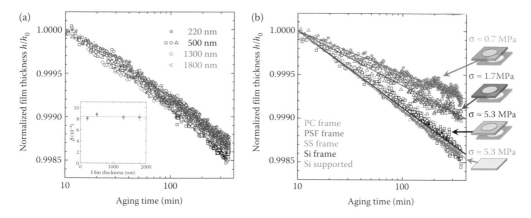

Figure 5.8 Physical aging characterized by the decrease in normalized film thickness h/h_0 as a function of logarithmic time after a temperature quench to 65°C from above T_g for free-standing PS films supported on rigid frames where the stress imparted to the PS film from thermal expansion mismatch between film and frame material can be explicitly calculated. (a) For free-standing PS films of different thicknesses above 200 nm, where no $T_g(h)$ reductions are observed, the physical aging rate β is independent of film thickness when held on stainless-steel (SS) frames that impart a stress of 4.5 MPa on cooling that is independent of film thickness. (b) For 500 nm thick free-standing PS films held by different frame materials, the physical aging rate β increases with increasing stress: PS held by polycarbonate (PC) frames ($\sigma \approx 0.7$ MPa), PS held by polysulfone (PSF) frames ($\sigma \approx 1.7$ MPa), PS held by silicon frames ($\sigma \approx 5.3$ MPa), and PS supported on silicon wafers ($\sigma \approx 5.3$ MPa). (Reprinted with permission from J. E. Pye and C. B. Roth, *Macromolecules*, 46, 9455–9463, 2013. Copyright 2013, American Chemical Society.)

by factors that alter cooperative motion [28,98,139,141]. Thus, it may be possible to use studies of glassy dynamics near interfaces to learn more about the glass transition and its length scales [148,151,153,201,202]. In many cases, property changes such as physical aging [24,28,113], glassy modulus [30], and low molecular weight viscosity [137] can be correlated with local changes in T_g near these interfaces. In addition, high molecular weight free-standing films have a small (10%–20%) fraction of the material that stays mobile to much lower temperatures that appears to be controlled by a second mechanism related to chain connectivity [171,173].

- Yet, despite this general picture, there are still many issues and questions we do not yet understand: *What exactly are these mechanisms that transfer mobility from a boundary into the material?* Do we need a viable framework for the glass transition in bulk systems before this can be answered, or can studies of such mechanisms be used to elucidate the glass transition in general? How are other material properties such as fragility [119,120], diffusion [203], and yielding [204,205] related or not to changes in the glass transition?

- *What makes a free surface "free"? What is the nature of a free surface and how is it similar or different from a liquid interface or that with another polymer?* An interface with a lower T_g polymer can sometimes lead to larger T_g reductions than a free surface [150,151], while other low T_g polymers [155,156] and some liquids [38,39] behave more like an interface with a solid substrate. What factors [154] control how dynamics are coupled across an interface?

- *What role does chain connectivity play in affecting material properties in polymer films?* Chain architecture [111–115], chain stiffness [33,117], and chain entanglements [69,105,107,108,206,207] appear to have additional facets we have yet to decipher. Chain adsorption [72,208] and grafting [209,210] also strongly alter the behavior of an interface, along with more subtler effects [211].

ACKNOWLEDGMENTS

The authors would like to thank Dr. Laura A.G. Gray for useful and informative discussions. CBR gratefully acknowledges support from National Science Foundation, Division of Materials Research, Polymers Program in the form of a CAREER award (Grant No. DMR-1151646), the ACS Petroleum Research Fund (PRF-48927-DNI7), and Emory University.

REFERENCES

1. C. L. Jackson and G. B. McKenna, *J. Non-Cryst. Solids*, 131, 221, 1991.
2. J. L. Keddie, R. A. L. Jones, and R. A. Cory, *Europhys. Lett.*, 27, 59, 1994.
3. J. L. Keddie, R. A. L. Jones, and R. A. Cory, *Faraday Discuss.*, 98, 219, 1994.
4. J. A. Forrest, K. Dalnoki-Veress, J. R. Stevens, and J. R. Dutcher, *Phys. Rev. Lett.*, 77, 2002, 1996.
5. J. A. Forrest and K. Dalnoki-Veress, *Adv. Colloid Interface Sci.*, 94, 167, 2001.
6. C. B. Roth and J. R. Dutcher, *J. Electroanal. Chem.*, 584, 13, 2005.
7. J. Baschnagel and F. Varnik, *J. Phys. Condens. Matter*, 17, R851, 2005.
8. M. Alcoutlabi and G. B. McKenna, *J. Phys. Condens. Matter*, 17, R461, 2005.
9. O. K. C. Tsui, Anomalous dynamics of polymer films, in *Polymer Thin Films*, edited by O. K. C. Tsui and T. P. Russell (World Scientific, Singapore), 267–294, 2008.
10. M. D. Ediger and J. A. Forrest, *Macromolecules*, 47, 471, 2014.
11. J. S. Sharp and J. A. Forrest, *Phys. Rev. Lett.*, 91, 235701, 2003.
12. C. J. Ellison and J. M. Torkelson, *Nat. Mater.*, 2, 695, 2003.
13. C. Zhang, Y. Guo, and R. D. Priestley, *Macromolecules*, 44, 4001, 2011.
14. S. F. Swallen, K. L. Kearns, M. K. Mapes, Y. S. Kim, R. J. McMahon, M. D. Ediger, T. Wu, L. Yu, and S. Satija, *Science*, 315, 353, 2007.
15. C. Kim, A. Facchetti, and T. J. Marks, *Science*, 318, 76, 2007.
16. C. R. Nugent, K. V. Edmond, H. N. Patel, and E. R. Weeks, *Phys. Rev. Lett.*, 99, 025702, 2007.
17. K. V. Edmond, C. R. Nugent, and E. R. Weeks, *Phys. Rev. E*, 85, 041401, 2012.
18. G. L. Hunter, K. V. Edmond, and E. R. Weeks, *Phys. Rev. Lett.*, 112, 218302, 2014.
19. P. S. Sarangapani and Y. Zhu, *Phys. Rev. E*, 77, 010501, 2008.
20. P. S. Sarangapani, A. B. Schofield, and Y. Zhu, *Phys. Rev. E*, 83, 030502, 2011.
21. P. S. Sarangapani, A. B. Schofield, and Y. Zhu, *Soft Matter*, 8, 814, 2012.
22. S. Kawana and R. A. L. Jones, *Eur. Phys. J. E*, 10, 223, 2003.
23. R. D. Priestley, L. J. Broadbelt, and J. M. Torkelson, *Macromolecules*, 38, 654, 2005.
24. R. D. Priestley, C. J. Ellison, L. J. Broadbelt, and J. M. Torkelson, *Science*, 309, 456, 2005.
25. Y. P. Koh and S. L. Simon, *J. Polym. Sci., Part B: Polym. Phys.*, 46, 2741, 2008.
26. K. Fukao and A. Sakamoto, *Phys. Rev. E*, 71, 041803, 2005.
27. K. Fukao and H. Koizumi, *Phys. Rev. E*, 77, 021503, 2008.
28. J. E. Pye, K. A. Rohald, E. A. Baker, and C. B. Roth, *Macromolecules*, 43, 8296, 2010.
29. B. W. Rowe, B. D. Freeman, and D. R. Paul, *Polymer*, 50, 5565, 2009.
30. C. M. Stafford, B. D. Vogt, C. Harrison, D. Julthongpiput, and R. Huang, *Macromolecules*, 39, 5095, 2006.
31. J. M. Torres, C. M. Stafford, and B. D. Vogt, *ACS Nano*, 3, 2677, 2009.
32. J. M. Torres, C. M. Stafford, and B. D. Vogt, *Polymer*, 51, 4211, 2010.
33. J. M. Torres, C. Wang, E. B. Coughlin, J. P. Bishop, R. A. Register, R. A. Riggleman, C. M. Stafford, and B. D. Vogt, *Macromolecules*, 44, 9040, 2011.
34. J. M. Torres, C. M. Stafford, D. Uhrig, and B. D. Vogt, *J. Polym. Sci., Part B: Polym. Phys.*, 50, 370, 2012.
35. P. A. O'Connell and G. B. McKenna, *Science*, 307, 1760, 2005.
36. P. A. O'Connell, S. A. Hutcheson, and G. B. McKenna, *J. Polym. Sci., Part B: Polym. Phys.*, 46, 1952, 2008.
37. P. A. O'Connell, J. Wang, T. A. Ishola, and G. B. McKenna, *Macromolecules*, 45, 2453, 2012.
38. J. Wang and G. B. McKenna, *Macromolecules*, 46, 2485, 2013.
39. J. Wang and G. B. McKenna, *J. Polym. Sci., Part B: Polym. Phys.*, 51, 1343, 2013.
40. T. B. Karim and G. B. McKenna, *Polymer*, 54, 5928, 2013.
41. D. S. Fryer, P. F. Nealey, and J. J. de Pablo, *J. Vac. Sci. Technol. B*, 18, 3376, 2000.

42. D. S. Fryer, P. F. Nealey, and J. J. de Pablo, *Macromolecules*, 33, 6439, 2000.

43. S. P. Delcambre, R. A. Riggleman, J. J. de Pablo, and P. F. Nealey, *Soft Matter*, 6, 2475, 2010.

44. M. K. Mundra, S. K. Donthu, V. P. Dravid, and J. M. Torkelson, *Nano Lett.*, 7, 713, 2007.

45. Y. Huang and D. R. Paul, *Ind. Eng. Chem. Res.*, 46, 2342, 2007.

46. C. A. Angell, *Science*, 319, 582, 2008.

47. D. Ortiz-Young, H.-C. Chiu, S. Kim, K. Voïtchovsky, and E. Riedo, *Nat. Commun.*, 4, 2482, 2013.

48. G. B. McKenna, *Eur. Phys. J. Spec. Top.*, 141, 291, 2007.

49. G. B. McKenna, *Eur. Phys. J. Spec. Top.*, 189, 285, 2010.

50. M. Tress, M. Erber, E. U. Mapesa, H. Huth, J. Mueller, A. Serghei, C. Schick, K.-J. Eichhorn, B. Volt, and F. Kremer, *Macromolecules*, 43, 9937, 2010.

51. F. Kremer, M. Tress, and E. U. Mapesa, *J. Non-Cryst. Solids*, 407, 277, 2015.

52. A. N. Raegen, M. V. Massa, J. A. Forrest, and K. Dalnoki-Veress, *Eur. Phys. J. E*, 27, 375, 2008.

53. O. Bäumchen, J. D. McGraw, J. A. Forrest, and K. Dalnoki-Veress, *Phys. Rev. Lett.*, 109, 055701, 2012.

54. M. D. Ediger and P. Harrowell, *J. Chem. Phys.*, 137, 080901, 2012.

55. L. Berthier and G. Biroli, *Rev. Mod. Phys.*, 83, 587, 2011.

56. L. Berthier, G. Biroli, J. P. Bouchaud, L. Cipelletti, and W. van Saarloos, editors, *Dynamical Heterogeneities in Glasses, Colloids, and Granular Media* (Oxford University Press, Oxford, UK), 2011.

57. J. C. Dyre, *Rev. Mod. Phys.*, 78, 953, 2006.

58. M. D. Ediger, C. A. Angell, and S. R. Nagel, *J. Phys. Chem.*, 100, 13200, 1996.

59. C. A. Angell, K. L. Ngai, G. B. McKenna, P. F. McMillan, and S. W. Martin, *J. Appl. Phys.*, 88, 3113, 2000.

60. C. B. Roth, K. L. McNerny, W. F. Jager, and J. M. Torkelson, *Macromolecules*, 40, 2568, 2007.

61. K. I. Winey and R. A. Vaia, *MRS Bull.*, 32, 314, 2007.

62. P. Rittigstein, R. D. Priestley, L. J. Broadbelt, and J. M. Torkelson, *Nat. Mater.*, 6, 278, 2007.

63. J. M. Kropka, V. Pryamitsyn, and V. Ganesan, *Phys. Rev. Lett.*, 101, 075702, 2008.

64. J. F. Moll, P. Akcora, A. Rungta, S. Gong, R. H. Colby, B. C. Benicewicz, and S. K. Kumar, *Macromolecules*, 44, 7473, 2011.

65. S. Sen, Y. Xie, A. Bansal, H. Yang, K. Cho, L. S. Schadler, and S. K. Kumar, *Eur. Phys. J. Spec. Top.*, 141, 161, 2007.

66. M. K. Mundra, C. J. Ellison, R. E. Behling, and J. M. Torkelson, *Polymer*, 47, 7747, 2006.

67. M. K. Mundra, C. J. Ellison, P. Rittigstein, and J. M. Torkelson, *Eur. Phys. J. Spec. Top.*, 141, 143–151, 2007.

68. G. Reiter, *Macromolecules*, 27, 3046, 1994.

69. Z. Yang, A. Clough, C.-H. Lam, and O. K. C. Tsui, *Macromolecules*, 44, 8294, 2011.

70. K. R. Thomas, A. Chenneviere, G. Reiter, and U. Steiner, *Phys. Rev. E*, 83, 021804, 2011.

71. R. D. Priestley, *Soft Matter*, 5, 919, 2009.

72. S. Napolitano, S. Capponi, and B. Vanroy, *Eur. Phys. J. E*, 36, 61, 2013.

73. R. D. Priestley, D. Cangialosi, and S. Napolitano, *J. Non-Cryst. Solids*, 407, 288, 2015.

74. C. L. Jackson and G. B. McKenna, *J. Chem. Phys.*, 93, 9002, 1990.

75. S. Herminghaus, K. Jacobs, and R. Seemann, *Eur. Phys. J. E*, 5, 531, 2001.

76. S. Herminghaus, R. Seemann, and K. Landfester, *Phys. Rev. Lett*, 93, 017801, 2004.

77. S. Peter, H. Meyer, and J. Baschnagel, *J. Polym. Sci., Part B: Polym. Phys.*, 44, 2951, 2006.

78. J. L. Keddie and R. A. L. Jones, *Israel J. Chem.*, 35, 21, 1995.

79. J. A. Forrest, K. Dalnoki-Veress, and J. R. Dutcher, *Phys. Rev. E*, 56, 5705, 1997.

80. O. K. C. Tsui and H. F. Zhang, *Macromolecules*, 34, 9139, 2001.

81. S. Kawana and R. A. L. Jones, *Phys. Rev. E*, 63, 021501, 2001.

82. K. Fukao and Y. Miyamoto, *Europhys. Lett.*, 46, 649, 1999.

83. O. K. C. Tsui, T. P. Russell, and C. J. Hawker, *Macromolecules*, 34, 5535, 2001.

84. G. B. DeMaggio, W. E. Frieze, D. W. Gidley, M. Zhu, H. A. Hristov, and A. F. Yee, *Phys. Rev. Lett.*, 78, 1524, 1997.

85. S. Kim, S. A. Hewlett, C. B. Roth, and J. M. Torkelson, *Eur. Phys. J. E*, 30, 83, 2009.

86. K. Fukao and Y. Miyamoto, *Phys. Rev. E*, 61, 1743, 2000.

87. Z. Fakhraai and J. A. Forrest, *Phys. Rev. Lett*, 95, 025701, 2005.

88. S. Gao, Y. P. Koh, and S. L. Simon, *Macromolecules*, 46, 562, 2013.

89. H. Huth, A. A. Minakov, and C. Schick, *J. Polym. Sci., Part B: Polym. Phys.*, 44, 2996, 2006.

90. H. Huth, A. A. Minakov, A. Serghei, F. Kremer, and C. Schick, *Eur. Phys. J. Spec. Top.*, 141, 153, 2007.

91. M. Y. Efremov, E. A. Olson, M. Zhang, Z. Zhang, and L. H. Allen, *Phys. Rev. Lett.*, 91, 085703, 2003.

92. K. Fukao and Y. Miyamoto, *Phys. Rev. E*, 64, 011803, 2001.

93. K. Fukao, T. Terasawa, Y. Oda, K. Nakamura, and D. Tahara, *Phys. Rev. E*, 84, 041808, 2011.

94. C. J. Ellison, R. L. Ruszkowski, N. J. Fredin, and J. M. Torkelson, *Phys. Rev. Lett.*, 92, 095702, 2004.

95. R. A. Riggleman, K. Yoshimoto, J. F. Douglas, and J. J. de Pablo, *Phys. Rev. Lett.*, 97, 045502, 2006.

96. R. A. Riggleman, J. F. Douglas, and J. J. de Pablo, *Phys. Rev. E*, 76, 011504, 2007.

97. S. Kim, M. K. Mundra, C. B. Roth, and J. M. Torkelson, *Macromolecules*, 43, 5158, 2010.

98. C. J. Ellison, M. K. Mundra, and J. M. Torkelson, *Macromolecules*, 38, 1767, 2005.

99. C. M. Evans, H. Deng, W. F. Jager, and J. M. Torkelson, *Macromolecules*, 46, 6091, 2013.

100. R. Seemann, K. Jacobs, K. Landfester, and S. Herminghaus, *J. Polym. Sci., Part B: Polym. Phys.*, 44, 2968, 2006.

101. P. Bernazzani, S. L. Simon, D. J. Plazek, and K. L. Ngai, *Eur. Phys. J. E*, 8, 201, 2002.

102. T. Lan and J. M. Torkelson, *Polymer*, 55, 1249, 2014.

103. K. Geng, F. Chen, and O. K. C. Tsui, *J. Non-Cryst. Solids*, 407, 296, 2015.

104. E. C. Glor and Z. Fakhraai, *J. Chem. Phys.*, 141, 194505, 2014.

105. H. R. Brown and T. P. Russell, *Macromolecules*, 29, 798, 1996.

106. W. Bisbee, J. Qin, and S. T. Milner, *Macromolecules*, 44, 8972, 2011.

107. D. M. Sussman, W.-S. Tung, K. I. Winey, K. S. Schweizer, and R. A. Riggleman, *Macromolecules*, 47, 6462, 2014.

108. L. Si, M. V. Massa, K. Dalnoki-Veress, H. R. Brown, and R. A. L. Jones, *Phys. Rev. Lett.*, 94, 127801, 2005.

109. J. D. McGraw, P. D. Fowler, M. L. Ferrari, and K. Dalnoki-Veress, *Eur. Phys. J. E*, 36, 7, 2013.

110. S. Xu, P. A. O'Connell, and G. B. McKenna, *J. Chem. Phys.*, 132, 184902, 2010.

111. E. Glynos, B. Frieberg, H. Oh, M. Liu, D. W. Gidley, and P. F. Green, *Phys. Rev. Lett.*, 106, 128301, 2011.

112. B. Frieberg, E. Glynos, G. Sakellariou, and P. F. Green, *ACS Macro Lett.*, 1, 636, 2012.

113. B. Frieberg, E. Glynos, and P. F. Green, *Phys. Rev. Lett.*, 108, 268304, 2012.

114. S.-F. Wang, S. Yang, J. Lee, B. Akgun, D. T. Wu, and M. D. Foster, *Phys. Rev. Lett.*, 111, 068303, 2013.

115. J. S. Lee, N.-H. Lee, S. Peri, M. D. Foster, C. F. Majkrzak, R. Hu, and D. T. Wu, *Phys. Rev. Lett.*, 113, 225702, 2014.

116. J. H. Kim, J. Jang, and W. C. Zin, *Langmuir*, 16, 4064, 2000.

117. A. Shavit and R. A. Riggleman, *Macromolecules*, 46, 5044, 2013.

118. C. G. Campbell and B. D. Vogt, *Polymer*, 48, 7169, 2007.

119. P. Z. Hanakata, J. F. Douglas, and F. W. Starr, *J. Chem. Phys.*, 137, 244901, 2012.

120. C. Zhang and R. D. Priestley, *Soft Matter*, 9, 7076, 2013.

121. C. Zhang, Y. Guo, K. B. Shepard, and R. D. Priestley, *J. Phys. Chem. Lett.*, 4, 431, 2013.

122. A. K. Torres Arellano and G. B. McKenna, *J. Polym. Sci., Part B: Polym. Phys.*, 53, 1261, 2015.

123. C. A. Angell, *Science*, 267, 1924, 1995.

124. K. Kunal, C. G. Robertson, S. Pawlus, S. F. Hahn, and A. P. Sokolov, *Macromolecules*, 41, 7232, 2008.

125. Q. Qin and G. B. McKenna, *J. Non-Cryst. Solids.*, 352, 2977, 2006.

126. L. Hong, V. N. Novikov, and A. P. Sokolov, *J. Non-Cryst. Solids.*, 357, 351, 2011.

127. J. van Zanten, W. Wallace, and W.-L. Wu, *Phys. Rev. E*, 53, R2053, 1996.

128. C. H. Park, J. H. Kim, M. Ree, B. H. Sohn, J. C. Jung, and W. C. Zin, *Polymer*, 45, 4507, 2004.

129. R. D. Priestley, M. K. Mundra, N. J. Barnett, L. J. Broadbelt, and J. M. Torkelson, *Aust. J. Chem.*, 60, 765, 2007.

130. C. B. Roth and J. R. Dutcher, *Eur. Phys. J. E*, 12, S103, 2003.

131. C. B. Roth, A. Pound, S. W. Kamp, C. A. Murray, and J. R. Dutcher, *Eur. Phys. J. E*, 20, 441, 2006.

132. C. J. Ellison and J. M. Torkelson, *J. Polym. Sci., Part B: Polym. Phys.*, 40, 2745, 2002.

133. K. Paeng and M. D. Ediger, *Macromolecules*, 44, 7034, 2011.

134. G. C. Berry and T. G. Fox, *Adv. Polym. Sci.*, 5, 261, 1968.

135. R. P. White, C. C. Price, and J. E. G. Lipson, *Macromolecules*, 48, 4132, 2015.

136. Z. Fakhraai and J. A. Forrest, *Science*, 319, 600, 2008.

137. Z. Yang, Y. Fujii, F. K. Lee, C.-H. Lam, and O. K. C. Tsui, *Science*, 328, 1676, 2010.

138. K. Paeng, S. F. Swallen, and M. D. Ediger, *J. Am. Chem. Soc.*, 133, 8444, 2011.

139. R. J. Lang and D. S. Simmons, *Macromolecules*, 46, 9818, 2013.

140. J. Mattsson, J. A. Forrest, and L. Borjesson, *Phys. Rev. E*, 62, 5187, 2000.

141. P. Z. Hanakata, J. F. Douglas, and F. W. Starr, *Nat. Commun.*, 5, 4163, 2014.

142. C. R. Daley, Z. Fakhraai, M. D. Ediger, and J. A. Forrest, *Soft Matter*, 8, 2206, 2012.

143. D. Qi, C. R. Daley, Y. Chai, and J. A. Forrest, *Soft Matter*, 9, 8958, 2013.

144. J. E. G. Lipson and S. T. Milner, *Eur. Phys. J. B*, 72, 133, 2009.

145. D. Long and F. Lequeux, *Eur. Phys. J. E*, 4, 371, 2001.

146. J. D. Stevenson and P. G. Wolynes, *J. Chem. Phys.*, 129, 234514, 2008.

147. R. P. White and J. E. G. Lipson, *Phys. Rev. E*, 84, 041801, 2011.

148. S. Mirigian and K. S. Schweizer, *J. Chem. Phys.*, 141, 161103, 2014.

149. C. B. Roth and J. M. Torkelson, *Macromolecules*, 40, 3328, 2007.

150. P. M. Rauscher, J. E. Pye, R. R. Baglay, and C. B. Roth, *Macromolecules*, 46, 9806, 2013.

151. R. R. Baglay and C. B. Roth, *J. Chem. Phys.*, 143, 111101, 2015.

152. H. Yoon and G. B. McKenna, *Macromolecules*, 47, 8808, 2014.

153. N. B. Tito, J. E. G. Lipson, and S. T. Milner, *Soft Matter*, 9, 9403, 2013.

154. R. J. Lang, W. L. Merling, and D. S. Simmons, *ACS Macro Lett.*, 3, 758, 2014.

155. C. M. Evans, S. Narayanan, Z. Jiang, and J. M. Torkelson, *Phys. Rev. Lett.*, 109, 038302, 2012.

156. S. Askar, C. M. Evans, and J. M. Torkelson, *Polymer*, 76, 113, 2015.

157. D. F. Siqueira, D. W. Schubert, V. Erb, M. Stamm, and J. P. Amato, *Colloid Polym. Sci.*, 273, 1041, 1995.

158. S. Mirigian and K. S. Schweizer, *J. Phys. Chem. Lett.*, 4, 3648, 2013.

159. S. Mirigian and K. S. Schweizer, *J. Chem. Phys.*, 140, 194506, 2014.

160. K. Dalnoki-Veress, J. A. Forrest, C. Murray, C. Gigault, and J. R. Dutcher, *Phys. Rev. E*, 63, 031801, 2001.

161. J. A. Forrest and J. Mattsson, *Phys. Rev. E*, 61, R53, 2000.

162. H. Liem, J. Cabanillas-Gonzalez, P. Etchegoin, and D. D. C. Bradley, *J. Phys. Condens. Matter*, 16, 721, 2004.

163. T. Miyazaki, R. Inoue, K. Nishida, and T. Kanaya, *Eur. Phys. J. Spec. Top.*, 141, 203, 2007.

164. C. Rotella, S. Napolitano, and M. Wubbenhorst, *Macromolecules*, 42, 1415, 2009.

165. S. Kim, C. B. Roth, and J. M. Torkelson, *J. Polym. Sci., Part B: Polym. Phys.*, 46, 2754, 2008.

166. S. Kim and J. M. Torkelson, *Macromolecules*, 44, 4546, 2011.

167. P. G. de Gennes, *C. R. Acad. Sci. IV Phys.*, 1, 1179, 2000.

168. P. G. de Gennes, *Eur. Phys. J. E*, 2, 201, 2000.

169. S. T. Milner and J. E. G. Lipson, *Macromolecules*, 43, 9865, 2010.

170. J. E. G. Lipson and S. T. Milner, *Macromolecules*, 43, 9874, 2010.

171. J. E. Pye and C. B. Roth, *Phys. Rev. Lett.*, 107, 235701, 2011.

172. P. A. O'Connell and G. B. McKenna, *Eur. Phys. J. E*, 20, 143, 2006.

173. J. E. Pye and C. B. Roth, *J. Polym. Sci., Part B: Polym. Phys.*, 53, 64, 2015.

174. C. B. Roth and J. R. Dutcher, *Phys. Rev. E*, 72, 021803, 2005.

175. C. B. Roth and J. R. Dutcher, *J. Polym. Sci., Part B: Polym. Phys.*, 44, 3011, 2006.

176. L. C. E. Struik, *Physical Aging in Amorphous Polymers and Other Materials* (Elsevier Scientific Publishing Company, Amsterdam), 1978.

177. J. M. Hutchinson, *Prog. Polym. Sci.*, 20, 703, 1995.

178. R. Greiner and F. R. Schwarzl, *Rheol. Acta*, 23, 378, 1984.

179. M. R. Tant and G. L. Wilkes, *Polym. Eng. Sci.*, 21, 874, 1981.

180. G. B. McKenna, Physical aging in glasses and composites, in *Long-Term Durability of Polymeric Matrix Composites*, edited by K. V. Pochiraju, G. P. Tandon, and G. A. Schoeppner (Springer US, Boston, MA), 237–309, 2012.

181. S. L. Simon, J. W. Sobieski, and D. J. Plazek, *Polymer*, 42, 2555, 2001.

182. Y. P. Koh and S. L. Simon, *Macromolecules*, 46, 5815, 2013.

183. I. M. Hodge, *J. Non-Cryst. Solids*, 169, 211, 1994.

184. Y. P. Koh, L. Grassia, and S. L. Simon, *Thermochim. Acta*, 603, 135, 2015.

185. E. A. Baker, P. Rittigstein, J. M. Torkelson, and C. B. Roth, *J. Polym. Sci., Part B: Polym. Phys.*, 47, 2509, 2009.

186. Y. C. Jean, R. W. Zhang, H. Cao, J. P. Yuan, C. M. Huang, B. Nielsen, and P. AsokaKumar, *Phys. Rev. B*, 56, R8459, 1997.

187. G. F. Meyers, B. M. DeKoven, and J. T. Seitz, *Langmuir*, 8, 2330, 1992.

188. V. M. Boucher, D. Cangialosi, H. Yin, A. Schönhals, A. Alegria, and J. Colmenero, *Soft Matter*, 8, 5119, 2012.

189. K. Paeng, R. Richert, and M. D. Ediger, *Soft Matter*, 8, 819, 2012.

190. Y. Chai, T. Salez, J. D. McGraw, M. Benzaquen, K. Dalnoki-Veress, E. Raphael, and J. A. Forrest, *Science*, 343, 994, 2014.

191. P. H. Pfromm and W. J. Koros, *Polymer*, 36, 2379, 1995.

192. K. D. Dorkenoo and P. H. Pfromm, *Macromolecules*, 33, 3747, 2000.

193. M. S. McCaig and D. R. Paul, *Polymer*, 41, 629, 2000.

194. Y. Huang and D. R. Paul, *Polymer*, 45, 8377, 2004.

195. Y. Huang and D. R. Paul, *J. Membrane Sci.*, 244, 167, 2004.

196. Y. Huang and D. R. Paul, *Macromolecules*, 39, 1554, 2006.

197. V. M. Boucher, D. Cangialosi, A. Alegria, and J. Colmenero, *Macromolecules*, 45, 5296, 2012.

198. L. A. G. Gray, S. W. Yoon, W. A. Pahner, J. E. Davidheiser, and C. B. Roth, *Macromolecules*, 45, 1701, 2012.

199. J. E. Pye and C. B. Roth, *Macromolecules*, 46, 9455, 2013.

200. L. A. G. Gray and C. B. Roth, *Soft Matter*, 10, 1572, 2014.

201. P. Z. Hanakata, B. A. Pazmiño Betancourt, J. F. Douglas, and F. W. Starr, *J. Chem. Phys.*, 142, 234907, 2015.

202. S. Butler and P. Harrowell, *J. Chem. Phys.*, 95, 4466, 1991.

203. J. M. Katzenstein, D. W. Janes, H. E. Hocker, J. K. Chandler, and C. J. Ellison, *Macromolecules*, 45, 1544, 2012.

204. B. J. Gurmessa and A. B. Croll, *Phys. Rev. Lett.*, 110, 074301, 2013.

205. B. Gurmessa and A. B. Croll, *Macromolecules*, 48, 5670, 2015.

206. F. Chen, D. Peng, C.-H. Lam, and O. K. C. Tsui, *Macromolecules*, 48, 5034, 2015.

207. C.-H. Lam and O. K. C. Tsui, *Phys. Rev. E*, 88, 042604, 2013.

208. S. Napolitano and M. Wubbenhorst, *Nat. Commun.*, 2, 260, 2011.

209. R. S. Tate, D. S. Fryer, S. Pasqualini, M. F. Montague, J. J. de Pablo, and P. F. Nealey, *J. Chem. Phys.*, 115, 9982, 2001.

210. A. Clough, D. Peng, Z. Yang, and O. K. C. Tsui, *Macromolecules*, 44, 1649, 2011.

211. C. Zhang, Y. Fujii, and K. Tanaka, *ACS Macro Lett.*, 1, 1317, 2012.

6

Mechanical and viscoelastic properties of polymer thin films and surfaces

GREGORY B. McKENNA AND MEIYU ZHAI

6.1 INTRODUCTION

There is a major interest in the behavior of amorphous polymers at the nanometer-size scale primarily due to the discovery [1–3] that glass-forming liquids, including polymers [4–9], when constrained to dimensions of approximately 100 nm or less, exhibit strong deviations in the glass transition temperature from the properties observed in the bulk material. There have been different reports concerning the glass transition temperature in confined systems, and it can increase, decrease, or remain the same depending upon the details of the experiment. In the case of liquids in porous media or polymers supported on a substrate, the strength of the interactions between the confining and confined media seems to have a very strong influence. Increases in the T_g relative to the bulk value are related to strong interactions while decreases are related to weaker interactions, though it has been reported that both an increase and a decrease can occur in the same film but from different measurements, for example, thickness dilatometry versus temperature and inelastic neutron scattering [10,11] on the same samples. In the case of freely standing polymer films, it seems that nearly all, but not all, reports to date suggest a reduction in the glass transition temperature [7–9,12–19], with such reductions being reported [20] to be as much as 122°C, for extremely thin films of polycarbonate.

In the case of amorphous, glassy polymers, an especially important aspect of their application is that they are normally used at a high fraction [21] of their T_g, with the result that large changes in the T_g can dramatically impact the mechanical response of the material. Therefore, it is of interest to be able to make measurements of the mechanical response of polymers when they are confined at the nanometer-size scale. Because polymer viscoelasticity is important and reflects how close one is to the T_g, it is generally better in polymeric materials to make measurements of viscoelastic properties than simply static ones, though the static properties can reflect changing T_g as well. Furthermore, because the behavior of polymers at the nanometer-size scale seems to also have a component that is related to the surface dynamics [22–25], the investigation of these by viscoelastic methods is also of interest. We emphasize that viscoelastic measurements are dynamic measurements, but not identical to dielectric measurements of dynamics. Dielectric spectroscopy is dependent on local dipole motions in the viscoelastic medium, while the viscoelastic measurements probe a more global response [26]. In what follows, we first describe the properties of interest from a macroscopic perspective and then describe a novel bubble inflation nanomechanical measurement method of determining the viscoelastic response of ultrathin (nanometric thickness) polymer films. This will be followed by the description of a spontaneous particle embedment method of measuring the surface viscoelastic properties of polymers that promises novel insights into surface dynamics that differs from classical nanoindentation methods that probe a larger length scale.

6.2 VISCOELASTIC RESPONSE OF POLYMERS AT THE MACROSCOPIC-SIZE SCALE

There are multiple excellent treatises [27–37] on the viscoelastic response of polymers at the macroscopic-size scale and the behaviors both below and above the glass transition temperature continue to be major areas of research. In this section, we describe only the main features of behavior with particular emphasis on the aspects that are related to the glass transition regime. The reader is referred to the cited literature for more details of the major questions that remain open in the understanding of the viscoelastic behavior of polymers at the macroscopic-size scale.

6.2.1 STRESS RELAXATION MODULUS

The stress relaxation modulus of a material is simply the ratio $G(t)$ of the response in stress $\sigma(t)$ to a step in strain γ, that is, $G(t) = \sigma(t)/\gamma$. If one looks simply at the response in linear time, as is frequently done, it is not very interesting simply because the short time response is lost in the long time behavior, which can extend for, literally, years or millennia. Therefore, it is common practice to represent the data on a double logarithmic representation of $G(t)$ versus t. Figure 6.1a shows the idealized response for a polymer (tested near to the glass transition temperature) in a linear representation and Figure 6.1b shows a double logarithmic representation of the same data. In Figure 6.1a, it appears that the material has virtually relaxed to zero by 1000 s. But, in Figure 6.1b, we see that the material is still relaxing and for much longer times. In fact, the value of the modulus at 1000 s is 3.1×10^6 Pa, which is more than 10 times the rubbery plateau in Figure 6.1b of 2×10^5 Pa. We will discuss subsequently how the extreme time range of information shown in Figure 6.1b can be obtained through the principles of time–temperature superposition, but it shows the important relaxations in the polymer. Below the glass transition regime, the glassy modulus is in the vicinity of 1 GPa, depending somewhat on the polymer, and this sort of regime is seen in all glass-forming systems, with different magnitudes of glassy modulus and with a relaxation time in the vicinity of 10–1000 s. In such systems, in fact this first "dispersion" would relax to zero rather than to the rubbery plateau. But, because of their long-chain nature,

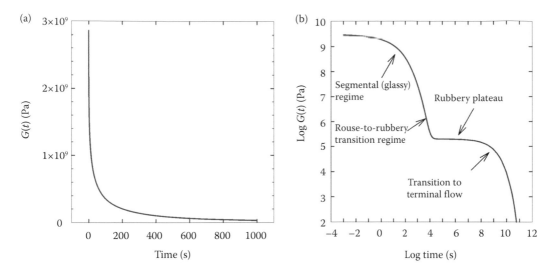

Figure 6.1 (a) Relaxation modulus as a function of time on a linear scale for a model polymer. (b) Same data as in (a) but now in a double logarithmic representation. This shows the importance of the double logarithmic representation of the viscoelastic response of polymeric materials because the spectrum of relaxation times (the response function) is extremely broad. Remark that in (a), the time runs to 1000 s and G(t) appears to be approaching zero without any features. This corresponds only to the short time response represented in (b), where we see that relaxation continues far beyond 1000 s.

polymers exhibit a rubbery or entanglement plateau [27] that has features of behavior similar to those of a cross-linked rubber. But, because the chains only entangle rather than being chemically linked together, they eventually slip free of each other and flow in a process that is currently thought to be similar to that of a snake moving (either backward or forward) along its principal axis referred to as reptation [38,39]. In the deep glassy regime, the material is used as a hard, near-solid structural element (such as resins for composite materials, thermoplastics for automotive parts, or thin dielectric layers in electronic devices). The regime from the rubbery plateau and into the terminal flow regime is important in the processing of the materials.

Of course, it is impossible to make measurements on a material for the time range of Figure 6.1b, but the evidence available suggests that this is, indeed, the range of relaxations seen in polymeric materials. So, if mechanical experiments are generally limited to times of about 0.1 s (a typical instrumental response time) to 10^6 s (about 12 days) or less, how does one obtain the range from 10^{-4} to 10^{12} s shown in the figure? The principles of time–temperature superposition make this possible. Without going into great detail, imagine that the glassy dispersion is represented by a relaxation function that looks like

$$G(t) = (G_g - G_r)e^{-(t/\tau_1)^{\beta_1}} \tag{6.1}$$

And the rubbery dispersion looks like

$$G(t) = G_r e^{-(t/\tau_2)^{\beta_2}} \tag{6.2}$$

where the τ_i are related to the relaxation time of the system, the G_g and G_r are the glassy and rubbery limiting moduli, respectively, and the β_i relate to the breadth of each of the relaxation processes, with a single relaxation time process represented if $\beta_i = 1$ and no relaxation occurring if $\beta_i = 0$. Though in detail it is now well known that there are more than two dispersions representing the relaxation processes in polymers, and that even for the case just depicted the temperature dependences of the two dispersions are different, for simplicity we assume that they are the same. In that case, time–temperature superposition tells us that the entire

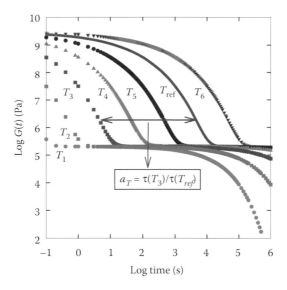

Figure 6.2 Schematic of how a series of experiments at a range of temperatures from T_6 to T_1 are used to define the time–temperature shift factor a_T. See text for discussion. Relevant master curve is shown in Figure 6.1 and time–temperature shift factor behavior is shown in Figure 6.3.

spectrum of relaxation times shifts uniformly with temperature, that is, a temperature shift factor a_T exists and it can be defined by

$$a_T = \frac{\tau_i(T)}{\tau_i(T_{ref})}$$

(6.3)

where T and T_{ref} are the temperature of interest and a reference temperature, respectively. If temperature increases, the τ_i decrease, and when the temperature decreases, the τ_i increase. In polymers (and other glass-forming materials), such shifting of relaxation times can be many orders of magnitude. Thus, by carrying out relaxation experiments at different temperatures and for relatively short times, one can create a "master curve" that covers both extremely short times and extremely long times at the temperatures of interest. Figure 6.2 shows a schematic of the procedure and we will refer to this subsequently. We see that seven "tests" were run and each was from 0.1 s for 10^6 s. The shifting between each temperature and the reference temperature in the present example is such that, relative to T_{ref}, T_1 is shifted by 5 decades in time (five orders of magnitude) to shorter times; T_2 by 4 decades; T_3 by 3 decades; T_4 by 2 decades; T_5 by 1 decade; and T_6 is shifted to longer times by 1 decade. These would be typical near the glass transition temperature and represent approximately 15–25°C in temperature range for such shift magnitudes. We see the glassy regime disappear almost completely at the highest temperature while the rubbery and terminal flow regimes are barely visible at the lower temperatures as the glassy response appears. The shifting of these segments by the amounts indicated would give the master curve at T_{ref} and is the master curve that is represented in Figure 6.1b.

Associated with the time–temperature shifting is the observation that the shift factors, in general, follow a very non-Arrhenius temperature dependence above the glass transition temperature. Figure 6.3a and b shows plots typical of the temperature dependence for the shift factors for the segmental relaxation for an amorphous polymer that follows the Williams, Landel, and Ferry [40] (WLF) or, equivalently, Vogel [41], Fulcher [42], Tammann [43] (VFT) [44],* type of response. The figures show the extremely rapid increase in relaxation

---

* It should be remarked that, in reality, Tammann (or Tammann and Hesse) did not propose this equation independently; rather, they cited the work of Fulcher [41] and claimed that they had used it before becoming aware of the Fulcher work. This has been accepted, though perhaps today one would think that they had been "scooped." Note that historically, the H is not generally used, that is, we use VFT not VFTH, though Ferry [26] has suggested using VFTH.

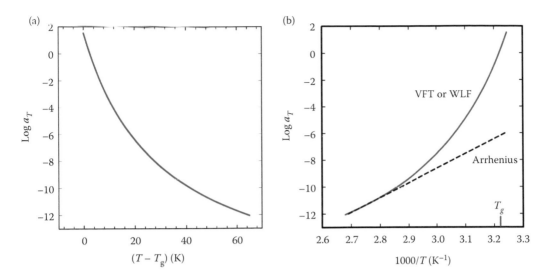

Figure 6.3 Temperature dependence of temperature shift factors versus (a) $T - T_g$ and (b) $1000/T$ showing how time scales apparently diverge as temperature is lowered toward the glass transition temperature.

times (scaled by the reference relaxation time) either with decreasing $T - T_g$ (Figure 6.3a) or with increasing $1/T$ (Figure 6.3b), the latter showing the strong non-Arrhenius or super-Arrhenius behavior. Once one has made the master curve and knows the shift factors, the response of the material at any temperature and time can be calculated—this is the importance of time–temperature superposition in both understanding and characterizing the behavior of polymers. One other aspect of the time–temperature superposition that is more difficult than what was done above is determining the behavior below the glass transition temperature. This is an area of intense study that began in earnest with the pioneering work of Struik on physical aging behavior. Below the glass transition temperature, because the material is out of equilibrium [21,45–49], the relaxation times deviate from the extrapolated VFT or WLF curves shown in Figure 6.3. How the equilibrium response, at presumably extremely long times, behaves is itself still a matter of discussion and the reader is directed to the relevant references for further information [44,50–61]. We now turn to the other viscoelastic properties of interest.

6.2.2 DYNAMIC MODULI AND CREEP COMPLIANCE

An important aspect of the linear viscoelastic response of materials is that all of the linear functions can be related through appropriate equations [27–36]. Hence, once $G(t)$ is known, the dynamic moduli $G'(\omega)$ and $G''(\omega)$ can be determined and, then, the time–temperature superposition is followed but now in terms of frequency. While the dynamic moduli are frequently measured directly at the macroscale because of the readily available instrumentation for their determination, at the nanoscale, which is of eventual interest in this chapter, they have been little investigated, except in indentation types of experiments. Hence we leave out a discussion of these properties. The other important and frequently used property in macroscopic measurements is the creep compliance of the material, an area especially developed in the works of D.J. Plazek [62–66]. In this instance, the measurements involve applying a constant stress to the sample and measuring the strain response. The time-dependent compliance (in shear) is denoted $J(t) = \gamma(t)/\sigma$. It is important to remind the reader that the creep compliance is not the reciprocal of the stress relaxation modulus, but is related either through the frequency dependent responses as $J^*(\omega) = 1/G^*(\omega)$ or through the Laplace transforms of the time-dependent responses as $s\hat{G}(s) = 1/s\hat{J}(s)$, where $\hat{G}(s)$ and $\hat{J}(s)$ are the Laplace transforms of the shear relaxation modulus and shear creep compliance, respectively. Creep measurements have much the same problem as the relaxation measurements, that is, they can only be performed over a limited time window similar to the above of about 10^{-1} s to approximately 10^6 s. Hence, time–temperature superposition is used to create a

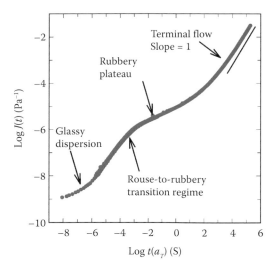

Figure 6.4 Double logarithmic representation of the creep compliance master curve for a poly(vinyl acetate) polymer showing the segmental, rubbery, and terminal flow regimes. The reference temperature is 84.3°C. (Data are from D. J. Plazek, *Polym. J.*, 12, 43–53, 1980.)

master curve of the creep response, as was done for the relaxation modulus. Figure 6.4 shows such a master curve, and the shift factors used to create the creep master curve are the same as those used to create the stress relaxation master curve for the same material. This is the case, as just noted, because all of the viscoelastic material properties are interrelated. Hence, the information obtained from the different experiments is the same if transformed properly. Importantly, there are often reasons that one chooses to carry out different types of experiment because the measurements themselves may be more or less sensitive to different aspects of the viscoelastic response of the material. A simple example is that the creep experiment can give the viscosity directly, while a flow experiment needs to be performed if one has the moduli because the viscosity makes up part of the relaxation function while it is a separable and additive part of the creep function. As seen in Figure 6.4, the short time response of the polymer is the glassy to segmental regime, followed at compliances several orders of magnitude higher by the rubbery plateau and, finally, by a terminal flow regime where the double logarithmic response goes to a slope of unity where $J(t \to \infty) \approx (\eta/t)$, where η is the shear viscosity. The data are for poly(vinyl acetate) and the reference temperature is 84.3°C. We also remark that between the segmental relaxation and the rubbery plateau is a regime related to the chain motions between the entanglement points and this is referred to as the Rouse regime [27], that is, the regime in which the chain dynamics are similar to those of a Gaussian chain before the effects of the constraints of the entanglements are felt.

The above discussion was for the macroscopic polymer behavior. What is the response at the nanometer-size scale and how does one make measurements on extremely small quantities of material such as ultrathin films? One way to address this problem is shown in work by McKenna and coworkers in which a membrane inflation method was developed [12–14,17,20,67,68] that permits one to determine the creep compliance of films having thicknesses in the nanometer-size regime. The method is now described and results of measurements on films having thicknesses ranging from below 10 nm to above 100 nm are shown.

6.3 MEMBRANE INFLATION

An important tool in the arsenal of many a rubber or polymer melt rheologist has been the membrane inflation experiment [69–73]. The method in thin metallic films has been referred to as a bulge test [74–77]. In this experiment, one takes a sheet of material, clamps it over a channel, generally of circular orifice and applies a pressure across the sheet. Once the sheet displacement is approximately three times the sheet thickness, one

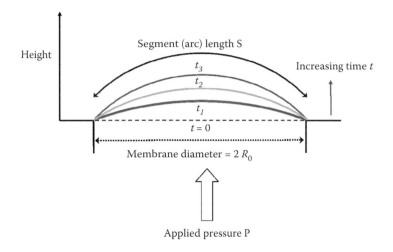

Figure 6.5 Schematic of membrane inflation experiment for a viscoelastic material.

is in the membrane limit, and the problem becomes equivalent to the biaxial stretching of a sheet, at least at the pole of the inflated membrane or bubble [78,79]. Figure 6.5 shows a schematic of a single membrane that is suspended across a channel and subjected to a pressure P. In the case of a viscoelastic material, the bubble will change shape as a function of time. In the case of the measurement of ultrathin films, the challenge is to make measurements of the bubble shape and to have pressures that are reasonable for measurement. One can take multiple approaches to this, but here we describe a method that uses templates that have "channels" of micron size so that the inflated bubbles or membranes are able to be imaged with an atomic force microscope (AFM). Figure 6.6 [12] shows a schematic of the concept of the bubble inflation measurement in an AFM. Figure 6.7 [80] shows a typical AFM image of a set of 5 μm diameter membranes that have been inflated. The film is 56 nm thick and subjected to a pressure of 5.2 kPa. By tracking the bubble size as a function of time, the viscoelastic properties can be extracted using the Boltzmann [81] linear superposition principle. The properties can be compared as a function of film thickness with the macroscopic properties. We will see that there are significant changes in properties as one reduces the film thickness to the nanometer-size range. First, we present the equations for the membrane inflation.

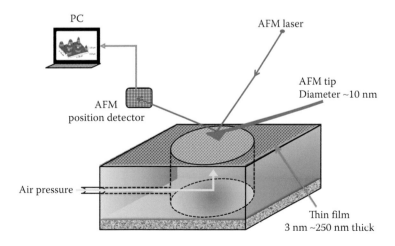

Figure 6.6 Schematic of membrane inflation experiment in an atomic force microscope.

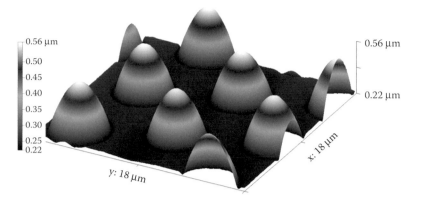

Figure 6.7 AFM image of a bubble inflation experiment for a polyurethane elastomer. The film was 56 nm thick, the bubble (through-channel) diameter was 5 μm, and the inflating pressure was 5.2 kPa. (Reprinted from *Polymer*, 55, M. Zhai and G. B. McKenna, Elastic modulus and surface tension of a polyurethane rubber in nanometer thick films, 2725–2733, Copyright 2014, with permission from Elsevier.)

6.3.1 MEMBRANE EQUATIONS

The inflation of a membrane is a combination of a bending problem and stretching [67–72]. When the bending term can be neglected, one can treat the membrane as, essentially, a biaxial stretching problem. In this case, the time-dependent properties of the material are amenable to a simple treatment as an approximation to a homogeneous deformation using Boltzmann [81] superposition procedures. In what follows, we use the approach of O'Connell and McKenna [67]. For the inflation of a membrane, the membrane stresses are determined from the bubble radius of curvature κ, the pressure P, and the membrane thickness h as

$$\sigma_{11} = \sigma_{22} = \frac{P\kappa}{2h} \tag{6.4}$$

and the stresses σ_{11} and σ_{22} are the equibiaxial stresses at the pole of the inflated (hemispherical shaped) bubble. The equal biaxial strains ε_{11} and ε_{22} at the pole of the bubble are

$$\varepsilon_{11} = \varepsilon_{22} = \frac{S}{2R_0} - 1 \tag{6.5}$$

where S is the segment or arc length of the deformed bubble and R_0 is the initial radius of the membrane (hole diameters of $2R_0$ in Figures 6.4 through 6.7. We further can write that the segment length is

$$S = 2\kappa \sin^{-1}\left[\frac{R_0}{\kappa}\right] \tag{6.6}$$

Since the pressure is applied and the parameters of film thickness and hole diameter are known, the strain can be obtained by making a measurement of the bubble profile and calculating the radius of curvature by fitting the shape to a circle*:

$$\kappa^2 = (x - a)^2 + (z - b)^2 \tag{6.7}$$

---

* The full profile can also be fit to a spherical cap in a similar fashion, but the time would not be as well defined as in the case of a single line profile across the center of the membrane due to the finite scan rate of the AFM.

where x and z are the coordinate axes along the hole diameter or radius and the height direction, respectively, and a and b are offset values for the fact that the circle is not centered on the coordinate axes. From the segment length calculation using Equations 6.6 and 6.7, one can then use Equation 6.5 to determine the biaxial strain in the inflated membrane. The apparent creep compliance $D_{app}(t)$ at any time t can then be determined from the ratio of the measured strain and the applied stress at time t. But this is not the desired compliance; rather, we want the actual creep compliance for the viscoelastic material and need to use the strain history in response to the applied stress with the Boltzmann [81] superposition integral to determine this [14,82].

$$\varepsilon_{11}(t) = \int_0^t D(t-\xi)\left\{\frac{d\sigma_{11}(\xi)}{d\xi}\right\}d\xi \tag{6.8}$$

which can be rearranged as [14,82]

$$\varepsilon_{11}(t) = \int_0^t \sigma_{11}(t-\xi)\left\{\frac{dD(\xi)}{d\xi}\right\}d\xi + D_g\sigma_{11}(t) \tag{6.9}$$

where the strain is the biaxial strain defined above, and the stress is the biaxial stress defined above. The creep compliance $D(t)$ is the viscoelastic response function that would be obtained were one able to perform a constant stress experiment following an instantaneous step in stress. The parameter D_g is the zero time glassy compliance (biaxial). Numerical solution of Equation 6.9 for known strain history and known stress history permits us to extract the creep compliance of the membrane. In the next section, we first describe results for the elastic response of a rubbery material. This is followed by a section on results from time-dependent measurements.

6.3.2 ELASTIC ANALYSIS

Although the bubble inflation method was developed originally for the purposes of making measurements of the properties of viscoelastic materials, the results that came from those studies also indicated a need to examine the elasticity of the membranes, in particular to consider surface tension effects as well as the modulus of the membrane. We now describe the equations relevant to this problem before moving to the problem of the time-dependent membrane.

6.3.2.1 STRESS–STRAIN APPROACH

If one only considers the stretching of the membrane, the stress–strain response is made up of two terms for the total stress $\sigma = \sigma_{tot}$ in the membrane [83,84]:

$$\sigma = \sigma_{11} = \sigma_{22} = \sigma_{tot} = E_{biax}\varepsilon + \sigma_{int} \tag{6.10}$$

where the stress σ_{int} is due to internal stresses and this is the sum of residual stresses σ_{res} and surface tension σ_s induced stresses.

$$\sigma_{int} = \sigma_{res} + \sigma_s \tag{6.11}$$

The surface tension induced stresses are written as

$$\sigma_s = \frac{2\gamma}{h} \tag{6.12}$$

where γ is the surface tension and h is, again, the membrane thickness. The residual stress is not known, in general, and the data can be treated to either obtain the residual stress, or if there is reason to believe there is no residual stress, to obtain the surface tension of the membrane material. Hence, a plot of stress versus

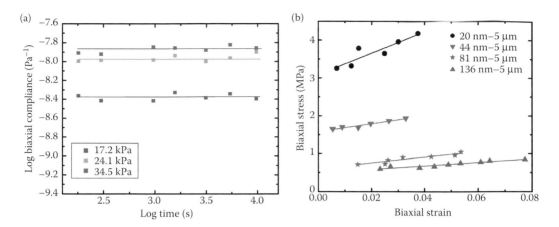

Figure 6.8 (a) Logarithm of compliance versus logarithm of time for a polyurethane rubber membrane of 56 nm thickness tested at room temperature. (Data from M. Zhai and G. B. McKenna, *Polymer*, 55, 2725–2733, 2014.) The figure shows that behavior is elastic. (b) Stress–strain plots for a poly(n-butyl methacrylate) polymer in the rubber regime for different thickness membranes showing slope and finite intercept. (Reprinted with permission from S. Xu, P. A. O'Connell, and G. B. McKenna, Unusual elastic behavior of ultrathin polymer films: Confinement-induced/molecular stiffening and surface tension effects, *J. Chem. Phys.*, 132, 184902. Copyright 2010, American Institute of Physics.)

strain, rather than going through [0,0], will have an intercept related to the internal stresses and a slope related to the biaxial modulus.

Figure 6.8 shows some typical results for the inflation of a polymer in the rubbery plateau regime that is behaving elastically. In Figure 6.8a, one sees that there is no time dependence to the membrane deformation at different stresses, thus justifying the elastic analysis. Figure 6.8b shows the stress–strain response and there it can be seen that there is an intercept at a positive value of stress and there is a finite slope related to the modulus.

Analysis of the data of Figure 6.8b using Equations 6.10 through 6.12, with the assumption that $\sigma_{res} = 0$, gives results that are of great interest in the understanding of the response of nanometric-scale polymer films. First, Figure 6.9 shows the biaxial modulus for the poly(n-butyl methacrylate) (PBMA) rubbery polymer as a

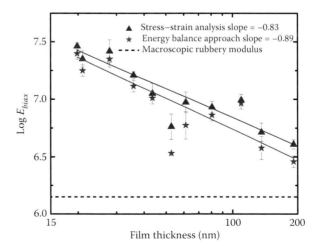

Figure 6.9 Rubbery modulus of a PBMA polymer as a function of film thickness. (Reprinted with permission from S. Xu, P. A. O'Connell, and G. B. McKenna, Unusual elastic behavior of ultrathin polymer films: Confinement-induced/molecular stiffening and surface tension effects, *J. Chem. Phys.*, 132, 184902. Copyright 2010, American Institute of Physics.)

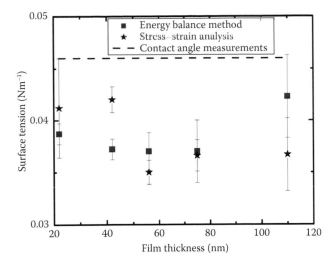

Figure 6.10 Surface tension for films of a polyurethane rubber obtained by membrane inflation compared with macroscopic measurements using contact angle determinations. (Reprinted from *Polymer*, 55, M. Zhai and G. B. McKenna, Elastic modulus and surface tension of a polyurethane rubber in nanometer thick films, 2725–2733, Copyright 2014, with permission from Elsevier.)

function of film thickness calculated both from the stress–strain analysis and from an energy balance analysis, described subsequently. Of significant interest in the figure is that the modulus of the film in this rubbery state is significantly greater than the macroscopic modulus, exceeding it by over 10-fold for the thinnest film. In fact, PBMA and the polyurethane rubber, while showing significant stiffening, show less than is seen in, for example, polystyrene (PS), polycarbonate (PC), or poly(vinyl acetate) (PVAc) for which the enhanced modulus can be up to 1000 times the macroscopic modulus when the films get to sub-10 nm in thickness [12–14,17,20,67,68]. The dependence on the thickness is just beginning to be considered from theoretical perspectives and much remains to be understood [80,85,86]. Of interest in the data of Figure 6.8b (and the published results for other materials) is the additional observation that, if there were no stiffening, then the macroscopic rubbery modulus would determine the slope of the stress–strain plots and in cases where it is very low such as in the rubbery regime of the macroscopic material, the slope would go toward zero and only the internal stresses (and surface tension) would govern the curves. We turn now to these stresses.

If one considers that in the polymeric membrane there should be no residual stress because it has been annealed above the glass transition temperature of the thin film, then the intercepts for the data in Figure 6.8 are related to the surface tension only (Equations 6.10 through 6.12). As seen in Figure 6.10, for the polyurethane rubber, the surface tensions are very close to the macroscopic values obtained from contact angle measurements [80]. Similar results have been found in other materials [83,84]. The results are important for two reasons. First, they show that the test itself is a means to determine the surface energies of ultrathin films. In addition, for the present conditions of a polymeric material above its glass transition temperature, the values are independent of film thickness and very close to the macroscopic surface tension. As indicated above, the data for the membrane inflation can also be treated by using an energy balance approach to the problem. We now look at this type of analysis.

6.3.2.2 ENERGY BALANCE APPROACH

If we think about the PdV energy of the gas used to inflate the bubble in the membrane problem, then the membrane resists that energy through bending of the membrane plate, stretching of the membrane, and surface energy resistances. The balance of the energies is then

$$PdV = \Delta E_{total} = \Delta E_{bending} + \Delta E_{stretching} + \Delta E_{surface} \tag{6.13}$$

where P is the applied pressure, V is the volume of the bubble created upon deforming the membrane, and the $\Delta E_{bending}$, $\Delta E_{stretching}$, $\Delta E_{surface}$ refer to the bending, stretching, and surface contributions to the elastic energy of the deformed membrane. Without going into the details (the reader is referred to Reference 84), and recognizing that the bending term is small, we can write the total energy, the stretching energy, and the surface energy terms as

$$\Delta E_{total} = PdV = \pi \left(\frac{PR_0^2}{2} + \frac{P\delta^2}{2} \right) d\delta \qquad (6.14)$$

$$\Delta E_{stretching} = \frac{2.168\pi Eh\delta^3}{R_0^2} d\delta \qquad (6.15)$$

$$\Delta E_{surface} = 4\pi\gamma\delta d\delta \qquad (6.16)$$

where γ is the surface energy of the polymer membrane, δ is the deflection of the membrane at the pole, h is the film thickness, and R_0 is the radius of the bubble.

Substituting Equations 6.14 through 6.16 into Equation 6.13 with $\Delta E_{bending} = 0$, we obtain

$$\frac{PR_0^2}{2\delta} + \frac{P\delta}{2} = \frac{2.168\, Eh\delta^2}{R_0^2} + 4\gamma \qquad (6.17)$$

and a plot of $(PR_0^2/2\delta) + (P\delta/2)$ versus $(2.168h\delta^2/R_0^2)$ should give a straight line with slope related to the Young's modulus and the intercept to the surface energy. Figure 6.11 shows such a plot for a set of individual bubbles in a single 56 nm film of a polyurethane elastomer [80] and we see there the reproducibility of the data, as well. Figures 6.9 and 6.10 show that the results from the energy balance analysis and the direct stress strain analysis are similar.

So far, the analysis has concerned the elastic membrane. We now turn to the viscoelastic case.

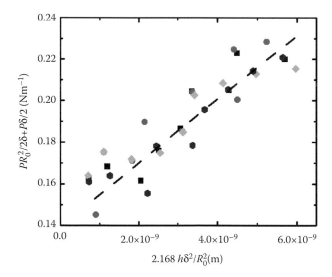

Figure 6.11 Plot of $(PR_0^2/2\delta) + (P\delta/2)$ versus $(2.168h\delta^2/R_0^2)$ as suggested by energy balance analysis of membrane inflation experiment. Data are for four individual bubbles in a single 56 nm thick polyurethane film. Slope is related to modulus and intercept to the surface energy. Line is as aid to eye. (Reprinted from *Polymer*, 55, M. Zhai and G. B. McKenna, Elastic modulus and surface tension of a polyurethane rubber in nanometer thick films, 2725–2733, Copyright 2014, with permission from Elsevier.)

6.3.3 VISCOELASTIC ANALYSIS

The original purpose [12–14,67] of developing the bubble inflation test for ultrathin films was to have a means to interrogate the dynamics of polymers at the nanometer-size scale in a fashion similar to that developed by Ferry [27]* and his students using viscoelasticity as a tool to study molecular physics of materials. Thus, the developments given at the beginning of the chapter are important. In the following paragraphs, we develop the bubble inflation method for the case of the viscoelastic membrane.

The first point of importance in the bubble inflation experiment is that it is a creep experiment because a load (pressure) is applied and the deformation response is measured. Hence, the property measured is related to the biaxial creep compliance. However, unlike the simple experiment of creep in shear or in uniaxial stretching, the stresses in the sample at constant applied pressure change because the bubble radius of curvature κ changes (see Equation 6.4). As a result, the treatment of the data requires that a time-dependent load and deformation be considered and that the Boltzmann [81] superposition be used to extract the actual material compliance. In general, one uses the following equation to solve for the creep compliance from the stress versus time and strain versus time data [27,35]:

$$\varepsilon(t) = \int_0^t D(t-t')\left[\frac{d\sigma(t')}{dt'}\right]dt' \tag{6.18}$$

which can be rearranged to a form developed by Riande et al. [82] that is used to circumvent discontinuities in the derivative of the stress history:

$$\varepsilon(t) = \int_0^t \sigma(t-t')\left[\frac{dD(t')}{dt'}\right]dt' + D_g\sigma(t) \tag{6.19}$$

where $\varepsilon(t)$ is the time-varying creep (biaxial) strain, $\sigma(t)$ is the time-varying biaxial stress, and $D(t)$ is the biaxial creep compliance. D_g is the zero-time glassy compliance and is time independent. (*Remark*: this comes from the observation [27] that the creep compliance can be represented by a function $D(t) = D_g + \Phi(t) + t/\eta$ where the function $\Phi(t)$ captures the time dependence and η is the material viscosity.)

Typical stress versus time data and strain versus time data for an inflating bubble are shown in Figure 6.12. The data are for a 36 nm thick polystyrene at a temperature of 69°C. The data were fitted to an analytical function for ease of analysis (in the present instances), modified stretched exponential functions of the following forms were used for the time-dependent stress $\sigma(t)$ and strain $\varepsilon(t)$ [14]:

$$\sigma(t) = \sigma_0 + \sigma_1 \exp\left(-\left[\frac{t}{\lambda_\sigma}\right]^{\beta_\sigma}\right) \tag{6.20}$$

$$\varepsilon(t) = \varepsilon_0 + \varepsilon_1\left(1 - \exp\left(-\left[\frac{t}{\lambda_\varepsilon}\right]^{\beta_\varepsilon}\right)\right) \tag{6.21}$$

---

* J. D. Ferry was one of the most influential scientists of the generation who began their careers in the 1930s, and his extensive development of molecular viscoelasticity remains a touchstone to students of macromolecular materials today. A full set of references to his works would be far beyond the scope of this chapter, therefore, his book is referenced in this context and the author acknowledges the intellectual framework he developed with his students that forms the basis for the view espoused here that molecular dynamics can be characterized using viscoelastic measurements and, in many instances, is the method of choice over, for example, methods such as dielectric spectroscopy, which has a broader frequency or time domain range of measurement but also has a smaller change of properties over the same frequency range (sometimes only a factor of two or three while the viscoelastic response changes by orders of magnitude, especially in the retardation domain) and sometimes the dielectric measurement probes a more local response than does the mechanical, which leads to different apparent outcome). In general, though, the methods are very complementary.

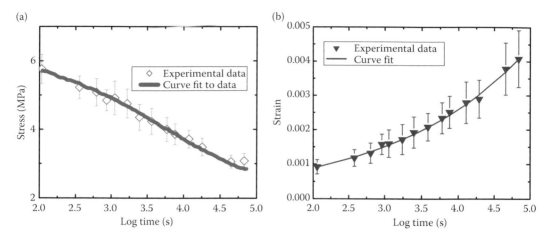

Figure 6.12 Plots of stress–time data (a) and strain–time data (b) for a 36 nm polystyrene film at 69°C along with the curve fits used to obtain analytical forms for implementation in Equation 6.19. (P. A. O'Connell, S. A. Hutcheson, and G. B. McKenna, Creep behavior of ultrathin polymer films. *J. Polym. Sci., Part B: Polym. Phys.,* 2008, 46, 1952–1965. Copyright Wiley-VCH Verlag GmbH & Co. KGaA. Reproduced with permission.)

These are used only to fit the stress versus time and strain versus time data so that we have analytical and smooth functions to put into Equation 6.19 and no physical interpretation is made of the parameters in the equations, though the λ values have the nature of retardation or relaxation times and the β values are related to the breadth of the viscoelastic response functions. Using the raw experimental data is problematic as it leads to noise propagating into the numerical solution for $D(t)$. This situation could be improved by taking more dense data, but because the bubble inflation measurements are carried out manually, this would be extremely tedious. We also found that the data treatment can lead to issues for estimates of the response at short times, where there is no data (the shortest experimental times are 100–200 s). But, for the purposes of the work to create master curves by, for example, time–temperature superposition, the short times were not used. Ideally, it would be possible to do the film inflation more rapidly, but the AFM imaging process did not allow this.

Equation 6.19 can then be solved numerically to determine the creep compliance for each experiment from the stress–time and strain–time data. Figure 6.13a shows the creep responses for a 36 nm polystyrene

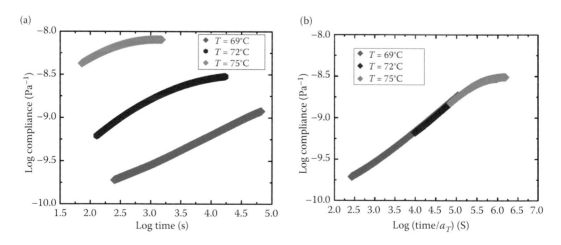

Figure 6.13 (a) Double logarithmic representation of the creep response of a 36 nm polystyrene film at three temperatures, as indicated. (b) Time–temperature superposition master curve created from data of Figure 6.13a. (P. A. O'Connell, S. A. Hutcheson, and G. B. McKenna, Creep behavior of ultrathin polymer films. *J. Polym. Sci., Part B: Polym. Phys.,* 2008, 46, 1952–1965. Copyright Wiley-VCH Verlag GmbH & Co. KGaA. Reproduced with permission.)

film at several temperatures and Figure 6.13b shows the created master curve from performing a combination of vertical and horizontal shifts of the data. Some vertical shifting was required to create the master curves, and though somewhat larger than what is found in macroscopic measurements, in general, they are nonsystematic and not large enough to affect the conclusions of stiffening in either glassy or rubbery states.

The bubble inflation experiments have proven highly useful in examining the film thickness dependence of the mechanical properties. Data such as those of Figure 6.13b can be readily obtained for other film thicknesses and compared both to each other and to macroscopic responses. The first thing to remark upon, though, is the observation in Figure 6.13 of the range of temperatures in which the material creeps from the glassy regime through a transition regime and onto a plateau. Recalling Figure 6.4, this is typical of polymeric materials, though in Figure 6.13 we do not reach a terminal flow regime. Of interest here is that the material does this, not at the macroscopic glass transition temperature, but at temperatures near to 70°C, which is approximately 30°C below the macroscopic glass transition temperature. Hence, for the thin films, the material seems to have a reduced T_g, as alluded to in Section 6.1. The other point of interest simply from the comparison of Figures 6.4 and 6.13 is that the difference between the glassy compliance and the compliance in the rubbery plateau regime in Figure 6.4 is approximately 3–4 orders of magnitude, typical for polymers [27], while the same apparent regime in the 36 nm film is approximately 1–1.5 orders of magnitude. This is related to the rubber stiffening seen in the elastic membranes described above. Both the film thickness dependence of the T_g and the stiffening of the rubbery regime can be obtained as a function of film thickness. We show this in the next paragraphs.

Once the master curves for different film thicknesses are determined, the time–temperature shift factors used to create them can be compared and the master curves themselves can be compared. Figure 6.14 shows the time–temperature shift factors for different film thicknesses from 4.2 to 22 nm for a polycarbonate material as a function of temperature. We see that these are offset significantly such that the reference temperatures, which are close to the T_g, are significantly lower than the macroscopic T_g for this material, which is approximately [20,51] 137–141°C. The lines in Figure 6.14 represent the VFT or WLF curve for the macroscopic material but shifted to the reference temperature so that the changing T_g effect can be determined quantitatively. This is done simply by assuming the macroscopic T_g and taking the temperature difference

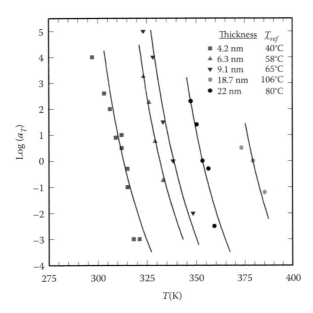

Figure 6.14 Logarithm of the time–temperature shift factors versus temperature for ultrathin films of polycarbonate. Thicknesses indicated in figure. Lines are for WLF or VFT temperature dependences adjusted for the reduced glass transition temperature of the thin films. (Reprinted with permission from P. A. O'Connell et al., Exceptional property changes in ultrathin films of polycarbonate: Glass temperature, rubbery stiffening, and flow, *Macromolecules*, 45, 2453–2459. Copyright 2012, American Chemical Society.)

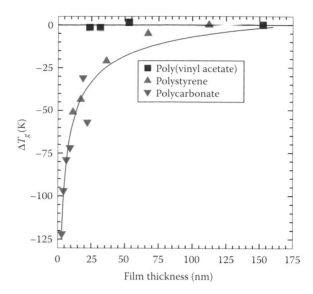

Figure 6.15 Glass transition temperature change relative to the macroscopic material for three different polymers as obtained from the membrane inflation experiments. The line is a "guide for the eye" and suggests that the polystyrene and polycarbonate may have the same thickness dependence. Further data required to substantiate this. (Data from P. A. O'Connell et al., *Macromolecules*, 45, 2453–2459, 2012.)

with the reference temperature of the thin films. Figure 6.15 shows the interpretation of these data as a ΔT_g relative to the bulk material T_g and the results are compared with results for a poly(vinyl acetate) polymer and with the polystyrene material of Figures 6.12 and 6.13. Of interest, beyond the fact of very large T_g reductions in both the polystyrene and the polycarbonate, is the observation that the reductions depend on the material investigated. The line in the figure suggests that the polycarbonate and the polystyrene may have the same T_g dependence on film thickness. However, this has not been verified because we have not succeeded in performing bubble inflation experiments on extremely thin (<10 nm) films of polystyrene as the spin cast PS films below this thickness seem to rupture during handling or testing, while the polycarbonate we could work down to the 3–4 nm range. On the other hand, in the case of the polycarbonate, the high T_g of the macroscopic material results in even the 18–22 nm thick films having to be tested at the upper temperature limits of the membrane inflation system as it is currently configured ($T \approx 120°C$). Hence, further work is required to complete the range of film thickness and T_g effects that may be seen.

We now compare the master curves for different film thicknesses. As shown in Figure 6.16, for polystyrene, we see that the master curves for each individual film thickness can be brought together and it appears that in the segmental regime (short times), the curves come close to overlapping. This would suggest that the glassy response, while being shifted to shorter times, would have the same shape, that is, that time–thickness superposition is valid. This particular aspect of behavior remains to be firmly validated simply because of the data uncertainty and the fact that breakdown of rheological simplicity ideas is often subtle. Hence, this question remains a challenge for future work. Also, of interest, particularly since there is significant work in the literature that suggests that there is a liquid surface layer on the films, there is no evidence from the bubble inflation creep experiments of an increase in compliance in the segmental regime as the films become thinner. Hence, either the glassy regime is becoming stiffer than in the macroscopic material, or the liquid layer is very thin. While there is uncertainty in the data, if the liquid layer were, for example, 3 nm in thickness [22,87], this would imply that the compliance of a 10 nm film would increase by a factor of 2.5, whereas the data actually suggest stiffening of this order of magnitude [14].

The last point of importance from the bubble inflation experimental method is the discovery that the "rubbery plateau regime" of the polymers seems to be greatly stiffened, that is, the rubbery compliance decreases dramatically relative to the macroscopic compliance or, as shown above for the polyurethane elastomer, the

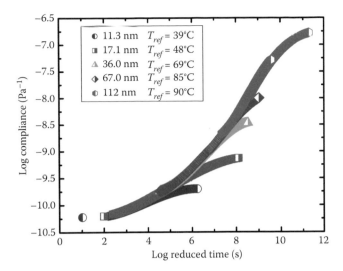

Figure 6.16 Logarithm of the compliance versus logarithm of reduced time for polystyrene films of different thickness, as indicated. (P. A. O'Connell, S. A. Hutcheson, and G. B. McKenna, Creep behavior of ultrathin polymer films. *J. Polym. Sci., Part B: Polym. Phys.*, 2008, 46, 1952–1965. Copyright Wiley-VCH Verlag GmbH & Co. KGaA. Reproduced with permission.)

modulus increases as the films get thicker. This is clearly seen in Figure 6.16 where the responses at long (reduced) times deviate from a single curve. The explanation for this is unclear, though it has been suggested by Ngai et al. [85] that the observed stiffening is related to the differences in the effects of the confinement of the molecules at the nanometer-size scale on the Rouse modes (being weak) and on the segmental modes (being strong). Hence, as the glass transition temperature decreases, the segmental modes become very rapid while the Rouse modes remain unchanged with thickness, and hence are "cut off" at low temperatures where the experiments are carried out. The magnitude of the stiffening is clear in Figure 6.16. Figure 6.17 shows a

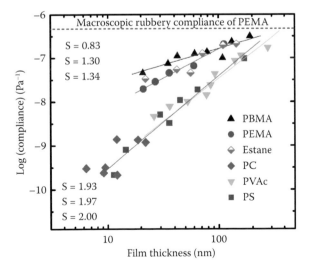

Figure 6.17 Plateau compliance versus film thickness for different polymeric ultrathin membranes showing that PVAc, PC, and PS have similar strong stiffening while the PBMA, PEMA, and estane polyurethane show weaker behaviors. S is the stiffening index. (Data from P. A. O'Connell et al., *Macromolecules*, 45, 2453–2459, 2012; M. Zhai and G. B. McKenna, *Polymer*, 55, 2725–2733, 2014; S. Xu, P. A. O'Connell, and G. B. McKenna, *J. Chem. Phys.*, 132, 184902, 2010; and X. Li and G. B. McKenna, *Macromolecules*, 48, 6329–6336, 2015; Reprinted with permission from X. Li and G. B. McKenna, Ultrathin polymer films: Rubbery stiffening, fragility, and Tg reduction, *Macromolecules*, 48, 6329–6336. Copyright 2015, American Chemical Society.)

compilation of the rubbery stiffening for polystyrene [13,14,84], poly(vinyl acetate) [12,13], polycarbonate [20], poly(n-butyl methacrylate) [83], poly(ethyl methacrylate) [88], and for the polyurethane rubber [80] described above. The first thing to notice in the figure is that the compliance for the first three of these materials seems to scale with the square of the film thickness. On the other hand, the stiffening of the PBMA, the PEMA, and the polyurethane is much weaker and the reasons for these differences are currently unknown. As just noted, Ngai et al. [85] suggest that it should vary with the shape of the segmental relaxation, which is related to the sensitivity of the Rouse and segmental modes to the chain confinement. In this case, the stiffening index S (the double logarithmic slope of the compliance versus film thickness) should correlate with the Ngai "coupling" parameter n_α and, as shown in Figure 6.18a, the results are consistent with this hypothesis. The figure also shows that the coupling parameter and the dynamic fragility index m of the glass-forming material correlate and Li and McKenna [88] extended the Ngai et al. [85] idea to the possible correlation between the stiffening index and the fragility index. As seen in Figure 6.18b, when the choice of fragility index is based on mechanical determinations, the correlation is very good. These findings merit further exploration.

On the other hand, Page et al. [86] have found similar large stiffening in film buckling experiments on a Nafion® material and propose that the polymer chains in confinement begin to lose their entropic character and the stiffening observed is due to tendency of the forces to be resisted by the chains acting as stiff reinforcing elements due to the confinement. Though Li and McKenna [88] concluded that the full set of stiffening data are not consistent with this model, it seems further exploration is warranted. In any case, understanding the origins of such behavior is important to the use of polymeric materials at the nanometer-size scale. The nanomechanical testing using the bubble inflation method is one important new means of addressing the fundamental questions (in addition to having identified some of the new phenomena).

In addition to the direct mechanics of measurement on ultrathin films by the membrane inflation method, we discuss in the following a spontaneous particle embedment method, built upon early works by Tabor [89] and Rimai and coworkers [90–92] with microparticles, followed by Forrest and coworkers [22,23] who used nanometer-sized particles, and that we have been exploiting in our own labs [24,25,93–95] to investigate surface viscoelastic properties at a scale of size that is smaller than normally obtained with nanoindentation methods. We remark here that there is also a method for characterizing the viscoelastic behavior based on a film dewetting method that was developed by Bodiguel and Fretigny [96–98] to investigate ultrathin films. The method is based on using the surface energy differences between the film and a supporting liquid substrate to mechanically force the film to shrink. A viscoelastic analysis can be used to extract the material

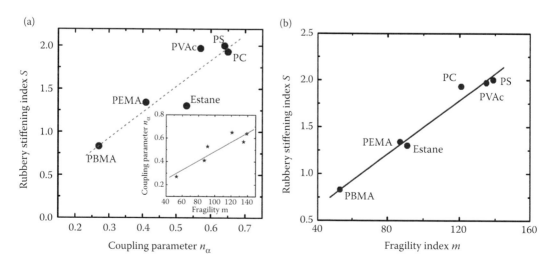

Figure 6.18 (a) Rubbery stiffening index versus coupling parameter. Inset: Coupling parameter versus fragility index. (Reprinted with permission from X. Li and G. B. McKenna, Ultrathin polymer films: Rubbery stiffening, fragility, and Tg reduction, *Macromolecules*, 48, 6329–6336. Copyright 2015, American Chemical Society.) (b) Rubbery stiffening index versus fragility index. (Data from X. Li and G. B. McKenna, *Macromolecules*, 48, 6329–6336, 2015.)

properties of the ultrathin films [99,100]. We do not elaborate on this method here as its limitations make it only marginally suitable for polymers in the glassy state itself, though it is readily used in the regime from times close to the Rouse regime through the terminal regime (see Figures 6.1 and 6.4).

6.4 SPONTANEOUS PARTICLE EMBEDMENT

Frequently in making nanomechanical measurements, one uses either a nanoindenter or an atomic force microscope to force a spherical or other shaped tip into a surface and to then use a contact mechanics analysis to extract the material properties. When the material is viscoelastic, the problem can be complicated by issues of the combination of the nonlinearity of the force–displacement curve with the viscoelasticity of the material interacting [101]. Methods have been developed to deal with these problems [101–106], though sometimes approaches more relevant to elastic plastic materials [107–111] have been used and these can be problematic for polymers, which are viscoelastic. Here, we look at the related problem that stems from the observation that a hard particle placed onto a soft surface will spontaneously embed into the surface due to the work of adhesion [112,113] between the surface and the particle. If the surface being probed does not flow to surround the particle, one can readily use the viscoelastic form of the JKR (Johnson, Kendall, and Roberts) [114] model to determine the surface properties. In what follows, we show results using, first, an elastic analysis and, then, following a viscoelastic approach.

6.4.1 ELASTIC ANALYSIS

Figure 6.19 shows the problem of a sphere on a surface before embedment and after embedment. We also show the idea that the sphere embeds with time when the material is viscoelastic, such as a polymer surface. In the case of an elastic material, one can relate the force on the particle to the work of adhesion through the JKR approach (we remark that it is in some instances going to be required to use an approach that differs, such as the Maugis model, but the differences are of detail and not of substance. Of course derived/extracted properties would differ somewhat). The difference in the approach here from that used for, for example, nanoindentation, is that the force function is determined by the particle/material couple rather than by the external forces applied by the indenter. First, we take the Hertzian elastic solution for the indentation of a rigid sphere of radius R indenting on a flat surface having a shear modulus G and a Poisson's ratio of v:

$$h^{3/2} = \frac{3P(1-v)}{8G\sqrt{R}} \tag{6.22}$$

and h is the embedment depth and P is the load required to create the indentation. The shear modulus is then given by

$$G = \frac{3P(1-v)}{8h^{3/2}R^{1/2}} \tag{6.23}$$

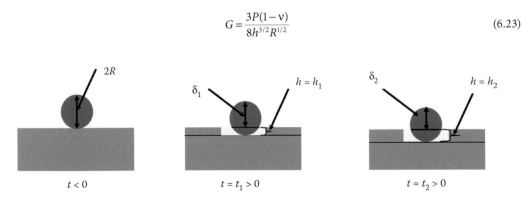

Figure 6.19 Schematic of embedding of a rigid particle into a viscoelastic substrate at different times.

The fact that the Poisson's ratio is not known can be a problem, but generally in the case of polymers is taken to be near to 0.3 in the glassy state and 0.5 in the rubbery state. As just noted, in the indentation experiment, the load P is prescribed by the instrument while in the case of the spontaneous particle embedment, the load results from the interfacial forces acting between the particle and the material surface of interest, that is, the work of adhesion W_a, which is defined as

$$W_a = \gamma_{SV} + \gamma_{LV} - \gamma_{SL} \tag{6.24}$$

where the surface energy contributions to the work of adhesion are γ_{SV}, the solid–vapor contribution (particle), γ_{LV}, the liquid–vapor contribution (substrate polymer), and γ_{SL}, the interfacial tension between the polymer substrate and the particle. The determination of the surface energy and work of adhesion is beyond the scope of this chapter and the reader is referred to the literature for its determination [24,93,115,116]. Then, the force between the particle and the substrate of interest is written as

$$P(t) = 2\pi W_a \sqrt{R^2 - (R - h(t) - h_m)^2} \tag{6.25}$$

The meniscus height h_m is formed upon initial contact between particle and flat material surface. For each particle–surface system, meniscus height h_m is a constant and it can be described by the following equation developed by Tabor [89] and Rimai and coworkers [90–92]:

$$h_m = \left\{ \frac{[R(\omega_A/2)^2]}{E^2} \right\}^{1/3} \tag{6.26}$$

Here, E is the Young's modulus (zero time modulus) of the surface. The meniscus height is used to determine the initial (zero time) force for the problem for the viscoelastic case. We first illustrate essentially elastic embedment experiments and show some of the promise and pitfalls of the method before moving on to consider the viscoelastic case.

6.4.2 PARTICLE CHARACTERIZATION

One of the most important factors in the particle embedment experiment is the particle itself. Hence, it is important to characterize size and shape. Figure 6.20 shows some micrographs of different submicron- or nanometer-sized particles and illustrate visually that one has both spherical particles (the polystyrene and silica) that are made from amorphous materials as well as faceted particles made from a crystalline material (gold). Perhaps more importantly, since even gold has a reasonable "sphericity," these particles upon close examination are polydisperse in diameter. Figure 6.21 illustrates the polydispersity of a set of 200 nm silica particles and that of a set of 200 nm gold particles. The polydispersity is an issue because it reduces the ability to make spatial measurements of the material properties (property mapping), and one can really only make measurements of averages, as we show subsequently. Also, it is of interest that the particle sizes are not the same as those specified by the particle suppliers. The AFM images used for the particle size distributions are much closer to the transmission electron micrograph estimated sizes, suggesting that the light scattering generally used by the manufacturers provides a larger number. The reasons for this are beyond the scope of the present work, but need to be considered in any estimate of properties as the differences are not negligible.

6.4.3 PARTICLE EMBEDMENT MEASUREMENTS AND MODULUS DETERMINATIONS

The particle embedment can be determined by taking atomic force microscope topographic images similar to those shown in Figure 6.22 where we see an AFM image of 200 nm polystyrene particles embedded into a lipid multilayer material, which is much softer than the polystyrene. A good point to keep in mind is that the

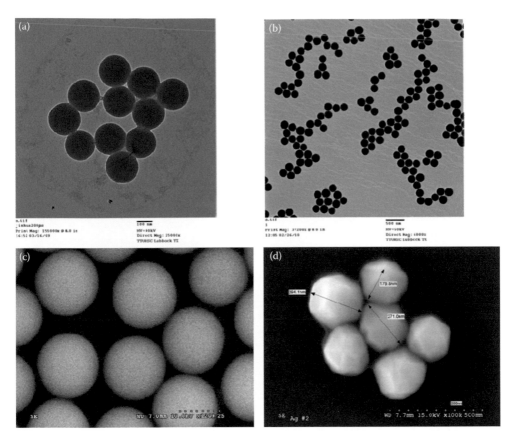

Figure 6.20 Electron micrograph images of different particles used for spontaneous particle embedment types of experiment. (a) and (b) show transmission electron micrograph images of 200 nm polystyrene and 235 nm silica particles, respectively. (c) and (d) show scanning electron micrograph images of the 200 nm polystyrene particles and 250 nm gold particles, respectively. (Reprinted with permission from J. Wang, K. Deshpande, and G. B. McKenna, Determination of the shear modulus of spin-coated lipid multibilayer films by the spontaneous embedment of submicrometer-sized particles, *Langmuir*, 27, 6846–6854. Copyright 2011 American Chemical Society. Images taken at Texas Tech University Imaging Center.)

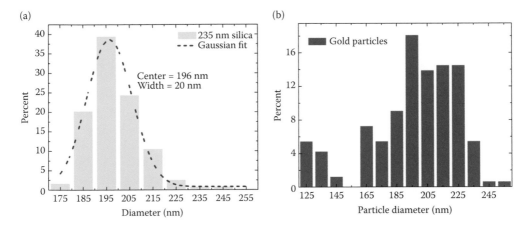

Figure 6.21 Typical particle size distributions for silica (a) and gold particles (b). The silica particles are reasonably narrowly distributed with a near-Gaussian distribution of diameters. The gold particles show a very non-Gaussian size distribution. (Reprinted from *Polymer*, 52, T. B. Karim and G. B. McKenna, Evidence of surface softening in polymers and their nanocomposites as determined by spontaneous particle embedment, 6134–6145, Copyright 2011, with permission from Elsevier.)

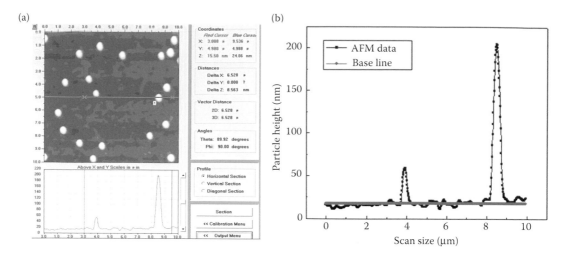

Figure 6.22 (a) Atomic force microscope image and line profile for 200 nm polystyrene particles embedding into a lipid multilayer material. (b) Line profile showing how the baseline to determine the particle height above the lipid surface is provided. (Reprinted with permission from J. Wang, K. Deshpande, and G. B. McKenna, Determination of the shear modulus of spin-coated lipid multibilayer films by the spontaneous embedment of submicrometer-sized particles, *Langmuir*, 27, 6846–6854. Copyright 2011, American Chemical Society.)

particle height is relatively easy to obtain from the line profile across the particle (line is shown in the image of Figure 6.22a) while convolution with the tip makes the lateral dimensions very imprecise, particularly as the particles get smaller [117,118].

Figure 6.22b shows the definition of a baseline and the particle heights so obtained. The lower height is actually a scan over the side of a particle. From measurements on many particles, one can obtain a distribution of particle heights and the average can be compared to the average from the particle size distribution obtained, as in Figure 6.21, from the particles on a hard surface that is weakly interacting so that the particles themselves do not deform.

Figure 6.23 shows the distribution of particle sizes for the above 200 nm silica particles embedded into a POSS-reinforced epoxy system. Of interest here is not only that the particles embed, as seen by the mean particle height going from approximately 196 nm in Figure 6.21 to approximately 191 nm in Figure 6.23, but also the distribution of embedments is described by simply applying Equation 6.22 to obtain the average modulus of the material and then using the known distribution of the particle sizes to calculate the distribution of particle embedment depths. At least for the system investigated, there is no evidence of a large heterogeneity of the material properties estimated from the embedment experiments. Of interest, of course, is the modulus estimated from the particle embedment itself. Table 6.1 shows some estimates [93,115,119] of modulus determined from several materials and with different particle types. There are several things to take from the table. First, for the elastic cases investigated, it appears that the surface modulus in soft materials is very close to that of the macroscopic material (this is most clear with the poly(dimethyl siloxane) rubber. And the said result is also consistent with work on nanoindentation with an interface force microscope as reported by Houston and coworkers [120,121] also for a poly(dimethyl siloxane) material. As one goes to the harder materials (polymer glasses), it is clear that the particle embedment measurements give a softer response than one expects from the macroscopic moduli. This specific response is consistent with reports of high surface mobility in polymer glasses [22,23,122], but it is at odds with reports from some nanoindentation measurements [109,123] in which it is reported that the modulus of the surface at shallow indentation depths is higher than the macroscopic modulus, an effect that is partially accounted for by high hydrostatic pressures beneath the probe tip. While this is of interest and clearly requires further investigation, the last material shown in Table 6.1 is polystyrene for which the particle embedment was observed for two different times at room temperature, 30 min and 27 h. As can be seen, the response is softer after 27 h than at 30 min. This is due to the viscoelasticity of the surface and is the subject to which we now turn.

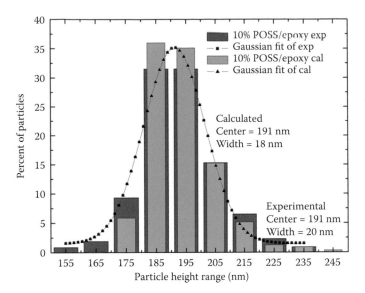

Figure 6.23 Distribution of particle heights after embedment of 199 nm diameter silica particles into a POSS/epoxy nanocomposite surface. The distributions are calculated using the average particle embedment depth to determine the surface modulus, or from the measured distribution itself. (Reprinted from *Polymer*, 52, T. B. Karim and G. B. McKenna, Evidence of surface softening in polymers and their nanocomposites as determined by spontaneous particle embedment, 6134–6145, Copyright 2011, with permission from Elsevier.)

Table 6.1 Particle embedment results for different, nominally, elastic materials

Material	Particle type	Particle diameter (nm)	Surface shear modulus (Pa)	Macroscopic shear modulus (Pa)	Reference
Cross-linked poly(dimethyl siloxane)	Silica	300	0.78×10^6	0.52×10^6	[119]
Cross-linked poly(dimethyl siloxane)	Silica	500	0.52×10^6	0.52×10^6	[119]
DPPC[a] lipid multi-bilayers	Polystyrene	203	$5.0\text{–}6.4 \times 10^6$	$0.2\text{–}200 \times 10^6$ [b]	[115]
DPPC lipid multi-bilayers	Silica	199	$29.2\text{–}62.7 \times 10^6$	$0.2\text{–}200 \times 10^6$ [b]	[115]
Epoxy	Silica	199	276×10^6		[93]
Epoxy/POSS	Silica	199	116×10^6	477×10^6 [c]	[93]
Epoxy	Gold	195	172×10^6	526×10^6 [c]	[93]
Epoxy/POSS	Gold	195	129×10^6 [b]	477×10^6 [c]	[93]
Cross-linked poly(dimethyl siloxane)	Gold	20	Particle imbibed[c]	0.52×10^6	[119]
DPPC lipid multi-bilayers	Gold	20	Particle imbibed[d]	$0.2\text{–}200 \times 10^6$ [b]	[115]
Polystyrene at RT, 30 min	Silica	199	37×10^6	1×10^9	[93]
Polystyrene at RT, 27 h	Silica	199	27×10^6	1×10^9	[93]

Note: Embedment modulus compared with macroscopic modulus.

[a] DPPC is 1, 2-dipalmitoyl-Sn-glycero-3-phosphotidylcholine.

[b] These are not macroscopic values but those from measurements on lipid bilayer by other methods [124,125].

[c] From Reference 126.

[d] Here, it was reported that the particles of very small diameter were observed to disappear into the surface of interest, suggesting that the material had yielded or failed under the contact stresses. For the lipid multi bilayers, one might imagine full wetting of the particles. For the cross-linked poly(dimethyl siloxane) material, however, the solid nature of the material would preclude such wetting.

6.4.4 VISCOELASTIC ANALYSIS

The idea of using spontaneous particle embedment to investigate the dynamics of a polymer surface was initially developed by Teichroeb and Forrest [22] and, while it was used in a fairly qualitative way, subsequent developments have provided a means to use the approach to extract quantitative information about surface viscoelastic properties. As just developed above, for the elastic case, the surface mechanical properties can be obtained by using the elastic model along with the embedment force–depth information. However, in the viscoelastic case, the problem becomes one of solving the viscoelastic problem and, if this is not done properly, results that are incorrect or difficult to interpret can be obtained. A good example of how this can be problematic is the conventional nanoindentation experiment on a viscoelastic surface (this is similar to the particle embedment method in that it uses the contact information between rigid tip [rigid particle] and material surface to obtain the surface properties). Sneddon's [108] method and the Oliver and Pharr [107] method are widely used to describe the elastic and plastic behaviors using the load–displacement curves in the indentation test. In particular, the unloading portion of the curve is used to obtain the stiffness. But when this method is used to test viscoelastic materials, creep occurs during the indentation process and this can even lead to the initial unloading curve showing a negative load–displacement relationship. The result is unphysical values of the obtained modulus, and it is not completely clear that one can overcome this difficulty readily. Cheng and Cheng [110,111] have shown that the instantaneous modulus of a viscoelastic material can be obtained from the initial unloading curve with a sufficiently fast unloading rate. However, how to decide the proper unloading rate is still a question. Furthermore, the approach gives a modulus that is "elastic," which means the method is not readily able to provide the viscoelastic response of the material, something fundamental for polymeric materials in particular. For the nanoindentation force–displacement curve itself, the viscoelastic problem involves both the time-dependent response and the changing tip geometry during the indentation test [102–106,123,127], and the effects of the two mix in the response that is measured. Furthermore, when the contact area is reduced to the nano or micro size, size effects have been reported. For example, Tweedie et al. [109] have found a stiffening of the surface of polymer glasses when the indentation depth ranges from 5 to 200 nm, and they argued that the contact load or pressure reduce molecular mobility (increase the glass transition temperature), which induces the stiffening behavior by shifting, for example, the relaxation modulus to effectively lower temperatures (see Figure 6.2 but now consider that the low temperatures are equivalent to high pressures). Contrary to this result, the findings of Teichroeb and Forrest [22] were interpreted to suggest surface softening within a few nanometers of the polymer surface based on spontaneous particle embedment experiments. They interpreted their findings as being due to the existence of a liquid-like layer at the surface, which is one of the popular explanations for the surface softening response [22,23,122,128] and one that has been supported by theoretical works [129–131]. Therefore, having a viscoelastic model that can be used to determine quantitative material properties is important and we turn now to the viscoelastic solution to the spontaneous particle embedment problem.

From the experimental perspective, the test is straightforward, if not necessarily simple. As in the elastic case discussed above, the particle dispersion in water is placed on the surface of the sample for which one wants to know the viscoelastic properties. Then the solution on the surface is dried at ambient conditions for 2–3 h. The particles spontaneously embed into the material surface due to the surface energy difference between the particles and the probed material, that is, the work of adhesion acts as the driving force for embedment. The heights of individual particles above the surface are acquired by AFM. Figure 6.24 depicts typical data for particle height versus time, for both 10 and 20 nm diameter gold particles embedding into polystyrene as reported by Teichroeb and Forrest [22,23]. Two different temperatures near to the glass transition temperature of polystyrene are shown ($T_g = 100°C = 373.2$ K for PS). The challenge is to analyze these data to quantitatively extract the viscoelastic properties of interest. Below, we describe how one does this for both the full viscoelastic response as well as for the isochronal (constant time) cases. One possible reason for using the isochronal analysis is that in many instances, the temperature capabilities of atomic force microscopes are insufficient to carry out continuous tests. In such cases, there may be advantages to making single-point measurements at room temperature after letting a set of samples sit in the oven for a given time (isochrone) and then making the particle height measurement at room temperature.

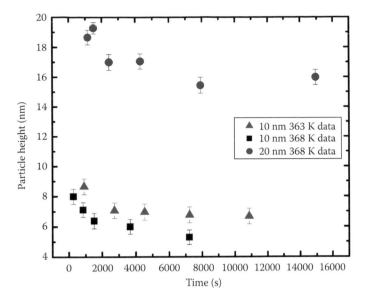

Figure 6.24 Measured apparent height of gold particles on polystyrene surface. (Data from J. H. Teichroeb and J. A. Forrest, *Phys. Rev. Lett.*, 91, 016104, 2003.)

Building on the elastic contact mechanics problem discussed above, we can develop the appropriate viscoelastic solution. The contact mechanics problem of the viscoelastic response with moving boundary deformation was solved by Lee and Radok [132] and Ting [133]. For an arbitrary particle indentation with a spherical contact, the relationship between the time-dependent embedment depth $h(t)$, the applied load $P(t)$, and viscoelastic function (creep compliance $J(t)$) can be expressed as

$$[h(t)]^{3/2} = \frac{3}{8\sqrt{R}} \int_0^t (1-\nu)J(t-\xi)\left(\frac{dP(\xi)}{d\xi}\right)d\xi \tag{6.27}$$

where R is relative radius of curvature, ν is the Poisson's ratio, $J(t-\xi)$ is the creep compliance, and $P(\xi)$ is the force. Based on the Lee and Radok equation, both isochronal compliance and time-dependent creep compliance can be extracted.

For the isochronal case, a Hertz-like equation can be obtained by simplifying Equation 6.27 to [93]

$$J_{app} \approx J(t) = \frac{8R^{1/2}\Delta h^{3/2}(t)}{3(1-\nu)P(t)} \tag{6.28}$$

And the force value $P(t)$ is obtained from Equations 6.25 and 6.26 above.

6.4.4.1 ISOCHRONAL RESULTS

Karim and McKenna [93,94] recently used Equation 6.28 to calculate the compliance using isochronal (constant time) values of embedment depth in different polymers. Because not all AFMs are capable of obtaining real-time images at high temperatures, using the isochronal solution to get the embedment depth at a specific temperature provides a good solution for exploring the mechanical behaviors at different temperatures. Previously, in order to obtain results at high temperature, Teichroeb and Forrest [22,23] had placed the sample on a hot stage and held the experimental temperature T for a predetermined time, then cooled the sample to 300 K and scanned the particle heights. They then placed the sample back into the oven at the same temperature T and let the particles embed for the next increment in time. Hence, near-continuous data

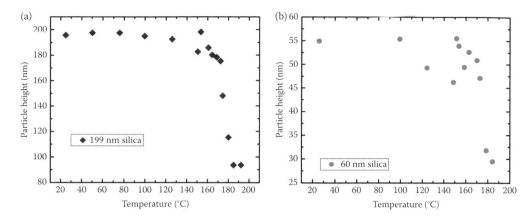

Figure 6.25 Embedment data for silica particles on PαMS at different temperatures, creep time 30 min. (a) 199 nm silica particles. (b) 60 nm particles. (Data from T. B. Karim and G. B. McKenna, *Polymer*, 52, 6134–6145, 2011.)

were obtained. Although this procedure could overcome the limitation of AFM operation problem at high temperature, it induces experimental variability. In the work by Karim and McKenna, only isochronal results were used in their analysis and avoided some of this uncertainty by only considering the isochronal aspects of behavior over the range of temperatures of interest. In that work, the sample was held at the constant temperature for 30 min, then cooled to room temperature over a time of approximately 5–6 h. Then AFM was used to scan the particle heights. Figure 6.25 shows isochronal particle heights on a poly(α-methyl styrene) (PαMS) ($T_g = 171°C$) as a function of temperature. As shown in Figure 6.25a for 199 nm diameter particles, after 30 min of creep (30 min isochrones), the particle height remained approximately constant and close to the initial height until the creep temperature increased to 150°C. Then the particle height changed rapidly as the temperature increased above 150°C, indicating increased surface mobility above 150°C. This is of interest as the glass transition of the PαMS is close to 189°C. Considering that the change of particle height at room temperature is small compared with the change at high temperature (experimental temperature), it was assumed that in the experiment, the embedment depth was frozen at the quench point and the scanned results at room temperature represent the particle heights after 30 min of creep at the relevant temperature. The Lee and Radok viscoelastic model is valid for the contact problem for the embedment depth up to half of the particle diameter, which also then determines the highest achievable experimental temperature in the test (for the 30 min isochrones). Particle size effects were also considered and tests were performed using 60 nm silica particles. As shown in Figure 6.25b, the 60 nm particles show a similar trend of particle height versus temperature as the 199 nm silica particles.

Then the compliance at each temperature can be calculated from Equation 6.28 using the corresponding isochronal embedment depth, and the results are shown in Figure 6.26. In the figure, the compliance extracted from the 60 and 199 nm silica particles are similar to each other, hence the surface properties are independent of the particle size, at least to 60 nm in particle diameter. The surface properties shown in Figure 6.26 are also compared with the macroscopic properties of PαMS obtained from conventional rheometry [134] and, surprisingly, while at low temperatures one sees a surface softening for the PαMS (room temperature to $T_g - 21°C$), a surface stiffening behavior was found at higher temperatures ($T_g - 21°C$ to $T_g + 21°C$). Similar crossover surface mechanical responses were observed for polystyrene samples using the particle embedment experiment and the same analysis method [94], though they also observed that an 8-arm, star-branched polystyrene only exhibited softening behavior [94] over the full temperature range that could be probed.

6.4.4.2 VISCOELASTIC RESULTS

We now consider the full viscoelastic case using the Lee and Radok [132] model to determine the time-dependent compliance. In the particle embedment experiment, the particle heights can be obtained at different

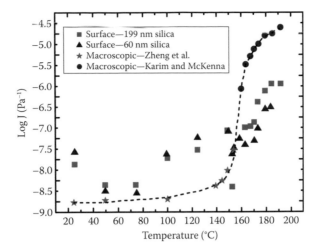

Figure 6.26 Plot of log J versus temperature for 199 nm silica particles and 60 nm silica particles embedment data, and the macroscopic compliance for PαMS obtained from Zheng et al. [134]. (Reprinted with permission from W. Zheng, G. B. McKenna, and S. L. Simon, The viscoelastic behavior of polymer/oligomer blends, *Polymer*, 51, 4899–4906. Copyright 2010, American Chemical Society.)

times as shown in Figure 6.24. Such data can be used to calculate the embedment force and, with the depth (height) information, extract the viscoelastic function. Before going further, two issues need to be pointed out here for special attention. The first is the lack of data for short times that leads to issues for estimating the short time response (similar to the problem discussed for the bubble inflation method) and the second is the scattered data points that can cause large errors in subsequent numerical solution. For the particle embedment experiment, the shortest material response times that can be easily achieved are 100–200 s due to the time required to obtain images of individual particles once temperature becomes stable. The longest time of experiment depends on the time when the particles sink more than half way into the surface where the viscoelastic contact model is no longer valid. Because of the difficulties for carrying out the particle embedment experiment, that is, good dispersion of particles on the surface, finding single particles in the AFM, the experimental data can be scattered as seen in the different figures of this section. One way to smooth the particle height (embedment) versus time data is by fitting it to a stretched exponential function with the form as

$$h(t) = a + b\exp\left[-\left(\frac{t}{c}\right)^d\right]$$

(6.29)

where a, b, c, and d are fitting parameters. The parameters shown in Equation 6.29 have no physical meaning but the analytical function provides a means to improve the numerical calculations needed to extract the viscoelastic function $J(t)$ from the data. As an example, the measured data from Teichroeb and Forrest [22], shown as the symbols in Figure 6.27a, were fitted using Equation 6.29, and the fitting result along with fitting parameters is shown in the figure. Then, using the meniscus height calculation of Hutcheson and McKenna [24] (h_m for a 10 nm gold particle on polystyrene is 0.549 nm) and using Equation 6.27 with the analytical function of Equation 6.29, one calculates the load history on the particle, as shown in Figure 6.27b.

Because of the generally scattered data, Hutcheson and McKenna [24,25] assumed that the surface creep compliance has a stretched exponential form [135,136] commonly used in polymer glass dynamics descriptions [137]:

$$J(t) = J_G + J_N\left\{1 - \exp\left[-\left[\frac{t}{\tau}\right]^\beta\right]\right\}$$

(6.30)

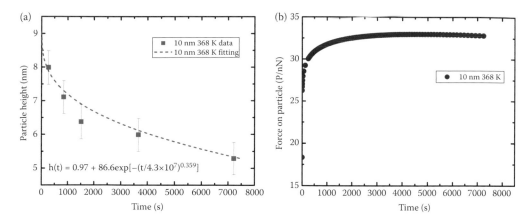

Figure 6.27 (a) Measured apparent height of 10 nm gold particles on polystyrene surface at 368 K and the fitting curve. The experimental data are from Teichroeb and Forrest [22] and the fitting curve is generated using the stretched exponential model shown in the figure. (b) Force profile calculated from apparent height experimental data shown in (a).

where $J(t)$ is the time-dependent creep compliance, J_G is the glassy compliance, J_N is the rubbery plateau compliance, τ is the retardation time, and β is the shape parameter. With this assumed form of the material function and the force profile $P(t)$, the creep compliance can be extracted by solving Equation 6.27 numerically. One reason for choosing this approach is the difficulty of getting accurate data with limited single particle embedment data. The problem is that fitting to the single embedment curve does not provide enough information to obtain accurate data for, for example, a sum of exponentials or even the stretched exponential model of Equation 6.30 without some constraints. Even for the function of Equation 6.30, the numerical analysis requires that one integrate from a finite time rather than zero because at zero time the derivative of the stretched exponential term is noncontinuous and the lower limit of the integral needs to be changed. In the Hutcheson and McKenna [24,25] work, Equation 6.30 was used to fit the macroscopic creep compliance data for a polystyrene sample from Plazek [64], which was of similar molecular weight to the material used by Teichroeb and Forrest [22]. Furthermore, it was assumed that the Poisson's ratio of the polymer changed from 0.3 to 0.5 as the polymer film moved from glass region to rubbery region and they provided the solution for the time-dependent Poisson's ratio using a similar stretched exponential form as that for the creep compliance itself of Equation 6.30:

$$\upsilon(t) = 0.3 + 0.2 \left\{ 1 - \exp\left[-\left[\frac{t}{\tau} \right]^{\beta} \right] \right\} \tag{6.31}$$

where τ is the retardation time and β is the shape parameter. These parameters were assumed to be the same as those for the creep compliance.

Combining Equation 6.27 with Equations 6.25, 6.30, and 6.31, the embedment depth can be calculated. Hutcheson and McKenna varied the retardation time τ in Equations 6.30 and 6.31 to get a good representation to the embedment data as shown in Figure 6.28. From the retardation time, one can then calculate a shift factor $a_T = \tau_{Plazek}(T)/\tau_{exp} = \tau_{Rheol}(T)/\tau_{exp}$ and use the reported time–temperature shifting parameters for polystyrene to obtain a "rheological" T_{rheol} temperature for the embedment response at an experimental temperature T_{exp}. This difference is essentially the shift in the glass transition for the specific measurement conditions.

Figure 6.29a shows the shift factors for the embedment experiments at different temperatures and the data from Plazek [64] for bulk polystyrene, which was fitted by the VFT [41–43] equation. As shown in this figure, the shift factors for the embedment tests lay close to the VFT fit, and the corresponding temperature for each overlap data point is referred to as the rheological temperature T_{rheol}. For the embedment data of Figure 6.29a, the rheological temperatures T_{rheol} are compared with experimental temperatures T_{exp} at which

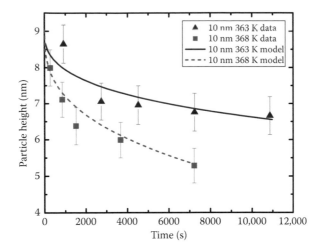

Figure 6.28 Comparison of the measured experimental results (symbols) and the fitting results using viscoelastic model (lines) for 10 nm gold particles embedding into a polystyrene surface. The retardation times extracted from the viscoelastic model for 363 and 368 K are 1.0×10^9 and 2.6×10^8 s, respectively. (Reprinted with permission from S. A. Hutcheson and G. B. McKenna, *Phys. Rev. Lett.*, 94, 076103-1–076103-4, Erratum, *Phys. Rev. Lett.*, 94, 189902. Copyright 2005 by the American Physical Society.)

the particle embedment tests were performed in Teichroeb and Forrest's work. The comparisons are shown in Figure 6.29b, in which a linear relationship between T_{rheol} and T_{exp} is observed. T_{exp} is smaller than T_{rheol} in the region of temperature below the glass transition temperature, while the opposite is the case for T_{rheol} and T_{exp} when the temperature goes above the glass transition temperature. The latter finding is in accordance with the rubbery stiffing behaviors observed from the bubble inflation test. This is also consistent with the findings of Karim and McKenna [95] in which surface stiffening of as much as one order of magnitude was observed above the glass transition temperature. The origins of the stiffening behavior of surfaces above the T_g as well as for the thin films are not currently known, though both Ngai et al. [85] and Page et al. [86] have made relevant suggestions. Furthermore, the combination of stiffening above T_g but softening below it is not understood.

Of interest here is that Teichroeb and Forrest [22,23] argued that their data supported the existence of a liquid-like layer on the free surfaces leading to unexpected surface properties, as well as providing a possible

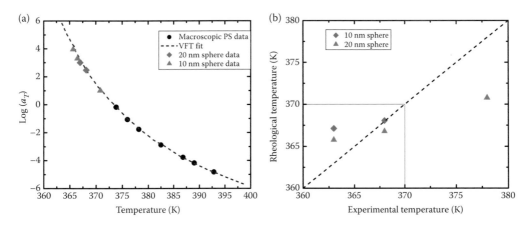

Figure 6.29 (a) Shift factors for bulk polystyrene (sphere) and the VFT fit (dashed line) obtained from Plazek's work [11]. Also, the shift factors for the surface rheology calculated for 10 nm sphere (square) and 20 nm sphere (triangle) are shown. (b) Comparison of experimental temperature with the corresponding surface rheological temperature. (With kind permission from Springer Nature: *Eur. Phys. J. E.*, Comment on "the properties of free polymer surfaces and their influence on the glass transition temperature of thin polystyrene films" by J. S. Sharp, J. H. Teichroeb, and J. A. Forrest, 22, 2007, 281–286, S. A. Hutcheson and G. B. McKenna.)

explanation for the reduction of the T_g in nanometer-thick films. However, from the work done by Hutcheson and McKenna [24,25], in which a major effort was made to build up the analysis method to extract the viscoelastic response from the particle embedment experiments, they found that the original data of Teichroeb and Forrest [22] were not consistent with a liquid layer and the response of the material was still within the segmental or glassy response regime (see Figure 6.4). One possible reason for the absence of such a liquid-like region in this case is the high surface energy of the gold particles and the resulting strong interactions between the polymer surface and the gold particles. Furthermore, as shown in Figure 6.30a, comparison of shear compliance obtained from gold particle embedment and silica particle embedment from Karim and McKenna [95] shows that the gold particles seem to give a stiffer isochronal compliance than do the silica particles. The higher surface energy for the gold (see Table 6.1), combined with their smaller size (giving a smaller contact area) induces a high pressure beneath the particles. Karim and McKenna [95] considered time–pressure superposition [27] and it was estimated that the glass transition temperature T_g changed much more for the gold particle embedment conditions than for the silica particle embedment conditions. Therefore, the isochronal response for the gold particles was stiffer (lower compliance) than that for the silica particles. Taking into account the T_g change with pressure beneath the gold particles and correcting the value of $T - T_g$, the results obtained from the different particle types and sizes agree with each other, as shown in Figure 6.30b.

Recently, Yoon and McKenna [138] used the particle embedment test with silica particles to probe underlayer effects in ultrathin polymer films. With an AFM equipped with good temperature control capabilities, the sample temperature could be set in the AFM and the particle heights constantly monitored, avoiding the error issues caused by removing the sample from the temperature chamber, quenching to measure particle height and then replacing in the oven, hence providing more accurate time-dependent particle height determinations. Then they used the Hutcheson and McKenna [24,25] method to extract the creep compliance at different temperatures and compare the surface rheological behavior with the macroscopic response. In their work, the surface dynamic behaviors were different from the macroscopic behavior and faster dynamics was observed below the macroscopic T_g while slower dynamics was observed above the macroscopic T_g. Their conclusion is in accordance with the previous results reported by Karim and McKenna [95] using the isochronal tests.

As mentioned at the very beginning of the section, the particle embedment experiment and the nanoindentation technique, applied for surface properties test, are similar to each other, and they both encounter the issues surrounding accurate determination of the viscoelastic information from the load–depth curves or

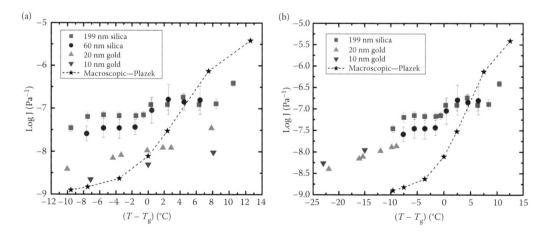

Figure 6.30 (a) Shear compliance calculated from particle height on linear polystyrene after 30 min creep versus $T - T_g$ for 199 nm silica particles, 60 nm silica particles, 20 nm gold particles, and 10 nm gold particles and the comparison of surface properties with the macroscopic response of polystyrene. (b) After correction for the pressure dependence of the glass transition temperature, the shear compliance data of (a) at different values of $T - T_g$. (Reprinted from *Polymer*, 54, T. B. Karim and G. B. McKenna, Comparison of surface mechanical properties among linear and star polystyrenes: Surface softening and stiffening at different temperatures, 5928–5935, Copyright 2013, with permission from Elsevier.)

depth–time curves. One possibility is to follow on to some recent work in which multiple indentation curves at different rates of indentation were fitted simultaneously to improve viscoelastic property determination [101]. In the future, it is our intention to provide a similar analysis for particle embedment by using multiple data sets at a single temperature in which particles of different sizes are treated simultaneously. Finally, one should be able to expand the range of properties investigated for polymers by using time–temperature super-position principle.

6.5 SUMMARY AND PERSPECTIVES

The previous paragraphs have summarized the major macroscopic properties of interest in the characterization of the linear viscoelastic properties of polymers. We then examined two different methods of making mechanical measurements of the viscoelastic properties of either polymer membranes or polymer surfaces. In the former case, we showed the TTU (Texas Tech University) "nanobubble inflation" test method that is a reduction of the classic membrane inflation experiment to a size scale that permits the inflation and imaging of bubbles having diameters of 1.2–5 μm and film thicknesses from 3 to 250 nm. Ongoing work promises to expand this range to below a μm and above 10 μm diameter. This ability also expands the range of material stiffnesses that could be investigated beyond the typical polymer properties to potentially metallic systems for larger-diameter bubbles and to softer systems with smaller bubbles. The method also permits the determination of the surface energy of ultrathin polymer films. Findings included reductions of the glass transition temperature in the ultrathin films, with the greatest observed T_g reduction being over 120°C for a 3 nm thick polycarbonate film. The T_g reduction was also found to depend on the material. In addition to the T_g reduction that was observed, a very strong rubbery stiffening was also observed and ultrathin films of polymers such as polycarbonate, polystyrene, and poly(vinyl acetate) were found to be as much as three orders of magnitude stiffer than the macroscopic rubbery material when thicknesses were less than 10 nm. At the same time, in other materials such as a polyurethane rubber and poly(n-butyl methacrylate), while showing stiffening, the amount of stiffening was more modest being approximately an order of magnitude for 20 nm in thickness. The method has the potential for other types of testing, such as measuring yield response of ultrathin polymer films, but has not yet been used extensively for such measurements. Future work should provide such investigations and other laboratories should make similar measurements to broaden the applications that have been achieved to date.

In the case of the measurement of surface mechanical properties, we have shown measurements of nanoindentation using a spontaneous particle embedment method that has advantages over either a classical nanoindentation apparatus or using an AFM as a nanoindenter. The method relies on the work of adhesion between the nanoparticles and the surface of the material of interest to create the driving force for indentation or embedment. We showed results for particles ranging in size from 10 nm to approximately 200 nm and the results for both elastic and viscoelastic conditions. Of great interest here is that with the very small particles, one performs nanoindentation measurements that are both nanometric in depth and in lateral dimensions. The results are at variance with reported results from nanoindentation and we were able to show that the observed stiffening of the surface of glassy polymers seems to be due, not to the surface properties themselves, but to the fact of high pressures beneath a nanoindenter tip. The general observation for the nanoparticle embedment experiments for materials below the glass transition temperature was an observed softening. On the other hand, there is also evidence that materials above the glass transition temperature may stiffen. It is recommended that further investigation in regard to both of these issues would be worth the effort.

As a last set of comments, the methods described are not the only ones available to characterize the nanomechanical response of polymeric materials. For example, as mentioned earlier, the liquid dewetting method can be used to determine the viscoelastic response and glass transition reduction of ultrathin films. The method though is limited for two reasons. The first is simply that the surface forces involved in dewetting are such that the glassy response, which are frequently of great interest mechanically, cannot be obtained with current imaging resolutions. Additionally, the method is limited by the couples of polymer and liquid from which it might dewet because most systems prefer to wet due to a favorable spreading parameter. When the polymer film wets the liquid, the deformation is a nonhomogeneous thinning as the film spreads and this is much more difficult to measure and analyze than the uniform shrinkage that occurs upon dewetting from the liquid. Hence, the dewetting measurement as developed originally by Bodiguel and Fretigny [96–98] is extremely clever and provides great insight to material behavior; unfortunately, it does so for a limited number of materials systems.

We mention another method here that has shown some very interesting nanomechanical results and that is the buckling method [86,139–142] in which a film of material is placed on a soft underlayer that can be easily deformed. This underlayer can then be used to apply a compression to the film and buckling takes place. The wavelength of the buckling combined with the film thickness can be used to extract an effective elastic modulus. While the method is relatively straightforward, the primary limitation is that it cannot be used at this point to make viscoelastic measurements, hence is limited to an elastic modulus, perhaps as a function of temperature. The method does not readily provide the time-dependent response [143]. But its ease of use has made it a good method for multiple investigations of ultrathin film mechanical modulus.

Last but not least, it is worthwhile for the reader to keep in mind that as a technological development nanoindentation has become an increasingly popular method to study the polymer and material surfaces [101–111], and both the apparent modulus and time-dependent behavior of the surface of interest can be characterized. However, several issues still remain to be solved before this method can be used universally. One problem arises due to the fact that going from the elastic case to the viscoelastic one is made more difficult because of the mixing of the shape parameters involved in the simple fact of the indentation tip geometry and its modification due to the time-dependent response of the material, even for a linear analysis. Viscoelastic analysis methods are needed to address this moving boundary case and the extraction of accurate material parameters can be challenging. Furthermore, uncertainties in the indentation experiment itself to which are added issues such as surface to tip boundary conditions can be different from those assumed in the analyses, for example, pile-up during the deformation, can lead to further difficulties in the measurement. Additionally, an indentation system with high data acquisition rate and good long-term stability at high temperature are necessary to adequately determine polymer properties over a wide time or temperature range, as suggested in Figures 6.1, 6.2, and 6.4.

The importance of nanotechnologies and the corresponding mechanical measurements at the nanometer-size scale remain challenges to the mechanics and materials physics communities. The above provides some novel methods and a potential for future investigations of material mechanical properties at small size and length scales.

ACKNOWLEDGMENTS

The authors thank the Office of Naval Research under grant N00014-11-1-0424, the National Science Foundation under grant DMR-DMR-1207070, and the John R. Bradford Endowment at Texas Tech University, each for partial support of this work. GBM is also grateful to the opportunity to work in the Laboratoire Sciences et Ingénierie de la Matière Molle Physico-chimie des Polymères et Milieux Dispersés, UMR 7615, E.S.P.C.I. in Paris, during a sabbatical leave opportunity where he prepared much of this manuscript.

REFERENCES

1. C. L. Jackson and G. B. McKenna, The glass transition of organic liquids confined to small pores, *J. Non-Cryst. Solids*, 131, 221–224, 1991.
2. J. Zhang, G. Liu, and J. Jonas, Effects of confinement on the glass transition temperature of molecular liquids, *J. Phys. Chem.*, 96, 3478–3480, 1992.
3. C. L. Jackson and G. B. McKenna, Vitrification and crystallization of organic liquids confined to nanoscale pores, *Chem. Mater.*, 8, 2128–1237, 1996.
4. J. L. Keddie, R. A. L. Jones, and R. A. Cory, Interface and surface effects on the glass transition temperature in thin polymer films, *Farad. Discuss.*, 98, 219–230, 1994.

5. J. L. Keddie, R. A. L. Jones, and R. A. Cory, Size-dependent depression of the glass transition temperature in polymer films, *Europhys. Lett.*, 27, 59–64, 1994.

6. P. Pissis, A. Kyritsis, G. Barut, R. Pelster, and G. Nimtz, Glass transition in 2- and 3-dimensionally confined liquids, *J. Non-Cryst. Solids*, 235, 444–449, 1998.

7. J. A. Forrest, K. Dalnoki-Veress, J. R. Stevens, and J. R. Dutcher, Effect of free surfaces on the glass transition temperature of thin polymer films, *Phys. Rev. Lett.*, 77, 2002–2005, 1996.

8. K. Dalnoki-Veress, J. A. Forrest, C. Murray, C. Gigault, and J. R. Dutcher, Molecular weight dependence of reductions in the glass transition temperature of thin freely standing polymer films, *Phys. Rev. E*, 63, 031801, 2001.

9. M. Alcoutlabi and G. B. McKenna, Effects of confinement on material behaviour at the nanometre size scale, *J. Phys. Condens. Matter*, 17, R461–R524, 2005.

10. C. L. Soles, J. F. Douglas, R. L. Jones, W.-L. Wu, H. Peng, and D. W. Gidley, Comparative specular x-ray reflectivity, positron annihilation lifetime spectroscopy, and incoherent neutron scattering measurements of the dynamics in thin polycarbonate films, *Macromolecules*, 37, 2890–2900, 2004.

11. R. Inoue, T. Kanaya, K. Nishida, I. Tsukushi, M. T. F. Telling, B. J. Gabrys, M. Tyagi, C. Soles, and W.-L. Wu, Glass transition and molecular mobility in polymer thin films, *Phys. Rev. E*, 80, 031802, 2009.

12. P. A. O'Connell and G. B. McKenna, Rheological measurements of the thermoviscoelastic response of ultrathin polymer films, *Science*, 307, 1760–1763, 2005.

13. P. A. O'Connell and G. B. McKenna, Dramatic stiffening of ultrathin polymer films in the rubbery regime, *Eur. Phys. J. E*, 20, 143–150, 2006.

14. P. A. O'Connell, S. A. Hutcheson, and G. B. McKenna, Creep behavior of ultrathin polymer films, *J. Polym. Sci., Part B: Polym. Phys.*, 46, 1952–1965, 2008.

15. A. N. Raegen, M. V. Massa, J. A. Forrest, and K. Dalnoki-Veress, Effect of atmosphere on reductions in the glass transition of thin polystyrene films, *Eur. Phys. J. E*, 27, 375–377, 2008.

16. C. B. Roth and J. R. Dutcher, Mobility on different length scales in thin polymer films, in *Soft Materials: Structure and Dynamics*, edited by J. R. Dutcher and A. G. Marangoni (Decker, New York), 2004.

17. S. H. Xu, P. A. O'Connell, G. B. McKenna, and S. Castagnet, Nanomechanical properties in ultrathin polymer films: Measurement on rectangular versus circular bubbles, *J. Polym. Sci., Part B: Polym. Phys.*, 50, 466–476, 2012.

18. J. E. Pye and C. B. Roth, Two simultaneous mechanisms causing glass transition temperature reductions in high molecular weight freestanding polymer films as measured by transmission ellipsometry, *Phys. Rev. Lett.*, 107, 235701, 2011.

19. J. E. Pye and C. B. Roth, Above, below, and in-between the two glass transitions of ultrathin free-standing polystyrene films: Thermal expansion coefficient and physical aging, *J. Polym. Sci., Part B: Polym. Phys.*, 53, 64–75, 2015.

20. P. A. O'Connell, J. Wang, T. A. Ishola, and G. B. McKenna, Exceptional property changes in ultrathin films of polycarbonate: Glass temperature, rubbery stiffening, and flow, *Macromolecules*, 45, 2453–2459, 2012.

21. G. B. McKenna, Physical aging in glasses and composites, in *Long-Term Durability of Polymeric Matrix Composites*, edited by K. V. Pochiraju, G. P. Tandon, and G. A. Schoeppner (Springer, New York), 237–309, 2011.

22. J. H. Teichroeb and J. A. Forrest, Direct imaging of nanoparticle embedding to probe viscoelasticity of polymer surfaces, *Phys. Rev. Lett.*, 91, 016104, 2003.

23. J. S. Sharp, J. A. Forrest, Z. Fakhraai, M. Khomenko, J. H. Teichroeb, and K. Dalnoki-Veress, Reply to comment on "the properties of free polymer surfaces and their effect upon the glass transition temperature of thin polystyrene films" by S. A. Hutcheson and G. B. McKenna, *E. Phys. J. E*, 22, 287–291, 2007.

24. S. A. Hutcheson and G. B. McKenna, Nanosphere embedding into polymer surfaces: A viscoelastic contact mechanics analysis, *Phys. Rev. Lett.*, 94, 076103-1–076103-4, 2005; Erratum, *Phys. Rev. Lett.*, 94, 189902, 2005.

25. S. A. Hutcheson and G. B. McKenna, Comment on "the properties of free polymer surfaces and their influence on the glass transition temperature of thin polystyrene films" by J. S. Sharp, J. H. Teichroeb, and J. A. Forrest, *Eur. Phys. J. E*, 22, 281–286, 2007.

26. R. Zorn, F. I. Mopsik, G. B. McKenna, L. Wilner, and D. Richter, Dynamics of polybutadienes with different microstructures. 2. Dielectric response and comparisons with rheological behavior, *J. Chem. Phys.*, 107, 3645–3655, 1997.

27. J. D. Ferry, *Viscoelastic Behavior of Polymers* (Wiley, New York), 3rd edn., 1980.

28. R. G. Larson, *The Structure and Rheology of Complex Fluids* (Oxford University Press, New York), 1999.

29. R. G. Larson, *Constitutive Equations for Polymer Melts and Solutions* (Butterworths, London), 1988.

30. C. W. Macosko, *Rheology: Principles, Measurements, and Applications* (VCH Publishers, New York), 1994.

31. G. B. McKenna, Viscoelasticity, in *Encyclopedia of Polymer Science and Technology,* edited by Herman Francis Mark, Jacqueline I. Kroschwitz (John Wiley, New York), 3rd edn., 1–144, 2002.

32. H. Leaderman, *Elastic and Creep Properties of Filamentous Materials and Other High Polymers* (The Textile Foundation, Washington, DC), 1943.

33. A. J. Staverman and F. Schwarzl, Linear deformation behaviour of high polymers, in *Die Physik der Hochpolymeren. Vierter Band. Theorie und Molekulare Deutung Technologischer Eigenschaften von Hochpolymeren Werkstoffen*, edited by H. A. Stuart (Springer-Verlag, Berlin), 1–125, 1956 (in English).

34. H. Leaderman, Viscoelasticity phenomena in amorphous high polymeric systems, in *Rheology. Vol. 2. Theory and Applications*, edited by F. R. Eirich (Academic Press, New York), 1–62, 1958.

35. A. S. Wineman and K. R. Rajagopal, *Mechanical Response of Polymers* (Cambridge University Press, New York), 2000.

36. W. N. Findley, J. S. Lai, and K. Onaran, *Creep and Relaxation of Nonlinear Viscoelastic Materials. With an Introduction to Linear Viscoelasticity* (North Holland, New York), 1976.

37. A. V. Tobolsky, *Properties and Structure of Polymers* (Wiley, New York), 1967.

38. P. G. de Gennes, Reptation of a polymer chain in the presence of fixed obstacles, *J. Chem. Phys.*, 55, 572–579, 1971.

39. M. Doi and S. F. Edwards, *The Theory of Polymer Dynamics* (Oxford Science Publishers, Clarendon Press, Oxford), 1986.

40. M. L. Williams, R. F. Landel, and J. D. Ferry, The temperature dependence of relaxation mechanisms in amorphous polymers and other glass-forming liquids, *J. Am. Chem. Soc.*, 77, 3701–3706, 1955.

41. H. Vogel, Das Temperaaturabhängigkeitsgesetz der Viskosität Flüssigkeiten, *Phys. Z.*, 22, 645–646, 1921.

42. G. S. Fulcher, Analysis of recent measurements of the viscosity of glasses, *J. Am. Ceram. Soc.*, 8, 339–355, 1925.

43. G. Tammann, W. Hesse, Abhängigkeit der Viscosität von der Temperatur bei unterkühlten Flüssigkeiten, *Z. Anorg. Allg. Chem.*, 156, 245–257, 1926.

44. G. B. McKenna and J. Zhao, Accumulating evidence for non-diverging time-scales in glass-forming fluids, *J. Non-Cryst. Solids*, 407, 3–13, 2014.

45. A. J. Kovacs, Transition Vitreuse dans les Polyméres Amorphes. Etude Phénoménologique, *Fortschr. Hochpolym.-Forsch.*, 3, 394–507, 1963.

46. C. T. Moynihan, P. B. Macedo, C. J. Montrose, P. K. Gupta, M. A. DeBolt, J. F. Dill et al., Structural relaxation in vitreous materials, *Ann. N. Y. Acad. Sci.*, 279, 15–35, 1976.

47. L. C. E. Struik, *Physical Aging in Polymers and Other Amorphous Materials* (Elsevier, Amsterdam), 1978.

48. G. B. McKenna, Glass formation and glassy behavior, in *Comprehensive Polymer Science, Vol. 2: Polymer Properties*, edited by C. Booth and C. Price (Pergamon Press, Oxford), 311–362, 1989.

49. G. W. Scherer, *Relaxation in Glass and Composites* (Krieger Publishing Co., Malabar, FL), 1992.

50. J. Zhao, S. L. Simon, and G. B. McKenna, Using 20-million-year-old amber to test the super-Arrhenius behavior of glass-forming systems, *Nat. Commun.*, 4, 1783, 2013.

51. P. A. O'Connell and G. B. McKenna, Arrhenius like temperature dependence of the segmental relaxation below T_g, *J. Chem. Phys.*, 110, 11054–11060, 1999.

52. S. L. Simon, J. W. Sobieski, and D. J. Plazek, Volume and enthalpy recovery of polystyrene, *Polymer*, 42, 2555–2567, 2001.

53. P. Badinarayanan and S. L. Simon, Origin of the divergence of the timescales for volume and enthalpy recovery, *Polymer*, 48, 1464–1470, 2007.

54. J. C. Dyre, Colloquium: The glass transition and elastic models of glass-forming liquids, *Rev. Mod. Phys.*, 78, 953–972, 2006.

55. T. Hecksher, A. I. Nielsen, N. B. Olsen, and J. C. Dyre, Little evidence for dynamic divergences in ultra-viscous molecular liquids, *Nat. Phys.*, 4, 737–741, 2008.

56. G. B. McKenna, Diverging views on glass transition, *Nat. Phys.*, 4, 673–674, 2008.

57. J. Zhao and G. B. McKenna, Temperature divergence of the dynamics of a poly(vinyl acetate) glass: Dielectric vs. mechanical behaviors, *J. Chem. Phys.*, 136, 154901-1–154901-8, 2012.

58. J. Zhao and G. B. McKenna, Response to comment on "temperature divergence of the dynamics of a poly(vinyl acetate) glass: Dielectric vs. mechanical behaviors" [*J. Chem. Phys.* 139, 137101 2013], *J. Chem. Phys.*, 139, 137102, 2013.

59. H. Wagner and R. Richert, Thermally stimulated modulus relaxation in polymers: Method and interpretation, *Polymer*, 38, 5801–5806, 1997.

60. R. Richert and H. Wagner, The dielectric modulus: Relaxation versus retardation, *Solid State Ionics*, 105, 167–173, 1998.

61. R. Richert, Comment on "temperature divergence of the dynamics of a poly(vinyl acetate) glass: Dielectric vs. mechanical behaviors" [*J. Chem. Phys.* 136, 1549012012], *J. Chem. Phys.*, 139, 137101, 2013.

62. D. J. Plazek, The temperature dependence of the viscoelastic behavior of poly(vinyl acetate), *Polym. J.*, 12, 43–53, 1980.

63. D. J. Plazek, Magnetic bearing torsional creep apparatus, *J. Polym. Sci. A-2*, 6, 621–638, 1968.

64. D. J. Plazek and V. M. O'Rourke, Viscoelastic behavior of low molecular weight polystyrene, *J. Polym. Sci. A-2*, 9, 209–243, 1971.

65. D. J. Plazek, C. A. Bero, and I. C. Chay, The recoverable compliance of amorphous materials, *J. Non-Cryst. Solids*, 172–174, 1816190, 1994.

66. D. J. Plazek and I. Echeverria, Don't cry for me Charlie Brown, or with compliance comes comprehension, *J. Rheol.*, 44, 831–841, 2000.

67. P. A. O'Connell and G. B. McKenna, Novel nanobubble inflation method for determining the viscoelastic properties of ultrathin polymer films, *Rev. Sci. Inst.*, 78, 013901-1–013901-12, 2007.

68. P. A. O'Connell and G. B. McKenna, A novel nano-bubble inflation method for determining the viscoelastic properties of ultrathin polymer films, *Scanning*, 30, 184–196, 2008.

69. R. S. Rivlin and D. W. Saunders, Large elastic deformations of isotropic materials. VII. Experiments on the deformation of rubber, *Philos. Trans. R. Soc. Lond. A*, 243, 251–288, 1951.

70. D. D. Joye, G. W. Poehlein, and C. D. Denson, A bubble inflation technique for the measurement of viscoelastic properties in equal biaxial extensional flow, II, *Trans. Soc. Rheol.*, 16, 421–445, 1972.

71. P. H. Mott, C. M. Roland, and S. E. Hassan, Strains in an inflated rubber sheet, *Rubber Chem. Technol.*, 76, 326–333, 2003.

72. G. B. McKenna and R. W. Penn, Time dependent failure of a polyolefin candidate material for blood pump applications, *J. Biomed. Mater. Res.*, 14, 689–703, 1980.

73. L. R. G. Treloar, *The Physics of Rubber Elasticity* (Clarendon, Oxford), 3rd edn., 1975.

74. J. W. Beams, Mechanical properties of thin films of gold and silver. In *Structure and Properties of Thin Films*, edited by D. A. N. Neugebauer and D. A. Vermilyea (Wiley, New York), 183–192, 1959.

75. J. J. Vlasek and W. D. Nix, A new bulge test technique for the determination of young's modulus and Poisson's ratio of thin films, *J. Mater. Res.*, 7, 3242–3249, 1992.

76. M. K. Small and W. D. Nix, Analysis of the accuracy of the bulge test in determining the mechanical properties of thin films, *J. Mater. Res.*, 7, 1553–1563, 1992.

77. J. Neggers, J. P. M. Hoefnagels, and M. G. D. Geers, On the validity regime of the bulge equations, *J. Mater. Res.*, 27, 1245–1250.

78. S. Gao, K.-T. Wan, and D. A. Dillard, A bending-to-stretching analysis of the blister test in the presence of tensile residual stress, *Int. J. Solids Struc.*, 42, 2771–2784, 2005.

79. S. P. Timoshenko and S. Woinowsky-Krieger, *Theory of Plates and Shells* (McGraw-Hill, New York), 2nd edn., 1969.

80. M. Zhai and G. B. McKenna, Elastic modulus and surface tension of a polyurethane rubber in nanometer thick films, *Polymer*, 55, 2725–2733, 2014.

81. L. Boltzmann, Zur Theorie der Elastischen Nachwirkung, *Sitzungsberichte der Mathematisch-Naturwissenschaftlichen Klass der Akademie Gemeinnutiger Wissenschaften zu Erfurt.*, 70, 275–300, 1874.

82. E. Riande, R. Diaz-Calleja, M. Prolongo, R. Masegosa, and C. Salom, editors, *Polymer Viscoelasticity: Stress and Strain in Practice* (CRC Press/Marcel Dekker, New York), 2000.

83. S. Xu, P. A. O'Connell, and G. B. McKenna, Unusual elastic behavior of ultrathin polymer films: Confinement-induced/molecular stiffening and surface tension effects, *J. Chem. Phys.*, 132, 184902, 2010.

84. P. A. O'Connell and G. B. McKenna, The stiffening of ultrathin polymer films in the rubbery regime: The relative contributions of membrane stress and surface tension, *J. Polym. Sci., Part B: Polym. Phys.*, 47, 2441–2448, 2009.

85. K. L. Ngai, D. Prevosto, and L. Grassia, Viscoelasticity of nanobubble-inflated ultrathin polymer films: Justification by the coupling model, *J. Polym. Sci., Part B: Polym. Phys.*, 51,214–224, 2013.

86. K. A. Page, A. Kusoglu, C. M. Stafford, S. Kim, R. J. Kline, and A. Z. Weber, Confinement-driven increase in ionomer thin-film modulus, *Nano Lett.*, 14, 2299–2304, 2014.

87. Y. Chai, T. Salez, J. D. McGraw, M. Benzaquen, K. Dalnoki-Veress, E. Raphael, and J. A. Forrest, A direct quantitative measure of surface mobility in a glassy polymer, *Science*, 343, 994–999, 2014.

88. X. Li and G. B. McKenna, Ultrathin polymer films: Rubbery stiffening, fragility, and Tg reduction, *Macromolecules*, 48, 6329–6336, 2015.

89. D. Tabor, Surface forces and surface interactions, *J. Colloid Interface Sci.*, 58, 2–13, 1977.

90. D. S. Rimai, L. P. Demejo, and R. C. Bowen, Mechanics of particle adhesion, *J. Adhes. Sci. Technol.*, 8, 1333–1355, 1994.

91. L. P. Demejo, D. S. Rimai, and J. H. Chen, Time dependent adhesion induced phenomena: The flow of a compliant silicone-polyester copolymer substrate over rigid micrometer size gold and polystyrene particles, *J. Adhes.*, 48, 47–56, 1995.

92. D. S. Rimai, L. P. Demejo, J. H. Chen, R. C. Bowen, and T. H. Mourey, Time-dependent adhesion-induced phenomena: Viscoelastic creep of a substrate polymer over rigid particles, *J. Adhes.*, 62, 151–168, 1997.

93. T. B. Karim and G. B. McKenna, Evidence of surface softening in Polymers and their nanocomposites as determined by spontaneous particle embedment, *Polymer*, 52, 6134–6145, 2011.

94. T. B. Karim and G. B. McKenna, Unusual surface mechanical properties of poly(α-methylstyrene): Surface softening and stiffening at different temperatures, *Macromolecules*, 45, 9697–9706, 2012.

95. T. B. Karim and G. B. McKenna, Comparison of surface mechanical properties among linear and star polystyrenes: Surface softening and stiffening at different temperatures, *Polymer*, 54, 5928–5935, 2013.

96. H. Bodiguel and C. Fretigny, Reduced viscosity in thin polymer films, *Phys. Rev. Lett.*, 97, 266105, 2006.

97. H. Bodiguel and C. Fretigny, Viscoelastic dewetting of a polymer film on a liquid substrate, *Eur. Phys. J. E: Soft Matter Biol. Phys.*, 19, 185–193, 2006.

98. H. Bodiguel and C. Fretigny, Viscoelastic properties of ultrathin polystyrene films, *Macromolecules*, 40, 7291–7298, 2007.

99. J. Wang and G. B. McKenna, Viscoelastic and glass transition properties of ultrathin polystyrene films by dewetting from liquid glycerol, *Macromolecules*, 46, 2485–2495, 2013.

100. J. Wang and G. B. McKenna, A novel temperature-step method to determine the glass transition temperature of ultrathin polymer films by liquid dewetting, *J. Polym. Sci., Part B: Polym. Phys.*, 51, 1343–1349, 2013.

101. M. Zhai and G. B. McKenna, Viscoelastic modeling of nanoindentation experiments: A multicurve method, *J. Polym. Sci., Part B: Polym. Phys.*, 9, 633–639, 2014.

102. M. R. VanLandingham, J. S. Villarrubia, W. F. Guthrie, and G. F. Meyers, *Nanoindentation of Polymers: An Overview* (Wiley-Blackwell, New Jersey), 15–44, 2001.

103. M. R. VanLandingham, N. K. Chang, P. L. Drzal, C. C. White, and S. H. Chang, Viscoelastic characterization of polymers using instrumented indentation. I. Quasi-static testing, *J. Polym. Sci., Part B: Polym. Phys.*, 43, 1794–1811, 2005.

104. G. Huang and H. Lu, Measurement of young's relaxation modulus using nanoindentation, *Mech. Time-Depend. Mater.*, 10, 229–243, 2006.

105. Z. Zhou and H. Lu, On the measurements of viscoelastic functions of a sphere by nanoindentation, *Mech. Time-Depend. Mater.*, 14, 1–24, 2010.

106. N. P. Daphalapurkar, C. Dai, R. Z. Gan, and H. Lu, Characterization of the linearly viscoelastic behavior of human tympanic membrane by nanoindentation, *J. Mech. Behav. Biomed.*, 2, 82–92, 2009.

107. W. C. Oliver and G. M. Pharr, An improved technique for determining hardness and elastic modulus using load and displacement sensing indentation experiments, *J. Mater. Res.*, 7, 1564–1583, 1992.

108. I. N. Sneddon, The relation between load and penetration in the axisymmetric Boussinesq problem for a punch of arbitrary profile, *Int. J. Eng. Sci.*, 3, 47–57, 1965.

109. C. A. Tweedie, G. Constantinides, K. E. Lehman, D. J. Brill, G. S. Blackman, and K. J. Van Vliet, Enhanced stiffness of amorphous polymer surfaces under confinement of localized contact loads, *Adv. Mater.*, 19, 2540–2546, 2007.

110. Y.-T. Cheng and C.-M. Cheng, General relationship between contact stiffness, contact depth, and mechanical properties for indentation in linear viscoelastic solids using axisymmetric indenters of arbitrary profiles, *Appl. Phys. Lett.*, 87, 111914, 2005.

111. Y.-T. Cheng, Obtaining viscoelastic properties from instrumented indentation, in *Time Dependent Constitutive Behavior and Fracture/Failure Processes*, Volume 3, edited by T. Proulx (Springer, New York), 119–120, 2011.

112. Z. Cao, M. J. Stevens, and A. V. Dobrynin, Adhesion and wetting of nanoparticles on soft surfaces, *Macromolecules*, 47, 3203–3209, 2014.

113. J. M. Y. Carrillo and A. V. Dobrynin, Dynamics of nanoparticle adhesion, *J. Chem. Phys.*, 137, 214902, 2012.

114. K. L. Johnson, K. Kendall, and A. D. Roberts, Surface energy and the contact of elastic solids, *Philos. Roy. Soc. Lond. A*, 324, 301–313, 1971.

115. J. Wang, K. Deshpande, and G. B. McKenna, Determination of the shear modulus of spin-coated lipid multibilayer films by the spontaneous embedment of submicrometer-sized particles, *Langmuir*, 27, 6846–6854, 2011.

116. H. Butt, K. Graf, and M. Kappl, *Physics and Chemistry of Interfaces* (Wiley-VCH Verlag GmbH & Co KGaA, Weinheim), 127–130, 2006.

117. D. Tranchida, S. Piccarolo, and R. A. C. Deblieck, Some experimental issues of AFM tip blind estimation: The effect of noise and resolution, *Measure. Sci. Technol.*, 17, 2630–2636, 2006.

118. N. Rana, A. Chas, and J. Foucher, Reconciling measurements in AFM reference metrology when using different probing techniques, in *Metrology, Inspection and Process Control for Microlithography XXV, Pt. 1 and Pt. 2*, edited by C. J. Raymond, Book Series: Proc. SPIE, 7971, 79117, 2011.

119. S. A. Hutcheson, Evaluation of Viscoelastic Materials: The Study of Nanosphere Embedment into Polymer Surfaces and Rheology of Simple Glass Formers Using a Compliant Rheometer, Ph.D. thesis, Department of Chemical Engineering, Texas Tech University, August 2008.

120. J. E. Houston, A local-probe analysis of the rheology of a "solid-liquid," *J. Polym. Sci. Polym. Phys.*, 43, 2993–2999, 2005.

121. M. P. Goertz, X.-Y. Zhu, and J. E. Houston, Temperature dependent relaxation of a "solid-liquid," *J. Polym. Sci. Polym. Phys.*, 47, 1285–1290, 2009.

122. Z. Fakhraai and J. A. Forrest, Measuring the surface dynamics of glassy polymers, *Science*, 319, 600–604, 2008.

123. R. K. Abu Al-Rub and A. N. M. Faruk, Prediction of micro and nano indentation size effects from spherical indenters, *Mech. Adv. Mater. Struc.*, 19, 119–128, 2012.

124. T. Hianik and V. I. Passechnik, *Bilayer Lipid Membranes: Structures and Mechanical Properties* (Kluwer Academic Publishers, Dordrecht, The Netherlands), 1995.

125. Y. L. Chen, C. A. Helm, and J. N. Israelachvili, Measurements of the elastic properties of surfactant and lipid monolayers, *Langmuir*, 7, 2694–2699, 1991.

126. Q. X. Li, S. A. Hutcheson, G. B. McKenna, and S. I. Simon, Viscoelastic properties and residual stresses in polyhedral oligomeric silsesquioxane-reinforced epoxy matrices, *J. Polym. Sci., Part B: Polym. Phys. Ed.*, 46, 2719–2732, 2008.

127. G. Peng, T. Zhang, Y. Feng, and R. Yang, Determination of shear creep compliance of linear viscoelastic solids by instrumented indentation when the contact area has a single maximum, *J. Mater. Res.*, 27, 1565–1572, 2012.

128. D. Qi, Z. Fakhraai, and J. A. Forrest, Substrate and chain size dependence of near surface dynamics of glassy polymers, *Phys. Rev. Lett.*, 101, 096101, 2008.

129. S. Peter, H. Meyer, J. Baschnagel, and R. Seemann, Slow dynamics and glass transition in simulated free-standing polymer films: A possible relation between global and local glass transition temperatures, *J. Phys. Condens. Matter*, 19, 205119, 2007.

130. K. F. Mansfield and D. N. Theodorou, Molecular dynamics simulation of a glassy polymer surface, *Macromolecules*, 24, 6283–6294, 1991.

131. T. R. Böhme and J. J. de Pablo, Evidence for size-dependent mechanical properties from simulations of nanoscopic polymeric structures, *J. Chem. Phys.*, 116, 9939–9951, 2002.

132. E. H. Lee and J. R. M. Radok, The contact problem for viscoelastic bodies, *J. Appl. Mech.-T. ASME*, 27, 438–444, 1960.

133. T. C. T. Ting, The contact stresses between a rigid indenter and a viscoelastic half-space, *J. Appl. Mech.-T. ASME*, 33, 845–854, 1966.

134. W. Zheng, G. B. McKenna, and S. L. Simon, The viscoelastic behavior of polymer/oligomer blends, *Polymer*, 51, 4899–4906, 2010.

135. R. Kohlrausch, Theorie des Elektrischen Rückstandes in der Leidener Flasche, *Annalen der Physik und Chemie von J.C. Poggendorff*, 91, 179–214, 1854.

136. G. Williams and D. C. Watts, Non-symmetrical dielectric relaxation behaviour arising from a simple empirical decay function, *Trans. Farad. Soc.*, 66, 80–85, 1970.

137. M. Alcoutlabi, F. Briatico-Vangosa, and G. B. McKenna, Effect of chemical activity jumps on the viscoelastic behavior of an epoxy resin: The physical aging response in carbon dioxide pressure-jumps, *J. Polym. Sci., Part B: Polym. Phys.*, 40, 2050–2064, 2002.

138. H. D. Yoon and G. B. McKenna, Substrate effects on glass transition and free surface viscoelasticity of ultrathin polystyrene films, *Macromolecules*, 47, 8808–8818, 2014.

139. C. M. Stafford, C. Harrison, K. L. Beers, A. Karim, E. J. Amis, M. R. VanLandingham, H.-C. Kim, W. Volksen, R. D. Miller, and E. E. Simonyi, A buckling-based metrology for measuring the elastic moduli of polymeric thin films, *Nat. Mater.*, 3, 545–550, 2004.

140. C. M. Stafford, S. Guo, C. Harrison, and M. Y. M. Chiang, Combinatorial and high-throughput measurements of the modulus of thin polymer films, *Rev. Sci. Instrum.*, 76, 062207, 2005.

141. C. M. Stafford, B. D. Vogt, C. Harrison, D. Julthongpiput, and R. Huang, Elastic moduli of ultrathin amorphous polymer films, *Macromolecules*, 39, 5095–5099, 2006.

142. J. M. Torres, C. M. Stafford, D. Uhrig, and B. D. Vogt, Impact of chain architecture (branching) on the thermal and mechanical behavior of polystyrene thin films, *J. Polym. Sci., Part B: Polym. Phys.*, 50, 370–377, 2012.

143. P. A. O'Connell, G. B. McKenna, E. P. Chan, and C. M. Stafford, Comment on "viscoelastic properties of confined polymer films measured via thermal wrinkling" by E. P. Chan, K. A. Page, S. H. Im, D. L. Patton, R. Huang, and C. M. Stafford, *Soft Matter*, 2009, 5, 4638–4641, *Soft Matter*, 7, 788–790 2011.

Glassy and aging dynamics in polymer films investigated by dielectric relaxation spectroscopy

KOJI FUKAO

7.1 INTRODUCTION

At high temperatures, polymeric material is in the liquid state. When the material is cooled down from a high temperature to a lower temperature at an appropriate cooling rate, it assumes a glassy state at the glass transition temperature, T_g [1]. For polymeric systems, the α-process, known as cooperative segmental motion, becomes slower with decreasing temperature [2]. The characteristic time of the α-process changes from a microscopic time scale to a macroscopic one during a temperature change of only several tens of degrees. At T_g, this characteristic time is equal to 10^2–10^3 s; hence, the segmental motion is almost frozen below T_g. As the mechanism of the glass transition is not yet fully understood, understanding it is one of the most important unsolved problems in condensed matter physics [3]. Dynamical heterogeneity is the most important concept for understanding the glass transition [4,5]. Many experiments and simulations have shown its existence [6–8]. That is, molecular motion related to the glass transition has shown cooperative behavior with respect to spatial correlation in dynamics [9]. Thus, the existence of dynamical heterogeneity may be directly related to the characteristic length scale, which governs the mechanism of the glass transition.

Intensive investigations of the glass transition in confined geometries, such as small molecules in nanopores and thin polymer films, have been performed with the expectation that the glass transition dynamics change when the system size is smaller than the characteristic length scale of the glass transition [10,11]. Since the 1990s, many experiments on confined systems have clearly shown the drastic deviation of T_g and related dynamics from those in the bulk, with a few exceptions [12,13]. In Part 2 of this book, there are several interesting chapters related to this topic.

In the glassy state, molecular mobility due to segmental motion is almost frozen because the characteristic time scale of molecular motions is extremely long. However, even in the glassy state, several physical quantities such as density and volume have very slow time evolution, and their behavior depends strongly on their thermal history. This is known as physical aging or aging. These phenomena are believed to correspond to a time evolution from the nonequilibrium glassy state to an equilibrium state caused by the local motion or the lower-frequency tail of the segmental motion. Several curious phenomena related to physical aging, including memory and rejuvenation effects, have been observed [14,15].

Dielectric relaxation spectroscopy (DRS) is a powerful tool that is especially used for slow dynamics [16]. A temporal change in permanent dipole moments attached to or included in polymer chains can be observed. In principle, it is possible to obtain experimental data on the dynamics over a very wide time (frequency) range from 10^{-12} to 10^6 s with the DRS technique. Compared with DRS, other dynamical techniques such as neutron scattering, Brillouin light scattering, and infrared absorption spectroscopy cover a narrow time range. Furthermore, the sample shape suitable for DRS measurements resembles a thin film because the strength of the signal obtained by DRS increases with decreasing film thickness d. Hence, the sensitivity of DRS measurements increases with decreasing d, while other dynamical techniques are usually less sensitive when d decreases and the sample mass decreases. For this reason, DRS is a very suitable and powerful experimental technique for the glass transition of thin polymer films, which must be investigated over a very wide time range.

Because the characteristic time of the α-process at the glass transition is located within the frequency range covered by DRS, the dynamics of the α-process in many systems could be well measured by using DRS alone. However, the sensitivity of DRS decreases, as the frequency of the applied electric field decreases and/ or the polarity of polymers decreases. In the low-frequency region, there are also some contributions from dc conductivity, which might disturb the observation of the α-process. Therefore, in some polymeric systems without strong polarity, such as polystyrene (PS), it is better to use other experimental techniques for the measurement of the dynamics near the glass transition in addition to DRS. In this chapter, thermal expansion spectroscopy (TES) will be introduced as a complementary technique, especially for measurements of the dynamics of the α-process in the very-low-frequency ranges.

In this chapter, we discuss how DRS and related techniques have contributed to the understanding of the dynamical nature of the glass transition in thin polymer films by introducing recent experimental works of ours and other groups. Discussions regarding the glass transition and dynamics in thin polymer films investigated by DRS, and aging dynamics in polymer glasses are provided in Sections 7.2 and 7.3, respectively. Section 7.4 deals with the glass transition dynamics of stacked thin polymer films. Finally, a summary is presented in Section 7.5.

7.2 GLASS TRANSITION AND DYNAMICS IN THIN POLYMER FILMS INVESTIGATED BY DRS

Using ellipsometric measurements in the early 1990s, Keddie et al. observed a large reduction in T_g in thin polymer films supported on Si substrate [17,18]. Following this seminal work, there have been many reports on the deviation of T_g in confined geometries such as thin polymer films. There have also been many reports on the reduction of T_g for small molecules in nanopores, observed through various experimental methods, including differential scanning calorimetry (DSC) [10]. A much larger reduction in T_g for freestanding thin polymer films (without substrate) has been reported [19,20]. There have also been several reports on the deviation of the dynamics of the α-process from the bulk with decreasing film thickness d in relation to the reduction in T_g [21–29]. It is widely accepted that the glass transition occurs if the time scale of the α-process becomes extraordinarily large during cooling from a liquid state and that T_g is strongly associated with the α-process. Hence, it is natural to consider whether the thickness dependence of T_g may be the same or similar to that of the dynamics of the α-process. To test the validity of this assumption, simultaneous measurements on the same sample under exactly the same condition are desirable.

In 1999, we performed simultaneous measurements of T_g and the α-process for thin films of polystyrene supported on glass substrates [22,23]. For this observation, we adopted dielectric measurements. Thin films of PS were prepared by spin-coating onto aluminum deposited on a glass substrate. The aluminum was vacuum-deposited again to form an upper electrode. Hence, thin films of PS used for our dielectric measurements were covered with aluminum layers on both sides. These films are termed "capped (supported) films." Capacitance and/or impedance measurements on the capped PS thin films supported on glass substrate were performed to evaluate the T_g and the dynamics of the α-process (segmental motion) as a function of d.

7.2.1 DIELECTRIC RELAXATION SPECTROSCOPY AND CAPACITIVE DILATOMETRY

In DRS measurements, we usually measure the complex electric capacitance of the sample condenser $C^*(\equiv C' - iC'')$, where C' and C'' are the real and imaginary parts of the electric capacitance. If the shape of the sample condenser is a parallel-plate type, then the value of C^* can be described by $C^*(\omega, T) = \varepsilon^*(\omega, T)C_0(T)$, where $\varepsilon^*(\omega, T)$ is the complex dielectric permittivity at ω and T, $\omega = 2\pi f$, f is the frequency of the applied electric field, and C_0 is the geometrical capacitance, which depends on the shape of the condenser. In this case [23],

$$C_0(T) = \varepsilon_0 \frac{S}{d} \sim \varepsilon_0 \frac{S_0}{d_0}[1 + (2\alpha_t - \alpha_n)\Delta T], \quad (7.1)$$

where ε_0 is the dielectric permittivity in vacuo; S and d are the area of the electrode and the thickness of the sample at temperature T, respectively; and S_0 and d_0 are, respectively, the area and thickness of the sample condenser at a standard temperature T_0. Furthermore, $\Delta T = T - T_0$, and α_t and α_n are, respectively, the linear thermal expansion coefficients of the sample along the direction parallel and normal to the film surface. The complex dielectric permittivity $\varepsilon^*(\equiv \varepsilon' - i\varepsilon'')$ is expressed as follows:

$$\varepsilon'(\omega, T) = \varepsilon_\infty(T) + \varepsilon'_{disp}(\omega, T), \ \varepsilon''(\omega, T) = \varepsilon''_{disp}(\omega, T). \quad (7.2)$$

Here, ε_∞ is the dielectric permittivity at very high frequency, and ε'_{disp} and ε''_{disp} are the real and imaginary parts of dielectric permittivity dependent on ω, which are associated with the change in orientational polarization with the electric field. Since the change in orientation of electric dipole moments attached to polymers is directly related to their molecular motions, DRS allows observation of the molecular motion of polymers through measurements of the dependence of ε'_{disp} and ε''_{disp} on frequency, temperature, and pressure [16].

On the other hand, if C' is observed in the frequency and temperature range in which the contributions from ε'_{disp} and ε''_{disp}, that is, the contributions from the molecular motion, are not appreciable, then the value of $C_0 \cdot \varepsilon_\infty$ can be mainly observed from capacitance measurements. When the temperature dependences of C_0 and ε_∞ are taken into account, the temperature coefficient of the capacitance $\tilde{\alpha}$ is obtained as follows [23,30]:

$$\tilde{\alpha} \equiv -\frac{1}{C'(T_0)} \frac{dC'(T)}{dT}$$
$$= -\left(\frac{1}{\varepsilon_\infty(T_0)} \frac{d\varepsilon_\infty}{dT} + \frac{1}{C_0(T_0)} \frac{dC_0}{dT}\right) \approx \zeta \alpha_n. \quad (7.3)$$

Here, ζ is a positive constant depending on the polymer and $\zeta \sim 2$ for PS. This relation suggests that the thermal expansion coefficient for the heating or cooling process can be obtained from the measurement of the real part of the capacitance (dielectric permittivity) at an appropriate frequency. This method is known as the capacitive dilatometry (CD) method. Therefore, a single capacitance measurement on the same sample can provide information on both the molecular motion of polymer chains and the volume change during a heating and/or cooling process.

7.2.2 GLASS TRANSITION IN PS THIN FILMS

Figure 7.1 shows the T_g of thin films of atactic PS with various d values and molecular weights observed by the CD method [23]. The value $\tilde{\alpha}$ changes from a larger value to a smaller value when the temperature decreases across T_g. Hence, the value of T_g can be determined as a crossover temperature of $\tilde{\alpha}$. The observed values of T_g are almost independent of d for $d > 200$ nm for PS thin films for a given molecular weight, although there is a molecular weight dependence of T_g of the bulk sample due to the effect of the chain ends. In Figure 7.1, T_g values for thin PS films of four molecular weights decrease as d decreases from 200 to 7 nm. The amount of the depression of T_g is about 30 K when $d = 7$ nm. This thickness dependence of T_g is consistent with that observed through ellipsometry by Keddie et al. [17,18]. The thickness dependence of T_g can be described well by

$$T_g(d) = T_g^\infty \left(1 - \frac{a}{d} \right), \tag{7.4}$$

where a is a positive constant and T_g^∞ is the T_g of the bulk sample. This thickness dependence of T_g suggests that there are regions with lower T_g and/or higher T_g in addition to the region with bulk T_g within the thin films [17,23,31].

With exactly the same PS thin films, the relaxation rate f_α (denoted by f_{max} in Figure 7.2) at which the dielectric loss ε'' exhibits a maximum due to the α-process at a given temperature was observed, and then a

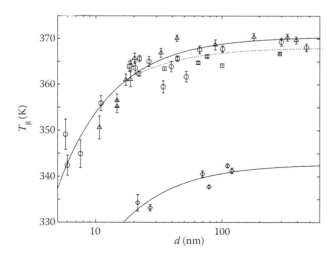

Figure 7.1 Thickness dependence of T_g of PS films obtained during the heating process (Δ corresponds to $M_w = 1.8 \times 10^6$, O to $M_w = 2.8 \times 10^5$, $\square$ to $M_w = 3.6 \times 10^4$, and $\Diamond$ to $M_w = 3.6 \times 10^3$, where M_w is the weight-averaged molecular weight). T_g values are determined as the crossover temperatures between the lines characterizing $C'(T)$ at 10 kHz below and above T_g. The curves were obtained from Equation 7.4. The upper solid curve is for $M_w = 1.8 \times 10^6$ and 2.8×10^5, the lower solid curve is for $M_w = 3.6 \times 10^3$, and the dotted curve is for $M_w = 3.6 \times 10^4$. (Reprinted with permission from K. Fukao and Y. Miyamoto, *Phys. Rev. E*, 61, 1743. Copyright 2000 by the American Physical Society.)

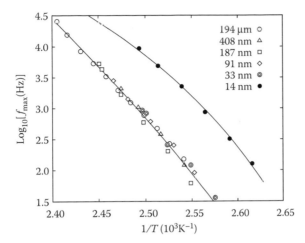

Figure 7.2 Peak frequency of dielectric loss due to the α-process as a function of the inverse of temperature for thin films of PS with various d values ($M_w = 1.8 \times 10^6$). Solid curves were obtained by fitting the data to the Vogel–Fulcher–Tammann equation 7.11. (Reprinted with permission from K. Fukao and Y. Miyamoto, *Phys. Rev. E*, 61, 1743. Copyright 2000 by the American Physical Society.)

dispersion map for the α-process in PS thin films was obtained, as shown in Figure 7.2. Evidently, the temperature dependence of f_α of the α-process for PS thin films with various d values down to 33 nm substantially overlapped with each other. However, when $d = 14$ nm, f_α is larger than that for thicker PS films at a frequency range of 20 Hz to 100 kHz. If the dynamics of the α-process in thin polymer films are also assumed to be associated with the value of T_g, analogous to that in the bulk system, then the lower (higher) value of T_g may be expected to be accompanied with faster (slower) dynamics of the α-process with decreasing d. Within the limited frequency range in Figure 7.2, the thickness dependence of f_α of the α-process appears to be different from that of T_g.

7.2.3 THERMAL EXPANSION SPECTROSCOPY

To determine whether there is a positive correlation between the change in T_g and the dynamics of the α-process in PS thin films as d decreases from the bulk to about 10 nm, we measured the temperature dependence of f_α of the α-process in the temperature and frequency regions down to the glass transition region, where f_α (denoted by f_m in Figure 7.3) is of the order of 10^{-3}–10^{-2} Hz. In principle, DRS is applicable to such low-frequency regions. However, PS is very weakly polar; hence, obtaining a meaningful dielectric signal near its T_g region is difficult. To avoid this problem, we adopted an alternative relaxation method, thermal expansion spectroscopy, to measure the dynamical response against an external disturbance [30,32–34]. In this method, we modulated the temperature, an external stimulus to the sample condenser, at an angular frequency ω_T as follows:

$$T(t) = \langle T \rangle + T_{\omega_T} e^{i\omega_T t}. \tag{7.5}$$

We measured the response, which is the change in C' at the same angular frequency ω_T:

$$C'(t) = \langle C' \rangle - C'_{\omega_T} e^{i(\omega_T t + \delta)}. \tag{7.6}$$

Here, T_{ω_T} and C'_{ω_T} are, respectively, the amplitudes of T and C' for the modulation at an angular frequency ω_T, and δ is the phase difference between them. $\langle T \rangle$ is the average temperature and $\langle C' \rangle$ is the average capacitance.

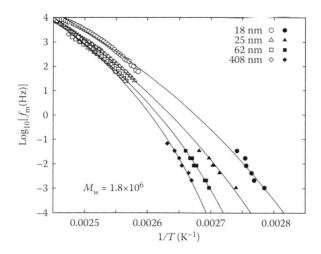

Figure 7.3 Dispersion map for PS thin films obtained from peak positions of the loss component α_n'' or ε'' showing peak values for various d values. f_m values satisfy the relation $2\pi f_m \tau_\alpha = 1$, where τ_α is the relaxation time of the α-process. Full and open symbols stand for the values measured by TES and DRS, respectively. The thicknesses are shown in the figure. The solid curve is calculated using the Vogel–Fulcher–Tammann equation. This figure shows the results for $M_w = 1.8 \times 10^6$. (Reprinted with permission from K. Fukao and Y. Miyamoto, *Phys. Rev. E*, 64, 011803. Copyright 2001 by the American Physical Society.)

If we select the appropriate angular frequency of the applied electric field, $\omega(\equiv\omega_E)$, then we can obtain the dynamic thermal expansion coefficient α_n^* as follows:

$$\alpha_n^* = -\zeta \frac{1}{C'(T_0)} \frac{dC'}{dT} = \zeta \frac{1}{C'(T_0)} \frac{C'_{\omega_T}}{T_{\omega_T}} e^{i\delta}. \tag{7.7}$$

The real and imaginary parts of α_n^* are defined as follows:

$$\alpha_n^* = \alpha_n' - i\alpha_n''. \tag{7.8}$$

In our study, TES was applied to measure the dynamics of the α-process for PS thin films at very low frequencies. Since changing the frequency of applied temperature modulation over a wide frequency region was difficult, we measured the temperature dependence of α_n' and α_n'' at frequencies ($f_T \equiv \omega_T/2\pi$) between 2.1 and 16.7 mHz. In these measurements, $\langle T \rangle$ was increased at a controlled constant rate of 0.1 or 0.5 K/min. The amplitude T_{ω_T} was set between 0.2 and 0.6 K. For capacitance measurements in TES, the frequency of the applied electric field was 100 kHz to avoid interference with dielectric relaxations. In these temperature-domain measurements, we could determine the temperature T_α at which the imaginary part of α_n^* shows a peak due to the α-process. We found that the distinct thickness dependence of f_α of the α-process is in the frequency range of 10^{-3}–10^{-2} Hz (Figure 7.3). In this range, the characteristic time of the α-process, which is the segmental motion for polymeric systems, corresponds to a macroscopic time scale. On the other hand, there is low or almost no thickness dependence in the frequency range of 20–10^5 Hz for films that are thicker than the critical thickness ($d_c \sim 20$ nm). Data obtained by TES and DRS (Figure 7.3) are used for low- and high-frequency ranges, respectively. Therefore, we could confirm the existence of a positive correlation between T_g and the dynamics of the α-process for thin polymer films.

Fakhraai and Forrest found that the thickness dependence of the T_g reduction in thin polymer films decreases with increasing cooling rate in ellipsometry measurements, and that there is almost no confinement effect at high cooling rates, which correspond to high frequencies for DRS measurements. The thickness and temperature dependence of the α-process observed by TES and DRS is consistent with the observed results by Fakhraai and Forrest [35].

7.2.4 RECENT DEVELOPMENTS IN GLASS TRANSITION IN PS THIN FILMS

After our studies, several experimental results have been reported in the literature. Lupascu et al. showed that the dynamics of the α-process has low thickness dependence, while T_g for PS thin films evaluated by CD decreases with decreasing d [36]. Recently, Boucher et al. showed that T_g evaluated by CD decreases with decreasing d in a manner similar to that observed through other experimental methods, while f_α of the α-process and the width of the dielectric loss curves remain independent of d for PS thin films [37]. The sample geometry and experimental methods in three different measurements, including ours, were the same. Despite this, the thickness dependence of T_g and the dynamics of the α-process differ. There are several possible origins of the difference in thickness dependence. First, the sample preparation method might affect the observed thickness dependence of T_g and the α-process. Recent studies have reported that the annealing process may affect the dynamics of the α-process [38]. The difference in thermal history may result in differences in thickness dependence. Second, Boucher et al. measured the dynamics of the α-process down to 1 Hz. However, the frequency range for the glass transition is usually located in the lower-frequency region, that is, 10^{-2}–10^{-3} Hz. Hence, the thickness dependence of the dynamics of the α-process may be observed with DRS measurements in this region, although meaningful dielectric loss signals are hard to obtain because of the weak polarity of PS. Third, the dielectric signal of the α-process is usually observed with contributions of dc conductivity, especially at the high-temperature, low-frequency region. Boucher et al. subtracted the contributions from dc conductivity to obtain the shape of the α-process. When the dielectric loss signals were large enough for PS, there would be no difference in the results before and after subtraction of the dc component. However, the polarity of PS is very weak and, hence, a small difference arising from subtraction may lead to a large difference in the experimental results after subtraction.

Similar simultaneous measurements of T_g and the dynamics of the α-process on thin films of labeled PS produced similar results, that is, a large T_g shift in thin films in accordance with the change in segmental mobility and broadening of dielectric loss spectra of the α-process with decreasing d [39]. The labeled PS has a very high polarity; hence, the dielectric signal due to the α-process is 70 times as large as that of neat PS. Thus, the uncertainty induced by subtraction of the dc component for this system could be substantially suppressed. As clearly shown later in this chapter, f_α values of the α-process of stacked thin films of poly(2-chlorostyrene) (P2CS) [40] and of poly(methyl methacrylate) (PMMA) [41] based on T_g measurements by CD nicely fit f_α values of the segmental motions at high temperatures and high frequencies measured by DRS, through the Vogel–Fulcher–Tammann (VFT) law [42–45].

Contrary to the results for labeled PS and stacked P2CS thin films, simultaneous measurements of DRS and CD on thin films of PS nanospheres with diameters less than 400 nm showed that T_g is depressed by confinement for the diameter range from 130 to 400 nm, while cooperative segmental mobility related to the α-process remains unchanged [46]. This result, together with observations of Boucher et al., suggests unexpected decoupling between T_g and segmental motion under confinement. Nevertheless, explaining the difference between these experimental results still poses a significant challenge [47].

Many recent studies showed that, in addition to a region with bulk mobility, a region with enhanced mobility exists near the free surface of polymer films, and that the mobility can exceed the bulk mobility by several orders of magnitude [48–51]. The mobile region extends several nanometers into the bulk polymer. The temperature dependence of the dynamics of the surface region with enhanced mobility is much weaker than that of the bulk region; above $T_g + 10$ K, the surface process merges with the bulk α-process. T_g and the average relaxation time τ_α of the α-process in thin-film geometry are affected by the dynamics of the surface mobile region. As the size of the surface mobile region is not highly thickness dependent, the influence of the surface mobile region on the average dynamics of the α-process is more evident in thin films than in the bulk systems. This is a scenario proposed by Ediger and Forrest [51].

The thin films used for DRS and CD measurements have no free surface and are supported on glass substrate. Both sides of the films are covered with electrodes (aluminum) [23,36,37]. These films are commonly known as capped supported films and show depression of T_g similar to that observed with uncapped supported thin films. Therefore, the above scenario could be used to explain the observed depression of T_g in thin films with a region with enhanced mobility. In fact, several DRS measurements have shown the existence of an additional relaxation process at a frequency higher than f_α of the α-process at a temperature near or below T_g [23,36,39,46]. This additional process merges with the α-process at high temperatures. This additional process can be associated with surface mobile regions.

Here, there might be a question whether the penetration of Al into the polymer layer occurs during the evaporation of Al onto the surface of the polymer films and affects the dynamics of the polymer thin layer. Recent experimental results clearly showed that these concerns are not the case. For example, x-ray photo-electron spectroscopy (XPS) measurement done by Bébin and Prud'homme proved that for coating the polymer surface with Al, the polymer substrate below the dense Al layer remains unperturbed, leading to a narrow Al/polymer interface [52]. Furthermore, a study using local dielectric spectroscopy has verified that the impact of confinement of polymer chains within thin supported films with one free surface is similar to that of capped supported films without any free surface [53].

7.3 AGING DYNAMICS IN THIN FILMS OF POLYMER GLASSES

In dielectric measurements, the capacitance of the sample condenser is a measured physical quantity and is the product of the dielectric permittivity and the area-to-thickness ratio, as mentioned in Section 7.2. For some polymeric systems such as PS, labeled PS, and P2CS, the contributions from both components can be observed separately, although only the change in dielectric permittivity can be observed mainly with PMMA during its isothermal aging process. In this section, we discuss the slow change in dielectric permittivity and/or volume of polymer-glass thin films based on dielectric measurements during aging below T_g. Through such measurements, the nature of the aging dynamics in polymer glasses is clarified. Here, it should be noted that the dielectric permittivity and volume could be measured very precisely through dielectric measurements of the thin-film geometry, although the confinement effect does not necessarily play an important role.

When polymers in the liquid state are cooled down to T_g, the characteristic time of segmental motions increases with decreasing temperature through a supercooled liquid state and reaches macroscopic time at T_g. Below T_g, the polymers form a glassy state, and the mobility due to the segmental motion is almost lost in the glassy state. And hence, the polymer in the glassy state seems to be solid and have a stable structure macroscopically. Nevertheless, it is well known that many physical quantities such as volume, elasticity, specific heat and dielectric permittivity change with a very large characteristic time even below T_g and several fascinating phenomena could be observed in the glassy state [14]. Although there is no appreciable contribution from the segmental motion in the glassy state, the segmental motion might be still activated with a very large but finite characteristic time down to a critical temperature, which is located by about 50 K lower than T_g and is usually called the Vogel temperature or the ideal glass transition temperature. Furthermore, several local motions might be activated even below T_g, depending on the structures of the polymeric systems. The mobility due to such contributions could be a driving force to induce several phenomena observed in the glassy state.

The glassy state, which can be obtained by an appropriate cooling process through the supercooled state from high-temperature liquid state, has a disordered structure like the structure of the liquid state. On the other hand, the mobility that can induce the glassy state to flow is lost like the solid state. Hence, the glassy state can be regarded as a frozen state of the liquid. It is well known that the glassy state is no thermodynamical equilibrium state, and hence, there are several different glassy states at a given temperature, depending on the thermal history before the present state [14,15].

7.3.1 KOVACS EFFECTS

One of the most well-known experiments on the dependence of glassy properties of polymers on the thermal history was done by Kovacs et al. [54]. In 1979, they performed an interesting experiment in which they elucidated the existence of memory effects in polymer glasses. They measured the volume relaxation at a point in the glassy state located on the line extrapolated from an equilibrium line in the liquid state, but arrived at via different thermal histories. As this can be regarded as a point on the equilibrium line, no further change in volume is expected. Nevertheless, their result revealed that the volume relaxation strongly depends on the thermal history of the polymer before reaching the equilibrium point. In other words, the thermal history could be memorized through the glassy state of the polymers.

An example of the Kovacs effect is shown in Figure 7.4 [55]. In Figure 7.4a, the vertical axis is the inverse of C' measured at 100 kHz for P2CS thin films with 22 nm thickness, and the horizontal axis is the temperature. Since the frequency of the applied electric field is 100 kHz, which is high enough to avoid the overlap of the dipolar relaxation with the volume change, we can expect that $1/C'$ is proportional to the volume change with temperature for P2CS thin films. The thin film was first at a thermal equilibrium state at a temperature above $T_g(T_0 = 413\ K)$, from which the thin film was cooled down to T_a. The thin film was isothermally aged at T_a for a period of t_a. In this figure, we selected two different thermal histories, that is, $T_a = 347.8$ and 357.6 K, for which $t_a = 80$ and 40 h, respectively. The thin film was then heated once more to T_f ($T_f = 367.1$ K). The temperature T_f is located on the line extrapolated from the equilibrium line above T_g. The thin film was then isothermally aged at T_f. If the thin film had already attained equilibrium at T_f, one would not expect the volume of the thin film would change with aging time t_a when aged isothermally at T_f. Nevertheless, our experimental results

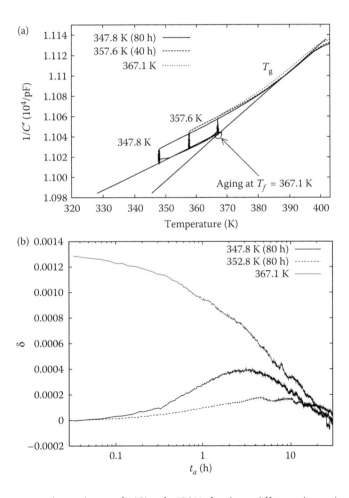

Figure 7.4 (a) Temperature dependence of $1/C'$ at $f = 10$ kHz for three different thermal treatments of 22 nm thick P2CS films. From $T_0 = 413$ K to $T_f = 367.1$ K (1) broken curves: via isothermal aging at $T_a = 357.6$ K for $t_a = 40$ h; (2) solid curve: via isothermal aging at $T_a = 347.8$ K for $t_a = 80$ h; (3) dotted curves: A direct temperature quench to 367.1 K. (b) The departure from equilibrium $\delta(t_a)$ as measured by C'^{-1} based on Equation 7.9 versus the the logarithm of the aging time for isothermal aging at $T_f = 367.1$ K. (Reprinted with permission from K. Fukao and D. Tahara, *Phys. Rev. E*, 80, 051802. Copyright 2009 by the American Physical Society.)

show an interesting dependence of $1/C'$ on t_a, as shown in Figure 7.4b. The dependence of the relative change in volume δ on t_a is plotted against the elapsed time at T_f in Figure 7.4b, where $t_a = 0$ corresponds to the time at which the temperature of the thin film reached T_f. $\delta(t_a)$ is defined by the departure from equilibrium as follows:

$$\delta(t_a) = \frac{v(t_a) - v(0)}{v(0)} = \frac{C'^{-1}(t_a) - C'^{-1}(0)}{C'^{-1}(0)}, \quad (7.9)$$

where $v(t_a)$ and $C'^{-1}(t_a)$ are the volume and the inverse of C' at time t_a. As shown in Figure 7.4b, $\delta(t_a)$ increases with increasing t_a and then decreases back to its initial value after reaching a peak. Furthermore, the time at which C'^{-1} shows a maximum strongly depends on the aging temperature T_a before the thin film reached T_f. This result suggests that the polymer glass can save memory of its thermal history even after heating up to T_f. As the aforementioned measurements could closely reproduce the Kovacs experiment, dielectric measurements may be very useful methods for investigating aging phenomena in polymer glasses.

7.3.2 Memory and rejuvenation effects

Next, we show more examples of interesting aging phenomena during the aging of thin films of PMMA, specifically, the memory and rejuvenation effects observed [56–58]. To do this, two different types of thermal histories are introduced as follows:

1. Constant-rate (CR) mode: After thermal equilibration at 403 K, the sample in a high-temperature liquid state is cooled down to a lower temperature below T_g at a constant rate (0.5 K/min). On the way to the final temperature, cooling is stopped at a temperature T_a (below T_g) and the sample is isothermally aged at T_a for a period τ_a. The sample is subsequently cooled down to room temperature at the same rate. Finally, the sample is heated again from room temperature to a temperature above T_g.

2. Temperature-cycling (TC) mode: The temperature of the sample is changed from a high temperature (above T_g) down to T_1 (below T_g) and is then maintained at T_1 for a period of τ_1 (stage 1). The temperature is then changed to $T_2(\equiv T_1 + \Delta T, \Delta T < 0)$ and is kept at T_2 for τ_2 (stage 2). Finally, the temperature is increased back to T_1 and is kept at T_1 for τ_3 (stage 3).

The aging phenomena of polymer glasses were observed through the dielectric permittivity, which is associated with thermal fluctuations of permanent dipole moments attached to polymer chains. The glassy state of polymers is a nonequilibrium state; hence, the energy level of the state tends to decrease, approaching the thermodynamically most stable state, which is usually a crystalline state. Thus, thermal fluctuation becomes smaller as the energy of the polymeric state decreases. Therefore, the state becomes more stable as the dielectric permittivity decreases.

First, we show the results obtained through the CR mode. Figure 7.5 shows the results for PMMA thin films observed through the CR mode with various aging temperatures T_a (374.2, 364.5, and 354.5 K) when $\tau_a = 10$ h. The frequency of the applied electric field was 20 Hz. In addition to the segmental motion, the β-process for PMMA, which is associated with the local motion of the side branch attached to the backbone chain, could also be observed. Hence, the dielectric permittivity in the temperature domain at a given frequency has a characteristic temperature dependence. To determine the effect of aging at T_a on the dielectric permittivity, we also measured the temperature dependence of the dielectric permittivity during the cooling and heating processes at the same rate without any isothermal aging at T_a. To obtain the contribution of isothermal aging only, data obtained without isothermal aging could be used as reference data ($\varepsilon'_{\text{ref}}$), which were subtracted from the observed values (ε') for the CR mode, including isothermal aging.

Figure 7.5 shows the change in ε' relative to $\varepsilon'_{\text{ref}}$ as a function of temperature. Although only the results on the real part of dielectric permittivity are shown, almost similar results were obtained for the imaginary part of dielectric permittivity. Open symbols represent results for the dielectric permittivity ($\varepsilon' - \varepsilon'_{\text{ref}}$) observed during the cooling process, including isothermal aging at T_a, and full symbols represent those observed in the subsequent heating process. Figure 7.5 shows that the dielectric permittivity $\varepsilon' - \varepsilon'_{\text{ref}}$ remained at zero for

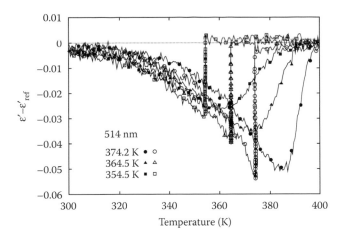

Figure 7.5 Temperature dependence of the real part of the complex dielectric permittivity for PMMA films with $d = 514$ nm observed by CR mode at three aging temperatures ($T_a = 374.2$, 364.5, and 354.5 K). The frequency of the applied electric field f was 20 Hz. The reference value ε'_{ref} was subtracted. The heating and cooling rates were 0.5 K/min and the aging times at T_a was 10 h. Open symbols display the data observed during the cooling process, while full symbols display those observed during the heating process. (Reprinted with permission from K. Fukao and A. Sakamoto, *Phys. Rev. E*, 71, 041803. Copyright 2005 by the American Physical Society.)

the cooling process down to the isothermal aging temperature T_a. As the temperature was kept at T_a for 10 h, the dielectric permittivity decreases with increasing aging time, deviating from the reference curves. This decrease in dielectric permittivity corresponds to the approach of the glassy system to an equilibrium stable state. If the equilibrium state is an inactive state, that is, a dead state, then this behavior in dielectric permittivity is considered an "aging" process.

After isothermal aging at T_a and restarting the cooling process at the same rate, the dielectric permittivity increases with decreasing temperature, approaching the reference curve at room temperature. The increase in dielectric permittivity indicates reversion to an active nonequilibrium state; that is, the glassy system experiences "rejuvenation" from the aged state during the cooling process after isothermal aging. Here, it should be noted that in some literatures [59,60], rejuvenation is defined so that all memories of the previous aging history are lost as if the sample had been reheated above T_g. In this chapter, however, "rejuvenation" is defined as the increase in dielectric permittivity with increasing aging time below T_g, as mentioned above.

Since the dielectric permittivity of the room-temperature glassy state after the first cooling process is on the reference curve, the glassy state is the same as that after the cooling process *without isothermal aging*. Nevertheless, during the subsequent heating process, the dielectric permittivity decreases again from the reference curve and reaches a minimum at a temperature about 10 K above T_a, as if the polymer glass could keep a memory on the isothermal aging during the preceding cooling process. This is the "memory effect" observed in polymer glass. A similar temperature change in $\varepsilon' - \varepsilon'_{ref}$ has been observed in PMMA films of 0.3 nm thickness [61] and in gels [62].

We show the results obtained by the TC mode. Figure 7.6a describes the dependence of the imaginary part of the dielectric permittivity (at 20 Hz relative to the reference curve) on the aging time for the thermal treatment as follows: $T_1 = 375.3$ K and $\tau_1 = 5$ h (stage 1), $T_2 = 355.8$ K ($\Delta T = -19.5$ K) and $\tau_2 = 5$ h (stage 2), and $T_1 = 375.3$ K and $\tau_3 = 17$ h (stage 3). In stage 1, the polymer glass starts to age isothermally at T_1 at $t_w = 0$ h. The origin of the time axis is defined by the moment at which the temperature reaches T_1. The relative dielectric permittivity decreases with increasing aging time. This decrease corresponds to the aging of polymer glass toward an equilibrium state. When the temperature decreases to T_2 at $t_w = \tau_1$, the dielectric permittivity quickly returns to the value on the reference curve and then starts to decrease with increasing aging time (stage 2). The quick return to the reference curve suggests that cooling from T_1 to T_2 rejuvenated the polymer glass aged for τ_1 during stage 1. At $t_w = \tau_1 + \tau_2$, the temperature is increased again to T_1 and then kept at T_1.

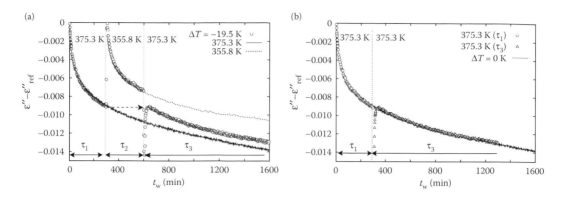

Figure 7.6 (a) Difference between ε'' and ε''_{ref} observed by the TC mode with $T_1 = 375.3$ K and $T_2 = 355.8$ K ($\Delta T = -19.5$ K) for PMMA thin films with $d = 20$ nm. (b) The difference $\varepsilon'' - \varepsilon''_{ref}$ obtained by shifting the data points in the third stage in the negative direction of the time axis by τ_2 after removing the data points in the second stage. Aging times at the first and second stages are $\tau_1 = \tau_2 = 5$ h. The horizontal axis of Figure 7.6b is the total aging time at T_1. The solid (—) and dotted curves ($\cdots$) are standard relaxation ones obtained by isothermal aging at $T_1 = 375.3$ K and $T_2 = 355.8$ K, respectively. It should be noted that the time origin of the dotted curve for T_2 is shifted from $t_w = 0$ to $t_w = \tau_1$. (Reprinted with permission from K. Fukao and A. Sakamoto, *Phys. Rev. E*, 71, 041803. Copyright 2005 by the American Physical Society.)

The dielectric permittivity consequently returns to the value that $\varepsilon'' - \varepsilon''_{ref}$ had reached at $t_w = \tau_1$ and begins to decrease with aging time (stage 3). To see only the aging dynamics at T_1, data on stage 2 were removed and those on stage 3 were shifted to the origin of the time axis by the amount of τ_2. Data obtained in this manner are plotted in Figure 7.6b. The horizontal axis of Figure 7.6b may be regarded as the total elapsed time at T_1. Figure 7.6b suggests that except for a small region just after the temperature jump from T_2 to T_1, the curve in stage 3 is perfectly connected to the curve on stage 1. The connected curve agrees very well with the solid curve, as apparent during the isothermal aging at T_1. This experimental result suggests that polymer glass can keep the full memory on the state of the polymer glass just after aging at T_1 for a period of τ_1, whereas the polymer glass experiences isothermal aging at T_2 for a period of τ_2. Aging at T_2 does not affect aging at T_1. As summarized by Figure 7.6, the polymer glass shows "rejuvenation," "memory effect," and "independence" between the aging at different temperatures (T_1 and T_2). If the amount of the temperature jump $|\Delta T|$ becomes smaller than 19.5 K, then the independence between aging at T_1 and T_2 becomes weaker; thus, aging at T_2 can affect aging at T_1 with an effective time τ_{eff}, which is less than τ_2 [56,58].

The interesting properties of aging in polymer glass in polymeric materials were discovered a long time ago [14]. Several useful protocols for spin glasses have been developed to examine the peculiar properties of aging phenomena in glassy systems. Several measurements on spin glass have clearly shown that the aging phenomena in spin glass exhibit behavior almost similar to those in polymer glass, including memory, rejuvenation, and independence of aging [63–65]. Considering the large difference in structure between polymers and spin glass, the similarity of the aging phenomena is unexpected. This implies the same physics behind the aging phenomena for both polymer glass and spin glass.

Here, we introduce a hierarchical model that explains the phenomena observed in the TC mode, which was originally developed for aging in spin glass [66]. We assume that the glass phase has a multivalley structure of a temperature-dependent free-energy surface. At temperature T_1, the system relaxes over the many valleys formed at T_1. When the system cools from T_1 to $T_2(T_2 < T_1)$, each valley of free energy splits into new smaller subvalleys. At large values of $|\Delta T|(\equiv |T_1 - T_2|)$, energy barriers separating the initial valleys are too high to coexist with the different initial valleys within the period of τ_2 at T_2. As only relaxations within the initial valleys can be activated, the occupation number of each initial valley is maintained during aging at T_2. When the system is heated again from T_2 to T_1, the small subvalleys merge back with their predecessors. Therefore, the memory at the end of the first isothermal aging process at T_1 can be memorized and then recalled at $\tau_1 + \tau_2$. The rejuvenation effects may also be explained in a similar way.

7.3.3 Aging in dielectric permittivity and volume

In the preceding sections, we showed peculiar aging phenomena of volume and dielectric permittivity of polymer glass, including memory and rejuvenation effects. As shown in Section 7.2, a change in capacitance with temperature and time for the aging process is associated with the change in dielectric permittivity and volume; hence, simultaneous measurements of both quantities may be achieved through a single capacitance measurement. Here, we present a result obtained from such measurements.

Figure 7.7 displays the temperature dependence of C' and C'' for P2CS thin films with 20 nm thickness relative to the values C'_{ref} and C''_{ref} observed during cooling and heating processes at a rate of 1 K/min. Values of C' and C'' were obtained during the ramping process with isothermal aging at T_a for 30 h in the cooling process, and those of C'_{ref} and C''_{ref} were obtained during the same ramping process without isothermal aging.

Figure 7.7a shows the temperature dependence of C' within the frequency range of 20 Hz to 100 kHz. Observations were done in CR mode with $T_a = 357.4$ K and $\tau_a = 30$ h, which is the same temperature protocol as that used for Figure 7.5. In Figure 7.7a, $C' - C'_{ref}$ at 100 kHz increases monotonically with increasing aging time during isothermal aging at T_a. During the subsequent cooling process after isothermal aging, the value of $C' - C'_{ref}$ decreases slightly with decreasing temperature, but most of the deviation induced during isothermal aging still occurs at 273 K. At frequencies above 40 kHz, the frequency dependence is very weak (Figure 7.7a). As discussed in Section 7.2, the frequency-independent value of C' at the high-frequency region may be mainly associated with the change in volume of the polymer glass. Hence, the increase in C' above 40 kHz with aging time might be due to the volume change in the isothermal aging process at T_a. A similar increase

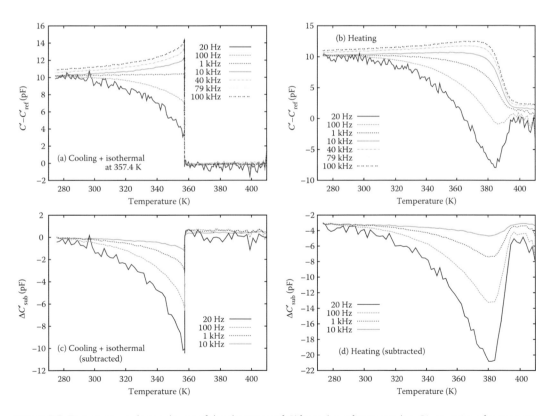

Figure 7.7 Temperature dependence of the deviation of C' from the reference value C'_{ref} at various frequencies between 20 Hz and 100 kHz, (a) during cooling with isothermal aging at 357.4 K and (b) during the subsequent heating cycle. The lower figures show the temperature dependence of $C' - C'_{ref}$ after subtracting the volume change contribution $C' - C'_{ref}$ at 100 kHz, $\Delta C'_{sub}$, at various frequencies between 20 Hz and 100 kHz, (c) during cooling with isothermal aging at 357.4 K and (d) during the subsequent heating cycle. (Reprinted with permission from K. Fukao and D. Tahara, *Phys. Rev. E*, 80, 051802. Copyright 2009 by the American Physical Society.)

in C' for PS thin films has also been observed [67]. The volume of polymer glass decreases and the density increases with aging time during aging below T_g. This phenomenon is commonly known as densification of glassy polymer [68]. It should be noted that the volume change induced during isothermal aging still survives even at 273 K. This suggests that the volume change *does not show full rejuvenation*. During the subsequent heating process, $C' - C'_{ref}$ remains almost constant up to about 380 K and then decreases rapidly to zero, as shown in Figure 7.7b. This suggests that the volume also retains the *memory* of the temperature at which isothermal aging was done.

In Figure 7.7a and b, it is clear that the temperature change in $C' - C'_{ref}$ strongly depends on the frequency for both cooling and heating processes when the frequency is reduced from 40 kHz to 20 Hz. This frequency dependence may be attributed to the overlap of the frequency dependence of dielectric permittivity due to molecular motion and volume change. Therefore, we may expect that $C' - C'_{ref}$ shows the same temperature dependence as that for the dielectric permittivity after subtracting the contribution from the volume change.

For this purpose, we define the value of $\Delta C'_{sub}(T, f)$ as follows:

$$\Delta C'_{sub}(T, f) \equiv [C' - C'_{ref}](T, f) - [C' - C'_{ref}](T, 100\ \text{kHz}). \tag{7.10}$$

As shown in Figure 7.7c and d, $\Delta C'_{sub}$ corresponds to the real part of the dielectric permittivity, ε'_{disp}. This temperature dependence of $\Delta C'_{sub}$ is very similar to that for the dielectric permittivity of PMMA (Figure 7.5) and for C'' of P2CS thin films (see Figure 3 in Fukao and Tahara [55]). Therefore, Figure 7.7c and d shows that ε'_{disp} exhibits *full rejuvenation and memory effects* in the CR mode, although volume does not show full rejuvenation. The above capacitance measurements reveal a clear difference in the aging behaviors of the volume and dielectric permittivity.

It should be noted that simultaneous measurement of the change in dielectric permittivity and volume of P2CS thin films during isothermal aging could be done through a single capacitance measurement. This simultaneous measurement was not possible in the case of PMMA because of a relaxation process related to molecular motion even at high frequency and because of the strong overlap of the change in dielectric permittivity with a possible volume change during isothermal aging.

7.3.4 CONFINEMENT EFFECT OF AGING IN POLYMER GLASS

In the preceding sections, we showed interesting aging phenomena observed through capacitance measurements. In ultrathin films of P2CS ($d < 10$ nm), an increase in ε'' (not in ε') with aging time has been observed. This anomalous increase in ε'' has been found to be strongly correlated with the presence of a mobile region within the thin film [69].

Many studies have attempted to characterize and understand the change in the rate of physical aging under confinement in molecular glasses in nanopores, supported polymer films, and nanocomposites [70–72]. The effect of confinement on physical aging depends on various factors, including interfacial interactions, confinement type, and aging conditions. Therefore, the confinement effect on the rate of physical aging is directly associated with that on T_g, because physical aging is only possible below T_g and because the driving force for physical aging at T_a is a function of a quench depth given by the equation $\Delta T_g \equiv T_g - T_a$. The characteristic time of physical aging (τ_{pa}) cannot be described by a simple linear relationship with τ_α, but the proportionality constant must be a function of the confinement condition, such as area/volume ratio A/V as in $\tau_{pa} = f(A/V)\tau_\alpha$. The relation between τ_{pa} and τ_α is still under debate [72].

7.4 GLASS TRANSITION DYNAMICS OF STACKED THIN POLYMER FILMS

As discussed in Section 7.2, T_g of thin polymer films substantially deviates from the bulk value. Several measurements have revealed that in (thin) polymer films, there is a surface region where molecular mobility is highly enhanced. However, there are various experimental data on the thickness dependence of the α-process (segmental motion) in PS thin films that do not necessarily agree with each other. One result shows that the relaxation rate of the α-process, f_α, has thickness dependence corresponding to the change in T_g [36]. Another shows that f_α is independent of d, while T_g of PS thin films in both cases decreases relative to the bulk T_g [37].

Here, we adopt another type of thin polymer film, a stacked thin polymer film, and show the results for the glass transition and the dynamics of the α-process observed by DRS and DSC. In stacked thin polymer films, we expect that there are many interfaces between thin polymer layers. Through the thermal and dielectric measurements on stacked thin polymer films, the nature of the glass transition in thin polymer films is discussed in this section.

The stacked thin polymer films were prepared as follows. A thin polymer film prepared on a glass substrate was floated onto a water surface, and then the floating thin film was gathered on the substrate on which thin polymer layers had been stacked previously. This procedure was repeated until the number of thin polymer layers reached the number required for measurements. Even if the layer thickness was 10 nm, the thickness of a stack consisting of 100 sheets of the polymer reached 1 μm, which corresponds to the thickness of thick films. Our question in this case is whether this stack had thin-film-like glass transition behavior or its bulk-like counterpart. There is an answer to this simple question in the literature [73,74]. The T_g of as-stacked PS thin films determined by DSC was found to be lower than that of single PS thin films supported on a substrate and higher than that of freestanding PS thin films. In this case, d of each thin layer in the stack, that of the single thin film supported on a substrate, and that of the freestanding film are equal. Furthermore, the T_g of as-stacked PS thin films has been reported to increase with increasing annealing time, approaching the bulk T_g during isothermal annealing at a temperature above the bulk T_g. Here, we have two further questions: (1) Does the T_g of as-stacked PS thin films truly reach the bulk T_g if the films are annealed for long enough? (2) Does the dynamics of the α-process of as-stacked PS thin films change continuously from thin-film-like dynamics to bulk-like dynamics upon annealing? To answer these questions, we performed DSC and DRS measurements on stacked thin films of PS, P2CS, and PMMA [40,41,75–77].

7.4.1 THERMAL PROPERTIES

Figure 7.8 shows the temperature dependence of the total heat flow observed at a rate of 10 K/min by DSC for as-stacked PS thin films and stacked thin films after annealing at 523 K for 12 h. Here, we prepared five stacked PS thin films with single layers of different thicknesses. The numbers of stacked single layers of 70, 55, 40, 20, and 13 nm thicknesses were 73, 100, 130, 224, and 400, respectively. The number of stacked thin layers had to be large enough for the total weight of stacked thin films to reach 1 mg to obtain a meaningful DSC signal. In Figure 7.8a, the T_g is about 350 K for as-stacked films of 13 nm thick PS layers. T_g increases with increasing d of single layers, approaching the bulk T_g. Furthermore, the bulk T_g could be restored after annealing of the as-stacked PS thin films at 523 K for 12 h, as shown in Figure 7.8b. This suggests that annealing can induce an increase in T_g up to the bulk T_g. The effect on as-stacked thin films during the annealing process is apparent. As-stacked PS thin films include many distinct interfaces between thin polymer layers, and each polymer chain is confined within a thin polymer layer. Hence, there is a density gap at the interface. However, the density gap at the interface can be reduced and eventually eliminated by annealing above T_g. At this final stage, the stacked thin film is identical to the bulk film. Therefore, the above result suggests the possibility of controllably lowering T_g from the bulk T_g by changing the annealing conditions for the stacked thin films.

7.4.2 DIELECTRIC PROPERTIES

Next, we show the results on the dynamics of stacked thin polymer films measured by DRS to elucidate how the dynamics of the α-process of as-stacked thin polymer films changes during annealing above T_g, where T_g increases with annealing time toward the bulk T_g. For this measurement, we chose stacked P2CS thin films. The chemical structure of P2CS is similar to that of PS, but P2CS has a much higher polarity. Hence, this polymer can be regarded as a model system for dynamics investigation of PS-like polymer by DRS.

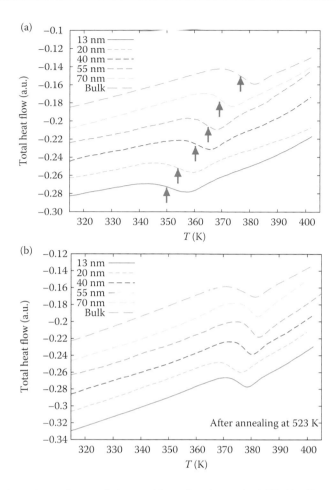

Figure 7.8 Temperature dependence of the total heat flow in stacked PS thin films of single-layer thicknesses of 13 nm to the bulk value during heating at a rate of 10 K/min: (a) as-stacked PS thin films, and (b) films after annealing at 523 K for 12 h. Each curve is shifted slightly along the vertical axis for ease of comparison. (Reprinted with permission from K. Fukao et al., *Phys. Rev. E*, 84, 041808. Copyright 2011 by the American Physical Society.)

Figure 7.9 shows the value of T_g obtained by CD method for stacked P2CS thin films and single thin films of P2CS with various d values. In Figure 7.9, T_g of single P2CS thin films monotonically decreases with decreasing d, and this thickness dependence of T_g is similar to that observed with PS thin films [17]. On the other hand, the T_g for stacked P2CS thin films annealed at 393 K for 2 h is located far below the bulk T_g and about 10 K below the T_g of the single thin film with the same thickness. Here, the number of layers in the stacked thin films is 10. This result suggests that even after stacking of 10 thin layers, the films had a thin-film-like T_g. The magnitude of the depression of T_g of the freestanding PS thin films is ca. 50 K larger than that of the supported thin films [19,78]. If the magnitude of T_g lowering of the freestanding thin films of P2CS could obey the same law for PS, then the T_g of the stacked P2CS thin films is between the T_g values of the supported single thin films and of the freestanding films. This result is consistent with the observed results for stacked PS thin films [73]. Furthermore, the T_g of stacked P2CS thin films annealed at 413 K for 2 h is higher than that of the stacked films annealed at 393 K.

From temperature-domain DRS measurements, we obtained the complex dielectric permittivity as a function of temperature at a given frequency. A dynamical process such as the α-process within the observed temperature and frequency window induces a signal related to the dynamical process. Consequently, ε″ reaches a peak in the temperature domain at a given frequency. We measured the peak temperature T_α of the dielectric

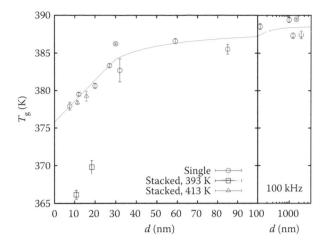

Figure 7.9 Thickness dependence of T_g for thin films of P2CS determined as a crossover temperature in the electric capacitance. o represents for T_g of single thin films; □ and Δ represent T_g of stacked thin films annealed at 393 and 413 K, respectively. In the latter case, the thickness of the horizontal axis is the averaged thickness of each single layer in 10 stacked thin films. (Reproduced from K. Fukao et al., *Eur. Phys. J. Special Topics*, 189, 165, 2010, with kind permission of The European Physical Journal (EPJ).)

loss for P2CS thin films due to the α-process at 20 Hz, 100 Hz, and 1 kHz. The value of T_α decreases with decreasing thickness in a manner similar to that of T_g, as shown in Figure 7.10. T_α values of the stacked P2CS thin films annealed at 393 and 413 K for 2 h are plotted in Figure 7.10 (see □, O and Δ, ∇). It is clear that T_α of the stacked thin films annealed at 393 K is almost equal to that of P2CS single thin films with thicknesses of 12 and 18 nm at 20 and 100 Hz. However, T_α of the stacked thin films annealed at 413 K is higher than that of the P2CS single thin films of the same thickness. These results suggest that the dynamics of the α-process changes when the annealing temperature increases from 393 to 413 K.

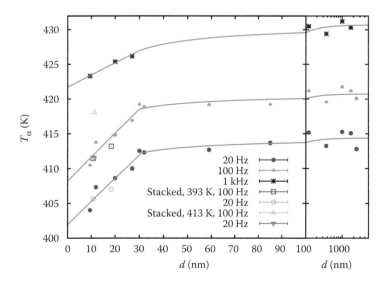

Figure 7.10 Thickness dependence of the temperature T_α determined as the peak temperature of the dielectric loss for single thin films of P2CS at $f = 20$ Hz (•), 100 Hz (Δ), and 1 kHz (*). The T_α for stacked thin films of P2CS is also plotted. □ and o, as well as Δ and ∇, represent T_α for stacked thin films annealed at 393 and 413 K, respectively. (Reproduced from K. Fukao., *Eur. Phys. J. Special Topics*, 189, 165, 2010, with kind permission of The European Physical Journal (EPJ).)

Here, we performed in situ dielectric measurements on stacked P2CS thin films during isothermal annealing process to investigate the time evolution in the dynamics of the α-process. Dielectric loss spectra of as-stacked P2CS thin films observed at 425 K show a gradual change with annealing time. The peak frequency due to the α-process of as-stacked P2CS thin films is shifted to the higher-frequency side from that for the bulk system. However, annealing at high temperature induces f_α of the α-process to move to the lower-frequency side, which corresponds to the slowing down of the dynamics of the α-process as it approaches that of the bulk. The characteristic time τ of the time evolution of f_α of the α-process is 30–45 h at annealing temperatures between 412 and 425 K. This result implies that there is a very slow process occurring during the isothermal annealing process [40]. The temperature dependence of τ can be well described by the Arrhenius law. The activation energy related to this time evolution is ca. 6 kcal/mol, which is much smaller than the activation energy of the α-process and/or β-process of PS [79].

The density gap at the interface between layers within the stacked thin polymer films can decrease during the isothermal annealing process of the stacked films. This reduction of the density gap is usually caused by mutual diffusion of polymer chains at the interface [80–82]. To determine whether the change in structure of the interface is directly associated with the change in T_g and the dynamics of the α-process, we recently performed neutron reflectivity (NR) measurements on stacked PMMA thin films during an isothermal annealing process above T_g. The time evolution of T_g and T_α during annealing of stacked PMMA thin films, as observed by DRS, was almost similar to that observed for stacked PS (P2CS) thin films [41]. For the NR measurements, we prepared alternately stacked thin films of hydrogenated PMMA (h-PMMA) and deuterated PMMA (d-PMMA) because a contrast of the scattering length density between thin polymer layers is necessary to obtain meaningful signals. From the NR measurements, we confirmed the existence of the time evolution of the interface structure between d-PMMA and h-PMMA layers during the annealing process. The characteristic time is almost the same as that of time evolution of the glassy dynamics observed by DRS [83].

7.4.3 T_g AND THE RELAXATION RATE OF THE α-PROCESS

In this section, we discuss in more detail the dependence of f_α of the α-process on the annealing time during isothermal annealing above T_g. Figure 7.11 shows the Arrhenius plot of f_α of the α-process at various annealing times for stacked P2CS thin films consisting of 10 single layers with 18 nm thickness. Data in

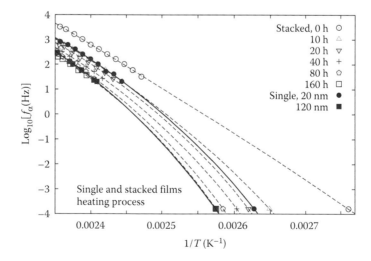

Figure 7.11 Dispersion map for the α-process of both single thin films and stacked thin films of P2CS. For stacked thin films of 18 nm thick P2CS layers, the temperature dependence of f_α at various annealing times $t_a = 0$ to 160 h are plotted. Results for single thin films with 20 and 120 nm thicknesses are also plotted. (Reprinted with permission from K. Fukao et al., *Phys. Rev. E*, 84, 041808. Copyright 2011 by the American Physical Society.)

the high-frequency region were obtained by DRS. T_g measured by CD on exactly the same sample was also plotted as a point $(1/T_g, \log_{10}(1/(2\pi\tau_g)))$ on the assumption that f_α of the α-process is equal to $1/2\pi\tau_g$ at T_g, where τ_g is a macroscopic time scale (for example, 10^3 s). In these measurements, as-stacked P2CS thin films were isothermally annealed at 425 K. During the annealing process, two DRS measurements at a rate of 1 K/min over a temperature range between 425 and 273 K intervened in DRS measurements under isothermal annealing every 10 h. We confirm that the results of the two measurements in between isothermal annealing processes agree well. Hence, the annealing effect due to the temperature change during ramping between 425 and 273 K was negligible compared with the effect of isothermal annealing at 425 K. The total annealing time under isothermal annealing may be regarded as the effective annealing time at 425 K without any correction.

Figure 7.11 shows the temperature dependence of f_α of the α-process of single thin films and stacked thin films of P2CS. The stacked thin films consist of 10 thin layers of 18 nm thickness. The temperature dependence of f_α at various annealing times is plotted. As shown in Figure 7.11, f_α of 20 nm thick and 120 nm thick single films of P2CS can be described well by the VFT law:

$$f_\alpha = f_{\alpha,0} \exp\left(-\frac{U}{T - T_0}\right), \tag{7.11}$$

where U is a positive constant, $f_{\alpha,0}$ is the relaxation rate of the α-process at very high temperature, and T_0 is the Vogel temperature at which f_α^{-1} diverges. f_α of the α-process at a given temperature increases with decreasing d. On the other hand, the temperature dependence of f_α of as-stacked P2CS thin films is a nearly straight line, suggesting a nearly Arrhenius behavior. As the annealing time increased from 0 to 160 h, f_α of the α-process decreased at a given temperature and the Arrhenius-type temperature dependence of f_α approached the VFT law, which is usually observed in bulk systems.

To classify the glassy dynamics and to explain the temperature dependence of f_α of the α-process, the fragility index m is commonly used. m is defined by the equation [84]

$$m = \left(\frac{d\log_{10}\tau_\alpha(T)}{d(T_g/T)}\right)_{T=T_g}, \tag{7.12}$$

where $\tau_\alpha = 1/(2\pi f_\alpha)$. In the concept of "fragility," which was introduced by Angell, glassy materials are classified as strong or fragile [85]. According to the definition of m in Equation 7.12, m is equal to the slope at T_g of the curve of the logarithm of the α relaxation time (τ_α) versus the inverse temperature normalized with T_g. As m decreases, the temperature dependence of τ_α approaches the Arrhenius law. It is well known that SiO_2 approximately exhibits the Arrhenius-type temperature dependence of τ_α. The structure of SiO_2 is self-reinforcing because every silicon atom is connected to four nearest-neighbor silicon atoms through bridging bonds. Such a network structure is resistant to thermal degradation and to rupture of any single bond. This resistance is reflected in the very small change in heat capacity that accompanies passage through the glass transition. Thus, this glassy material is a "strong" glass. In contrast, for materials that show strongly non-Arrhenius-type temperature dependence of τ_α, m is larger; such materials are examples of "fragile" glass.

We evaluated m as a function of annealing time. m increases from 50 to 120 as the annealing time increased from 0 to 160 h during isothermal annealing of stacked P2CS thin films at 425 K. In other words, the glassy dynamics of the α-process of stacked P2CS thin films changed from strong to fragile. This change suggests the possibility of controlling the fragility by changing the annealing conditions for stacked polymer thin films.

The experimental results shown in Figure 7.11 can be explained well by the strong correlation between T_g evaluated by CD and f_α of the α-process via the VFT law, and the changes in T_g and f_α with annealing are induced by the change in the strength of interfacial interaction, that is, enhanced surface/interfacial mobility. Figure 7.11 suggests strong coupling between T_g and the α-process (segmental motion) even in thin-film geometry, which is identical to that in the bulk.

7.5 SUMMARY

Since the discovery of the drastic reduction of T_g in thin polymer films, DRS has been a powerful tool for the investigation of the dynamics related to the glass transition. As discussed in this chapter, we showed how DRS and CD methods have contributed to the understanding of the dynamical nature of the glass transition in thin polymer films. In Section 7.2, we showed the results on T_g and the dynamics of segmental motion (the α-process) in thin polymer films obtained by DRS and CD. A comparison between our results and others was performed to show recent trends and controversial issues in this area.

In Section 7.3, aging phenomena observed in polymer glass, including memory and rejuvenation effects, were presented. In some polymeric systems such as PS and P2CS, the contributions of changes in dielectric permittivity and volume during the aging process could be observed separately through a single capacitance measurement. The nature of the aging dynamics in polymer glasses based on such measurements was discussed.

In Section 7.4, the glass transition dynamics of stacked thin polymer films was discussed to elucidate the effect of interfacial interaction on the dynamics in thin polymer films. Thermal and dielectric measurements clearly showed the thin-film-like glass transition behavior for as-stacked thin polymer films. However, the glass transition dynamics of the stacked thin polymer films changed from thin-film-like to the bulk-like dynamics with increasing annealing time during an isothermal annealing process. These results clearly suggest that the interfacial mobile region plays a crucial role in determining the nature of the glass transition dynamics.

There remain important issues in this area that could not be addressed in this chapter. One is the distribution of T_g and the relaxation time of the α-process in thin polymer films. Ellison and Torkelson measured the T_g through fluorescence measurements at different positions in multilayered polymer films consisting of neat polymer layers and a dye-labeled polymer layer [86]. Rotella et al. investigated the distribution of T_g and dielectric relaxation strength inside thin polymer films by using labeled PS [87]. Takaki et al. measured the time evolution of f_α of the α-process at various positions within thin polymer films during the annealing process by using multilayered films of PS and P2CS [77]. These measurements clearly showed the existence of the positional dependence of the glass transition and related dynamics in thin polymer films.

Another important topic is the decoupling between different dynamical modes. It is well known that decoupling between translational motion and rotational motion occurs near the glass transition region in bulk amorphous materials [88]. Recently, it has been shown through dielectric measurements that diffusion of small molecules and cold crystallization kinetics in thin polymer films slow down considerably, while segmental relaxation remains almost unchanged even for thicknesses several times larger than the polymer size [89,90]. Several studies using tracer diffusivity measurements report a large deviation of T_g from the bulk T_g and the nearly unchanged or suppressed diffusion of small molecules within the polymer matrix in thin supported films and freestanding films of PS [78,91]. Very recently, we performed dielectric measurements on thin films of amorphous polyamide random copolymers [92]. We observed the thickness dependence of the electrode polarization process in addition to the α-process in the thin films. We could also observe the accelerated diffusion of charge carriers within polymer matrix in thin films, while the α-process does not show an appreciable thickness dependence.

DRS and related techniques that were developed a long time ago seem to be considered classical experimental methods compared with other experimental techniques for measuring the dynamics. However, DRS still has potential to produce new and unexpected results, especially regarding the dynamical behavior of the glass transition in confinement.

ACKNOWLEDGMENTS

Most of the experimental results in this chapters came from collaborative works with T. Hayashi, A. Hoshino, H. Koizumi, D. Long, H. Miyaji, Y. Miyamoto, K. Nakamura, Y. Oda, R.D. Priestley, A. Sakamoto, Y. Saruyama,

P. Sotta, D. Tahara, H. Takaki, N. Taniguchi, T. Terasawa, J.M. Torkelson, S. Uno, and S. Yamawaki. The author would like to thank all collaborators for fruitful collaborations. These works were supported by a Grant-in-Aid for Scientific Research (B) (Nos. 25287108, 21340121, 16340122), Scientific Research (C) (Nos. 14540377, 11640395), and Exploratory Research (Nos. 25610127, 23654154, 19654064, 16654068) from the Japan Society for the Promotion of Science, Scientific Research in the Priority Area "Soft Matter Physics" from the Ministry of Education, Culture, Sports, Science and Technology of Japan.

REFERENCES

1. K. L. Ngai, The glass transition and the glassy state, in *Physical Properties of Polymers*, edited by J. Mark (Cambridge University Press), 72–152, 2004.
2. G. R. Strobl, Mechanical and dielectric response, in *The Physics of Polymers* (Springer-Verlag, Berlin, Heidelberg), 223–286, 2007.
3. P. W. Anderson, *Science*, 267, 1615, 1995.
4. H. Sillescu, *J. Non-Cryst. Solids*, 243, 81, 1999.
5. L. Berthier, G. Biroli, J.-P. Bouchaud, L. Cipelletti, and W. van Saarloos, *Dynamical Heterogeneities in Glasses, Colloids, and Granular Media* (Oxford University Press, New York), 1–450, 2011.
6. T. Muranaka and Y. Hiwatari, *Phys. Rev. E*, 51, R2735, 1995.
7. R. Yamamoto and A. Onuki, *Phys. Rev. E*, 58, 3515, 1998.
8. E. R. Weeks, J. C. Crocker, A. C. Levitt, A. Schofield, and D. A. Weitz, *Science*, 287, 627, 2000.
9. G. Adam and J. H. Gibbs, *J. Chem. Phys.*, 43, 139, 1965.
10. M. Alcoutlabi and G. B. McKenna, *J. Phys. Condens. Matter*, 17, R461, 2005.
11. S. Napolitano, S. Capponi, and B. Vanroy, *Eur. Phys. J. E*, 36, 61, 2013.
12. M. Y. Efremov, A. V. Kiyanova, J. Last, S. S. Soofi, C. Thode, and P. F. Nealey, *Phys. Rev. E*, 86, 021501, 2012.
13. M. Tress, M. Erber, E. U. Mapesa, H. Huth, J. Müller, A. Serghei, C. Schick, K.-J. Eichhorn, B. Voit, and F. Kremer, *Macromolecules*, 43, 9937, 2010.
14. L. C. Struick, in *Physical Aging in Amorphous Polymers and Other Materials* (Elsevier, Amsterdam), 1–244, 1978.
15. J.-P. Bouchaud, Aging in glassy systems: New experiments, simple models, and open questions, in *Soft and Fragile Matter: Nonequilibrium Dynamics, Metastability and Flow*, edited by M. Cates and M. Evans (IOP publishing, London), 285–304, 2000.
16. F. Kremer and A. Schönhals, in *Broadband Dielectric Spectroscopy* (Springer-Verlag, Berlin, Heidelberg), 1–729, 2000.
17. J. L. Keddie, R. A. L. Jones, and R. A. Cory, *EPL (Europhys. Lett.)* 27, 59, 1994.
18. J. L. Keddie, R. A. L. Jones, and R. A. Cory, *Faraday Discuss.*, 98, 219, 1994.
19. J. A. Forrest, K. Dalnoki-Veress, J. R. Stevens, and J. R. Dutcher, *Phys. Rev. Lett.*, 77, 2002, 1996.
20. J. A. Forrest, K. Dalnoki-Veress, and J. R. Dutcher, *Phys. Rev. E*, 56, 5705, 1997.
21. J. A. Forrest, C. Svanberg, K. Révész, M. Rodahl, L. M. Torell, and B. Kasemo, *Phys. Rev. E*, 58, R1226, 1998.
22. K. Fukao and Y. Miyamoto, *EPL (Europhys. Lett.)*, 46, 649, 1999.
23. K. Fukao and Y. Miyamoto, *Phys. Rev. E*, 61, 1743, 2000.
24. K. Fukao, S. Uno, Y. Miyamoto, A. Hoshino, and H. Miyaji, *Phys. Rev. E*, 64, 051807, 2001.
25. K. Fukao, S. Uno, Y. Miyamoto, A. Hoshino, and H. Miyaji, *J. Non-Cryst. Solids*, 307–310, 517, 2002.
26. L. Hartmann, W. Gorbatschow, J. Hauwede, and F. Kremer, *Eur. Phys. J. E*, 8, 145, 2002.
27. J. S. Sharp and J. A. Forrest, *Phys. Rev. E*, 67, 031805, 2003.
28. J. Sharp and J. Forrest, *Eur. Phys. J. E*, 12, 97, 2003.
29. C. Svanberg, *Macromolecules*, 40, 312, 2007.
30. C. Bauer, R. Böhmer, S. Moreno-Flores, R. Richert, H. Sillescu, and D. Neher, *Phys. Rev. E*, 61, 1755, 2000.
31. G. B. DeMaggio, W. E. Frieze, D. W. Gidley, M. Zhu, H. A. Hristov, and A. F. Yee, *Phys. Rev. Lett.*, 78, 1524, 1997.

32. K. Fukao and Y. Miyamoto, *J. Phys. IV France*, 10, 243, 2000.
33. K. Fukao and Y. Miyamoto, *Phys. Rev. E*, 64, 011803, 2001.
34. C. Bauer, R. Richert, R. Böhmer, and T. Christensen, *J. Non-Cryst. Solids*, 262, 276, 2000.
35. Z. Fakhraai and J. A. Forrest, *Phys. Rev. Lett.*, 95, 025701, 2005.
36. V. Lupascu, S. J. Picken, and M. Wübbenhorst, *J. Non-Cryst. Solids*, 352, 5594, 2006.
37. V. M. Boucher, D. Cangialosi, H. Yin, A. Schonhals, A. Alegria, and J. Colmenero, *Soft Matter*, 8, 5119, 2012.
38. A. Serghei and F. Kremer, Unexpected preparative effects on the properties of thin polymer films, in *Characterization of Polymer Surfaces and Thin Films*, vol. 132 of Progress in Colloid and Polymer Science, edited by K. Grundke, M. Stamm, and H.-J. Adler (Springer, Berlin, Heidelberg), 33–40, 2006.
39. R. D. Priestley, L. J. Broadbelt, J. M. Torkelson, and K. Fukao, *Phys. Rev. E*, 75, 061806, 2007.
40. K. Fukao, T. Terasawa, Y. Oda, K. Nakamura, and D. Tahara, *Phys. Rev. E*, 84, 041808, 2011.
41. T. Hayashi and K. Fukao, *Phys. Rev. E*, 89, 022602, 2014.
42. G. S. Fulcher, *J. Am. Ceram. Soc.*, 8, 339, 1925.
43. G. S. Fulcher, *J. Am. Ceram. Soc.*, 8, 789, 1925.
44. H. Vogel, *Phys. Z.*, 22, 645, 1921.
45. G. Tammann and W. Hesse, *Z. Anorg. Allg. Chem.*, 156, 245, 1926.
46. C. Zhang, V. M. Boucher, D. Cangialosi, and R. D. Priestley, *Polymer*, 54, 230, 2013.
47. R. D. Priestley, D. Cangialosi, and S. Napolitano, *J. Non-Cryst. Solids*, 407, 288, 2015.
48. J. E. Pye and C. B. Roth, *Phys. Rev. Lett.*, 107, 235701, 2011.
49. K. Paeng, S. F. Swallen, and M. D. Ediger, *J. Am. Chem. Soc.*, 133, 8444, 2011.
50. K. Paeng and M. D. Ediger, *Macromolecules*, 44, 7034, 2011.
51. M. D. Ediger and J. A. Forrest, *Macromolecules*, 47, 471, 2014.
52. P. Bébin, and R. E. Prud'homme, *Chem. Mater.*, 15, 965, 2003.
53. H. K. Nguyen, M. Labardi, S. Capaccioli, M. Lucchesi, P. Rolla, and D. Prevosto, *Macromolecules*, 45, 2138, 2012.
54. A. J. Kovacs, J. J. Aklonis, J. M. Hutchinson, and A. R. Ramos, *J. Polym. Sci. Polym. Phys. Ed.*, 17, 1097, 1979.
55. K. Fukao and D. Tahara, *Phys. Rev. E*, 80, 051802, 2009.
56. K. Fukao and A. Sakamoto, *Phys. Rev. E*, 71, 041803, 2005.
57. K. Fukao, A. Sakamoto, Y. Kubota, and Y. Saruyama, *J. Non-Cryst. Solids*, 351, 2678, 2005.
58. K. Fukao and S. Yamawaki, *Phys. Rev. E*, 76, 021507, 2007.
59. G. B. McKenna, *J. Phys. Condens. Matter*, 15, S737, 2003.
60. H.-N. Lee and M. D. Ediger, *Macromolecules*, 43, 5863, 2010.
61. L. Bellon, S. Ciliberto, and C. Laroche, *Eur. Phys. J. B*, 25, 223, 2002.
62. A. Parker and V. Normand, *Soft Matter*, 6, 4916, 2010.
63. F. Lefloch, J. Hammann, M. Ocio, and E. Vincent, *EPL (Europhys. Lett.)*, 18, 647, 1992.
64. K. Jonason, E. Vincent, J. Hammann, J. P. Bouchaud, and P. Nordblad, *Phys. Rev. Lett.*, 81, 3243, 1998.
65. E. Vincent, Ageing, rejuvenation and memory: The example of spin-glasses, in *Ageing and the Glass Transition*, vol. 716 of Lecture Notes in Physics, edited by M. Henkel, M. Pleimling, and R. Sanctuary (Springer, Berlin, Heidelberg), 7–60, 2007.
66. E. Vincent, J. P. Bouchaud, J. Hammann, and F. Lefloch, *Philos. Mag. B*, 71, 489, 1995.
67. K. Fukao and H. Koizumi, *Phys. Rev. E*, 77, 021503, 2008.
68. R. Greiner and F. Schwarzl, *Rheol. Acta*, 23, 378, 1984.
69. D. Tahara and K. Fukao, *Phys. Rev. E*, 82, 051801, 2010.
70. R. D. Priestley, C. J. Ellison, L. J. Broadbelt, and J. M. Torkelson, *Science*, 309, 456, 2005.
71. R. D. Priestley, *Soft Matter*, 5, 919, 2009.
72. D. Cangialosi, V. M. Boucher, A. Alegria, and J. Colmenero, *Soft Matter*, 9, 8619, 2013.
73. Y. P. Koh, G. B. McKenna, and S. L. Simon, *J. Polym. Sci., Part B: Polym. Phys.*, 44, 3518, 2006.
74. Y. P. Koh and S. L. Simon, *J. Polym. Sci., Part B: Polym. Phys.*, 46, 2741, 2008.
75. K. Fukao, Y. Oda, K. Nakamura, and D. Tahara, *Eur. Phys. J. Special Topics*, 189, 165, 2010.

76. K. Fukao, T. Terasawa, K. Nakamura, and D. Tahara, Heterogeneous and aging dynamics in single and stacked thin polymer films, in *Glass Transition, Dynamics and Heterogeneity of Polymer Thin Films*, vol. 252 of *Advances in Polymer Science*, edited by T. Kanaya (Springer, Berlin, Heidelberg), pp. 65–106, 2013.

77. K. Fukao, H. Takaki, and T. Hayashi, Heterogeneous dynamics of multilayered thin polymer films, in *Dynamics in Geometrical Confinement*, Advances in Dielectrics, edited by F. Kremer (Springer International Publishing, Heidelberg), 179–212, 2014.

78. C. Rotella, S. Napolitano, and M. Wübbenhorst, *Macromolecules*, 42, 1415, 2009.

79. N. McCrum, B. Read, and G. Williams, in *Anelastic and Dielectric Effects in Polymeric Solids* (Dover Publications, New York), Dover Books on Engineering, 414, 1967.

80. A. Karim, A. Mansour, G. P. Felcher, and T. P. Russell, *Phys. Rev. B*, 42, 6846, 1990.

81. M. Stamm, S. Hüttenbach, G. Reiter, and T. Springer, *Europhys. Lett.*, 14, 451, 1991.

82. S. J. Whitlow and R. P. Wool, *Macromolecules*, 24, 5926, 1991.

83. T. Hayashi, K. Segawa, K. Sadakane, K. Fukao, and N. L. Yamada, *J. Chem. Phys.*, submitted.

84. R. Böhmer and C. A. Angell, *Phys. Rev. B*, 45, 10091, 1992.

85. C. Angell, *J. Non-Cryst. Solids*, 73, 1, 1985.

86. C. J. Ellison and J. M. Torkelson, *Nat. Mater.*, 2, 695, 2003.

87. C. Rotella, S. Napolitano, L. De Cremer, G. Koeckelberghs, and M. Wübbenhorst, *Macromolecules*, 43, 8686, 2010.

88. F. Fujara, B. Geil, H. Sillescu, and G. Fleischer, *Z. Phys. B*, 88, 195, 1992.

89. S. Napolitano, D. Prevosto, M. Lucchesi, P. Pingue, M. D'Acunto, and P. Rolla, *Langmuir*, 23, 2103, 2007.

90. B. Vanroy, M. Wübbenhorst, and S. Napolitano, *ACS Macro Lett.*, 2, 168, 2013.

91. X. Zheng, M. H. Rafailovich, J. Sokolov, Y. Strzhemechny, S. A. Schwarz, B. B. Sauer, and M. Rubinstein, *Phys. Rev. Lett.*, 79, 241, 1997.

92. N. Taniguchi, K. Fukao, P. Sotta, and D. R. Long, *Phys. Rev. E*, 91, 052605, 2015.

Cooperative motion as an organizing principle for understanding relaxation in supported thin polymer films

PAUL Z. HANAKATA, BEATRIZ A. PAZMIÑO BETANCOURT,
JACK F. DOUGLAS, AND FRANCIS W. STARR

8.1 INTRODUCTION

Solidification by the formation of a glass below the glass transition temperature T_g is one of the most important material properties for the processing and applications of polymeric systems. In particular, T_g can be strongly altered in ultrathin polymer films, and there are many reviews emphasizing experimental aspects of the dynamics of thin polymer films [1–5]. While much progress has been made toward understanding dynamical changes in thin films, a theoretical framework to describe and predict the effects of confinement on glass formation remains a significant scientific challenge. A substantial body of research has established that glass-forming (GF) liquids in general are "dynamically heterogeneous" [6–8], exhibiting both spatial and temporal fluctuations in local mobility. The notion of heterogeneity, in the form of cooperative rearragement, underlies the Adam–Gibbs (AG) theory [9] and the conceptually related random first-order transition theory (RFOT)

[10,11], and these theories provide a useful framework for understanding the dynamics of GF liquids in general. However, both the AG and RFOT models involve many heuristic assumptions that require validation by either simulations or measurements. In this regard, simulations can play a leading role, since fluid properties such as the scale of cooperative rearrangements and the configurational entropy are not readily estimated from experiments. In particular, the actual form of cooperative motion in glass-forming liquids, the hypothetical "cooperatively rearranging regions" (CRR) of AG, or the "entropic droplets" of RFOT, has been an object of much speculation. The central goal of the present review is to develop and investigate a model for glass formation in which the CRR are quantified by string-like cooperative motions [8–23]. Specifically, we consider the ability of this "string model" to rationalize the diverse changes to glass formation in thin supported polymer films. We systematically explore the influence of film thickness, polymer–substrate interaction strength, and substrate stiffness, since all of these parameters can significantly alter thin film dynamics.

The primary outcome of the string model is that it condenses the seemingly intractable range of control parameters that affect the glass formation of polymer films into just the activation parameters for molecular rearrangement. In developing this description, we address a variety of questions about liquid dynamics, for example: "What is the nature of cooperative rearranging motions in glass-forming fluids? How do the activation barriers for molecular rearragement at high temperature relate to the barriers approaching T_g, where the motion is highly collective? How does confinement alter the scale of molecular rearrangements and the associated activation free energy? Are the scales of heterogenous motion and gradients in film relaxation related?" The answers to these questions lead to a theory of the dynamics of glass-forming liquids based on a physical description of cooperative dynamics that we test for a broad range of conditions by molecular dynamics simulations. Our description of relaxation in glass-forming liquids preserves the concepts of the AG model, while providing a molecular interpretation for their origin. We review results showing that the string-like cooperative rearrangements have a size that is directly linked to the growth of the activation free energy, giving molecular form to the CRR of AG. By considering the possible relation of these dynamical strings to existing descriptions of equilibrium polymerization, we realize important extensions to the AG approach, such as the propensity of CRR to saturate to a finite scale at low temperature, preventing the divergence of the relaxation or viscosity at low temperature. We also find that the neglect of the activation entropy in the AG model is unwarranted, and without this term, the correct dynamics cannot be recovered. Changes in the activation enthalpy and entropy of the standard Arrhenius parameters governing the high-temperature relaxation dynamics are found to play an essential role in the dynamics of thin polymer films. Thus, the string model of relaxation offers a significant extension of the original heuristic AG theory. This string model provides a basis for recent analytic work by Freed [24], who developed an extension of transition state theory to account for multiparticle barrier crossing events, thereby establishing a firmer theoretical foundation for this model of the dynamics of glass formation.

While the string model describes the fluid or viscous relaxation through a relation to cooperative displacements of highly mobile particles, it is important to recognize that regions of *low-mobility* particles also play an important role in the dynamics of glass-forming materials [23,25–37]. The immobile particles are crucial for understanding aspects of elastic response [26,27], the stretching exponent β of relaxation functions (like the density–density correlation function), and the decoupling phenomenon, in which the structural relaxation time scales differently with temperature than the diffusive relaxation time. The effect of the immobile particles is exhibited in the dynamics of superheated crystalline materials [38], where string-like collective motion and non-Arrhenius relaxation are observed, but where there are no finite immobile clusters, no decoupling phenomenon, and β = 1. There is clearly more than one type of dynamic heterogeneity arising in glass-forming liquids, the mobile and immobile particle clusters each contribute significantly to the overall viscoelastic relaxation process. Evidently, these mobility subsets play complementary roles in understanding the complete dynamics of glass-forming liquids. This chapter focuses on viscous relaxation, so we emphasize the mobile particles that facilitate the viscous relaxation response.

In the following sections, we first review the basic phenomenology of incipient glass formation in simulations of thin polymer films, including how these changes are manifest in terms of the local variations in structure and dynamics. Based on these data, we examine the empirical validity of the AG-inspired relation, taking the CRR to be defined by the scale of string-like cooperative motion. We then consider the theoretical description of these strings, which enables us to make predictions for relaxation behavior at lower temperatures than can be studied via equilibrium molecular simulations. Finally, given the challenges to experimentally probing these cooperative motions, we consider their possible relation to the scale of interfacial changes of mobility in thin films. While our results are developed in the context of polymer-based materials, the approaches we use should be general to small-molecule glass formers as well.

8.2 INFLUENCE OF FILM THICKNESS AND SUBSTRATE PROPERTIES ON GLASS FORMATION

8.2.1 SUMMARY OF T_g AND FRAGILITY CHANGES

Although the formation of a glass involves changes in a wide array of thermodynamic, dynamic, and structural features, the two most frequently tabulated features are the glass transition temperature T_g and fragility m. Accordingly, we first report the effects of varying film thickness as well as the substrate properties on these key metrics. Figure 8.1a shows that, relative to the bulk, T_g of polymer films on a rough surface increases with decreasing film thickness, while for the smooth surfaces, T_g decreases with decreasing film thickness. This is consistent with many previous studies [39–44]. For the rough substrate, we also study a variety of substrate interaction strengths and substrate rigidities. This allows us to tune the T_g changes, as decreasing substrate

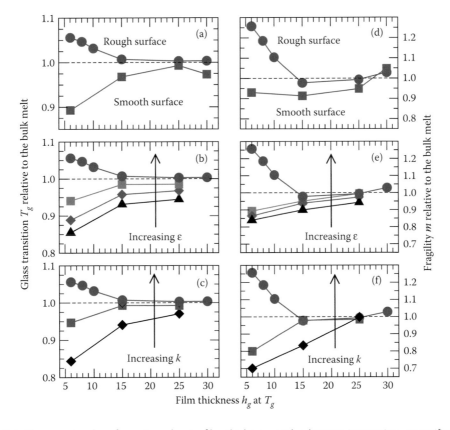

Figure 8.1 Changes to glass formation due to film thickness and substrate interaction, quantified by the changes to the glass transition temperature T_g and fragility m. (a) T_g relative to the bulk as a function of film thickness. T_g of films supported on the rough surface increases, while the opposite occurs for a smooth substrate. (b) For the rough substrate, T_g increases with increasing substrate attraction and (c) increasing surface rigidity. The interaction strengths in (b) are (from bottom to top) $c_{mw} = 0.477, 0.6, 0.8,$ and 1.0. The surface rigidity parameters in (c) are (from bottom to top) $k = 10, 50,$ and 100. The top T_g curve in (a), (b), and (c) is for the same parameters $\epsilon = 1.0$ and $k = 100$. (d)–(f) show the relative fragility m for the same systems shown in (a)–(c). (Figure adapted from Hanakata, P. Z., Douglas, J. F., and Starr, F. W., *Nat. Commun.*, 5, 256207, 2014.)

interaction or rigidity decreases T_g (Figure 8.1b and c). Along with T_g changes, we expect changes to the fragility, m, which quantifies the T dependence of relaxation near T_g. Formally, m is defined as the logarithmic slope of τ near T_g:

$$m(T_g) = \frac{\partial \ln \tau}{\partial (T/T_g)}\bigg|_{T_g}. \tag{8.1}$$

We evaluate m from the same fit used to estimate T_g. Experimentally, m is often found to vary in proportion to T_g [45]. Figure 8.1d–f shows a correlation between T_g and m for all films, but this relation is not strictly proportional, as expected from recent studies [40,46]. Note that films supported on a smooth substrate may have a nonmonotonic thickness dependence of T_g and m on thickness. Specifically, we showed that T_g or m decreases with decreasing film thickness on the smooth substrate up to some critical thickness, but that T_g increases for very thin films when interfacial effects become dominant [40].

8.2.2 GIBBS–THOMSON-INSPIRED MODEL FOR THE THICKNESS DEPENDENCE OF T_g

We base our understanding of the general trend in T_g as a function of film thickness on similar observations that have already been established to describe the shift of the melting point T_m of crystalline materials of finite dimensions [47–50]. The melting temperature T_m and T_g in materials that crystallize normally exhibit a constant proportionality relationship, $T_g \approx (2/3)T_m$ in bulk materials [51,52]. Thus, it is natural to think that T_g will exhibit a comparable behavior, even though the character of the ordering process is different in these two types of transitions; clearly, these phenomena share common driving forces and interactions. In the case of the melting point, the Gibbs–Thomson relation is defined as,

$$\frac{T_m(h)}{T_m(h \to \infty)} = 1 - C/h, \tag{8.2}$$

where T_m is the melting temperature of the material, h is the lamellar thickness of the material and it corresponds to film thickness or nanoparticle diameter in "confined" inorganic crystals, and the constant C depends on the interfacial energies involved [49]. Equation 8.2 has been highly successful in describing melting point in polymers, and we refer the reader the following literature for a more detailed discussion of models predicting the variation of T_m [53–59]. The important point for the present discussion is that the sign of C depends on the interfacial properties (of both interfaces) so that melting points shift are observed to be up or down [57,60] depending on the substrate interactions. The usual situation is that interactions are such that T_m shifts downward with confinement and these shifts can be very large as in the case of the observed T_g shifts in polymer films. Jackson and McKenna invoked this analogy between melting point depression due to finite sizes to describe their pioneering paper on T_g depression of small-molecule glass-forming liquids in controlled pore glasses [47]. Similarly, this point of view has been advocated by other authors [61–63]. In particular, the change in T_g can be expressed as,

$$\frac{T_g(h)}{T_g(h \to \infty)} = 1 - a(l_p/h), \tag{8.3}$$

where l_p is the polymer persistence length, $l_p = \sigma = 1.0$ (in our model), and a is a constant on the order of unity if the polymer–substrate interaction is not highly attractive, so that atomic motions at this boundary are not greatly inhibited. In the case that the polymer–polymer attractive interaction is very strong, as for poly(methyl methacrylate) (PMMA) on a silica wafer or when $\varepsilon \geq 0.88$ in our polymer model film, we may expect that interfacial motions are suppressed, causing an increase in T_g. This is the same mechanism that has been demonstrated to cause T_m increase in crystals under confinement. In the highly attractive surface interaction case, we expect the magnitude of $a \sim O$ (1), but its sign is to be reversed as already observed when $\varepsilon = 1.0$, where $a = -0.25$ (see inset in Figure 8.2b). The origin of the $1/h$ scaling on both T_g and T_m shifts is the high interfacial surface area of the film, since the $1/h$ scaling derives from the surface–volume ratio of the material. Exactly the same behavior shown in Figure 8.2 has been reported in both experimental studies of T_g shifts in thin polymer films [64–67] and simulations [63]. Figure 8.2b shows that this simple scaling argument provides a good qualitative explanation of our own T_g shifts data when ε is varied, and that the parameter a can be well approximated to be linear with ε, suggesting a crossover (at $a = 0$) from decreased to increased T_g at $\varepsilon \approx 0.88$ for fixed $k = 100$. Likewise, varying the substrate stiffness k leads to a linear dependence of $a = -0.015\,k + 1.03$ at fixed $\varepsilon = 1$, indicating a crossover in T_g changes near $k \approx 69$.

Notice that while the crossover scale in Equation 8.3 is set by the statistical segment length, l_p. Figure 8.2a shows that there is a scale at which T_g shifts become noticeable, that is, $\Delta T_g \geq 0.02$. In our model, this onset scale is $\xi_o = 50l_p$ approximately, which for polymers that have a moderate mass and that have a typical

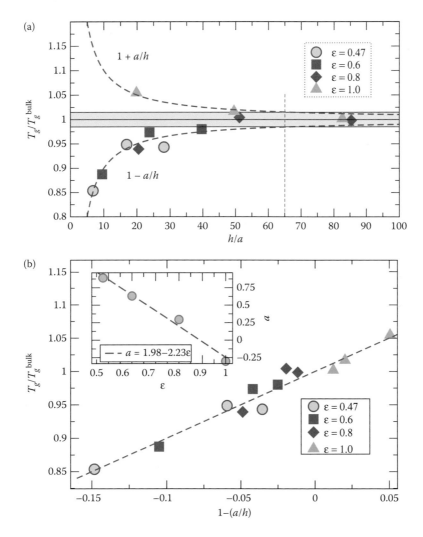

Figure 8.2 (a) Normalized T_g shift for thin films when ε is varied and $k = 100$. The symbols represent values obtained by simulations, the dashed lines illustrate the normalized behavior for positive and negative shifts. The highlighted region indicates the length scale at which the changes in T_g are negligible. (b) Reduction of simulated data using the Thomson relation. The inset shows the variation of a as a function of ε. Note that a follows a linear relation with ε, with $a = 1.98 - 2.23\varepsilon$. (Figure reproduced with permission from Hanakata, P. Z. et al., *J. Chem. Phys.*, 142, 234907, 2015.)

persistence length on the order of 1–2 nm gives $\xi_o \approx 100$ nm, generally consistent with experimental findings [66–70]. By making the polymer have bulky side groups, it should be possible to increase this onset scale. Torkelson and coworkers have shown that by making the chain side groups bulky, it is indeed possible to push the onset scale for T_g shifts up to 400 nm, consistent with the current discussion [68]. A vanishing of finite-size shifts of transition temperatures at some "special" value at the surface interaction is commonly reported in thermodynamic transitions, and below we find this effect when we vary the strength of the polymer–substrate interaction. A "critical condition" describing the changeover from being predominantly an attractive to a repulsive interaction is commonly encountered in polymer chromatography [71], where this condition is crucial to the chromatographic separation process. The observation of the θ point [72] in the solution properties of polymers has a similar compensation condition, so it should not be surprising that this compensation effect arises in the context of glass formation in thin films. Finite-size scaling of the thermodynamic properties was also found in early observations by Keddie and coworkers at a similar thickness range where changes in T_g occur [64].

8.2.3 LOCAL STRUCTURE AND DYNAMICS OF SUPPORTED FILMS

It is widely appreciated that the observed changes to the overall T_g and fragility are due in large part to changes in the interfacial properties, both near the substrate and the free surface, qualitatively illustrated in Figure 8.3. In simulation, we can readily resolve both structure and dynamics locally. We first contrast the local dynamics and monomer density as function of distance z from the substrate boundary of rough or smooth substrates with monomer–substrate interaction strength $\varepsilon = 1$. We evaluate both $\rho(z)$ and $\tau_s(z)$ with a bin size $\delta z = 0.875$.

In Figure 8.4a, we observe that the monomer density near either the smooth or rough substrate increases weakly, and has a steady value through most of the film. The density drops to zero over a narrow window at the free boundary region. At the center of the film, the density has a value close to the bulk. The density profile of the film on a smooth substrate is essentially identical to that of a film on a rough substrate.

In addition, we contrast the local structure parallel to the substrate by evaluating the density pair correlation function $g(r_\parallel)$ (see Figure 8.4b and c). Far from the substrate, $g(r_\parallel)$ of both systems is indistinguishable, as the monomers are completely unperturbed by the substrate. Near the substrate, we see that there is a slight

Figure 8.3 Visualization of the polymer film relaxation gradient. The tubes represent the segments between bonded monomers of the simulated polymer film of thickness $h_g = 15$, supported on a rough substrate. The color gradient describes the local relaxation time as a function of distance from the substrate: the slowest relaxing chains are colored blue, while the most rapidly relaxing chains are shown in red. This illustration makes visually apparent the existence of a substrate, central, and free-surface region of the film. (Figure reproduced with permission from Hanakata, P. Z., Douglas, J. F., and Starr, F. W., *Nat. Commun.*, 5, 256207, 2014.)

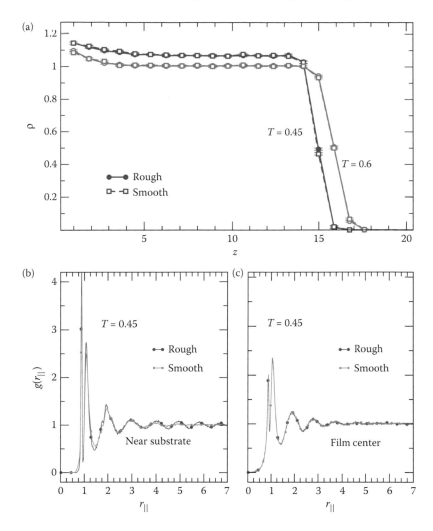

Figure 8.4 (a) Monomer density profile $\rho(z)$ of a film, $h_g = 15$, supported on a rough or smooth substrate. The values of ρ represent the average and the error bars represent the uncertainty calculations obtained form the standard deviation of 100 configurations. (b) Pair correlation function in the direction parallel to the substrate $g(r_\parallel)$ near the substrate. (c) $g(r_\parallel)$ at the film center. Monomers near a rough substrate are slightly more densely packed and have better local ordering in comparison to those near the smooth substrate. (Figure reproduced with permission from Hanakata, P. Z. et al., *J. Chem. Phys.*, 142, 234907, 2015.)

difference in the local structure. In particular, near the substrate, $g(r_\parallel)$ of the rough substrate has a somewhat larger first peak, indicating that the monomers near the rough substrate are more ordered than those near the smooth substrate. In addition, there is a weak long-range ordering of monomers for the rough substrate, potentially induced by the periodicity of the substrate atoms.

We next examine to what degree the local film dynamics reflect the changes in the density described above. To do so, we evaluate the self-part of the dynamical density correlation function [73], and condition it by the distance z from the substrate in intervals $\delta z = 0.875$. From this, we obtain the distance-dependent self-relaxation time $\tau_s(z)$. To provide context for the time scale probed by these calculations, consider that for a simple polymer (like polystyrene), the reduced time units that we use can be approximately mapped to 1 ps. Figure 8.5a shows that the dynamics of the film on a rough or smooth substrate at the same T are nearly identical over the range from the center of the film to the free boundary region. However, there are large differences of relaxation time near the substrate. The local relaxation time τ_s increases close to the rough substrate, but decreases near the smooth substrate. The enhanced dynamics near the smooth substrate are in part a consequence of the fact that monomers can "slide" along the substrate due to the substrate smoothness (see References 40 and 44). This effect disappears for a rough substrate. An increasing relaxation time approaching the rough substrate has also been observed in a computational study of a binary Lennard-Jones (LJ) liquids, as well as in a bead–spring model of polymer melts with a relatively strong interaction [74–77]. Evidently, substrate roughness is highly relevant for the polymer film dynamics, and this factor must be controlled for consistent results.

Frequently, it is presumed that the interior of the polymer film can be considered to have bulk-like behavior. For sufficiently thick films, this is probably a valid assumption, but it must break for thin-enough films. To illustrate this concern, Figure 8.5b shows the position-dependent relaxation $\tau_s(z)$, normalized by the bulk value for many temperatures for one film thickness. Significantly, these data show that for a given thickness, whether or not the interior appears bulk-like is also temperature dependent. In other words, as the film cools toward T_g, the scale of interfacial effects is growing, and as a consequence, whether or not the interior is bulk-like depends on both thickness and temperature. Additionally, the temperature dependence of the suface and interior relaxation can differ substantially, such that the interior relaxation can be as much as 10^7 slower than surface relaxation near T_g [78–82]. Later, we discuss the relationship of this interfacial scale to that of the scale of cooperative rearrangements.

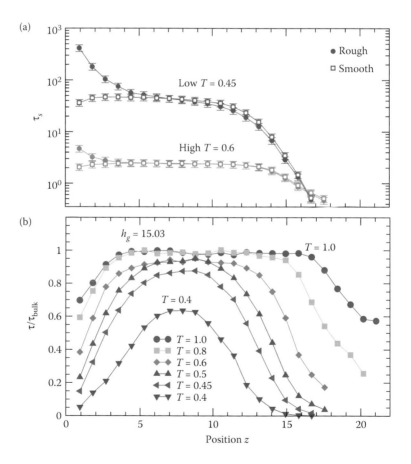

Figure 8.5 (a) Relaxation time τ_s as a function of distance z from the substrate Although the averaged densities of two systems are identical, the local dynamics are clearly distinct from one another, particularly near the substrate. (b) Position-dependent relaxation time $\tau_s(z)$ relative to the bulk polymer for films on a smooth substrate, demonstrating that $\tau_s(z)$ is altered from the bulk, even near the center of the film where the density is the same as the bulk. The deviation of τ_s from the bulk becomes more significant at low temperature. (Figure adapted from Hanakata, P. Z., Douglas, J. F., and Starr, F. W., J. Chem. Phys., 137, 244901, 2012; Hanakata, P. Z. et al., J. Chem. Phys., 142, 234907, 2015.)

A convenient way to parameterize local dynamical changes is by considering the local dependence of T_g and m as a function of distance z from the substrate. This provides a way of summarizing the behavior of $\tau_a(z, T)$, shown in Figure 8.5. Figure 8.6a shows that T_g increases near the rough substrate, reflecting the observed increase of τ_s near the attractive substrate. Near the free boundary region, T_g decreases due to the enhanced mobility of monomers at the free boundary region. For relatively thick films, we find that there is a substantial film region where monomers have a T_g close to the bulk value. This is a situation in which the film thickness is large compared to the perturbing scales of the interfaces [40]. T_g is often found to be proportional to m, as observed in the overall dynamics. However, we do not see this proportionality between the local T_g and m. Specifically, m *decreases* approaching the rough substrate while T_g increases. This opposing trend has also been observed in polymer–nanoparticle composites [83].

Figure 8.6c and d contrasts the local variation of T_g and m for rough and smooth substrates of a relatively thick film, $h_g = 15$. In contrast to the increasing T_g of polymer films near the rough substrate, T_g of smooth substrate decreases close to the smooth substrate, which is consistent with variation of τ_s (Figure 8.5). Note

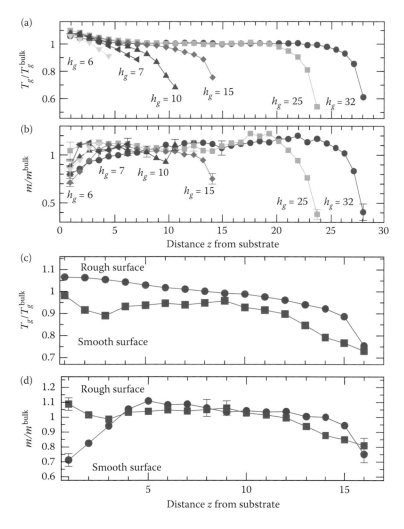

Figure 8.6 Local T_g and fragility. (a) Relative T_g and (b) fragility m of polymer films supported on a rough substrate as a function of distance z from the substrate for many thicknesses. (c) Relative T_g and (d) m of a polymer film ($h_g = 15$) supported on a rough or a smooth substrate. Near the free substrate, T_g decreases for both systems. Near the substrate, T_g increases for film supported on a rough substrate, but decreases for a smooth substrate. Notice that the uncertainties for T_g are smaller than the symbols. (Figure reproduced with permission from Hanakata, P. Z. et al., *J. Chem. Phys.*, 142, 234907, 2015.)

that T_g and m are slightly depressed for films on the smooth substrate, even at the middle of the film, a scenario where the perturbing scales of both interfaces become comparable to film thickness.

Substrate roughness is relevant to the film dynamics, but there are other physically relevant variables. We next investigate the dependence of dynamics on the interaction strength as well as rigidity of the rough substrate. First, we examine the role of substrate interaction strength. Figure 8.7a and b shows how relaxation time τ for two representative film thicknesses changes as we vary the interaction ε between the rough substrate and the polymers. The overall changes in dynamics result from the competing effects of the substrate and free interface, so that τ can be higher or lower relative to the bulk. As we have established, the free boundary region decreases τ while a substrate with a relatively strong interaction increases τ. Thus, for a given thickness, τ decreases with decreasing substrate interaction strength.

We find a similar effect by varying the stiffness k of the bonds describing the substrate stiffness. Specifically, increasing the flexibility of the substrate (decreasing k) results in a smaller τ (Figure 8.7c and d). Evidently, monomers of the chains near the substrate are less constrained, since the substrate atoms are more flexible. The complete local analysis of the dependence of dynamics on flexibility of the substrate will be discussed in

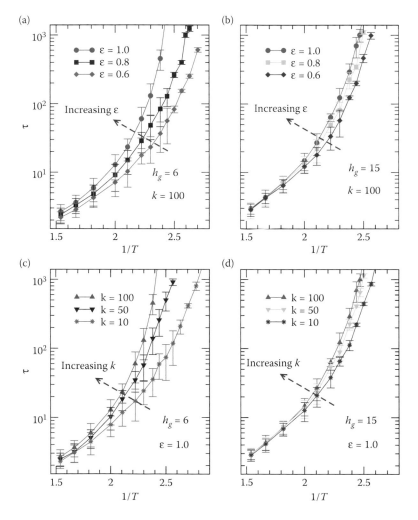

Figure 8.7 The T dependence of the relaxation time τ for two film thicknesses, (a) $h_g = 6$ and (b) $h_g = 15$, with various interfacial strength ε, and fixed rigidity $k = 100$. (c) and (d) show the effect of variable rigidity k for fixed $\varepsilon = 1$. In general, τ is decreased as we decrease the substrate strength or the molecular substrate stiffness. (Figure reproduced with permission from Hanakata, P. Z. et al., *J. Chem. Phys.*, 142, 234907, 2015.)

the next section. By comparing Figure 8.7a and b, as well as Figure 8.7c and d, we can see that the substrate interaction or the flexibility of the substrate has greater influences on the thinner film, expected since the thinner film has a larger surface-to-volume ratio.

We next evaluate the resulting dependence of T_g and fragility on the substrate interaction strength and rigidity of the rough substrate films. Figure 8.8a and b shows how T_g of three representative thicknesses change as a function of substrate interaction strength ε at fixed rigidity $k = 100$. Generally, increasing the polymer–substrate interaction increases both T_g and fragility as monomer dynamics near the substrate presumably become progressively slower. This general trend of a decreasing T_g with decreasing substrate interaction has also been observed in both experiments and computational works [39,41,42,84]. This depression of fragility is also consistent with the findings in a free-standing film [19], which formally corresponds to taking the limit ε → 0. Evidently, the dependence of substrate–polymer interaction of T_g or m becomes more significant for thinner films, as indicated by a steeper variation of T_g or m with ε.

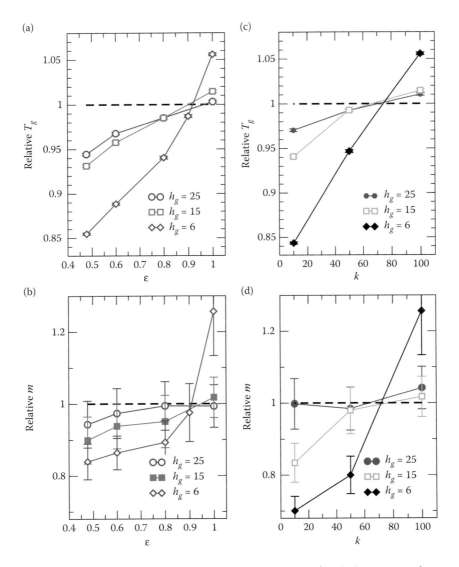

Figure 8.8 Dependence of the relative T_g and m of three representative film thicknesses as a function of substrate strength ε (panels a and b) or substrate rigidity k (panels c and d). For thinner films, the range of T_g and m is wide due to the larger substrate-to-volume ratio. (Figure reproduced with permission from Hanakata, P. Z. et al., *J. Chem. Phys.*, 142, 234907, 2015.)

We found similar trends for T_g and m by varying the substrate rigidity. That is, increasing substrate rigidity k at fixed substrate interaction strength ($\varepsilon = 1$) increases both T_g and m. It is interesting to note that there appears to be a nearly fixed point for T_g and m as a function of ε. Specifically, T_g and m are independent of film thickness for $\varepsilon \approx 0.9$ ($k = 100$) or $k \approx 70$ ($\varepsilon = 1$). These values are consistent with the crossover values of ε and k estimated using the Gibbs–Thomson-inspired model (Figure 8.2). We emphasize that the lack of T_g and m changes at the crossover point does not mean there are no changes in local dynamics, but rather that there is a balance between the dynamic enhancement at the free boundary region and the slowing down of the dynamics near the substrate. In fact, the increasing behavior of m with decreasing thickness is only observed for values $k > 70$ and $\varepsilon > 0.9$. This compensation effect is reminiscent of the self-excluded volume interactions of polymers in solution near their θ point [85,86], and the compensation point for isolated polymers interacting with surfaces [87].

Both results potentially offer us insights into how T_g changes in multilayer films, which are "stacks" of polymer films with different species characterized by different flexibility, interpolymer interaction, or molecular weight. Multilayer film experiments by Torkelson and coworkers have shown that a given layer of the multilayer film may have different T_g depending on the properties of neighboring layers [88]. Here, we emphasize that changes in dynamics do not necessarily arise from the substrate interaction strength alone; changes in the rigidity of the interface (for example, polymer films placed on a polymer substrate with the same substrate interaction strength, but having different molecular flexibility) and substrate roughness are also relevant.

8.3 STRING-LIKE COLLECTIVE REARRANGEMENTS AS AN ORGANIZING PRINCIPLE FOR DYNAMICS

From the discussion of the preceding sections, it is clear that there are a variety of factors that can alter the dynamics of thin polymer films, including film thickness, roughness, polymer–substrate interaction, and stiffness of the substrate. These effects are all significant and the observed changes of the film dynamics involve the convolution of all these variables. Accordingly, we must address the central question: Is there a unifying description to explain the broad array of effects on the polymer dynamics? There has been much speculation that these changes revolve around changes in the collective dynamics of the polymer molecules, where the Adam–Gibbs theory is often discussed without a specific definition of the hypothetical "cooperatively rearranging regions" that are relevant to understanding these property changes. Simulations have identified cooperative rearrangements that are quantitatively linked to the structural relaxation time for bulk polymer materials [23], and a similar connection has also been established in model polymer nanocomposites [83,89]. These string-like motions therefore offer a molecular realization of the abstract CRR. In the following, we examine the changes of cooperative string-like motions in thin films and demonstrate an impressive reduction of all our simulation data for structural relaxation in thin polymer films, as well as polymer–nanoparticle composites, based on this unifying framework.

8.3.1 STRING-LIKE COLLECTIVE MOTION IN THIN FILMS

On cooling, molecular motion becomes increasingly cooperative on time scales between the collision time and the relaxation time of the intermediate scattering function, and associated with this phenomenon there are particles of high and low mobility that form clusters. Within clusters of relatively high mobility, both simulations and experiments [8,12–23] have revealed the existence of cooperative molecular motion involving particle exchange motion in a string-like form.

To identify string-like displacements of highly mobile monomers, previous works have provided a tested procedure [12,13,23]. First, one identifies subset of monomers with relatively enhanced mobility. Since the distribution of particle mobilities varies continuously, the first challenge is how to distinguish these monomers. For a variety of systems [25,90–92], it has been shown that choosing the subset of particles that have moved farther than is expected from the Gaussian approximation at the characteristic time t^* offers a useful metric to identify the highly mobile particles. Depending on the system, these mobile particles typically account for

5–7% of the particles. We follow the choice of Reference 92 where the same model was examined, and select mobile particles as the 6.5% of particles with the greatest displacement over any chosen interval t. This allows us to see the evolution of mobile particle properties over all t, in addition to the characteristic time t^*.

Among these mobile monomers, strings can be distinguished by a simple condition that, when one of the two mobile monomers has taken the place of the other, they are then part of a string, namely

$$\min[|\vec{r}_i(t^*) - \vec{r}_j(0)|, |\vec{r}_i(0) - \vec{r}_j(t^*)|] \leq \delta, \tag{8.4}$$

where $\delta = 0.55^{13}$ is smaller than the monomer core radius.

We show a representative example of the time t and temperature T dependence of the average string size $L(t)$ in Figure 8.9. For both very short (ballistic) and very long (diffusive) time intervals, monomer motion is characterized by uncorrelated Gaussian displacements. Consequently, $L(t)$ exhibits a characteristic maximum size L at intermediate times when displacements are most strongly correlated. The characteristic peak size and time both increase as temperature decreases [12]. The growing string size (inset of Figure 8.9) directly embodies the increasing cooperativity of molecular motion upon cooling that is at the foundation of the heuristic Adam and Gibbs description [9]. The relationship of the strings to the temperature-dependent activation energy is discussed in the next section.

Notably, this cooperative motion is largely insensitive to chain connectivity, and this type of collective motion should not be confused with reptation where the chains are thought to move preferentially along their backbone coordinates [93].

Before we examine the quantitative relation between the size of string-like cooperative motion and relaxation time, we first catalog the behavior of the characteristic peak string size $L(T)$ for the range of film thicknesses and substrate interactions we have discussed. Figure 8.10a and b compares the $L(T)$ for two film thicknesses with a rough or smooth substrate at two representative thicknesses. For a given film thickness, $L(T)$ is larger and grows faster for a film with rough substrate, which is qualitatively consistent with $\tau(T)$. The variation of $L(T)$ with substrate interaction strength ε or substrate rigidity k, shown in Figure 8.10c and d, and $L(T)$ in these cases also qualitatively capture the variation of $\tau(T)$ (Figure 8.7b and d). The similarities

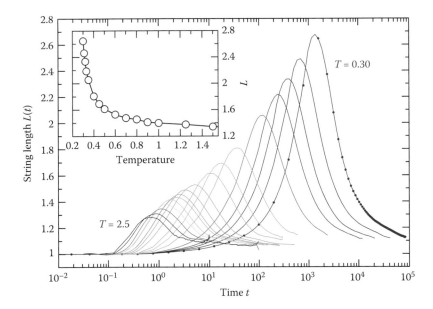

Figure 8.9 The dynamical string length $L(T)$ for all T studied. The inset shows the T dependence of the characteristic peak value, simply denoted as L. The color gradient goes from yellow at highest T to blue at lowest T. (Figure reproduced with permission from Starr, F. W., Douglas, J. F., and Sastry, S., *J. Chem. Phys.*, 138, 12A541 (page 18), 2013.)

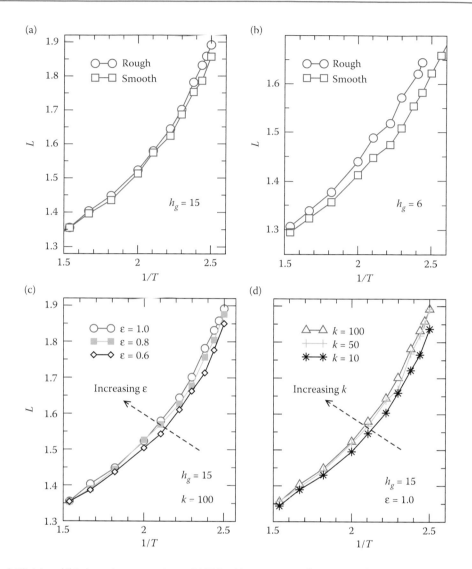

Figure 8.10 (a) and (b) show the comparison of $L(T)$ for films supported on a smooth or rough substrate. (c) and (d) show the variation of $L(T)$ with varying polymer–substrate interaction ε or substrate rigidity k. The variation of $L(T)$ with h_g, ε, and k mimics that of $\tau(T)$ shown in Figure 8.7 This qualitative consistency suggests the applicability of the string model of relaxation [97] to quantify $\tau(T)$. (Figure adapted from Hanakata, P. Z. et al., *J. Chem. Phys.*, 142, 234907, 2015.)

between the variation of $L(T)$ and $\tau(T)$ suggest that we may be able to predict changes in fragility from the variation of $L(T)$, as found in previous works [19,83,89,94].

It is tempting to associate the changes in relaxation time τ directly to the changes in L. However, it is apparent that the changes in L are rather minute, and we shall soon see that equally important, if not more so, are the changes to the activation parameters (with varying thickness and substrate interactions) that relate L to τ.

8.3.2 ADAM–GIBBS-INSPIRED STRING MODEL OF RELAXATION

It has long been argued that polymer relaxation is governed by the scale of cooperative motion. From a theoretical perspective, the classical arguments of Adam–Gibbs [9], and our extension of this model based on numerical simulation evidence and thermodynamic modeling [95], provides a theoretical perspective for testing this proposition. Specifically, according to AG theory, the activation Gibbs free energy $\Delta G_a(T)$ is extensive of the size z^* of "cooperatively rearranging regions", so that τ can be formally written in terms of the general transition state theory relation. Classical transition state theory [96–98] implies that the structural relaxation time can be described by an Arrhenius T dependence:

$$\tau(T) = \tau_0 \exp[\Delta G_a(T)/k_B T], \Delta G_a(T) \equiv z^* \Delta\mu, \tag{8.5}$$

where $\Delta\mu$ is the activation free energy at high temperatures when particle motion does not involve a significant cooperative motion, so that z^* equals a constant. AG originally assumed that $z^* \approx 1$ at high temperatures, corresponding to completely uncooperative motion, but a constant value of z^* at high T is all that is required to recover Arrhenius dynamics. Recent simulations have shown that, despite the rather heuristic nature of the original arguments of AG, Equation 8.5 with z^* identified specifically with the average size L of the cooperative string-like particle exchange motion provides a good description for $\tau(T)$ in polymer melt simulations, even in the case when nanoparticles have been added to tune the fragility over a wide range [23,40,83,94]. In their original work, AG identified z^* with the minimal scale of cooperative movement while we identify z with the average over a highly polydisperse distribution of collective movements observed in our simulations. This is an important conceptual difference between the string model and the AG model, but both models lead to Equation 8.5 as their final result, so that the string model certainly preserves the spirit of the AG relation. Recently, we have established a quantitative correspondence between the $L(T)$ and a living polymerization theory [95], and inspired by these results, Freed [24] has systematically derived Equation 8.5 from transition state theory assuming that the transition states involve many-body transition events in the form of equilibrium polymers with z as the average string length.

To provide a quantitative test of the relation between cooperative motion and relaxation, we follow References 23,83,89,99, and identify CRR size z^* with the relative size L/L_A of string-like cooperative particle arrangements. $L_A \equiv L(T_A)$ is the value of the string size at the temperature T_A, above which an Arrhenius law for $\tau(T)$ holds. To determine T_A, we find consistent estimates of two independent procedures. One method uses $\tau(T)$ to determine the temperature at which $\tau(T)$ departs from its Arrhenius behavior, $\tau(T) = \tau_0 \exp[\Delta E/k_B T]$. The other approach is to identify the temperature below which particle caging first emerges, where caging is found by the logarithmic derivative of the mean square displacements $\partial \ln\langle r^2(t)\rangle/\partial \ln t$. This derivative develops a minimum, which is typically near 1 ps in molecular fluids. The temperature at which this minimum disappears is T_A [23,95,100].

We have proposed a formal extension to the Adam–Gibbs description based on the string model for the dynamics of glass-forming liquids [95,100], in which τ is described by the AG-inspired relation

$$\tau(h,T) = \tau_0(h) \exp\left[\frac{L(T)}{L_A(h)} \frac{\Delta\mu(h,T)}{k_B T}\right], \tag{8.6}$$

where $\Delta\mu(h, T)$ is the high-temperature activation free energy for $T > T_A$:

$$\Delta\mu(h,T) = \Delta H_a(h) - T\Delta S_a(h), \tag{8.7}$$

where $\Delta H_a(h)$ and $\Delta S_a(h)$ are the enthalpic and the entropic contributions of the high T activation free energy, respectively. These basic energetic parameters vary with film thickness h_g and type of interaction (ε and k). Note that Equation 8.6 for $T \geq T_A$ becomes $\tau(h, T) = \tau_0 \exp[(\Delta\mu(h, T)/k_B T]$, the typical activation form of transition state theory. In fact, Equation 8.6 at T_A implies that $\tau_0(h)$ is not a free parameter, but instead is determined by the relation

$$\tau_0(h) = \tau_A(h) \exp[-\Delta\mu(h,T_A(h))/k_B T_A(h)], \tag{8.8}$$

where $\tau_A \equiv \tau(T_A)$ so that ΔH_a and ΔS_a are the only undetermined parameters in Equation 8.6. This relation was noted and tested in Starr et al. [102]. The string model prediction for the structural relaxation time of a film of thickness h can then be formally written:

$$\tau(h,T) = \tau_A(h)\exp\left[\frac{L(T)}{L_A(h)}\frac{\Delta\mu(h,T)}{k_BT} - \frac{\Delta\mu(h,T_A)}{k_BT_A}\right], \qquad (8.9)$$

where $\Delta H_a(h)$ and $\Delta S_a(h)$ are the only parameters on which τ depends, just as in ordinary transition state theory for homogeneous fluids.

We now demonstrate the applicability of Equation 8.6 in quantitatively describing the dynamics of all the films we have thus far studied. Figure 8.11 shows the linear relationship between $\ln(\tau)$ and $\Delta\mu L/k_BT$ for different films thickness in Figure 8.11a, and for different substrate rigidity and strength of interactions in Figure 8.11b. The universal collapse of τ in terms of string size was noted recently in a brief communication [94], but the variation of the relaxation time prefactor $\tau_0(h)$ was considered as a free parameter in that work; τ_0 is *not* a free parameter in the string model of glass formation. This characteristic time is entirely determined

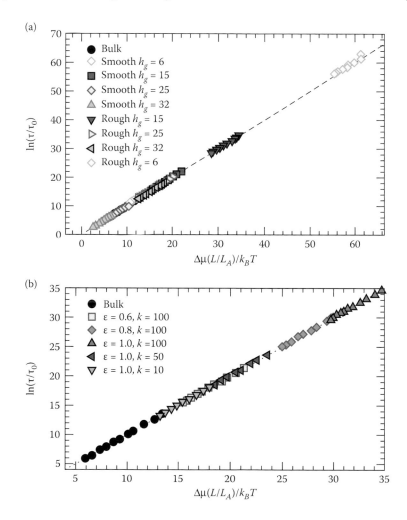

Figure 8.11 Structural relaxation time τ in terms of the average strings size L for (a) various thicknesses, and (b) various polymer–substrate interactions or substrate rigidities. τ is scaled by $\tau_0 = \tau_A\exp[-\Delta\mu(T_A)/k_BT_A]$ where $\Delta\mu(T_A) = \Delta H_a - T_A\Delta S_a$, and ΔH_a and ΔS_a are determined by fitting to Equation 8.6 over a broad T range. (Figure reproduced with permission from Hanakata, P. Z. et al., *J. Chem. Phys.*, 142, 234907, 2015.)

by τ_A, ΔH_a, and ΔS_a (Equation 8.8). As we discussed in Hanakata et al. [101], τ_0 varies strongly with confinement, while τ_A varies only weakly. The changes τ_0 are then due almost entirely to changes of ΔH_a, and ΔS_a. Evidently, the significant changes in the relaxation time of glass-forming films and nanocomposites derive in large part from the high-temperature activation parameters, which are typically considered in relation to glass formation.

The data reduction of τ in terms of string size holds for all film thicknesses supported on a rough or smooth substrates, and applies as well as to the bulk polymer material. The same reduced variable description describes a representative film supported on a rough substrate for various substrate interaction or supporting substrate rigidity (see Figure 8.11b). Although the method of data reduction is identical between these figures, we separate them for clarity. This remarkable data reduction shows that we can quantitatively describe the film dynamics of all these films based on the string model relation (Equation 8.6), despite a wide range of dynamical changes due to film thickness, polymer–substrate interaction, or substrate rigidity. For instance note that, Figure 8.10c and d shows that the average extent of cooperative motion L does not significantly change with ε or k, but that the structural relaxation time does change considerably. Therefore, the changes in τ must result from the variations of ΔH_a and ΔS_a, which we discuss next.

The successful reduction of the relaxation data using string size is not unique to polymer films. In particular, it has been argued by several authors [102–104] that there should be a correspondence between the changes in the glass formation of polymer thin films and polymer–nanoparticle composites. Indeed, we can likewise describe polymer nanocomposite relaxation time data to the same precision based on the same model as the present work, over a range of particle concentrations and for an attractive and repulsive polymer–particle interaction [100].

8.3.3 RELATIONSHIP OF THE ENTHALPY AND ENTROPY OF ACTIVATION

Before we consider how ΔH_a and ΔS_a depend on film properties, we consider the origin and relationship of these parameters in the high-temperature limit, where cooperative motion does not complicate our discussion.

The basic physical picture is that the free energy of activation is related to the free energy cost of removing a molecule from its local molecular environment [98,105,106]. This perspective implies that ΔH_a should scale in approximate proportion to the heat of vaporization H_{vap} or the cohesive interaction energy of the fluid. Although this argument is simple, its experimental validity has been established for hundreds of fluids [107]. More recently, simulations of simple Lennard-Jones fluids in two and three dimensions have shown that ΔH_a scales in proportion to the interaction parameter ε [108,109,110], the natural measure of intermolecular interaction strength in simple pair potential models such as LJ fluids and also our polymer model.

Next, we consider the entropic contribution ΔS_a, which generally arises from entropy changes needed to surmount complex multidimensional potential energy barriers in condensed materials [111–114]. The variation of ΔS_a with molecular parameters is less well understood, and the factor $\exp[-\Delta S_a/k_B]$ is often just absorbed into the measured prefactor τ_0 as a practical matter; however, this is not an option for glass-forming liquids. In many small-molecule fluids, the intermolecular potential is weak, and therefore the variation of ΔS_a can be reasonably neglected. For molecules with many internal degrees of freedom, such as polymers, there can be a considerable variation in ΔS_a. In particular, a survey by Bondi [111] revealed that $\Delta S_a/k_B$ could vary over a 100 units, and can even change sign, so that variations of ΔS_a cannot be ignored.

Following the logic that free energy of activation is determined by the free energy cost of removing a molecule from its local molecular environment suggests a proportionate contribution to ΔS_a from the cohesive intermolecular interaction [114,115]. This effect is evident in Trouton's rule [116,117], which relates the heat of vaporization H_{vap} and the entropy of vaporization S_{vap} of gases and the Barclay–Butler phenomenological relation linking enthalpies and entropies of solvation in many mixtures [118–121]. Indeed, many studies have established the specific relation [112–114]

$$\Delta S_a = \Delta S_0 + \Delta H_a/k_B T_{comp} \tag{8.10}$$

supported by observations on diverse materials, where ΔS_0 captures a background contribution associated with the internal configurational degrees of freedom of fluid molecules. This linear relation has long been established for the Arrhenius activation parameters of bulk polymer fluids [122,123]. In glass-forming materials, the entropy–enthalpy "compensation temperature" T_{comp} is often found to be near the glass transition temperature of the fluid, sometimes termed the "melting temperature of the glass" [124]; in crystalline solid materials, T_{comp} is often found to be near the melting temperature T_m of the solid [125,126]. We indeed find "entropy–enthalpy" compensation in our simulated glass-forming films (shown in Figure 8.12) where the compensation temperature, $T_{comp} = 0.18$, a temperature similar to the estimated VFT temperature. These observations suggest that T_{comp} is determined by a physical condition at which the intermolecular cohesive interaction is insufficient to keep the material in the solid state, so that the fluid then begins to explore liquid-like configurations, but a quantitative understanding of how T_{comp} relates to the structure of the potential energy substrate remains to be determined.

For the thinnest film, we have considered, $h_g = 6$, there is a different relation between ΔH_a and ΔS_a for film thickness. Apparently, there is a limiting thickness where boundary effects become the dominant contribution. This interesting effect requires further study, and may be a consequence of the strong deviation of ρ from its bulk value very near the substrate.

We now specifically confront the issue of how the parameters ΔH_a and ΔS_a depend on film thickness; the effects of boundary geometry and interaction strength are discussed in Hanakata et al. [101]. At the free surface, the activation barrier for molecular rearrangements is reduced. Thus, we can generally expect that reducing thickness (increasing surface-to-volume ratio) will reduce the overall activation enthalpy $\Delta H_a(h)$. This is similar to the argument for shifts of T_m and T_g with thickness discussed in Section 8.2.2. Accordingly, we expect that $\Delta H_a(h)$ scales with film thickness

$$\Delta H_a(h) - \Delta H_a(\text{bulk}) \sim 1/h. \tag{8.11}$$

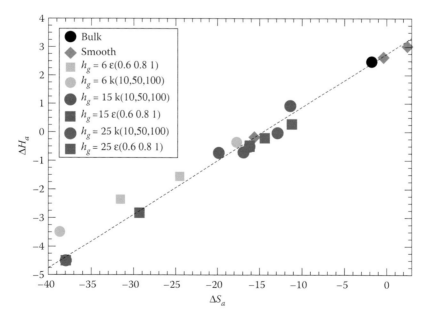

Figure 8.12 Variation of ΔH_a and ΔS_a for different film thicknesses, substrate roughnesses, substrate interactions, and substrate rigidities. The slope defines compensation temperature $T_{comp} = 0.18$. The data for the thinnest film ($h_g = 6$) show a deviation from the thicker film data, which is discussed in the main text. (Figure reproduced with permission from Hanakata, P. Z. et al., *J. Chem. Phys.*, 142, 234907, 2015.)

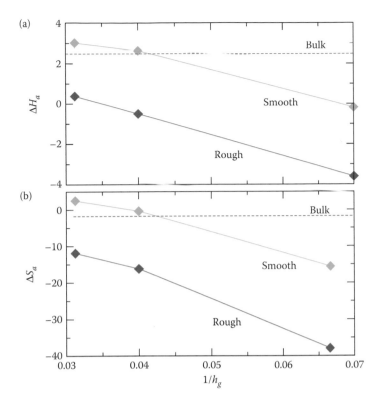

Figure 8.13 Changes in the enthalpic contribution ΔH_a (a) and entropic contribution ΔS_a (b) of the activation free energy with changing confinement on smooth or rough substrates. (Figure adapted from Hanakata, P. Z. et al., *J. Chem. Phys.*, 142, 234907, 2015.)

Figure 8.13a is consistent with a linear variation of $\Delta H_a(h) - \Delta H_a$(bulk) with $1/h_g$. Interestingly, the sign of ΔH_a depends on whether substrate–polymer interactions are stronger or weaker than the polymer–fluid interactions. The component of ΔS_a that is linked to ΔH_a (via the compensation relation 8.10) is expected to follow the same dependence as ΔH_a, suggesting a similar inverse dependence on thickness, and shown in Figure 8.13.

8.4 LIVING POLYMERIZATION MODEL OF STRINGS

While the results of the previous section show the promise of the size and activation barrier for string-like motions to explain the large changes in polymer film relaxation time, they do not provide a theoretical prediction for the T dependence of dynamics. This requires a theoretical model for the variation of L with temperature. Furthermore, the string model has so far only been validated in the limited temperature range accessible to equilibrium molecular dynamics simulation, making it difficult to directly relate these findings to experimental measurements of polymer films in the glass state. To address these issues, in this section we discuss a statistical mechanical model for string formation based on a living polymerization theory to quantitatively describe and predict the scale of cooperative string-like particle rearrangement clusters, as well as the relation of these dynamic clusters to the thermodynamics of the glass-forming liquids [95]. The theory quantitatively describes the interrelation between the average string length L, configurational entropy S_{conf}, and the order parameter Φ for string assembly, without free parameters. Moreover, by combining this theory with the Adam–Gibbs-inspired string model, we can predict the equilibrium relaxation time τ to arbitrarily low temperatures, where the model parameters are precisely determined from comparatively high-temperature simulation results. In doing so, we gain insight into experimentally measured changes to the activation barrier for rearrangement for polymer films in the glass state, a regime in which equilibrium simulations are not possible. We remind the reader that the cooperative string-like motion is largely insensitive to chain connectivity; thus, the dynamic polymerization model we use to describe the strings should not be confused with the permanent bonds that form the backbone of polymer chains. In this regard, the living polymerization approach should be transferable to glass formation in general, including small-molecule systems.

8.4.1 THEORETICAL FORMULATION

We begin by describing the model for strings as dynamic "polymers" (independent from permanent chain connectivity) that form and disintegrate in equilibrium. We consider a model for initiated equilibrium polymerization by Dudowicz et al. [127], which assumes that chain formation is governed by two simple reactions:

$$2M + 2I \rightarrow M_2 I_2,$$ (8.12)

$$M_2 I_2 + M \rightleftharpoons M_{2+i} I_2 \; i = 2,3,\ldots,\infty,$$ (8.13)

where the reaction of the monomer species M, mobile particles in the context of glass-forming liquids, requires an energetically excited initiating species I. The fraction of linked mobile particles Φ serves as the order parameter for the string self-assembly process, and L and Φ of living polymer solutions are related by

$$L = \frac{1}{1 - \Phi + \dfrac{r}{2}},$$ (8.14)

where r is the ratio of the initiator to the monomer concentration ϕ_0. Φ is limited to the range $r \leq \Phi \leq 1$. Consequently, L has an upper bound determined by r, a fact that has important implications for glass formation at low T. Note that the high-temperature limit of L (when $\Phi \rightarrow r$) from Equation 8.14 is larger than 1, consistent with our molecular dynamics simulations observations on GF liquids.

To test the applicability of this model of string formation for GF fluids, we need to map observable quantities from the simulation to the input variables of the theory. For initiated or "living" polymerization, $r \approx \Phi$ for T above the "onset" temperature in the theory (see Figure 7 of Reference 127). Similarly, we know that string-like motion in GF liquids is only prevalent below an onset temperature T_A where relaxation becomes non-Arrhenius and caging becomes prevalent [128]. Thus, we restrict the application of the polymerization model to $T \lesssim T_A$. For our model, $T_A \approx 0.75$ [95]. Identifying the onset condition in the simulation and in living polymers, we have $L(T_A) \equiv L_A$. The approximations $r \approx \Phi_A$ and Equation 8.14 give rise to the useful closed analytic form

$$L \approx \frac{L_A (1 - \dfrac{\Phi_A}{2})}{1 - \Phi + \dfrac{\Phi_A}{2}}, T \leq T_A$$ (8.15)

expressed in terms of observable properties, L_A, Φ_A, Φ. In the living polymerization model, L_A and Φ_A are not independent, but we take these as independent observables in our application of Equation 8.15 because of the approximation $\Phi_A \approx r$. Equations 8.14 and 8.15 imply that L saturates to a constant value at low T, a point extensively discussed below. In living polymer solutions, the magnitude of r (and thus Φ_A, by analogy) links the string mass at high and low T and also governs the cooperativity of the polymerization transition [129], as measured by the rate of change of $L(T)$, and the magnitude of the change in the specific heat; similar definitions of cooperativity have been applied to GF liquids [89,130].

In the living polymerization model, the order parameter that controls this process is the extent of polymerization Φ, given by the fraction of monomers that are forming the string-like structures. In our case, this is simply the fraction of those highly mobile monomers that are forming strings. The behavior of $\Phi(T)$ is analytically described by the polymerization model in Dudowicz et al. [127]:

$$\Phi = 1 - \frac{\phi(T)}{\phi_0}, \tag{8.16}$$

where $\phi(T)$ is the fraction of monomers, given by

$$\phi(T) = \frac{1}{2} e^{\frac{\Delta G_p}{k_B T}} \left[1 + \phi_0 \left(1 - \frac{\Phi_A}{2} \right) e^{-\frac{\Delta G_p}{k_B T}} - \right.$$

$$\left. \sqrt{\left(1 + \phi_0 (1 - \frac{\Phi_A}{2}) e^{-\frac{\Delta G_p}{k_B T}} \right)^2 - 4\phi_0 (1 - \Phi_A) e^{-\frac{\Delta G_p}{k_B T}}} \right]. \tag{8.17}$$

The free energy $\Delta G_p = H_p - T\Delta S_p$ describes the thermodynamics of polymerization (string formation) where ΔH_p and ΔS_p are the enthalpy and entropy of chain assembly, respectively. Accordingly, the association into a strings is governed by an equilibrium constant $k = \exp(\Delta G_p/k_B T)$. The volume fraction of mobile particles $\phi_0 = f_0(\pi/6) = 0.034$, using a mobile particle fraction $f_0 = 0.065$ determined in an earlier work [92].

8.4.2 Validation of the Living Polymerization Model

We validate this living polymerization model by comparing its predictions to the known T dependence of $L(T)$ from our simulation results; a more comprehensive test of the model is described in Pazmiño Betancourt et al. [95]. Combining Equations 8.15 through 8.17 provides the predicted $L(T)$, which we show accounts for the molecular simulation results Figure 8.14a, from which we fix the parameters $\Delta H_p = -1.55$ and $\Delta S_p/k_B = -0.46$. We also show a high-T expansion of the theoretical result, which predicts an approximately Arrhenius T dependence of $L(T)$ at elevated temperatures [95]. This full theory and the approximate expansion result show dramatically different behavior at low T. The high-T expansion predicts unbounded string growth, but by the nature of the approximations required, this estimate is not reliable at low T. In contrast, the complete model predicts that $L(T)$ reaches a plateau $L(T \rightarrow 0) = L_A[(2/\Phi_A) - 1] \approx 4.1$, with a radius of gyration $R_g = 1.6$ $\sigma \approx 1.6$ nm to 3.2 nm [95], a length scale consistent with empirical CRR size estimates obtained from specific heat measurements in small-molecule liquids and estimates of the change in the activation energy in GF liquids [131–133]. This range of length scales is also in accord with the typical "cooperative motion scale" from Boson peak and nuclear magnetic resonance (NMR) measurements [134].

Given the success of the theory to describe the available string data, we consider the predictions of the string model for relaxation at much lower T than accessible by equilibrium simulation. Combining the predicted form for $L(T)$ (Equations 8.15 through 8.17) with the AG-inspired string model (Equation 8.9), we demonstrate the validity of the approach to describe relaxation time $\tau(T)$ in the region where simulation data are available, as well as predict $\tau(T)$ at T below temperatures accessible by equilibrium simulation (Figure 8.14b). The predicted plateau of $L(T)$ at low T has a significant consequence for the low-T behavior of τ. Specifically, such a plateau implies a return to an Arrhenius temperature dependence, as illustrated in the inset of Figure 8.14b. Consequently, fragility estimates based on the string model are clearly smaller than values estimated from the VFT fit. This non-Arrhenius to Arrhenius crossover in the string model occurs rather close to the expected T_g for the system. Given the extrapolation required to assign a precise value for T_g, it is plausible that this crossover behavior occurs roughly on entering the glass state. Such Arrhenius dependence near T_g has been observed in a variety of GF fluids [135], and recent experimental aging results in a 20 million-year-old amber sample show Arrhenius behavior continuing significantly below T_g [136]. Thus, the predicted plateau in L at low T is physically plausible.

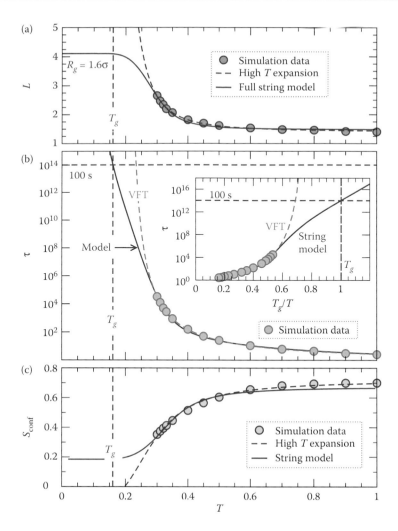

Figure 8.14 (a) Temperature dependence of $L(T)$ from numerical simulations in comparison with the full model for string T dependence, as well as the high-T expansion. (b) Temperature dependence of $\tau(T)$ from simulation data, as well as the string model prediction when combined with the AG expression (Equation 8.9). The VFT fit is also shown for comparison. The horizontal dashed blue line illustrates the value where $\tau \approx 100$ s, which typically defines the glass transition temperature, $T_g = 0.16$. The inset shows τ as a function of T_g/T in order to highlight the predicted return to Arrhenius (strong) behavior in the vicinity of T_g. (c) Configurational entropy $S_{conf}(T)$ from numerical simulations [23] in comparison with the predicted variation from the string model. (Figure adapted from Pazmiño Betancourt, B. A., Douglas, J. F., and Starr, F. W., *J. Chem. Phys.*, 140, 204509, 2014.)

Based on the inverse scaling relation between L and S_{conf}, we can also use L from the polymerization model to anticipate the low-T behavior of S_{conf}, shown in Figure 8.14c. The low-T plateau of L corresponds to the saturation of S_{conf} to a residual low-T residual value that is about one-quarter of its value near T_A. Consequently, the Kauzmann entropy "catastrophe" [135], in which S_{conf} of the fluid would be negative for $T \to 0$, is naturally avoided in the string polymerization model. The residual entropy is determined by Φ_A and L_A, and in the living polymerization model, these parameters control the sharpness of the polymerization transition. If we consider the extrapolation of the high-T expansion to estimate S_{conf} at low T (Figure 8.14c), we see the extrapolation predicts a Kauzmann temperature $T_0 = 0.2$ (i.e., where $S_{conf} = 0$). This is consistent with the VFT extrapolated divergence temperature, and an earlier extrapolation of S_{conf} based on a simple polynomial fit [23]. A similar vanishing of S_{conf} due to using the high-temperature expansion outside the range of validity also occurs in a recent entropy theory for polymer glass formation [137].

8.4.3 PREDICTIONS FOR GLASSY POLYMER FILMS

One of the values of the string model, apart from a general conceptual understanding of the temperature dependence of relaxation in glass-forming fluids, is that the model provides predictions for relaxation at temperatures below those accessible in simulation. This even allows us to address the behavior of thin polymer films below T_g, where a variety of experiments have observed an Arrhenius T dependence of viscosity or relaxation time below T_g with an activation energy of the Arrhenius relaxation that decreases with decreasing film thickness [67,138]. As described in the previous subsection, the nearly constant value for L below T_g predicted by our theory leads to an expectation of an Arrhenius T dependence when combined with our string model, Equation 8.9. Moreover, our simulation results for the dependence of the activation parameters on film thickness indicate that the activation barrier is a decreasing function of film thickness, as observed experimentally for films in the glass state.

We can test this expected trend for the variation of the enthalpy and entropy of activation with film thickness using the determined activation parameters from molecular simulation at comparatively high temperatures, where simple Arrhenius relaxation should apply. Figure 8.15 shows the predicted behavior of τ in the glass regime of the polymer film, reported for the first time in this chapter. We see that the slopes of the Arrhenius curves become progressively smaller with film confinement. which reproduces a characteristic fixed point, here defined by simply scaling the temperature by the value where $\tau = 100$ s—a canonical definition for T_g. These predictions accord with experimental findings [67,138], and the Arrhenius nature is a consequence of a saturation in the asymptotic size for cooperative rearrangement at low T. The decrease in activation barrier with decreasing thickness is readily rationalized for supported films: the relaxation near the free surface has an intrinsically lower activation barrier and makes a progressively larger contribution to the overall activation barrier as films become thinner. We caution that our predictions are for fully equilibrated relaxation, while experiments below T_g are necessarily susceptible to nonequilibrium effects. At the same time, experiments [136] examining highly aged samples support the possibility that equilibrium relaxation below T_g follows an Arrhenius T dependence.

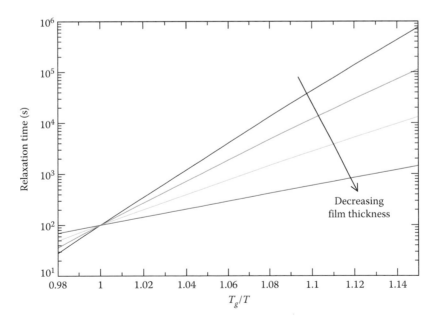

Figure 8.15 Predicted behavior of the equilibrium relaxation time τ crossing into the regime that would be glassy on experimental time scales. The combined polymerization and string models predict an Arrhenius T dependence of τ at low T, where the activation decreases with decreasing thickness.

8.5 RELATION BETWEEN THE SCALE OF INTERFACIAL MOBILITY AND STRING SIZE

As discussed earlier, the changes of T_g are due in part to the enhancement or suppression of dynamics in the interfacial regions. It is precisely the scale of this interfacial region on which we now focus our attention. This interfacial region represents another form of dynamic heterogeneity, which is static, rather than transient, like the strings we previously discussed. A natural question that arises is: to what degree does this interfacial heterogeneity scale relate to the intrinsic scale of dynamic heterogeneity of the fluid? In this section, we find that the interfacial scale grows on cooling in a quantitatively similar way to the string size. These findings are consistent with the experiments by Roth and coworkers [88,139], described in Chapter 5.

To quantify interfacial dynamics and the corresponding interfacial length scale, we utilize the local dynamics as a function of distance z from the substrate boundary, as shown, for example, in Figure 8.16a. We quantify the dynamical length scale ξ of the free-interface (mobile-layer scale) based on local relaxation time τ_s near the free surface. To evaluate the layer size, we identify the positions where $\tau_s(z)$ and $\rho(z)$ deviate from the near-constant behavior in the middle of the film. Specifically, we define the scale ξ as the length where $\rho(z)$ or $\tau_s(z)$ deviates by 30% from the mean value of the central region of the film. This fraction is chosen to minimize artifacts from noise in our data, and other reasonable choices do not affect our findings, demonstrated in detail in Hanakata et al. [94]. Note that, for thin films, there may be no middle layer, especially at low T. In this case, we define ξ from the extremum that separates substrate and free surface layers. In the latter case, ξ is dominated by the overall film thickness. A similar choice was examined by Simmons and coworkers [140].

An advantage of this definition for the interfacial scale is that it depends on the behavior near the interface as well as the film interior. We find below that both contributions are important for our results. Note that the bulk material does not provide a useful reference to define ξ, since the dynamics near the center of the film often differs from the bulk [40]. We limit our analysis to films in which the interfacial scale is no larger than half of the film thickness [40]. The inset of Figure 8.16a shows the growth of ξ with decreasing T for four representative films supported on a rough or smooth surface. The growth of this dynamical length scale has been observed in previous studies, where the scale is inferred in a similar fashion [40,63,74,140]. We emphasize that this interfacial mobility scale is *distinct* from the scale of interfacial density changes, so there is no simple relation between mobility and local density [40]. Note that ξ is normalized by its value ξ_A at the onset temperature T_A for "slow" dynamics. In glass-forming systems, the appearance of non-Arrhenius T dependence and cooperative motion only occurs below an "onset" temperature T_A. For our films, $T_A \approx 0.6$, with a variation of not more than $\approx$10% for all the substrates and film thicknesses we examine. Given the small changes in T_A with film type, we adopt the approximation $T_A \approx 0.6$ for all our films.

We next consider the possible relation of the interfacial scale to the scale of cooperative string-like motion we have already discussed. Figure 8.16b shows that L and ξ grow proportionally, relative to their values ξ_A and L_A, at the onset temperature T_A:

$$[L/L_A - 1] = A[\xi/\xi_A - 1]. \tag{8.18}$$

The proportionality constant A fluctuates with no apparent trend around a mean $A = 0.32$. For reference, the values of $L(T_A)$ and $\xi(T_A)$ are relatively constant for all systems; specifically, $L_A \approx 1.41$ and $\xi_A \approx 3.6$. Moreover, the precise values used for ξ_A and L_A are not critical for our findings. Given that L varies inversely proportional to the configurational entropy S_{conf} [23], ξ should also vary inversely proportional to S_{conf} (with the factors of ξ_A and L_A included). Indeed, Stephenson and Wolynes [141] predict a direct inverse proportionality, but our findings show this relation must include an additive term accounting for the scales of collective motion at T_A. This provides our first direct indication that the interfacial scale might indeed reflect a fundamental dynamical heterogeneity scale. The robustness of this relation between interfacial and heterogeneity

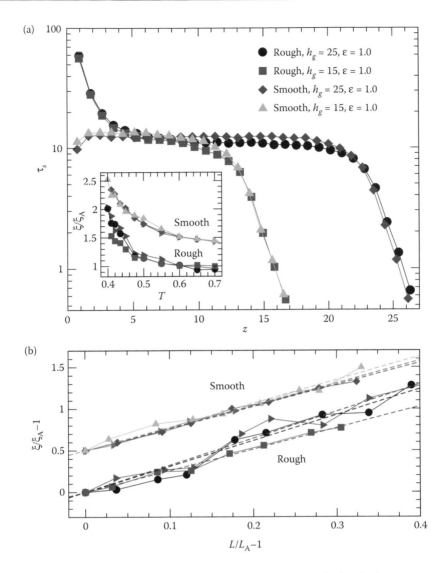

Figure 8.16 Interfacial scale of films with smooth or rough substrates. (a) The local relaxation time τ_s as function of distance z from the substrate for films supported on a rough or smooth surface. The inset shows the resulting free surface mobile layer scale, ξ, for films supported on a rough or smooth substrate as function of temperature T; ξ is normalized by its value at T_A. We also show ξ data for thickness 32 (triangle right) in the inset. The values of ξ for the smooth surface are shifted by 0.5 for clarity. (b) $\xi/\xi_A - 1$ for films supported on a rough or smooth surface as a function of $L/L_A - 1$, showing proportionality between these quantities. The dashed lines indicate a linear proportionality. The normalized $\xi/\xi_A - 1$ values of films on a smooth surface are shifted by 0.5 for clarity of the figure. (Figure reproduced with permission from Hanakata, P. Z., Douglas, J. F., and Starr, F. W., *Nat. Commun.*, 5, 256207, 2014.)

scales is demonstrated by examining variations in surface–polymer interaction strength ϵ_{sm} and surface rigidity k [94].

The linear relation between ξ and L implies a direct link between the interfacial scale and relaxation time. Specifically, combining Equations 8.18 and 8.5 yields

$$\tau = \tau_\infty \exp\left[\frac{(1-A)\Delta\mu}{k_B T}\right] \exp\left[\frac{\xi(T)}{\xi_A} \frac{A\Delta\mu}{k_B T}\right]. \tag{8.19}$$

Our data indicate that the T dependence of the leading term $\exp[(1 - A)\Delta\mu/k_B T]$ is weak in comparison to the term involving ξ, both because $1 - A$ is relatively small and since ξ grows on cooling. Consequently, if one approximates the first exponential term as constant, the expression simplifies to

$$\tau \approx \tau_\xi \exp\left[\frac{\xi(T)}{\xi_A}\frac{A\Delta\mu}{k_B T}\right]. \qquad (8.20)$$

More generally, without approximation $\tau_\xi = \tau_\infty \exp[(1 - A)\Delta\mu/k_B T]$. The approximation that τ_ξ is constant is important for possible experimental application, since L (and hence A) is not easily measured. Using this approximation, we find that ξ works nearly as well as L to describe $\tau(T)$ using the same ΔH and ΔS values already determined from the relation of τ with L. In other words, we do not allow these parameters to be refit. This parameterization results in a collapse of all data shown in Figure 8.17. This provides direct evidence that

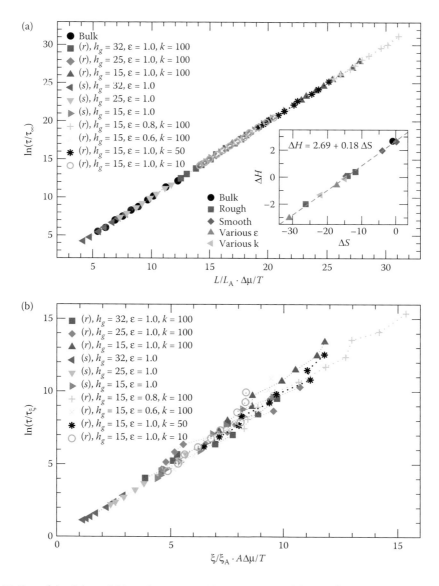

Figure 8.17 Test of the Adam–Gibbs relation using (a) string size L and (b) interfacial scale ξ. Collapse of $\log(\tau/\tau_\xi)$ versus $\xi/\xi_A \cdot A\Delta\mu/T$ for all systems. (Figure adapted from Hanakata, P. Z., Douglas, J. F., and Starr, F. W., *Nat. Commun.*, 5, 256207, 2014.)

the interfacial dynamical scale contains the information on changes to CRR size that are fundamental to the AG and RFOT approaches.

An important aspect of this finding is that it opens a possible route to indirectly access the cooperativity scale in experiments. A similar relation between the scale of cooperative motion and the interfacial mobility scale has been found in the interfacial zone of Ni nanoparticles [142,143]. Tanaka and coworkers [144] have also found a relationship between the interfacial mobility scale as we have defined it and a correlation length associated with immobile particle clusters having high local order. The observations of the present work suggests that there might be complementarity between these different types of clusters of extreme mobility, and future work should explore this relationship quantitatively.

8.6 CONCLUSIONS AND PROSPECTS

The dynamics of supported thin polymer films are highly sensitive to changes in film thickness, substrate interaction strength and rigidity, as well as any other effect that influences the relative contribution of the interfacial regions. The dynamics of the film interior can also deviate from bulk behavior, further complicating the interpretation of dynamical changes. Despite the apparent nonuniversal and significant shifts of the dynamics of thin polymer films in comparison to bulk melts, we find that the relaxation of all simulated films can be organized in a universal way based on the string model of glass formation, an extension of the Adam–Gibbs model that is grounded in computational evidence. The same universal reduction of the relaxation data also applies to polymer nanocomposite data for systems with attractive and nonattractive polymer–nanoparticle interactions, over a wide range of nanoparticle concentrations [83,100].

In our view, this universal description of the dynamics of both thin polymer films and nanocomposites provides a practical organizational principle for these systems, and should also be applicable to understanding the dynamics of glass formation in liquids broadly. In particular, these findings support the physical significance of cooperative motions, postulated by Adam and Gibbs, as the physical origin of the growing relaxation time of glass-forming liquids. Our work takes this classic approach further by actually identifying the geometric form of the cooperative motion, and then describes these dynamical strings in terms of a statistical mechanical model of equilibrium polymerization. This model of dynamic chain self-assembly has been applied to many other condensed matter systems (the formation of worm-like micelles, the thermal polymerization of sulfur and other atomic species, actin polymerization, aggregation of antibody proteins, etc.), and thus links glass formation to other well-studied thermodynamic phenomena. Many of the heuristic assumptions of the original Adam–Gibbs work can be derived from this statistical mechanical model, placing the Adam–Gibbs framework on sounder theoretical grounds and making the results of this model more physically intuitive and the predictions concrete. Indeed, Freed [24] recently formulated an extension of transition state theory to incorporate collective barrier crossing events, consistent with our simulation results. By modeling the strings as equilibrium polymers, we extend the predictions of the string model to cover the whole glass transition range, including the glass state, where equilibrium simulations are not possible. Combining the predicted string size with string theory of relaxation suggests that there is no divergence in the relaxation time or viscosity at a finite temperature; instead, the theory predicts that relaxation should return to Arrhenius temperature dependence at low temperatures (in the limit of equilibrium relaxation), due to the saturation of the scale of cooperative rearrangements. This also implies that configurational entropy does not vanish at finite temperature, but similarly saturates to a small low-temperature value. These predictions, combined with the observed changes to the activation free energy for rearrangements in our simulations, offer an explanation for recent observations on polymer films that indicate Arrhenius relaxation in thin films in the glass state where the activation energy progressively decreases with film thickness [67,138,145]. Finally, our examination of the scale of the interfacial region of supported films indicates a linear relationship of the interfacial scale to the string size, offering a more experimentally accessible metric to estimate the scale of cooperative motions.

Our findings suggest that an important contribution to T_g changes arises from changes in the high-temperature activation barriers, where cooperativity is not required to facilitate molecular rearragement. Specifically, the activation energy and entropy in the high-temperature Arrhenius relaxation regime are simply scaled by the scale of string-like cooperative motions at low temperature to determine the overall scale of barriers. Thus, almost paradoxically, changes to the high-temperature activation parameters can dramatically alter the temperature dependence of the dynamics of glass-forming polymer films and polymer nanocomposites, even if the scale of cooperative motions is unchanged. Indeed, in our simulations, we find that the primary effect of changing film thickness, substrate interaction, or substrate rigidity is to alter these activation parameters, rather than the scale of cooperative motions. This suggests that the preoccupation with changes to T_g, while practically useful, may be distracting from the more fundamental origin of dynamical changes. Accordingly, we hope to see experimental efforts to determine how the activation parameters depend on molecular structure (for example, chain length, monomer structure, bond rotation potentials), confinement, additives, pressure, or interfacial interactions. Recent computational work [146] has begun to address this problem by systematically considering how the activation energy parameters of alkanes depend on chain length. Such computational studies would provide a useful library if extended to many polymers. There is also a need to develop accessible and reliable methods to estimate the extent of collective motion in glass-forming polymer materials. Recent work relating the extent of collective motion to colored noise in mobility fluctuations [142,143] and the thickness of the interfacial mobility layer of the polymer–air interface of supported polymer [94,140] and crystalline films [143] are promising starts, but this work is also in its infancy. We are thus at the beginning of the development of a truly predictive molecular theory of the dynamics of glass-forming liquids.

8.7 APPENDIX: MODELING AND SIMULATION

In the simulations presented in this chapter, we modeled polymers as unentangled chains of beads linked by harmonic springs. The film substrate is modeled either as a collection of substrate atoms or by a perfectly smooth substrate. Nonbonded monomers or atoms of the substrate interact with each other via the Lennard-Jones potential, and we use a shifted-force implementation to ensure continuity of the potential and forces at the cutoff distance r_c. We choose $r_c = 2.5\sigma_{ij}$ to include interparticle attractions, where σ_{ij} is the monomer "diameter" in the LJ potential. The index pair ij distinguishes interactions between monomer–monomer (mm), substrate–monomer (sm), and substrate–substrate (ss) particles. The LJ interaction is not included for the nearest neighbors along the chain. These monomers are connected by a harmonic spring potential $U_{bond} = (k_{chain}/2)(r - r_0)^2$ with bond length $r_0 = 0.9$ (equilibrium distance) and spring constant $k_{chain} = (1111)\epsilon_{mm}/\sigma^2_{mm} \cdot r_0$. The spring constant is chosen as in Peter et al. [63], but we choose r_0 smaller than in Peter et al. [63] because we found that crystallization occurs readily in the films for the value used in Reference 63.

The interaction between monomers and the smooth substrate is given by

$$V_{smooth} = \frac{2\pi}{3}\epsilon_{sm}\rho_s\sigma^3_{ss}\left[\frac{2}{15}\left(\frac{\sigma_{sm}}{z}\right)^9 - \left(\frac{\sigma_{sm}}{z}\right)^3\right], \tag{8.21}$$

where z is the distance of a monomer from the substrate. This is the same smooth substrate model that we studied in our previous work [40]. To model the rough substrate, we tether the substrate atoms to the sites of triangular lattice (the 111 face of a face-centered cubic lattice) with harmonic potential

$$U_s(r_i) = \frac{k_s}{2}(|\vec{r}_i - \vec{r}_{ieq}|)^2, \tag{8.22}$$

where $\vec{r}_{eq}$ denotes an equilibrium position on the triangular lattice and k_s is the harmonic spring constant [75]. We choose the lattice spacing to be $2^{1/6}\sigma_{ss}$, where $\sigma_{ss} = 0.80\sigma_{mm}$ and $\sigma_{sm} = \sigma_{mm}$, where $\sigma_{mm} - 1$, and $\epsilon_{mm} - 1$; thus, varying k_s allows us to examine the role of substrate rigidity on the polymer dynamics. We note that, if the parameters for the substrate particle size and interaction strength are not chosen wisely, the resulting film may (partially) crystallize [147], precluding an analysis of the properties of glass formation.

All values reported are in reduced units, where T is given by ε/k_B, where k_B is Boltzmann's constant, and time is given in units of $(m\sigma^2/\epsilon_{mm})^{1/2}$. For a simple polymer (like polystyrene) with $T_g \approx 100°C$, the reduced units can be mapped to physical units relevant to real polymer materials, where the size of a chain segments σ is typically about 1–2 nm, time is measured in ps, and $\epsilon \approx 1\,kJ/mol$.

We simulate films of variable thicknesses with $N_c = 200$, 300, 400, 600, or 1000 chains of 10 monomers each. These sizes correspond to thicknesses with value of roughly 6–25 monomer diameters. We use various interaction strengths ($\epsilon_{sm} \equiv \varepsilon$) between the rough substrate and polymers, ranging from 0.4 to 1.0 ϵ_{mm} with a fixed surface rigidity $k_s = 100$; we vary the strength of the substrate rigidity ($k_s \equiv k$) over the range from 10 to 100 with a fixed $\varepsilon = 1$. For this range of model parameters, we find that T_g of the film can be higher or lower than the bulk value. Additionally, we simulate a pure bulk system of 400 chains of $M = 10$ monomers each at zero pressure for the purpose of comparison.

We define film thickness $h(T)$ as a distance from the substrate where the density profile along the z direction, perpendicular to the substrate, $\rho(z)$ decreases to 0.10. Other reasonable criteria do not affect our qualitative findings. The resulting $h(T)$ is well described by an Arrhenius form, which we use to extrapolate the thickness value $h_g \equiv h(T_g)$ at the glass transition.

To quantify the overall dynamics of the films and bulk system, we evaluate the coherent intermediate scattering function:

$$F(q,t) \equiv \frac{1}{NS(q)}\left\langle \sum_{j,k=1}^{N} e^{-iq.[r_k(t)-r_j(0)]} \right\rangle, \tag{8.23}$$

where r_j is the position of monomer j and $S(q)$ is the static structure factor, as shown in Figure 8.18. We define the characteristic time τ by $F(q_0,\tau) = 0.2$, where q_0 is the location of the first peak in of $S(q)$. To quantify dynamics locally within the film, we use the self (or incoherent) $F_{self}(z, q, t)$ part (i.e., $j = k$) of Equation 8.23 on the basis of the position z of a monomer at $t = 0$. We define the relaxation time $\tau_s(z)$ by $F_{self}(z, q_0,\tau_s) = 0.2$.

We estimate T_g by fitting our data to the Vogel–Fulcher–Tammann (VFT) equation:

$$\tau(T) = \tau_\infty e^{DT_0/(T-T_0)}, \tag{8.24}$$

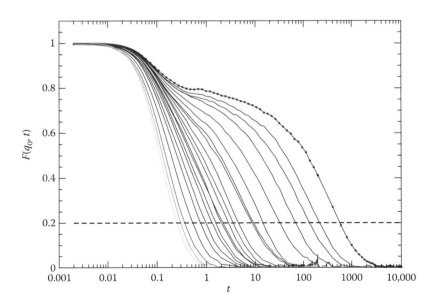

Figure 8.18 Coherent scattering function $F(q_0,t)$. Temperature is indicated by the color gradient, which goes from blue at the lowest T, where relaxation is highly nonexponential, to yellow at a high T.

where τ_∞ is an empirical prefactor normally on the order of a molecular vibrational time (10^{-14} to 10^{-13} s) [148], D is a measure of "fragility," and T_0 is a temperature at which τ extrapolates to infinity. Equation 8.24 should only be applied above the glass transition temperature. In a lab setting, T_g is often defined as T at which the relaxation time reaches 100 s [149], and we adopt this simple criterion.

REFERENCES

1. J. A. Forrest and K. Dalnoki-Veress, *Adv. Colloid Interface Sci.*, 94, 167, 2001.
2. C. B. Roth and J. R. Dutcher, *J. Electroanal. Chem.*, 584, 13, 2005.
3. M. Alcoutlabi and G. McKenna, *J. Phys. Condens. Matter*, 17, R461, 2005.
4. G. McKenna, *Eur. Phys. J. Spec. Top.*, 189, 285, 2010.
5. M. D. Ediger and J. A. Forrest, *Macromolecules*, 47, 471, 2014.
6. M. D. Ediger, *Annu. Rev. Phys. Chem.*, 51, 99, 2000.
7. H. Sillescu, *J. Non-Cryst. Solids*, 243, 81, 1999.
8. S. C. Glotzer, *J. Non-Cryst. Solids*, 274, 342, 2000.
9. G. Adam and J. H. Gibbs, *J. Chem. Phys.* 43, 139, 1965.
10. T. R. Kirkpatrick, D. Thirumalai, and P. G. Wolynes, *Phys. Rev. A*, 40, 1045, 1989.
11. V. Lubchenko and P. G. Wolynes, *Annu. Rev. Phys. Chem.*, 58, 235, 2007.
12. C. Donati, J. F. Douglas, W. Kob, S. J. Plimpton, P. H. Poole, and S. C. Glotzer, *Phys. Rev. Lett.*, 80, 2338, 1998.
13. M. Aichele, Y. Gebremichael, F. W. Starr, J. Baschnagel, and S. C. Glotzer, *J. Chem. Phys.*, 119, 5290, 2003.
14. A. H. Marcus, J. Schofield, and S. A. Rice, *Phys. Rev. E*, 60, 5725, 1999.
15. Z. Zhang, P. J. Yunker, P. Habdas, and A. G. Yodh, *Phys. Rev. Lett.*, 107, 208303, 2011.
16. E. R. Weeks, J. C. Crocker, A. C. Levitt, A. Schofield, and D. A. Weitz, *Science*, 287, 627, 2000.
17. Z. Zheng, F. Wang, and Y. Han, *Phys. Rev. Lett.*, 107, 065702, 2011.
18. Y. Gebremichael, M. Vogel, and S. Glotzer, *J. Chem. Phys.*, 120, 4415, 2004.
19. R. A. Riggleman, K. Yoshimoto, J. F. Douglas, and J. J. de Pablo, *Phys. Rev. Lett.*, 97, 2006.
20. M. Vogel, B. Doliwa, A. Heuer, and S. Glotzer, *J. Chem. Phys.*, 120, 4404, 2004.
21. T. Schroder, S. Sastry, J. Dyre, and S. Glotzer, *J. Chem. Phys.*, 112, 9834, 2000.
22. N. Giovambattista, F. W. Starr, F. Sciortino, S. V. Buldyrev, and H. E. Stanley, *Phys. Rev. E*, 65, 2002.
23. F. W. Starr, J. F. Douglas, and S. Sastry, *J. Chem. Phys.*, 138, 12A541 (pages 18) 2013.
24. K. F. Freed, *J. Chem. Phys.*, 141, 141102, 2014.
25. C. Donati, S. C. Glotzer, P. H. Poole, W. Kob, and S. J. Plimpton, *Phys. Rev. E*, 60, 3107, 1999.
26. D. Long and F. Lequeux, *Eur. Phys. J. E*, 4, 371, 2001.
27. A. Dequidt, D. Long, P. Sotta, and O. Sanséau, *Eur. Phys. J. E*, 35, 61, 2012.
28. K. Vollmayr-Lee and A. Zippelius, *Phys. Rev. E*, 72, 041507, 2005.
29. K. Vollmayr-Lee, W. Kob, K. Binder, and A. Zippelius, *J. Chem. Phys.*, 116, 5158, 2002.
30. M. Dzugutov, S. I. Simdyankin, and F. H. M. Zetterling, *Phys. Rev. Lett.*, 89, 195701, 2002.
31. H. Tanaka, *J. Non-Cryst. Solids*, 351, 3385, 2005.
32. H. Tanaka, T. Kawasaki, H. Shintani, and K. Watanabe, *Nat. Mater.*, 9, 324, 2010.
33. J. C. Conrad, P. P. Dhillon, E. R. Weeks, D. R. Reichman, and D. A. Weitz, *Phys. Rev. Lett.*, 97, 265701, 2006.
34. A. V. Anikeenko and N. N. Medvedev, *Phys. Rev. Lett.*, 98, 235504, 2007.
35. U. R. Pedersen, T. B. Schrøder, J. C. Dyre, and P. Harrowell, *Phys. Rev. Lett.*, 104, 105701, 2010.
36. A. J. Dunleavy, K. Wiesner, R. Yamamoto, and C. P. Royall, *Nat. Commun.*, 6, 6089, 2015.
37. R. Pinney, T. B. Liverpool, and C. P. Royall, *J. Chem. Phys.*, 143, 244507, 2015.
38. H. Zhang, M. Khalkhali, Q. Liu, and J. F. Douglas, *J. Chem. Phys.*, 138, 12A538, 2013.
39. J. A. Torres, P. F. Nealey, and J. J. de Pablo, *Phys. Rev. Lett.*, 85, 3221, 2000.
40. P. Z. Hanakata, J. F. Douglas, and F. W. Starr, *J. Chem. Phys.*, 137, 244901, 2012.
41. D. Hudzinskyy, A. V. Lyulin, A. R. C. Baljon, N. K. Balabaev, and M. A. J. Michels, *Macromolecules*, 44, 2299, 2011.

42. C. Batistakis, A. V. Lyulin, and M. A. J. Michels, *Macromolecules*, 45, 7282, 2012.

43. S. Peter, H. Meyer, and J. Baschnagel, *J. Chem. Phys.*, 131, 2009.

44. K. Binder, J. Baschnagel, and W. Paul, *Progr. Polym. Sci.*, 28, 115, 2003.

45. Q. Qin and G. B. McKenna, *J. Non-Cryst. Solids*, 352, 2977, 2006.

46. M. D. Marvin, R. J. Lang, and D. S. Simmons, *Soft Matter*, 10, 3166, 2014.

47. C. L. Jackson and G. B. McKenna, *J. Chem. Phys.*, 93, 9002, 1990.

48. C. L. Jackson and G. B. McKenna, *Chem. Mater.*, 8, 2128, 1996.

49. Q. Jiang, Z. Zhang, and J. Li, *Chem. Phys. Lett.*, 322, 549, 2000.

50. F. G. Shi, *J. Mater. Res.*, 9, 1307, 1994.

51. R. G. Beaman, *J. Polym. Sci.*, 9, 470, 1952.

52. R. F. Boyer, *Rubber Chem. Technol.*, 36, 1303, 1963.

53. M. Zhao and Q. Jiang, *Solid State Commun.*, 130, 37, 2004.

54. H. Li, M. Paczuski, M. Kardar, and K. Huang, *Phys. Rev. B*, 44, 8274, 1991.

55. Q. Jiang and F. Shi, *Mater. Lett.*, 37, 79, 1998.

56. Q. Jiang, H. Tong, D. Hsu, K. Okuyama, and F. Shi, *Thin Solid Films*, 312, 357, 1998.

57. Z. Zhang, J. C. Li, and Q. Jiang, *J. Phys. D Appl. Phys.*, 33, 2653, 2000.

58. M. E. Fisher and A. E. Ferdinand, *Phys. Rev. Lett.*, 19, 169, 1967.

59. G. A. T. Allan, *Phys. Rev. B*, 1, 352, 1970.

60. M. Zhao, X. H. Zhou, and Q. Jiang, *J. Mater. Res.*, 16, 3304, 2001.

61. Q. Jiang, C. C. Yang, and J. C. Li, *Macromol. Theory Simul.*, 12, 57, 2003.

62. Z. Zhang, M. Zhao, and Q. Jiang, *Phys. B Condens. Matter*, 293, 232, 2001.

63. S. Peter, H. Meyer, and J. Baschnagel, *J. Polym. Sci., Part B: Polym. Phys.*, 44, 2951, 2006.

64. J. L. Keddie, R. A. L. Jones, and R. A. Cory, *EPL (Europhys. Lett.)*, 27, 59, 1994.

65. C. M. Evans, H. Deng, W. F. Jager, and J. M. Torkelson, *Macromolecules*, 46, 6091, 2013.

66. R. D. Priestley, C. J. Ellison, L. J. Broadbelt, and J. M. Torkelson, *Science*, 309, 456, 2005.

67. Z. Fakhraai and J. A. Forrest, *Phys. Rev. Lett.*, 95, 025701, 2005.

68. C. J. Ellison, M. K. Mundra, and J. M. Torkelson, *Macromolecules*, 38, 1767, 2005.

69. J. H. Kim, J. Jang, and W.-C. Zin, *Langmuir*, 17, 2703, 2001.

70. K. Fukao and Y. Miyamoto, *Phys. Rev. E*, 61, 1743, 2000.

71. C. M. Guttman, E. A. Di Marzio, and J. F. Douglas, *Macromolecules*, 29, 5723, 1996.

72. K. F. Freed, *Renormalization Group Theory of Macromolecules* (Wiley-Interscience Publication, New York), 1987.

73. J.-P. Hansen and I. R. Mcdonald, *Theory of Simple Liquids* (Academic Press, London), 3rd edn., 2006.

74. P. Scheidler, W. Kob, and K. Binder, *Europhys. Lett.*, 59, 701, 2002.

75. J. Baschnagel and F. Varnick, *Condens. Matter*, 17, 852, 2005.

76. G. D. Smith, D. Bedrov, and O. Borodin, *Phys. Rev. Lett.*, 90, 226103, 2003.

77. A. Virgiliis, A. Milchev, V. Rostiashvili, and T. Vilgis, *Eur. Phys. J. E*, 35, 1, 2012.

78. Y. Sun, L. Zhu, K. L. Kearns, M. D. Ediger, and L. Yu, *Proc. Natl. Acad. Sci. U. S. A.*, 108, 5990, 2011.

79. L. Zhu, C. W. Brian, S. F. Swallen, P. T. Straus, M. D. Ediger, and L. Yu, *Phys. Rev. Lett.*, 106, 256103, 2011.

80. K. Paeng, R. Richert, and M. D. Ediger, *Soft Matter*, 8, 819, 2012.

81. C. W. Brian and L. Yu, *J. Phys. Chem.*, A 117, 13303, 2013.

82. W. Zhang, J. F. Douglas, and F. W. Starr, *J. Chem. Phys.* (submitted) 2016.

83. B. A. Pazmiño Betancourt, J. F. Douglas, and F. W. Starr, *Soft Matter*, 9, 241, 2013.

84. D. S. Fryer, R. D. Peters, E. J. Kim, J. E. Tomaszewski, J. J. de Pablo, P. F. Nealey, C. C. White, and W. L. Wu, *Macromolecules*, 34, 5627, 2001.

85. H. Yamakawa, *Modern Theory of Polymer Solutions* (Harper and Row Publishers, London), 1971.

86. K. F. Freed, *J. Phys. A Math. Gen.*, 18, 871, 1985.

87. J. F. Douglas, *Macromolecules*, 22, 3707, 1989.

88. C. B. Roth, K. L. McNerny, W. F. Jager, and J. M. Torkelson, *Macromolecules*, 40, 2568, 2007.

89. F. W. Starr and J. F. Douglas, *Phys. Rev. Lett.*, 106, 115702, 2011.

90. D. Thirumalai and R. D. Mountain, *Phys. Rev. E*, 47, 479, 1993.

91. W. Kob, C. Donati, S. J. Plimpton, P. H. Poole, and S. C. Glotzer, *Phys. Rev. Lett.*, 79, 2827, 1997.

92. Y. Gebremichael, T. B. Schrøder, F. W. Starr, and S. C. Glotzer, *Phys. Rev. E*, 64, 051503, 2001.

93. D. Masao and S. F. Edwards, *The Theory of Polymer Dynamics* (Oxford University Press, Oxford), 1988.

94. P. Z. Hanakata, J. F. Douglas, and F. W. Starr, *Nat. Commun.*, 5, 256207, 2014.

95. B. A. Pazmiño Betancourt, J. F. Douglas, and F. W. Starr, *J. Chem. Phys.*, 140, 204509, 2014.

96. H. Eyring, *J. Chem. Phys.*, 4, 283, 1936.

97. R. H. Ewell, *J. Appl. Phys.*, 9, 252, 1938.

98. S. Glasstone, K. Laidler, and H. Eyring, *The Theory of Rate Processes: The Kinetics of Chemical Reactions, Viscosity, Diffusion and Electrochemical Phenomena, International Chemical Series* (McGraw-Hill Book Company, Incorporated, New York), 1941.

99. F. W. Starr, P. Z. Hanakata, B. A. Pazmiño Betancourt, S. Sastry, and J. F. Douglas, Fragility and cooperative motion in polymer glass formation, in *Fragility of Glass Forming Liquids*, edited by A. L. Greer, K. F. Kelton, and S. Sastry (Hindustan, New Delhi, India), 337–361, 2013.

100. B. A. Pazmiño Betancourt, P. Z. Hanakata, F. W. Starr, and J. F. Douglas, *Proc. Natl. Acad. Sci.*, 112, 2966, 2015.

101. P. Z. Hanakata, B. A. Pazmiño Betancourt, J. F. Douglas, and F. W. Starr, *J. Chem. Phys.*, 142, 234907, 2015.

102. F. W. Starr, T. B. Schrøder, and S. C. Glotzer, *Phys. Rev. E*, 64, 021802, 2001.

103. B. Amitabh, L. C. Yang Hoichang, and S. L. S. Cho Kilwon, Benicewicz Brian C. and Kumar Sanat K., *Nat. Mater.*, 4, 693, 2005.

104. P. Rittigstein, R. D. Priestley, L. J. Broadbelt, and J. M. Torkelson, *Nat. Mater.*, 6, 278, 2007.

105. E. W. Madge, *J. Appl. Phys.*, 5, 39, 1934.

106. F. Eirich and R. Simha, *J. Chem. Phys.*, 7, 116, 1939.

107. L. Qun-Fang, H. Yu-Chun, and L. Rui-Sen, *Fluid Phase Equilibria*, 140, 221, 1997.

108. R. Speedy, F. Prielmeier, T. Vardag, E. Lang, and H.-D. Lüdemann, *Mol. Phys.*, 66, 577, 1989.

109. H. G. E. Hentschel, S. Karmakar, I. Procaccia, and J. Zylberg, *Phys. Rev. E*, 85, 061501, 2012.

110. T. Iwashita, D. M. Nicholson, and T. Egami, *Phys. Rev. Lett.*, 110, 205504, 2013.

111. A. Bondi, *J. Chem. Phys.*, 14, 591, 1946.

112. A. Yelon, B. Movaghar, and R. S. Crandall, *Rep. Prog. Phys.*, 69, 1145, 2006.

113. A. Yelon, E. Sacher, and W. Linert, *Catal. Lett.*, 141, 954, 2011.

114. A. Yelon and B. Movaghar, *Phys. Rev. Lett.*, 65, 618, 1990.

115. G. Boisvert, N. Mousseau, and L. J. Lewis, *Phys. Rev. Lett.*, 80, 203, 1998.

116. R. M. Digilov and M. Reiner, *Eur. J. Phys.* 25, 15, 2004.

117. L. K. Nash, *J. Chem. Ed.* 61, 981, 1984.

118. I. M. Barclay and J. A. V. Butler, *Trans. Faraday Soc.*, 34, 1445, 1938.

119. R. P. Bell, *Trans. Faraday Soc.*, 33, 496, 1937.

120. M. G. Evans and M. Polanyi, *Trans. Faraday Soc.*, 32, 1333, 1936.

121. H. S. Frank, *J. Chem. Phys.*, 13, 493, 1945.

122. R. M. Barrer, *Trans. Faraday Soc.*, 39, 48, 1943.

123. C. E. Waring and P. Becher, *J. Chem. Phys.*, 15, 488, 1947.

124. J. C. Dyre, *J. Phys. C*, 19, 5655, 1986.

125. R. K. Eby, *J. Chem. Phys.*, 37, 2785, 1962.

126. G. J. Dienes, *J. Appl. Phys.*, 21, 1189, 1950.

127. J. Dudowicz, K. F. Freed, and J. F. Douglas, *J. Chem. Phys.*, 111, 7116, 1999.

128. S. Sastry, P. G. Debenedetti, and F. H. Stillinger, *Nature*, 393, 554, 1998.

129. A. J. Rahedi, J. F. Douglas, and F. W. Starr, *J. Chem. Phys.*, 128, 024902, 2008.

130. J. F. Douglas, J. Dudowicz, and K. F. Freed, *J. Chem. Phys.*, 125, 144907, 2006.

131. U. Tracht, M. Wilhelm, A. Heuer, H. Feng, K. Schmidt-Rohr, and H. W. Spiess, *Phys. Rev. Lett.*, 81, 2727, 1998.

132. P. B. Macedo and A. Napolitano, *J. Chem. Phys.*, 49, 1887, 1968.

133. O. Yamamuro, I. Tsukushi, A. Lindqvist, S. Takahara, M. Ishikawa, and T. Matsuo, *J. Phys. Chem. B*, 102, 1605, 1998.

134. L. Hong, P. D. Gujrati, V. N. Novikov, and A. P. Sokolov, *J. Chem. Phys.*, 131, 194511, 2009.

135. P. G. Debenedetti, *Metastable Liquids* (Princeton University Press, Princeton), 1996.

136. J. Zhao, S. L. Simon, and G. B. McKenna, *Nat. Commun.*, 4, 1783, 2013.

137. J. Dudowicz, K. Freed, and J. Douglas, *Adv. Chem. Phys.*, 137, 125, 2008.

138. E. C. Glor and Z. Fakhraai, *J. Chem. Phys.*, 141, 194505, 2014.

139. J. E. Pye, K. A. Rohald, E. A. Baker, and C. B. Roth, *Macromolecules*, 43, 8296, 2010.

140. R. J. Lang and D. S. Simmons, *Macromolecules*, 46, 9818, 2013.

141. J. D. Stevenson and P. G. Wolynes, *J. Chem. Phys.*, 129, 234514, 2008.

142. H. Zhang and J. F. Douglas, *Soft Matter*, 9, 1254, 2013.

143. H. Zhang, Y. Yang, and J. F. Douglas, *J. Chem. Phys.*, 142, 084704, 2015.

144. K. Watanabe, T. Kawasaki, and H. Tanaka, *Nat. Mater.*, 10, 512, 2011.

145. A. Schönhals, H. Goering, C. Schick, B. Frick, and R. Zorn, *J. Non-Cryst. Solids*, 351, 2668, 2005.

146. C. Jeong and J. F. Douglas, *J. Chem. Phys.* 143, 144905, 2015.

147. M. E. Mackura and D. S. Simmons, *J. Polym. Sci. B*, 52, 134, 2014.

148. C. Angell, *Polymer*, 38, 6261, 1997.

149. C. A. Angell, *Science*, 267, 1924, 1995.

Mechanical properties of polymers and nanocomposites close to the glass transition

ALAIN DEQUIDT, DIDIER R. LONG, SAMY MERABIA, AND PAUL SOTTA

9.1 INTRODUCTION

The most striking feature of glass-forming liquids, either simple or polymeric liquids, is a steep increase of their viscosity as temperature decreases [1]. Usually, the glass transition temperature T_g is arbitrarily defined as the temperature at which the viscosity reaches 10^{12} Pa.s in simple liquids, or at which the dominant relaxation time τ_α becomes larger than a macroscopic time scale, for example, $\tau_g \sim 1$ s or 10^2 s [2]. Another important feature of glass transition is the strongly heterogeneous nature of the dynamics close to T_g [3,4]. This strongly heterogeneous nature has been demonstrated experimentally over the past 20 years using nuclear magnetic resonance (NMR) [5–7], fluorescence recovery after photobleaching (FRAP) [8–12], dielectric hole burning [13], or solvation dynamics [14]. These studies have demonstrated the coexistence of domains with relaxation time distributions spread over more than four decades at temperatures typically 20 K above T_g. The diffusion dynamics of molecular probes about 1 nm in diameter has also been measured, showing that the Stokes' law is essentially valid in the high-temperature range, but breaks down below $T_g + 20$ K [8]. This was one of the earlier indications of the spatial nature of dynamical heterogeneities and provided an estimate of their size ξ. The characteristic size ξ has been estimated by NMR [6] to be 3–4 nm at $T_g + 20$ K (in the case of van der Waals liquids), whereas it is as small as 1 nm in glycerol [7].

An important issue has been to know whether relaxation mechanisms take place over an increasing length scale as temperature decreases [15]. This issue has led to a great interest in studying the dynamics in thin liquid films over the past 15 years. Since the work by Keddie and coworkers [16] who observed a decrease of T_g for deposited thin films, many other researchers have confirmed the effect of confinement on thin-film dynamics (see, for example, [17–25]). In some instances, an increase of T_g is observed when the interaction with the substrate is strong [22,23,26–31]. In addition to understanding the origin of T_g shifts measured in thin films, the link between T_g shifts at interfaces and mechanical properties of reinforced elastomers has also been an issue of importance. Indeed, experiments performed on silica-filled elastomers have shown that mechanical properties are changed in a way that is consistent with the T_g shifts measured by NMR on the same samples [32–36], or with other experiments performed on thin deposited films such as those mentioned above. This point of view is consistent with some other studies performed over the past few years [37–43].

Of high importance also are the plastic properties of glassy polymers. When applying a deformation below T_g, one observes a yield stress at deformation amplitude of a few percents, which corresponds to a peak in the stress–strain curve. Beyond the yield stress, the stress displays a plateau at a value slightly lower than the peak: the polymer undergoes plastic flow. At deformations larger than about a few tens of percents, strain hardening is observed in some cases, in particular, for high molecular weight entangled or cross-linked polymers. These properties have been studied for many years due their high practical importance [44]. The oldest model for describing plastic deformation is the Eyring model [45,46], according to which free energy barriers are reduced under the applied stress. This effect is controlled by the so-called activation volume V^*, which is an adjustable parameter without clear interpretation [47,48]. The idea that plastic flow results from a stress-induced acceleration of the dynamics at the molecular level has been supported by several recent experiments [49–53]. Loo et al. [49] have studied the dynamics in the amorphous phase of polyamide under stress by NMR. Kalfus et al. [50] have studied dielectric relaxation during plastic flow. They observed that secondary relaxations are not significantly affected during deformation, whereas the α-relaxation is modified: an increase of tan δ is indeed observed in the low-frequency domain. Lee et al. [51,53] and Bending et al. [52] have shown that small probe diffusion is accelerated under applied deformation, and that the so-called stretching exponent β increases. They interpreted this result by the fact that the dynamics becomes faster and more homogeneous during plastic deformation.

Mechanical properties of glasses have been also studied by molecular dynamics (MD) simulations [54–60]. In particular, Leonforte et al. [56] and Riggleman et al. [54,57] have shown that mechanical properties are heterogeneous. When submitted to an applied strain, the deformation field is nonaffine. Riggleman et al. have shown that the elastic modulus is heterogeneous on a scale of order 1 nm, with some regions having negative moduli. The possible link between dynamical heterogeneities and these mechanical heterogeneities has been discussed in Dequidt et al. [61]. The idea that stress enhances molecular mobility is supported by these simulations. This idea has also been considered by Chen and Schweizer within the nonlinear Langevin equation model (NLE) [47,48,62–64]. Their model reproduced many features of plastic deformation. Though these studies support that the dynamics is enhanced during plastic deformation, a detailed description of the dynamical behavior from the the molecular level up to the scale of a few tens of nanometer during plastic deformation is still lacking. Some of these approaches are mean field theories, which lack a spatial description.

We describe in this chapter a model for the dynamics in the bulk developed recently, which proposes an interpretation both for thin-film experiments and the heterogeneous dynamics in the bulk [65–69], in the case of van der Waals liquids. This model is a mesoscale model, with spatial resolution typically the size of dynamical heterogeneities. This so-called percolation of free volume distribution (PFVD) model [70] allows to reach larger spatial scale and time scales as compared to molecular dynamics simulations [54–60]. It requires the incorporation of specific data regarding the considered polymers, which may be obtained experimentally. The details of the physics on smaller spatial scales are included in a few adjustable parameters, related to thermodynamical properties of the considered polymers and phenomenological laws such as the Williams–Landel–Ferry (WLF) law.

The PFVD model can be considered as an extension of the free volume model. The idea that the free model is a relevant parameter for describing relaxation processes has been challenged by several authors (see, for example, [71–74]). In particular, a series of experiments led to the conclusion by some authors that the free volume is not a relevant parameter by considering the evolution of the dynamics as a function of both the pressure and the temperature, or equivalently as a function of both the density and the temperature. By comparing the respective contributions of each of these macroscopic parameters, Tarjus et al. concluded that, as a universal rule, the relevant parameter for controlling the dynamics is temperature, the role of the density being secondary [72,73]. For instance, at a given density, the considered liquids were found to have faster dynamics at higher temperature (and thus higher pressure). The average density cannot thus be sufficient for describing the dynamics. Long and Merabia [75] argued that, in the case of nonpolar liquids, "the temperature and pressure dependence of the dynamics can be explained quantitatively by taking into account the whole spectrum of density fluctuations" [69,75]. The *facilitation mechanism* proposed by Merabia and Long [66] and discussed in Section 9.2.3 is key in this regard, as well as for calculating the evolution of the dynamics in out-of-equilibrium conditions (aging and rejuvenating). It must be noted also that the case of liquids with a high density of hydrogen bonds (as it is the case for glycerol) is very different from that of nonpolar liquids. Their dynamics is controlled by a network of hydrogen bonds and the concept of free volume is not relevant for them as it has been discussed in Merabia and Long [75]. The difference between liquids with strong specific interactions and nonpolar liquids is discussed in Sections 9.2.6 and 9.2.11.

The point of view from which this model derives is that van der Waals liquids exhibit generic features regarding the glass transition. These features are the consequences of generic thermodynamic properties (see [76,77]), and of generic relaxation mechanisms on molecular scale. Relaxation mechanisms are best qualitatively interpreted within the framework of the free volume model which assumes that the slowing down is the consequence of the reduction of free volume when cooling the liquids. The aim of this chapter is to discuss how these generic features allows for describing many different properties within a unified picture, and with a few adjustable parameters: (a) the heterogeneous nature of the dynamics [65], (b) the violation of the Stokes' law observed for small probes [66], (c) aging and rejuvenation phenomena [69], (d) the shift of glass transition temperature at interfaces [61,65,67], (e) the dependence of the dynamics as a function on both the temperature and the pressure [75], (f) mechanical properties of glassy polymers (in the linear regime), in the bulk and in thin films [61], and (g) reinforcement mechanisms in nano-filled elastomers [35,36,61,78]. The microscopic behavior during plastic deformation is not discussed here and is part of current extensions of the present model [79].

The generic thermodynamical properties have been described for instance in Long and Lequeux [65] and Masnada [76]. These features can be summarized by the existence of critical temperature that depends on the strength of the molecular interactions and by a close packing density. The generic features of the glass transition are contained in the Williams–Landel–Ferry law, which describes the evolution of the relaxation time as a function of temperature. As we discuss in this chapter, many detailed aspects regarding the glass transition in van der Waals liquids can be quantitatively described thanks to the parameters contained in these laws.

In Section 9.2, we introduce and discuss the physical basis of the PFVD model. The mechanical properties are discussed at the level of a percolation model. In Section 9.3, we introduce a more quantitative extension of the model, in which the mechanical properties are obtained from the spatial distribution of relaxation times. The model is solved numerically in three-dimensional (3D), in the bulk, and in confinement. T_g shifts are obtained, and we show that the glass transition in confinement is broadened as compared to the bulk. These results may allow for understanding reinforcement properties in filled elastomers, both in the linear regime and in the plastic regime of deformation. Note that more direct experiments aimed at measuring mechanical properties of confined polymers are very scarce and only partial results have been obtained yet [80]. The physical properties of filled elastomers are discussed in Section 9.4. Then, we introduce and discuss a model that allows for calculating mechanical properties of filled elastomers and which is a coarse-grained version of the model regarding mechanical properties in confinement presented in Section 9.3.

9.2 DESCRIPTION OF THE GLASS TRANSITION MODEL

9.2.1 THERMODYNAMICAL MODEL FOR VAN DER WAALS LIQUIDS

It has been discussed in Long and Lequeux [65] and Masnada [76] (see also the quoted references) that the thermodynamics of compressible van der Waals liquids depends in an universal way on a few parameters.

Essentially, the equilibrium density results from a balance between van der Waals attractive forces and confinement entropy. This point of view is consistent with that of Prigogine [77,81–83]. The attractive energy is directly related to the Hamaker constant C of the considered polymer [84], whereas the confinement entropy per monomer is taken to be $\ln(\rho_0 - \rho)$, where ρ is the monomer number density, and ρ_0 is analogous to a close packing density of monomers and would correspond to the density extrapolated at $T = 0$. Expressions for the equilibrium density $\rho_{eq}(T)$, dimensionless thermal expansion coefficient χ_T, and bulk (compressibility) modulus K of the polymer melt were obtained as a function of the ratio T/T_c, where the temperature $T_c \sim C\rho_0^2$ (when expressed in joule) is the only energy scale in the system and is proportional to the strength of van der Waals attractions between molecules or monomers. T_c is of order 1000 K for usual polymers. The vicinity of T_g corresponds to the low-temperature regime defined by $T \ll T_c$ [65]. In this regime, the following approximations are conceptually important, even though not perfectly accurate numerically:

$$\chi_T \sim \epsilon$$
$$K \sim \frac{k_B T \rho_0}{\epsilon^2} \tag{9.1}$$

where the dimensionless quantity

$$\epsilon = \frac{\rho_0 - \rho_{eq}}{\rho_0} \approx \frac{T}{4T_c} \tag{9.2}$$

can be interpreted as the free volume fraction. k_B is the Boltzmann constant. Typical values for the free volume ϵ are about 0.2–0.15 in the regimes of interest corresponding to the WLF regime. Typically, $\rho_0 \sim 10^{28}$ m^{-3}. The parameters of the model can be determined for each polymer by fitting the pressure–volume–temperature (PVT) data [76].

9.2.2 WLF LAW AND FREE VOLUME MODEL

Throughout the chapter, τ_α denotes the time that dominates the mechanical behavior. The variation of this relaxation time or of the viscosity in the bulk is given by the empirical WLF law (or Vogel–Fülcher–Tammann [VFT] law in the context of simple liquids) [2,85]:

$$\log\left(\frac{\tau_\alpha(T)}{\tau_\alpha(T_0)}\right) = \frac{-C_1(T - T_0)}{C_2 + T - T_0} \tag{9.3}$$

where T_0 is a reference temperature. Here, typically, T_0 is close to the glass transition temperature T_g, and $\tau_\alpha(T_0)$, the relaxation time at temperature T_0, is a macroscopic time scale, for example, comparable to 100 s. C_1 and C_2 are constants that depend on the considered polymers. The relaxation time would diverge at the Vogel temperature $T_\infty = T_0 - C_2$, with $T_\infty \sim T_g - 50$ K typically. In practice, however, the relaxation times become too long to be measurable below $T_g - 30$ K, which means that Equation 9.3 is in general checked at temperatures above $T_g - 20$ K only.

The so-called free volume model [2] interprets phenomenologically the WLF law. The basic idea is that for one molecule (or for one monomer in the context of polymers) to move, one needs to pack a number $n \propto v_m/v(T)$ other molecules, where v_m and $v(T)$ are, respectively, the volume of one molecule and the average free volume per molecule. That results in an entropic barrier for one molecule to move, and a jump time given by $\tau_\alpha = \tau_0 \exp\left(\Theta v_m/v(T)\right)$, where Θ is a number of order one and τ_0 is a microscopic time scale comparable to 10^{-12} s. Linearization of $v(T)$ allows qualitatively to recover the WLF law. The issue we consider now is to write the WLF law as a function of the density $\rho_{eq}(T)$ instead of the temperature as is usually the case. Then, one would expect that the free volume $v(T)$ should be given by $v(T) = 1/\rho_{eq} - 1/\rho_0$ while the volume v_m is taken to be, for example, $1/\rho_0$. One expects then a dependence $\tau_\alpha = \tau_0 \exp\left(\Theta \rho_0/(\rho_0 - \rho_{eq})\right)$. However, it turns out not to be the case: it is not possible to obtain the sharp increase of the relaxation time τ_η between $T_g + 100\ K$ and T_g

(about 12 orders of magnitude) using this value of the parameter ρ_0. We therefore need to introduce another adjustable parameter $\rho_0 < \rho_0$, which represents the density at which the relaxation time is supposed to diverge. It was shown in Merabia and Long [66] that Equation 9.3 may be expressed in the equivalent form:

$$\tau_\alpha(T) = \tau_0 \exp \frac{\Theta}{\tilde{\epsilon}(T)} \tag{9.4}$$

where τ_0 is a vibrational time of order 10^{-12} s, Θ is a number of order unity, and

$$\tilde{\epsilon} = \frac{\tilde{\rho}_0 - \rho_{eq}}{\tilde{\rho}_0} \tag{9.5}$$

is the "dynamical" free volume fraction, with ρ_{eq} the average (equilibrium) number density at temperature T. The density $\tilde{\rho}_0$ is significantly smaller than ρ_0. This is a fact that has been recognized many years ago [86] for which there is no satisfactory interpretation in the literature yet. Typically, $(\tilde{\rho}_0 - \rho_{eq})/\tilde{\rho}_0 \sim 0.03$, whereas $(\rho_0 - \rho_{eq})/\rho_0 \sim 0.15$ close to T_g [66]. Close to T_g, the free energy barriers in the liquids, that is, the argument in the exponential of Equation 9.4 are of order $30k_BT$ typically, corresponding to macroscopic relaxation times of order 10^2 s.

9.2.3 SCALE OF DYNAMICAL HETEROGENEITIES: FACILITATION MECHANISM

The underlying assumption of the PFVD model is that density fluctuations at scale $\xi \sim 3 - 5$ nm [66,69] create a wide distribution of relaxation times. Dynamical heterogeneities might also be related to some (yet unspecified) spatial organization on a microscopic scale. However, no structural transition has been observed when lowering temperature or increasing the pressure in van der Waals glass formers, over the whole WLF regime of these liquids [87]. Such a hypothetical transition could be observed by x-ray or neutron scattering, whereas no particular change in the structure factor has ever been observed close to T_g. A structural transition would also result in a discontinuity of thermodynamical quantities such as the bulk modulus (or its derivative) as a function of temperature or pressure, which has not been observed yet. By contrast, all experimental evidence indicates that no structural change occurs when cooling a glass former down to T_g.

To account for the large spatial scales of dynamical heterogeneities, one needs to consider a conserved quantity, which involves α-relaxation processes for relaxing. In van der Waals liquids, both decreasing the temperature and increasing the density lead to a slowing down of the dynamics [75]. Thus, two different kinds of fluctuations, of temperature or of density, may be considered on a microscopic scale for explaining dynamical heterogeneities. However, thanks to high-frequency phonons unrelated to α-relaxation, temperature fluctuations can relax in less than 1 ns typically on the nanometer scale. These fluctuations are thus short-lived compared to the glass transition time scales and cannot be related to the spatial nature of dynamical heterogeneities. Thus, there is no other clear candidate than long-lived density fluctuations for explaining the spatial nature of dynamical heterogeneities.

Experimentally, two kinds of density fluctuations are observed: long-lived density fluctuations, which can be associated with α-relaxation, that is, elementary monomer jumps, and short-lived ones associated to fast vibrations such as acoustic vibrations, which takes place at a fixed molecular environment [88]. The coexistence of density fluctuations associated to very different relaxation mechanisms can be inferred by considering, for example, dilatational compliance measurements of liquids [88]. Liquids close to T_g exhibit a short time scale compliance $1/3K_0$, that is, immediately after a hydrostatic stress is applied. Then, at long times, the dilatational compliance saturates to a higher value $1/3K > 1/3K_0$. The compliance at short time scales can be associated to fast relaxation mechanisms, of elementary time scale of order 1 ps, whereas the long-time behavior is associated with the much longer α-relaxation, that is, the elementary monomer jumps, which are the relevant mechanisms. Therefore, following Robertson [89], the density fluctuations that we consider in this chapter are those associated with long relaxation processes. Note that the fast density fluctuations discussed here are the fast phonons through which thermal fluctuations relax. We just present here a different viewpoint. Note that it is experimentally observed that both moduli K and K_0 are comparable and that the ratio K/K_0 is of order 0.3 typically.

We propose thus that slow (respectively fast) subunits correspond to upward (respectively downward) density fluctuations at a characteristic scale $N = \rho\xi^3$ monomers to be determined [65,66,67,69]. At the length scale ξ, the distribution of density fluctuations $\delta\rho = \rho - \rho_{eq}$ follows the Boltzmann law:

$$p(\delta\rho) \propto \exp\left(-\frac{K\xi^3}{2k_BT}\frac{\delta\rho^2}{\rho_{eq}^2}\right) \tag{9.6}$$

Typical fluctuations of the density are thus of order

$$\frac{\delta\rho}{\rho_{eq}} \approx \pm\sqrt{\frac{k_BT}{K\xi^3}} \tag{9.7}$$

which yields $\delta\rho/\rho_{eq} \approx 1\%$ with $K = 1$ GPa, $\xi = 3$ nm, and $T = 300$ K. Combining Equations 9.1 and 9.7 with $N = \rho\xi^3$, it is useful to define

$$\frac{\delta\rho}{\tilde{\rho}_0} = \frac{\alpha\epsilon}{N^{1/2}} \tag{9.8}$$

where α is a dimensionless number that characterizes the relative amplitude of the fluctuations and follows the Gaussian statistics $Q(\alpha)$ with standard deviation 1: $Q(\alpha) \sim \exp(-\alpha^2/2)$.

Then, the characteristic scale N_c must be determined. According to Equation 9.6, on very small scales, one can find very dense subunits with probability of order one. Therefore, the relaxation time τ_{slow} of these subunits should be very long. Equation 9.4 is valid on a macroscopic scale. We assume that we can use it on a microscopic scale, with the local density. One obtains then

$$\tau_{slow} \sim \tau_0 \exp\left(\frac{\Theta}{\tilde{\epsilon} - \alpha\epsilon/N^{1/2}}\right) \tag{9.9}$$

where α is of order +1. This relaxation time would be observed if relaxation took place within the considered volume *at fixed density*, by local reorganization of free volume within the considered subunit. However, another relaxation process takes place and competes with the local reorganization. This is the relaxation of density fluctuations, which takes place through diffusion that occurs at the boundary of the subunit into the lower-density neighboring subunits. This process is schematized in Figure 9.1. The local relaxation time in the fast surrounding subunits is much smaller than the local one within the slow subunit. The local relaxation time in the fast subunit is given by

$$\tau_{fast} \sim \tau_0 \exp\left(\frac{\Theta}{\tilde{\epsilon} + \alpha\epsilon/N^{1/2}}\right) \tag{9.10}$$

where α is of order +1. The lifetime of a considered density fluctuation on scale N is then

$$\tau_{life} \sim \tau_0 N^{2/3} \exp\left(\frac{\Theta}{\tilde{\epsilon} + \alpha\epsilon/N^{1/2}}\right) \tag{9.11}$$

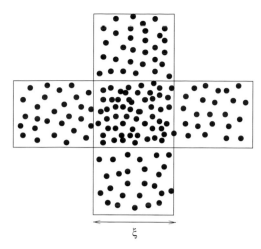

Figure 9.1 Two mechanisms compete for determining the α-relaxation: (1) local reorganization of free volume, at fixed local density (central subunit), and (2) diffusion of monomers from the densest subunit in the more mobile neighboring subunits, or equivalently, free volume diffusion from the fast surrounding subunits toward the densest one. A dense subunit can melt through this process. A subunit of large density will melt and relax in a longer time if it is surrounded only by subunits that are themselves relatively slow. The lifetime of a dense subunit surrounded by much faster subunits with internal relaxation time τ_{fast} is $\tau_{life} = \tau_{fast} N^{2/3}$. This time is frequently shorter than the relaxation time a dense subunit would have if its overall density was maintained fixed by imaginary walls. The latter would be τ_{slow} as given in the text by Equation 9.9. (Reprinted with permission from S. Merabia and D. Long, *J. Chem. Phys.*, 125, 234901. Copyright 2006, American Institute of Physics.)

The factor $N^{2/3}$ comes from the fact that free volume has to diffuse on a scale $\xi = aN^{1/3}$, with elementary jump time τ_{fast} on the monomer scale a. At small scale (i.e., small N), the diffusion process is faster than the individual monomer jump time, which would be observed if the density of the slow subunit was maintained fixed. From the discussion above, the largest relaxation times in the system are determined by the relation

$$\tau_{life} = \tau_{slow} \qquad (9.12)$$

Indeed, for obtaining long relaxation times, one needs large density fluctuations, and also long lifetimes of density fluctuations (see Figure 9.2). From the relation (9.12), one obtains an expression for the scale N_c of dynamical heterogeneities:

$$N_c \approx \frac{\Theta^2 \epsilon^2}{\bar{\epsilon}^4} \frac{1}{\left(\ln \left(\frac{\Theta^2 \epsilon^2}{\bar{\epsilon}^4} \right) \right)^2} \qquad (9.13)$$

The scale $\xi = aN_c^{1/3}$ ($a \approx 0.45$ nm is one monomer length [2]) thereby is the smallest scale at which the lifetime of density fluctuations can be equal or larger than τ_α (see Figure 9.2). Density fluctuations on smaller scales are irrelevant for the α-relaxation process, being too short lived.

The physical mechanism that we have introduced corresponds to the melting of dense subunits by less dense surrounding ones, which are fast subunits. We call this process a facilitation mechanism by which very dense subunits relax by diffusion of free volume from the surrounding environment [66]. The term "facilitation" was coined by Chandler et al. [90,91], and by Chen et al. [70] in the context of our model. Note that the Merabia and Long facilitation mechanism has been discussed and used by Tito et al. [92] for studying T_g shifts in suspended films.

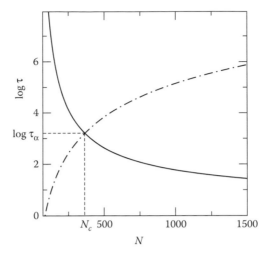

Figure 9.2 Determination of the length scale of dynamical heterogeneities. The black solid line represents the relaxation time τ_{slow} given by Equation 9.9. The dashed-dotted line represents the lifetime of density fluctuations τ_{life} (see Equation 9.11). On large scale, density fluctuations are long lived. On shorter length scales, one observes comparatively large density fluctuations, which should lead to large relaxation times τ_{slow}. On the other hand, on short length scales, density fluctuations are short lived. There exists an intermediate length scale on which one has large density fluctuations that are long lived. That corresponds to the scale of dynamical heterogeneities N_c. (Reprinted with permission from S. Merabia and D. Long, *J. Chem. Phys.*, 125, 234901. Copyright 2006, American Institute of Physics.)

From the discussion regarding the facilitation mechanism, we have seen that two processes compete for allowing individual molecular jumps: The first one corresponds to a jump, at fixed local free volume. By that, we mean that there is no reallocation of free volume on a scale larger than $aN_c^{1/3}$ on the considered time scale. This is the dominant process in relatively fast subunits, because there is a large amount of local free volume. The second one is the dominant one in relatively slow subunits. The local fraction of free volume is too small for allowing a molecular jump at fixed local fraction of free volume. Local reorganization of the free volume would be too slow for allowing an individual molecular jump on the corresponding time scale. The process of dissolution by diffusion in faster neighboring subunits is much faster. Then, once the local fraction of free volume has increased through the free volume diffusion process, an individual molecule can jump. This facilitation process happens in a time that is smaller than the time that would be required if the initial density was maintained fixed by imaginary walls. The characteristic times of these two competing processes are schematized in Figure 9.2.

By using the WLF law on the scale N_c, and by using the statistics of density fluctuations, one can obtain the distribution of relaxation time as exemplified in Figure 9.3. The corresponding parameters are given in Merabia and Long [69]. See more detailed discussion later in the text.

As a consequence of the facilitation mechanism, the relaxation time τ_α satisfies the relation

$$\tau_\alpha = N_c^{2/3}\tau_{fast} \tag{9.14}$$

where τ_{fast} corresponds to a fraction q_c (to be determined later) of the fast subunits. The significance of this relation is that the relaxation of the slowest subunits corresponds both to internal processes and to the melting by the faster environment. The relaxation time τ_{fast} satisfies the relation

$$\int_0^{\tau_{fast}} Q(\tau)d\tau = q_c \tag{9.15}$$

where q_c is an adjustable parameter, typically a few tens of percents.

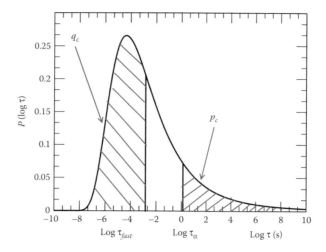

Figure 9.3 Equilibrium relaxation time distribution at temperature T_g. τ_α is the relaxation time of dense subunits that percolate (fraction p_c) while τ_{fast} is the typical relaxation time of mobile units (fraction q_c). At equilibrium, one has $\tau_\alpha = \tau_{fast}N_c^{2/3}$ (see [69] for the parameters used). (Reprinted with permission from S. Merabia and D. Long, J. Chem. Phys., 125, 234901. Copyright 2006, American Institute of Physics.)

One obtains an expression for the scale N_c of dynamical heterogeneities (Equation 9.13). The only adjustable parameters are those required for the WLF law (Equation 9.3) and for describing the thermodynamics of the considered liquid. It was obtained in Merabia and Long [66] that $N_c \approx 1000, 700, 100, 100, 400, 300$ for *ortho*-terphenyl (OTP), poly(methyl methacrylate) (PMMA), poly(*n*-butyl methacrylate) (PBMA), polyisobutylene (PIB), poly(vinyl acetate) (PVAc), and polystyrene (PS) at their respective T_g, which corresponds to ξ between 2 nm (PIB, PBMA) and 3–4 nm (PS, OTP, PVAc, PMMA). N_c is plotted as a function of temperature for OTP, PS, and PVAc in Figure 9.4. N_c is a decreasing function of temperature. At $T_g + 80$ K, it is of order 10, which is very small and corresponds to one monomer and its close neighbors. In this high-temperature regime, the description we just proposed breaks down. The associated scale for dynamical heterogeneities

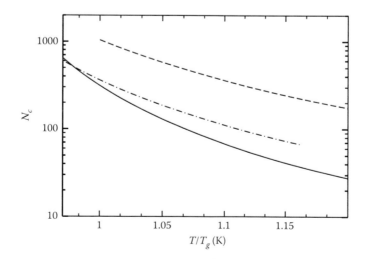

Figure 9.4 Variation of N_c as a function of T/T_g, given by Equation 9.28, for PS (continuous line), PVAc (dashed-dotted line), and OTP (dashed line). At T_g, we find that $N_c \approx 3–400$ for PS and PVAc, and $N_c \approx 1000$ for OTP, which amount to characteristic sizes for the heterogeneities of about $\xi \approx 3–4$ nm for these liquids. The parameters with which N_c have been calculated are given in Merabia and Long [66]. (Reproduced from S. Merabia and D. Long, Eur. Phys. J. E, 9, 195, 2002. With kind permission of The European Physical Journal [EPJ].)

is so small that dynamics is close to being homogeneous. The values obtained here are comparable to those measured by NMR by Spiess and coworkers regarding PVAc (3 ± 1 nm) and OTP (3 ± 1 nm) [6,93].

9.2.4 DYNAMICAL HETEROGENEITIES AND MECHANICAL RELAXATION

As we have just discussed, dynamics is strongly heterogeneous in supercooled liquids, with relaxation times spanning many decades, up to 8 as measured by dielectric spectroscopy (DSC) [14]. Slow regions coexist with fast ones on a scale of a few nanometers. Then, the question is: what is the dominant relaxation time that is measured in a given experiment? We will see indeed that the probed time scale depends on the considered experiment. In particular, in a mechanical experiment, which relaxation times are probed? The latter may be defined in the following way. Let's apply a step strain to a polymer liquid. The dominant relaxation time will be obtained by considering the final relaxation of the stress. The issue is: can we identify in Figure 9.3 this dominant relaxation time?

Our point of view is the following. In this kind of experiment, very fast relaxation times are not relevant because their contribution to the stress relaxes very fast. They contribute to the short time scales response. In other words, their contribution to the viscosity for instance is very small, being proportional to their own relaxation times. Very slow subunits are very rare and surrounded by a sea of faster subunits. Final relaxation takes place before their own internal relaxation has taken place because the stress they bear is transmitted by the surrounding environment, which relaxes in shorter time scale than the internal relaxation time of these slow subunits. Therefore, their internal relaxation time is not probed in mechanical experiments as the one just mentioned. Accordingly, it has been proposed [65,67] that the longest relaxation time probed in such experiments corresponds to a percolation threshold, p_c. Subunits slower than those corresponding to the percolation threshold cannot support stress on longer time scales because the surrounding environment has relaxed. The final relaxation mechanism in dynamically heterogeneous liquids is schematized in Figure 9.5. The time scale corresponding to the percolation threshold p_c thus appears as a cutoff in the relaxation time distribution plotted in Figure 9.3. As a consequence, the effective distribution of relaxation times is given by the "bare distribution" (Figure 9.3), with an effective cutoff at long times τ_α defined by

$$\int_{\tau_\alpha}^{\infty} Q(\tau)d\tau = p_c \tag{9.16}$$

Figure 9.5 The final relaxation mechanism in a van der Waals liquid, in the WLF (or VFT) regime. Dynamical subunits have been divided into three types: fast subunits (light gray), subunits of internal relaxation time τ_α (gray), and subunits with internal relaxation time $\tau \gg \tau_\alpha$ (dark gray). The ensemble of subunits with internal relaxation time larger or equal to τ_α percolate. The dominant relaxation time is τ_α since the slowest subunits can rotate or diffuse in those with dominant time τ_α. (Reproduced from S. Merabia, P. Sotta, and D. Long, *Eur. Phys. J. E*, 15, 189, 2004. With kind permission of *The European Physical Journal* [EPJ].)

This is the time scale that controls the final relaxation of the stress. Shorter time scales contribute to the relaxation on short time scales; longer time scales are not relevant because they are too rare. By rewriting Equation 9.4, the bare internal relaxation time of subunits corresponding to a fluctuation α is then

$$\tau(\alpha) = \tau_0 \exp\left(\frac{\Theta}{\tilde{\epsilon} - (\alpha - \alpha_\eta)\epsilon/N^{1/2}} \right)$$

$$Q(\alpha) \propto \exp\left(\frac{-\alpha^2}{2} \right)$$

(9.17)

α_η is defined by

$$\int_{\alpha_\eta}^{\infty} Q(\alpha)d\alpha = p_c$$

(9.18)

where p_c is a 3D percolation threshold, typically close to 0.1 [67]. The glass transition temperature then corresponds to the temperature at which subunits with relaxation time larger or equal to $\tau_g \sim 100$ s percolate. Final relaxation in dynamically heterogeneous liquids is schematized in Figure 9.5.

9.2.5 GLASS TRANSITION IN THIN FILMS

In the same way as for the glass transition in the bulk, for the thin suspended polymer films to be in the glassy state, we assume that subunits of relaxation time τ larger than τ_g percolate. Note that the percolation we require is not in the direction normal to the film but in the direction of the plane parallel to the film so as to build macroscopic clusters of dynamics slower than τ_g. On the scale ξ, the film has a finite number $n = h/\xi$ of layers in the direction normal to the film (see Figure 9.6). For $\xi \approx 3-5$ nm (which amounts to taking $N_c \approx 300-1000$), n is typically of order 10. Then the situation, as far as percolation is concerned, is neither a 3D nor a two-dimensional (2D) problem but in the cross-over between the two regimes. The situation is schematized in Figure 9.6. The percolation threshold $p_c(n)$ [68] for a film of thickness n in the direction normal to the plane is

$$p_c(n) - p_c = \mu_1 n^{-1/\nu}$$

(9.19)

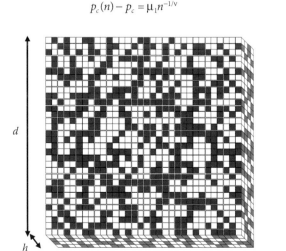

Figure 9.6 Percolation in a system of dimensions $d \times d \times h$: this is a situation of cross-over between 2D and 3D percolation. (Reproduced from P. Sotta and D. Long, *Eur. Phys. J. E*, 11, 375, 2003, with kind permission of *The European Physical Journal* [EPJ].)

where the equality holds asymptotically, and where μ_1 is a constant of order 1. The exponent v is the critical exponent for the 3D correlation length $v \approx 0.88$ [94].

A larger fraction of slow subunits is required for percolating as compared to the bulk: glass transition takes place at a lower temperature. The glass transition temperature $T_g(n)$ of a film of thickness $n = h/\xi$ is then determined by requiring that the fraction of slow domains is equal to $p_c(n)$. Following the free volume model, we assume that subunits of relaxation time τ_g correspond to subunits of density ρ_c, that is, $\tau(\rho_c) = \tau_g$. For the film being glassy, it needs to have a fraction of subunits of density larger than ρ_c equal to $p_c(n)$. The average (i.e., equilibrium) density of the film at its glass transition, $\rho_g(n)$, thus satisfies the relation

$$\frac{\rho_c - \rho_g(n)}{\rho_g(n)} = \left(\frac{T}{K\xi^3}\right)^{1/2} F(p_c(n))$$

$F(x)$ is the reciprocal function of $Erf(x)$, where Erf defined by

$$Erf(x) = \frac{1}{(2\pi)^{1/2}} \int_x^{+\infty} \exp(\frac{-u^2}{2}) du$$

is analogous to the error function $erf(x)$ [95]. Note that $F(x)$ is positive for $x < 0.5$ and negative for $x > 0.5$. It thus leads to a new glass transition temperature for the film, $T_g(n)$, given by the relation

$$\frac{\Delta T_g(n)}{T_g} = (F(p_c(n)) - F(p_c^{3D})) \frac{1}{N_c^{1/2}} \tag{9.20}$$

where we have used the relations $(K/(T\rho_0))^{1/2}\chi_T \approx 1$ (Equation 9.1). One obtains then

$$\frac{\Delta T_g(n)}{T_g} \approx \frac{-1.3}{N_c^{1/2}} n^{-1/v} = \frac{-1.3}{N_c^{(1/2 - 1/(3v))}} \left(\frac{a}{h}\right)^{1/v} \tag{9.21}$$

where we have used $n = h/\xi = h/aN_c^{1/3}$. The prefactor has been obtained by calculation on a cubic lattice [68]. Note that the dependence in h of the variation of T_g is independent of the considered type of percolation, since the correlation length exponent is universal. Only the prefactor indeed can be affected by considering another model of percolation. Note also that Equation 9.21 is valid for all polymers for which the dominant intermolecular interactions are van der Waals interactions. Indeed, we have used universal thermodynamical and dynamical properties. The exponent regarding the dependence in N_c is approximately equal to -0.13, which is very small. For a film 10 nm thick, this model predicts a T_g shift of order a few tens of kelvins.

Let's consider now the glass transition in the case of strongly adsorbed polymer films. As mentioned in the introduction, increases of T_g by up to 60 K for films of thickness 80 Å, has been observed by several groups [26–28,31]. For explaining the increase of the glass transition temperature, let's note that the substrate induces some kind of percolation. Indeed, even at temperatures larger than the bulk glass transition temperature, continuous paths of slow domains (i.e., of densities larger than ρ_c) can connect the two interfaces of the film. Indeed, the probability that both substrates are connected by slow subunits is never zero, but simply decreases exponentially with the distance d between the substrates, as $p \sim \exp(-d/\zeta)$, where ζ is the size of the slow aggregates. In 3D percolation, for a fraction of occupied sites $p < p_c$, the size ζ of the aggregates is

$$\zeta \approx \xi\mu_2(p_c - p)^{-v}$$

μ_2 is a prefactor of order unity [94]. Expressed as a function of temperature, one has $\zeta \approx \xi|(T - T_g)/T_g|^{-v}$ [65,67,68].

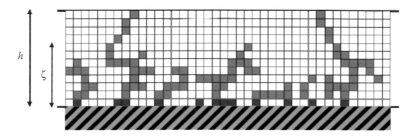

Figure 9.7 Schematics of a film of thickness h in the regime $\zeta < h$. Only percolation clusters connected to the lower surface (i.e., to the substrate) are shown. Very few clusters reach the upper surface. These slow clusters may dominate the mechanical response of the confined polymer. (Reproduced from P. Sotta and D. Long, *Eur. Phys. J. E*, 11, 375, 2003. With kind permission of *The European Physical Journal* [EPJ].)

We will see in the next section how we can calculate the elastic modulus of confined films by using this model, as a function of temperature. At this stage, we assume that a thin film is glassy when the size ζ of the aggregates of slow domains is comparable to the thickness of the film (see Figure 9.7). The probability of aggregates of size $r > \zeta$ is exponentially decreasing but nonzero. It means that at any temperature, the fraction of sites on one interface connected to the other interface is nonzero, though small at high temperature. This effect is schematized in Figure 9.7. Because the glassy modulus is 4 orders of magnitude larger than the one in the rubbery regime, these rare events may dominate the elastic modulus in thin films. As a consequence, we expect the glass transition of strongly adsorbed films to be broadened as compared to the glass transition in the bulk. Hence, the very slow decrease of the elastic modulus measured experimentally [33,34,96–99] in filled elastomers as we shall see below.

Note that Lipson and Milner have extended the model for the glass transition presented here to get further insight into the glass transition in suspended films (see, for example, [92]). Essentially, their approach takes into account free volume diffusion from the free surface in a way that is analogous to the facilitation mechanism discussed above. In a further extension, they take into account also the de Gennes sliding mechanism for explaining the molecular weight dependence of the glass transition temperature in thin suspended films [100].

To summarize, in the case of a suspended film, we proposed that the glass transition occurs when the slow domains percolate in the direction parallel to the film, which requires a larger fraction of slow domains than in the bulk and results in a decrease of the glass transition temperature [65]. In the case of films deposited on a substrate with which the interactions are strong, glass transition occurs when both interfaces are connected with continuous paths of slow domains, that is, when the correlation length of the 3D percolation problem is comparable to the thickness of the film, which requires a smaller fraction of slow subunits as compared to the bulk glass transition. In both cases—suspended films and strongly interacting films—we predict a shift of T_g of the form (see Figure 9.8)

$$\frac{\Delta T_g}{T_g} = \beta \left(\frac{a}{h} \right)^{1/\nu} \tag{9.22}$$

where β is a number of order unity, a is one monomer length (typically 5 Å), and h is the thickness of the film. The sign of β is positive for strongly interacting films (SI films) and negative for suspended or weakly interacting films (WI films). The corresponding percolation thresholds, as functions of the thickness of the films, from which the T_g shift is derived, are plotted in Figure 9.8. A consequence of our model is that the change of the dynamics induced by an interface is long ranged: dominant relaxation times can be changed by several orders of magnitude at distances a few tens of nanometers, as it has been demonstrated in Berriot et al. [32,34].

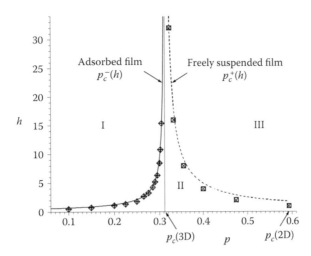

Figure 9.8 Diagram of the percolation thresholds $p_c^-(h)$ and $p_c^+(h)$ as a function of the film thickness h. $p_c^-(h)$ is the percolation threshold in an adsorbed film. It is defined here as the ensemble of points for which the size of aggregates satisfies $\zeta(p) \equiv h$. Diamond symbols correspond to results from numerical simulation described in Sotta and Long [68]. The continuous line is the fit with Equation 9.19 (with a negative sign of μ). $p_c^+(h)$ is the percolation threshold in a freely suspended film. Square symbols correspond to the data from Reference 68. The dashed line is the fit with Equation 9.19 and a positive μ. (Reproduced from P. Sotta and D. Long, *Eur. Phys. J. E*, 11, 375, 2003. With kind permission of *The European Physical Journal* [EPJ].)

9.2.6 DYNAMICAL HETEROGENEITIES: POLAR VERSUS NONPOLAR LIQUIDS

Ediger and coworkers have measured by NMR the length scale ξ of dynamical heterogeneities in the case of glycerol [101]. They obtained $\xi \sim 1$ nm, which is barely larger than one molecular size. We have seen that in the case of van der Waals (fragile) liquids, relaxation mechanisms involve a few hundreds of molecules or monomers, which sets the scale—3–5 nm—for dynamical heterogeneities. In the case of glycerol, a molecule can build up to three hydrogen bonds with its neighbors. Given the energy of one bond ($\sim 10^{-20}$ J [84]), hydrogen bonds control the glass transition. Indeed, the energy involved with three bonds is of order 5×10^{-20} J, which is sufficient for creating relaxation times of order 100 s at temperatures ~ 200 K. Thus, relaxation processes take place on the molecular scale, by hydrogen bond breaking for one molecule. In that case, the heterogeneities could indeed take place on the scale of one molecule, between bonded and nonbonded molecules, or as a function of the statistics of active bonds. The mechanism of the glass transition are therefore very different from those discussed above for nonpolar liquids, for which a large number of molecules need to be packed for creating long relaxation times, with relaxation processes taking place thanks to free volume diffusion. The case of such strong glass former such as glycerol is beyond the scope of the model described here and would require a specific description.

As a consequence, in the case of strongly polar liquids, that is, for liquids with a high density of strong polar interactions, one expects that viscosity may be determined by the presence of a hydrogen bond network if the density of polar groups and their strength are high enough. This is certainly the case for glycerol. Then the associated length scale for dynamical heterogeneities is much smaller, comparable to the molecular scale. In particular, we expect no Stokes' law violation for optical probes during diffusion in glycerol similar to the ones observed in nonpolar polymers. We may also expect that negligible T_g shifts should be observed in confinement for strongly polar polymers. This prediction of the PFVD model regarding T_g at interfaces in strongly polar liquids [61,66,67,69,75] may be confirmed by recent experiments with nanoparticles embedded in glycerol for which the authors observed no modification of the dynamics, or changes that are not correlated with local modification of the density in the vicinity of the nanoparticles [74].

9.2.7 MEASURING τ_α IN DIELECTRIC AND IN MECHANICAL EXPERIMENTS

A dominant relaxation time in a supercooled liquid can be measured using a great diversity of techniques. Usually, what is called a dominant relaxation time, or τ_α, is the time scale that dominates properties at long times. In mechanical experiments, and in simple liquids, this long-time property is the viscosity η that is related to the α-relaxation time by $\eta \approx K\tau_\alpha$, where K is the high-frequency shear modulus, which is comparable to the bulk modulus. It may also be measured for instance by applying a static electric field, letting the system be polarized and equilibrated in the presence of the electric field with a polarization $\mathbf{P}_\infty$ [102]. After the system has been equilibrated, one can remove the electric field and measure the relaxation of the polarization $\mathbf{P}(t)$. The dominant relaxation time is then defined by

$$\tau_\alpha = \int_0^\infty F(t)dt \tag{9.23}$$

where $F(t) = P(t)/P_\infty$. Equivalently, the dominant relaxation time can be calculated as

$$\tau_\alpha - \lim_{\omega \to 0} \frac{\Im F^*(\omega)}{\omega} \tag{9.24}$$

where

$$F^*(\omega) = i\omega \int_0^\infty F(t)\exp(-i\omega t)dt \tag{9.25}$$

and $\Im F^*$ is the imaginary part of F^*. Equation 9.24 reads

$$\tau_\alpha = \lim_{\omega \to 0} \frac{\epsilon''(\omega)}{\omega \Delta \epsilon} \tag{9.26}$$

where $\Delta \epsilon$ is the amplitude of the relaxation process, which is the difference between the low-frequency and high-frequency polarization. Examples of relaxation function $F(t)$ are schematized in Figure 9.9.

The relation between Equation 9.23 and the distribution of relaxation times $Q(\tau)$ reads

$$\tau_\alpha = \int_0^\infty F(t)dt \approx \int^{\tau_{cutoff}} \tau Q(\tau)d\tau \approx \tau_{cutoff} \tag{9.27}$$

Thus, in this experiment, one measures and quantifies the long-time part of the relaxation process. Equation 9.27 shows that τ_α is given by the first moment of the relaxation time distribution as given in Figure 9.3 for instance. However, one does not integrate up to ∞ and there is an upper bound in the integral that plays an essential role, τ_{cutoff}, according to the discussion regarding final relaxation. These measurements are thus dominated by the cutoff at long times. One does not measure internal processes of very slow and very rare subunits with relaxation times larger than τ_{cutoff}: their local polarization relaxes not by internal processes that are very long but by their random reorientation on a time scale τ_{cutoff} because these rare and slow subunits are surrounded by faster subunits. On time scales larger than τ_{cutoff}, the macroscopic polarization has relaxed because very slow subunits have reoriented randomly on this time scale. This long-time cutoff corresponds to a percolation threshold, with $p_c \approx 0.1$. This cutoff is the same in mechanical or dielectric experiments, which is the reason why the dominant relaxation time τ_α measured in dielectric spectroscopy or in mechanical experiments is essentially the same. Note that at long times, $Q(\tau) \sim \tau^{-1}$. $\tau Q(\tau)$ is thus essentially a constant. The integral in Equation 9.27 is thus of order τ_{cutoff}.

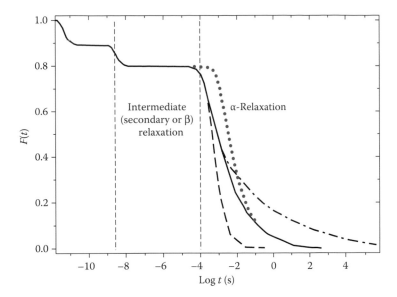

Figure 9.9 Schematic of typical relaxation function in a supercooled liquid. The quantity $F(t)$ may be the normalized low q structure factor $S(q,t)$, the relaxation of the polarization $P(t,T)$ (normalized by the equilibrium polarization at long times P_∞), the time-dependent part of the dilatational bulk compliance $J_2(t,T)$, or the transient part of the shear compliance $J_1(t,T)$ as defined in the text. Over the whole VFT (or WLF) temperature range, the α-relaxation is well separated from other relaxation modes. The temperature considered here is close to the glass transition temperature. The width of the final relaxation (about four decades here) is related to the width of the relaxation time distribution (see Figure 9.4). The dominant relaxation time is defined as $\tau_\alpha = \int_0^\infty F(t)dt$ and is thus essentially the long-time cutoff of the relaxation time distribution (see Figure 9.4). The continuous curve shows the relaxation in the bulk. The same relaxation function at the same temperature is slower in thin films with strong interactions with the substrate (dashed-dotted curve) and faster in a thin suspended film (dashed curve). The dotted curve illustrates the possible difference at short times between dielectric (continuous curve) and bulk dilatational compliance. However, their long-time behavior is the same. (Reproduced from S. Merabia, P. Sotta, and D. Long, *Eur. Phys. J. E*, 15, 189, 2004. With king permission of *The European Physical Journal* [EPJ].)

The dominant relaxation time can also be measured by mechanical experiments. A mechanical compliance is equivalent to a polarization in dielectric experiment. For instance, one can measure the relaxation of the deformation of a rubber after an instantaneous and small variation of pressure or after a small shear stress is applied. This relaxation is quantified by the time-dependent dilatational compliance and the transient shear compliance of the liquid, which we denote $J_2(t,T)$ and $J_1(t,T)$, respectively [88]. These functions may be defined in the following way: a small stress σ is applied to the system, which acquires a deformation $\epsilon(t)$ that relaxes toward the equilibrium values $\epsilon_\infty = J_{1,\infty}\sigma$ or $\epsilon_\infty = J_{2,\infty}\sigma$. After the system has been equilibrated, the stress is removed and the time-dependent compliances $J_1(t,T) = \epsilon(t)/\sigma$ and $J_2(t,T) = \epsilon(t)/\sigma$ (which relaxes to zero by using this definition) may be measured. Here also, τ_α can be obtained by Equation 9.27 where $F(t)$ is defined by $F(t) = J_1(t, T)/J_{1,\infty}$ or $J_2(t, T)/J_{2,\infty}$. An example of this relaxation function $F(t)$ is schematized in Figure 9.9.

The dominant relaxation time may also be defined by measuring the peak at low frequencies of the imaginary part of $J_1^*(\omega,T)$ or of $J_2^*(\omega,T)$, which are defined according to Equation 9.25. One can, for example, look at $J^*(\omega,T)$ at fixed temperature, as a function of frequency. Then, the α-relaxation corresponds to the peak at low frequency. When the corresponding peak is located at $\omega \sim 10^{-2}$ s, the temperature is the glass transition temperature.

9.2.8 PROBING THE LONG-TIME PART OF THE DISTRIBUTION OF RELAXATION TIMES

The question then is the following: Would it be possible to devise experiments that give access to the long-time part of the relaxation time distribution, that is, access to the relaxation time spectrum for times larger

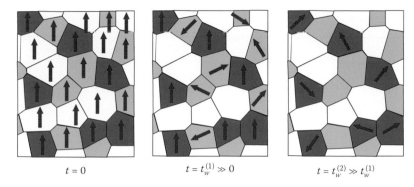

$t = 0$ $t = t_w^{(1)} \gg 0$ $t = t_w^{(2)} \gg t_w^{(1)}$

Figure 9.10 Schematics of a dielectric hole-burning experiment. At $t = 0$, an electric field is applied to polarize the material. Red arrows denote the polarization of each dynamical subunit. The polarization is then much larger as compared to the statistical fluctuation of the polarization on this scale, or the polarization in the linear regime (small applied electric field). After a waiting time $t_w^{(1)}$, dynamical subunits with relaxation times larger than $t_w^{(1)}$ (dark green) have preserved their initial polarization, while dynamical subunits with intermediate relaxation time comparable to $t_w^{(1)}$ have started to reorient and subunits with relaxation times shorter than $t_w^{(1)}$ have fully relaxed. After a waiting time $t_w^{(2)} \gg t_w^{(1)}$ comparable to the slowest relaxation times in the material, only slowest subunits have not yet fully relaxed their internal polarization, while the material as a whole is no longer polarized. When applying again an electric field at time $t_w^{(2)}$, the electric field is coupled to these randomly oriented strong dipoles: the dielectric response is different as compared to the dielectric response of a fully equilibrated system. Performing this experiment for different waiting times $t_w^{(2)}$ may give access to the full relaxation time distribution of the polymer, beyond the relaxation time τ_α.

than τ_α? Our point of view is that this is the case regarding mechanical properties in confinement, such as in filled elastomers where the mechanical response is dominated by the mechanical properties of polymer confined between two neighboring fillers typically 5–10 nm apart, as we shall see later in this chapter. We will see that these mechanical properties are controlled by the long-time part, beyond the bulk τ_α, of the distribution of relaxation times. As we shall discuss, mechanical experiments in confinement probe the long-time part of the relaxation time distribution. In dielectric experiments, a way to probe long relaxation time subunits would be the following. Let's apply a very strong electric field, much stronger than the one in linear dielectric spectroscopy. Then, the polymer chains may get a polarization stronger than the one in linear experiments, which is essentially a biased orientation of fluctuating dipoles. In this nonlinear experiment, the local polarization may be much larger. Then, after removing the applied electric field, the local polarization may relax by an internal process on the monomer scale (fast subunits), or by random reorientation of the subunits. On time scales larger than τ_α, the latter process dominates for the very slow subunits. They still keep the memory of the very high electric field because of their polarization $\mathbf{P}_{slow}$, but their individual polarization becomes a random orientation, on a time scale τ_α. This effect is schematized in Figure 9.10. On much larger time scales, these randomly oriented polarizations relax through monomer-scale processes. However, on a time scale larger than τ_α, but not too long, these large dipoles, though randomly oriented, are still present. By applying again an electric field, the dielectric response is modified as compared to the initially equilibrated system because of the coupling between the applied electric field and these strong, randomly oriented dipoles. The fraction of these strong dipoles relax according to the bare distribution of relaxation times. Thus, this kind of experiment—called dielectric hole burning [103]—may provide access to the long-time part of the distribution. This nonlinear dielectric spectroscopy relaxation process is schematized in Figure 9.10. Performed systematically for various waiting times t_w, this experiment may give access to the whole relaxation time spectrum.

9.2.9 SMALL PROBE DIFFUSION

Because the dynamics are strongly heterogeneous, the issue of the relaxation time probed by a given technique versus the time probed by another one is relevant. Ediger et al. and Fujara et al., for instance [104–107], have shown that the variation with temperature of the time scale associated to probe diffusion is different from that of the dominant time measured in mechanical or dielectric spectroscopy, and that the corresponding discrepancy can reach up to 3 orders of magnitude when lowering the temperature. This effect can be quantified for instance by considering the evolution of the quantity $D\eta/T$ as a function of temperature. If the diffusion is controlled by the same relaxation time as the viscosity, this quantity is a constant, independent of the temperature. This is the case of large (colloidal) particles for which the Stokes–Einstein relation holds. However, for very small probes, there is a fundamental difference between the diffusion coefficient, and the dominant relaxation time in, for example, mechanical or dielectric experiments. When a very small probe (for example, fluorescent molecule) diffuses, it spends a fraction of order 1 of its time in slow subunits. Then, it diffuses very slowly. But it also spends a fraction of order 1 of its time in rapid regions, where it diffuses rapidly. Then, the diffusion coefficient is controlled by the rapid dynamics regions. Note that this is true for molecules smaller than the characteristic size ξ of the dynamical heterogeneities.

Thus, the diffusion coefficient depends on the relaxation time distribution $Q(\tau)$ (Equation 9.17) by a relation of the type

$$D \sim \int_{\tau_{fast}} \frac{1}{\tau} Q(\tau) d\tau \approx \tau_{fast}^{-1}$$

(9.28)

and is therefore dominated by the fastest relevant time τ_{fast} accessible to the probe in the system [66]. It is a short-time cutoff in Equation 9.28, which depends on the relative size d/ξ of the probe and of the scale of the heterogeneities [66]. When the scale of the heterogeneities is smaller than the size of the probe, the latter interact at any time with a statistically averaged environment, and diffusion is controlled by τ_α. Then, $\tau_{fast} \approx \tau_\alpha$ and the ratio $D\eta/T$ is a constant in this regime [66]. In the lower-temperature regime, the scale of heterogeneities might become larger than the probe diameter, and τ_{fast} becomes smaller than τ_α. The discrepancy then increases when lowering the temperature, up to several orders of magnitudes, depending on the system [66,104,105], as a consequence of the increase of the scale ξ of dynamical heterogeneities.

According to this discussion, the viscosity of the supercooled liquid and the diffusion coefficient of the tagged molecule vary, respectively, as

$$\eta \propto \tau_{slow} = \tau_\alpha = \tau_0 \exp\left(\frac{\Theta}{\tilde{\epsilon}}\right)$$

and

$$D^{-1} \propto \tau_{fast} = \tau_0 \exp\left(\frac{\Theta}{\tilde{\epsilon} + \alpha\epsilon/N_c^{1/2}}\right)$$

(9.29)

where α is a number of order one, function of the ratio $r = d/\xi$, where d is the diameter of the probe and ξ the scale of dynamical heterogeneities. When $r \geq 1$, $\alpha = 0$. α starts increasing when $r \lesssim 1$: Stokes' law violation starts to appear. An example of the evolution of the quantity $D\eta/T$ is given in Figure 9.11 with both experimental results from Fujara et al. [11], and the theoretical curve derived from the PFVD model.

9.2.10 Aging and rejuvenating

Up to now, we have discussed results regarding systems at equilibrium. However, when cooling below T_g at sufficiently low temperatures, the evolution toward equilibrium is very slow. It may even not be possible to reach equilibrium in an accessible time scale. This evolution toward equilibrium is usually called aging. The issue we consider here is to describe the role of dynamical heterogeneities, their evolution during the aging process, as well as after the system has been heated up (rejuvenation).

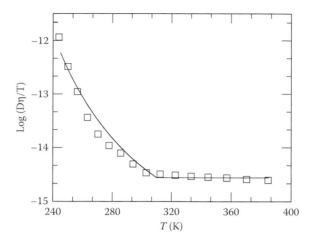

Figure 9.11 Comparison of the slowest relevant times (associated to viscosity) to the fastest ones (associated to molecular diffusion) in the case of a small fluorescent probe (TTI) in OTP [66]. The experimental data are from Reference 104. The parameters used for the theoretical calculation are given in Merabia and Long [66]. (Reproduced from S. Merabia and D. Long, *Eur. Phys. J. E, 9, 195, 2002*. With kind permission of *The European Physical Journal* [EPJ].)

During aging, various quantities evolve with time. The volume of the sample decreases with time, with a decreasing but never vanishing rate [108–111]. For describing the evolution of the compliance J during aging, Struik proposed the following heuristic relation:

$$J(t,t_w) = \mathcal{J}\left(\frac{t}{t_w^\mu}\right) \tag{9.30}$$

where μ is an exponent close to 1 at temperature a few tens of kelvins below T_g. t_w is the aging time, that is, the elapsed time after the system has been cooled down [112]. The Struik relation implies that the evolution of the dominant relaxation time satisfies the relation $\tau_\alpha \sim t_w^\mu$. Calorimetric studies have also shown that the enthalpy of samples evolves slowly with time during aging [113,114].

Though aging involves huge time scales, up to years in Struik's experiments, its effect can be erased upon heating on a much shorter time scale. This effect, studied in particular by Kovacs and coworkers [110] is called temporal asymmetry. Another effect, also described by Kovacs and other authors, is the so-called Kovacs memory effect [115–118]. The experiment is the following: let's consider a system cooled at a temperature well below T_g, which is then allowed to age. Then, if it is reheated, the evolution of the volume can be nonmonotonous, and exhibits an overshoot, for particular aging times and reheating temperatures. The amplitude of the effect can be of order 0.1%.

As we shall discuss now, dynamical heterogeneities are key for understanding these effects. In particular, the facilitation mechanism discussed above plays a key role in understanding temporal asymmetry between cooling and rejuvenating.

9.2.10.1 BASIC EQUATIONS FOR AGING AND REJUVENATING

As discussed, we assume that aging corresponds to the relaxation toward equilibrium, even though this equilibrium cannot be reached on experimentally accessible time scales. We assume that aging corresponds to the evolution of the distribution of density fluctuations toward the equilibrium one. let's thus introduce the out-of-equilibrium excess of free energy per monomer [69,119,120]:

$$F = T\int p(\rho)\ln\left(\frac{p}{p_{eq}}(\rho)\right)d\rho$$

where $p_{eq}(\rho)$ is the temperature-dependent equilibrium density fluctuation distribution on the scale N_c. Following an Onsager-like picture, the evolution of the distribution of density fluctuations is thus given by

$$\frac{\partial p}{\partial t} - \frac{\partial}{\partial \rho}\left(\gamma(\{\rho\})p_{eq}(\rho)\frac{\partial}{\partial \rho}\left(\frac{p}{p_{eq}}\right)\right) = 0 \tag{9.31}$$

which is a Fokker–Planck equation, where $\gamma(\{\rho\})$ in Equation 9.31 can be understood as a diffusion coefficient in density space. It is related to the relation time in the liquid $\tau(\rho)$ (in first reading, one may assume $\gamma(\{\rho\}) = \tau^{-1}(\rho)$), but does not depend only on ρ but on the whole density distribution $p(\rho)$ as we shall discuss below.

9.2.10.2 EQUILIBRIUM AND OUT-OF-EQUILIBRIUM SITUATIONS

As discussed above, the relaxation process for the slow part of the relaxation spectrum is controlled by a facilitation mechanism: the slowest subunits melt thanks to a diffusion process of free volume from the faster environment. That results in the relation, valid at equilibrium:

$$\tau_\alpha = N_c^{2/3}\tau_{fast} = \tau_{p_c} \tag{9.32}$$

where τ_{fast} corresponds to a fraction q_c of the fastest subunits (see Equation 9.16). This fraction needs to be determined. As we shall see, rejuvenating experiments allow for an estimate of q_c (see Equation 9.15).

In out-of-equilibrium conditions, the width of the bare relaxation time distribution changes. Equation 9.32 is no longer valid. In aging situations, the distribution of relaxation times is narrower as compared to the equilibrium one (see an example in Figure 9.12). One has then typically $N_c^{2/3}\tau_{fast} > \tau_{p_c}$. Then, the dominant relaxation time is τ_{p_c}. In rejuvenating experiments, the distribution of relaxation times is broader than the one at equilibrium (see an example in Figure 9.13) and one has typically $N_c^{2/3}\tau_{fast} < \tau_{p_c}$. Then, the dominant relaxation τ_α is $N_c^{2/3}\tau_{fast}$. In the general situation, the dominant relaxation time is given by

$$\tau_\alpha = \min\left[N_c^{2/3}\tau_{fast}, \tau_{p_c}\right] \tag{9.33}$$

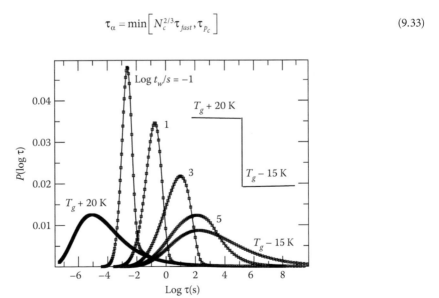

Figure 9.12 Evolution of the relaxation time distributions, as a function of time, during aging. They are calculated by solving the Fokker–Planck equation 9.31 for PVAc (parameters are given in Merabia and Long [69]). The polymer has been cooled from $T_g + 20$ K to $T_g - 15$ K at time 0^+. The waiting times are log $t_w(s) = -1$; 1; 3; and 5. We also plotted both equilibrium distributions at $T_g + 20$ K and at $T_g - 15$ K. Note that during aging, the distribution of relaxation times is narrower as compared to the distributions at equilibrium. (Reprinted with permission from S. Merabia and D. Long, *J. Chem. Phys.*, 125, 234901. Copyright 2006, American Institute of Physics.)

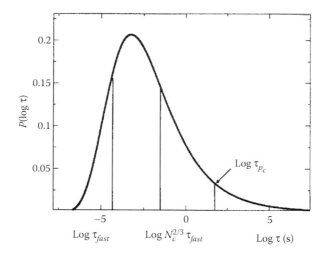

Figure 9.13 When the distribution of relaxation times is broader than that at equilibrium (which is the case during rejuvenating), the dominant relaxation time, that is, the effective cutoff in the distribution is given by $\tau_\alpha = N_c^{2/3}\tau_{fast}$, which is then smaller than τ_{p_c}. (Reprinted with permission from S. Merabia and D. Long, *J. Chem. Phys.*, 125, 234901. Copyright 2006, American Institute of Physics.)

This relation expresses in a concise way the facilitation mechanism. The spectrum of relaxation times $\gamma^{-1}(\{\rho\})$ may be calculated from the bare distribution of relaxation times, which is a function of the density, with the following condition:

$$\gamma(\{\rho\}) \sim \tau^{-1}(\rho) \quad \text{when} \quad \tau < \tau_\alpha(t)$$
$$\gamma(\{\rho\}) \sim \tau_\alpha^{-1}(t) \quad \text{otherwise}$$

(9.34)

where $\tau_\alpha(t)$ is given by Equation 9.33 and $\tau(\rho)$ is given by Equation 9.4.

9.2.10.3 AGING AFTER A QUENCH: RESULTS OF SIMULATIONS

In Figure 9.12, we plotted the distribution of relaxation times calculated for PVAc (see [69] for specific details) following a temperature quench from $T_i = T_g + 20$ K to $T_f = T_g - 15$ K. Distributions of relaxation times are plotted after various waiting times t_w. Both equilibrium distributions at $T_g + 20$ K and at $T_g - 15$ K are also plotted. The distribution translates toward longer times as the system ages. The distribution of relaxation times is narrower indeed than that at equilibrium. The dominant relaxation time $\tau_\alpha(t)$ is then equal to τ_{p_c}. Note that the melting time $\tau_{fast}N_c^{2/3}$ would be larger than τ_{p_c} and thus cannot be the dominant relaxation time. The evolution of the whole distribution is slowed by the evolution of $\tau_\alpha(t)$. The evolution of the dominant relaxation time $\tau_\alpha = \tau_{p_c}(t_w)$, which is a function of the aging time t_w is plotted in Figure 9.14. The following features may be observed (see [69]). In a first step, and on logarithmic scale, the dominant relaxation time remains equal to that of the initial temperature T_i:

$$\tau_\alpha(t_w) = \tau_\alpha(T_i) \quad \text{for } t_w < \tau_\alpha(T_i)$$

Then, after an elapsed time comparable to the latter, it starts increasing with an exponent close to one, up to the equilibrium value at the final temperature T_f. After a time, $t_w > \tau_\alpha(T_i)$:

$$\tau_\alpha(t_w) = Ct_w^\mu \quad \text{for } \tau_\alpha(T_i) < t_w < \tau_\alpha(T_f)$$

with $\mu \approx 1$: the relaxation time evolves proportionally with the elapsed time t_w. This regime lasts until a waiting time t_f comparable to $\tau_\alpha(T_f)$, where $T_f = T_g - 15$ K.

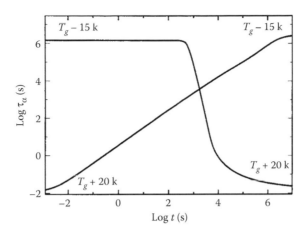

Figure 9.14 Dominant relaxation time of PVAc, during a temperature down-jump between $T_g + 20$ K and $T_g - 15$ K, and during a temperature up-jump at $T_g + 20$ K of a polymer sample initially at equilibrium at $T_g - 15$ K (data from [69]). These curves are calculated using the Fokker–Planck equation without coupling for the dynamics between dense subunits and mobile subunits ($q_c = 1.0$). The time scale that controls both evolutions is the same and is the dominant equilibrium relaxation time $\tau_\alpha(T)$ at the lowest temperature $T = T_g - 15$ K. There is no temporal asymmetry. (Reprinted with permission from S. Merabia and D. Long, *J. Chem. Phys.*, 125, 234901. Copyright 2006, American Institute of Physics.)

9.2.10.4 REHEATING (ANNEALING) AND TEMPORAL ASYMMETRY

We can calculate the evolution of the dominant relaxation time as given by Equation 9.33 for various temperature up-jumps. In particular, we will consider the time it takes for the system to return to equilibrium, after it is heated from a temperature below T_g up to a temperature above T_g. We will see indeed that dynamical heterogeneities are key for understanding temporal asymmetry. To do so, we solve first the Fokker–Planck equation 9.31 without coupling between the dynamics of dense and less dense subunits, which is without the facilitation mechanism. Then, the longest relaxation time is always equal to τ_{p_c} defined by Equation 9.16.

9.2.10.4.1 Fokker–Planck equation without dynamic coupling between slow and fast subunits

In Figure 9.14, we plotted the evolution of the dominant relaxation time τ_α after a temperature up-jump from $T_g - 15$ K up to $T_g + 20$ K. In the first time, the dominant relaxation time is stationary and equal to that at the initial temperature. After an elapsed time of about $t \simeq 10^3$ s, the dominant relaxation time starts decreasing with time. At first, the decrease is rapid, then the variation slows down. Finally, the dominant relaxation time becomes equal to that at equilibrium at the final temperature. However, the time it takes for the dominant relaxation time to reach the equilibrium value is comparable to the initial relaxation time $\tau_\alpha(T_i)$: this is a very long process, comparable in duration to the time it takes for equilibrating the system at low temperature T_i: there is no temporal asymmetry. In the same simulations, it was observed that the volume relaxes very slowly toward the equilibrium value, on a time scale comparable to $\tau_\alpha(T_i)$. This is not what is observed experimentally. Indeed, the time to reach equilibrium when heating a sample is much smaller than the initial dominant relaxation time.

The problem of the present model without coupling (facilitation mechanism) can be observed in Figure 9.15 in which we plotted the evolution of the relaxation time distribution. We see that even at long heating times, there remains long relaxation times in the system. Without the facilitation mechanism, slow subunits melt with a characteristic time corresponding to their own relaxation time. Very long relaxation times disappear upon heating after a time $\tau_\alpha(T_i)$. With the implementation of the facilitation mechanism, we will see that very slow subunits melt thanks to the fast surrounding subunits in a much shorter time than $\tau_\alpha(T_i)$.

9.2.10.4.2 Fokker–Planck equation with dynamic coupling between slow and fast subunits

The dynamic coupling between slow and fast subunits is quantitatively determined by the value of the parameter q_c (see Equation 9.15). To determine the correct value of this parameter, we consider the evolution of the

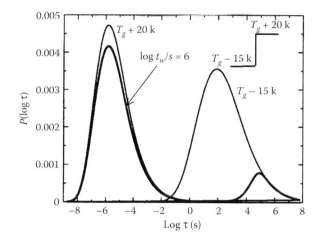

Figure 9.15 Distribution of relaxation times during a temperature up-jump at $T_g + 20$ K of a polymer sample initially at equilibrium at $T_g - 15$ K. The evolution is calculated using the Fokker–Planck equation without coupling ($q_c = 1.0$). We see that long relaxation times persist a long time ($t = 10^6$ s) after the temperature has been increased. This persistence explains the absence of temporal asymmetry observed in the absence of coupling. We have also displayed the initial distribution of relaxation times (equilibrium at $T_g - 15$ K) and the final one (equilibrium at $T_g + 20$ K). (Reprinted with permission from S. Merabia and D. Long, *J. Chem. Phys.*, 125, 234901. Copyright 2006, American Institute of Physics.)

volume as a function of time during a heating process, with various values of q_c and compare the resulting dynamics to experimental data. In Figure 9.16, we plotted the evolution of the volume after heating a sample for three different values of q_c: 20%, 30%, and 40%. The evolution that is the closest to the data obtained by Kovacs [108] corresponds to the value $q_c = 30\%$, which we retain in the following. The corresponding evolution of the dominant relaxation time is plotted in Figure 9.17. We can observe that the sample melts in a time that is much shorter than its aging time: we recover temporal asymmetry.

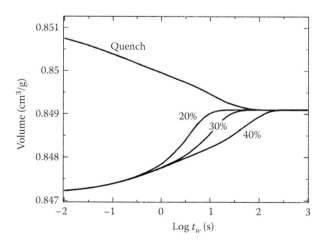

Figure 9.16 Volume during a temperature down-jump from $T_g + 5$ K down to T_g and during a temperature up-jump from $T_g - 5$ K to T_g calculated by using the Fokker–Planck equation. The different temperature up-jump curves correspond to different choices of the parameter q_c (see Equation 9.15). From left to right: $q_c = 20\%$; 30%, and 40%. The larger the value of q_c, the longer it takes for melting slow subunits. We found the value of $q_c = 30\%$ to provide reasonable agreement with known experimental melting time scales [108]. (Reprinted with permission from S. Merabia and D. Long, *J. Chem. Phys.*, 125, 234901. Copyright 2006, American Institute of Physics.)

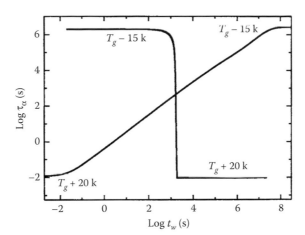

Figure 9.17 Comparison of the evolution of the dominant relaxation time during a temperature down-jump between $T_g + 20$ K and $T_g - 15$ K, and during a temperature up-jump between the same temperatures. The dominant relaxation time is calculated solving the Fokker–Planck equation 9.31 with coupling and with a parameter $q_c = 30\%$. With this coupling, a polymer initially at equilibrium at a temperature below T_g can melt in a time that is much shorter than the dominant relaxation time in the initial glassy state. (Reprinted with permission from S. Merabia and D. Long, *J. Chem. Phys.*, 125, 234901. Copyright 2006, American Institute of Physics.)

The evolution of the distribution of relaxation times after reheating a sample initially at a temperature $T_g - 15$ K to a temperature $T_g + 20$ K (see [69]) is the following. The distribution is broader than the equilibrium one. The dominant relaxation time here is thus

$$\tau_\alpha(t) = N_c^{2/3} \tau_{fast}(t)$$

which acts as a cutoff in the relaxation time distribution at any time during the reequilibration after heating. Physically, it means that fast subunits melt denser ones in a time that is at most $N_c^{2/3} \tau_{fast}(q_c)$. This process is responsible for the temporal asymmetry. At short times, the distribution is nearly equal to that at the initial temperature. In this regime, the dominant relaxation time and the volume have not evolved. This regime lasts for a time comparable to the shortest times present in the initial distribution, here about 10 s. After this initial period, fast regions appear within the system: the time $N_c^{2/3} \tau_{fast}$ becomes shorter than the time corresponding to the 10% densest regions. Then, the fastest regions will melt the slowest ones (see Figure 9.2). The time that controls the kinetics of melting is linked to the creation of fast regions in the neighborhood of slow regions, which melt them. That explains the temporal asymmetry. As a consequence, studying rejuvenating dynamics allows for an estimate of q_c. This mechanism has also been shown to be able to explain Kovacs memory effects [110,115,116,118], which is the nonmonotonous evolution of the volume after heating up a sample that has been aged for times t_w smaller than the time required for equilibrating them, and typically times t_w comparable to the new equilibrium relaxation time after the polymer has been heated up. We display such an evolution in Figure 9.18. Finally, note that a similar approach as that described here [69] regarding volume relaxation has been proposed recently by Medvedev and Caruthers [121], which also emphasizes the role of dynamical heterogeneities in these relaxation mechanisms.

9.2.11 DEPENDENCE OF THE DYNAMICS ON BOTH TEMPERATURE AND PRESSURE

The issue to know whether the free volume is the proper parameter for describing dynamics in supercooled liquids is still debated in the literature. As discussed in the introduction, experiments that led to the conclusion that this is not the case [72,73] prove in fact only that *the average free volume* is not sufficient for calculating the dynamics [75]. Merabia and Long have considered this issue by studying data from the literature regarding

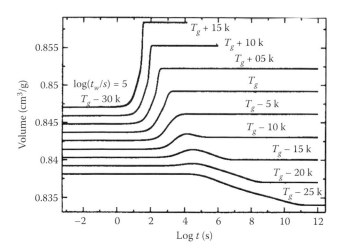

Figure 9.18 Variation of the volume for a PVAc sample after heating. The initial states are nonequilibrium ones: the samples have aged for log t_w(s) = 5 at temperature T_g – 30 K. The final temperatures are, respectively, T_g – 20 K; T_g – 15 K; T_g – 10 K; T_g – 5 K; T_g and T_g + 5 K; T_g + 10 K; T_g + 15 K. The evolution of the volume may be nonmonotonous depending on the difference between the initial and the final volumes. The difference for the volumes of the different samples at short time is due to the contribution of the fast modes, which is proportional to $\alpha_g \, \Delta T$. (Reprinted with permission from S. Merabia and D. Long, *J. Chem. Phys.*, 125, 234901. Copyright 2006, American Institute of Physics.)

nonpolar molecular liquids such as 1,1'-di(4-methoxy-5-methylphenyl)cyclohexane (BMMPC); 1,1'-bis(*p*-methoxyphenyl)cyclohexane (BMPC), and flexible nonpolar polymers such as poly(methyltolylsiloxane) (PMTS) and poly(methylphenylsiloxane) (PMPS) [122–125].

They have shown that taking into account the whole spectrum of density fluctuations allows for almost quantitatively describing the dependence of the dynamics on pressure and temperature for the nonpolar molecular liquids and flexible polymers just mentioned. In the case of PVAc, which displays weak polar interactions, the PFVD model is also close to account quantitatively for the intial slope $d\log(\tau_\alpha)/dP$ of the pressure dependence of the dynamics. Key is the facilitation mechanism introduced in Merabia and Long [66], which has been used also for describing rejuvenating [69]. Consider, for instance, increasing the temperature for a sample at constant density, which is obtained by applying a pressure. Then, because of the decrease of the bulk modulus, the density fluctuations, which are proportional to $(T/K)^{1/2}$ increase: the spectrum of density fluctuations broadens and the systems populate faster subunits of lower densities. Thanks to the facilitation mechanism discussed above, this process results in a shortening of the lifetime of density fluctuations, and thus in a reduction of the dominant relaxation time that was calculated in Merabia and Long [75]. In particular, this model shows that the dynamics can be changed even at fixed density, in agreement with experiments. The main point here is that *what matters is not the average free volume (or density of the system), but its distribution.* According to the PFVD model, a broader distribution of free volume, even at fixed average density, thus results in an acceleration of the dynamics.

9.3 MECHANICAL RELAXATION OF GLASSY POLYMERS: QUANTITATIVE DESCRIPTION

9.3.1 MODELING THE MECHANICAL BEHAVIOR

9.3.1.1 VISCOELASTIC BEHAVIOR CLOSE TO GLASS TRANSITION

In this section, it will be shown how mechanical properties of polymers close to T_g, both in the bulk and in confined films, can be described with the model introduced above. As will be described below in the next section, in polymer nanocomposites, numerous studies demonstrate that the mechanical behavior of the polymer in between filler particles is strongly affected by confinement. This phenomenon plays a fundamental role in the mechanical reinforcement observed in these materials. Therefore, understanding and modeling mechanical properties in the vicinity of the glass transition, specifically in confined polymer film, is a crucial issue.

The mechanical behavior of polymers in the linear regime of deformation is viscoelastic. On cooling a polymer down to T_g, its mechanical behavior changes from that of a viscous fluid (with a viscosity depending on the chain length) or an elastomer (elastic modulus of order 0.5 MPa), toward an effectively elastic solid with a very high storage modulus of several GPa. The response is elastic at high frequency, with a high-modulus G_g corresponding to the glassy state. At lower frequency, the stress progressively relaxes, the polymer flows, and its response becomes viscous. We will assume that the polymer is cross-linked, or that it is entangled with a shear period shorter than the disentanglement time, so that the mechanical behavior at low frequency is again elastic, with a low modulus G_r corresponding to the rubbery state.

Up to this point, the PFVD model has been discussed only within the percolation picture [65,67]. The mechanical behavior was not modeled explicitly in a quantitative way. Thus, the PFVD model has been extended in order to describe and calculate the linear mechanical properties of polymers in the vicinity of the glass transition, both in the bulk and in thin confined films subjected to shear in the linear regime of deformation [61].

Our point of view is that the mechanical response of polymers in the vicinity of the glass transition is essentially driven by the local dynamics at the scale of the subunits of dynamical heterogeneities ($\xi \sim 3 - 5$ nm) (see Figures 9.19 and 9.20). The mechanical response can be described by the low-frequency, rubbery modulus, the high-frequency, glassy modulus, and by the distribution of relaxation times associated to the glassy subunits. The stress field, the strain field, and the dynamical state have to be described at the scale ξ of dynamical heterogeneities. Their dynamical evolution (aging/rejuvenating) is coupled to the stress field, which results from

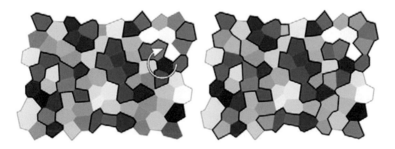

Figure 9.19 Schematic view of the heterogeneous dynamics of bulk polymer above T_g (left) and below T_g (right). The green heterogeneities have a relaxation time longer than 10^2 s and behave essentially like a rigid solid. Gray heterogeneities have a faster relaxation time shorter than 10^2 s. They behave essentially like a fluid or a rubber. Glass transition is viewed as the percolation of slow heterogeneities. (Reproduced from A. Dequidt et al., *Eur. Phys. J. E,* 35, 61, 2012. With kind permission of *The European Physical Journal* [EPJ].)

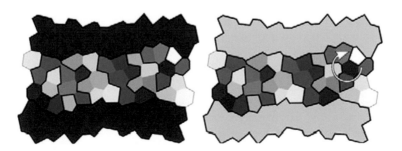

Figure 9.20 Percolation across a thin slab is easier than across a bulk sample. With strong substrate–polymer interactions (left), rigid clusters can be immobilized by a connection to one or both of the substrates. If the substrate–polymer interaction is weak (right), the clusters are freer than in the bulk, so dynamics is accelerated under confinement. (Reproduced from A. Dequidt et al., *Eur. Phys. J. E,* 35, 61, 2012, with kind permission of *The European Physical Journal* [EPJ].)

the imposed deformation. Dynamical subunits transmit the stress through couplings with their neighbors. Thus, in the physical model, the elementary objects (degrees of freedom) are the dynamical subunits of size $\xi \sim 3$ nm, which interact with the neighboring subunits. The couplings between neighboring subunits are supposed to have finite relaxation times, with a distribution given by Equation 9.17. When a local coupling relaxes, the local stress vanishes, or equivalently, the elastic energy stored in the contact between the neighboring subunits is released and dissipated. This process is the main source of viscous dissipation in the real system close to T_g. Macroscopic relaxation occurs as a result of random microscopic relaxation of subunits.

9.3.1.2 MODELING THE MECHANICAL RESPONSE

In order to calculate physical quantities according to the physical model described above, it is necessary to translate this general theoretical picture into a numerical model. The distinction between the assumptions of the theoretical and numerical models should remain very clear. For calculating the mechanical response in the numerical model, dynamical subunits are represented by nodes and the couplings between subunits are described by glassy springs connecting nodes, which have finite relaxation times, with a relaxation time distribution given by Equation 9.17 (see Figure 9.21). Then, the high-frequency contribution to the force (glassy force) between two neighboring subunits is given by

$$\vec{F}_g \sim -G_0'\xi(\vec{R}_{ij} - \vec{R}_{ij}^{ref})$$ (9.35)

where G_0' is the high-frequency (glassy) modulus ($G_0' \sim 10^9$ Pa). $\vec{R}_{ij} = \vec{R}_i - \vec{R}_j$ with $\vec{R}_i$ the vector position of subunit i.

Microscopic relaxation is modeled by breaking of glassy springs. The reference state $\vec{R}_{ij}^{ref}$ is reset to the actual relative positions of the subunits every time the elastic link breaks, which means that, when a link (spring) breaks, the local glassy force vanishes instantaneously.

In addition to glassy forces, repulsive interactions between nodes accounts for incompressibility and very loose, permanent springs model the rubbery elasticity of cross-linked or entangled polymers. The forces corresponding to these additional interactions are, respectively,

$$\vec{f}_{rep} = 12\, u_0 \left(\frac{\sigma}{r}\right)^{12} \frac{\vec{r}}{r^2}$$ (9.36)

$$\vec{f}_{el} = -k_\infty \left(r - l_0\right)\vec{u}_r$$ (9.37)

with $\vec{u}_r = \vec{r}/r$. $\sigma, l_0 \sim \xi$ while $k_\infty \ll k_0$ and u_0 is small enough not to interfere in shear measurements. Repulsion forces are cut off and regularized at long distance.

In this picture, the mechanical response is essentially controlled by the couplings between glassy subunits that are able to transmit the stress over the duration of the mechanical loading, that is, with a relaxation time longer than the period of the applied strain or the inverse strain rate. In other words, the mechanical response

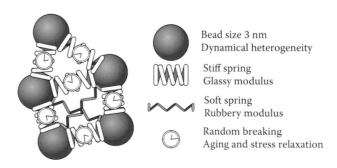

Bead size 3 nm
Dynamical heterogeneity

Stiff spring
Glassy modulus

Soft spring
Rubbery modulus

Random breaking
Aging and stress relaxation

Figure 9.21 Numerical model.

is driven by the long-time part of the distribution of relaxation times. For deformations of small amplitude, the applied strain does not change the distribution of relaxation times (linear response): random microscopic relaxations are spontaneous and uncorrelated to macroscopic loading. This linear regime is the focus of this section.

Let's now describe how the heterogeneity of relaxation times is modeled. Once a spring breaks, it is immediately rebuilt with a new reference state $\vec{R}_{ij}^{ref} = \vec{R}_{ij}$ corresponding to an initially vanishing force, and its lifetime (age) is correspondingly reset to zero. Breaking of glassy springs occurs randomly. The probability dP_{break} that a spring of age t (the time elapsed since last rupture) breaks within the interval $[t, t + dt]$ is given by

$$dP_{break} \sim \frac{dt}{\tau_\alpha(t)} \tag{9.38}$$

where $\tau_\alpha(t)$ is the instantaneous relaxation time of the glassy link. When quenched slightly below T_g, a polymer undergoes *aging*, that is, its average relaxation time depends on its age and increases progressively toward an equilibrium value. Equilibrium is reached after a waiting time longer than the longest typical relaxation time of the distribution. Before reaching equilibrium, the systems usually exhibit an aging behavior, during which the overall relaxation time τ_α follows the Struik law (see, for example, [69,111])

$$\tau_\alpha \sim t_w^\mu \tag{9.39}$$

with $\mu \sim 1$. Aging is taken into account at the microscopic scale. In Equation 9.38, the breaking probability per unit time dP_{break}/dt depends on the spring age t, through the dependence of the instantaneous relaxation time $\tau_\alpha(t)$ on the age. At a rupture, the instantaneous relaxation time of the glassy link is reset to τ_{min}, the lower limit of the relaxation time distribution covered in the simulations. It may then progressively recover a large value. At equilibrium (i.e., after very long aging), the spring relaxation times must be distributed according to the equilibrium distribution $p_{eq}(\tau)$ expected at the experiment temperature given by Equation 9.17. Technically, to insure that this condtion is satisfied, the breaking probability per unit time is chosen to be

$$\frac{dP_{break}}{dt} = \frac{dt}{\tau_\alpha(t)} = -\frac{d\log p_{eq}(t)}{dt} \tag{9.40}$$

Note that, for a distribution $p_{eq}(\log(\tau))$ extending over several decades, $p_{eq}(\tau) = (1/\tau)p_{eq}(\log(\tau)) \sim 1/\tau$, so that $\tau(t) \sim t$. Note that dP_{break} is very close to (but lower than) dt/t. Young springs are thus more likely to break than old ones and aging should approximately follow Struik's macroscopic law [69,111], that is, $\tau(t_w) \propto t_w^\beta$, where t_w is the aging time and $\beta \lesssim 1$. The aging behavior at rest corresponding to this breaking dynamics is illustrated in Figures 9.22 and 9.23. The ensemble average of rupture times is plotted as a function of the time passed since a given rupture in Figure 9.23. This is a measure of the relaxation time as a function of waiting time (after a quench for instance) and is to be compared to the Struik aging behavior. A power law is found with exponent $\mu \approx 0.6$. The short-time plateau is due to the discrete time step, whereas the long-time plateau in principle coincides with the mean of the relaxation time distribution. In practice, the plateau value is lower, due to the finite duration of the simulation.

The simulated 3D systems correspond to $10\xi \times 10\xi$ (about 30 nm × 30 nm with $\xi \simeq 3$ nm) with periodic boundary conditions in the directions parallel to the substrates and a thickness adjusted between about 2ξ and 18ξ (6–54 nm). The substrates are modeled by repulsive walls akin to the excluded volume potential, invariant in the plane of the surface for a perfectly smooth substrate. Adhesion between the polymer and the surface is due to the adsorption of the monomers at the surface. In the simulations, the nodes within a distance $z \lesssim \xi$ from the substrate may either adsorb or desorb randomly. While a node is adsorbed, its velocity is equal to that of the substrate. Let's denote by τ_{ads}^{-1} and τ_{des}^{-1} the adsorption and desorption rates. At equilibrium, the ratio between the desorption and adsorption rates τ_{des}^{-1} and τ_{ads}^{-1} must be equal to the

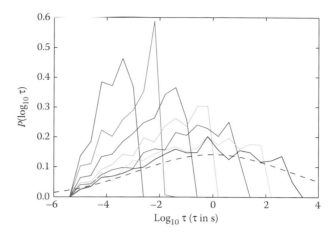

Figure 9.22 Evolution of the simulated distribution of relaxation times during aging after a quench from high temperature. The dotted black line is the equilibrium Gaussian distribution. Full lines correspond to distributions of relaxation time after having aged 10^{-3} to 10^3 s.

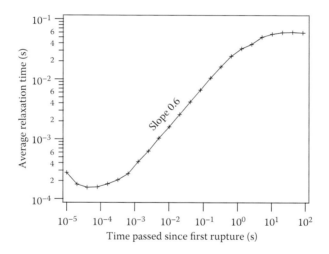

Figure 9.23 Evolution of the simulated average relaxation time during aging after a quench from high temperature. The simulation approximately follows Struik's power law, with an exponent smaller than 1. (Reproduced from A. Dequidt et al., *Eur. Phys. J. E*, 35, 61, 2012. With kind permission of *The European Physical Journal* [EPJ].)

Boltzmann factor

$$\frac{\tau_{\text{ads}}}{\tau_{\text{des}}} = e^{\frac{-U}{k_B T}} \tag{9.41}$$

where $U > 0$ is a parameter of the model that represents the adsorption energy of a domain of size ξ. The adsorption rate is of the form

$$\tau_{\text{ads}}^{-1} = \tau_0^{-1} e^{\frac{-E}{k_B T}} \tag{9.42}$$

where τ_0 is the microscopic time and E is a free energy barrier. Accordingly, the desorption rate is

$$\tau_{\text{des}}^{-1} = \tau_0^{-1} e^{\frac{-(U+E)}{k_B T}} \tag{9.43}$$

Table 9.1 Equivalence between the properties of the real system and the parameters of the numerical model

Physical property	Numerical equivalent
Glassy modulus at high-frequency G_g	Glassy spring stiffness k_g
Rubbery modulus at low-frequency G_r	Soft spring stiffness k_r
Stress relaxation	Spring breaking
Viscous dissipation	Energy release during breaking
Heterogeneous dynamics	Distribution of spring relaxation times
Size of the dynamical heterogeneities ξ	Typical distance between the nodes ξ
Aging	Relaxation time dependence on the spring age
Incompressibility	Excluded volume σ^3
Strength of the interaction with the substrate	Adsorption energy E_{ads}

The desorption energy, $U + E$, corresponds to the free energy barrier that an entire subunit has to overcome for desorbing from the substrate. The corresponding desorption energy per monomer is thus $E_{des} = (U + E)$ $(a/\xi)^2$. As we shall see below, the relevant desorption energies per monomer correspond to a few tenth of $k_B T$. Assuming $(\xi/a)^2 = N_c^{2/3} \approx 50$, a free energy barrier of $0.2 k_B T$ per monomer corresponds to a free energy of $10 k_B T$ per subunit.

The values of the parameters used in the simulations are given in Table 9.1, with their corresponding physical quantities.

9.3.2 HOW MECHANICAL PROPERTIES ARE MODIFIED IN CONFINEMENT AS COMPARED TO THE BULK

9.3.2.1 TEMPERATURE DEPENDENCE OF THE MECHANICAL RESPONSE

Shear experiments at constant shear rate are simulated by displacing the substrates at constant velocity with respect to each other. The adsorbed nodes are carried along with the surfaces and exert back a force on the substrate. The average force per unit area on the substrate is the stress. Stress can thus be recorded together with the strain as a function of time and stress–strain curves can be simulated in this way.

Examples of simulated stress–strain curves are shown in Figure 9.24, which illustrates a generic behavior. In the shown example, simulation parameters correspond to polystyrene. At temperatures far above T_g (or at low shear rate), the glassy springs break very often compared to the strain rate. They have no time to accumulate stress. The viscoelastic behavior is dominated by the elastic contribution of the rubbery springs, and the stress–strain curve is essentially linear with a slope corresponding to the rubbery modulus G_∞. At temperatures close to or below T_g (or at a higher shear rate), the glassy springs dominate the mechanical response because they have a long lifetime. Stress–strain curves start with a much steeper slope corresponding to the glassy modulus G_g. After a time comparable to the average relaxation time τ, the springs have independently broken and been deformed again and a stress plateau is reached corresponding to the viscous regime. The rubbery springs contribute as a small slope on this plateau. From the mechanical point of view, the glass transition can be defined from the storage modulus G', that is the slope of stress–strain curves at the origin. Below T_g, $G' \approx G_g$, while G' approaches G_∞ as T increases above T_g. Note that the temperature variations of G_g and G_∞ were not implemented in the model here.

9.3.2.2 T_g SHIFTS

The impact of confinement on $G'(T)$ with strong polymer–substrate interactions is shown in Figure 9.25. The glass transition clearly occurs at higher temperatures as the thickness is reduced. The simulated T_g shift is up to about 20 K at the smallest thickness (9 nm) and is still significant at about 20 nm. See Figure 9.26. The transition also appears to be broader and the curves $G'(T)$ are noisy at intermediate temperatures in the thinnest simulated films. This will be discussed in the subsequent sections.

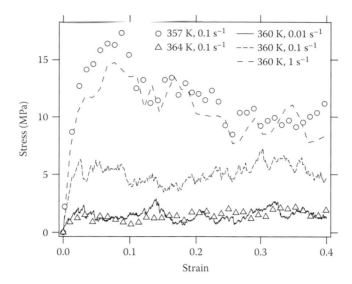

Figure 9.24 Simulated stress–strain curves of polystyrene at various temperatures and shear rates around T_g, showing an approximate time–temperature equivalence. At high temperature (low shear rate), the behavior is elastic with a low modulus. At low temperature (high shear rate), the response to shear is first elastic with a high modulus, and then viscous with a plateau after a time of order τ_α. G' is defined as the initial slope. (Reproduced from A. Dequidt et al., *Eur. Phys. J. E*, 35, 61, 2012. With kind permission of *The European Physical Journal* [EPJ].)

9.3.2.3 BROADENING OF THE GLASS TRANSITION

Confinement does not produce a simple shift in the $G'(T)$ curves. By comparing the $G'(T)$ curves obtained at different thicknesses (Figure 9.25), we observe that the most striking evolution of the curves for the thinnest films is the broadening of the mechanical glass transition. The transition is relatively abrupt in the bulk whereas it is more and more gradual under confinement. As the polymer becomes more confined, slow heterogeneities retain a significant contribution to the mechanical response at higher temperature. This is consistent with the qualitative interpretation based on percolation across a slab.

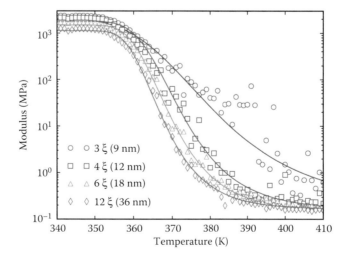

Figure 9.25 Simulated elastic modulus G' of polystyrene as a function of temperature for several confinement thicknesses in the case of strong polymer–substrate interactions. Glass transition corresponds to the transition from a high modulus at low temperature to a low modulus at high temperature. Under confinement, T_g is shifted to higher temperatures. Glass transition also appears broader and more noisy at smaller thicknesses. The lines are fits of an empirical expression. (Reproduced from A. Dequidt et al., *Eur. Phys. J. E*, 35, 61, 2012. With kind permission of *The European Physical Journal* [EPJ].)

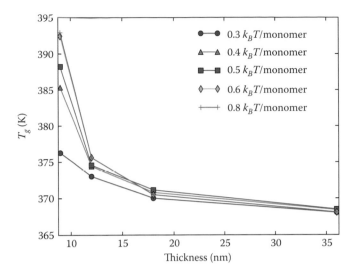

Figure 9.26 Simulated T_g shifts of polystyrene as a function of thicknesses at various polymer–substrate interactions characterized by the desorption energy barrier. T_g is defined here as the temperature at which $G' = 10$ MPa. The simulated mechanical properties are modified under confinement up to more than 20 nm. T_g shifts are more important in the case of strong substrate–polymer interactions and can reach more than 20 K in 9 nm films. (Reproduced from A. Dequidt et al., *Eur. Phys. J. E*, 35, 61, 2012. With kind permission of *The European Physical Journal* [EPJ].)

Thus the simulations show that under confinement with strong polymer–substrate interactions, the glass transition is not simply shifted to higher temperatures but also broadened toward high temperatures.

9.3.2.4 EFFECT OF POLYMER–SUBSTRATE INTERACTION

The polymer–substrate interaction can be varied in the simulation by varying the adsorption energy E_{ads}. However, since the mechanical deformation is imposed from the surfaces, only relatively strong interactions can be simulated. Freely suspended films, which experimentally exhibit negative T_g shifts, cannot be simulated using this method. In order to quantify the T_g shift, one can arbitrarily define T_g as the temperature at which $G' = 10$ MPa. Simulations show that, in agreement with experiments and with qualitative interpretation, T_g is shifted to higher temperatures when the polymer–substrate interaction is stronger, as illustrated in Figure 9.26 and in Table 9.2. With a low desorption energy ($0.3k_BT$ per monomer), the shift is about 7 K at 9 nm, whereas with an energy barrier corresponding to strong adsorption ($0.6k_BT$ per monomer and higher), the shift is about 20 K at 9 nm. In the limit of high adsorption energies, desorption of a whole dynamical heterogeneity almost never occurs and the results become insensitive to the actual value of E_{ads}. In the simulations, this occurs beyond about $0.6k_BT$ per monomer.

9.3.2.5 HOW RARE EVENTS CONTROL THE MECHANICAL PROPERTIES OF CONFINED POLYMERS

Since the distribution of relaxation times is several decades broad, a few subunits with long relaxation time may survive relatively far above T_g in the equilibrated system, while the average relaxation τ_α time is fast.

Table 9.2 Simulated T_g shifts

| Thickness (nm) | Desorption energy barrier (k_BT monomer) | | | | |
	0.3	0.4	0.5	0.6	0.8
9	372 K	381 K	383 K	387 K	387 K
12	370 K	371 K	3/1 K	372 K	372 K
18	367 K	368 K	368 K	368 K	368 K
36	365 K	366 K	366 K	366 K	366 K

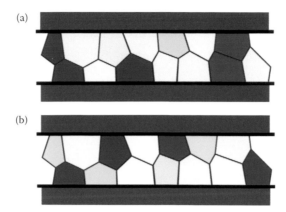

Figure 9.27 Two possible situations in thin polymer films with the same distribution of dynamic heterogeneities. (a) Clusters of slow heterogeneities connect both substrates together. The film has a high elastic modulus. (b) Heterogeneities are dispersed and no cluster connects both substrates. The film has a low elastic modulus. The two situations may correspond to the same sample at different times, when dynamical heterogeneities evolve following thermal fluctuations. (Reproduced from A. Dequidt et al., *Eur. Phys. J. E*, 35, 61, 2012. With kind permission of *The European Physical Journal* [EPJ].)

These may form rigid clusters connecting the substrates, so that the thin film resists shear and has a higher shear modulus than bulk polymer at the same temperature. It is essentially this mechanism that explains both the T_g shifts and the broadening of the glass transition on increasing temperature.

In finite-size simulations, particular configurations with small modulus G', comparable to the rubbery modulus G_r (no rigid cluster connecting both substrates), or with high modulus, comparable to a fraction of the glassy modulus G_g (a rigid cluster connects both substrates), may occur with finite probability. These two situations are illustrated in Figure 9.27. The resulting scattering of instantaneous stress values from several simulations of the same system above bulk T_g at the same level of deformation is shown in Figure 9.28. This distribution is peaked at nearly zero stress, which means that most of the time, the elastic modulus is small,

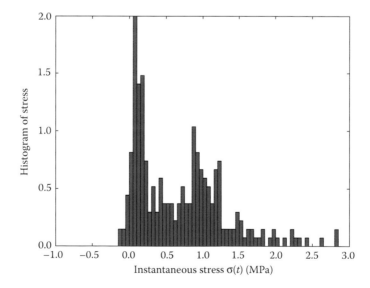

Figure 9.28 Histogram of simulated stress at 390 K, for several random outcomes in a polystyrene film of thickness 3ξ, after a deformation of 10^{-2} at shear rate 0.1 s^{-1}. Most of the time, stress is low. The histogram has a peak around 0. But sometimes, a cluster connects both substrates and confers a high elastic modulus on the film. The histogram has a tail reaching high stress values. (Reproduced from A. Dequidt et al., *Eur. Phys. J. E,* 35, 61, 2012. With kind permission of *The European Physical Journal* [EPJ].)

with a tail that extends to higher stress values. This corresponds to the presence of one or more rigid connecting clusters. The tail significantly increases the average stress, and so the average modulus. The increase in G' in confined polymer films is thus the consequence of events with a low probability, but with a high contribution.

9.3.2.6 ARE THE EFFECTS OF CONFINEMENT THE SAME AS REGARDS DIFFERENT EXPERIMENTAL TECHNIQUES: MECHANICAL SPECTROSCOPY, DIELECTRIC SPECTROSCOPY, NMR?

Positive T_g shifts at the interfaces with strongly interacting substrates may be interpreted as the contribution of slow clusters connected to one substrate. The broadening of the mechanical glass transition observed in the simulated thin films is interpreted as the contributions of slow, rigid clusters connecting both surfaces together. These contributions have an impact on mechanical properties due to the very large contrast in elastic modulus (3–4 orders of magnitude) below and above T_g. In other experimental techniques, such as dielectric spectroscopy or DSC, the contrast is much smaller. The dielectric permittivity and the heat capacity keep the same order of magnitude below and above T_g, so the contributions of rare slow clusters is probably too small to be detected. With these techniques, confinement may induce a small T_g shift, but no measurable broadening of the glass transition. Therefore, the effect of confinement above $T_g + 20$ K might not be detectable with these techniques, whereas it might be using dynamic mechanical analysis. NMR may be used to estimate the fraction of polymer with a slow relaxation time. However, in this technique as well, it may be difficult to detect very small fractions of the material (say, of the order 1%), which may still have a significant effect on the mechanical response due to the large contrast in modulus.

The closer it is to the interface, the higher the fraction of polymer belonging to rigid clusters of slow subunits immobilized by the substrate. Thus, on average, there is a gradient of relaxation times at the interface. The contribution of this slow fraction becomes more sensitive as the confinement becomes stronger. The corresponding signal can in principle be detected using all the mentioned experimental techniques. But since mechanical measurements are so much sensitive to slow subunits, they are impacted at much higher temperatures above T_g by confinement, than using other techniques.

9.4 FILLED ELASTOMERS

9.4.1 INTRODUCTION

In this section, we discuss the behavior of elastomers filled with particles or aggregates of submicrometric, typically 10–100 nm, sizes. Filled elastomers exhibit specific properties qualitatively different from those of pure elastomers, first of all a strong increase of the elastic modulus [44]. Our point of view is that a major (if not predominent in most cases) contribution to reinforcement is due to the presence of T_g gradients in the polymer matrix close to filler interfaces. This assumption is still the subject of an active debate, if not controversy. It is based on an analogy with the behavior of thin polymer films deposited on a substrate. Global mechanical properties, which have been characterized extensively over the past 50 years or so, are presented in Section 9.4.2. Some interpretations that have been proposed for these properties are presented in Section 9.4.3. We shall then present a (perhaps nonexhaustive) set of experimental results that are in favor of the hypothesis of T_g gradients in Section 9.4.4. The way in which mechanical properties can be modeled is introduced in Section 9.4.5.

9.4.2 MECHANICAL PROPERTIES OF REINFORCED ELASTOMERS

9.4.2.1 LINEAR RESPONSE

The linear shear modulus is strongly enhanced, with a temperature dependence different from that of the pure matrix, as exemplified by the data reported by Wang (Figure 9.29) [99]. The reinforcement factor R, defined as the ratio of the modulus of the filled material over that of the pure elastomer matrix, depends

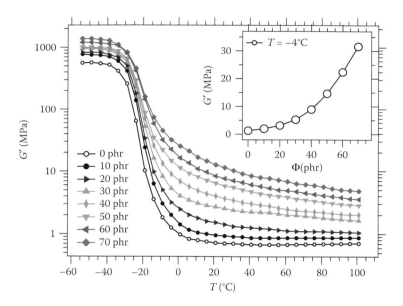

Figure 9.29 The elastic modulus G' measured in oscillatory strain in the linear regime as a function of temperature for a series of styrene butadiene rubber filled with various amounts of carbon black (in phr [g per hundred g of rubber] units). Inset: G' as a function of the filler amount at $T = -4°C$. (Reprinted with permission from M.-J. Wang, *Rubber Chem. Technol.*, 71, 520. Copyright 1998, Rubber Division, American Chemical Society, Inc.)

strongly on temperature, with a maximum at typically 20 K above the T_g of the pure matrix. At maximum, R can reach 30, as in Figure 9.29, and up to about 200 in some cases [126]. Often, R remains large with respect to unity over a very broad temperature range (more than 140 K above T_g), as in Figure 9.29, while in some cases, and specifically in model systems with well-dispersed spherical particles, R instead exhibits a relatively narrow peak and steep decrease as T increases (Figure 9.30) [34]. Note also that the modulus does not increase linearly with the volume fraction Φ.

With regard to the glass transition, the mechanical response of filled elastomers does not reduce to a master curve when using shift factors a_T from the unfilled matrix only. Instead, a master curve can be constructed

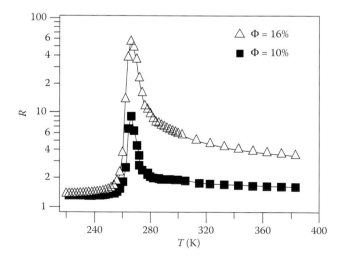

Figure 9.30 The reinforcement factor $R = G'/G_0'$ as a function of temperature for model poly(ethyl acrylate) (PEA) elastomers filled with well-dispersed spherical silica particles at two different volume fractions Φ. (From J. Berriot et al., *Europhys. Lett.*, 64, 50, 2003. IOP Publishing is acknowledged.)

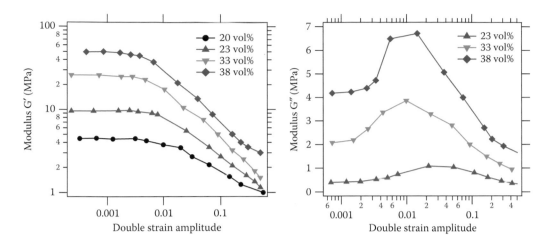

Figure 9.31 The storage G' and loss moduli G'' (in MPa) measured in carbon black–filled butyl rubbers as a function of the deformation amplitude and three different filler volume fraction: □: $\Phi = 38.6\%$; O: $\Phi = 33.6\%$; ●: $\Phi = 23.2\%$. (A. R. Payne, *J. Appl. Polym. Sci.*, 1963, 7, 873. Copyright Wiley-VCH Verlag GmbH & Co. KGaA. Reproduced with permission.)

with a different shift factor (solid circles in inset), which describes the filled elastomer system with a distribution of α-relaxation times, suggesting the relaxatory (or viscoelastic) nature of the filler-induced reinforcement [127].

9.4.2.2 NONLINEAR RESPONSE

Under oscillatory strain (at frequencies typically between 0.1 and 10 Hz), the elastic modulus reversibly decreases down to values comparable to that of the unfilled matrix as the strain amplitude increases between a few percents and a few tens of percents, while the modulus stays roughly constant in the unfilled matrix [126,128–131]. The loss modulus correlatively often shows a maximum. Data reported by Payne examplify this specific nonlinear behavior, denoted as Payne effect (Figure 9.31) [130]. The amplitude of this decrease is generally correlated to the amplitude of reinforcement. This nonlinear mechanical response under oscillatory strain has been recently analyzed in more detail [132,133]. It actually corresponds to two distinct types of nonlinear response, one related to the decrease of the modulus (the slope at the origin in stress–strain cycles) as the amplitude of the cycle increases, the other one related to nonlinearity of the shape of a large amplitude cycle [132]. After the system has been put at rest and during subsequent deformations, the elastic modulus in the linear regime is smaller than that of the initial system, but recovers progressively—at least partially—the initial value [128,134–136]. This is the so-called Mullins effect, as examplified in Figure 9.32. These features are essential for damping materials, impact absorbers, or tires, as regards rolling resistance, grip, and durability [97]. Understanding this unique behavior is thus an issue of major importance in order to be able to control and tune the properties of these systems. [44,97,126,130,131,134,136–139].

9.4.3 PHYSICAL MECHANISMS AT THE ORIGIN OF REINFORCEMENT

9.4.3.1 LOCAL STRAIN AMPLIFICATION

Different physical mechanisms have been proposed to explain reinforcement. The first one is the so-called hydrodynamic effect of Guth and Gold [140], inspired by the approach of Einstein for the viscosity of diluted colloidal suspensions, in which the reinforcement R is expressed as a linear function of the volume fraction Φ only as $R = 1 + 2.5\Phi$ and which was later extended to more complex forms to account for both higher volume fractions and correlations between fillers. Note that local strain amplification in a filled matrix is the same physical concept as the effect described by Guth and Gold and followers, even though the dependences of local strain amplification and modulus enhancement on Φ may be different, as shown recently [141].

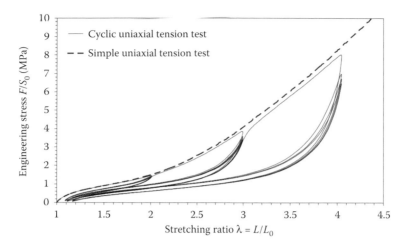

Figure 9.32 Stress–strain response of a carbon black–filled styrene-butadiene rubbers (SBR) elastomer submitted to simple uniaxial tension and to cyclic uniaxial tension with increasing maximum stretching ratios. (Reprinted from *Eur. Polym. J.*, 45, J. Diani, B. Fayolle and P. Gilormini, 601, Copyright 2009, with permission from Elsevier.)

9.4.3.2 OTHER MECHANISMS?

It has long been recognized that the properties of filled elastomers, specifically the very high values of modulus reached in some cases and its strong temperature variation, together with the peculiar nonlinear effects, involve physical mechanisms that are more complex than the above essentially geometrical effect.

It has been proposed that trapping effects associated with chain adsorption on the filler surfaces would induce an increased density of effectively active constraints (cross-links and entanglements) in the vicinity of filler particles, which would be responsible for the enhanced modulus (beyond the pure hydrodynamic effect) [142–144]. These constraints would then be released due to chain desorption under large amplitude strain, which would explain the drop of the modulus and dissipative properties [145,146]. Local strain amplification might also contribute to modulus enhancement due to finite chain extensibility [144]. Indeed, some studies have shown that reinforcement is affected by the polymer molecular weight and filler surface treatment, and is particularly pronounced when distances between fillers are comparable to chain sizes [142,143]. Interpretations were based on careful analysis of mechanical experiments. However, there has been no direct experimental evidence of an increased entanglement density close to filler particles, nor of disentanglement under large amplitude strain. On the other hand, molecular dynamics simulations show that the entanglement density may be increased in the vicinity of an adsorbing surface [147].

9.4.3.3 FILLER NETWORKING

Strong reinforcement is commonly understood to result from the presence of a mechanical network formed by filler percolation, which would mostly transmit the stress. Indeed, it is generally observed that the modulus deviates from the pure hydrodynamic prediction and increases quite steeply above a threshold fraction, which depends on the type of fillers [148]. Destruction/reformation/reorganization of the filler network would then explain the observed drop of the modulus under large amplitude strain. [97,130,149,150]. To elucidate the nature of this network, it is thus of crucial importance to study its relationship to both the dispersion state of the fillers and the filler–matrix interactions.

9.4.3.4 GRADIENT OF GLASS TRANSITION TEMPERATURE AT INTERFACES, GLASSY BRIDGES

The above mechanisms certainly cannot explain the very large values of the modulus that can be reached a little above the elastomer matrix T_g, suggesting an approach in which the behavior of the elastomer matrix itself is drastically modified. Indeed, it has been recognized that reinforcement must involve the modification of the polymer dynamics in the presence of fillers [97,99,151]. In this context, a T_g shift of the matrix in the vicinity of fillers, creating glassy bridges, has been proposed as a possible physical explanation [32,33,35]. In thin polymer films with attractive polymer–substrate interactions, an increase of T_g has been reported as the film thickness decreases [152]. This T_g shift was modeled by a T_g gradient according to [65,67]

$$T_g(z,\omega) = T_g(\omega)\left(1+\left(\frac{\delta}{z}\right)^{1/\nu}\right) \tag{9.44}$$

where z is the distance from the solid surface. This equation is directly derived from Equation 9.22, where $T_g(\omega)$ is the frequency-dependent glass transition temperature in the bulk and δ the characteristic length of the gradient.

The relative contributions of the various mechanisms mentioned above may vary, depending on material parameters (matrix cross-linking, reinforcing system) and on test conditions (strain amplitude, strain rate, and temperature). Indeed, it has been claimed that glassy bridges may not be the dominant mechanism in little reinforced systems and/or very far above the matrix T_g [153,154]. Therefore, it is still a major issue to gain selective information on the various mechanisms contributing to reinforcement in filled polymer melts and elastomers.

9.4.4 Experimental evidence for a T_g gradient

Direct experimental evidence for T_g gradients or glassy layers are very difficult to obtain in nanocomposites, if not out of reach in many cases, mostly because a small fraction of the polymer is generally affected, which claims for very clean model systems. Based on aging experiments, Struik has inferred the presence of a fraction of glassy polymer above the matrix T_g in filled materials [155].

9.4.4.1 NMR EXPERIMENTS

Time-domain proton nuclear magnetic resonance provides information on the elastomer dynamics in quite a direct way. NMR experiments have shown the existence of different polymer mobility domains in filled elastomers [156,157], based on the very different transverse relaxation times for the rigid (glassy) fraction and the elastomeric (bulk) part of the matrix. The whole NMR signal of an elastomer filled with well-dispersed spherical particles may be computed by taking into account the T_g gradient around filler particles, according to

$$M(T,t) = M^{graft} + \frac{1}{V_{tot}} \int\limits_{V_{tot}} M^{PEA}(T - T_g(z),t)dV(z) \tag{9.45}$$

where $T_g(z)$ follows Equation 9.44. $M^{PEA}(T,t)$ describes the NMR signal of the pure, unfilled matrix, measured and/or modeled at temperature T (between T_g and $T_g + 120$ K). V_{tot} corresponds to the average polymer volume per particle. A rigid contribution M^{graft} from immobilized protons in the grafters must generally be added [158].

The immobilized (rigid) polymer fraction has been evaluated systematically using Equation 9.45 in a series of silica-filled model poly(ethyl acrylate) (PEA) elastomers with different grafters (with or without covalent filler/silica bonds) and at different temperatures, with the gradient parameter δ (see Equation 9.44) and grafted thickness as adjustable parameters [158,159], as exemplified in Figure 9.33a. This shows explicitly that a T_g gradient describes the polymer dynamics in filled elastomers very well. The gradient parameter δ was found to be between 0.1 and 0.2 nm for all samples [18,19]. Accordingly, considering as "glassy" the polymer below $T_g + 20$ K, glassy layer thicknesses ranging from 2 to 6 nm (respectively 1 to 3 nm) for samples with (respectively without) covalent grafting were obtained in the considered temperature range, which is also in agreement with literature results [33,160,161].

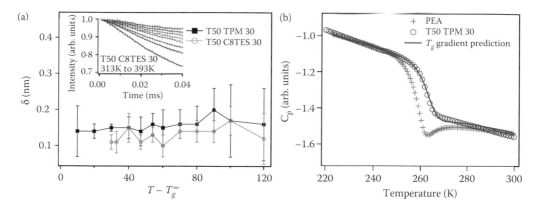

Figure 9.33 (a) Inset: Proton NMR signal at various temperatures in a filled PEA sample (crosses), fitted with Equation 9.45 (curves) at all temperatures with the same M^{graft} contribution and gradient parameter δ (see Equation 9.44) as the only adjustable parameter for each T. (b) Heat capacity for the pure (crosses) and filled (circles) PEA samples at cooling and heating rates 10 K/min. The sample is the same as in Figure 9.2a. The curve is the prediction from the T_g gradient model. The curve is the fit with the T_g gradient model. The parameter e_0 was fixed at the value obtained from NMR and δ was free. Values found for δ are very close to those obtained from NMR. (Reprinted with permission from A. Papon et al., *Phys. Rev. Lett.*, 108, 065702. Copyright 2012 by the American Physical Society.)

9.4.4.2 DSC EXPERIMENTS

A classical signature of the glass transition in polymers is the DSC response. The DSC response of a filled sample is compared with that of the pure matrix in Figure 9.33b. The apparent T_g in the filled sample is clearly shifted to higher temperature and broader than in the pure matrix. Using the same approach as for NMR, the heat capacity of the filled sample was integrated using the T_g gradient and pure PEA matrix $c_p^{PEA}(T)$ signal, according to

$$c_p(T) = \frac{1}{V_{tot}} \int_{V_{tot}} c_p^{PEA}(T - T_g(z))dV(z) \tag{9.46}$$

The model is in very good agreement with the experimental DSC curve, using the same parameters (grafted thickness and δ) as determined previously from NMR (Figure 9.33) [159].

9.4.4.3 A NEW MASTER CURVE FOR FILLED ELASTOMERS

The effects of a T_g gradient around filler particles on the mechanical behavior of filled elastomers can be tested experimentally. According to Equation 9.44, at a temperature $T > T_g$, the polymer is glassy, with a high elastic modulus ($\approx 10^9$ Pa), within a shell of thickness e_g, which depends on T and the measurement frequency ω, given by

$$e_g(T, \omega) = \delta \left(\frac{T_g(\omega)}{T - T_g(\omega)} \right)^\nu \tag{9.47}$$

Beyond this shell, the elastic modulus remains nearly that of the bulk matrix ($\approx 10^6$ Pa). A simple approach is to assume that reinforcement is a function of the volume fraction $\Phi_{eff}(T, \omega)$ of solid material—particles plus glassy shell—only, given by

$$\Phi_{eff}(T, \omega) = \Phi \left(1 + \frac{e_g(T, \omega)}{a} \right)^3 = \Phi \left(1 + \frac{\delta}{a} \left(\frac{T_g(\omega)}{T - T_g(\omega)} \right)^\nu \right)^3 \tag{9.48}$$

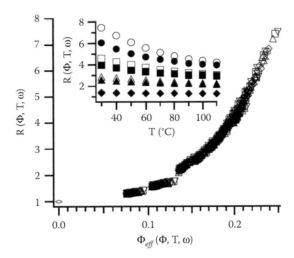

Figure 9.34 Master curve of the reinforcement in poly(ethyl acrylate) (PEA) matrices filled with well-dispersed spherical silica particles (radius 25 nm) grafted with a silane, which copolymerizes with ethyl acrylate monomers (about two grafters per nm^2) as a function of the effective volume fraction Φ_{eff}, using Equation 9.48, for frequencies from 0.01 to 10 Hz, volume fraction Φ from 6.7% to 17.7%, and temperature from $T_g + 50$ K to $T_g + 130$ K. Inset: Reinforcement as a function of temperature for volume fraction and frequencies, respectively: ○ 17.7%, 10 Hz, ● 17.7%, 0.01 Hz, □ 15%, 10 Hz, ■ 15%, 0.01 Hz, △ 12%, 10 Hz, ▲ 12%, 0.01 Hz, ◇ 6.7%, 10 Hz, ◆ 6.7%, 0.01 Hz. (Reproduced from J. Berriot et al., *Europhys. Lett.*, 64, 50, 2003.)

In this approximation, typically valid for $T > T_g + 50$ K, the reinforcement R for a given sample is a function of $T_g(\omega)/(T - T_g(\omega))$. This new time–temperature superposition specific to filled elastomers has been validated in a series of well-controlled model samples, as shown in Figure 9.34 [34]. However, at lower temperature, a core–shell description is not sufficient anymore and a more complex behavior is observed [158].

Finally, the T_g gradient model in nanocomposites is illustrated in Figure 9.35 [34,35]. The glassy polymer layer around a filler particle becomes broader and the interface more gradual as T comes closer to the bulk T_g. The robustness of the gradient model has been shown through its ability to describe both DSC and NMR measurements with and without solvent.

9.4.4.4 BREAKDOWN OF STRESS-OPTICAL LAW IN REINFORCED ELASTOMERS

The constitutive equation of an elastomer is based on the very general stress-optical law, which writes, for uniaxial extension:

$$\sigma = \frac{k_B T}{b^3} \langle P_2(\cos\theta) \rangle \approx k_B T \nu \left(\lambda^2 - \lambda^{-1} \right) \tag{9.49}$$

where σ is the stress, b^3 the volume of a statistical segment, ν the density of cross-links, λ the elongation ratio, and $\langle P_2(\cos\theta) \rangle$ the average of the segmental orientation parameter $(3\cos^2\theta - 1)/2$, with θ the angle between any chain segment and the direction of the strain. This expresses that the physical origin of the stress in an elastomer is related to chain configurational statistics, that is, to chain segment ordering under an applied strain.

The behavior of the matrix in a filled elastomer may be probed by measuring the average segmental orientation $\langle P_2(\cos\theta) \rangle$ and comparing it to the stress value [162]. By analyzing the deviations with respect to the behavior of the pure, unfilled elastomer, the contribution due to strain amplification effects in the elastomer matrix can be selectively distinguished. Figure 9.36 shows the moduli, measured at 50% (G'_{50}) and 0.1% (G'_0) strain amplitude, plotted versus the ratio (denoted S_2) of $\langle P_2(\cos\theta) \rangle$ to the strain function $\lambda^2 - \lambda^{-1}$, for unfilled and filled natural rubber elastomers with various cross-link densities. S_2 was measured by x-ray scattering. While the modulus decreases as the strain increases (Payne effect), as seen from the difference between G'_{50} and G'_0 values in Figure 9.36, the response of the elastomer matrix in terms of the ratio S_2 (equivalent to modulus as regards segmental orientation in response to strain) stays constant over the whole strain range. Therefore,

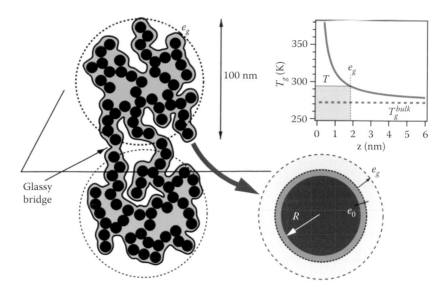

Figure 9.35 Schematics of a glassy bridge between two filler aggregates. The glassy layer is in gray. The T_g gradient around a filler particle is schematized on the right. The layer of thickness e_0 corresponds to immobilized protons at the particle surface, and e_g is the glassy layer thickness at temperature T. The fraction of glassy polymer in a plane normal to the applied stress is $\Sigma \sim 1\%$, while the macroscopic strain is amplified in between fillers by a factor typically $\lambda \sim 10$, which results in a macroscopic modulus $G' \approx G'_g \lambda \Sigma \approx 10^7$ to 10^8 Pa. (Reprinted with permission from S. Merabia, P. Sotta, and D. R. Long, *Macromolecules*, 41, 8252. Copyright 2008, American Institute of Physics.)

Equation 9.49 breaks down in reinforced materials. Data for G'_{50} almost collapse with data of the unfilled samples. This shows that, at medium/large strain amplitudes, mechanical reinforcement observed in the filled samples is mostly due to strain amplification and entropic elasticity in the bulk elastomer matrix. At small strain amplitudes, the increased modulus is related to additional, nonentropic reinforcement mechanisms, which are active even at $T_g + 120$ K, and is not proportional to the cross-link density of the matrix, as it is the case for unfilled systems. Therefore, G'_0 cannot be expressed as a separable formula of the form $G' = G'_m f(\Phi)$, with G'_m the elastic modulus of the pure matrix and $f(\Phi)$ a reinforcement function depending only on the reinforcing filler system, namely, the volume fraction, structure, morphology, and dispersion state of fillers.

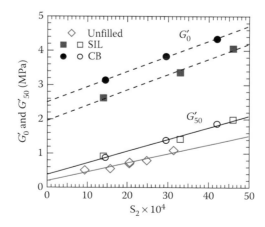

Figure 9.36 Storage shear moduli G'_0 (full symbols) and G'_{50} (empty symbols) for vulcanized natural rubber elastomers, as a function of the slope S_2 of $\langle P_2 \rangle$ measured by x-ray scattering: unfilled (◊); SIL (○): silica-filled; CB (□): carbon black–filled samples. (Reprinted with permission from R. Pérez Aparicio et al., *Macromolecules*, 46, 2013, 8964. Copyright 2013, American Chemical Society.)

9.4.4.5 DEPENDENCE ON THE DISPERSION STATE AND FILLER–MATRIX INTERACTIONS: NUMBER OF GLASSY BRIDGES

The dispersion state of the fillers and filler–matrix interactions play a predominant role in reinforcement. Since changing the interactions have an effect on the dispersion, it is a challenge to control *independently* the dispersion state and the interactions at filler–matrix interfaces. In Papon et al. [132], model-filled rubbers with the same chemical structure, including same filler–matrix interfaces, but with different filler dispersion states, have been elaborated, in an attempt to quantify their respective roles [98,158,163]. From these model systems, it was shown that the particle arrangement controls the strain softening at small strain amplitude (Payne effect), as well as the temperature dependence of the elastic modulus. The Payne effect vanishes and the elastic modulus depends only weakly on T when particles are well separated. Conversely, samples with aggregated particles show large Payne effect and their elastic modulus drops markedly as T increases, as shown in Figure 9.37. These effects provide further evidence that glassy bridges play a key role on the mechanical properties of filled rubbers. Indeed, the relative amplitude of the Payne effect is well correlated to the number of glassy bridges in the same series of samples [163].

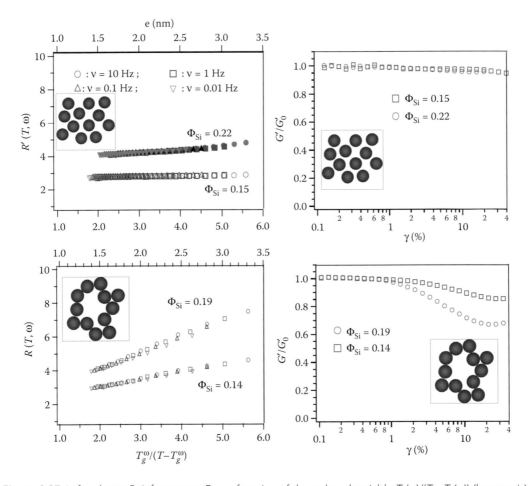

Figure 9.37 Left column: Reinforcement R as a function of the reduced variable $T_g(\omega)/(T - T_g(\omega))$ (lower axis), proportional to the glassy layer thickness e (upper axis) (see Equation 9.47), in PEA elastomers filled with well-dispersed (top) or partially aggregated (bottom) spherical silica particles with covalent grafting (TPM). Right column: Shear modulus as a function of the strain amplitude (Payne effect), normalized to the modulus in the linear regime, for well-dispersed (top) or partially aggregated (bottom) systems. The dispersion state of silica particles was characterized by small-angle neutron scattering. (Reproduced from H. Montes et al., *Eur. Phys. J. E*, 31, 263, 2010. With kind permission of *The European Physical Journal* [EPJ].)

9.4.5 MODELING REINFORCEMENT MECHANISMS IN FILLED ELASTOMERS

Theoretical modeling of filled elastomers is faced with a number of difficulties of various origins. First, the behavior of the elastomer matrix itself is very complex [164,165], involving excluded volume interactions, nonaffine displacements of cross-links, trapped entanglements, and network defects. Introducing filler particles induces additional issues. Mechanical properties depend in a complex way on parameters such as temperature and filler volume fraction, dispersion state, structure and size, as well as the nature and strength of filler–matrix interactions.

Experimental results reviewed above have shown that the presence of glassy layers around fillers plays an essential role in reinforcement [32–34,166]. It has long been proposed that maximum reinforcement occurs when the elastomer matrix is effectively glassy in the vicinity of the fillers [99,155,167–171]. In fact, the whole set of mechanical properties of filled elastomers, in particular, as a function of the polymer–filler interaction, filler volume fraction, and temperature, can be rationalized based on the presence of glassy layers/glassy bridges. This picture is consistent with experimental reports of an increase of T_g in adsorbed thin polymer films with strong polymer–substrate interactions. [16–18,26–28,61,65,67]. Indeed, the relationship between glass transition in thin films and physical properties of nanocomposites has been emphasized by many authors [38–43,172–176].

The purpose of this section is to introduce the mesoscale approach that has been proposed to model large-scale elastic and plastic properties of filled elastomers [177–179]. Given the relevant spatial and temporal scales of the problem, molecular simulations are inappropriate [180] and a coarse-grained approach at the scale of the fillers is unavoidable. Among recent related works, let's mention the finite element mapping with spring network representations by Gusev [181] and the multiscale model of Bauerle et al. [182,183]. The elementary length scale of the model is typically of the order 10–100 nm, depending on the filler diameter and/or on the typical distance between fillers. It incorporates physical features relevant to the behavior of the elastomer matrix in the presence of a T_g gradient, and predicts semiquantitatively the mechanical behavior of filled elastomers, in the linear (reinforcement) as well as nonlinear (Payne and Mullins effects) regimes. All model parameters are based on physical, measurable quantities. They must, however, be mapped to each real physical system. In particular, for simulation purposes, the perhaps complex structure of the fillers has been simplified into spherical filler particles.

Fillers are modeled by solid spheres that interact with their neighbors through hardcore repulsions, "rubbery" permanent elastic springs that mimic the elastic response of the cross-linked, bulk polymer matrix and "glassy" elastic springs that mimic the behavior of glassy polymer layers. The local T_g is given by [35,166]

$$T_g(z,\sigma) \approx T_g\left(1+\left(\frac{\delta}{z}\right)^{1/\nu}\right) - \frac{\sigma}{K} \approx T_g\left(1+\frac{\delta}{z}\right) - \frac{\sigma}{K} \tag{9.50}$$

where z is half the distance between nearest filler interfaces, T_g is the bulk glass transition temperature of the pure elastomer, and $\nu \approx 0.88$. The first term in the right-hand side of Equation 9.50 is the T_g gradient already described in Equation 9.44, with δ depending on matrix–filler interactions. For strong interactions, $\delta \sim 1 - 2$ nm [35]. The decrease of T_g due to the local stress σ in the second term describes the plasticizing effect of an applied stress, typical in a glassy polymer. K relates the yield stress σ_y to temperature T and the polymer T_g by $\sigma_y = K(T_g - T)$, with K of the order 10^6 Pa/K typically [184]. Hence, glassy bridges may yield under stress when their local T_g decreases below the temperature of the experiment.

9.4.5.1 LIFETIME OF GLASSY BRIDGES: AGING

Within glassy layers, polymer relaxation is modeled by breaking of glassy springs, with breaking time identified to the local relaxation time τ_α of the polymer given by the William–Landel–Ferry law of the polymer [2], referenced to the local T_g:

$$\log\left(\frac{\tau_\alpha(z,\sigma)}{\tau_g}\right) = -\frac{C_1(T - T_g(z,\sigma))}{C_2 + (T - T_g(z,\sigma))} \tag{9.51}$$

where $T_g(z,\sigma)$ is given by Equation 9.50 with $\tau_g = 100$ s (relaxation time at T_g) and C_1 and C_2 the WLF parameters of the considered polymer [2]. Equation 9.51 gives *the equilibrium value of the breaking time*, obtained when the distance z, local stress σ, and temperature T have been maintained over a long time. In general, the breaking time, denoted $\tau_\alpha(t)$, depends on the history of the glassy bridge. It evolves linearly with time toward its equilibrium value with a factor (aging coefficient) $\alpha \lesssim 1$ [69].

9.4.5.2 IMPLEMENTATION OF THE MESOSCALE MODELING

The numerical implementation of the model has been described in detail in Long and Sotta [177–179]. Strain- or strain rate-controlled experiments are simulated. The degrees of freedom of the numerical model are the positions of the fillers. The filler diameter d sets the dimensionless unit length scale. Assuming $d \simeq 20$ nm, typical dimensionless δ values corresponding to $\delta = 1$ nm are of order 0.05. The dynamics are noninertial. Each particle i interacts with typically 10 neighboring particles. The equation of motion for particle i is

$$\sum_j \vec{F}_{el}^{ij} + \vec{F}_{hs}^{ij} + \vec{F}_{Hydro}^{ij} = \vec{0} \tag{9.52}$$

where $\vec{F}_{hs}^{ij}$ is the hardcore repulsion, $\vec{F}_{Hydro}^{ij} = -\zeta\left(d\vec{R}_{ij}/dt\right)$ is the viscous friction force, and $\vec{F}_{el}^{ij}$, the elastic force between particles i and j, is the sum of the glassy bridge and rubbery matrix contributions:

$$\vec{F}_{el}^{ij} = -k_\infty(\vec{R}_{ij} - l_0\vec{u}) - k_0(\vec{R}_{ij} - \vec{R}_{ij}^{ref}) \tag{9.53}$$

$\vec{R}_{ij} = \vec{R}_i - \vec{R}_j$ is the vector between the centers of particles i and j ($\vec{R}_i$ is the position of the center of particle i), l_0 the equilibrium length of rubbery springs, and $\vec{u} = \vec{R}_{ij}/\|\vec{R}_{ij}\|$. The rubbery spring constant is set to $k_\infty = 1$, which maps to an elastic modulus of typically 0.5–1 MPa. The glassy modulus is typically 3 orders of magnitude larger than the rubbery modulus. The glassy spring constant is thus set to $k_0 \sim 10 - 100$ (k_0 is smaller than 10^3 because geometrical aspects regarding glassy bridges have to be taken into account, as schematized in Figure 9.35) [35]. All simulation parameters can be mapped to physical parameters of a real system [35].

9.4.5.3 SIMULATION RESULTS: REINFORCEMENT IN THE LINEAR AND NONLINEAR REGIMES, STRAIN SOFTENING

Figure 9.38 illustrates simulated reinforcement curves in the linear regime, as a function of the volume fraction or of the interaction parameter δ. Filler particles were distributed randomly in the simulation box. The

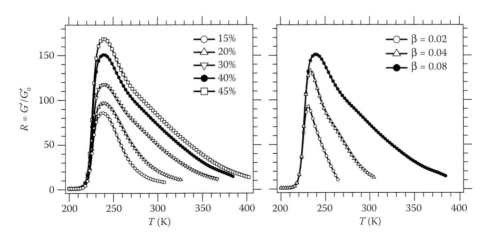

Figure 9.38 Reinforcement as a function of temperature. Left: For different filler volume fractions Φ as indicated, with interaction parameter $\beta = 0.08$. Right: For different values of the interaction parameter β as indicated, $\Phi = 0.4$. Compare with the curves in –ith. (Reprinted with permission from S. Merabia, P. Sotta, and D. R. Long, *Macromolecules*, 41, 2008, 8252. Copyright 2008, American Chemical Society.)

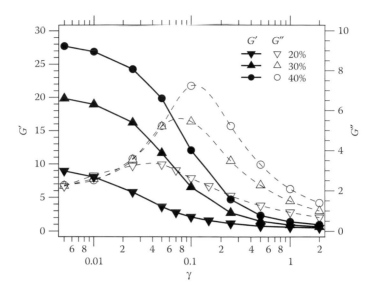

Figure 9.39 Simulated elastic moduli G' and G'' (in MPa) in oscillatory shear at a pulsation $\omega = 1$ rad/s, as a function of the shear amplitude γ, at $T = 283$ K for different volume fractions Φ as indicated. Parameter $\beta = 0.08$.

variation of reinforcement as a function of T is very similar to experimentally observed ones (see Figure 9.30). In particular, reinforcement extends over a very broad temperature range, over which glassy bridges are still active due to the presence of small distances between particles. The importance of the interaction parameter δ (which drives the thickness of glassy layers) is emphasized.

The simulated Payne effect in samples with various filler volume fractions is illustrated in Figure 9.39. The kinetics of yield, birth, and aging of glassy bridges is a key feature for explaining the dependence of $G'(\omega, \gamma_0)$ and $G''(\omega, \gamma_0)$ on the strain amplitude (Payne effect), and especially the presence of the peak of $G''(\omega, \gamma_0)$ at intermediate deformation amplitudes for strongly reinforced elastomers [35,126]. Here, the ratio $\tan \delta = G''/G'$ is overestimated as compared to real systems, which might be due to the simplified simulated geometry (spherical beads instead of fractal aggregates). With fractal aggregates, glassy bridges might break over a broader deformation range, resulting in a less abrupt G' decrease and a smaller G'' amplitude.

The G' value measured at an amplitude γ is directly related to the fraction of bridges with breaking times comparable to or larger than the experimental time scale (the period of the oscillatory strain). The strain-softening mechanisms consists of a lowering of the local T_g, or equivalently of a sharp decrease of the breaking times of glassy bridges as the local stress increases under strain, according to Equation 9.50. The integrated distributions of breaking times (i.e., the number $P(\tau)$ of glassy bridges with a breaking time equal to or larger than τ) in a system, which has been sheared at various amplitudes γ are plotted in Figure 9.40. Note that these distributions result from birth and death of glassy bridges measured in the steady state of oscillatory strain.

9.4.5.4 AGING OR REBIRTH OF GLASSY BRIDGES WITHIN A STRAIN CYCLE

The kinetics of birth and death of glassy bridges are also essential for understanding the plastic behavior of filled elastomers [36,142]. As we shall see, this kinetics are also key for understanding the higher harmonic response of filled elastomers.

In fact, the whole nonlinear response may be interpreted by the kinetics of birth and death of glassy bridges. As explained above, the decrease of the elastic modulus (Payne effect), that is, the decrease of the slope of $\sigma'(\gamma)$ at zero strain as the strain amplitude γ_0 increases, may be related to a decrease of the average number of glassy bridges as the average stress increases. On the other hand, the nonlinearity of the response (the presence of higher harmonics in the response), and more specifically, the strain hardening at larger strain, can be interpreted as the consequence of the slowing down of the strain rate at the extremities of the cycle, by considering the aging—or rebirth—of glassy bridges within a strain cycle.

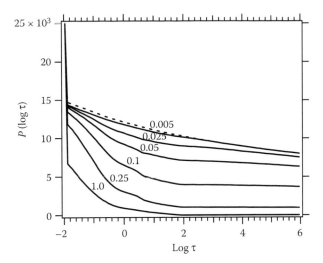

Figure 9.40 Distributions of relaxation times P(log τ) after oscillatory shear of various amplitudes γ as indicated, at pulsation ω = 1.0 rad/s, at T = 263 K in the system Φ = 0.40, β = 0.04, K = 0.3. Dashed curve: Distribution at equilibrium. The sharp drop at log τ ≈ −2 corresponds to the short-time cutoff in our simulations and is thus an artifact of the simulation. At very small γ, the distribution of breaking times is only slightly altered. (Reprinted with permission from S. Merabia and D. Long, *J. Chem. Phys.*, 125, 234901. Copyright 2006, American Institute of Physics.)

Results of simulations are shown in Figure 9.41, in comparison to experiments. Oscillatory cycles of three different strain amplitudes have been applied. The stress–strain curves are analyzed so as to separate the elastic and dissipative (viscous) contributions to the total stress, as described in Papon et al. [133].

When the strain rate slows down (at maximum strain), a fraction of glassy bridges may age and recover relatively long relaxation times, and thus contribute to the elastic response. Conversely, at high values of the strain rate (at zero strain), glassy bridges break more frequently. Therefore, the observed hardening would not correspond to strain hardening per se, but to what might be denoted as *slowing down hardening*, a property that should be observed in self-healing systems in general.

The proposed mechanism, related to the modulation of the number of effectively active glassy bridges within one strain cycle, is illustrated in Figure 9.42. Compared to the initial state of the system at rest (schematized in a), the average number of active glassy bridges (with long relaxation times) decreases under stress. Furthermore, this number is modulated within one cycle, as explained above. This provides an interpretation for the observed behavior, which is well reproduced by simulations.

9.4.5.5 SIMULATION RESULTS: RECOVERY BEHAVIOR (MULLINS EFFECT)

The same breaking–aging mechanism may explain the Mullins effect. After applying large amplitude oscillatory strain, breaking of a large fraction of glassy bridges results in a drastic drop of the elastic modulus, which recovers progressively as a result of aging, as illustrated in Figure 9.43. When a static strain is applied for a long time, glassy bridges recover long relaxation times in this new imposed deformation state, thus creating a new reference state corresponding to a plastic deformation with respect to the initial state at rest.

9.4.5.6 ARE PHYSICAL INTERACTIONS SUFFICIENT FOR OBTAINING REINFORCEMENT, OR IS GRAFTING NECESSARY?

It results from the proposed interpretation of reinforcement properties that grafting is not a necessary condition for obtaining reinforcement. Physical interactions are sufficient to induce a T_g gradient in the vicinity of filler particles and thus to induce the presence of glassy bridges, which is the main reinforcement mechanism. A further consequence is that the distribution of distances between filler particles or aggregates is a key

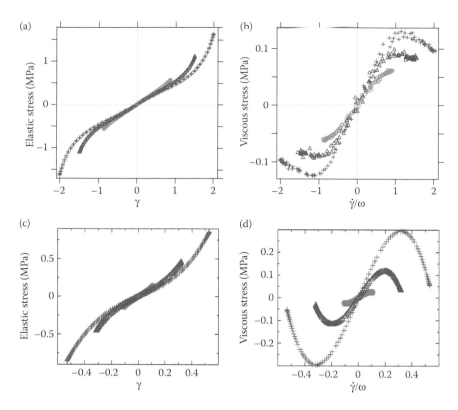

Figure 9.41 Stress versus strain or strain rate in cycles of three different amplitudes in a PEA sample filled with silica: (a) Elastic stress; (b) viscous stress. The signal analysis is described in Papon et al. [133]. (c and d) Corresponding simulated stress versus strain or strain rate curves in cycles of three different amplitudes. Values of simulation parameters are given in Papon et al. [133]. The general nonlinear behavior is qualitatively reproduced. (Reprinted with permission from A. Papon et al., *Macromolecules*, 45, 2891. Copyright 2012 by the American Physical Society.)

feature of a given system, because it determines the number of glassy bridges that are effectively active, at a given volume fraction, temperature, and strength of the surface interactions.

9.4.6 Conclusion

In conclusion, we have shown how the properties of elastomers filled with nanometric particles or aggregates can be rationalized in a unified way, based on the assumption of T_g gradients in the polymer matrix close to filler interfaces, which lead to the presence of glassy bridges between filler particles or aggregates. The increase of T_g is the part of the matrix that is confined in between filler particles, as well as the broadening of the glass transition, as it was described in Section 9.3, is the essential physical mechanism that may explain the very high values of the modulus characterizing these materials. Mesoscale modeling at the scale of filler particles and aggregates, that is, typically 100 nm or more, has been developed based on this idea. It was shown that the generic behavior of reinforced elastomers can be modeled. Adjusting simulation results quantitatively to data obtained in real systems, however, would involve further work. Specifically, the number of effective glassy bridges in real systems depends in a crucial way on the distribution of distances between nearest-neighbor particles, that is, in turn, to the dispersion state. The dispersion state of the fillers depends in a complex way on both the processing conditions and on filler–elastomer interactions. Modeling complex dispersion states is a further development of the model. Note further that systems relevant for applications are often reinforced by aggregates that themselves have complex, fractal morphologies and sometimes rough or porous surfaces.

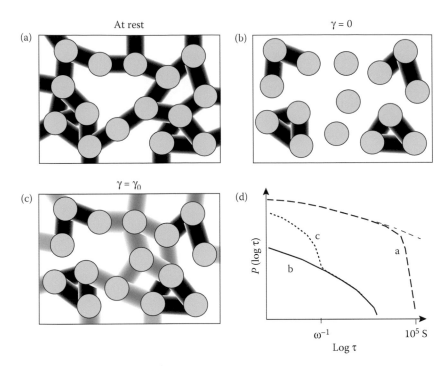

Figure 9.42 Schematic snapshots of the glassy bridge network in a filled elastomer. Glassy bridges with relaxation times much longer than (respectively comparable to) the loading period ω^{-1} are shown in black (respectively in gray). White background: Rubbery matrix or glassy bridges with relaxation times much shorter than ω^{-1}: (a) in a sample at rest after an aging time of 10^5 s; (b) the same sample under large amplitude oscillatory shear, taken at maximum shear rate, that is, at zero instantaneous strain; (c) taken at maximum instantaneous strain. Since the strain rate is zero at this point, glassy bridges with longer relaxation times have time to form as compared to (b). (d) integrated distributions of relaxation times corresponding to cases (a) (dashed curve), (b) (full curve), (c) (dotted curve), and at equilibrium (thin dashed curve). (Reprinted with permission from A. Papon et al., *Macromolecules*, 45, 2891. Copyright 2012 by the American Physical Society.)

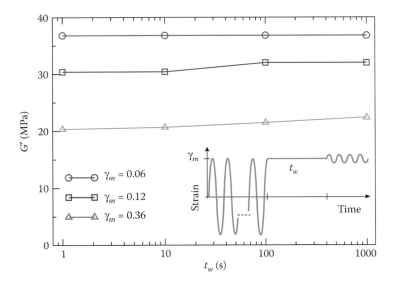

Figure 9.43 Small amplitude storage modulus G' after oscillatory shearing at large amplitude γ_m and aging in a deformed state, for different values of γ_m, as a function of the aging time t_w. The whole simulated experiment is schematized in the inset (pulsation $\omega = 1$ s^{-1}). (Reprinted with permission from S. Merabia and D. Long, *J. Chem. Phys.*, 125, 234901. Copyright 2006, American Institute of Physics.)

REFERENCES

1. M. D. Ediger, C. A. Angell, and S. R. Nagel, *J. Phys. Chem.*, 100, 13200, 1996.
2. J. D. Ferry, *Viscoelastic Properties of Polymers* (John Wiley & Sons Inc., New York), 1980.
3. M. D. Ediger, *Annu. Rev. Chem.*, 51, 99, 2000.
4. R. Richert, *J. Phys. Condens. Matter*, 14, R703, 2002.
5. K. Schmidt-Rohr and H. W. Spiess, *Phys. Rev. Lett.*, 66, 3020, 1991.
6. U. Tracht, M. Wilhelm, A. Heuer, H. Feng, K. Schmidt-Rohr, and H. W. Spiess, *Phys. Rev. Lett.*, 81, 2727, 1998.
7. S. A. Reinsberg, X. H. Qiu, M. Wilhelm, H. W. Spiess, and M. D. Ediger, *J. Chem. Phys.*, 114, 7299, 2001.
8. M. T. Cicerone, F. R. Blackburn, and M. D. Ediger, *Macromolecules*, 28, 8224, 1995.
9. C.-Y. Wang and M. D. Ediger, *Macromolecules*, 30, 4770, 1997.
10. M. T. Cicerone, P. A. Wagner, and M. D. Ediger, *J. Phys. Chem. B*, 101, 8727, 1997.
11. F. Fujara, G. Geil, H. Sillescu, and G. Fleischer, *Z. Phys. B.*, 88, 195, 1992.
12. Y. Hwang, T. Inoue, P. A. Wagner, and M. D. Ediger, *J. Polym. Sci., Part B: Polym. Phys.*, 38, 68, 2000.
13. B. Schiener, R. Bohmer, A. Loidl, and R. V. Chamberlin, *Science*, 274, 752, 1996.
14. R. Richert, *J. Chem. Phys.*, 113, 8404, 2000.
15. G. Adam and J. H. Gibbs, *J. Chem. Phys.*, 43, 139, 1965.
16. J. L. Keddie, R. A. L. Jones, and R. A. Cory, *Europhys. Lett.*, 27, 59, 1994.
17. D. B. Hall, A. Dhinojwala, and J. M. Torkelson, *Phys. Rev. Lett.*, 79, 103, 1997.
18. J. Mattsson, J. A. Forrest, and J. Borjesson, *Phys. Rev. E*, 62, 5187, 2000.
19. S. Kawana and R. A. L. Jones, *Phys. Rev. E*, 63, 021501, 2001,
20. J. Q. Pham and P. F. Green, *J. Chem. Phys.*, 116, 5801, 2002.
21. L. Hartmann, W. Gorbatschow, J. Hauwede, and F. Kremer, *Eur. Phys. J. E*, 8, 145, 2002.
22. C. J. Ellison and J. M. Torkelson, *J. Polym. Sci., Part B: Polym. Phys.*, 40, 2745, 2002.
23. C. J. Ellison and J. M. Torkelso, *Nat. Mater.*, 2, 695, 2003.
24. K. Fukao and Y. Miyamoto, *Phys. Rev. E*, 61, 1743, 2000.
25. J. Q. Pham and P. F. Green, *Macromolecules*, 36, 1665, 2003.
26. W. E. Wallace, J. H. van Zanten, and W. L. Wu, *Phys. Rev. E*, 52, R3329. 1995.
27. J. H. van Zanten, W. E. Wallace, and W. L. Wu, *Phys. Rev. E*, 53, R2053, 1996.
28. Y. Grohens, M. Brogly, C. Labbe, M.-O. David, and J. Schultz, *Langmuir*, 14, 2929, 1998.
29. Carriere P., Grohens Y., Spevacek J., and Schultz J., *Langmuir*, 16, 5051, 2000.
30. Y. Grohens, L. Hamon, G. Reiter, A. Soldera, and Y. Holl, *EPJ E*, 8, 217, 2002.
31. W. E. Wallace, N. C. Beck Tan, W.-L. Wu, and S. Satija, *J. Chem. Phys.*, 108, 3798, 1998.
32. J. Berriot, F. Lequeux, H. Montes, L. Monnerie, D. Long, and P. Sotta, *J. Non-Cryst. Solids*, 307, 719, 2002.
33. J. Berriot, H. Montes, F. Lequeux, D. Long, and P. Sotta, *Macromolecules*, 35, 9756, 2002.
34. J. Berriot, H. Montes, F. Lequeux, D. Long, and P. Sotta, *Europhys. Lett.*, 64, 50, 2003.
35. S. Merabia, P. Sotta, and D. R. Long, *Macromolecules*, 41, 8252, 2008.
36. S. Merabia, P. Sotta, and D. R. Long, *J. Polym. Sci., Part B: Polym. Phys.*, 48, 1495, 2010.
37. B. Metin and F. D. Blum, *Langmuir*, 26, 5226, 2010.
38. P. Rittigstein and J. M. Torkelson, *J. Polym. Sci., Part B: Polym. Phys.*, 44, 2935. 2006.
39. D. Ciprari, K. Jacob, and R. Tannenbaum, *Macromolecules*, 39, 6565, 2006.
40. K. Putz, R. Krishnamoorti, and P. F. Green, *Polymer*, 48, 3540, 2007.
41. J. M. Kropka, K. W. Putz, V. Pryamitsyn, V. Ganesan, and P. F. Green, *Macromolecules*, 40, 5424, 2007.
42. Y. -Q. Rao and J. M. Pochan, *Macromolecules*, 40, 290, 2007.
43. P. Rittigstein, R. D. Priestley, L. J. Broadbelt, and J. M. Torkelson, *Nat. Mater.*, 6, 278, 2007.
44. L. E. Nielsen and R. F. Landel, *Mechanical Properties of Polymers and Composites* (Marcel Dekker, New York), 1994.
45. H. Eyring, *J. Chem. Phys.*, 3, 107, 1935.

46. H. Eyring, *J. Chem. Phys.*, 4, 283, 1936.
47. K. Chen and K. S. Schweizer, *Europhys. Lett.*, 79, 26006, 2007.
48. K. Chen and K. S. Schweizer, *Macromolecules*, 41, 5908, 2008.
49. L. S. Loo, R. E. Cohen, and K. K. Gleason, *Science*, 288, 116119, 2000.
50. J. Kalfus, A. Detwiler, A. J. Lesser, *Macromolecules*, 45, 4839, 2012.
51. H.-N. Lee, R. A. Riggleman, J. J. de Pablo, and M. D. Ediger, *Macromolecules*, 42, 4328, 2009.
52. B. Bending, K. Christison, J. Ricci, and M. D. Ediger, *Macromolecules*, 47, 800, 2014.
53. H. -N. Lee, K. Paeng, S. F. Swallen, and M. D. Ediger, *Science*, 323, 231, 2009.
54. R. A. Riggleman, H.-N. Lee, M. D. Ediger, and J. J. de Pablo, *Soft Matter*, 6, 287, 2010.
55. K. Yoshimoto, T. S. Jain, K. van Workum, P. F. Nealey, and J. J. de Pablo, *Phys. Rev. Lett.*, 93, 175501, 2004.
56. F. Leonforte, R. Boissiere, A. Tanguy, J. P. Wittmer, and J.-L. Barrat, *Phys. Rev. B*, 72, 224206, 2005.
57. R. A. Riggleman, H.-N. Lee, M. D. Ediger, and J. J. de Pablo, *Phys. Rev. Lett.*, 99, 215501, 2007.
58. M. Tsamados, A. Tanguy, C. Goldenberg, and J.-L. Barrat, *Phys. Rev. E*, 80, 026112, 2009.
59. G. J. Papakonstantopoulos, R. A. Riggleman, J. L. Barrat, and J. J. de Pablo, *Phys. Rev. E*, 77, 041502, 2008.
60. R. A. Riggleman, K. S. Schweizer, and J. J. de Pablo, *Macromolecules*, 41, 4969, 2008.
61. A. Dequidt, D. R. Long, P. Sotta, and O. Sanseau, *Eur. Phys. J. E*, 35, 61, 2012.
62. K. S. Schweizer and E. J. Saltzman, *J. Chem. Phys.*, 119, 1181, 2003.
63. K. S. Schweizer and E. J. Saltzman, *J. Chem. Phys.*, 55, 241, 2004.
64. K. Chen and K. S. Schweitzer, *Phys. Rev. Lett.*, 102, 038301, 2009.
65. D. Long and F. Lequeux, *Eur. Phys. J. E*, 4, 371, 2001.
66. S. Merabia and D. Long, *Eur. Phys. J. E*, 9, 195, 2002.
67. S. Merabia, P. Sotta, and D. Long, *Eur. Phys. J. E*, 15, 189, 2004.
68. P. Sotta and D. Long, *Eur. Phys. J. E*, 11, 375, 2003.
69. S. Merabia and D. Long, *J. Chem. Phys.*, 125, 234901, 2006.
70. K. Chen, E. J. Saltzman, and K. S. Schweizer, *J. Phys. Condens. Matter*, 21, 503101, 2009.
71. J. C. Dyre, *Rev. Mod. Phys.*, 78, 953, 2006.
72. M. L. Ferrer, C. Lawrence, B. G. Demirjian, D. Kivelson, C. Alba-Simionesco, and G. Tarjus, *J. Chem. Phys.*, 109, 8010, 1998.
73. C. Alba-Simionesco, D. Kivelson, and G. Tarjus, *J. Chem. Phys.*, 116, 5033, 2002.
74. S. Cheng, S. Mirigian, J.-M. Y. Carrillo, V. Bocharova, B. G. Sumpter, K. S. Schweizer, and A. P. Sokolov, *J. Chem. Phys.*, 143, 194704, 2015.
75. S. Merabia and D. Long, *Macromolecules*, 41, 3284, 2008.
76. E. M. Masnada, G. Julien, and D. R. Long, *J. Polym. Sci., Part B: Polym. Phys. Ed.*, 52, 419, 2014.
77. I. Prigogine, *The Molecular Theory of Solutions* (North-Holland Publishing Co, Amsterdam), 1957.
78. A. Papon, S. Merabia, L. Guy, H. Montes, F. Lequeux, P. Sotta, and D. R. Long, *Macromolecules*, 45, 2891, 2012.
79. Dequidt et al., submitted to *Macromolecules*, 2016.
80. J. A. Hammerschmidt, W. L. Gladfelter, and G. Haugstad, *Macromolecules*, 32, 3360, 1999.
81. I. Prigogine, N. Trappeniers, and V. Mathot, *Discuss. Faraday Soc.*, 15, 93, 1953.
82. I. Prigogine, N. Trappeniers, and V. Mathot, *J. Chem. Phys.*, 21, 559, 1953.
83. I. Prigogine, V. Mathot, and N. Trappeniers, *J. Chem. Phys.*, 21, 560, 1953.
84. J. N. Israelachvili, *Intermolecular and Surface Forces* (Academic Press, London), 1992.
85. J. E. Mark, *Physical Properties of Polymers Handbook* (American Institute of Physics, Woodbury, New York), 1996.
86. R. N. Haward, *The Physics of Glassy Polymers* (Applied Science Publishers, London), 1973.
87. C. A. Angell, *Science*, 267, 1924–1935, 1995; F. H. Stillinger, *Science*, 267, 1935–1939, 1995; D. Richter, *Science*, 267, 1939–1945, 1995.
88. G. W. Scherer, *Relaxation in Glass and Composites* (John Wiley & Sons, New York),1986.

89. R. E. Robertson, *J. Polym. Sci. Polym. Symp.*, 63, 173, 1978.

90. Y.-J. Jung, J. P. Garrahan, and D. Chandler, *Phys. Rev. E*, 69, 061205, 2004.

91. J. P. Garrahan and D. Chander, *Phys. Rev. Lett.*, 89, 035704, 2002.

92. N. B. Tito, J. E. G Lipson, and S. T. Milner, *Soft Matter*, 9, 3173, 2013.

93. S. A. Reinsberg, A. Heuer, B. Doliwa, H. Zimmermann, and H. W. Spiess, *J. Non-Cryst. Solids*, 307, 208, 2002.

94. D. Stauffer and A. Aharony, *Introduction to Percolation Theory* (Taylor & Francis, London), 1994.

95. I. S. Gradshteyn and I. M. Ryzhik, *Table of Integrals, Series, and Products* (Academic Press, New York), 1980.

96. J. Frohlich, W. Niedermeier, and H.-D. Luginsland, *Composites A*, 36, 449, 2005.

97. G. Heinrich and M. Kluppel, *Adv. Polym. Sci.*, 160, 1, 2002.

98. H. Montes, T. Chaussée, A. Papon, F. Lequeux, and L. Guy, *Eur. Phys. J. E*, 31, 263, 2010.

99. M.-J. Wang, *Rubber Chem. Technol.*, 71, 520, 1998.

100. P.-G. De Gennes, *Eur. Phys. J. E*, 2, 367, 2000.

101. S. A Reinsberg, X. H. Qiu, M. Wilhelm, H. W. Spiess, and M. D. Ediger, *J. Chem. Phys.*, 114, 7299, 2001.

102. N. G. McCrum, B. E. Read, and G. Williams, *Anelastic and Dielectric Effects in Polymeric Solids* (Dover Publications, Inc., New York), 1967.

103. G. Diezemann, *J. Chem. Phys.*, 123, 204510, 2005.

104. F. Fujara, B. Geil, H. Sillescu, and G. Fleischer, *Z. Phys. B.*, 88, 195, 1992.

105. M. T. Cicerone, F. R. Blackburn, and M. D. Ediger, *Macromolecules*, 28, 8224, 1995.

106. C.-Y. Wang and M. D. Ediger, *Macromolecules*, 30, 4770, 1997.

107. C.-Y. Wang and M. D. Ediger, *J. Chem. Phys.*, 112, 6933, 2000.

108. A. J. Kovacs, *J. Polym. Sci*, 30, 131, 1958.

109. G. Braun and A. J. Kovacs, *Phys. Chem. Glasses*, 4, 152, 1963.

110. A. J. Kovacs, *Fortschr. Hochpolym. Forsch.*, 3, 394, 1963.

111. A. J. Kovacs, J. J. Aklonis, J. M. Hutchinson, and A. R. Ramos, *J. Polym. Sci. Polym. Phys. Ed.*, 17, 1097, 1979.

112. L. C. E. Struik, *Physical Aging in Amorphous Polymers and Other Materials* (Elsevier, Amsterdam), 1978.

113. S. L. Simon, J. W. Sobieski, and D. J. Plazek, *Polymer*, 42, 2555, 2001.

114. P. Bernazzani and S. L. Simon, *J. Non-Cryst. Solids*, 307–310, 470, 2002.

115. K. Adachi and T. Kotaka, *Polym. J.*, 14, 959, 1982.

116. S. Hozumi, T. Wakabayashi, and K. Sugihara, *Polym. J.*, 1, 632, 1970.

117. H. H.-D. Lee and F. J. McGarry, *Polymer*, 34, 4267, 1993.

118. R. Greiner and F. R. Schwartl, *Rheol. Acta*, 23, 378, 1984.

119. M. Doi and S. F. Edwards, *The Theory of Polymer Dynamics* (Oxford Science Publications, Oxford), 1986.

120. S. R. de Groot and P. Mazur, *Non-Equilibrium Thermodynamics* (Dover Publications, New York), 1984.

121. G. Medvedev and J. Caruters, *Macromolecules*, 48, 788, 2015.

122. C. M. Roland, M. Paluch, T. Pakula, and R. Casalini, *Philos. Mag.*, 84, 1573, 2004.

123. C. M. Roland, S. Bair, and R. Casalini, *J. Chem. Phys.*, 125, 124508, 2006.

124. M. Paluch, R. Casalini, A. Patkowski, T. Pakula, and C. M. Roland, *Phys. Rev. E*, 68, 031802, 2003.

125. M. Paluch, C. M. Roland, R. Casalini, G. Meier, and A. Patkowski, *J. Chem. Phys.*, 118, 4578, 2003.

126. A. R. Payne, *J. Appl. Polym. Sci.*, 7, 873, 1963.

127. A. Mujtaba, M. Keller, S. Ilisch, H.-J. Radusch, M. Beiner, T. Thurn-Albrecht, and K. Saalwächter, *ACS Macro Lett.*, 3, 481, 2014.

128. F. Bueche, *J. Appl. Polym. Sci.*, 15, 271, 1961.

129. A. R. Payne, *J. Appl. Polym. Sci.*, 21, 368, 1962.

130. A. R. Payne, *J. Appl. Polym. Sci.*, 9, 1073, 1965.

131. G. Kraus, *J. Appl. Polym. Sci. Appl. Polym. Symp.*, 39, 75, 1984.

132. A. Papon, H. Montes, F. Lequeux, and L. Guy, *J. Polym. Sci., Part B: Polym. Phys.*, 48, 2490, 2010.

133. A. Papon, S. Merabia, L. Guy, F. Lequeux, H. Montes, P. Sotta, and DR. R. Long, *Macromolecules*, 45, 2891, 2012.

134. G. Kraus, *Rubber Chem. Tech.*, 51, 297, 1978.

135. J. Diani, B. Fayolle, and P. Gilormini, *Eur. Polym. J.*, 45, 601, 2009.

136. J. A. C. Harwood, L. Mullins, and A. R. Payne, *J. Appl. Polym. Sci.*, 9, 3011, 1965.

137. A. I. Medalia, *Rubber Chem. Technol.*, 60, 45, 1986.

138. D. C. Edwards, *J. Mater. Sci.*, 25, 4175, 1990.

139. A. I. Medalia, *Rubber Chem. Tech*, 51, 437, 1978.

140. E. Guth and O. Gold, *Phys. Rev.*, 53, 322, 1938.

141. J. Domurath, M. Saphiannikova, G. Ausias, and G. Heinrich, *J. Non-Newtonian Fluid Mech.*, 171–172, 8, 2012.

142. S. S. Sternstein and A.-J. Zhu, *Macromolecules*, 35, 7262, 2002.

143. R. B. Bogoslovov, C. M. Roland, A. R. Ellis, A. M. Randall, and C. G. Robertson, *Macromolecules*, 41, 1289, 2008.

144. W. F. Reichert, D. Göritz, and E. J. Duschl, *Polymer*, 34, 1216, 1993.

145. V. M. Litvinov, R. A. Orza, M. Klüppel, M. van Duin, and P. Magusin, *Macromolecules*, 44, 4887, 2011.

146. P. G. Maier and D. Goeritz, *Kautsch. Gummi Kunstst.*, 49, 18, 1996.

147. M. Vladkov and J. L. Barrat, *Macromolecules*, 40, 3797, 2007.

148. A. Mujtaba, M. Keller, S. Ilisch, H.-J. Radusch, T. Thurn-Albrecht, K. Saalwächter, and M. Beiner, *Macromolecules*, 45, 6504, 2012.

149. K. Akutagawa, K. Yamaguchi, A. Yamamoto, H. Heguri, H. Jinnai, and Y. Shinbori, *Rubber Chem. Technol.*, 81, 182, 2008.

150. M. Klüppel, R. H. Schuster, and G. Heinrich, *Rubber Chem. Technol.*, 70, 243, 1997.

151. G. Heinrich, M. Klüppel, and T. A. Vilgis, *Reinf. Elastomers*, 6, 195, 2002.

152. D. S. Fryer, P. F. Nealey, and J. J. de Pablo, *Macromolecules*, 33, 6439, 2000.

153. C. G. Robertson, C. J. Lin, M. Rackaitis, and C. M. Roland, *Macromolecules*, 41, 2727, 2008.

154. C. G. Robertson and M. Rackaitis, *Macromolecules*, 44, 1177, 2011.

155. L. C. E. Struik, *Polymer*, 28, 1521, 1987.

156. S. Kaufman, W. Slichter, and D. Davis, *J. Polym. Sci. A-2 Polym. Phys.*, 9, 829, 1971.

157. V. M. Litvinov and P. Steeman, *Macromolecules*, 32, 8476, 1999.

158. A. Papon, K. Saalwächter, K. Schäler, L. Guy, F. Lequeux, and H. Montes, *Macromolecules*, 44, 913, 2011.

159. A. Papon, H. Montes, M. Hanafi, F. Lequeux, L. Guy, and K. Saalwächter, *Phys. Rev. Lett.*, 108, 065702, 2012.

160. W. Zheng and S. L. Simon, *J. Chem. Phys.*, 127, 194501, 2007.

161. S. Napolitano, V. Lupascu, and M. Wübbenhorst, *Macromolecules*, 41, 1061, 2008.

162. R. Pérez Aparicio, A. Vieyres, P.-A. Albouy, O. Sanséau, L. Vanel, D. R. Long, and P. Sotta, *Macromolecules*, 46, 8964, 2013.

163. A. Papon, H. Montes, F. Lequeux, J. Oberdisse, K. Saalwächter, and L. Guy, *Soft Matter*, 8, 4090, 2012.

164. J. Bastide and L. Leibler, *Macromolecules*, 21, 2647, 1988.

165. M. Rubinstein and S. Panyukov, *Macromolecules*, 35, 6670, 2002.

166. H. Montès, F. Lequeux, and J. Berriot, *Macromolecules*, 36, 8107, 2003.

167. S. Kaufmann, W. P. Slichter, and D. D. Davis, *J. Polym. Sci. A2*, 9, 829, 1971.

168. B. Haidar, H. Salah Deradji, A. Vidal, and E. Papirer, *Macromol. Symp.*, 108, 147, 1996.

169. G. Tsagaropoulos and A. Eisenberg, *Macromolecules*, 28, 6067, 1995.

170. G. Tsagaropoulos and A. Eisenberg, *Macromolecules*, 28, 396, 1995.

171. G. Tsagaropoulos, J.-S. Kim, and A. Eisenberg, *Macromolecules*, 29, 2222, 1996.

172. J.-Y. Lee, K. E. Su, E. P. Chan, Q. Zhang, T. Emrick, and A. J. Crosby, *Macromolecules*, 40, 7755, 2007.

173. A. Bansal, H. Yang, C. Li, K. Cho, B. C. Benicewicz, S. K. Kumar, and L. S. Schadler, *Nat. Mater.*, 4, 693, 2005.

174. L. S. Schadler, *Nat. Mater.* 6, 257, 2007.

175. A. A. Gusev, *Macromolecules*, 39, 5960, 2006.

176. J. Kalfus and J. Jancar, *Polymer*, 48, 3935, 2007.

177. D. Long, D. and P. Sotta, IMA Volume in mathematics and its applications, in *Modeling of Soft Matter*, edited by M.-C. T. Calderer and E. M. Terentjev (Springer, New York), 141, 205, 2005.

178. D. Long and P. Sotta, *Macromolecules*, 39, 6282, 2006.

179. D. Long and P. Sotta, *Rheol. Acta*, 44, 1029, 2007.

180. F. W. Starr, B. T. Schroder, and S. C. Glotzer, *Macromolecules*, 35, 4481, 2002.

181. A. A. Gusev, *Phys. Rev. Lett.*, 93, 0304302, 2004.

182. S. A. Baeurle, T. Usami, and A. A. Gusev, *Polymer*, 47, 8604, 2006.

183. S. A. Baeurle, A. Hotta, and A. A. Gusev, *Polymer*, 46, 4344, 2005.

184. S. Wu, *Polym. Int.*, 29, 229, 1992.

POLYMER GLASSES UNDER DEFORMATION

A molecular perspective on the yield and flow of polymer glasses
The role of enhanced segmental dynamics during active deformation

MARK D. EDIGER AND KELLY HEBERT

The mechanical properties of polymer glasses are often critical in determining the best material for a particular application. Extremely stiff materials (high modulus) may be important for some applications, while avoiding catastrophic failure due to fracture (high toughness) may be more important for others. The mechanical properties of a polymer glass will depend upon both molecular structure and many experimental variables, including temperature and the mode of deformation (tension, compression, or shear).

In this chapter, we discuss the mechanical response of polymer glasses from a molecular perspective. In particular, we consider how deformation changes the rate at which polymer segments rearrange and how this in turn influences the mechanical response of the material. It will be shown that this focus on the changes in *dynamics* provides an understanding of many important features of polymer glass deformation. Of course, it is also true that the *structure* of a polymer glass must be altered by nonlinear deformation. Although not a major focus, we will make some comments about these structural changes at the end of this chapter (Section 10.8).

10.1 INTRODUCTION TO POLYMER GLASS DEFORMATION

In order to start our discussion, let's consider one type of deformation experiment used to characterize polymer glasses. Figure 10.1 shows "stress–strain" curves that describe the response of a polymer glass to

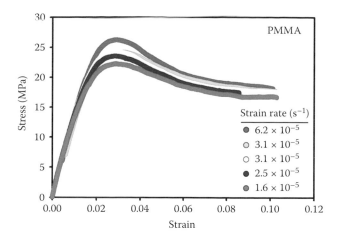

Figure 10.1 Mechanical response of PMMA deformed at constant strain rates as listed in the legend. Engineering stress (extension force divided by cross-sectional area) is plotted against global strain (fractional increase in sample length). These tests were performed at $T_g - 19$ K ($T_g = 392$ K). (Data for the four highest strain rates from Bending, B. et al., *Macromolecules*, 47, 800, 2014.)

a constant strain rate deformation. For these data, the sample was deformed uniaxially in tension although qualitatively similar data are also obtained in compression. These particular experiments were performed at $T_g - 19$ K on a lightly cross-linked poly(methyl methacrylate) (PMMA) glass after annealing the sample at the testing temperature for about 30 min; here, T_g denotes the glass transition temperature obtained with differential scanning calorimetry at heating rate of 10 K/min. These details are specified for completeness; qualitatively similar data could be obtained for almost any polymer glass (including uncross-linked samples) at a similar temperature below T_g. The different curves shown in Figure 10.1 come from experiments performed at different strain rates as indicated.

Figure 10.1 is useful for identifying several characteristic features of polymer glass deformation. At extremely small strains (perhaps up to 0.002), the stress increases linearly with strain. If the stress is released after such a deformation, the sample returns very nearly to its original length; these small deformations are essentially reversible. The slope of the stress–strain curve at such small deformations defines the Young's modulus E. The modulus is essentially independent of strain for such small strains and this mechanical experiment can be considered to be in the linear response regime. Larger deformations are nonreversible and are not in the linear response regime. As shown in Figure 10.1, a maximum in the stress is reached (the yield stress) at a strain of about 0.03. Larger strains result in a decrease of the stress (strain softening). At even larger strains than those shown in the figure, the stress increases again (strain hardening). The area underneath the stress–strain curve up to the point where the sample breaks is a measure of toughness. As discussed in Section 10.8, strain hardening plays a critical role in delocalizing the deformation response of a polymer glass; such delocalization is necessary for a tough material (see Chapter 13 for a more detailed discussion of strain hardening).

Figure 10.1 also shows the influence of strain rate on polymer glass deformation and this provides an interesting method to quantify the nonlinearity of these experiments. A useful comparison can be made with the predictions of the Maxwell model. The Maxwell model is the simplest model for the mechanical response of a viscoelastic liquid; it can be represented by a purely elastic spring (with modulus E) in series with a purely viscous dashpot (with viscosity η). Even though the Maxwell model has only a single relaxation time ($\tau = \eta/E$), it is a useful starting point for understanding linear viscoelastic behavior. Using this model, calculations for three constant strain rate experiments are displayed in Figure 10.2, with the range of strain rates matching the range in Figure 10.1. The Maxwell model predicts that the stress will initially rise linearly with strain and then reach a steady-state value (the flow stress σ_{flow}), with σ_{flow} depending linearly upon the strain rate (a factor of 6 for the strain rate range displayed). In contrast, note that the stresses at a strain of 0.10 in Figure 10.1 differ by only 10%. The near-independence of the post-yield stress data with respect to strain

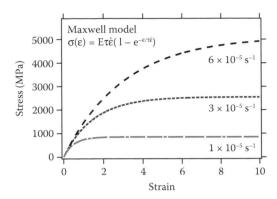

Figure 10.2 Maxwell model calculations for three constant strain rate deformations using parameters for the PMMA glass of Figure 10.1. After an initial linear regime at small strains, individual curves reach a steady-state flow stress, σ_{flow}, at higher strains. For the Maxwell model, the flow stress depends linearly on the strain rate in contrast to the data shown in Figure 10.1.

rate, in contrast to the prediction of linear viscoelasticity, indicates that these deformations are deeply in a nonlinear regime.

10.2 THE ROLE OF ENHANCED SEGMENTAL DYNAMICS DURING DEFORMATION

We can use the Maxwell model to provide further insight into why the response of a polymer glass to deformation is so highly nonlinear. The Maxwell model predicts the following value for the flow stress: $\sigma_{flow} = E\tau\dot{\varepsilon}$. Here, τ is the relaxation time, which we identify as the time for segmental rearrangements of the polymer, and $\dot{\varepsilon}$ is the strain rate. For PMMA at $T_g - 19$ K, we can estimate that $\tau = 50,000$ s in the absence of deformation [1]. Using a strain rate of $\dot{\varepsilon} = 3.1 \times 10^{-5}$ s^{-1} and the measured value of E, we can calculate from the Maxwell model that σ_{flow} should be about 2500 MPa, a value about 150 times larger than the flow stress shown in Figure 10.1. If we turn this calculation around and use the observed flow stress during deformation as an input, we can calculate that the *effective* segmental relaxation time during deformation is ~350 s, or about 150 times shorter than the segmental relaxation time in the absence of deformation. As we will show below, this estimate is in semiquantitative agreement with experiments that directly measure segmental dynamics during deformation.

The calculation in the previous paragraph gives us a reasonable qualitative picture of the deformation of polymer glasses: something about the deformation causes the segmental dynamics to speed up enormously and this allows the polymer glass to flow at much lower stresses than otherwise would have been possible. This is actually an old idea that goes back at least as far as Eyring. Eyring [2] predicted that stress accelerates the rate at which barriers will be overcome in a solid. This can be interpreted as "landscape tilting." The multidimensional potential energy landscape (PEL) governs the thermodynamics and dynamics of a system at constant volume. At any given instant, the configuration of a system is specified by its position on the PEL. Eyring calculated that stress decreases the heights of some barriers on the PEL and increases the heights of others. The system evolves in the direction of the decreased barriers and thus stress accelerates dynamics.

The idea that deformation enhances the rate of segmental dynamics has been included in models of polymer glass deformation for the last 50 years (see Chapter 14). Some models follow Eyring [2] in using stress [3–10] or strain rate [11,12] as the control variable. Others use strain [9], configurational entropy [13,14], configurational internal energy [15,16], the amplitude of density fluctuations [17–19], or free volume [20,21]. Many of these models make use of a *material* or *effective time* formalism [8,22,23]. In this approach, a linear mechanical model is used to predict the nonlinear deformation of a polymer glass with one modification: the relaxation time specifying the polymer segmental dynamics is allowed to change during the deformation in response to stress or some other control variable. As a preview, we warn the reader that experiments show that no simple mechanical variable (such as stress or strain rate) can predict how segmental relaxation times evolve during all phases of deformation.

While the above paragraphs have emphasized the nonlinearity of the mechanical response as a key feature of polymer glass deformation, we wish to briefly mention other aspects of this problem that are particularly challenging. *Polymer glasses are out of equilibrium* from a thermodynamic perspective and thus these systems are constantly evolving due to structural relaxation (physical aging) [24–26] (see Chapter 2). Structural relaxation certainly influences the material response during and immediately following deformation as well. In addition, it is known that *dynamics in polymer glasses are spatially heterogeneous*, that is, segments in one region of the sample can relax on time scales orders of magnitude faster than segments in another region only a few nanometers away [27–29]. There is no reason that these regions of differing dynamics should respond uniformly to deformation. The experiments shown below indicate that there are substantial changes in the heterogeneous dynamics during deformation; slower regions experience dynamics that are enhanced by a greater extent than are the faster regions.

Several of the ideas presented above in Sections 10.1 and 10.2 will be revisited in the remainder of this chapter. Section 10.3 will discuss how segmental dynamics are monitored during deformation, and will also introduce the optical probe technique featured in this chapter. The behavior of segmental mobility during constant strain rate and constant stress mechanical protocols, as measured using the optical probe technique, will be illustrated in Sections 10.4 and 10.7, respectively. Section 10.5 includes a discussion of experiments that compare mechanical and optical measures of segmental dynamics. The effect of temperature on segmental dynamics during deformation is discussed in Section 10.6. Comments about changes in structure during deformation will be made in Section 10.8.

10.3 EXPERIMENTAL METHODS FOR MEASURING SEGMENTAL DYNAMICS DURING DEFORMATION

Prior to work in our lab, which began about 7 years ago, there were experimental results showing indications that segmental dynamics were enhanced during the deformation of a polymer glass. Solid-state nuclear magnetic resonance experiments were used to study deformation induced segmental mobility in the amorphous regions of a semicrystalline nylon 6 sample [30]. Other experiments demonstrated that the diffusion of a plasticizer into poly(ether imide) glass under compression far below T_g was similar to diffusion at T_g in the absence of deformation [31]. While both of these approaches showed enhanced dynamics as a result of deformation, neither quantified the change in the average segmental relaxation time during active deformation.

The optical technique [1,32–39] developed in our lab to monitor segmental dynamics during deformation relies on the observation that the reorientation of a molecular probe can be a good reporter of segmental dynamics. We typically utilize the fluorescent probe DPPC (*N,N'*-dipentyl-3,4,9,10-perylenedicarboximide, shown in Figure 10.4), which is dissolved in the polymer matrix at a concentration of ~10^{-6} M. In polymer melts above T_g, this probe and other similar probes have ensemble-average reorientation times that closely track the polymer segmental relaxation times [32]; the probe reorientation time is typically longer than dielectric measurements of the segmental relaxation process, but the ratio of these two times is essentially independent of temperature. There is also good evidence that the probes reasonably track segmental dynamics in the glassy state during physical aging [37]. We assume that the close correspondence between probe reorientation and polymer segmental dynamics is also valid during deformation and the results below support this view.

To measure probe reorientation and mechanical deformation at the same time, we need optical access to the polymer glass sample. Figure 10.3 shows how this is accomplished. Figure 10.3a provides the sample dimensions; the sample is thin enough to ensure that deformation-induced birefringence does not interfere with the optical polarization measurements described below. Figure 10.3b shows a deformation apparatus that fits on top of a confocal optical microscope [1]. A computer controls the linear actuator and reads the load cell, allowing experiments at constant strain rate, constant stress, or combinations of these experiments

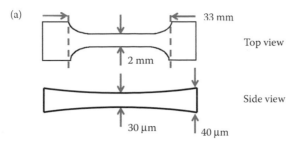

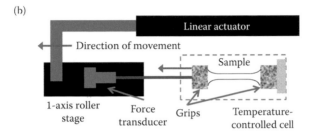

Figure 10.3 Sample geometry and schematic of deformation apparatus. (a) shows the dimensions of a typical sample prior to deformation. (b) is a schematic of the deformation apparatus that sits atop a confocal optical microscope; a top view is shown. A programmable linear actuator drives the deformation. (Reprinted with permission from B. Bending et al., *Macromolecules*, 47, 2014, 800–806. Copyright 2014, American Chemical Society.)

with stress or strain relaxation. The temperature-controlled cell that houses the polymer glass sample has an optical window on the bottom to allow optical access prior to, during, and after deformation.

Probe reorientation is measured with an optical photobleaching method. At the start of the experiment, probe molecules in the glass are orientationally isotropic. A linearly polarized laser excites a subset of probe molecules and a small fraction of these are permanently photobleached; the photobleached molecules are not orientationally isotropic but preferentially have orientations that allow them to efficiently absorb the polarized laser excitation. This step, which requires a fraction of a second, establishes an anisotropic distribution of *unbleached* probe molecules. As probe molecules reorient, this anisotropic orientational distribution will evolve into an isotropic distribution. This evolution is monitored by illuminating the sample with a weak reading beam of circularly polarized light. At the start of a measurement, the fluorescence that results from the reading beam displays the highest intensity in the polarization orthogonal to the excitation beam polarization. The unbleached probes attain an isotropic distribution on a time scale that defines the probe relaxation time τ_{probe}; as a result, the fluorescence from the reading beam becomes unpolarized. From these experiments, we obtain a time-dependent anisotropy decay function r(t) that describes the reorientation of the ensemble of probe molecules.

Figure 10.4 shows that segmental dynamics are enhanced during deformation. The figure shows anisotropy decay functions for DPPC in lightly cross-linked PMMA obtained just prior to deformation (in black) and during various stages of a constant strain rate deformation experiment; this particular data was acquired at a global strain rate of 3.1×10^{-5} s^{-1}, as shown in Figure 10.1. The more rapid decay curves obtained during deformation indicate that the ensemble of probes reorients more rapidly during deformation and thus segmental relaxation is also occurring more rapidly. Each of these anisotropy decay curves is fit to the Kohlrausch–Williams–Watts function (KWW):

$$r(t) = r(0)e^{-(t/\tau_{probe})^{\beta}}$$

Here, τ_{probe} is a characteristic reorientation time, β is a parameter that characterizes the shape of the decay process ($\beta < 1$ is a stretched exponential decay), and r(0) is the initial anisotropy. Another characteristic

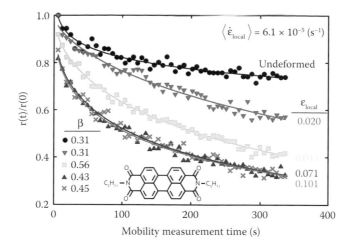

Figure 10.4 Anisotropy decay data at various strains during constant strain rate deformation of a PMMA glass at $T_g - 19$ K. Local strains and the value of the KWW β parameter are shown in the legends. Solid lines through the data are fits to the KWW function. The structure of the probe DPPC is shown. (Reprinted with permission from B. Bending et al., *Macromolecules*, 47, 2014, 800–806. Copyright 2014, American Chemical Society.)

reorientation time discussed in this chapter is τ_c, which is obtained by integrating r(t)/r(0). Although τ_c and τ_{probe} both track the segmental dynamics of the polymer, τ_{probe} emphasizes a faster portion of the relaxing segments as compared to τ_c [34].

In order to interpret these results, we need to track the evolution of the local strain during deformation. Because these samples neck during extension and do not deform homogeneously, the local strain is not equal to the global strain (which is controlled by the linear actuator). Local strain is determined by tracking the evolution of the distance between lines that are photobleached on the sample within a few hundred microns of the spot where mobility is measured. Prior to yield, the local strain very nearly equals the global strain. After yield, the local strain generally exceeds the global strain by a factor of 2–3 (because we perform our experiments close to the point where necking originates) [1].

Before discussing our results in more detail, we briefly describe another approach to measuring segmental dynamics during deformation. Subsequent to our development of the optical method described above, Lesser and coworkers [40] have developed a dielectric technique for measuring segmental relaxation during deformation. Electrodes are brought into contact with the two opposite sides of the polymer glass sample to form the capacitor needed for this method. The results obtained indicate that dynamics are enhanced during deformation and qualitatively match the optical experiments in other aspects as well. While this approach has the advantage that the polymer segments are being monitored directly (rather than probe molecules), the extent to which the average segmental dynamics are changed during deformation has not yet been quantified with this method.

10.4 SEGMENTAL MOBILITY DURING CONSTANT STRAIN RATE DEFORMATION

From data such as those shown in Figure 10.4, we can track the evolution of the average segmental relaxation time during a constant strain rate deformation experiment. Figure 10.5 shows these changes for experiments at $T_g - 19$ K at five different strain rates; the data set indicated by circles corresponds to the experiment shown in Figure 10.4. Each experiment shows a significant shortening of the probe reorientation time as yield is approached followed by a post-yield regime where the probe time is almost constant. At the lowest strain rates, the mobility increases about a factor of 25 during deformation while at the highest strain rates, mobility

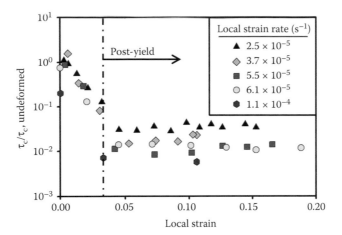

Figure 10.5 Evolution of segmental mobility with increasing local strain during constant strain rate deformation of a PMMA glass at T_g – 19 K. Average post-yield local strain rates are displayed in the legend. The y-axis scales τ_c during deformation to its value for an undeformed glass with the same thermal history. (Data for the four highest local strain rates were reported in B. Bending et al., *Macromolecules*, 47, 800, 2014.)

increases of more than a factor of 100 are observed. Figure 10.6 shows another important feature of these experiments; the KWW β parameter increases during deformation, indicating a narrowing of the distribution of relaxation times. A change of the KWW β parameter from 0.3 to 0.6 is very substantial, indicating that the distribution of relaxation times narrows from more than 3 decades to about 1 decade [41]. We note that the KWW β parameter is sensitive to the size of the probe molecule used in these experiments. While the qualitative trends shown in Figure 10.6 are expected to apply to the distribution of segmental relaxation times, the absolute values of the KWW β parameter may be shifted to some extent [34].

Prior to detailed analysis of this data, we return briefly to the Maxwell model described earlier. We noted that if we use the Maxwell model to calculate an effective relaxation time during deformation, that relaxation time would need to be shortened by more than a factor of 100 in response to deformation. The data in Figure 10.5 shows changes of this magnitude. This indicates that the qualitative interpretation of the stress/strain experiment in terms of enhanced segmental dynamics is consistent with our direct measurements of probe relaxation times during deformation.

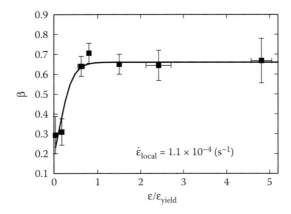

Figure 10.6 Evolution of the KWW β parameter during constant strain rate deformation of a PMMA glass deformed at T_g – 19 K. The x-axis scales local strain by the strain at yield. The KWW β parameter increases from its pre-deformation value of ~0.31 up until yield. After yield, β levels off. The average post-yield local strain rate is displayed in the legend. Guides to the eye are shown as solid lines.

The qualitative features of Figure 10.5 are precisely those expected in light of recent computer simulations. Riggleman and coworkers used molecular dynamics computer simulations of coarse-grained polymer chains to study deformation in the glassy state [42]. Their results show a strong decrease (up to a factor of 1000) in segmental relaxation times prior to yield followed by a nearly constant post-yield relaxation time; the simulations indicate that the post-yield relaxation time depends strongly on strain rate. While these simulations did not detect large changes in the KWW β parameter during deformation, they did detect dynamics that were more spatially homogeneous in the flow regime [38]. This greater spatial uniformity is consistent with the narrowing of the distribution of relaxation times observed in Figure 10.6 and provides a useful spatial interpretation of these results. Recent simulations by Rottler and coworkers [43,44] are also consistent with these experimental results (see Chapter 11).

A particularly useful feature of the simulations by Riggleman and de Pablo [42] is that they were performed both in tension and compression. One possible interpretation of the data in Figure 10.5 is that segmental mobility increases with strain because free volume increases; it is known that the density of glassy polymers decreases during extension. The simulations observed essentially identical values of the segmental relaxation time during tension and compression as a function of strain, as long as the strain rates were the same. This result argues against a free volume interpretation of the segmental relaxation time changes that occur in a stress/strain experiment, since the simulated polymer glass increased in volume during extension but densified during compression [45]. This result is particularly valuable since it has not yet been possible to perform optical measurements of probe mobility in polymer glasses during compression.

The features in Figure 10.5 also match predictions based upon a molecular theory of Chen and Schweizer [17]. In response to a constant strain rate deformation, they predict that the segmental relaxation time shortens dramatically in the pre-yield regime but remains essentially constant post-yield. They further predict that at constant temperature, the post-yield value of the segmental relaxation time scales with strain rate to the power −0.86. Figure 10.7 shows a plot of our experimental data in this format. The data obtained at T_g − 11 K shows a slope of −0.80 ± 0.06, which is compatible with the theory. A more detailed comparison can be found in Hebert et al. [39]. The theory of Chen and Schweizer does not consider spatially heterogeneous dynamics or a distribution of relaxation times, and thus their approach cannot account for the results shown in Figure 10.6. A recent mesoscopic constitutive approach of Medvedev and Caruthers [16] does predict changes in the distribution of relaxation times and is qualitatively consistent with the results shown in Figure 10.6.

We now return to the question of what controls the segmental relaxation time during deformation. Clearly, the answer cannot simply be *strain rate*, as can be seen in the pre-yield regime of Figure 10.5; here, the strain

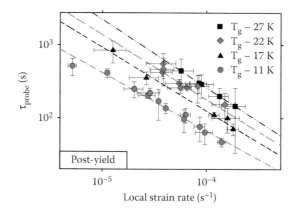

Figure 10.7 Dependence of average post-yield τ_{probe} values on local strain rate for PMMA glasses at four different temperatures undergoing constant strain rate deformation. Within each temperature series, τ_{probe} decreases with increasing strain rate. At a fixed strain rate, τ_{probe} decreases with increasing temperature. Each data point corresponds to the average post-yield data for one deformation; error bars represent one standard deviation in the data. Dashed lines represent power law fits to the data. (Data from K. Hebert et al., *Macromolecules*, 48, 6736, 2015.)

rate is constant while the relaxation time is changing significantly. The answer cannot simply be *stress* either, for the relaxation time is roughly constant in the post-yield regime while the stress is decreasing (see Figure 10.1). The theory of Chen and Schweizer [17] allows a molecular interpretation of the results shown in Figure 10.5. In this theory, the segmental relaxation time is controlled by two factors. Stress decreases the relaxation time through an Eyring-like mechanism (landscape tilting). In addition, the amplitude of local density fluctuations influences the relaxation time; we interpret this variable to indicate the position of the system on the PEL. The Chen/Schweizer approach indicates that landscape tilting is responsible for the pre-yield decrease in the relaxation time; below we show experimental data consistent with this view. In the post-yield regime, the two mechanisms operate together to maintain a constant relaxation time; as the deformation progresses into the strain softening regime, the decrease in the stress means that landscape tilting contributes less to the enhanced mobility but the system is being pulled up higher on the PEL (where energy barriers are lower) and this effect increases mobility enough to make up the difference.

10.5 CAN SEGMENTAL MOBILITY DURING DEFORMATION BE MEASURED FROM A PURELY MECHANICAL EXPERIMENT?

The experiments described above make a compelling case that changes in segmental dynamics are intimately connected with the nonlinear mechanical deformation of polymer glasses. As such, it is reasonable to ask if there is not some other (easier!) way to obtain these segmental relaxation times than the optical measurements described above. For example, if these relaxation times could be obtained from a purely mechanical measurement, this would allow much broader access to this fundamentally important information. There is a significant history of efforts along these lines, including the work of Yee [46], Martinez-Vega [47], and their coworkers. There are also cautionary notes in the literature about the difficulty of obtaining molecular relaxation times during a nonlinear mechanical measurement [48]. Inspired by recent work from Caruthers, Medvedev, and coworkers [49–51], we have tested one particular idea for obtaining segmental relaxation times from a purely mechanical measurement. Our approach was to perform the optical measurement of probe reorientation during the proposed mechanical measurement so that the relaxation time derived from mechanical measurements could be directly compared to the probe reorientation time. As we describe below, we found only a partial correspondence between the mechanical and probe relaxation times.

Figure 10.8 shows mechanical measurements performed on a lightly cross-linked PMMA glass in which a constant strain rate deformation is followed by a stress relaxation experiment (during stress relaxation, the strain is fixed and decay of stress is measured as a function of time). The four experiments shown in the figure all share a common strain rate and differ only in the strain at which stress relaxation was initiated. It

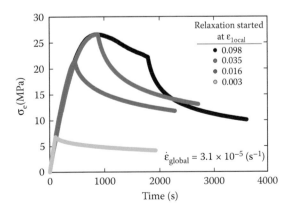

Figure 10.8 Stress as a function of time for a PMMA glass at $T_g - 19$ K deformed at a constant strain rate of 3.1×10^{-5} s^{-1}, followed by stress relaxation at various strains, as displayed in the legend. For clarity, only part of the stress relaxation data is shown for each test.

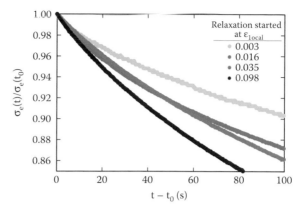

Figure 10.9 Stress relaxation response for the data in Figure 10.8 normalized to the initial stress and shifted to the starting time of stress relaxation. The local strain at which stress relaxation was initiated is provided in the legend. For clarity, only the first 100 s of stress relaxation data is shown.

has been proposed that the initial rate of stress relaxation is an accurate measurement of the rate of segmental dynamics [50,51]. Figure 10.9 shows an overlay of the stress relaxation experiments in a format that normalizes the data to the initial stress and shifts time to overlap the starting times of the stress relaxation portion of the experiments. Following References 50 and 51, we define a mechanical relaxation time from the inverse of the initial decay rates of these curves (determined from a linear fit); thus, τ_{mech} is about a factor of two shorter at a strain of 0.098 than at a strain of 0.003.

Figure 10.10 compares the mechanical relaxation times extracted as described in the previous paragraph to optical measurements of probe reorientation that were obtained during the same experiments. For simplicity, we present only the probe measurements obtained during the constant strain rate portion of the experiment; we make this comparison because τ_{mech} was proposed as measure of mobility during constant strain rate deformation. As expected, the probe relaxation times have behavior consistent with Figure 10.5; τ_{probe} changes by about a factor of 30 in the pre-yield regime. In contrast, τ_{mech} changes by only a factor of 2 in the pre-yield regime. In the post-yield regime, τ_{mech} is similar to τ_{probe}; we have preliminary results that confirm that this agreement in the post-yield regime is also obtained at other strain rates.

Unfortunately, Figure 10.10 indicates that we have not yet succeeded in our effort to identify a purely mechanical experiment that tracks the segmental dynamics throughout all stages of a constant strain rate

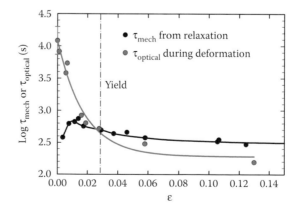

Figure 10.10 Evolution of τ_{mech}, as determined by stress relaxation, and optically measured τ_{probe} with strain during the constant strain rate deformations of Figures 10.8 and 10.9. During deformation, τ_{probe} experiences almost a 100-fold decrease, which is not observed in τ_{mech}. The x-axis scales the local strain to the strain at yield. Solid lines are guides to the eye.

deformation. Our understanding is that τ_{probe} accurately tracks the average segmental dynamics. In contrast, τ_{mech} is also influenced by the width of the distribution of relaxation times. In a linear viscoelastic experiment, τ_{mech} would increase as a function of strain merely because there is a distribution of relaxation times, even though these relaxation times would not be changing during the deformation. Because the actual mechanical experiment shown is highly nonlinear, we speculate that the values of τ_{mech} shown in Figure 10.10 are the combination of two effects. As the strain increases, the distribution of relaxation times tends to increase τ_{mech} while the nonlinear deformation tends to decrease τ_{mech}. Apparently, these two effects that influence τ_{mech} nearly cancel under these conditions. It remains to be established if any purely mechanical experiment can directly track changes of segmental relaxation times during all phases of nonlinear mechanical deformation of a polymer glass.

Although τ_{mech} does not provide the same information as τ_{probe} and the KWW β parameter, it joins these parameters as observables that characterize (in various ways) the influence of a constant strain rate deformation on the dynamics of polymer glasses. We anticipate that these three observables in concert provide quite a demanding test of theories, models, and simulations.

10.6 EFFECT OF TEMPERATURE ON SEGMENTAL DYNAMICS DURING DEFORMATION

Temperature has a very strong influence on the deformation properties of polymer glasses. Taking polycarbonate as an example, we note that the yield stress doubles from 30 to 60 MPa as the temperature is dropped from T_g − 20 K to T_g − 100 K (for this comparison, we use a series of experiments in extension with the strain rate of 4×10^{-3} s^{-1} [52]). We can think about this increase in yield stress with decreasing temperature in qualitative terms as follows. At a lower temperature, the time scale for segmental relaxation of a given glass will be longer, since less thermal energy is available to surmount energy barriers in the PEL. On the other hand, at yield, we anticipate that the segmental relaxation time will be approximately the same no matter what the deformation temperature, since the segmental dynamics need to be fast enough to enable flow at the strain rate imposed by the experiment (see next paragraph for more details on this point). Thus we anticipate that, at low temperature, the segmental relaxation time must change by a much larger factor in the pre-yield regime than at high temperature. According to the work of Chen and Schweizer [17], the landscape tilting mechanism dominates in the pre-yield regime, so higher levels of stress will be required at low temperature in order to drive the system to yield, in general agreement with the experimental data discussed above.

We have performed a series of experiments to quantitatively test the effect of temperature on the segmental relaxation time during constant strain rate deformation [39]. The results are qualitatively consistent with the scenario from the previous paragraph but differ in one important detail. The post-yield segmental dynamics of the same polymer glass deformed at the same strain rate, but at different temperatures, are somewhat faster at higher temperatures. This effect is shown in Figure 10.7 for lightly cross-linked glasses of PMMA. We interpret the faster dynamics at higher temperature in Figure 10.7, even for the same strain rate, to mean that thermally activated barrier crossing is quite important in the post-yield regime for these polymer glasses. Under the conditions of our experiments, we estimate [39] that the free energy barriers that are crossed in the flow state are still very substantial (with barrier heights of about 39 kT during flow versus 45 kT in the absence of deformation).

Our conclusion that thermal barrier crossing is very important for segmental dynamics during post-yield deformation is somewhat different than recent theoretical and simulation work. As discussed in Hebert et al. [39], the temperature dependence shown in Figure 10.7 is larger than anticipated by the theory of Chen and Schweizer. The role of thermal barrier crossing has also been investigated in computer simulations by Chung and Lacks [53,54]; these authors came to the conclusion that barrier hopping plays a relatively minor role during deformation. There is certainly an opportunity for additional computer simulations on this topic, particularly those which calculate the instantaneous segmental relaxation time during deformation across a wide range of temperature.

10.7 SEGMENTAL MOBILITY DURING CREEP DEFORMATION

There are two reasons for investigating changes in segmental relaxation times during different types of deformations. First, a wide variety of deformation schemes are utilized to characterize polymeric materials; measurements of molecular motion during these experiments will be useful for understanding the observed mechanical response. Second, different deformation schemes provide an opportunity to check our understanding of the fundamental factors that control changes in segmental dynamics. In creep deformation, stress is held constant and this provides a way to test the extent to which stress controls mobility during deformation.

Figure 10.11 shows results for two creep experiments on lightly cross-linked PMMA samples [37]. In a creep experiment, the stress is held constant while the strain is the dependent variable. Figure 10.11a shows the changes in the local strain during these creep experiments; the increase in the strain corresponds to the creep portion of the measurement while the subsequent decrease occurs after the stress is set to zero (i.e., strain recovery). Both of these experiments show a large permanent set, as indicated by the persistent strain even after long times in the absence of stress. Figure 10.11b shows the changes in segmental relaxation times during each of these experiments. One important feature of these experiments is that these samples were aged to equilibrium prior to deformation; the experiment temperature was sufficiently close to T_g that

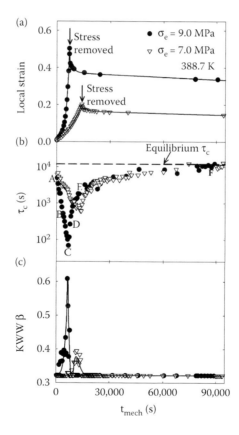

Figure 10.11 Local strain, τ_c, and the KWW β parameter during creep and recovery of PMMA aged to equilibrium at $T_g - 6$ K. Engineering stress in each of the two deformations is shown in the legend. In (a), it is observed that strain is not fully recovered after the deformation, despite the recovery of the equilibrium τ_c (b). (b) shows enhancement of segmental mobility by up to a factor of ~100 during deformation, and the evolution of τ_c into equilibrium after deformation. During deformation, the KWW β parameter (c) drastically increases while stress is applied but rapidly recovers its pre-deformation value after stress is released. (Reprinted with permission from H. N. Lee and M. D. Ediger, *Macromolecules* 43, 2010, 5863–5873. Copyright 2010, American Chemical Society.)

equilibrium could be achieved in one day of aging. Thus, in a technical sense, these deformation experiments started in the equilibrium supercooled liquid state. The equilibrium relaxation time at the temperature of the mechanical experiment is marked in Figure 10.11b. It is important to note the segmental relaxation time has nearly returned to its predeformation (equilibrium) value by the end of the experiment; in contrast, the strain remains at a large value. This result indicates that segmental dynamics are not directly controlled by strain. In Figure 10.11c, the changes in the KWW β parameter during these experiments are presented. The β parameter is highly correlated with instantaneous segmental relaxation times such that large β parameters are observed when the relaxation times are the shortest. We interpret this to indicate that the distribution of segmental relaxation times narrows as the average segmental relaxation time decreases.

The behavior shown in Figure 10.11 is consistent with the interpretation presented earlier for constant strain rate experiments. In order to connect the behavior shown in Figure 10.11 to the constant strain rate experiments, we note that almost all the creep data shown in this figure was obtained in the flow regime. In a creep experiment, the onset of flow is analogous to yield in a constant strain rate experiment. All the data shown for the flow regime in Figure 10.11 demonstrate the same correlation between strain rate and segmental relaxation time as was indicated for the constant strain rate data in Figure 10.7. That is, when the strain is increasing most rapidly (the highest strain rate), the segmental relaxation time is the shortest. The KWW β parameter behavior shown in Figure 10.11c is consistent with the behavior shown for constant strain rate experiments in Figure 10.6; in the post-yield flow regime, larger values of β are associated with higher strain rates. The qualitative features shown in (a) and (b) of Figure 10.11 have also been observed in molecular dynamics computer simulations of polymer glasses subjected to creep and strain relaxation [35,38].

For the purposes of this chapter, the most insightful creep deformation experiments are those in which the stress is too low to cause flow. In such an experiment, the strain changes very little and the relaxation time can be measured very accurately at constant engineering stress. The results of many such experiments are shown in Figure 10.12 [34]. These experiments were performed at three temperatures below T_g for lightly cross-linked PMMA glasses. The three dashed lines are fits to the Eyring model $\tau \propto \sigma / \sinh((\sigma \cdot V)/(2 \cdot k_b T))$, where $k_b T$ is the thermal energy, with one fitting parameter (the activation volume, V). For each temperature, the Eyring model accurately describes the effect of stress on the segmental relaxation time up to true stress levels of about 10 MPa. As the samples enter the flow regime at higher true stresses, deviations from the Eyring behavior are observed.

The results shown in Figure 10.12 are completely consistent with our earlier interpretation of constant strain rate experiments. In the pre-yield regime of both creep and constant strain rate experiments, the landscape tilting mechanism described by the Eyring equation accurately describes the observed segmental

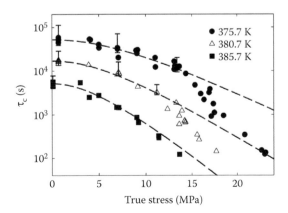

Figure 10.12 τ_c as a function of true stress during constant stress deformation of PMMA glasses at three temperatures, as shown in the legend. At low stress, the data agrees with the Eyring model (dashed lines). τ_c deviates from the Eyring model during flow at higher stresses. (H. N. Lee et al., *J. Polym. Sci. B Polym. Phys.*, 2009, 47, 1713–1727. Copyright Wiley-VCH Verlag GmbH & Co. KGaA. Reproduced with permission.)

relaxation times. Once flow occurs, segmental relaxation during creep is faster than would be expected on the basis of landscape tilting alone. During flow, the system is being pulled up the PEL into a regime in which energy barriers are lower; this second mechanism is then responsible for the deviations from Eyring behavior shown in Figure 10.12. Chen and Schweizer [18] have developed a theory for polymer glass deformation during creep and our experiments [32,33,36,37] are qualitatively consistent with the predictions of their theory. The interpretation provided in this paragraph is also consistent with their theory, if we identify the larger local density fluctuations discussed in the theory with higher regions of the PEL.

10.8 CHANGES IN POLYMER STRUCTURE DURING DEFORMATION

Structure/property relationships are at the heart of materials science and it is useful to address the role of structure changes in the deformation of polymer glasses. When a macroscopic polymer glass sample is deformed by 20% or 50%, it is clear that there is a significant change in the structure of individual chains. One manifestation of this is the birefringence developed by polymer glasses during deformation (and typically maintained after deformation ceases). The observed value of the birefringence depends upon both local and large-scale rearrangements of the polymer chains and this effect has been explored extensively by Osaki, Inoue, and their coworkers [55,56]. Neutron scattering experiments allow the changes in structure caused by deformation to be determined more directly. A recent experiment compared deformed and undeformed polymer glasses, and concluded that chains are deformed affinely on the scale of ~10 nm (roughly R_g for the chains studied) but are isotropically arranged on a much smaller length scale (~2 nm) [57]. At a qualitative level, we understand this result to mean that the segmental mobility during deformation is sufficient to extensively rearrange segments on a small length scale; on the length scale of the entire polymer chain, however, polymer chains do not have sufficient mobility to relax. Molecular dynamics computer simulations on nonpolymeric glassy systems (i.e., spherical particles) also indicate that large-scale deformation is affine while more local rearrangements are not [58].

Changes in polymer structure on large length scales are highly relevant for an understanding of the strain hardening that occurs at large deformations. Strain hardening determines the extent to which strain is localized during extension, with higher levels of strain hardening favoring delocalization of the strain, which in turn leads to higher ductility [59]. Strain hardening cannot occur if the molecular weight of the polymer chains is too low or if the entanglement density of the system is too low, and under these conditions polymer glasses fail via fracture at low strains [60,61].

We take the point of view that the yielding of polymer glasses (up to strains at which strain hardening remains insignificant) is a generic glass problem that can be understood without considering polymer molecular weight or large-scale changes in polymer structure. Yielding phenomena do not depend significantly upon polymer molecular weight as long as the molecular weight is high enough to ensure that yield occurs before failure. Consistent with this view, we have preliminary results indicating that glasses of uncross-linked polymer chains give rise to the same results as those shown for lightly cross-linked samples in Figures 10.5 and 10.7, if the temperature is adjusted to constant $T - T_g$. Furthermore, we expect that the detailed molecular structure of different polymers (polystyrene, polycarbonate, or PMMA) will not have a significant influence on the main features shown in Figures 10.5 through 10.7. We also have preliminary results that support this statement.

The very large changes in dynamics that occur during deformation in our experiments are ultimately connected to changes in the structure of the glass at a very local level. In the theory of Chen and Schweizer [17], the segmental relaxation time is controlled by two mechanisms. Prior to yield, the landscape tilting mechanism dominates; we interpret this to mean that very small changes in local structure, induced by the applied stress, lower some energy barriers and allow faster dynamics. At later stages of the deformation, the amplitude of local density fluctuations also influences the time scale of segmental relaxation in the glass. The amplitude of these density fluctuations increases in a typical deformation of a polymer glass and this is one of the factors that enhance dynamics in the post-yield regime of a constant strain rate experiment (see Figure 2 of Reference 16). In the context of the Chen and Schweizer theory, these local density fluctuations are the coarse-grained representation of the local structural changes that occur in the post-yield regime of the deformation of a polymer glass. As Chen and Schweizer have noted, the predicted changes in the amplitude of these local density fluctuations is quite small and only a few attempts have been made to measure these changes with x-ray scattering [62]. We should not be surprised that a very small change in structure can give rise to a large change in dynamics as this is a generic feature of the glass problem. For example, it is also difficult to pinpoint the changes in local structure that are responsible for the super-Arrhenius temperature dependence of many supercooled liquids.

As the yielding of polymer glasses appears to be part of a more generic yielding problem that includes other classes of amorphous materials, we wish to briefly comment on this connection. At a phenomenological level, the yielding of metallic and colloidal glasses is similar to that of polymer glasses [63,64]. Much of the literature on the yielding of metallic and colloidal glasses focuses on shear transformation zones, which are local rearrangements in structure that occur in response to deformation. A useful feature of colloidal glasses is that direct imaging of these local rearrangements is possible. For example, Schall et al. [65] showed that these rearrangements have the geometric features expected for shear transformation zones. In their experiments on colloids, local structural rearrangements were shown to occur via thermal barrier crossing, with barriers that are almost as high during deformation as they would be in the absence of deformation; our experiments on polymer glasses discussed above agree with their conclusions [39]. If the shear transformation zone models and the model of Chen and Schweizer are both correct, then we expect that shear transformation zones are the local rearrangements that are represented in a coarse-grained manner by the local density fluctuations of Chen and Schweizer. Of course, there are also important differences between colloidal glasses and polymer glasses. One manifestation of this is the difference in the characteristic barrier height during deformation near T_g (~18 kT for colloids [65] versus ~39 kT for polymer glasses [39]). In spite of these differences, we expect that studies of the deformation of other classes of amorphous materials will continue to deepen our understanding of polymer glass deformation.

10.9 CONCLUDING REMARKS

The deformation of polymer glasses is an important and challenging problem that is not yet sufficiently understood. In industry, the deformation properties of polymer glasses are generally predicted using models that must be extensively parameterized against experimental data. Because these models are not based upon a completely accurate fundamental understanding of the deformation process, such models often fail when applied outside the range of experiments from which they were parameterized; that is, a model that accurately predicts the results of a constant strain rate deformation may fail completely in describing a cyclic deformation. Our experiments, in combination with simulations and more fundamental theory/modeling, have the goal of advancing our understanding to the point where the right physics can be included in models used to predict the properties of polymer glasses in an industrial setting. One could be optimistic that such a model could accurately predict the mechanical response of a polymer glass to many different types of deformations, over time periods encompassing the 40-year lifespan of a product.

Our experiments indicate that enhanced segmental dynamics is a key feature of the deformation of polymer glasses. In future experiments, we hope to more critically examine the underlying mechanisms of enhanced segmental dynamics in theories. For example, the work of Chen and Schweizer indicates that both landscape tilting and the increased amplitude of density fluctuations contribute to enhanced mobility during a constant strain rate experiment. We expect that experiments that reverse the application of stress or strain during deformation can sensitively determine the separate contributions of each mechanism (since a tilted landscape can be untilted by removing stress).

Our experiments also indicate that changes in the width of the distribution of segmental relaxation times during deformation can be very substantial. It is unclear what role these changes play in the macroscopic mechanical properties of polymer glasses. At present, the Chen and Schweizer approach cannot describe this effect, although the theory predicts several features which are qualitatively consistent with our results. The model of Caruthers and Medvedev demonstrates changes in the width of segmental relaxation times and provides at least qualitative agreement with our experiments. Does a model need to accurately account for these changes in the width of relaxation time distribution in order to predict the real mechanical properties of polymer glasses? Or are these changes in the width of distribution a "detail" with little consequence for mechanical properties?

There is an opportunity to unify the description of the deformation of different types of glassy materials. While shear transformation zones feature prominently in the metallic glass literature, they are rarely mentioned in the description of polymer glass deformation. We expect that shear transformation zones are equally important for different classes of glassy materials. Simulations can play a critical role in testing theoretical approaches on different classes of glassy materials; we anticipate that some unification of the theoretical approaches taken in these two communities should be possible.

ACKNOWLEDGMENTS

MDE gratefully acknowledges extensive and useful collaboration with Juan de Pablo, Rob Riggleman, Jim Caruthers, Grisha Medvedev, Ken Schweizer, Hau-Nan Lee, Steve Swallen, Keewook Paeng, Ben Bending, and Josh Ricci. In particular, MDE thanks Jim Caruthers for providing the ideas that led to our work on polymer glass deformation. We gratefully acknowledge the support of the National Science Foundation, Division of Materials Research, Polymers Program for support of this work (1404614, 1104770, 0907607).

REFERENCES

1. B. Bending, K. Christison, J. Ricci, and M. D. Ediger, *Macromolecules*, 47, 800, 2014.
2. H. Eyring, *J. Chem. Phys.*, 4, 283, 1936.
3. R. N. Haward and G. Thackray, *Proc. R.l Soc. Lond. A Math. Phys. Sci.*, 302, 453, 1968.
4. C. P. Buckley and D. C. Jones, *Polymer*, 36, 3301, 1995.
5. D. S. A. De Focatiis, J. Embery, and C. P. Buckley, *J. Polym. Sci. B Polym. Phys.*, 48, 1449, 2010.
6. T. A. Tervoort, E. T. J. Klompen, and L. E. Govaert, *J. Rheol.*, 40, 779, 1996.
7. E. M. Arruda and M. C. Boyce, *Int. J. Plasticity*, 9, 697, 1993.
8. E. T. J. Klompen, T. A. P. Engels, L. E. Govaert, and H. E. H. Meijer, *Macromolecules*, 38, 6997, 2005.
9. R. A. Schapery, *Int. J. Solids Struct.*, 2, 407, 1966.
10. B. Bernstein and A. Shokooh, *J. Rheol.*, 24, 189, 1980.
11. S. M. Fielding, R. G. Larson, and M. E. Cates, *Phys. Rev. Lett.*, 108, 048301, 2012.
12. S. M. Fielding, R. L. Moorcroft, R. G. Larson, and M. E. Cates, *J. Chem. Phys.*, 138, 12A504, 2013.
13. R. M. Shay and J. M. Caruthers, *Polym. Eng. Sci.*, 30, 1266, 1990.
14. S. R. Lustig, R. M. Shay, and J. M. Caruthers, *J. Rheol.*, 40, 69, 1996.
15. J. M. Caruthers, D. B. Adolf, R. S. Chambers, and P. Shrikhande, *Polymer*, 45, 4577, 2004.
16. G. A. Medvedev and J. M. Caruthers, *J. Rheol.*, 57, 949, 2013.
17. K. Chen and K. S. Schweizer, *Macromolecules*, 44, 3988, 2011.
18. K. Chen and K. S. Schweizer, *Phys. Rev. E*, 82, 041804, 2010.
19. K. Chen and K. S. Schweizer, *Macromolecules*, 41, 5908, 2008.
20. W. G. Knauss and I. J. Emri, *Comput. Struct.*, 13, 123, 1981.
21. W. G. Knauss and I. Emri, *Polym. Eng. Sci.*, 27, 86, 1987.
22. I. L. Hopkins, *J. Polym. Sci.*, 28, 631, 1958.
23. L. W. Morland and E. H. Lee, *Trans. Soc. Rheol.*, 1957–1977, 4, 233, 1960.
24. L. C. E. Struik, *Polym. Eng. Sci.*, 17, 165, 1977.
25. C. A. Angell, K. L. Ngai, G. B. McKenna, P. F. McMillan, and S. W. Martin, *J. Appl. Phys.*, 88, 3113, 2000.
26. G. B. McKenna, *J. Phys. Condens. Matter*, 15, S737, 2003.
27. M. D. Ediger and P. Harrowell, *J. Chem. Phys.*, 137, 080901, 2012.
28. M. D. Ediger, *Annu. Rev. Phys. Chem.*, 51, 99, 2000.

29. R. Richert, *J. Phys. Condens. Matter*, 14, R703, 2002.

30. L. S. Loo, R. E. Cohen, and K. K. Gleason, *Science*, 288, 116, 2000.

31. Q. Y. Zhou, A. S. Argon, and R. E. Cohen, *Polymer*, 42, 613, 2001.

32. H. N. Lee, K. Paeng, S. F. Swallen, and M. D. Ediger, *J. Chem. Phys.*, 128, 134902, 2008.

33. H. N. Lee, K. Paeng, S. F. Swallen, and M. D. Ediger, *Science*, 323, 231, 2009.

34. H. N. Lee, K. Paeng, S. F. Swallen, M. D. Ediger, R. A. Stamm, G. A. Medvedev, and J. M. Caruthers, *J. Polym. Sci. B Polym. Phys.*, 47, 1713, 2009.

35. H. N. Lee, R. A. Riggleman, J. J. de Pablo, and M. D. Ediger, *Macromolecules*, 42, 4328, 2009.

36. H. N. Lee and M. D. Ediger, *J. Chem. Phys.*, 133, 014901, 2010.

37. H. N. Lee and M. D. Ediger, *Macromolecules*, 43, 5863, 2010.

38. R. A. Riggleman, H. N. Lee, M. D. Ediger, and J. J. de Pablo, *Soft Matter*, 6, 287, 2010.

39. K. Hebert, B. Bending, J. Ricci, and M. D. Ediger, *Macromolecules*, 48, 6736, 2015.

40. J. Kalfus, A. Detwiler, and A. J. Lesser, *Macromolecules*, 45, 4839, 2012.

41. C. P. Lindsey and G. D. Patterson, *J. Chem. Phys.*, 73, 3348, 1980.

42. R. A. Riggleman, G. N. Toepperwein, G. J. Papakonstantopoulos, and J. J. de Pablo, *Macromolecules*, 42, 3632, 2009.

43. M. Warren and J. Rottler, *Phys. Rev. Lett.*, 104, 205501, 2010.

44. M. Warren and J. Rottler, *J. Chem. Phys.*, 133, 164513, 2010.

45. R. A. Riggleman, H. N. Lee, M. D. Ediger, and J. J. de Pablo, *Phys. Rev. Lett.*, 99, 215501, 2007.

46. A. F. Yee, R. J. Bankert, K. L. Ngai, and R. W. Rendell, *J. Polym. Sci. B Polym. Phys.*, 26, 2463, 1988.

47. J. J. Martinez-Vega, H. Trumel, and J. L. Gacougnolle, *Polymer*, 43, 4979, 2002.

48. G. B. McKenna and L. J. Zapas, *Polym. Eng. Sci.*, 26, 725, 1986.

49. E. W. Lee, G. A. Medvedev, and J. M. Caruthers, *J. Polym. Sci. B Polym. Phys.*, 48, 2399, 2010.

50. J. W. Kim, G. A. Medvedev, and J. M. Caruthers, *Polymer*, 54, 3949, 2013.

51. G. A. Medvedev, J. W. Kim, and J. M. Caruthers, *Polymer*, 54, 6599, 2013.

52. C. Bauwens-Crowet, J. C. Bauwens, and G. Homès, *J. Polym. Sci. A-2 Polym. Phys.*, 7, 735, 1969.

53. Y. G. Chung and D. J. Lacks, *Macromolecules*, 45, 4416, 2012.

54. Y. G. Chung and D. J. Lacks, *J. Polym. Sci. B Polym. Phys.*, 50, 1733, 2012.

55. T. Inoue, H. Okamoto, and K. Osaki, *Macromolecules*, 24, 5670, 1991.

56. K. Osaki, H. Okamoto, T. Inoue, and E.-J. Hwang, *Macromolecules*, 28, 3625, 1995.

57. F. Casas, C. Alba-Simionesco, H. Montes, and F. Lequeux, *Macromolecules*, 41, 860, 2008.

58. A. Tanguy, J. P. Wittmer, F. Leonforte, and J. L. Barrat, *Phys. Rev. B*, 66, 174205, 2002.

59. H. E. H. Meijer and L. E. Govaert, *Progr. Polym. Sci.*, 30, 915, 2005.

60. R. S. Hoy and M. O. Robbins, *J. Polym. Sci. B Polym. Phys.*, 44, 3487, 2006.

61. J. Rottler, *J. Phys. Condens. Matter*, 21, 463101, 2009.

62. E. Munch, J.-M. Pelletier, B. Sixou, and G. Vigier, *Phys. Rev. Lett.*, 97, 207801, 2006.

63. C. P. Amann, M. Siebenbürger, M. Krüger, F. Weysser, M. Ballauff, and M. Fuchs, *J. Rheol. (1978–Present)*, 57, 149, 2013.

64. C. P. Amann et al., *arXiv:1302.2030* [cond-mat.soft], 2013.

65. P. Schall, D. A. Weitz, and F. Spaepen, *Science*, 318, 1895, 2007.

Local relaxation, aging, and memory of polymer glasses at rest and under stress

JÖRG ROTTLER

The process of physical aging or structural recovery is one of the defining features of glassy materials [1–4]. When a glass forms below the glass transition temperature, the dynamics are not completely frozen but the system continues to evolve slowly toward an eventual equilibrium state. The time for the glass to reach equilibrium depends on the details and depth of the temperature quench and can be very large. As a result, material properties change with the waiting or aging time elapsed since vitrification, which produces a fascinating array of nonequilibrium phenomena: nonlinear response, memory, and hysteresis effects.

Physical aging in polymers is also important because it influences the macroscopic failure mechanisms in load-bearing applications. Although older glasses initially develop a higher resistance to plastic flow, they also exhibit a stronger degree of softening after yielding, which promotes shear instabilities and strain localization. Understanding the molecular mechanism behind age-induced embrittlement [5] is therefore of fundamental importance to control the materials' lifetime to failure. This chapter first presents an overview of the various phenomena associated with physical aging in quiescent and deformed polymer glasses, and then discusses insights into the underlying molecular-level processes that have been gained recently through a combination of experiments and computer simulations.

11.1 GLASSY AGING PHENOMENOLOGY

During physical aging, molecules explore configurational space through thermal activation and find lower energy configurations that optimize local packing. Bulk thermodynamic quantities such as enthalpy and volume are generally found to decrease logarithmically with waiting time. In the simplest kind of aging experiment, the material experiences a temperature jump from a high-temperature state to a nonequilibrium glass state at lower temperature, from which it slowly evolves toward a new equilibrium state at that temperature.

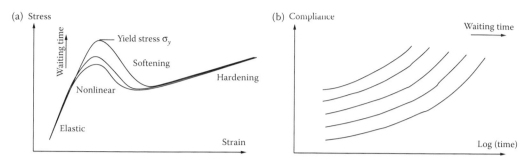

Figure 11.1 (a) Sketch of a typical stress–strain response of a glassy polymer in uniaxial deformation after a temperature quench followed by aging for different waiting times t_w. The upper yield stress σ_y increases with increasing age or waiting time, but the subsequent strain hardening is age independent. (b) Sketch of typical creep compliance curves of a glassy polymer in the linear regime for several increasing material ages.

We denote the time elapsed since the glass was formed by the waiting time t_w. A volume relaxation or aging rate may be defined as the logarithmic slope in plots of volume versus time:

$$r_V = -\frac{1}{V_0}\frac{dV(t_w)}{d\log(t_w)},\tag{11.1}$$

where V_0 is a reference volume. In thin films, physical aging manifests itself as a reduction of the film thickness $h(t_w)$, which may be quantified by an analogous expression where volume is replaced by thickness [6].

Since physical aging evolves the glass toward configurations with deeper energy minima in the free energy landscape, one may expect a reduction of molecular mobility and an increase in relaxation time. These have important effects on the mechanical properties of polymer glasses. The typical response of a polymer glass to deformation at constant strain rate is shown schematically in Figure 11.1a as a function of strain $\epsilon(t)$: The stress increases first elastically and then nonlinearly up to a maximum value, after which the material exhibits softening before the stress rises again in the strain hardening regime. Of particular importance for this chapter is the maximum or *upper yield stress* σ_y, which exhibits aging effects and increases logarithmically with waiting time t_w [7,8]. For glassy solids aged for a time t_w until deformation, one may write [9,10]

$$\sigma_y = \sigma_0 + s_0 \ln[(t_w + t_{\text{eff}})/t_0],\tag{11.2}$$

where s_0 is a parameter that describes the yield stress aging rate and t_0 is a microscopic time. An effective time $t_{\text{eff}} \propto \dot{\epsilon}^{-1}$ was included to reflect any additional contributions to the aging process during the time required to reach yield, and $\dot{\epsilon}$ is the strain rate. In terms of activated processes in an energy landscape, the increase of the yield stress with aging can be understood as an increase in the barriers for plastic deformation. During aging, both enthalpy and entropy decrease as a denser glass has fewer configurational degrees of freedom. Therefore, both energetic and entropic contributions contribute to changes in the free energy barrier for rearrangement [11]. The *lower yield stress* at the end of the strain softening regime usually exhibits a weaker or no age dependence at all, and the large-strain hardening behavior is found to be independent of waiting time.

Aging effects can be further probed by subjecting the polymer to a step stress σ_{ext} and measuring the compliance $J(t, t_w) = \epsilon(t, t_w)/\sigma_{\text{ext}}$ as a function of time t after the material has aged without loading for a time t_w. The total age of the glass is then $t_w + t$. In a series of pioneering experiments [1], Struik found that the plastic part of the compliance ($J_{\text{pl}}(t, t_w) = J(t, t_w) - 1/2E$, where $1/2E$ is the elastic part with E the modulus) exhibits a shift to longer time with increasing age, a behavior often referred to as *time-aging time superposition*. The compliance hence follows a scaling behavior:

$$J_{\text{pl}}(t,t_w) = F\left(t_0^{\mu-1}\frac{t}{t_w^\mu}\right),\tag{11.3}$$

where $F(.)$ is a scaling function with $F(0) = 0$, t_0 is a microscopic time, and μ is the apparent "aging exponent." In polymer glasses, values of $\mu = 1$ (normal aging) or $\mu < 1$ (subaging) are usually reported. Similar scaling behavior is also found for the magnetization of spin glasses. The form of the scaling function frequently follows a Kohlrausch-type law, $F(u) = \exp(cu^\beta) - 1$, with a stretching exponent $\beta < 1$, but depends on details of the system. The scaling behavior implies that there exists a characteristic relaxation time that increases with waiting time, $\tau_{\rm rel} \propto t_w^\mu$, a behavior called *power law aging*.

The above expression is valid for $t \ll t_w$, that is, when the testing time t is much smaller than the waiting time t_w so that aging effects during the testing interval can be neglected. The long time response can be described by replacing t with Struik's effective time [1] $t_{\rm eff} \leq t$:

$$t_{\rm eff}(t,t_w;\mu) = \int_0^t dt' \frac{t_w^\mu}{(t'+t_w)^\mu} = \frac{t_w}{1-\mu}\left[\left(1+\frac{t}{t_w}\right)^{1-\mu} - 1\right]$$
$$\sim \begin{cases} t, & t/t_w \ll 1, \\ \dfrac{t^{1-\mu}t_w^\mu}{(1-\mu)}, & t/t_w \gg 1, \end{cases} \tag{11.4}$$

which captures the slowing down of the dynamics. According to this expression, aging effects in the response (11.3) disappear for $t/t_w \gg 1$. The aging exponent $\mu = \mu(\sigma_{\rm ext},T)$ usually depends on the applied stress and the temperature T. It typically decreases with increasing $\sigma_{\rm ext}$, a behavior sometimes termed "rejuvenation" (see below), since in this case, larger stresses will lead to a decrease of the characteristic response time of the compliance in the aging regime $t_0 \ll t \lesssim t_w$. Additionally, the aging exponent typically decreases with decreasing $T < T_g$ and also drops to zero above the glass transition temperature.

Physical aging processes take place essentially on the monomer level and are not a polymer-specific effect. They have therefore been studied in many other glass formers, notably colloidal glasses (often polystyrene beads) [12], clay suspensions ("laponite") [13–16], and gels [17]. In these materials, density takes over the role of temperature as a control parameter. Many of the qualitative features of the rheological aging in these soft aging glasses mirror those observed in molecular glasses [18], notably the scaling behavior of the creep compliance [17] and the age dependence of the yield stress [19]. Since the size of the "monomers" is several orders of magnitude greater than in molecular glasses, their relaxation behavior can be probed more directly through light scattering and diffusive wave spectroscopy (DWS) [20–22] after a suitable starting state has been prepared via preshearing. The correlation function of the temporal fluctuations of the scattered light provides access to the self-intermediate (incoherent) scattering function (ISF):

$$F_s(\mathbf{q},t,t_w) = \langle\exp[i\mathbf{q}\Delta\mathbf{r}(t,t_w)]\rangle \approx \exp[-\mathbf{q}^2\langle\Delta\mathbf{r}^2(t,t_w)\rangle/6] + \dots \quad \text{if} \quad \mathbf{q}^2\langle\Delta\mathbf{r}^2(t,t_w)\rangle/6 \ll 1, \tag{11.5}$$

where $\Delta\mathbf{r}(t,t_w) = \mathbf{r}(t_w + t) - \mathbf{r}(t_w)$ is the displacement of a tagged particle during time t after a waiting time interval t_w. Accordingly, $F_s(\mathbf{q},t)$ measures the decorrelation of the density due to the motion of particles on length scales $\sim 1/q$. Figure 11.2a illustrates the typical behavior of the ISF with increasing waiting time if q corresponds to the first peak of the static structure factor $S(q)$. One usually observes an initial age-independent decay to a plateau that is associated with monomer vibrations around their local cages. These degrees of freedom are in equilibrium with the thermal bath and hence do not show aging effects. By contrast, the subsequent decreases from the plateau due to particles escaping their cages shifts to longer times with increasing age. If one analyzes instead the *mean-squared displacement* (MSD), $\langle\Delta\mathbf{r}^2(t, t_w)\rangle$, one also observes a caging plateau where the MSD changes little, followed by a rise when particles leave their trapped configurations (see also Figure 11.3). A structural relaxation time $\tau_{\rm rel}$, often also denoted α-relaxation time τ_α, is usually extracted from the decay of the ISF to a threshold value. Alternatively, the aging part of the ISFs can be collapsed with a scaling procedure analogous to the mechanical response, and the aging behavior of $\tau_{\rm rel}$ can be extracted from the shift factors. A plot of $\tau_{\rm rel}$ versus t_w (see Figure 11.2b) reveals the aforementioned power law behavior.

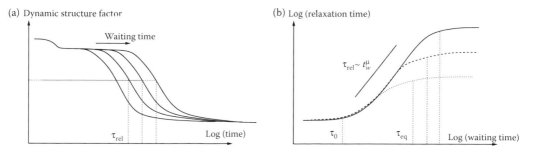

Figure 11.2 (a) Sketch of the self-intermediate scattering function (ISF) of an aging glass with increasing waiting time t_w. (b) Evolution of the structural relaxation time τ_{rel} extracted from the decay of the ISFs for three different temperatures, where each of which implies a different equilibration time.

It is important to realize that aging phenomena can only be observed in a limited time window [23]. First, the relaxation time cannot become smaller than some "initial age" τ_0 that corresponds to the relaxation time of the initial equilibrium (liquid) state. At large times, aging ceases when τ_{rel} reaches the equilibration time τ_{eq} of the material. These two times bracket the range of times in which aging effects occur. The dependence of τ_{rel} on t_w therefore has a sigmoidal shape and the power law regime occurs for $\tau_0 < t_w < \tau_{eq}$, see Figure 11.2b. Observing clear power law behavior and reliably obtaining the aging exponent μ therefore requires $\tau_0 \ll \tau_{eq}$.

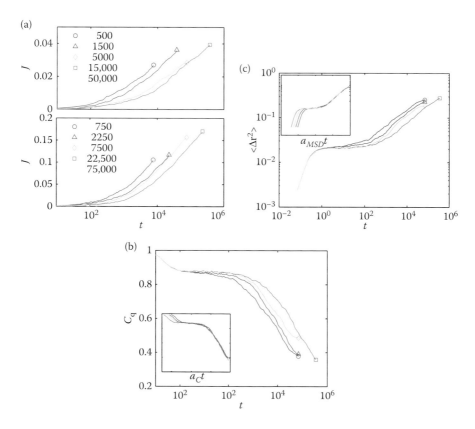

Figure 11.3 Molecular dynamics simulation of uniaxial deformation of a polymer glass after aging for five different waiting times (see legend). (a) Plastic creep compliance $J_{pl}(t, t_w)$ for applied stresses $\sigma_{ext} = 0.4$ (top) and $\sigma_{ext} = 0.5$ (bottom). (b) Decay of the ISF (here denoted C_q) for $\sigma_{ext} = 0.4$. (c) The MSD for the same simulation. Insets show time-aging time superposition with shift factors a_c and a_{msd}, respectively. Data shown in reduced simulation units. (Reprinted with permission from M. Warren and J. Rottler, *Phys. Rev. E*, 76, 031802. Copyright 2007 by the American Physical Society.)

Colloidal glass formers are thought to be a representative model systems for molecular glasses on larger length scales. Although their aging phenomenologies are qualitatively similar, some important differences have been pointed out recently. For instance, equilibration times τ_{eq} were found to vary much less strongly than in molecular glasses, where they strongly depend on temperature [22]. Moreover, aging exponents $\mu > 1(1.5 - 2)$ have been reported in light scattering studies on collodial glasses [20,24], while time-aging time superposition of the creep response always leads to $\mu \leq 1$ in both kinds of glasses [25]. These discrepancies suggest that differences exist in the nature of the structural rearrangements at the particle level during aging.

11.2 MECHANICAL RESPONSE, MOLECULAR MOBILITY, RELAXATION TIME DISTRIBUTIONS, AND DYNAMICAL HETEROGENEITY

Molecular simulations have played an important role in elucidating relevant aspects of the dynamics of supercooled liquids and glasses [26,27]. Simulations provide access to full trajectories for each monomer, from which virtually any microscopic observable of interest can be computed. This flexibility comes at the price of limitations in accessible time scale, which are generally much shorter than in experiments. In standard molecular dynamics simulations, deformation rates are therefore orders of magnitude faster than in experiments. One can exploit the fact, however, that in glasses many polymeric degrees of freedom are frozen out both on experimental and computational time scales, so that simulations are in fact operating in a comparable regime. An alternative simulation strategy is to deform the glass in the limit of zero temperature and strain rate by evolving atomic positions via potential energy minimization. This protocol removes all rate effects, and finite temperature vibrational contributions to the free energy can be included via the quasiharmonic approximation [28]. For both protocols, the preparation of glassy samples remains a challenge, as the temperature quenches do occur much faster in simulations than on laboratory time scales.

A polymer chain can be modeled with empirical force fields at varying levels of resolution that range from an explicit representation of every atom to coarser models that represent groups of atoms by a single site. In the context of polymer mechanics, so-called united atom force fields that subsume hydrogen atoms into a single site at the position of the carbon atom have been employed extensively to study polyethylene [29], polystyrene, and polycarbonate [30]. If one coarse-grains even further and represents multiple repeat units on the carbon backbone with a single interaction site, one arrives at bead–spring models. A standard choice for interactions between segments separated by a distance r is the Lennard-Jones potential appropriate for van der Waals forces:

$$U_{LJ}(r) = 4\epsilon \left[\left(\frac{\sigma}{r} \right)^{12} - \left(\frac{\sigma}{r} \right)^6 \right] - U_{LJ}(r_c), \tag{11.6}$$

which is usually truncated at a distance r_c and vertically adjusted for continuity. The parameters ϵ and σ set an energy and length scale of the model. Chain connectivity is imposed either by stiff harmonic springs, $U_b(r) = k(r - r_0)^2$, or a nonlinear (finitely extensible) spring, $U_{FENE}(r) = -0.5kr_0^2 \ln[1 - (r/r_0)^2]$, that is imposed in addition to the Lennard-Jones potential. Although chemical specificity is lost at that level of modeling, these coarse-grained models are ideal tools to explore generic effects in macromolecules over the widest possible simulation time window. With regard to mechanical deformation, they have been shown to reproduce the typical shape of stress–strain curves of glassy polymers as shown in Figure 11.1a, including the dependence of the upper yield stress on temperature, strain rate, and loading conditions. Logarithmic waiting time dependence of σ_y, volume, and potential energy was also reproduced in several studies based on monomeric binary Lennard-Jones mixtures [9,31], in accordance with the interpretation that physical aging is primarily driven by van der Waals interactions. Simulations of polystyrene [32] and polyethylene [33] with more detailed united-atom force fields confirm the importance of interchain interactions for the increase of the yield stress that is observed upon slower cooling from the melt.

An important advantage of simulations is that bulk mechanical properties can be studied simultaneously with processes at the monomer level. In this way, it could be proven that in polymer glasses, molecular mobility controls macroscopic response. Creep compliance curves obtained from molecular dynamics simulations of a bead–spring polymer glass composed of 320 chains of N = 100 beads are shown for two values of loading below the yield stress and five different waiting times after a rapid quench from the melt in Figure 11.3a [34], mimicking the Struik protocol. They exhibit the expected stiffening with increasing amount of aging and obey time-aging time superposition according to Equations 11.3 and 11.4. Figure 11.3b and c show the decay of the ISF and the evolution of the MSD (computed directly from the trajectories) as the glass deforms. The aging parts of these functions also obey a scaling behavior as can be seen in the insets. The entire aging phenomenology of polymer glasses is thus reproduced by such simple coarse-grained polymers. The shift factors and corresponding aging exponents for macroscopic compliance, ISF, and MSD were found to scale in the same way with waiting time, thereby establishing a direct link between the monomer relaxation time and the mechanical response function.

A deeper understanding of the origin of the power law aging behavior of the mean relaxation time requires an analysis of the relaxation time distribution. Stretched exponential behavior of the response functions already signals the presence of more than a single relaxation time. The existence of a plateau in the MSD suggests that individual particles spend long times trapped in local cages, from which they can escape only infrequently. Glassy relaxation is therefore appropriately described by a hopping dynamics, in which particle trajectories can be decomposed into stationary periods interrupted by rapid rearrangements or hops to new positions. The distribution of hopping times can then be identified with the monomeric relaxation time distribution.

A measurement of the hopping time distribution thus requires identification of such hops from particle trajectories acquired through simulations or video microscopy. This can be done by identifying periods when particle displacements significantly exceed the rms vibrational amplitude [35], or by identifying periods when a given particle trajectory fluctuates strongly through various thresholding criteria [36,37]. Once a given trajectory is decomposed into discrete hop events, it is straightforward to extract the distribution of times elapsed between two consecutive hopping events. An investigation by Warren and Rottler showed that the distribution of these residence or persistence times follows a broad power law or *Pareto* law, $p_\tau(\tau) \sim \tau^{-1-x}$, with an exponent $x < 1$ [36]. This form of the distribution occurs in both monomeric binary mixtures as well as polymer glasses; specific polymeric effects manifest themselves only in the monomer displacement distributions $p_\Delta(\mathbf{r})$. The existence of nonstationary dynamics can now be understood from the fact that the mean relaxation time $\bar{\tau} = \int_0^\infty d\tau \tau p_\tau(\tau)$ diverges for $x < 1$. An initially counterintuitive observation is that the distributions $p_\tau(\tau)$ do not show any dependence on the waiting time t_w, a result reported for both fragile (Lennard-Jones based) and strong (silica) glass formers [38]. Explicit age dependence is, however, encoded in the distributions of times $p_1(\tau_1)$ between the beginning of the measurement and the first hop, which is distinct from all other times since the time to the preceeding hop is unknown. These times increase indeed with increasing t_w as the ensemble ages as a whole, and therefore reflect the history of the sample.

The emergence of power law distributions of trapping times can be understood in a phenomenological energy landscape picture, where thermally activated relaxation events out of traps of depth $E \geq 0$ occur with an Arrhenius rate $\tau_0^{-1} \exp(-E/k_B T)$. One can now make the simplifying assumption that the energy E is chosen anew after each jump, which corresponds to a mean-field picture where correlations in the energies are neglected. In a seminal paper, Monthus and Bouchaud [39] suggested that for glasses, an appropriate distribution of trap energies is the exponential, $\rho(E) = \exp(-E/k_B T_0)$ and $T < T_0$, where T_0 is the "glass transition temperature." Glassy (nonergodic) behavior can now be understood from the fact that the equilibrium probability to find the system in a trap of energy E, $P(E)_{eq} = \rho(E)\exp(E/k_B T)$, ceases to be normalizable for $T < T_0$. In this case, the system settles into deeper and deeper traps with increasing age, and the probability distribution of trapping times $\tau = \tau_0 \exp[E/k_B T]$ has the above-mentioned form with $x = T/T_0$. The value of the aging exponent itself, however, is always $\mu = 1$ in the trap model in its mean-field or "annealed disorder" version. Equilibration effects are also not included unless cutoffs in the distribution of trap energies are introduced.

The trap model as described above is an abstract concept that does not specify the degrees of freedom that are described by the energy landscape. If one takes as those the individual monomers of the aging glass, and determines in addition to the persistence time distribution $p_\tau(\tau)$ also the distribution of hop distances $p_\Delta(\mathbf{r})$, one can describe the space–time trajectories of the monomers with the help of *continuous time random walks* (CTRWs). The distribution of times $p_1(\tau_1)$ to the first hop serves as an initial condition. Subsequently, the position of each walker in space and time is advanced by drawing time increments and displacements according to the distributions $p_\tau(\tau)$ and $p_\Delta(\mathbf{r})$. An important property of CTRWs is the fact that each hop is a renewal event, that is, the internal clock of each walker is reset after a hop and $p_\tau(\tau)$ has no explicit waiting time dependence. The validity of this concept was recently confirmed by computer simulations [40]. Warren and Rottler showed that CTRWs parameterized with distributions from molecular simulations could reproduce the aging behavior of particle trajectories, in particular, the MSD, the full distribution of displacements after various times (van Hove functions) [36], and the ISF [41]. This success indicates that essential components of the activated dynamics of physical aging can be captured on a mean-field level.

Since the trap model and by extension the aging CTRW strictly predict a simple aging scenario ($\mu = 1$), one may ask how they can be consistent with the frequently reported apparent subaging behavior. One possibility is that the only true scaling behavior in molecular glasses is indeed $\mu = 1$, but measurements might yield a lower value if the dynamical aging regime is influenced by crossover effects from onset and end of aging. If measurements could be carried out over a wider range, simple aging would emerge. Simulation work suggests that this may indeed be the case for the class of Lennard-Jones glass formers [41]. Similarly, a trap model cannot explain apparent "superaging" regimes ($\mu > 1$) that have been reported in light scattering experiments on some soft glasses. More work is needed to conclusively identify the origin of this behavior.

Despite its success, the CTRW model does not provide a complete coarse-grained picture of glassy dynamics. Although correlations in time exist due to the broad tail of the waiting time distributions, spatial correlations are entirely absent as each particle hops independently from each other. In real supercooled liquids and glasses, however, the distributions of the squared displacements possess higher-order moments, commonly quantified by a nonvanishing *non-Gaussian parameter* $\alpha_2 = 3\langle\Delta\mathbf{r}^4\rangle/5\langle\Delta\mathbf{r}^2\rangle^2 - 1$ [42]. The value of α_2 is maximal when the MSD begins to depart from the caging plateau (see Figure 11.3) and the monomers begin to diffuse. Moreover, one finds that upon decrease of temperature, the dynamics becomes increasingly cooperative: Fast particles tend to form clusters of activity, while other parts of the material remain essentially immobile. The partition into slow and fast regions, however, is not static but evolves dynamically so that slow regions can become fast at a later time. This *dynamical heterogeneity* (DH) has become a central component of our current understanding of glassy matter [43,44].

Dynamical heterogeneity arises because the motion of one particle at position $\mathbf{r}$ during a time interval t affects the motion of another particle at position $\mathbf{r}'$. Signatures of DH do not show up in static correlation functions such as the structure factor or even two-point dynamical correlations such as the ISF or MSD. They can only be observed through the study of multipoint correlation functions that measure the degree of "overlap" between two configurations separated by a time interval t, thereby revealing correlations in the mobility. The spatial extent of DH—the size of a correlated region—can be estimated from fluctuations of the total mobility.

As an alternative to correlation functions, DH can be visualized directly at the single monomer level by recording the monomers that have experienced a hop in a given time interval. Increasing the time interval then includes more and more hops. Figure 11.4 shows the gradual buildup of a cluster of hopping monomers in a molecular dynamics simulation of an aging polymer glass. These images show only those monomers that are spatially close to other monomers already in the cluster. Multiple clusters grow simultaneously nearby (not shown) and eventually merge to form few very large clusters. After $t \gg t_{rel}$, all monomers have relaxed throughout the entire simulation.

A more quantitative characterization of DH can be obtained by using the number of monomers $N_{caged}(t,t_w)$ that remain caged (i.e., have no yet hopped) up to a time t as a measure of mobility. The top panel of Figure 11.5 shows that N_{caged} exhibits aging effects and decays to zero at later times with increasing age. The curves shown above represent an average over 300 simulation runs. If one computes the variance of N_{caged} over these runs,

$$\chi_4(t,t_w) = \frac{V}{k_B T N^2}\left(\langle N_{caged}^2\rangle - \langle N_{caged}\rangle^2\right),\tag{11.7}$$

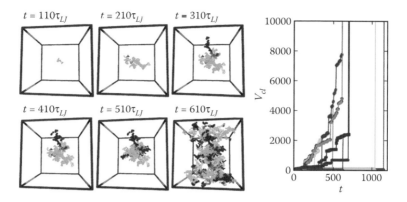

Figure 11.4 Snapshots of a growing cluster of hopping monomers in an aging polymer glass at constant temperature. Time labels indicate the observation interval and coloring indicates spatial depth. The simulation box size is approximately 20 monomer diameters. Panel on the right shows the growth of 15 example clusters as a function of time. (Reprinted with permission from A. Smessaert and J. Rottler, *Phys. Rev. E*, 88, 022314. Copyright 2013 by the American Physical Society.)

one obtains a dynamical susceptibility that can be interpreted as the volume of a correlated cluster. $\chi_4(t,t_w)$ exhibits rather rich behavior as can be seen in the bottom panel of Figure 11.5. In all cases, $\chi_4(t,t_w)$ reaches a maximum value at a time that increases with increasing age. Interestingly, this time corresponds very closely to the bulk relaxation time of the system. At this time, a dominating cluster has formed that spans essentially over the entire system. DH is therefore maximal at that time and drops to zero when all monomers have experienced a relaxation event. The growth of the peak value of $\chi_4(t,t_w)$ with waiting time t_w is consistent with both a slow logarithmic or weak power law growth [37,45]. Therefore, the older (or colder) the glass, the larger the size of the correlated regions.

Further characterization of DH can be performed via studying the size distributions of the clusters of hopping particles. These are shown in Figure 11.6 for multiple observation times t since the beginning of the measurement. As time progresses, the distributions become broader and develop a power law form with an

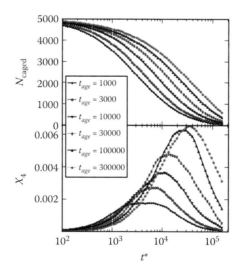

Figure 11.5 Top panel: Number of monomers N_{caged} that have not yet hopped as a function of time after having been aged for six different waiting times t_w. Bottom panel: Dynamical susceptibility (Equation 11.7) computed from the variance of N_{caged}. Values given in reduced simulation units. (Reprinted with permission from A. Smessaert and J. Rottler, *Phys. Rev. E*, 88, 022314. Copyright 2013 by the American Physical Society.)

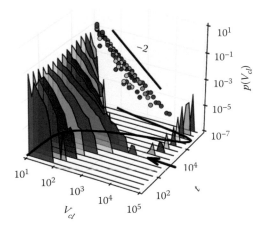

Figure 11.6 Volume distributions $p(V_{cl})$ of clusters of hopping monomers in an aging polymer glass measured at various times t. The back wall shows an overlay of distributions in a single plane, and the solid line indicates a power law with exponent -2. The black solid curve on the floor indicates the dynamical susceptibility $\chi_4(t)$ as a function of time. Values in reduced simulation units. (Reprinted with permission from A. Smessaert and J. Rottler, *Phys. Rev. E*, 88, 022314. Copyright 2013 by the American Physical Society.)

exponent close to -2. This fairly rapid decay of the cluster size distribution suggests that large-size fluctuations are rare enough so that the mean-field behavior of physical aging is not destroyed. At the time of the maximum of χ_4, only a few very large individual clusters remain. The distribution becomes bimodal and eventually contains only one system-spanning cluster. Very similar behavior can also be observed in simulations of the supercooled regime, where the polymer is still in equilibrium and the mean relaxation time does not change with waiting time [46].

The different heterogeneity parameters (non-Gaussianity, dynamic susceptibility [47], size of clusters) all illuminate closely related aspects of glassy dynamics. The times at which they reach their maximum values all exhibit similar dependence on temperature (above the glass transition) or waiting time (below the glass transition). Their mutual relationships and implications for the slowing down of the dynamics at the glass transition, however, is not yet fully understood [48].

11.3 ACTIVE DEFORMATION

We now turn our attention toward plastically deformed polymer glasses. Two questions have received particular attention in recent years: where do plastic events occur in the polymer glass, and how are the monomer dynamics changed by the application of stress or strain?

11.3.1 ELASTIC HETEROGENEITY AND LOCATION OF FAILURE

The mechanical response of solids is usually described by continuum mechanics of a homogeneous medium that can be characterized by a set of elastic moduli. It is now well accepted, however, that amorphous solids exhibit strong mechanical heterogeneity at the nanoscale. In molecular simulations of polymer glasses, local elastic moduli can be computed in small-volume elements containing several tens of monomers. The moduli contain a *Born* term from the change in potential energy due to particle displacements, a fluctuation term resulting from correlations of the local stress tensor, and a kinetic term [49]. While the first term describes affine deformation, the second term captures the effects of nonaffine strains that are particularly important in amorphous solids. A map of local shear moduli G in a bead–spring polymer glass computed from such simulations is shown in Figure 11.7 and reveals many soft regions embedded in a matrix of rigid regions. The distributions of the moduli was found to be well fitted by Gaussians with widths broadening with decreasing

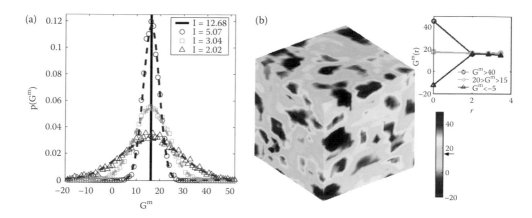

Figure 11.7 (a): Distributions of elastic shear moduli for three different domain sizes l. The solid line indicates the bulk value. (b): Map of local elastic shear moduli in a model polymer glass. The size of the volume elements is approximately $l = 2$ and contain about 8–10 monomers, while the size of the simulation box is approximately 12 monomer diameters. (Reprinted with permission from K. Yoshimoto et al., *Phys. Rev. Letters*, 93, 175501. Copyright 2004 by the American Physical Society.)

size of the volume elements [49]. Deviations from the bulk value become noticeable once the length scale drops to about five monomer diameters. In monomeric 2D glass formers, deviations from Hooke's law have been found on scales less than five molecular diameters [50]. It was also found that regions of low shear moduli do not correlate with local density or stress. It is important to note that some regions exhibit in fact negative values G. These are stabilized by nearby regions of positive stiffness, so that the bulk mean value is always positive as required for mechanical stability.

The emergence of structural heterogeneity has important consequences for the local plastic response of polymer glasses to deformation. Molecular simulations of a model polymer glass under tension show that irreversible monomer rearrangements correlate well with locations of small positive shear moduli [51]. In this study, the magnitude of the local irreversible or nonaffine residual displacement field was largest for the smallest values of the shear modulus. These regions arc also those where the mean-squared vibrational amplitude of the monomers, also called Debye–Waller factor, takes on very large values. They can therefore be considered particularly soft. Simulation studies of the athermal deformation of amorphous 2D binary mixtures give further evidence that the nonaffine displacement field resulting from the macroscopic shear deformations of a few percent strain is directly related to the spatial structure of the elastic moduli [50]. Again, local density was found not to correlate with the occurrence of plastic events. Local elastic moduli can therefore be invoked as a predictor of plastic activity.

A closely related description has been developed around the notion of soft vibrational modes. Amorphous solids generally exhibit an excess amount of low-energy modes (Boson peak), and a large number of them are quasilocalized, that is, they involve only very few particles. A superposition of the amplitudes of the lowest-frequency normal modes also yields a heterogeneous partitioning into hard and soft regions. Multiple simulation studies in supercooled liquids [52], aging polymer glasses [53], and athermal packings under shear [54] confirm that regions of large vibrational amplitude overlap with the loci of structural rearrangements. Moreover, the directions of particle displacements align preferentially with the polarization vectors of the soft modes [53]. This correlation is robust and insensitive to the specific model system or diagnostic of the rearrangement. Low-energy sound waves therefore scatter off flow defects and provide information about the particle displacements that are most easily excited. Both the elastic moduli approach and the soft mode approach are linear harmonic descriptions, but contain information about the nonlinear response of the material.

The correlation between small local shear moduli and shear deformation finds a counterpart in a relation between local bulk elastic moduli B and failure through cavitation [55,56]. Since the bulk modulus indicates the sensitivity of the material to volumetric deformation, it is not surprising to find the probability of cavitation enhanced near regions of low values of B. The distribution of these moduli is again well described by a Gaussian, and other quantities such as density of monomers or density of chain ends do not correlate with the appearance of cavities [55]. It is important to realize that while cavities nucleate systematically in zones of low B, it is not possible to predict precisely in which of these zones the cavity will nucleate.

11.3.2 ACCELERATED DYNAMICS

The notion that molecular dynamics must speed up in deforming materials is intuitive and has been the subject of frequent investigation. However, only recently has it become possible to directly measure molecular mobility at the segmental level in polymer glasses, and establish quantitative relationships to the deformation rate. Accelerated dynamics can be quantified with any observable that measures a structural relaxation process. One of the early molecular dynamics studies, for instance, reported an increase in the transition rate between dihedral angles in polyethylene under compressive deformation [29]. These transitions can be expected to have the lowest energy barriers for rearrangement. Similarly, enhanced monomer diffusion (inferred from mean-squared displacements) has been reported in atomistic simulation of polystyrene and polycarbonate [30].

In a series of pioneering experiments, Lee et al. developed an optical photobleaching technique to measure the local mobility in a poly(methyl methacrylate) glass during creep deformation [57]. These experiments measure the rate of reorientation of small dye molecules embedded in the polymer matrix. Motion of the surrounding polymers induces decorrelation and hence rapid decay in the corresponding correlation function $r(t)$ that can be fitted to a stretched exponential (Kohlrausch–Williams–Watts law, $r(t) = r_0 \exp(-(t/\tau)^\beta)$. The salient findings are a decrease of the mean relaxation time $\tau_c = \int dt\, r(t)/r_0$ up to three orders of magnitude accompanied by an increase of the stretching exponent β, which suggests a narrowing of the relaxation time spectrum under deformation. Moreover, a plot of strain rate versus inverse relaxation time τ_c reveals a near proportionality of flow rate and relaxation rate.

All of these observations have been successfully reproduced qualitatively in molecular dynamics simulations on the bead–spring level. Figure 11.8a, for instance, shows results from creep simulations of a polymer glass composed of short chains of length $N = 32$ monomers [58]. The authors investigate the decay of the ISF, Equation 11.5, as well as the decay of the bond autocorrelation function:

$$C_b(t) = \langle P_2(\hat{b}(t) \cdot \hat{b}(0)) \rangle, \tag{11.8}$$

where $P_2(x) = (3x^2 - 1)/2$ is the second Legendre polynomial and $\hat{b}(t)$ is a unit vector aligned along the bonds connecting monomers along the polymer chain. Again both functions exhibit stretched exponential decay, and bulk relaxation times extracted from fits and normalized by the times from undeformed samples are shown in Figure 11.8b of Figure 11.8 as a function of strain rate $\dot{\epsilon}$. Results from three different correlation functions roughly collapse onto the same curve, indicating that the relaxation times are all similarly affected by the deformation. For larger rates, one observes a regime where $\tau \sim |\dot{\epsilon}|^{-1}$ over roughly two decades. In this regime, the mean relaxation time is therefore reduced by up to two orders of magnitude. This acceleration was reported for both tensile and compressive deformations, suggesting that local free volume does not play a role [59]. One possible interpretation of this behavior is in terms of an energy landscape picture: The faster the glass flows, the higher its position on the energy landscape where barriers for rearrangement are lower [60]. This intuition was confirmed by Riggleman et al. [58] by computing the so-called inherent structure energies, which are found by minimizing the potential energy of a configuration. Accordingly, the smallest relaxation times are found when the inherent structure energy is highest.

A more detailed understanding of the phenomenon of increased mobility due to deformation can be obtained by studying the changes in the relaxation time distributions. These can be obtained by applying the hop analysis techniques described above to glasses that are subject to different deformation protocols. Molecular dynamics simulations were again performed with a bead–spring model consisting of $N = 10$ beads. Figure 11.9 provides an overview of the modifications that arise during creep (a,d), constant strain rate (b,e),

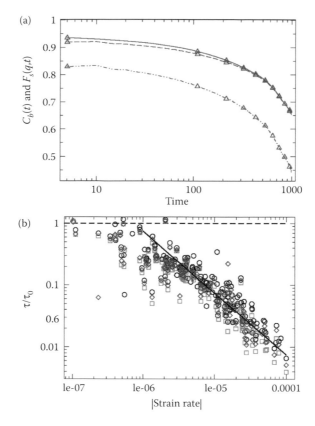

Figure 11.8 Measurement of segmental relaxation dynamics during creep of a polymer glass. (a) Shows the decay of bond autocorrelation $C_b(t)$ (solid line) and the ISF $F_s(q,t)$ for two wavevectors $q = 3.74$ (dashed line) and $q = 7.14$ (dash-dotted line). (b) Shows the characteristic relaxation times from stretched exponential fits versus strain rate to the three correlation functions. Solid line has slope −1. (Reprinted with permission from R. A. Riggleman, K. S. Schweizer, and J. J. de Pablo, *Macromolecules*, 41, 4969. Copyright 2008, American Chemical Society.)

and a step strain (c,f) deformation. In all cases, three different ages are studied and two different deformation amplitudes (dashed and dotted lines) are compared to an undeformed glass (solid lines), while the insets illustrate the mechanical response.

Focusing first on the times t_1 required to undergo a first rearrangement during creep deformation (Figure 11.9a), one observes a much more rapid decay of these distributions with increasing stress amplitude. The likelihood of finding short hop times is much enhanced while finding very large times is decreased, indicating accelerated dynamics and a narrowing of the spectrum as observed in experiments [57]. The distribution of persistence times (times between subsequent hops) shown in Figure 11.9d features the broad power law decay that induces physical aging. However, applying stress modifies the tail of the distributions such that it decays more rapidly for large times τ. This modification implies a truncation of physical aging as the likelihood of seeing very large relaxation time is decreased due to flow.

When the glass is instead deformed at constant strain rate $\dot{\epsilon}$, the primary effect on both hop time distributions is instead a truncation of the power law tails at times of order $\sim \epsilon_y/\dot{\epsilon}$ (Figure 11.9b and e), where $\epsilon_y \sim 0.1$ is the yield strain. In this deformation mode, the external drive imposes an upper bound on the longest possible relaxation times. As a result, mean relaxation times decrease and physical aging ceases after the cutoff time scale is reached, consistent with the observation that aging effects in the stress–strain curves disappear beyond ϵ_y (see Figure 11.1a). After a step strain perturbation, one observes again an increase in the probability of finding short first hop times t_1, but no futher changes to the persistence times τ.

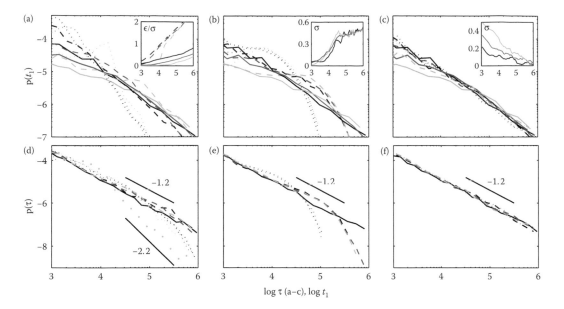

Figure 11.9 Distributions of first hop times t_1 (a)–(c) and persistence times τ (d)–(f) in an aging polymer glass subject to deformation after undergoing physical aging for three different waiting times $t_w = 750$ (black), 7500 (gray), and 75,000 (light gray). (a) and (d) correspond to a step stress with amplitude $\sigma = 0.4$ (dashed) and 0.5 (dotted). Solid dots show $p(\tau)$ for particles whose cages were created at $t < \tau_\alpha$ and $t_w = 75,000$. (b) and (e) refer to deformation at imposed strain rate $\dot{\epsilon} = 8.9 \times 10^{-7}$ (dashed) and 8.9×10^{-6} (dotted), while (c) and (f) show the effect of a step strain with amplitude $\epsilon = 0.02$ (dashed) and 0.04 (dotted). For comparison, the corresponding distributions in an undeformed glass that is only undergoing physical aging are also shown in every panel (solid lines). Straight lines indicate power law with the given slopes. All values are given in reduced simulation units. (Reprinted with permission from M. Warren and J. Rottler, *Phys. Rev. Lett.*, 104, 205501. Copyright 2010 by the American Physical Society.)

In order to develop a unified picture of accelerated dynamics, it is useful to describe the changes to the first hop time distributions through their cumulative distributions, $P_1(t,t_w) = \int_0^t p_1(t_l')dt_l'$. These give the fraction of particles that have already hopped after time t, and the function $1 - P_1(t,t_w)$ is qualitatively similar to the ISF and $N_{\text{caged}}(t,t_w)$. One can then quantify the degree of acceleration by comparing the times at which undeformed and deformed cumulative distributions reach the same value, that is, $P_{1,u}(t_u,t_w) = P_{1,d}(t_d,t_w)$. An "acceleration ratio" defined as t_u/t_d was found to increase by up to two orders of magnitude, which agrees quantitatively with the reduction in the mean relaxation time obtained from the decay of the ISF in Figure 11.8b. Remarkably, values of t_u/t_d for different deformation protocols, amplitudes, and waiting times appear to collapse onto a master curve when time is parameterized by the total strain $\epsilon(t)$ [61], while alternatives using other deformation variable were unsuccessful. This result might suggest that it is the total accumulated deformation rather than the instantaneous stress that is the quantity governing accelerated dynamics.

11.4 REJUVENATION EFFECTS

One of the most fascinating and unique aspects of glasses is the interaction between their intrinsic slow aging dynamics and external perturbations. The question whether and how precisely mechanical deformation alters the state and history of a glassy is of both fundamental and practical importance. Several different protocols have been devised to study memory effects in glasses, and each illuminates a different facet. Perhaps the earliest notion of "rejuvenation" in the sense of an erasure of aging dates back to Struik's experiments on polymers [1,62], in which he observed a reduction in the aging time shift factor in compliance measurements, and hence the aging exponent μ with increasing applied stress. As the applied stress approaches the yield stress from below, the material appears to age less quickly and even stops aging once μ = 0. This effect has since been confirmed in rheological experiments on pastes using the same protocol [17] and molecular dynamics simulations of creep in polymer glasses [34]. Similarly, plastic deformation can reverse the aging of the yield stress σ_y and reduce the subsequent yield drop, so that the material resembles a younger glass [8,10,11,63]. Both atomistic and coarse-grained simulations unequivocally reproduce the disappearance of aging effects in the post-yield, large-strain response [31–33].

An important question arising from these observations is then whether the time to reach equilibrium has also been changed. Results by McKenna from an experiment on an epoxy glass suggest that this is not the case [21]: While a reduction of the aging exponent is clearly seen, the end of the power law aging regime (see Figure 11.2b) is independent of applied stress. Moreover, volume recovery curves quickly returned to the volume of undeformed samples. These results point to a merely transient effect of the mechanical perturbation, and the "material clock" has not been altered. Direct molecular dynamics simulations of the equilibration times in aging polymers under load have not yet been carried out as reaching such ultralong time scales is very difficult. Simulations were able to confirm, however, that for perturbation with subyield stresses, the potential energy per particle returns to the aging trajectory of an unperturbed glass over a time scale of the structural relaxation time [64].

An alternative protocol to investigate erasure of memory is to subject the material to creep deformation, and then study the recovery regime after the stress is removed. Using optical probe techniques described previously, Lee and Ediger designed such an experiment and measured the molecular mobility in poly(methyl methacrylate) glasses after deformation with different stress amplitudes [65]. These experiments clearly establish two qualitatively different situations that depend on the degree of deformation: While accelerated dynamics but no erasure of aging is observed in the preflow regime, plastic deformation in the flow regime is required to take the sample into a state independent of the predeformation history. The flow regime is further characterized by increasingly homogeneous dynamics as evidenced by an increasing exponent β in stretched exponential fits, an effect reproduced in simulations using constant strain rate deformation [66]. Consistent with this view, simulations of glassy polymers under cyclic deformation reported maximal dynamical correlations for strain amplitudes of order the yield strain, while heterogeneity decreased for larger deformations [67].

Molecular dynamics simulations are again able to reproduce all these observations qualitatively and help develop a general scenario [68]. Figure 11.10 shows results for the structural relaxation time τ_α (here determined from the decay of the ISF) in the recovery regime following a deformation protocol closely modeled after that of Reference 65. Values of τ_α group according to the amount of strain ϵ that the sample experienced during the creep deformation. For unstressed glasses, the relaxation time is constant until the recovery time t_r (measured since the stress was released) reaches the total age of the system t_a, upon which aging sets in (solid line). Creep deformation lowers the relaxation time by up to two orders of magnitude in agreement with previous studies. The subsequent behavior can be classified according to idealized scenarios shown in the right panel. Curves (a) and (b) correspond to situations where the glass still remembers its original age, since it only merges with the original aging trajectory at a time where the power law regime is resumed. Since the amount of strain experienced by the polymer during creep is less than the yield strain of order 5%, this behavior corresponds to the pre-yield regime, where experiments report enhanced dynamics but no altering of the extent of memory [65]. In an energy landscape interpretation, this regime is dominated by the reduction of barriers due to deformation [60]. By contrast, curve (c) corresponds to a true erasure scenario, since the aging trajectory is reached at times sooner than the original age. The simulation results establish a gradual crossover from transient to permanent loss of memory, which appears to be controlled by the amount of strain. As in experiments, full erasure only occurs for strains larger than the yield strain. Remarkably, the quantity controlling the degree of acceleration and rejuvenation is the total strain. If one considers the relaxation time as the relevant state variable, then these results give strong support to the notion that plastic flow is required for full rejuvenation, while the effect of subyield stresses may be transient in nature.

Results from atomistic simulations of atactic polystyrene films subjected to oscillatory shear provide a similar picture of the rejuvenation effect [69]. Instead of the decorrelation of the bond vectors, the authors investigated segmental relaxation dynamics via the vectors connecting the phenyl side group with the backbone. Relaxation times extracted from stretched exponential fits to Equation 11.8 show power law aging in

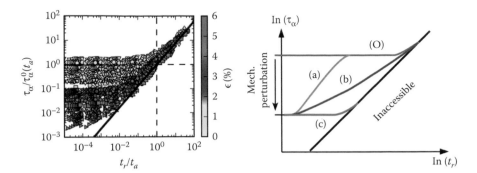

Figure 11.10 Left: Molecular dynamics simulation of a polymer glass in the recovery regime. Relaxation times τ_α are plotted as a function of recovery time t_r, which is normalized by the total age t_a of the sample. The color bar indicates the amount of strain the sample experienced during the deformation period. Right: Sketch of possible rejuvenation paths. Line (○) is the path of an unperturbed glass and lines (a)–(c) are possible paths of glasses after mechanically perturbation. The diagonal black line in both figures indicates the generic aging dynamics. (Reprinted with permission from A. Smessaert and J. Rottler, *Macromolecules*, 45, 2928. Copyright 2012, American Chemical Society.)

the unsheared glass. Immediately following cyclic shearing, the relaxation times are reduced but rise again in the recovery regime.

A third perspective on mechanical rejuvenation can be established using potential energy landscape concepts. The relevant observable here is the mean potential minimum or inherent structure energy E_{is}, which is computed easily from atomistic configuration using standard energy minimization techniques. In general, E_{is} decreases with temperature, decreasing cooling rate and increasing age (once in the glassy state), as the sample can reach deeper energy minima in each case. Monte Carlo simulations of binary Lennard-Jones mixtures established that shear deformation at constant strain rate can lift up values of E_{is} from their history-dependent glassy values to a higher history-independent value corresponding to a liquid structure [70], suggesting a rejuvenation effect if E_{is} is taken as the relevant state variable.

A particularly useful protocol consists of deformation cycles that return the material to a nominally undeformed state and monitor the passage of the system through the energy landscape. Lyulin and Michels considered such extension–compression loops in atomistic molecular dynamics simulations of polystyrene and polycarbonate [71]. Results from these calculations are shown in Figure 11.11 and reveal important differences due to the quench protocol. For well-annealed glasses (Figure 11.11c and d), both polymers exhibit the rejuvenation effect as a strain cycle with amplitude of approximately 25% repositions the sample at a *higher* inherent structure energy than before deformation. By contrast, when the same deformation loop is performed on more rapidly quenched glasses, the samples reach a *lower* inherent structure energy. These results closely mirror those of Lacks and Osborne, who studied cyclic deformation in monomeric Lennard-Jones glasses [72]. The quenched glasses start out much higher on the potential energy landscape than the annealed glasses, since the quench times are much shorter than thermal relaxation time scales. Mechanical stimulus can therefore speed up accessing lower energy configurations, which has motivated the term *overaging* [72]. Reference 71 suggests that rejuvenation (overaging) occurs when the cooling time is much longer (shorter) than the time it takes to strain to yield.

It is important to ask how closely the rejuvenated states really resemble those obtained through a simple quench followed by physical aging, even if they exhibit comparable relaxation times. The simulations of Reference 68 that were based on the bead–spring model investigated several different structural observables such as coordination number, the shape of Voronoi volumes, as well as inherent structure energy in the recovery regime. While these quantities show fairly good agreement with undeformed polymer glasses for strains up to the yield strain, differences become apparent for larger amounts plastic deformation. In atomistic simulations with more detailed force fields, it is possible to partition the total energy into contributions from bond stretching, bending, torsion, and van der Waals (Lennard-Jones) interactions. Significant differences were reported in the case of polystyrene between thermally cooled and mechanically perturbed glasses

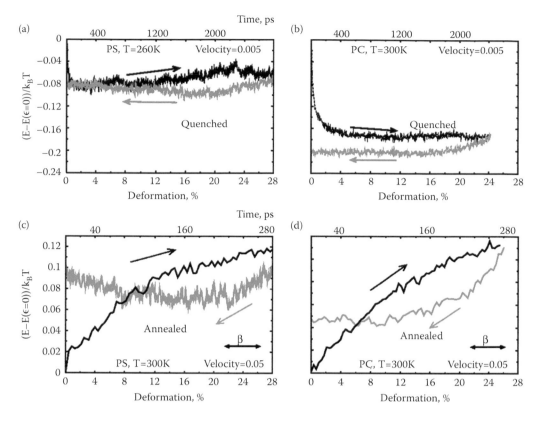

Figure 11.11 Inherent structure energy changes relative to an undeformed state during closed deformation–recompression loops for quenched (a),(b) and for annealed (c),(d) glassy polystyrene and polycarbonate. (Reprinted with permission from A. V. Lyulin and M. A. J. Michels, *Phys. Rev. Lett.*, 99, 085504. Copyright 2007 by the American Physical Society.)

[71], suggesting that these protocols place the polymer glasses into different parts of configuration space. The energy landscape studies on monomeric glasses further support the notion that plastically rejuvenated glasses resemble as-quenched glasses, but visit different regions of the energy landscape [72]. Consistent with this view, different distributions of the torsional angles in polystyrene are obtained in molecular dynamics simulations that compare shear deformation with temperature changes [73]. An emerging conclusion from this work is that mechanical rejuvenation does occur in terms of average relaxation times, but not in a literal sense of creating a material that is statistically indistiguishable from younger states.

11.5 OPEN QUESTIONS

This chapter reviewed recent advances in understanding the molecular mechanisms of physical aging and plastic deformation in polymer glasses with the help of molecular simulations. Much progress has been made in characterizing nanoscale elasticity, dynamical heterogeneity, accelerated dynamics, and the interplay between structural relaxation and mechanical perturbation. There are also outstanding challenges:

- What kind of structural changes occur during physical aging?
  The primary signature of physical aging is the power law increase of the relaxation time with time elapsed in the glassy state. What kind of structural changes occur during the aging process, and how can they be quantified? Simulations show robustly that the mean inherent structure energy

decreases logarithmically with the waiting time. However, standard structural order parameters such as radial distribution functions or static structure factors exhibit only extremely weak changes. Since the aging process locally optimizes the molecular packing, is it reasonable to look for an increased degree of local, short-range order. An investigation of structural order via the bond orientational parameter Q_6 that projects the bond angles of the nearest neighbors onto spherical harmonics and is maximized by a perfect fcc lattice exhibited no change in aging bead–spring polymer glasses [34]. However, a *surface order parameter* sensitive to quasicrystalline order was found to increase logarithmically with age [34]. Clearly, the highly constrained nature of polymer glasses efficiently prevents the formation of high-symmetry structural motifs. More detailed atomistic simulation of polyethylene, where the potential energy also depends on bond bending and torsion, reported a higher fraction of dihedrals in the trans-state upon slower cooling [33]. It would be important to identify those structural order parameters that are particularly relevant for characterizing the state of aging in polymeric systems.

A much stronger evolution toward crystalline order has been found in simulations of colloidal hard-sphere glasses that age after a volume quench [74]. Here, correlation functions based on the bond orientational parameter Q_6 exhibit distinct aging behavior and permit the extraction of a slowly growing *static* correlation length with inceasing age. Moreover, the relaxation time was found to increase exponentially with the static correlation length, suggesting a deeper connection between static structural order and aging dynamics. Finding an equivalent growing static correlation length in polymer glasses would be very interesting.

- Is there a structural origin of dynamic heterogeneity?
Given the existence of nanoscale heterogeneity of the elastic moduli and related quantities in amorphous solids, it is tempting to suspect a link to dynamical heterogeneity. The existence of such a correlation has been proved in computer simulations of monomeric glass formers with an elegant method known as the *isoconfigurational ensemble* [75]. Instead of averaging over many initial spatial configurations, one considers here always the same initial atomic positions, but averages instead over many different initial momenta. By following the mean-squared displacements of each particle up to the bulk relaxation time, one recovers dynamical heterogeneity as a decomposition into fast- and slow-moving particles or spatially varying *propensity* for motion. Since the momentum averaging restores isotropy, this heterogeneity must be encoded in the initial configuration. This result, however, does not answer which structural property is the best predictor of the propensity. Chaudury et al. showed that in supercooled liquids the memory of active sites lasts up to τ_{rel}, but correlates only weakly with the presence of empty sites [76].

Other obvious choices for local structural order parameters such as the instantaneous potential energy per particle or the coordination number around specific types of particles also correlate only weakly with clusters of fast-moving particles. However, rather good overlap with DH was reported for these quantities after they were averaged in the isoconfigurational ensemble [77,78]. Unfortunately, such simulations are extremely time consuming. Establishing robust correlations between locally preferred structures and particle mobility continues to be a focus of current research. A recent comparison across several monomeric supercooled liquids revealed that measures of local energy and structure correlate differently in different models [79]. A generalized structural metric across amorphous materials does not yet exist.

In the context of polymer glasses, the concept of soft vibrational modes [52] has been shown to capture a statistically significant fraction of the local mobility [53]. A comparison between maps of soft regions and clusters of particles that have rearranged since the maps were constructed does reveal correlations that peak at the time when DH is maximal [53]. These correlations are not perfect but can be further improved by applying isoconfigurational averaging. This result shows that substantial information about the ensuing dynamics can indeed be extracted from the structure alone. Exploring these correlations in different modes of deformation could yield important clues for constructing molecular-level theories of polymer deformation.

REFERENCES

1. L. C. E. Struik, *Physical Aging in Amorphous Polymers and Other Materials* (Elsevier Scientific Publishing Company, Amsterdam, the Netherlands), 1978.

2. I. M. Hodge, *Science*, 267, 1945, 1995.

3. J. M. Hutchinson, *Progr. Polym. Sci.*, 20, 703, 1995.

4. S. L. Simon, in *Encyclopedia of Polymer Science and Technology* (John Wiley & Sons, Inc., Hoboken, NJ), 2002.

5. H. A. Visser, T. C. Bor, M. Wolters, L. L. Warnet, and L. E. Govaert, Plastics, *Rubber Composites*, 40, 201, 2011.

6. J. E. Pye and C. B. Roth, *Macromolecules*, 23, 9455, 2013.

7. C. G.'Sell and G. B. McKenna, *Polymer*, 33, 2103, 1992.

8. H. G. H. van Melick, L. E. Govaert, B. Raas, W. J. Nauta, and H. E. H. Meijer, *Polymer*, 44, 1171, 2003.

9. J. Rottler and M. O. Robbins, *Phys. Rev. Lett.*, 95, 225504, 2005.

10. E. T. J. Klompen, T. A. P. Engels, L. E. Govaert, and H. E. H. Meijer, *Macromolecules*, 38, 6997, 2005.

11. D. J. A. Senden, J. A. W. van Dommelen, and L. E. Govaert, *J. Polym. Sci., Part B: Polym. Phys.*, 50, 1589, 2012.

12. V. Viasnoff and F. Lequeux, *Phys. Rev. Lett.*, 89, 065701, 2002.

13. B. Abou, D. Bonn, and J. Meunier, *Phys. Rev. E*, 64, 021510, 2001.

14. A. Knaebel, M. Bellour, J.-P. Munch, V. Viasnoff, F. Lequeux, and J. L. Harden, *EPL (Europhys. Lett.)*, 52, 73, 2000.

15. M. Bellour, A. Knaebel, J. L. Harden, F. Lequeux, and J.-P. Munch, *Phys. Rev. E*, 67, 031405, 2003.

16. S. Jabbari-Farouji, E. Eiser, G. H. Wegdam, and D. Bonn, *J. Phys. Condens. Matter*, 16, L471, 2004.

17. M. Cloitre, R. Borrega, and L. Leibler, *Phys. Rev. Lett.*, 85, 4819, 2000.

18. G. B. McKenna, T. Narita, and F. Lequeux, *J. Rheol. (1978–Present)*, 53, 489, 2009.

19. M. Siebenbürger, M. Ballauff, and T. Voigtmann, *Phys. Rev. Lett.*, 108, 255701, 2012.

20. R. Bandyopadhyay, D. Liang, J. L. Harden, and R. L. Leheny, *Solid State Commun. Soft Condens. Matter Soft Condens. Matter*, 139, 589, 2006.

21. V. A. Martinez, G. Bryant, and W. van Megen, *Phys. Rev. Lett.*, 101, 135702, 2008.

22. X. Di, K. Z. Win, G. B. McKenna, T. Narita, F. Lequeux, S. R. Pullela, and Z. Cheng, *Phys. Rev. Lett.*, 106, 095701, 2011.

23. G. B. McKenna, *J. Phys. Condens. Matter*, 15, S737, 2003.

24. X. Di, X. Peng, and G. B. McKenna, *J. Chem. Phys.*, 140, 054903, 2014.

25. X. Peng and G. B. McKenna, *Phys. Rev. E*, 90, 050301, 2014.

26. J. Rottler, *J. Phys. Condens. Matter*, 21, 463101, 2009.

27. J.-L. Barrat, J. Baschnagel, and A. Lyulin, *Soft Matter*, 6, 3430, 2010.

28. N. Lempesis, G. G. Vogiatzis, G. C. Boulougouris, L. C. A. v. Breemen, M. Hütter, and D. N. Theodorou, *Mol. Phys.*, 111, 3430, 2013.

29. F. M. Capaldi, M. C. Boyce, and G. C. Rutledge, *Phys. Rev. Lett.*, 89, 175505, 2002.

30. A. V. Lyulin, B. Vorselaars, M. A. Mazo, N. K. Balabaev, and M. A. J. Michels, *EPL (Europhys. Lett.)*, 71, 618, 2005.

31. F. Varnik, L. Bocquet, and J.-L. Barrat, *J. Chem. Phys.*, 120, 2788, 2004.

32. B. Vorselaars, A. V. Lyulin, and M. a. J. Michels, *J. Chem. Phys.*, 130, 074905, 2009.

33. D. K. Mahajan, R. Estevez, and S. Basu, *J. Mech. Phys. Solids*, 58, 1474, 2010.

34. M. Warren and J. Rottler, *Phys. Rev. E*, 76, 031802, 2007.

35. K. Vollmayr-Lee, *J. Chem. Phys.*, 121, 4781, 2004.

36. M. Warren and J. Rottler, *EPL (Europhys. Lett.)*, 88, 58005, 2009.

37. A. Smessaert and J. Rottler, *Phys. Rev. E*, 88, 022314, 2013.

38. K. Vollmayr-Lee, R. Bjorquist, and L. M. Chambers, *Phys. Rev. Lett.*, 110, 017801, 2013.

39. C. Monthus and J.-P. Bouchaud, *J. Phys. A Math. Gen.*, 29, 3847, 1996.

40. J. Helfferich, *Eur. Phys. J. E*, 37, 1, 2014.

41. M. Warren and J. Rottler, *Phys. Rev. Lett.*, 110, 025501, 2013.

42. W. Kob, C. Donati, S. J. Plimpton, P. H. Poole, and S. C. Glotzer, *Phys. Rev. Lett.*, 79, 2827, 1997.

43. M. D. Ediger, *Annu. Rev. Phys. Chem.*, 51, 99, 2000.

44. L. Berthier, G. Biroli, J.-P. Bouchaud, L. Cipelletti, and W. v. Saarloos, *Dynamical Heterogeneities in Glasses, Colloids, and Granular Media* (Oxford University Press, Oxford, UK), 2011.

45. A. Parsaeian and H. E. Castillo, *Phys. Rev. E*, 78, 060105, 2008.

46. F. W. Starr, J. F. Douglas, and S. Sastry, *J. Chem. Phys.*, 138, 12A541, 2013.

47. E. Flenner and G. Szamel, *Phys. Rev. Lett.*, 21, 217801, 2010.

48. X. Di and G. B. McKenna, *J. Chem. Phys.*, 138, 12A530, 2013.

49. K. Yoshimoto, T. S. Jain, K. V. Workum, P. F. Nealey, and J. J. de Pablo, *Phys. Rev. Lett.*, 93, 175501, 2004.

50. M. Tsamados, A. Tanguy, C. Goldenberg, and J.-L. Barrat, *Phys. Rev. E*, 80, 026112, 2009.

51. G. J. Papakonstantopoulos, R. A. Riggleman, J.-L. Barrat, and J. J. de Pablo, *Phys. Rev. E*, 77, 041502, 2008.

52. A. Widmer-Cooper, H. Perry, P. Harrowell, and D. R. Reichman, *Nat. Phys.*, 4, 711, 2008.

53. A. Smessaert and J. Rottler, *Soft Matter*, 10, 8533, 2014.

54. M. L. Manning and A. J. Liu, *Phys. Rev. Lett.*, 107, 108302, 2011.

55. A. Makke, M. Perez, J. Rottler, O. Lame, and J. Barrat, *Macromol. Theory Simul.*, 20, 826, 2011.

56. G. N. Toepperwein and J. J. de Pablo, *Macromolecules*, 44, 5498, 2011.

57. H.-N. Lee, K. Paeng, S. F. Swallen, and M. D. Ediger, *Science*, 323, 231, 2009.

58. R. A. Riggleman, K. S. Schweizer, and J. J. de Pablo, *Macromolecules*, 41, 4969, 2008.

59. R. A. Riggleman, H.-N. Lee, M. D. Ediger, and J. J. de Pablo, *Phys. Rev. Lett.*, 99, 215501, 2007.

60. Y. G. Chung and D. J. Lacks, *Macromolecules*, 45, 4416, 2012.

61. M. Warren and J. Rottler, *Phys. Rev. Lett.*, 104, 205501, 2010.

62. L. C. E. Struik, *Polymer*, 38, 4053, 1997.

63. L. E. Govaert, H. G. H. van Melick, and H. E. H. Meijer, *Polymer*, 42, 1271, 2001.

64. M. Warren and J. Rottler, *Phys. Rev. E*, 78, 041502, 2008.

65. H.-N. Lee and M. D. Ediger, *J. Chem. Phys.*, 133, 014901, 2010.

66. R. A. Riggleman, H.-N. Lee, M. D. Ediger, and J. J. de Pablo, *Soft Matter*, 6, 287, 2010.

67. N. V. Priezjev, *Phys. Rev. E*, 89, 012601, 2014.

68. A. Smessaert and J. Rottler, *Macromolecules* 45, 2928, 2012.

69. D. Hudzinskyy, M. A. J. Michels, and A. V. Lyulin, *Macromol. Theory Simul.*, 22, 71, 2013.

70. M. Utz, P. G. Debenedetti, and F. H. Stillinger, *Phys. Rev. Lett.*, 84, 1471, 2000.

71. A. V. Lyulin and M. A. J. Michels, *Phys. Rev. Lett.*, 99, 085504, 2007.

72. D. J. Lacks and M. J. Osborne, *Phys. Rev. Lett.*, 93, 255501, 2004.

73. Y. G. Chung and D. J. Lacks, *J. Chem. Phys.*, 136, 124907, 2012.

74. T. Kawasaki and H. Tanaka, *Phys. Rev. E*, 89, 062315, 2014.

75. A. Widmer-Cooper, P. Harrowell, and H. Fynewever, *Phys. Rev. Lett.*, 93, 135701, 2004.

76. P. Chaudhuri, S. Sastry, and W. Kob, *Phys. Rev. Lett.*, 101, 190601, 2008.

77. G. S. Matharoo, M. S. G. Razul, and P. H. Poole, *Phys. Rev. E*, 74, 050502, 2006.

78. M. S. G. Razul, G. S. Matharoo, and P. H. Poole, *J. Phys. Condens. Matter*, 23, 235103, 2011.

79. G. M. Hocky, D. Coslovich, A. Ikeda, and D. R. Reichman, *Phys. Rev. Lett.*, 113, 157801, 2014.

Experiments-inspired molecular modeling of yielding and failure of polymer glasses under large deformation

SHI-QING WANG AND SHIWANG CHENG

12.1 INTRODUCTION

There are many aspects regarding the physics of polymer glasses as summarized previously [1]. For the subject concerning mechanics of polymer glasses, different treatments have been put forward [2–5]. In the Introduction, besides a brief overview of what this chapter will focus on, which deals with yielding, crazing, and brittle-to-ductile transition (BDT), we use much of the space to discuss the background from new (molecular-level) perspectives and relationships between different methodologies, and to draw similarities as well as differences at a molecular level between mechanics of polymer glasses and melt rheology. For clarity, subtitles are used to highlight the key elements under discussion. Throughout the chapter, we adopt the widely used nomenclature in polymer science and engineering. For example, yielding denotes a ductile response in the sense that the load peaks before failure (i.e., macroscopic separation). Strain localization, for example, necking initiated by shear yielding, is not regarded here as failure, although crazing could be viewed as localized failure. Consistent with the description of Ward [5], being ductile means that the system is capable of yielding and appreciable drawing during post-yield uniaxial extension. Here, post-yield refers to the regime following the stress maximum, which is commonly called the yield point. Brittle failure takes place when the stress response to continuing deformation is monotonic up to the point of macroscopic fracture. These definitions are graphically expressed in (a) to (c) of Figure 12.13.

12.1.1 IMBALANCE OF RESEARCH ACTIVITIES ABOVE AND BELOW T_g

It is very challenging to establish a molecular-level understanding of leading characteristics (for example, yielding and failure) concerning mechanical responses of polymer glasses to large external deformation. Polymer giants de Gennes and Edwards did not invest their efforts on the subject although it closely resembles the subject of melt rheology where they made breakthroughs and lasting contributions. Obviously, nonequilibrium and nonergodic polymer glasses are not an ideal system to exercise statistical mechanics. The powerful scaling apparatus to treat polymer melt dynamics is not so useful here. Consequently, polymer physics has undergone rather asymmetric development. The past four decades have witnessed a spurt of energy and intensity in molecular-model-based research of polymers above the glass transition temperature [6–10]. For example, the subject of polymer melt rheology has consumed a great deal of efforts from generations of polymer scientists and engineers. In contrast, much less efforts have been spent to develop molecular-level knowledge about the mechanical behavior of polymer glasses. Thus, pioneering efforts from Kramer and coworker [11–15], early studies based on molecular dynamics (MD) simulations from several groups [16–51], and theoretical attempts of Schweizer and coworkers [52–55] are particularly notable.

In comparison with melt rheology that deals with mechanical behavior above the glass transition temperature T_g, mechanics of polymer glasses has extra degrees of complexity [1,4]: intersegmental van der Waals interactions cannot be simply modeled as friction as in melt rheology where effects of intermolecular uncrossability allow us to draw analogy with the successful theoretical treatment of rubber elasticity [7]. In the glassy state, intersegmental interactions immobilize individual segments and account for much of the plastic dissipation and appreciable elastic energy storage in large external deformation [17,56–58]. Intersegmental interactions can be significantly altered upon thermal and mechanical treatments that have no parallel in melt rheology. For example, physical aging allows a polymer glass to evolve toward the equilibrium state [59–62]. During the annealing, individual segments move around and settle down closer to their equilibrium rates, that is, moving along the energy landscape to deeper "valleys" [61,62]. Segmental dynamics slow down upon aging. On the other hand, mechanical "rejuvenation" has an "opposite" effect to aging [34,35,38,63]. Ductile (post-yield) deformation below T_g rearranges packing at the segmental level, bringing some segments over the potential barrier while driving others away from local minima of potential wells [34,35]. This is rather different from other (for example, thermal) means to rejuvenate a polymer glass. The system finds itself situated in a higher energy state after such mechanical rejuvenation [35,38]. Both physical aging and mechanical "rejuvenation" are the additional factors to consider in the description of mechanical behavior of polymer glasses. Finally, the intermolecular uncrossability that is responsible for entanglement above T_g also operates to retain molecular networking via covalent bonds in the glassy state.

12.1.2 CHEN–SCHWEIZER THEORY FOR DEFORMATION AND YIELDING OF POLYMER GLASSES

What are the structural and dynamic parameters that are experimentally observable and can be used to depict polymer glasses of different microstructures? In the glassy state, how does the alpha relaxation time τ_α (typically much longer than 100 s) change under large deformation? How do we describe the physical aging and mechanical "rejuvenation" in terms of segmental dynamics? Eyring's activation idea of lowering potential energy barriers by stress [64] is widely invoked to discuss the origin of plastic flow: Application of a mechanical stress σ reduces the barrier for molecular activation so that τ_α decreases exponentially with σ. Schweizer and coworkers have developed a nonlinear Langevin equation (NLE)-based microscopic theory for polymer glasses under deformation [54,55,65]. They identified the amplitude S_0 of thermal density fluctuations as a key structural variable that can be affected by external deformation. Specifically, segmental dynamics are described in terms of a "dynamic free energy" that depicts how intersegmental interactions cause segmental caging by erecting a potential barrier F_B. According to the theory, the alpha relaxation time $\tau_\alpha(S_0)$ depends explicitly on S_0 through F_B. In this language, physical aging leads to a decrease of S_0 [55]. Mechanically induced structural disordering rejuvenates the glassy state and causes S_0 to increase in proportion to the rate of plastic dissipation, consistent with small-angle x-ray scattering measurements [66]. Thus, the polymer NLE theory has gone beyond the phenomenological idea of Eyring [64] to describe the acceleration of relaxation, yielding, plastic flow, as well as strain hardening, and appears to be a first crucial step toward a microscopic description of polymer glasses under large deformation.

In more than one way, the Chen–Schweizer theory [55,65] bypassed the intractable challenge to formulate a many-body nonequilibrium statistical mechanical theory for molecular mechanics of polymer glasses by describing physical aging and mechanical "rejuvenation" in terms of their influences on S_0. According to Schweizer and coworkers, during strain-rate-controlled deformation, yielding occurs when the alpha relaxation time becomes sufficiently short as a result of the reduced barrier height so that plastic deformation can take place through segmental rearrangement. The Chen–Schweizer theory gives us the idea that the amplitude of density fluctuations may be an "order parameter." Analogous to S_0, we wish to introduce a state variable S that will be loosely referred to throughout the chapter as a quantity to depict the glassy state, for example, in terms of the system's location on the energy landscape or the amplitude of thermal density fluctuations. Physical aging drives S to go down and mechanical "rejuvenation" creates a higher S. In other words, a state with higher S should have higher segmental mobility and is more likely to respond to large deformation in a ductile manner. It is perhaps most appropriate to call S some kind of segmental packing parameter. In the absence of a microscopic theory, S is rather abstract and loosely defined. According to our understanding to be described in some detail in this chapter, it is through chain networking that high stress can emerge to lower the barrier height and speed up segmental mobility during large deformation.

12.1.3 POLYMER GLASSES UNDER LARGE DEFORMATION ARE STRUCTURAL HYBRID: PRIMARY STRUCTURE AND CHAIN NETWORK

Polymeric glasses of high molecular weight (MW) differ from small-molecule organic glasses in a fundamental way. They can be highly ductile, that is, can undergo yielding and cold drawing [1]. Glassy polymers of low molar mass are, however, as brittle as other organic glasses, even in compression [67]. The contrast suggests the necessity to take the chain networking into account when depicting yielding behavior [68,69]. The concept of chain entanglement has been applied to describe mechanics of polymer glasses since the beginning when Haward and Thackray [70] and Argon [71] assumed the stress increase in post-yield regime (known as "strain hardening") to arise from chain conformational entropy changes. While such an association may be questionable [72,73], as shown by Hoy in Chapter 13, it is reasonable to think of polymer glasses of sufficiently high molecular weight as a structural hybrid [68], made of a primary structure due to short-ranged van der Waals intersegmental interactions and a "mechanical" chain network arising from the interchain penetration

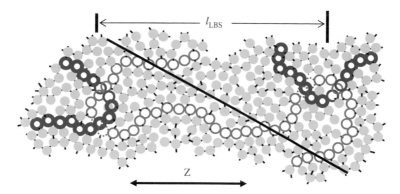

Figure 12.1 Perceived hybrid structure for polymer glasses undergoing large deformation (for example, extension along Z direction) that causes a load-bearing chain network to emerge, where dots and short bars represent polymer segments connected by van der Waals bonds. A subchain (denoted by thin rings) with two hairpins depicts a load-bearing strand (LBS) that defines the mesh size (i.e., l_{LBS}) of the chain network. Its minimum length is determined by the criterion that a plane (represented by the inclined line) can cut through the LBS three times. The junctions at the pairs of hairpins are obviously due to the intermolecular chain uncrossability. Only such junctions are strong enough: for them to be destroyed, that is, for mutual chain sliding to occur at the junctions, many segments would need to move collectively.

of Gaussian chains, as shown in Figure 12.1. Although this chain network and the familiar concept of entanglement network share the same origin, which is intermolecular uncrossability, we need to avoid phases such as "entanglement" and "entanglement network" to draw a distinction between the familiar definition of chain entanglement in melt rheology and the chain network concept for polymer glasses. During large deformation of polymer glasses, the chain network may be perceived as shown in Figure 12.1, made of load-bearing strands (LBSs) that form by pairing hairpins from different chains, whereas segments on non-load-bearing strands (nLBSs) constitute the primary structure. This hybrid emerges only beyond the initial elastic deformation regime within which all segments belong to the primary structure. In other words, beyond the elastic deformation regime, segments on some strands continue to undergo affine deformation and therefore become mobilized or activated. Such strands become load-bearing. Throughout the chapter, the *chain network* refers to a network made of only the LBSs. The fraction of segments belonging to the chain network increases with increasing deformation at the expense of the fraction belonging to the primary structure (associated with the nLBSs). For entangled polymer melts, there is no such conversion to speak of. The structure of the entanglement network is straightforwardly depicted by the packing model [74–77]. We perceive the chain network to have a different structure. Since the average length of an LBS is proportional to the Kuhn length l_K and that of an entanglement strand is proportional to the packing length p [78,79], the ratio of the number density v_{LBS} of LBSs to that (v_e) of entanglement strands is proportional to $(p/l_K)^2$. This indicates that (v_{LBS}/v_e) is lower for polymers with smaller p because most polymers have comparable l_K [79]. The ratio of the areal densities of these two types of strands can be shown in Figure 12.10b to be proportional to p/l_K. This difference will be explicitly described in Section 12.4 when we depict our molecular picture in detail.

12.1.4 WHY CAN POLYMER GLASSES UNDERGO DUCTILE, LARGE-STRAIN DEFORMATION?

The behavior of low-molar-mass organic glasses indicates that vitreous segments cannot uniformly hop out of deep confining potentials to undergo macroscopic yielding and global plastic deformation. Breakdown of intersegmental van der Waals (vdW) bonds anywhere in a macroscopic specimen terminates the ongoing deformation in such organic solids whose cohesion derives from these weak and short-ranged vdW interactions. Consequently, S does not have a chance to increase uniformly across a macroscopic system, and acceleration of segmental relaxation can never take place on a global scale. We have recently come to recognize [68,69] the critical role played by the chain network: An increase of S in the primary structure is either due to conversion of nLBSs to LBSs or to the activation by the displacement of LBSs in the chain network that is undergoing affine deformation on the length scale of LBSs and beyond. It is perhaps the chain network that drives a polymer glass to yield. This causality relationship does not change whether or not there is any structural heterogeneity. Section 12.4 discusses the physical picture in detail.

12.1.5 CONSTITUTIVE MODELING AND FRACTURE MECHANICS VERSUS MOLECULAR PHYSICS: BRIDGING A GAP OF 10^6

Phenomenologically motivated constitutive modeling was the first step in a quantitative description of solid mechanics as well as liquid rheology of polymers [70,80–91]. However, constitutive modeling that is not derived from first principle-based molecular theories lacks molecular ingredients necessary to explain why polymer glasses of high molecular weight can yield and organic glasses of low molar mass are only brittle. For example, when depicting elastic-like deformation such as strain hardening, it made a questionable connection with rubbery elasticity [70–72,86,87,89,92,93]. The problem with such an identification is centrally discussed in the Chapter 13.

Fracture takes place in glassy polymers. From the characteristics of fracture behavior, for example, the fracture stress as a function of crack length, the analysis based on fracture mechanics [2] allows one to evaluate the fracture toughness G_c, which measures the ability of polymer glasses to resist fracture in the presence of a crack. In polymer glasses, the fracture energy per unit area of the fracture surface, G_c, includes contributions from both surface energy and plastic dissipation. Fracture mechanical methods themselves do not deal with the molecular origin of toughness and do not describe the dissipative mechanism in terms of molecular characteristics. As pointed out by Williams [2,94], "the precise nature of the molecular mechanisms involved in the processes are not important to the methods." Therefore, although fracture mechanics helps us predict how cracks in glassy polymers propagate and when fracture may take place, it does not answer such a question as why atactic polystyrene (PS) is brittle at 30°C below its $T_g = 100°C$ but bisphenol-A polycarbonate (bpA-PC) is ductile even 200°C below its $T_g = 145°C$. Moreover, fracture mechanical measurements find G_c to be ca. 400, 1000, and 1700 J/m^2 for PS, poly(methyl methacrylate) (PMMA), and bpA-PC, respectively. However, fracture mechanics cannot explain why the G_c of bpA-PC is so much higher. To reiterate, fracture mechanics does not contain any insight to explain the fact that a brittle polymer glass, for example, PS that is typically brittle at room temperature, no longer suffers from fracture at large deformation and undergoes yielding and plastic deformation after melt stretching [68].

To explore molecular mechanisms for ductility in polymer glasses, we need to probe physical origins of yielding, strain localization such as necking, and brittle fracture. This is a daunting task for polymer physicists because we need to connect the macroscopically observable world (on 1 mm) to the microscopic kingdom (1 nm) that is hardly amenable to direct observations, spanning six orders of magnitude in length scale. In Sections 12.1.3 and 12.1.4, a first helpful hint about how to carry out such an undertaking was mentioned, and details are described in Section 12.4. To bridge the gap of six orders of magnitude between molecular behavior (nm scale) and macroscopic phenomena (mm scale), computer simulations may be indispensable. We should use molecular dynamics (MD) simulations to test the proposed speculation [69] of how chain networking are necessary for the observed yielding, crazing, and BDT and learn how molecular events lead to the observed mechanical behavior. If we can pinpoint the microscopic mechanisms for yielding and failure, guidelines can be developed to help molecular and engineering designs of the next-generation polymeric materials, and molecular mechanics of glassy polymers will have a sound foundation.

12.1.6 DRAWING ANALOGY OF MOLECULAR MECHANICS OF POLYMER GLASSES WITH MELT RHEOLOGY

There is a strong analogy between melt rheology and mechanics of amorphous polymeric solids because both subjects aim to delineate the mechanical response to external deformation and to identify the microscopic origin of mechanical stress. In melt rheology, the dominant and pertinent relaxation time (known as the terminal relaxation) τ is highly molecular weight dependent and can be explicitly characterized, whose smooth temperature dependence can be treated empirically [7,95]. To study yielding in polymer glasses, the alpha relaxation time τ_α for segmental dynamics should be a relevant parameter whose temperature dependence remains an important topic to investigate [96,97]. While it is conventional to take the glass transition as the point where $\tau_{\alpha g} = 100$ s, far below T_g, τ_α must be orders of magnitude higher than $\tau_{\alpha g}$. Defining Wi_α for polymer glasses as a product of the deformational rate and τ_α. When either Weissenberg number Wi (product of rate and relaxation time τ) or Wi_α are greater than unity, polymers of high molecular weight might undergo yielding, respectively, either above or below T_g. The yielding is either a transition from elastic stretching of the entanglement network to disentanglement and flow at $T > T_g$ or a transformation from a vitreous state to an activated plastic state well below T_g. For melts with significant chain entanglement, intrachain elastic retraction force grows to reach imbalance with the interchain grip force, leading to breakdown of the entanglement network, at the yield point [98]. For polymer glasses, yielding can take place during external deformation when the displacement of the chain network has sufficiently accelerated segmental dynamics. In other words, we proposed [69] that the yielding of polymer glasses is driven by the chain network. If such activation could proceed without the breakdown of the chain network until the effective $\tau_{\alpha(\text{eff})}$ becomes shorter than the experimental time scale set by the deformation rate, plastic deformation may take place on macroscopic scales in polymer glasses. Thus, although the same polymer shows macroscopic yielding both above and below T_g, the molecular physics behind yielding are entirely different, involving either irrecoverable collapse of the entanglement network or activation of the primary structure at the segmental level by the chain network. Well above T_g, entangled polymers yield at a large strain (over 100%) for $Wi \gg 1$, whereas the yielding occurs at a few percent of strain for $T < T_g$.

Under circumstances where a polymer can be regarded as elastic either above or below T_g, large fast deformation often produces strain localization [5,99–102]. Any description of strain localization in polymeric liquids is beyond the conventional scope and objective of experimental rheology that is developed on the premise of homogeneous deformation in typical rheometric setups. To answer in a satisfactory manner why strain localization such as wall slip and shear banding occur [103], a realistic molecular picture is required to explain why and how the entanglement structure (responsible for initial rubbery elastic deformation) fails [10,98]. Similarly, to understand why strain localization such as necking occurs upon shear yielding in uniaxial extension of polymer glasses, we need a molecular-level explanation of the yielding behavior. There is another superficial parallelism between the world of polymer glasses and that of polymer melts: Ductile polymer glasses become brittle and lose their ability to yield at low-enough temperatures. Entangled melts can also undergo a yield-to-rupture transition as temperature decreases (but still well above T_g) so that Wi well exceeds the ratio of reptation time to Rouse time [99,104]. Brittle fracture of a glassy polymer occurs at a few percent of strain [1] whereas the same polymer above T_g shows melt rupture after a great deal of stretching, with melt stretching ratio as high as 10. In both cases, the molecular networking suffers a massive breakdown.

In nonlinear melt rheology, the mission is to depict the effect of chain entanglement and how and why entanglement breaks down upon large deformation [10,98]. Analogously, molecular mechanics of polymer glasses does not regard it as its mission to predict the glass transition temperature based on specific microstructures. Instead, it aims to describe when and why yielding can or cannot take place and the condition for the BDT. For this mission to be feasible, we need to characterize the structure of the chain network in terms of relevant molecular parameters. In other words, the ductility of a polymer glass depends explicitly on the characteristics of the chain network [69]. The major task in molecular mechanics of glassy polymers is to delineate when and how the chain network fails under significant deformation.

Although there has been a widely known phenomenological assertion since the early 1980s that "a useful way to increase the 'brittle' fracture stress and decrease the ductile-to-brittle transition temperature of a glassy polymer is to decrease its entanglement contour length l_{ent}" [14,105], there has been insufficient theoretical understanding about why l_{ent} is a relevant structural parameter and what "entanglement" in polymer glasses really does. Indeed, "our understanding of the nature and role of entanglements in an organic glass is still rather limited" [106]. Similarly, the proposal [68] of a hybrid structure for polymer glasses (cf. Figure 12.1) alone does not constitute a theory for polymer mechanics below T_g. Not only do we need to identify and define the structure of the chain network as well as parameterize it according to basic molecular parameters, we also need to describe the interplay between the chain network and the primary structure.

12.1.7 OVERVIEW OF THIS CHAPTER

Besides yielding, two leading characteristic mechanical responses are BDT and crazing during continuous extension of polymer glasses. In Section 12.2, we will review the previous attempts to explain the origin of BDT and then list five factors that can affect BDT. Section 12.3 is on crazing. Apart from mentioning the debate concerning the initiation of crazing [107–110], we indicate how external procedures such as mechanical rejuvenation, pressurization, and melt stretching suppress crazing and physical aging brings back crazing (S. Cheng, P. Lin, X. Li, J. Liu, Y. Zheng, and S.-Q. Wang, unpublished, 2015). Section 12.4 is devoted to a brief description of the recent molecular model [69] for yielding and BDT that may also be applicable to provide a crude account on the cause for crazing. The brief discussion on how crazing occurs is not a theory and is simply to indicate the versatility of the model. Since the model was inspired by a collection of key phenomenology reviewed in Sections 12.2 and 12.3, it naturally explains why BDT and crazing can be affected by the various factors. We conclude in Section 12.5 with a brief outlook to indicate that we may be witnessing a renaissance in the subject of yielding and failure in polymer glasses. For simplicity and conciseness, we focus only on tensile extension behavior of polymer glasses, leaving the subject of polymer glasses under compression for future discussion although the same physical picture should apply independent of the deformation mode. In passing, we note that our molecular picture applies only to polymeric glasses and cannot be used to describe colloidal and metallic glasses since the microscopic physics in each type of glass is rather different.

12.2 PHENOMENA OF YIELDING AND BRITTLE FAILURE

In this section, we discuss yielding and lack of yielding (brittle response) of polymer glasses in terms of the available theoretical explanations and known phenomenology. Yielding is a rather ubiquitous mechanical characteristic of polymeric solids [5]. Yielding of polymer glasses is a bit more surprising since nonpolymeric organic glasses cannot do so. Unless we can appreciate and understand from molecular viewpoints why polymer glasses yield, there is little chance for us to explain how they lose their ductility with lowering temperature. Conversely, it appears that a full understanding of yielding depends on efforts to examine why the sample glass loses its ability to yield at a sufficiently low temperature and what causes brittle fracture. To the lowest order, the Eyring phenomenological model [1,52,64] has been widely applied to describe plastic deformation although the Eyring idea does not explain why yielding can take place in polymer glasses, that is, why high stress can emerge, and why organic glasses of low molar mass cannot undergo macroscopic plastic deformation at all. Plastic deformation of polymer glasses may or may not take place homogeneously, depending on the preparation of the glassy state. Both melt stretching and mechanical rejuvenation can remove shear yielding and allow homogeneous extension.

12.2.1 YIELDING OF THE PRIMARY STRUCTURE

Yielding and plastic deformation of polymer glasses have been modeled [80,84–87,89,111–113] at the constitutive level ever since Haward and Thackray [70] proposed to combine a Maxwell element in series with a spring. Empirically, it is known that the entanglement network density affects strain softening and post-yield hardening behavior. At different compositions of PS blended with poly(2,6-dimethyl-1,4-phenylene oxide) (PPE) [114], the degree of strain hardening, measured in term of how strongly mechanical stress grows with deformation in post-yield regime, is found to correlate with the entanglement network density ν_e. At the extremes, PPE shows the highest level of strain hardening, and PS shows the lowest.

It is important to note that the deformation modes influence mechanical responses. For example, at room temperature, PS and PMMA of high molecular weight that are brittle in uniaxial extension actually yield and undergo plastic deformation in uniaxial compression. Existing literature has not addressed such a remarkable difference between extension and compression from a molecular viewpoint. A tentative view will be discussed at the end of Section 12.4.1. In either extension or compression, the question is what drives the glassy state to undergo plastic deformation. In other words, how does external deformation cause a polymer glass to yield? Can the primary structure formed by short-ranged intersegmental attractions yield on its own?

12.2.2 EXPLANATION FOR BRITTLE-TO-DUCTILE TRANSITION: LDWO HYPOTHESIS AND OTHER PROPOSALS

All polymer melts are rheologically similar, given a similar level of entanglement and when examined at a similar value of Wi. In comparison, supercooled melts, that is, polymer glasses, clearly behave differently at a comparable distance below their glass transition temperatures T_g. At room temperature during tensile extension, bpA-PC is completely ductile, and PS and PMMA are brittle. In the early account of brittle behavior of glassy polymers [115], Vincent found that the critical tensile strength is correlated with the bond number density ϕ for a dozen polymer glasses as shown in Figure 12.2, revealing $f_c = 0.04$ nN as the slope. According to Vincent, the correlation implied the brittle fracture is facilitated by chain scission. Since it takes $f_b = 5$ nN of force to break a covalent bond, Vincent also realized that chain scission would amount to having a tensile strength more than a hundred times the experimental value. To reconcile, it is common to apply the Griffith's criterion [1], attributing the presence of defects for local stress concentration, which is a standard treatment in fracture mechanics of solids. Unfortunately, any discussion along this line ignores the possibility that a large fraction of backbone bonds are not load-bearing.

The standard textbook explanation concerning the nature of brittle-to-ductile transition is the Ludwig–Davidenkov–Wittman–Orowan (LDWO) hypothesis [5] that brittle fracture occurs when the yield stress exceeds a critical value known as the brittle stress σ_B [5,115–117]. In Figure 12.3 that summarizes typical experimental data, brittle fracture and plastic flow are depicted as two independent processes. With lowering temperature, the yield stress σ_y increases until it crosses over the brittle stress line σ_B. Figure 12.3 implies that at a given temperature, the glass would always turn from ductile to brittle as the applied rate increases. According to the hypothesis, BDT should shift to lower temperatures when the deformation rate is lowered. In applying the LDWO hypothesis, it is implicitly envisioned that chain scission causes the brittle fracture. The approach based on the LDWO hypothesis has been rather appealing as explained in Ward's book: "The influence of chemical and physical structure on the brittle–ductile transition can be analyzed by considering

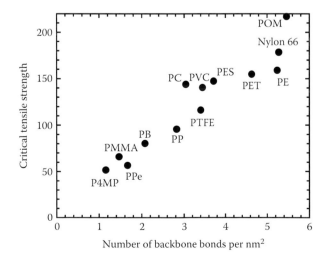

Figure 12.2 Critical stress at the BDT for 13 different polymers in terms of the number of backbone bonds per unit area of 1 nm². The number of backbone bonds per nm² is calculated from the crystallographic data. Data points are read and replot from Reference 115, which also provides the acronyms for these polymers.

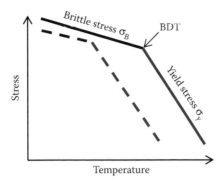

Figure 12.3 Depiction of the LDWO hypothesis where the brittle stress line meets the yield stress line at the BDT. Since the yield stress decreases faster than brittle stress with lowering rate and is lower than brittle stress at a lower temperature, the BDT moves to a lower temperature at a lower rate as shown by the dashed lines.

how these factors affect the brittle stress curve and the yield stress curve, respectively. As will be appreciated, this approach bypasses the relevance of fracture mechanics to brittle failure...." [5]. On the other hand, as a mere restatement of the experimental observation, the LDWO hypothesis does not have anything to say about the molecular mechanism leading to the BDT. Consequently, it cannot adequately explain why PS turns ductile at room temperature upon sufficient hydrostatic pressurization [118,119]. It cannot even explain why high molecular weight is necessary to ensure ductile behavior. Within the framework of the LDWO hypothesis, the observed ductility of PS and PMMA at room temperature under pressure has been interpreted to suggest that the pressure dependence of the yield stress is different from that of the brittle stress [119,120].

Since the pioneering work of Vincent motivated by the LDWO hypothesis, the concept of chain entanglement had been proposed, encouraging subsequent speculations to relate the entanglement density to the tensile strength of polymer glasses [117]. It is said that polymer glasses with higher entanglement density v_e tend to be more ductile. A well-known explanation for the difference between bpA-PC and PS is that bpA-PC has higher entanglement density than that of PS [14]. To avoid brittle fracture, one would need to have high v_e. Such a popular view is however not very helpful because whether a polymer glass of high molecular weight is brittle or not depends on the testing temperature. Even PS of high molecular weight turns ductile in a narrow range of temperatures below T_g. When we raise temperature, v_e does not change, yet a polymer glass can turn ductile from brittle within a few degrees. Clearly, focusing on the chain network structure alone is insufficient.

There have been explicit proposals to link the BDT to segmental [121,122] and secondary relaxations [123–126]. More specifically, phenyl rings of the backbone of bpA-PC have been claimed to be responsible for its remarkable ductility [125]. However, not all polymer glasses possess a beta transition or phenyl rings. The statement from Ward's book [5] best summarizes the state of mind in the field: "It was thought at first that the brittle–ductile transition was related to mechanical relaxation and in particular to the glass transition,.... Because the brittle–ductile transition occurs at fairly high strains, whereas the dynamic mechanical behavior is measured in the linear, low-strain region, it is unreasonable to expect that the two can be linked directly." On the other hand, it is known that the characteristic of crazing, for example, the draw ratio in the fibrils does correlate with the structure of polymer entanglement [13].

12.2.3 KEY PHENOMENOLOGY ON BDT

Treating the BDT as a leading phenomenon, we indicate several factors that are known to influence the characteristics of the BDT [68,69]. A microscopic model has to be able to provide a coherent and unified explanation for why each factor affects the BDT as shown below. Clearly, past understandings have difficulty in explaining the following five effects on BDT: (a) physical aging, (b) melt stretching, (c) blending, (d) pressurization, and (e) mechanical rejuvenation.

12.2.3.1 PHYSICAL AGING

Physical aging allows the system to evolve toward equilibrium [11,59]. Consequently, after adequate aging, many polymer segments are trapped into deeper potential wells so that the glass becomes more difficult to yield [34,61,62]. Such annealing, usually carried out just below T_g, involves no shape change and thus does not alter the structure of the chain network. Since aging affects the BDT behavior [68,69,127], the idea to only associate polymer ductility with "entanglement network density" is obviously not helpful. The mass density increase does not change the entanglement density appreciably but the effect of physical aging on S or density fluctuations can be expected to be significant.

It is straightforward to show that adequate physical aging can even make the most ductile bpA-PC lose its ability to undergo cold drawing. It was shown in Figure 12.4, for example, that aged PC loses its ability to extend [68]. Effects of physical aging on other characteristics of polymer glasses have been well documented in the literature [1] and need not be reviewed here.

12.2.3.2 MELT STRETCHING (GEOMETRIC CONDENSATION OF CHAIN NETWORK)

One intriguing way to change the mechanical behavior of polymer glasses is to perform melt extension [129–131]. It is noteworthy that pre-deformation of polymer melts involving simple shear is not effective. We assert that uniaxial, biaxial, and planar extension can each produce significant melt stretching effect although data involving biaxial and planar extension are rare to find in the academic literature. After sufficient melt stretching, the most brittle PS becomes ductile in tensile extension at room temperature [68], a striking phenomenon known since 1960s [129,131]. Figure 12.5 is a recent study of the phenomenon [68], showing the brittle PS becoming completely ductile at room temperature (RT). Such melt-stretched PS specimens remain ductile until a temperature well below RT. Specifically, for a melt stretching ratio of $\lambda_{ms} = 2.2$, a series of tensile tests show in Figure 12.6 that the BDT shifts to a temperature as low as −15°C. Without melt stretching, the BDT usually occurs around 70–80°C [120,132]. A truly satisfactory and realistic explanation of the physics behind the phenomenon had remained elusive until now, involving the recognition that melt stretching produces a geometric condensation [133] of the LBSs as illustrated in Figure 12.7.

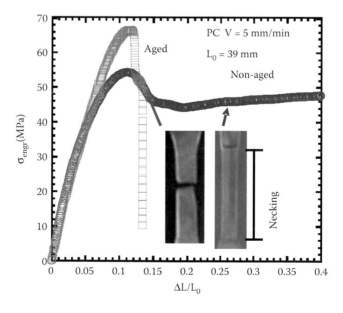

Figure 12.4 The comparison of failure behavior of bpA-PC before and after aging. After aging, bpA-PC becomes brittle when they are stretched at room temperature. Aging takes place under vacuum at 147°C for 10 days. (Reprinted from G. D. Zartman et al., *Macromolecules*, 45, 6719, 2012, with permission.)

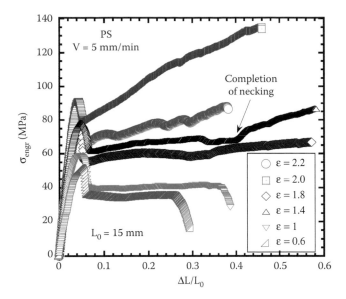

Figure 12.5 Mechanical response of melt-stretched PS along the stretching Z direction. Melt stretching took place at 130°C at $\dot{\varepsilon} = 0.2\,\mathrm{s}^{-1}$ to Hencky strains of $\varepsilon = 0.6$, 1.0, 1.4, 1.8, 2.0, and 2.2. Otherwise brittle PS undergoes shear yielding and necking propagation after melt stretching. (Reprinted from G. D. Zartman et al., *Macromolecules*, 45, 6719, 2012, with permission.)

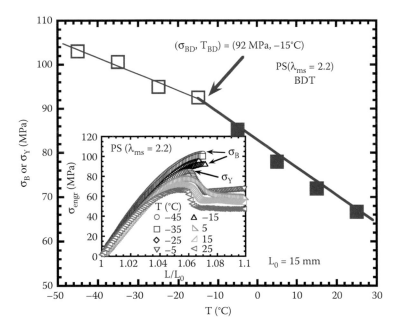

Figure 12.6 Brittle versus ductile behavior of melt-stretched PS in tensile deformation, at a constant cross-head speed of $V = 6$ mm/min, obtained with an Instron 5567 with an environment chamber. Melt stretching to $\lambda_{ms} = 2.2$ was carried at 120°C using a constant Hencky strain rate of $0.2\,\mathrm{s}^{-1}$ ($Wi = 633$) with an Sentmanat Extension Rheometer (SER) fixture on Advanced Rheometric Expansion System (ARES) with a temperature control chamber. The open symbols show the brittle stress and the filled symbols indicate the yielding stress at different temperatures. A BDT can be identified around −15°C. The inset shows the unpublished raw data at temperatures ranging from −45°C to 25°C.

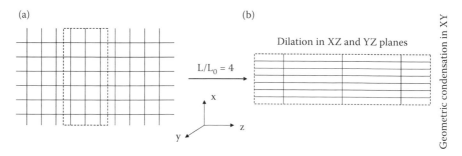

Figure 12.7 Illustration of geometric condensation of entanglement strands produced by (affine) uniaxial extension of a well-entangled polymer melt, a concept first discussed in Zartman et al. [68] as well as Liu et al. [133]. Here, the straight horizontal lines represent effective entanglement strands condensed by a factor of four given by the melt stretching ratio L/L_0. Since the cross-sectional area shrinks by four in the XY plane, the entanglement network is dilated when viewed in the plane of XZ or YZ.

12.2.3.3 BLENDING (DILUTION OF CHAIN NETWORK)

Another way to alter the structure of the chain network without affecting the glassy state is to simply dilute it with a low-molar-mass component that cannot form its own chain network but have the same T_g as that of the high-molar-mass component [68]. Such a (binary) mixture at different compositions would have different network densities. It has been demonstrated, as shown in Figure 12.8, that even the most ductile bpA-PC can be made brittle by such dilution. Conversely, the structure of the chain network in a polymer glass can also be boosted by blending with a second miscible polymer of network density. For example, blending PS with PPE improves the tensile response [114].

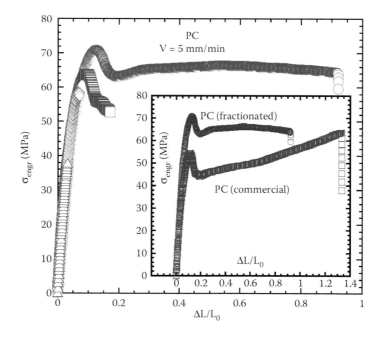

Figure 12.8 Stress versus elongation of PC blends with different blending ratios of a high-MW component 167 kg/mol (PC167 K) to a low-MW component 6.9 kg/mol (PC7 K): 100:0 (circles), 80:20 (squares), 60:40 (diamonds), and 40:60 (triangles). The inset shows the apparent dependence of the characteristic mechanical response on the molecular weight based on the comparison between PC167 K (fractionated) and PC63 K (commercial). (Reprinted from G. D. Zartman et al., *Macromolecules*, 45, 6719, 2012, with permission.)

12.2.3.4 PRESSURIZATION

It has been demonstrated that PS and PMMA become totally ductile under sufficient hydrostatic pressure P, that is, above 0.5 kbar [118,119]. What does pressurization do to the hybrid structure? Obviously it does not appreciably increase the density of LBSs in the chain network. But can it increase the strength of the chain network? If so, how? Since the yield stress is seen to be an increasing function of P, the pressure has strengthened intersegmental interactions due to the increased mass density. A molecular account for the pressure effect has been missing since the pioneering work in the mid-1970s until now [69].

12.2.3.5 MECHANICAL "REJUVENATION"

One of the most intriguing effects is that of mechanical "rejuvenation" [35,59,63]. Such rejuvenation involves post-yield deformation of polymer glasses well below T_g. The "cold" deformation can be realized through twisting [113] or roll-milling [128,134,135]. For example, after milling, brittle PMMA can become ductile at room temperature in tensile extension (P. Lin and S.-Q. Wang, unpublished, 2015). After either adequate twisting or milling, bpA-PC no longer undergoes shear yielding and necking, and can extend homogeneously at room temperature (S. Cheng, P. Lin, X. Li, J. Liu, Y. Zheng, and S.-Q. Wang, unpublished, 2015, 128). Mechanical rejuvenation may or may not involve any shape change. For example, twisting does not change the chain network density, whereas milling only slightly does. It is known through computer simulation that mechanical rejuvenation elevates the energy landscape of polymer glasses [34,37]. Thus, mechanical rejuvenation makes a polymer glass more prone to yielding and plastic deformation.

No coherent theoretical understanding is available to offer satisfactory explanations for how and why the five factors can each dramatically influence mechanical responses of polymer glasses. Conversely, any useful theoretical description must first meet the challenge to explain how these five effects can be understood collectively, coherently, and microscopically.

12.3 CRAZING: AN EXTREME FORM OF STRAIN LOCALIZATION

Crazing is another leading type of nonlinear response to sufficient deformation or high stress, apart from such phenomena as macroscopic yielding and BDT. It is well known and generally accepted that crazing is a surface phenomenon [107,108,110]. Crazes not only characteristically originate from free surfaces but may also remain on them without growing deep into the bulk in the case of extension of bpA-PC above room temperature. Consequently, in our view, crazing may not always be a precursor to brittle fracture in polymer glasses. Moreover, crazes are not unique to brittle polymers.

As demonstrated by many workers [12–14,109,136–139], particularly Kramer and his coworkers [12–14,104], crazes are isolated regions where fibril-like structures are observed that involve significant extension [14]. Voids are present in crazes because the transverse dimensions perpendicular to the direction of extension are constrained in the glassy state. According to Argon [110], "the basic understanding of the molecular level processes that govern craze initiation still remains murky" although numerous simulation studies have been carried out [136–138,140–142]. There are at least two schools of thought concerning how crazing is initiated. Bucknall regarded crazing as a frustrated fracture process due to the presence of small inclusions and surface imperfections [107]. But according to Argon [110], the study of Bucknall "does not contribute much to the fundamental understanding of craze initiation." Contrasting the idea of fracture for craze initiation, Argon advocated the view that craze takes place through "local dilatant plastic shear events" [110]. Neither was speaking of any molecular mechanism for the localized failure.

Below, we mention several observations that motivate us to look for an overall molecular picture. Some of these experimental observations have yet to be published (S. Cheng, P. Lin, X. Li, J. Liu, Y. Zheng, and S.-Q. Wang, unpublished, 2015). In short, we emphasize that a realistic and comprehensive molecular model for mechanics of polymer glasses should not only explain why five different factors affect the BDT but also offer some insights into the physics behind crazing.

12.3.1 EFFECT OF MELT STRETCHING

It can be shown that melt stretching not only suppresses BDT, e.g., pushing it below room temperature for PS and PMMA [68,130,143] but also suppresses crazing [144,145]. What is the mechanism that makes melt-stretched polymer glasses craze-free? The melt stretching is effective when it has caused the entanglement network (and thus the chain network) to deform essentially affinely to a sufficient degree. After quenching to cool the stretched melt down below T_g, the chain network has undergone geometric condensation as shown in Figure 12.7. Without removing any foreign inclusions, we need to explain why the geometric condensation suppresses crazing.

12.3.2 EFFECTS OF MECHANICAL REJUVENATION AND PHYSICAL AGING

Twisting, roll-milling, and cold drawing, all below T_g, are three useful protocols to mechanically "rejuvenate" polymer glasses. With sufficient mechanical rejuvenation using any of the three methods, crazing no longer emerges in bpA-PC even though no inclusions are removed during rejuvenation (S. Cheng, P. Lin, X. Li, J. Liu, Y. Zheng, and S.-Q. Wang, unpublished, 2015). Here, twisting (back and forth) is a particularly powerful method because the sample is left untouched with regard to its surface and geometrical shape. Unpublished preliminary results from X. X. Li show that the observed crazing in bpA-PC is induced by surface contamination: when the specimen was handled using latex gloves, crazing no longer appear under the same condition.

Adequate physical aging can restore crazing in the previously rejuvenated craze-free polymer glasses (S. Cheng, P. Lin, X. Li, J. Liu, Y. Zheng, and S.-Q. Wang, unpublished, 2015). The dominant effect of annealing should be to cause better local segmental packing and reduced S. Moreover, any surface defects could only diminish during annealing, making it harder to initiate crazing. Plastic events should be more difficult to take place after aging. It is unlikely that the role of inclusions would change upon either "mechanical rejuvenation" or physical aging. Thus, the effects of rejuvenation and aging on crazing seem at odds with both Bucknall's and Argon's accounts for craze initiation [110].

12.3.3 EFFECT OF PRESSURIZATION

Sufficient hydrostatic pressure not only makes PS ductile at room temperature in extension, but also considerably increases the tensile stress σ_{craze} at which crazing emerges. For example, σ_{craze} more than doubles when the applied pressure increases from 1 bar to 600 bar [120]. The observed pressure effect on crazing appears related to the suppression of the BDT by pressure mentioned in Section 12.2.3.4.

12.4 PLAUSIBLE MOLECULAR MODEL FOR YIELDING, CRAZING, AND BDT

Mechanical behavior of polymer glasses is extraordinarily complicated and difficult to understand at a molecular level. The physical system is obviously more intricate than what we find it to be in the melt state. Given such a reality, we need to build up penetrating insights through a careful examination of available phenomenology. If there is insufficient phenomenological understanding, we must more rigorously pursue new experimental observations. Our first attempt to formulate a theoretical picture is necessarily phenomenologically based. Fortunately, many key experimental findings are available. Specifically, the five effects on BDT listed in Section 12.2.3 and the three effects on crazing described in Section 12.3 have placed significant constraints on what the molecular picture may look like. Needless to say, the most straightforward and important clue is that polymer glasses of low molar mass have no strength and are always brittle even in compression [67].

To reiterate, it seems that we have to address how polymer glasses lose their ductility if we are to have a complete picture of yielding. It is our assumption that yielding is produced by the affinely deforming chain network. On the one hand, the chain networking is essential to provide structural integrity in polymer glasses during large deformation, that is, beyond the elastic deformation regime where the Young's modulus is defined and measured from the slope in the stress versus strain curve. On the other hand, it is clear from the observed effects of physical aging and mechanical rejuvenation on BDT and crazing that the characteristics of the primary structure are also important to include in the theoretical description of molecular mechanics of polymer glasses. In the absence of any other available account, we will use S to inform us about the change in the primary structure [55,65], while acknowledging that S may not be sufficient to capture the essence of the glassy state, which can be changed by physical aging and mechanical rejuvenation. Analogous to the friction coefficient in melt rheology, the dynamics of the glassy state change with temperature, that is, the alpha relaxation time τ_α is strongly temperature dependent [55]. Initial mechanical response at a small strain of ε_{el} is glassy and elastic, showing the Young's modulus to be on the order of 1 GPa, as long as the deformation time is significantly shorter than τ_α (i.e., $Wi_\alpha \gg \varepsilon_{el}$). Plastic deformation cannot take place unless segments are able to keep pace with the imposed deformation [52,54,64]. Like rheology of entangled polymeric liquids where the key is to depict when rubbery elastic deformation ceases due to force imbalance leading to chain disentanglement [10,98,102], we must answer how a polymer glass is able to achieve global mobilization of the primary structure before failure of the chain network. We need to ask what molecular events produce brittle failure.

Based on the hybrid structure of Figure 12.1 along with the available phenomenological information reviewed in Sections 12.2 and 12.3, it is possible to develop a rheology-like theoretical treatment that can take physical aging and mechanical rejuvenation into account. These different procedures have distinct effects on the characteristics of the primary structure and can influence how the chain network affects the primary structure during large deformation. To the zeroth order, the task is to figure out how non-load-bearing strands become either activated or converted to LBSs during continuous external deformation. We have recently come to realize that the integrity of the chain network may be required for the mechanical response of polymer glasses to be ductile during large deformation. This recognition asserts that the chain network plays a predominant role in the molecular mechanics of polymer glasses [69]. Let's imagine tensile extension. It is the chain network shown in Figure 12.9, starting from the boundaries and extending over the whole system (sample), which enables the bulk away from the ends of the sample to undergo large extension.

At low strains in the pre-yield regime, the glass is made entirely of the primary structure. Moving beyond the regime of initial elastic deformation, on average, the segmental mobility increases rather gradually from its quiescent value as most nLBSs are initially immobile. The chain network becomes denser as more LBSs emerge with growing deformation during the conversion from nLBSs to LBSs. At the same time, other nLBSs may be activated. Upon global yielding, the mobilized segments form a macroscopic continuous phase where the chain network is embedded. In the post-yield regime, the average segmental mobility reaches a level higher than the rate of external deformation. This viewpoint naturally is consistent with the experimental verification based on a series of optical photobleaching experiments [146–150]. If we were to apply the idea of Eyring to explain the sharply increasing molecular mobility at the yield point, then it is the displacement of LBSs that has caused

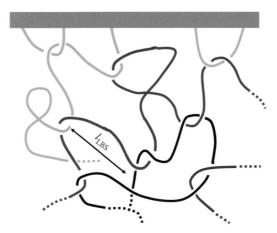

Figure 12.9 Conceptual picture of the chain network in a glassy polymer of high molecular weight, enabling it to undergo large extension imposed from the "ends," one of which is explicitly represented by the top surface.

the barrier heights to be lowered so that the surrounding segments can become delocalized from their confining cages. In the absence of a chain network, it is difficult for us to envision where the Eyring stress would come from and how yielding could take place. Arguing in a different way, any Eyring-like plastic deformation is expected to be extremely localized at the ends of the sample (cf. Figure 12.9) for any organic glasses of low molar mass, including polymer glasses, and consequently tensile drawing must only be brittle in such glasses.

Since the chain network plays a central role in our description of polymer glasses under large deformation, we need to investigate its basic properties. To contrast, we first briefly review the structure of the entanglement network as follows. The successful packing model prescribes the mesh size l_{ent} (i.e., the size of an entanglement strand) of the entanglement network to be linearly proportional to the packing length p [78,79]. Thus, the number density of entanglement strands is given by $v_e = (\rho/M_e)N_a \sim 1/p(l_{ent})^2 \sim p^{-3}$, where M_e is the entanglement molecular weight and N_a is the Avogadro constant. Similarly, the areal density of entanglement strands is given by $\psi_{ent} \sim v_e l_{ent} \sim 1/p^2$.

According to our analysis, the chain network has a different structure from that of the entanglement network, which can be also described in terms of known molecular parameters [69]. As depicted in Figure 12.1, the junctions of the chain network are made of pairs of hairpins. The minimum (critical) chain length l_{LBS} to make two hairpins is the mesh size of this chain network. As shown in Figure 12.1, an LBS of length l_{LBS} is long enough to return to a plane three times. We show below that l_{LBS} scales with the Kuhn length l_K.

Denoting the chain thickness in terms of s, we can determine how many times a Gaussian chain of molecular weight M (with n backbone bonds) has folded back within its pervaded volume $\Omega = (4\pi/3)R^3$, where $R = (C_\infty nl^2)^{1/2}$ is the chain's end-to-end distance. The volume of Ω accommodates Q chains of the same length R, where Q is given by the ratio of Ω to the chain's physical volume, that is, $1/v = M/\rho N_a = pR^2$, so that $Q = v\Omega \sim R/p$. On the length scale of backbone bond length l, we have $nsl = pR^2 = p(C_\infty nl^2)$. Therefore, the molecular thickness $s = pl_K$, where the Kuhn length $l_K = C_\infty l$. Since there are Q chains in Ω, qQs should be equal to $\pi R^2/6$, where q is the number of times a Gaussian chain of length R returns to a plane of area equal to $\pi R^2/6$. This means $q \sim R^2/sQ \sim pR/s = R/l_K$. At $q = 3$, R becomes l_{LBS} so that we have $l_{LBS} \sim l_K$.

The structure of the chain network can be characterized in terms of its density v_{LBS} and mesh size l_{LBS}. Specifically, the number density of LBSs is given by $v_{LBS} \sim 1/p(l_{LBS})^2$, and correspondingly the areal density ψ is given by

$$\psi \sim v_{LBS} l_{LBS} \sim 1/p l_K. \tag{12.1}$$

We have previously concluded based on available data on l_K that most of the commonly known linear polymers have rather comparable chain flexibility [79], that is, similar values of l_K. This means, to the leading order, ψ scales like $1/p$, contrasting ψ_{ent} of the entanglement network that scales like $1/p^2$. In other words, the entanglement network is doubly sensitive to the molecular thickness characterized by p. Figure 12.10a shows the structural dependence of the two networks on p for 12 different polymers [69]. The present scaling analysis leaves the prefactor in Equation 12.1 undetermined. Based on the available data on the values of p and l_K of common polymers [78,79], we can express the ratio of $\psi/\psi_{ent} \sim p/l_K$ for many polymers as shown in Figure 12.10b, which shows that ψ converges toward ψ_{ent} as p increases.

The preceding analysis applies in the limit of infinitely long chains. For chains of finite molecular weight M, we expect ψ to decrease as the ratio $h = M_{LBS}/M$ increases. A plausible formula would be $\psi(h) = \psi(0)(1 - 2h)$ for $h < 1/2$, where $\psi(0)$ is given by Equation 12.1.

Having described the structure of the chain network, we are ready to review a recently proposed molecular model by discussing its three essential elements in the following four paragraphs before presenting its interpretations of BDT and crazing in Sections 12.4.1 and 12.4.2. We need to emphasize from the beginning that the following theoretical considerations lack any direct proof because they are not derived from first principles. For example, although the concept of LBS is as natural for us to envision as Guth et al. [151,152] identified Gaussian chains between cross-links in their account of rubber elasticity, there is no statistical mechanical derivation of LBS's characteristics. It is in this sense that the development of a suitable molecular model is clearly still in its infancy.

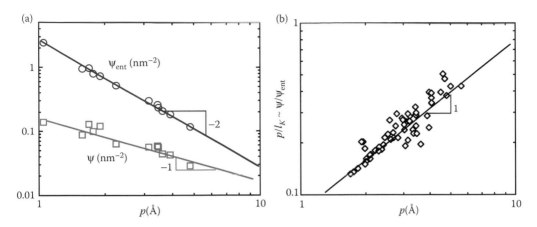

Figure 12.10 (a) Areal density ψ of load-bearing strands (LBSs) in the Gaussian chain network and areal density ψ_{ent} of entanglement strands of the different polymers. The unknown prefactor in Equation 12.1 is chosen in such a way to avoid overlaying the two quantities that show different scaling dependences on the packing length p. The data describe 12 polymers in the order of increasing p: bpA-PC, PEEK, PE, PET, POM, PPO, PVAc, PMA, PMMA, PαMS, PS, PtBS. (Reprinted from S.-Q. Wang et al., *J. Chem. Phys.*, 141, 094905, 2014, with permission.) (b) The ratio of ψ to ψ_{ent} for dozens of polymers studied in Fetters et al. [78] and Wang [79] as a function of the packing length p.

Since all polymer glasses can be brittle, any theoretical account for yielding of polymer glasses must not only explain how yielding takes place but also address when yielding cannot happen during continuous external deformation. One key element of the molecular picture [69] for yielding and failure of polymer glasses is the following assertion: Under external deformation, it is the affine deformation of the chain network on the scale of LBSs that displaces segments of nLBSs away from their potential minima, weakening and breaking intersegmental van der Waals bonds. Actually, there is evidence for affine deformation on scales smaller than LBSs and distortion at the level of covalent bonds [153], in support of the idea of chain tension in LBSs. The deforming chain network drives S to increase and τ_α to drop. Beyond the linear response (elastic deformation) regime, segments on each LBS may push around surrounding segments of nLBSs while keeping the deformation of LBS affine. It seems apparent that if ψ is higher, at a given degree of deformation, more segments of nLBSs can be displaced by LBSs, as shown by the comparison between (a)–(a')–(a") and (b)–(b')–(b") in Figure 12.11, where the areal density of LBSs differs by a factor of two. This means that the overall $S(\psi)$ may be an increasing function of ψ for a given degree of deformation. In Figure 12.11, we depict the activation zones (AZs) with irregular shapes in lighter gray color and include spatial heterogeneities, represented by the black patches where the segmental motions are more restricted because of denser packing of segments (corresponding to smaller S) relative to the surroundings. During deformation, the chain network tends to reduce the heterogeneity because of the emergence of AZs. Although the AZs seems analogous to the shear transformation zones (due to local density deficiency) perceived by Argon [3], the proposed physical origin of AZs is rather different.

Polymeric glasses are believed to be inherently heterogeneous [51], perceived by Long and Lequeux [154] to be made of a percolating set of glassy domains. We also note that the depiction of AZs as patches in Figure 12.11(b)–(b')–(b") resembles the report [30,155,156] of soft spots in small deformation of glasses that are more amenable to local structural rearrangement. Apparently, such an inhomogeneous structural response can also take place at very low strains in a nonpolymeric glass.

Chain tension f_{ct} on an LBS is a crucial concept in the model that is required to elucidate the mechanical interplay between the chain network and primary structure [69]. We expect f_{ct} to depend sensitively on the state of the primary structure, that is, being an explicit function of S, decreasing with increasing S for a given degree of external deformation. Unlike polymer melts or rubbers (above T_g) where segmental relaxation rate is usually much higher than the applied rate, the glassy surroundings allow the stretched LBSs to retain distortions at the covalent bond level [157,158]. Consequently, chain tension f_{ct} can build up orders of magnitude higher than the intrachain elastic retraction force $f_{ent} \sim k_B T/l_{ent}$ in the melt state. It is plausible that for a given

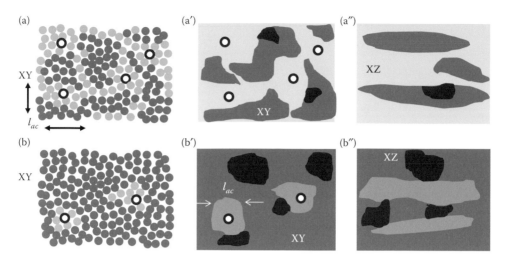

Figure 12.11 Two groups of illustration of (a) to (a″) for a polymer glass under extension along the Z axis whose ψ is twice that of the second polymer glass depicted in (b) to (b″). This figure combines with Figure 12.1 to depict a complete image of how high-molar-mass polymer glasses respond to large extension. In (a) and (b), the glasses, viewed along the Z axis, consist of the LBSs (open dots), the activation zones (light dots) of dimension l_{ac} and vitreous regions (dark dots). In (a′) and (b′), viewed along the Z axis, the states are represented as various continuum regions, including the activated region surrounding the LBSs and patches of more vitreous state to the show presence of heterogeneities. With reduced ψ, the primary structure is less activated, as shown by the darker "tones." In a side view, for example, in the XZ plane, the glasses may look like what is depicted in (a″) and (b″). It is useful to note that these sketches are slightly different from those presented in Wang et al. [69] where the activation zones were depicted as circular tubes surrounding the LBSs for simplicity.

extent of deformation, different amounts of chain tension occur in different LBSs, depending on the respective values of S and ψ.

The third and last component of the model is to delineate how and why macroscopic failure occurs. We assume that the mechanical integrity is provided by the chain network and specimen breakup can take place only if the chain network has fallen apart macroscopically. The chain network can break down in one of the two ways, that is, through either chain pullout or scission, depending on the magnitude of chain tension f_{ct}. The revealing observations [118,119] that pressurization makes PS and PMMA ductile in tensile extension at room temperature unambiguously assures us that the brittle fracture of PS at 1 bar at room temperature involves massive chain pullout. The reasoning goes as follows. Under sufficiently high pressure, PS was observed to exhibit a much higher yield stress σ_y, that is, $\sigma_y(1\ kbar) \sim 3\sigma_B(1\ bar)$, implying that the brittle stress $\sigma_B(1\ kbar)$ at 1 kbar would be more than three times $\sigma_B(1\ bar)$. Since the chain-breaking force f_b in $\sigma_B = \psi f_b$ does not change with P, we would need to conclude that ψ would have increased several times when P increases from 1 bar to 1 kbar, that is, ψ(1 bar) is only a little fraction of ψ(1 kbar). In other words, there must be a great deal of chain pullout at 1 bar since ψ at 1 bar is only a small fraction of its value at 1 kbar. Thus, it is pullout that leads to brittle fracture in PS and PMMA at room temperature.

In summary, all three components of the model have to do with the chain network—chain networking is a lead actor in the theater of glassy polymer mechanics. To complete the description of the model, we need to make some speculations about the chain network. We assert that the chain network fails through chain pullout. Pullout takes place when the chain tension f_{ct} exceeds a critical level denoted by f_{cp}. This pullout force f_{cp} should depend on temperature T and the state of the primary structure, characterized by S (in the absence of a quantitative depiction). Chain network failure occurs only when $f_{ct} > f_{cp}$. Clearly, f_{cp} might vary from one polymer to another. The following comparison between our model and available experimental data on brittle stresses for different polymers indicates that the magnitude of f_{cp} saturates to a common level for these polymer glasses. This saturation occurs perhaps because T_{BD} is sufficiently below T_g for most of these polymers.

Before addressing this issue of universality, let's consider a better defined situation. At a given temperature, we examine the same polymer glass and ask whether the chain tension f_{ct} would change with ψ. This question was left unanswered in the recently proposed model [69]. Upon melt stretching, PS can become ductile in cold drawing at room temperature. Owing to the geometric condensation effect, depicted in Figure 12.7, melt stretching has increased ψ, making it easier for the chain network to activate the primary structure [69]. But this is only half of an explanation on why increased ψ defers brittle fracture. The other half of the explanation must answer why the chain network does not fail at an increased ψ. A logical answer is that there is not enough chain tension buildup to reach the threshold f_{cp} because of the enhanced activation leading to a higher S. In other words, it is reasonable and plausible that chain tension $f_{ct}[S(\psi)]$ as a decreasing function of S becomes lower at a higher ψ. Until computer simulations are applied to explore such a relationship, we shall treat it a conjecture.

Having delineated our molecular picture concerning why yielding can take place in polymer glasses of high molecular weight, we are ready to depict two leading forms of mechanical behavior of glassy polymers. We shall demonstrate that the new molecular-level understanding can qualitatively account for the BDT and crazing phenomena.

12.4.1 BRITTLE-TO-DUCTILE TRANSITION

Any polymer glass of high molecular weight is ductile when it is not far below T_g. Even PS is ductile at 70°C and PMMA from capillary extrusion can be ductile at 35°C [132]. At high temperatures, albeit still dozens of degrees below T_g, polymer glasses can yield on a time scale many orders of magnitude shorter than τ_α as the displacement of LBSs drives surrounding segments to overcome the confining barriers. It is important to appreciate that our perceived chain network depicted in Figure 12.1 and Equation 12.1 is actually rather sparse. In other words, during large deformation, the emergent chain network only involves a fraction of segments that belong to LBSs. Most segments interact with one another via van der Waals bonds that make up the primary structure. Affine deformation on segmental scales only lasts for an initial period in the elastic deformation regime. External deformation cannot cause them to uniformly displace over any appreciable distance in the absence of the chain network. At sufficiently low temperatures, chain tension f_{ct} can build up to rather high levels as segmental activation by LBSs becomes more difficult. The primary structure may not get sufficiently mobilized by the chain network before pullout at $f_{ct} \sim f_{cp}$. The loss of LBSs self-feeds, leading to a macroscopic crack and brittle fracture. In other words, a polymer glass may not have a chance to reach global yielding before structural failure via chain pullout. For a given rate of deformation, there exists a temperature or a narrow window of temperature where the chain network barely accomplishes the mission of activating the primary structure when f_{ct} increases to f_{cp}. We have suggested [69] that the temperature T_{BD} at BDT is a Goldilocks temperature where the yielding of the primary structure occurs on the verge of the chain network breakdown. If this is true, the measured stress σ_{BD} at the BDT might directly reveal the structure of the chain network. Specifically, we might have

$$\sigma_{BD} = f_{cp}\psi. \tag{12.2}$$

The threshold value $f_{cp}(P,T_g/T)$ is expected to depend on temperature T and pressure P. If T_g has any meaning, it may be that T_g/T controls S and therefore $\tau_\alpha(S)$ in the absence of deformation. In other words, farther below T_g, that is, at a higher value of T_g/T, it should be more difficult to mobilize the glassy state by reducing τ_α. Since the junctions are made of pairs of hairpins and structurally the same for Gaussian chains, the magnitude of f_{cp} could be a relatively universal constant, independent of microstructural details when a polymer is sufficiently below T_g (i.e., at a "large" value of T_g/T). In other words, if T_{BD} is significantly lower than T_g, then f_{cp} in Equation 12.2 might approach a common value f* so that σ_{BD} explicitly reveals the structure of the chain network, that is, $\sigma_{BD} \sim \psi$. Conversely, since ψ is known for various polymers, apart from a prefactor in Equation 12.1, Equation 12.2 could be used to predict their σ_{BD}.

Such a speculation is rather bold and perhaps incredible. On the other hand, it might work provided that BDT would take place well below T_g. Polymer glasses with small enough p could have large enough ψ (cf. Equation 12.1) to be ductile well below T_g. It turns out that available literature data on the BDT, as shown in

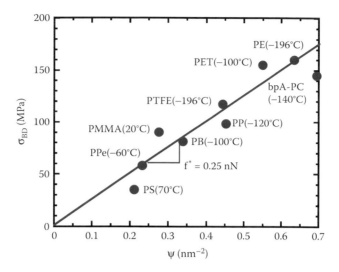

Figure 12.12 Breaking tensile stress σ_{BD} at the BDT for nine different polymer glasses in terms of the area density of the load-bearing strands, where the straight line is a fit of Equation 12.2 to the data. Here, PPe is poly(pentene-1). The slope shows a universal value of the average chain pullout force f* = 0.25 nN at the BDT. (Reprinted from S.-Q. Wang et al., *J. Chem. Phys.*, 141, 094905, 2014, with permission, where the data for PS and PMMA were updated.)

Figure 12.12, appears to support Equation 12.2. The agreement between Equation 12.2 and the available data is truly encouraging. Although it is not a direct validation of the molecular model, it has revealed that σ_{BD} is sensitive to the structure of the loading-bearing network. For example, if $\psi_{ent} \sim 1/p^2$ is used in Equation 12.2 instead of ψ of Equation 12.1, there will be no correlation between σ_{BD} and the entanglement network areal density ψ_{ent}. The resemblance between Figures 12.2 and 12.12 has been explained in Appendix B of Reference 69: The number of backbone bonds per unit area ϕ, which is the X axis in Figure 12.2, actually scales with molecular parameters such as packing and Kuhn lengths identically as ψ of Equation 12.1 does. Thus, our theoretical picture in terms of Equations 12.1 and 12.2 gives a first satisfactory explanation for why Figure 12.2 shows such a remarkable correlation between the breaking stress σ_{BD} and the molecular characteristics. Figure 12.12 indicates a plausible estimate of the absolute values of ψ for these polymers. If the areal ψ of LBSs was taken to be a factor of 20 lower, which appears implausible, the slope would have a value of 5 nN, comparable to the force required to break a covalent bond.

The molecular picture first proposed in a recent study [69] not only explains the well-known correlation of σ_{BD} with the molecular network structure as shown in Figure 12.12 but also contains rational and coherent explanations for most of the known effects summarized in Section 12.2.3. Let's review the list one by one. (a) *Aging* drives the state closer to the equilibrium state, making the primary structure harder to be activated on the one hand and allowing the more chain tension to build up by the external deformation on the other hand. Consequently, aging should move T_{BD} to higher temperature and reduce the ductility. Upon aging, more and stronger heterogeneities might emerge [159], so that the mechanical strength of the chain network may not be spatially uniform. (b) *Melt extension*, when done at sufficiently high rates to achieve near-affine deformation, changes the structure of the chain network. Specifically, LBSs become condensed in the cross section perpendicular to the direction of melt stretching; the chain network is similarly dilated when cold drawn in a direction perpendicular to the uniaxial melt stretching direction, as indicated in Figure 12.7. Thus, *melt stretching* (along the Z direction) can either suppress or prompt brittle fracture by changing ψ. When cold drawing is along the Z axis, the chain network of a higher ψ (due to the geometric condensation) more efficiently activates the glassy state so that chain tension does not get to build up to reach the point of chain pullout. Upon cold drawing perpendicular to the melt stretching direction, fewer LBSs (geometric dilation, cf. Figure 12.7) are available to activate their surrounding segments so that significantly high chain tension emerges beyond the threshold for pullout. (c) Incorporation of a low-molar-mass component into a polymer glass of high molecular weight degrades ductility. A reduction in ψ makes it more difficult to activate the glass as shown by the comparison between (a)–(a″) and (b)–(b″) in Figure 12.11. Moreover, in a less mobile surrounding, more chain tension can build up because $f_{ct}(\psi)$ increases with decreasing ψ. Consequently, such a mixture is less resistant against brittle failure. We further speculate that the presence of low-molar-mass components could also have an adverse effect on the strength of the junctions and lowers the value of f_{cp}. (d) *Pressure* may raise the effective T_g and therefore increases the alpha relaxation time [160]. The effect of pressurization can be understood to increase $f_{cp}(P)$ and decrease S. Thus, the reported suppression of brittle failure by pressure [118–120] most plausibly indicates that the critical level of chain tension $f_{cp}(P)$ for pullout increased significantly with P [69]. The condition for chain pullout is no longer met at a sufficiently high applied pressure although it is also more difficult for yielding to take place. (e) Finally, the *mechanical rejuvenation* is evidently able to drive a polymer glass to a higher energy state, making it easier for large deformation to cause yielding. In other words, the glassy state is less resistant to conversion into a state of plastic deformation after rejuvenation. This effect should make it easier to yield and harder to build up chain tension. Therefore, brittle failure can be deferred after such a pretreatment.

Before closing this section, we wish to emphasize that the molecular model depicted in Figure 12.11 and Equations 12.1 and 12.2 has nothing to say about any structural or dynamic heterogeneity. In other words, it can neither depict nor anticipate inherent heterogeneity [161–163]. Heterogeneities may cause chain tension to arise in a spatially inhomogeneous manner. Conversely, if heterogeneities are domains where intersegmental interactions are stronger because of closer local packing, they would be more strongly affected if LBSs pass through them. This could be one way to envision how deformation reduces heterogeneities in agreement with the optical photobleaching measurements [146–150]. Finally, we point out that this model allows us to contemplate the effect of deformation rate on mechanical responses: Polymer glasses are usually observed to turn brittle upon increasing the applied extensional rate, in agreement with the LDWO hypothesis in Figure 12.3. If the mechanical failure is due to breakdown of the chain network via chain pullout, then slower deformation could allow more time for pullout to "nucleate," leading to behavior opposite to the common intuition. A preliminary study has revealed that both PMMA and PS become much less ductile at a given temperature when the extensional rate is lowered for a range of rates in a sizable temperature window [132]. Such results can be summarized as shown in the right-hand-side panel of Figure 12.13, along with the effect of temperature on mechanical responses of polymer glasses, shown in the left-hand-side panel.

Now we are ready to propose a molecular-level explanation for why a brittle polymer glass in tensile extension is ductile in uniaxial compression. PS of high molecular weight is a typical example, brittle for extension at room temperature, but exhibiting a robust ductile response to compression. Compressible polymer glasses may have higher density under compression, making it more difficult for chain pullout to take place. This

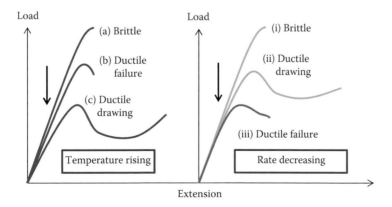

Figure 12.13 Load (tensile force) as a function of extension for uniaxial drawing of polymer glasses of high molecular weight as the temperature increases from (a) to (c) at a fixed rate (left) or at a fixed temperature as the drawing speed decreases from (i) to (iii) (right).

conclusion is consistent with the well-known effects [118,119] of hydrostatic pressure on tensile behavior of PS and PMMA. For example, PS turns ductile in extension when the applied pressure is at 0.4 kbar, that is, 40 MPa. Since the yield compressive stress for PS well exceeds 40 MPa, compression has the added "benefit," raising the level of f_{cp} well beyond the threshold f* ~ 0.25 nN. So that the chain network no longer suffer from a structural breakdown during compression.

12.4.2 CRAZING

Clearly, it is polymer physicists' task to explain why melt stretching, pressurization, mechanical rejuvenation, and physical aging affect the conditions for crazing. Strain localization is always challenging to theorize. Global strain localization on the scale of system's characteristic dimensions usually involves a "horse race" in tensile extension because the loading elements are in series. For example, initiation of shear yielding at one location across the specimen will directly affect the strain field elsewhere along the specimen. Crazing are extremely localized events that take place independently, on scales rather small relative to the dimensions of a specimen [12–14,107,109,110]. New crazes can emerge when the existing crazes do not grow to prevent craze-free regions from experiencing increasing strain. It is not obvious that the well-recognized heterogeneities are directly responsible for crazing. In other words, we can hardly claim that the emergence of crazing is evidence for or manifestation of inherent structural or dynamic heterogeneities [161,163].

Crazing is commonly viewed as a surface phenomenon [107,110]. Yet it is often regarded as a precursor to brittle fracture. Actually, crazing seems neither a necessary nor sufficient condition for brittle failure. PMMA can show brittle failure without visible crazing. Crazes can be present during ductile drawing.

There is a vast literature on initiation, growth, and characteristics of crazing [12–14,105]. Donald summarized [105] the comprehensive understanding on various aspects of crazing due to Kramer, Kambour, and others [13,14,105]. For several polymers, the degree of extension, λ_{craze}, in the crazing fibrils has been correlated with the maximum extension λ_{max} of the entanglement network to suggest the importance of entanglement. Moreover, craze initiation stress σ_{craze} is found to be independent of molecular weight, in the case of PS with sufficiently high molecular weight [164,165]. Thus, "scission was identified as the operative mechanism" and "historically, it was initially assumed that crazes formed via scission" [105]. The logic was that the disentanglement process leading to fracture should be molecular weight dependent and thus would not be a valid mechanism for crazing. For a "highly entangled polymer" such as bpA-PC that has an entanglement density v_e a factor of 10 higher than that of PS, scission would not occur since "bpA-PC so often deforms not by crazing, but readily forms shear bands" [105].

According to our model, the condition for chain pullout, which is a phrase we prefer to use in placement of "disentanglement" that we would like to reserve for rheology of polymers well above T_g, is determined by the magnitude of f_{cp}. In other words, the strength of the junctions (cf. Figure 12.1) depends on the structure of Gaussian chain hairpins that should be universal and independent of molecular weight. In the limit where the amount of chain ends that produce dangling strands is negligible, the condition for pullout is molecular weight independent just as scission is. Thus, the lack of molecular weight dependence for σ_{craze} is insufficient evidence to suggest that chain scission is a leading cause for crazing.

Since the strength of covalent bonds does not increase with pressure, the same logic applies equally well regarding the molecular mechanism for crazing. This is the logic introduced in the paragraph starting with "The third and last component of the model is…" to argue against chain scission as the responsible process for brittle fracture. As shown Figure 12.14, σ_{craze} increases significantly with the applied pressure P. This would mean that the applied pressure had altered the chain network structure, doubling the value of ψ at P = 0.6 kbar if chain scission was responsible. In other words, considerable chain pullout must have taken place at 1 bar leading to a value of ψ only half the value at 0.6 kbar. Therefore, crazing, like the brittle fracture, should be dominated by chain pullout. The chain pullout mechanism can also be applied to explain the molecular weight (MW) effect reported in De Focatiis et al. [165]: At a lower molecular weight, there are more non-load-bearing dangling strands so that ψ is lower, and $f_{ct}(\psi)$ is higher, causing pullout to occur more readily. On the other hand, early experiments show that the craze can grow at an angle to the principal stress in pre-orientated polymer glasses. Such behavior is at odds with the chain scission mechanism [139,166].

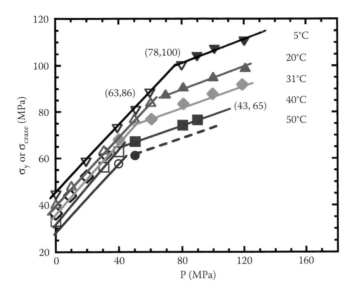

Figure 12.14 Craze initiation stress σ_{craze} (open symbols) and yield stress σ_y (filled symbols) at different temperatures during uniaxial extension of PS as a function of the applied hydrostatic pressure P. (Replotted from K. Matsushige, S. V. Radcliffe, and E. Baer, *J. Appl. Polym. Sci.*, 20, 1853, 1976.)

Crazing reflects severe strain localization of the chain network, involving great extension in the crazes and hardly any deformation elsewhere. Such inhomogeneous response to external deformation plausibly involves dynamic yielding of the chain network. The time required for an LBS to free itself at the junction (depicted in Figure 12.1) depends on the level of chain tension in the LBS that arises from deformation of the LBS. We conjecture that sufficient tension can induce over time LBSs to undergo pullout. Depending on the local packing, the strength of network junctions may vary widely. For example, junction (a) is expected to be stronger than junction (b) in Figure 12.15. There should also be a wide spectrum of rising chain tension during extension as well as a Gaussian length distribution of LBSs. Pullout starts at either weaker junctions or involves shorter LBSs that have higher tensions. Sliding of LBSs during pullout ceases at junctions that are strong enough to withstand the existing chain tension. Correspondingly, the chain network locally reaches a limiting level of extension. As the applied extension continues, this state of high network extension may spread into surrounding regions, and the width of observable crazes grows, which is defined by boundaries between highly extended network and intact parts of the network. In short, sufficient chain tension buildup is necessary for pullout to produce a localized failure of the chain network leading to crazing. Polymer glasses of low molecular weight cannot develop crazes because chain networking is a prerequisite for crazing. On a given experimental time scale if chain tension can increase to reach a threshold so that significant pullout can nucleate, then we could observe crazing. Our discussion on the BDT suggests that the magnitude of f_{cp}

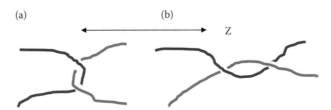

Figure 12.15 Two different junctions (a) and (b) in the chain network. For extension along the Z axis as indicated, junction (a) should be a much stronger one than junction (b) because fewer segments may need to be displaced around during pullout. With decreasing molecular weight, there would be fewer junctions of type (a).

is significantly lower than $f_b \sim 5$ nN. If $f_{cp} \ll f_b$ is true for most LBSs, then chain scission should not be the operative mechanism for either crazing or brittle fracture.

Within such a picture, it is straightforward to give a plausible explanation about the effects of mechanical rejuvenation, physical aging, and pressurization on crazing. The consequence of rejuvenation is to make it easier for segments to gain mobility during deformation. In other word, from the higher energy state, the primary structure can be more readily activated by the chain network. The more mobilized segments surrounding LBSs make it harder for chain tension to build up and reach the threshold for pullout. Crazing no longer takes place in such pretreated polymer glasses, as indicated in Section 12.3.2. If such rejuvenated polymer glasses are allowed to undergo sufficient physical aging, crazes reemerge because in the aged glass, segments settle down to lower energy states with lower mobility and are harder to get mobilized. Consequently, chain tension may grow until chain pullout takes place. The effect of pressure is to make the polymer glass more resistant against chain pullout. The suppression of pullout defers the onset of crazing as well as the point of the BDT.

To explain why melt stretching suppresses crazing is a bit challenging. It is more instructive to ask what microscopic changes melt stretching produces. We have already explained why melt stretching can make a brittle polymer glass ductile. From the geometric condensation of LBSs arises a higher value of ψ. The effects of melt stretching on crazing and BDT suggest that chain tension $f_{ct}(\psi)$ is a decreasing function of ψ. With increased ψ, the primary structure is more effectively activated. The increased segmental mobility makes it harder for chain tension to build up in the LBSs. As a result, pullout of LBSs leading to localized straining of the chain network may be avoided.

12.5 CONCLUSION

The purpose of this chapter is to indicate that the expedition has reached a new stage to identify molecular mechanisms and explore microscopic origins of a variety of mechanical behavior in large deformation of polymer glasses. Molecular mechanics of glassy polymers seems show rather promising progress although most of the theoretical discussions presented in this chapter are based only on plausible and logical speculations. The usefulness of such speculations lies in the fact that they can suggest many new experiments to interrogate the mechanical behavior of polymer glasses. For example, experiments to study melt stretching and rejuvenation effects on crazing have been motivated, designed carried out to learn how the recently available molecular picture might be used to rationalize the observations (S. Cheng, P. Lin, X. Li, J. Liu, Y. Zheng, and S.-Q. Wang, unpublished, 2015). Although the present theoretical picture has not explicitly taken into account any dynamical information at either segmental or chain level, it should be regarded as a valid starting point, to the best of our judgment. To proceed, we need to work diligently to collect more phenomenology necessary for an eventual establishment of a molecule-level theoretical description of yielding, brittle–ductile transition and crazing in polymeric glasses. We need to determine using molecular dynamics simulations whether these speculative ideas are realistic. Although we have only described extensional behavior regarding yielding, crazing, and BDT, a similar line of discussion can be carried out for compression. It is beyond the scope of this chapter to discuss the similarity and difference between extension and compression, although some preliminary comparisons have been made [67].

Evidently, polymer physics plays a pertinent role in the study of mechanical behavior of polymeric materials. Today, it seems feasible that chemical details, for example, microstructural differences of various glassy polymers can be accounted for in terms of several molecular parameters that characterize the structure of the chain network as shown in the discussions surrounding Equation 12.1. Physics-based molecular mechanics, depicted in Section 12.4, has enabled us to examine why bpA-PC is still ductile at 200°C below its T_g and PS is brittle at 30°C below its T_g. Neither constitutive modeling nor fracture mechanical depiction of failure of polymer glasses in large deformation can address why melt stretching, pretreatment by mechanical rejuvenation and pressurization, respectively, suppress crazing and brittle fracture. Neither of such methodologies seems to be able to satisfactorily explain the occurrence of a BDT in a narrow temperature window.

Building on the microscopic theory [54,55,65] of Schweizer and coworkers for polymer dynamics in the glassy state, we might hope to place on firmer grounds the concept of hybrid structure for polymer glasses under large deformation. In other words, if we can apply the theoretical description of Schweizer and coworkers to characterize the primary structure, a more comprehensive treatment may be made by depicting the interplay between the primary structure and chain network. It may be possible to develop a molecular-level account of nonlinear mechanical behavior of polymer glasses because the chain network structure can be straightforwardly parameterized in terms of well-known molecular variable such as packing length p and Kuhn length l_K. Although we cannot currently make quantitative predictions from the new molecular picture [69], many phenomena can be understood in terms of the central idea: the chain network drives the primary structure toward yielding during large deformation.

No polymer glasses of low molecular weight can undergo large post-yield deformation in either extension or compression. In our opinion, plastic deformation cannot take place in glassy polymers in the absence of a structurally and mechanically stable chain network. Most experimental systems contain various defects, leading to a stress concentration. Whenever an adequate chain network is present, any inherent imperfections do not necessarily lead to premature brittle failure. On the other hand, little is known about whether a defect-free polymer glass of low molar mass could exhibit adequate ductility, for example, capable of undergoing high extension. We do know [67], by increasing the molecular weight sufficiently, the same glassy polymer can change from a brittle glass to being entirely ductile, that is, able to undergo considerable plastic deformation without deliberately removing any impurities. In this chapter, we have avoided the phrase *plastic flow* in favor of *plastic deformation* since in the post-yield regime the chain network is still present and can undergo continuing elastic deformation, evidenced by the internal energy buildup [58].

It is encouraging that the recently available theoretical ideas can account for the five effects on BDT listed in Section 12.2.3 and the effects on crazing listed in Section 12.3. Both new experiments and computer simulations should be carried out to evaluate the basic characteristics of the chain network, by verifying the existence of LBSs emergent from large external deformation and determining the network structure in terms of the average chain length l_{LBS} of LBSs. In particular, we need to look for evidence of activation zones surrounding LBSs that might emerge during large deformation. On the other hand, it has remained a most formidable task to develop analytical tools to quantify the intermolecular chain uncrossability and derive l_{LBS} based on Gaussian chain statistics. Furthermore, it remains intractable to depict chain dynamics in the glassy state in the presence of external deformation. In other words, we do not know how segmental vitrification affects the relaxation dynamics and mechanical response of a glassy polymer chain to external deformation. Because of such profound theoretical difficulties, it is essential to use simple insights and intuitions to guide computer-simulation-directed research and to further accumulate new phenomenology through theoretically motivated experiments.

ACKNOWLEDGMENTS

The authors thank Xiaoxiao Li, Panpan Lin, and Jianning Liu for discussions and Ken Schweizer and Buck Crist for their helpful comments on the manuscript. This work is supported, in part, by the National Science Foundation through an EAGER grant (DMR-1444859) and ACS-PRF (54047-ND7).

REFERENCES

1. R. N. Haward and R. J. Young, *The Physics of Glassy Polymers* (Chapman & Hall, London), 2nd edn., 1997.
2. J. G. Williams, *Fracture Mechanics of Polymers* (Halsted Press [Ellis Horwood Limited], New York), 1987.
3. A. S. Argon, *The Physics of Deformation and Fracture of Polymers* (Cambridge University Press, New York), 2013.
4. A. J. Kinloch and R. J. Young, *Fracture Behavior of Polymers* (Elsevier, London), 1985.

5. I. M. Ward and J. Sweeney, *Mechanical Properties of Solid Polymers* (John Wiley & Sons, Ltd, Chichester, UK), 3rd edn., 2012.

6. P. G. De Gennes, *J. Chem. Phys.*, 55, 572, 1971.

7. M. Doi, and S.F. Edwards, *The Theory of Polymer Dynamics* (Oxford University Press, New York), 1986.

8. H. Watanabe, *Prog. Polym. Sci.*, 24, 1253, 1999.

9. T. C. B. McLeish, *Adv. Phys.*, 51, 1379, 2002.

10. S. Q. Wang, Y. Wang, S. Cheng, X. Li, X. Zhu, and H. Sun, *Macromolecules*, 46, 3147, 2013.

11. J. M. Hutchinson, *Prog. Polym. Sci.*, 20, 703, 1995.

12. R. P. Kambour, *J. Polym. Sci. Macromol. Rev.*, 7, 1, 1973.

13. E. J. Kramer, *Adv. Polym. Sci.*, 52–3, 1, 1983.

14. E. J. Kramer and L. Berger, in *Crazing in Polymers Vol. 2*, edited by H. Kausch (Springer, Berlin/Heidelberg), 1, 1990.

15. C. Creton, E. Kramer, H. Brown, and C.-Y. Hui, in *Molecular Simulation Fracture Gel Theory* (Springer, Berlin/Heidelberg), 53, 2001.

16. R. S. Hoy and M. O. Robbins, *J. Polym. Sci., Part B: Polym. Phys.*, 44, 3487, 2006.

17. R. S. Hoy and M. O. Robbins, *Phys. Rev. Lett.*, 99, 117801, 2007.

18. R. S. Hoy and M. O. Robbins, *Phys. Rev. E*, 77, 031801, 2008.

19. R. S. Hoy and M. O. Robbins, *J. Chem. Phys.*, 131, 244901, 2009.

20. R. S. Hoy and C. S. O'Hern, *Phys. Rev. E*, 82, 041803, 2010.

21. M. O. Robbins and R. S. Hoy, *J. Polym. Sci., Part B: Polym. Phys.*, 47, 1406, 2009.

22. J. Rottler and M. O. Robbins, *Phys. Rev. E*, 64, 051801, 2001.

23. J. Rottler and M. O. Robbins, *Phys. Rev. E*, 68, 011507, 2003.

24. J. Rottler and M. O. Robbins, *Comput. Phys. Commun.*, 169, 177, 2005.

25. M. Warren and J. Rottler, *Phys. Rev. E*, 76, 031802, 2007.

26. A. Y. H. Liu and J. Rottler, *Soft Matter*, 6, 4858, 2010.

27. M. Warren and J. Rottler, *Phys. Rev. Lett.*, 104, 205501, 2010.

28. M. Warren and J. Rottler, *J. Chem. Phys.*, 133, 164513, 2010.

29. A. Smessaert and J. Rottler, *Macromolecules*, 45, 2928, 2012.

30. S. S. Schoenholz, A. J. Liu, R. A. Riggleman, and J. Rottler, *Phys. Rev. X*, 4, 031014, 2014.

31. S. Jabbari-Farouji, J. Rottler, O. Lame, A. Makke, M. Perez, and J.-L. Barrat, *ACS Macro Lett.*, 4, 147, 2015.

32. D. L. Malandro and D. J. Lacks, *J. Chem. Phys.*, 107, 5804, 1997.

33. D. L. Malandro and D. J. Lacks, *J. Chem. Phys.*, 110, 4593, 1999.

34. D. J. Lacks and M. J. Osborne, *Phys. Rev. Lett.*, 93, 255501, 2004.

35. B. A. Isner and D. J. Lacks, *Phys. Rev. Lett.*, 96, 025506, 2006.

36. Y. G. Chung and D. J. Lacks, *J. Phys. Chem. B*, 116, 14201, 2012.

37. Y. G. Chung and D. J. Lacks, *Macromolecules*, 45, 4416, 2012.

38. Y. G. Chung and D. J. Lacks, *J. Chem. Phys.*, 136, 124907, 2012.

39. A. V. Lyulin, N. K. Balabaev, M. A. Mazo, and M. A. J. Michels, *Macromolecules*, 37, 8785, 2004.

40. A. V. Lyulin, B. Vorselaars, M. A. Mazo, N. K. Balabaev, and M. A. J. Michels, *Europhys. Lett.*, 71, 618, 2005.

41. A. V. Lyulin, J. Li, T. Mulder, B. Vorselaars, and M. A. J. Michels, *Macromol. Symp.*, 237, 108, 2006.

42. A. V. Lyulin and M. A. J. Michels, *J. Non-Cryst. Solids*, 352, 5008, 2006.

43. A. V. Lyulin and M. A. J. Michels, *Phys. Rev. Lett.*, 99, 085504, 2007.

44. T. Mulder, V. A. Harmandaris, A. V. Lyulin, N. F. A. van der Vegt, B. Vorselaars, and M. A. J. Michels, *Macromol. Theory Simul.*, 17, 290, 2008.

45. B. Vorselaars, A. V. Lyulin, and M. A. J. Michels, *Macromolecules*, 42, 5829, 2009.

46. B. Vorselaars, A. V. Lyulin, and M. A. J. Michels, *J. Chem. Phys.*, 130, 074905, 2009.

47. J.-L. Barrat, J. Baschnagel, and A. Lyulin, *Soft Matter*, 6, 3430, 2010.

48. D. Hudzinskyy, M. A. J. Michels, and A. V. Lyulin, *J. Chem. Phys.*, 137, 124902, 2012.

49. G. J. Papakonstantopoulos, R. A. Riggleman, J.-L. Barrat, and J. J. de Pablo, *Phys. Rev. E*, 77, 041502, 2008.

50. H. N. Lee, R. A. Riggleman, J. J. de Pablo, and M. D. Ediger, *Macromolecules*, 42, 4328, 2009.

51. R. A. Riggleman, J. F. Douglas, and J. J. de Pablo, *Soft Matter*, 6, 292, 2010.

52. K. Chen and K. S. Schweizer, *Macromolecules*, 41, 5908, 2008.

53. K. Chen, E. J. Saltzman, and K. S. Schweizer, *J. Phys. Condens. Matter*, 21, 503101, 2009.

54. K. Chen, E. J. Saltzman, and K. S. Schweizer, *Annu. Rev. Cond. Mat. Phys.*, 1, 277, 2010.

55. K. Chen and K. S. Schweizer, *Phys. Rev. E*, 82, 041804, 2010.

56. K. Chen and K. S. Schweizer, *Phys. Rev. Lett.*, 102, 038301, 2009.

57. P. Lin, S. Cheng, and S.-Q. Wang, *ACS Macro Lett.*, 3, 784, 2014.

58. P. Lin, J. Liu, and S.-Q. Wang, *Polymer* 89, 143, 2016.

59. L. C. E. Struik, *Physical Aging in Amorphous Polymers and Other Materials* (Elsevier Scientific Publishing Company, Amsterdam, Oxford, New York), 1978.

60. F. H. Stillinger, *J. Chem. Phys.*, 88, 7818, 1988.

61. F. H. Stillinger, *Science*, 267, 1935, 1995.

62. P. G. Debenedetti and F. H. Stillinger, *Nature*, 410, 259, 2001.

63. G. B. McKenna, *J. Phys. Condens. Matter*, 15, S737, 2003.

64. H. Eyring, *J. Chem. Phys.*, 4, 283, 1936.

65. K. Chen and K. S. Schweizer, *Macromolecules*, 44, 3988, 2011.

66. E. Munch, J.-M. Pelletier, B. Sixou, and G. Vigier, *Phys. Rev. Lett.*, 97, 207801, 2006.

67. J. Liu, P. Lin, S. Cheng, W. Wang, J. W. Mays, and S.-Q. Wang, *ACS Macro Lett.*, 4, 1072, 2015.

68. G. D. Zartman, S. Cheng, X. Li, F. Lin, M. L. Becker, and S. Q. Wang, *Macromolecules*, 45, 6719, 2012.

69. S.-Q. Wang, S. Cheng, P. Lin, and X. Li, *J. Chem. Phys.*, 141, 094905, 2014.

70. R. N. Haward and G. Thackray, *Proc. R. Soc. Lond. A*, 302, 453, 1968.

71. A. S. Argon, *Philos. Mag.*, 28, 839, 1973.

72. E. J. Kramer, *J. Polym. Sci., Part B: Polym. Phys.*, 43, 3369, 2005.

73. R. S. Hoy, *J. Polym. Sci., Part B: Polym. Phys.*, 49, 979, 2011.

74. T. A. Kavassalis and J. Noolandi, *Phys. Rev. Lett.*, 59, 2674, 1987.

75. Y. H. Lin, *Macromolecules*, 20, 3080, 1987.

76. N. Heymans, *J. Mater. Sci.*, 21, 1919, 1986.

77. J. Rault, *J. Non-Newtonian Fluid Mech.*, 23, 229, 1987.

78. L. J. Fetters, D. J. Lohse, D. Richter, T. A. Witten, and A. Zirkel, *Macromolecules*, 27, 4639, 1994.

79. S. Q. Wang, *Macromolecules*, 40, 8684, 2007.

80. C. P. Buckley and D. C. Jones, *Polymer* 36, 3301, 1995.

81. H. X. Li and C. P. Buckley, *Int. J. Plast.*, 26, 1726, 2010.

82. S. M. Fielding, R. G. Larson, and M. E. Cates, *Phys. Rev. Lett.*, 108, 048301, 2012.

83. G. A. Medvedev and J. M. Caruthers, *J. Rheol.*, 57, 949, 2013.

84. E. T. J. Klompen, T. A. P. Engels, L. E. Govaert, and H. E. H. Meijer, *Macromolecules*, 38, 6997, 2005.

85. L. C. A. van Breemen, E. T. J. Klompen, L. E. Govaert, and H. E. H. Meijer, *J. Mech. Phys. Solids*, 59, 2191, 2011.

86. M. C. Boyce, D. M. Parks, and A. S. Argon, *Mech. Mater.*, 7, 15, 1988.

87. E. M. Arruda and M. C. Boyce, *Int. J. Plast.*, 9, 697, 1993.

88. E. M. Arruda, M. C. Boyce, and H. Quintus-Bosz, *Int. J. Plast.*, 9, 783, 1993.

89. O. A. Hasan and M. C. Boyce, *Polym. Eng. Sci.*, 35, 331, 1995.

90. M. C. Boyce and E. M. Arruda, *Rubber Chem. Technol.*, 73, 504, 2000.

91. R. G. Larson, *Constitutive Equations for Polymer Melts and Solutions* (Butterworth-Heinemann, Boston, MA), 1988.

92. M. C. Boyce and R. N. Haward, The post-yield deformation of glassy polymers, in *The Physics of Glassy Polymers*, edited by R. N. Haward, and R. J. Young (Springer Science+Business Media, Dordrecht), 213, 1997.

93. E. M. Arruda, M. C. Boyce, and R. Jayachandran, *Mech. Mater.*, 19, 193, 1995.

94. J. G. Williams, Fracture mechanics, in *The Physics of Glassy Polymers*, edited by R. N. Haward, and R. J. Young (Chapman & Hall, London), 343–362, 1997.

95. J. D. Ferry, *Viscoelastic Properties of Polymers* (Wiley, New York), 1980.
96. C. A. Angell, K. L. Ngai, G. B. McKenna, P. F. McMillan, and S. W. Martin, *J. Appl. Phys.*, 88, 3113, 2000.
97. Q. Qin and G. B. McKenna, *J. Non-Cryst. Solids*, 352, 2977, 2006.
98. S. Q. Wang, S. Ravindranath, Y. Wang, and P. Boukany, *J. Chem. Phys.*, 127, 064903, 2007.
99. Y. Wang and S. Q. Wang, *Rheol. Acta*, 49, 1179, 2010.
100. A. Y. Malkin, A. Arinstein, and V. G. Kulichikhin, *Prog. Polym. Sci.*, 39, 959, 2014.
101. X. Zhu and S. Q. Wang, *J. Rheol.*, 57, 223, 2013.
102. S.-Q. Wang, *Soft Matter*, 11, 1454, 2015.
103. S. Q. Wang, S. Ravindranath, and P. E. Boukany, *Macromolecules*, 44, 183, 2011.
104. Y. Y. Wang and S. Q. Wang, *Macromolecules*, 44, 5427, 2011.
105. A. M. Donald, Crazing, in *The Physics of Glassy Polymers*, edited by R. N. Haward, and R. J. Young (Chapman & Hall, London), 295–342, 1997.
106. R. N. Haward and R. J. Young, Introduction, *The Physics of Glassy Polymers* (Chapman & Hall, London), 2nd edn., Chapter 1, 1997.
107. C. B. Bucknall, *Polymer*, 48, 1030, 2007.
108. C. B. Bucknall, *Polymer*, 53, 4778, 2012.
109. A. S. Argon, *Pure. Appl. Chem.*, 43, 247, 1975.
110. A. S. Argon, *Polymer*, 52, 2319, 2011.
111. P. J. Dooling, C. P. Buckley, and S. Hinduja, *Polym. Eng. Sci.*, 38, 892, 1998.
112. T. A. Tervoort, E. T. J. Klompen, and L. E. Govaert, *J. Rheol.*, 40, 779, 1996.
113. L. E. Govaert, P. H. M. Timmermans, and W. A. M. Brekelmans, *J. Eng. Mater T. ASME*, 122, 177, 2000.
114. H. G. H. van Melick, L. E. Govaert, and H. E. H. Meijer, *Polymer*, 44, 2493, 2003.
115. P. I. Vincent, *Polymer*, 13, 558, 1972.
116. P. I. Vincent, *Polymer*, 1, 425, 1960.
117. B. H. Bersted, *J. Appl. Polym. Sci.*, 24, 37, 1979.
118. K. Matsushige, S. V. Radcliffe, and E. Baer, *J. Appl. Polym. Sci.*, 20, 1853, 1976.
119. K. Matsushige, S. V. Radcliffe, and E. Baer, *J. Mater. Sci.*, 10, 833, 1975.
120. J. A. Sauer, *Polym. Eng. Sci.*, 17, 150, 1977.
121. S. M. Aharoni, *Macromolecules*, 18, 2624, 1985.
122. S. M. Aharoni, *J. Appl. Polym. Sci.*, 16, 3275, 1972.
123. L. P. Chen, A. F. Yee, and E. J. Moskala, *Macromolecules*, 32, 5944, 1999.
124. J. Y. Jho and A. F. Yee, *Macromolecules*, 24, 1905, 1991.
125. A. F. Yee and S. A. Smith, *Macromolecules*, 14, 54, 1981.
126. S. Wu, *J. Appl. Polym. Sci.*, 46, 619, 1992.
127. A. C. M. Yang, R. C. Wang, and J. H. Lin, *Polymer*, 37, 5751, 1996.
128. L. J. Broutman and R. S. Patil, *Polym. Eng. Sci.*, 11, 165, 1971.
129. L. J. Broutman and F. J. McGarry, *J. Appl. Polym. Sci.*, 9, 609, 1965.
130. Y. Tanabe and H. Kanetsuna, *J. Appl. Polym. Sci.*, 22, 1619, 1978.
131. D. H. Ender and R. D. Andrews, *J. Appl. Phys.*, 36, 3057, 1965.
132. X. Li and S.-Q. Wang, *ACS Macro Lett.*, 4, 1110, 2015.
133. G. Liu, H. Sun, S. Rangou, K. Ntetsikas, A. Avgeropoulos, and S. Q. Wang, *J. Rheol.*, 57, 89, 2013.
134. L. J. Broutman and S. M. Krishnakumar, *Polym. Eng. Sci.*, 14, 249, 1974.
135. H. G. H. van Melick, L. E. Govaert, B. Raas, W. J. Nauta, and H. E. H. Meijer, *Polymer*, 44, 1171, 2003.
136. A. R. C. Baljon and M. O. Robbins, *Science*, 271, 482, 1996.
137. A. R. C. Baljon and M. O. Robbins, *Macromolecules*, 34, 4200, 2001.
138. J. Rottler, S. Barsky, and M. O. Robbins, *Phys. Rev. Lett.*, 89, 148304, 2002.
139. P. Beardmore and S. Rabinowitz, *J. Mater. Sci.*, 10, 1763, 1975.
140. T. N. Krupenkin and G. H. Fredrickson, *Macromolecules*, 32, 5029, 1999.
141. T. N. Krupenkin and G. H. Fredrickson, *Macromolecules*, 32, 5036, 1999.
142. J. Rottler and M. O. Robbins, *Phys. Rev. E*, 68, 011801, 2003.
143. Y. Tanabe and H. Kanetsuna, *J. Appl. Polym. Sci.*, 22, 2707, 1978.

144. S. Cheng, L. Johnson, and S. Q. Wang, *Polymer*, 54, 3363, 2013.

145. D. S. A. De Focatiis and C. P. Buckley, *Polymer*, 52, 4045, 2011.

146. H. N. Lee, K. Paeng, S. F. Swallen, and M. D. Ediger, *Science*, 323, 231, 2009.

147. H.-N. Lee, K. Paeng, S. F. Swallen, and M. D. Ediger, *J. Chem. Phys.*, 128, 134902, 2008.

148. H.-N. Lee, K. Paeng, S. F. Swallen, M. D. Ediger, R. A. Stamm, G. A. Medvedev, and J. M. Caruthers, *J. Polym. Sci., Part B: Polym. Phys.*, 47, 1713, 2009.

149. H.-N. Lee and M. D. Ediger, *Macromolecules*, 43, 5863, 2010.

150. B. Bending, K. Christison, J. Ricci, and M. D. Ediger, *Macromolecules*, 47, 800, 2014.

151. E. Guth and H. M. James, *Indus. Eng. Chem.*, 33, 624, 1941.

152. H. M. James and E. Guth, *J. Chem. Phys.*, 11, 455, 1943.

153. F. Casas, C. Alba-Simionesco, H. Montes, and F. Lequeux, *Macromolecules*, 41, 860, 2008.

154. D. Long and F. Lequeux, *Eur Phys J E*, 4, 371, 2001.

155. M. L. Manning and A. J. Liu, *Phys. Rev. Lett.*, 107, 108302, 2011.

156. J. Rottler, S. S. Schoenholz, and A. J. Liu, *Phys. Rev. E*, 89, 042304, 2014.

157. S. Cheng and S. Q. Wang, *Phys. Rev. Lett.*, 110, 065506, 2013.

158. S. Cheng and S.-Q. Wang, *Macromolecules*, 47, 3661, 2014.

159. H.-N. Lee and M. D. Ediger, *J. Chem. Phys.*, 133, 014901, 2010.

160. C. M. Roland, S. Hensel-Bielowka, M. Paluch, and R. Casalini, *Rep. Prog. Phys.*, 68, 1405, 2005.

161. M. Goldstein, *J. Chem. Phys.*, 51, 3728, 1969.

162. M. D. Ediger, *Annu. Rev. Phys. Chem.*, 51, 99, 2000.

163. G. Adam and J. H. Gibbs, *J. Chem. Phys.*, 43, 139, 1965.

164. J. F. Fellers and B. F. Kee, *J. Appl. Polym. Sci.*, 18, 2355, 1974.

165. D. S. A. De Focatiis, C. P. Buckley, and L. R. Hutchings, *Macromolecules*, 41, 4484, 2008.

166. J. S. Harris and I. M. Ward, *J. Mater. Sci.*, 5, 573, 1970.

Modeling strain hardening in polymer glasses using molecular simulations

ROBERT S. HOY

This chapter reviews strain hardening in polymer glasses (GSH). After an introduction to GSH and a review of its industrial and scientific importance, the (rather poor) state of understanding of GSH that existed about 10 years ago is summarized, and then the dramatic progress made since is discussed. Focus is given to computational work, particularly that of the author, but key experimental developments are also discussed. The chapter concludes with a summary of open issues that remain in the field, and some plausible strategies for dealing with them.

13.1 WHY GLASSY-POLYMERIC STRAIN HARDENING IS INTERESTING

Massive, post-yield strain hardening is a mechanical response found only in bulk polymeric solids, melts, and networks. It arises from the uniquely polymeric properties of topological chain connectivity and random-walk-like large-scale chain structure. These features stabilize polymeric solids (and particularly glasses) against brittle failure and are critical for their use as lightweight, tough, load-bearing materials. The industrial importance of understanding strength and failure of polymeric materials has made quantitatively predicting their entire range of mechanical response one of the main goals of physical polymer science. Polymer scientists and engineers have worked toward this goal for many decades [1–4].

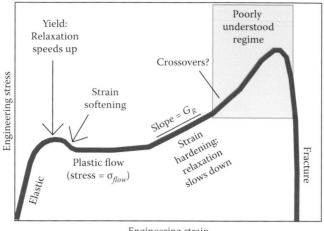

Figure 13.1 Mechanical response of ductlile polymer glasses.

Polymeric glasses are also of great intellectual interest due to the wide range of energy, length, and time (EL&T) scales controlling their properties. Figure 13.1, a schematic depiction of typical stress–strain curves for ductile polymer glasses, illustrates the mechanical consequences of this range. Undeformed systems occupy low-lying regions of a rugged free-energy landscape [5]. In the linear elastic regime, systems remain near their initial energy minima, and stress is controlled by local forces at the Kuhn scale or below. Yield occurs when energetic barriers to segmental rearrangements are overcome; the resulting increase in local mobility (see Chapter 10) [6–8] produces strain softening. In the plastic flow regime, the stress $\sigma = \partial W/\partial \epsilon$ is relatively constant, where W is the work done on the system and ϵ is its strain. Strain hardening begins when $\partial W/\partial \epsilon$ must increase to drive further segmental rearrangements while maintaining chain connectivity [9]. This increase becomes more dramatic as the scale over which chains are oriented approaches that of the entanglement mesh. Finally, fracture occurs when cohesive forces, either primary covalent bonds or secondary nonbonded interactions, no longer suffice to maintain material integrity (Chapter 12).

While great progress has been made in recent years toward understanding phenomena at strains up to and including the early stages of strain hardening within a single framework, a coherent theoretical picture including dramatic hardening and fracture (i.e., the "poorly understood regime" in Figure 13.1) remains elusive. The remainder of this chapter will cover the dramatic progress made toward obtaining such a picture over the past dozen years, and outline remaining "roadblocks" as well as potential strategies for overcoming them.

13.2 DOMINANT THEORETICAL VIEW OF STRAIN HARDENING PRIOR TO 2003

The first theory of strain hardening in polymer glasses, formulated by Haward [9], was based on rubber elasticity. It assumes that entanglements between polymer chains act like chemical cross-links in a rubber, and that strain hardening arises from the decrease in entropy of the associated, affinely deforming cross-linked network. The stress beyond the plastic flow regime is assumed to take the form

$$\sigma(\bar{\lambda}) \simeq \sigma_{flow} + G_R g(\bar{\lambda}) \tag{13.1}$$

where (following Figure 13.1) σ_{flow} is the plastic flow stress, G_R is the "strain hardening modulus," and $\bar{\lambda}$ is the stretch tensor describing the macroscopic deformation (Figure 13.2, Table 13.1). Here, $\tilde{g}$ is a dimensionless

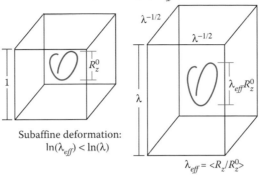

Stretches: λ (macroscopic), λ_{eff} (chain-scale)

$\lambda_{eff} = \langle R_z/R_z^0 \rangle$

Figure 13.2 Schematic comparision of the (in general subaffine) chain-scale stretch $\bar{\lambda}_{eff}$ to the macroscopic-sample-level stretch $\bar{\lambda}$. Here, the vertical direction is z and the transverse directions are x and y. A uniaxially applied constant-volume deformation with $\lambda_{zz} \equiv \lambda$ (and therefore $\lambda_{xx} = \lambda_{yy} = \lambda^{-1/2}$) will produce the depicted subaffine chain-scale deformation satisfying $|\ln(\lambda_{eff})| < |\ln(\lambda)|$.

Table 13.1 Functional forms for strain hardening assuming affine, constant-volume deformation by a stretch $\bar{\lambda}$

Deformation mode	$\bar{\lambda}$	$g(\lambda)$	$\tilde{g}(\lambda)$
Uniaxial	$\lambda_x = \lambda_y = \lambda_z^{-1/2} = \lambda$	$\lambda^2 - \dfrac{1}{\lambda}$	$\dfrac{1}{3}\left(\lambda^2 + \dfrac{2}{\lambda}\right)$
Plane strain	$\lambda_x = \lambda_z^{-1} = \lambda, \lambda_y = 1$	$\lambda^2 - \dfrac{1}{\lambda^2}$	$\dfrac{1}{3}\left(\lambda^2 + 1 + \dfrac{1}{\lambda^2}\right)$

Note: $g(\lambda)$ and $\tilde{g}(\lambda)$ are defined in Equations 13.1 and 13.2. Formulas for constant-volume shear (both pure and simple) deformations can be related to the below two cases via rotations of the Cartesian axes.

function that describes the reduction in the system's configurational entropy, for example, from the classical theory of rubber elasticity [10]:

$$S(\bar{\lambda}) = S_0 - \rho_x k_B \tilde{g}(\bar{\lambda}), \tag{13.2}$$

where ρ_x is cross-link density, or the classical theory's nonlinear variants. Then, $g(\bar{\lambda}) = \partial\tilde{g}/\partial\bar{\lambda}$, and from the rubber-elastic Helmholtz free energy $F(\bar{\lambda}) = -TS(\bar{\lambda})$, $\sigma(\bar{\lambda}) = \partial F/\partial\bar{\lambda}$ gives Equation 13.1. Haward reasonably assumed that entanglements in the glassy state are inherited from the melt, and that ρ_x could therefore be replaced by the entanglement density ρ_e from melt rheology, for example, from the melt plateau modulus $G_N^0 \simeq \rho_e k_B T$ [11] measured for the given polymer far above its T_g. Thus, the theory predicts [9] a glassy strain hardening modulus $G_R = \rho_e k_B T$.

Equation 13.1 and its nonlinear extensions correctly predict the *shape* of stress–strain curves for a wide variety of glassy polymers under uniaxial stress, plane strain, and shear [12–14]. The reason they do so is that $g(\bar{\lambda})$ has the same functional form as the increase in stress with λ, for each of these strain protocols (Table 13.1). For this reason, Haward's theoretical approach and others based upon it were widely adopted. A particularly well-developed theory of this type was the "eight-chain" model developed by Arruda and Boyce [15,16]. As was usual for models based on Equation 13.1, σ_{flow} was calculated separately and independently from the strain hardening contribution, using a modified version [17] of Argon's earlier phenomenological theories for polymeric plasticity [18,19]. The two major advances of the eight-chain model were its improved treatments of deformation-mode-dependence (for example, tension versus shear) and nonlinear, "dramatic" hardening. Both were captured by treating the entanglement network producing GSH as a uniformly cross-linked rubber of strand length N_e, and accounting for the finite extensibility of Gaussian-coil segments between cross-links. The model's prediction for the difference between principal stresses along axes i and j (in the strain hardening regime) is

$$\sigma_i - \sigma_j = \tau_{flow}^{ij} + G_R \frac{L^{-1}(h)}{3h}(\lambda_i^2 - \lambda_j^2),$$

$$\text{with } h = \left(\frac{\tilde{g}(\bar{\lambda})}{\sqrt{N_e/C_\infty}} \right).$$

(13.3)

Here, τ_{flow}^{ij} is an independently modeled, rate- and temperature-dependent plastic flow stress [17], L^{-1} is the inverse Langevin function, and C_∞ is the chain stiffness constant (characteristic ratio). The $(\lambda_i^2 - \lambda_j^2)$ term is equivalent to the $g(\bar{\lambda})$ in Table 13.1. The h term captures finite extensibility; $h \to 0$ for a random coil obtained by taking $N_e \to \infty$, and $h \to 1$ as chains pull taut ($\tilde{g}(\bar{\lambda}) \to \sqrt{N_e/C_\infty}$). As discussed in numerous review articles (for example, [20]), Equation 13.3's description of rubber elasticity well *above* T_g is excellent.

13.3 PROBLEMS WITH THE OLD VIEW

The Boyce group noted early on [16,21] that fitting experimental stress–strain curves for various polymer glasses to Equation 13.3 produced values of N_e that were far smaller than estimates from G_N^0, and that decreased with decreasing T. They attempted to explain these results by assuming that entanglement is strongly temperature dependent, and ρ_e increases rapidly below T_g. However, their fits to Equation 13.3 produce values of N_e (in the expression for h) that are inconsistent with the values of ρ_e in the prefactor, and show a dependence on deformation protocol (for example, tension versus compression [16]) that is not explained by the theory. More seriously, the assumption of strongly temperature-dependent entanglement density is in direct conflict with extensive experimental data that strongly support the notion that the same N_e measured in melts (from the standard [22] relation $G_N^0 = 4\rho k_B T/5N_e$) controls the ultimate mechanical properties of glasses. For example, the craze extension ratio Λ is excellently predicted by

$$\Lambda \simeq \left(\frac{N_e}{C_\infty} \right)^{1/2}$$

(13.4)

for essentially all glassy polymers and is nearly independent of temperature for $T \lesssim 0.9T_g$ [3,23,24]. Moreover, such T-dependent entanglements would necessarily also be highly time and σ dependent (i.e., they would have a characteristic lifetime $\tau(\sigma,T)$), and would therefore be subject to strain-rate-dependent, stress-activated annihilation. The physical validity of including such entities in a rubber-elasticity-based theory for GSH, like those of References 21 and 25, is, at best, unclear.

Another problem with entropic models is that in the glassy state, the conformational entropy of polymer chains is much less than its equilibrium value (because glasses are nonergodic and not all microstates of chains are equally likely to be occupied). Furthermore, its evolution under deformation is *a priori* unknown, and one can no longer safely assume (as was done in Equation 13.2) that deformation will simply decrease $(S - S_0)/\rho_x k_B$ by a factor of $\tilde{g}(\bar{\lambda})$. This casts serious doubt on the applicability of the approach used to derive the hardening terms in Equations 13.1 and 13.3.

A final problem with the entropic approach is its (non)treatment of energetic contributions. From the first law of thermodynamics ($dW = dQ + dU$, where W is the work done *on* the system, U is the internal energy of the system, and Q is the heat transfer *away from* the system), stress can be separated into an energetic component σ^U and a thermal component σ_z^Q:

$$\sigma = \frac{\partial W}{\partial \epsilon} \quad : \quad \sigma^U = \frac{\partial U}{\partial \epsilon} \quad : \quad \sigma^Q = \frac{\partial Q}{\partial \epsilon} = \sigma - \sigma^U, \tag{13.5}$$

where $\epsilon_{eff} \equiv \ln(\lambda)$ is the true strain along the strain-controlled direction. Internal energy increases during strain hardening (measured by differential scanning calorimetry (DSC) and deformation calorimetry experiments [26–28]) can account for up to ~20% of the stress in "tightly" entangled systems ($N_e \lesssim 10C_\infty$), particularly in the dramatic hardening regime that Equation 13.3 is designed to treat. Examples of relatively loosely and tightly entangled polymers are polystyrene (PS) and polycarbonate (PC), respectively [26]. Early simulations of GSH [29,30] yielded similar findings, and quantitatively associated σ^U with pair, bond, angle, and torsional contributions (i.e., the increases in U result from mechanically induced distortions of local structure).

Haward had noted the above-mentioned discrepancies in the magnitude of G_R, and pointed out that the same functional forms $g(\bar{\lambda})$ dominate the large-strain response of hyperelastic (neo-Hookean) materials, as early as 1987 [31]. He later noted [13] that the prefactors C obtained when fitting $G_R = C\rho_e k_B T$ to experimental data for G_R are nontrivially chemistry dependent, and suggested that G_R is primarily determined by polymer glasses' local, Kuhn-segment-scale structure, rather than entanglements. Buckley's popular "glass-rubber" model of GSH [32] included energetic terms modeling these distortions, but (as was accepted practice in the mechanical engineering community from the 1950–1990s) assumed that equilibrium thermodynamics can be used to calculate the free energy increase during deformation and therefore the stress. As noted above, this assumption is invalid.

Unfortunately, it seems that these important observations and ideas were largely ignored by most of the modeling community, and the flaws in the entropic picture of GSH were not seriously challenged (at least in any sustained fashion) until they were dramatically illustrated in experiments performed by Van Melick and colleagues [33] in 2003. They measured G_R in PS–polyphenylene oxide (PPO) blends over a wide range of T. Varying the blend composition from PS-rich to PPO-rich and applying further (variable) cross-linking allowed them to vary ρ_e over a large range. They measured G_R over a wide range of temperatures and showed it decreases linearly with T, and becomes small near the glass transition temperature T_g, scaling roughly as $(1 - T/T_g)$. Moreover, while their measured G_R were indeed linearly proportional to entanglement densities ρ_e^{rheo} obtained from independent measurements of G_N^0, the resulting prefactors $A(T)$ in $G_R = A(T)\rho_e^{rheo}$ were two orders of magnitude larger than $k_B T$, even near T_g; see Figure 13.3. Alternatively, fitting the same data to $G_R = \rho_e k_B T$ gave values of ρ_e that were similarly two orders of magnitude (or more) larger than their ρ_e^{rheo}. They noted that taking such values literally would imply the entanglement mesh size φ [22] approaches monomeric dimensions, which is of course nonsensical. These issues were nicely summarized by E. J. Kramer, who challenged [34] the theory/simulation community to produce better, more physically based models for GSH.

13.4 A DIGRESSION ON LENGTH AND TIME SCALES: CHOOSING A MOLECULAR MODEL

As mentioned above, GSH involves phenomena with wide ranges of controlling energy, length, and time scales. These scales are not generally well separated, and their couplings are both *a priori* unclear and generally strain dependent. For example, long-range stress transmission via covalent bond tensions that are correlated on the scale of entanglement length becomes important only in the "dramatic" hardening regime [35]. This complexity presents a challenge for both analytic theory and computational modeling: which scales to treat? Figure 13.4 depicts length scales relevant to polymer glasses. Under typical experimental conditions ($T \lesssim 300$ K, $|\sigma| \lesssim 100$ MPa), glasses remain nonconducting and no significant e^- transport occurs. Electrons mediate excluded volume, van der Waals, covalent bond, angle, and torsional interactions, but since these interactions do not change very much with T or σ, they can be safely assumed to be independent of T and

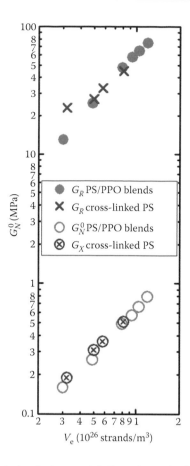

Figure 13.3 Comparison of the strain hardening moduli G_R to the melt plateau moduli G_N^0 measured for van Melick et al.'s [33] systems. If the entropic prediction $G_R = \rho_e k_B T$ were correct, the ratio G_R/G_N^0 would be of order unity for these systems since glasses were measured slightly below and melts slightly above T_g. The actual, measured ratio is of order 100, illustrating the serious problems with this prediction. (Figure taken from E. J. Kramer, *J. Polym. Sci., Part B: Polym. Phys.*, 43, 3369, 2005.)

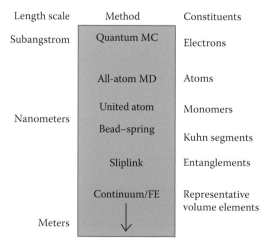

Figure 13.4 Modeling techniques for polymer glasses, associated with the smallest length scales/constituents they treat explicitly.

σ, and thus electrons can be coarse-grained out. Similar arguments apply for the interactions between small atoms within monomers.

The problem becomes more difficult on scales comparable to or above those of monomers. Chemical details are, of course, critical in determining the strength and functional form of the covalent bond, bond angle, and torsional interactions $U_{bond}(\ell)$, $U_{bend}(\theta)$, and $U_{tors}(\psi)$, where ℓ is bond length, and θ (ψ) are, respectively, the bending (torsion) angles formed by two (three) consecutive bond vectors. They are likewise critical in determining the structure and hence the mechanical properties of the solid state [36,37].

United-atom models are typically coarse-grained at the level of one to three "beads" per monomer, and include forms for U_{bond}, U_{bend}, and U_{tors} that are (at least ideally) obtained by matching to experimental data for specific polymer chemistries [36,37]. Consequences for mechanics can be large, for example, torsional interactions affect the balance of strain softening and hardening, which plays a major role in determining stability against fracture [3]. Energy storage and dissipation also occurs at the bending and torsion angle levels during GSH [29,30]. If one wishes to model GSH for a specific polymer chemistry [38–41], united-atom models are clearly the best choice (since atomistic models remain too computationally expensive).

However, the most essential features for those features of GSH common to all well-entangled ductile polymer glasses (Figure 13.1) seem to be excluded volume and van der Waals interactions, chain connectivity, and chain stiffness at the bond-angle level. The Kremer–Grest (KG) bead–spring model [42,43] includes all of these, and has proven ability to capture the behavior of real polymers to an extraordinary degree. Its utility for studies of the glass transition, aging, yield, crazing, and so on is described in Chapters 3 and 11. Notably, its nonlinear rubber-elastic response at high T [44] agrees excellently with predictions of the eight-chain model for rubber [15].

Since the KG model coarse-grains at the level of ~3 chemical monomers per CG bead [42], and has correspondingly softer $U_{bond}(\ell)$ and $U_{bend}(\theta)$ compared to united-atom models, its computational cost per (unit volume × unit time) is ~2 orders of magnitude lower. It is thus able to treat correspondingly longer length and time scales. For mechanics studies, simulations of ~$10^5 - 10^6$ beads (sufficient that finite-size effects are minimal for modeling of bulk systems) can now easily be conducted at strain rates that map to rates achievable in split-Hopkinson-bar-type experiments [45,46], that is, $10^4 - 10^5$ s^{-1}, up to strains producing fracture. Its larger bead diameter also makes it more readily adaptable to studies of glassy polymer *nanocomposite* mechanics; see, for example, [47–49]. Yet another advantage is that it is more readily comparable to the extensive, microphysics-based polymer mechanics theories of Chen and Schweizer [50–55] and Fielding et al. [56,57], which are coarse-grained at the Kuhn segment scale.

Here, we are interested primarily in those features of GSH common to all ductile polymer glasses (Figure 13.1), and so, below, will focus on KG studies except where noted otherwise. The simulations described below were conducted using protocols very similar to those described in Chapters 3 and 11; here, we will discuss only those aspects specifically relevant to GSH studies. Since we are focusing on bulk systems (i.e., not thin films or surface effects), we consider simulations employing three-dimensional periodic boundary conditions, with periods L_i along the Cartesian directions $i = x$, y, and z. Systems contain M polymer chains, each containing N monomers. M is chosen to be sufficiently large that no chains cross periodic boundaries more than once, and the total number of monomers employed, $MN \gtrsim 10^5$, is sufficiently large to be free of major finite-size effects.

All monomers have mass m and interact via a truncated and shifted Lennard-Jones (LJ) pair potential:

$$U_{LJ}(r) = 4u_0 \left[\left(\frac{a}{r} \right)^{12} - \left(\frac{a}{r} \right)^6 - \left(\frac{a}{r_c} \right)^{12} + \left(\frac{a}{r_c} \right)^6 \right], \tag{13.6}$$

where a is monomer diameter, u_0 is the intermonomer binding energy, and r_c is the cutoff radius; $U_{LJ} = 0$ for $r > r_c$. Covalent backbone bonds between neighboring beads along each chain are modeled using the standard finitely extensible nonlinear elastic (FENE) potential:

$$U_{FENE}(\ell) = -\frac{1}{2} k R_0^2 \ln \left[1 - \left(\frac{\ell}{R_0} \right)^2 \right], \tag{13.7}$$

Table 13.2 Values of the rheological entanglement lengths (N_e),[a] chain stiffness constants (C_∞), and tube diameters ($d_t = (N_e \ell_0 / k)^{1/2}$) for [60,62] bead–spring melts at $k_B T/u_0 = 1.0$

k_{bend}	N_e	C_∞	d_t/a
0	85	1.8	12.3
0.75	42	2.2	9.4
1.5	27	2.7	8.4
2.0	23	3.3	8.6

Source: Adapted from R. S. Hoy and M. O. Robbins, *Phys. Rev. E*, 72, 061802, 2005.

Note: Entanglement densities in glasses are within a few percent of melt values, and C_∞ drops by only ~5% over the range $1.0 > k_B T/u_0 \geq 0$.

[a] References 65–67 and other early works reported $N_e \simeq 70$ for $k_{bend} = 0$ chains. This value was shown in Hoy et al. [62] to be too low due to systematic errors in the formulas used to calculate N_e, and corrected to $N_e = 85$.

with the canonical parameters [24] $R_0 = 1.5a$ and $k = 30u_0 a^{-2}$. The equilibrium bond length is $\ell_0 = 0.96a$; note the condition $\ell_0 < a$ and the resulting length scale competition was shown in Abrams and Kremer [43], Bennemann et al. [58] and Mackura and Simmons [59] to completely suppress crystallization. This choice of parameters prevents chain crossing and is therefore suitable for simulations of entangled systems.

Entanglement density is controlled by varying the strength of the bending potential:

$$U_{bend}(\theta) = k_{bend}(1 - \cos(\theta)), \tag{13.8}$$

where θ is the angle formed by two consecutive bond vectors and is zero for straight trimers. Relevant values for chain and entanglement structure are given in Table 13.2. Topological analyses [60–62] show that N_e decreases from 85 to 23 as k_{bend}/u_0 is increased from 0 to 2.0, and that entanglements in the glassy state are indeed (to an excellent approximation) inherited from the melt. Note that a key parameter for GSH—N_e/C_∞— drops from 50 to 7 over this range, spanning the range from "loose" to "tight" entanglement [35,63].

For GSH studies, glasses are typically obtained by rapidly cooling melts from well above T_g to well below it. After cooling, systems are deformed to a stretch $\bar{\lambda}_{final}$ at a true strain rate $\dot{\epsilon}$. If the extension/compression direction is taken to be z (following Figure 13.2), then $\lambda = \lambda_z = L_z/L_z^0$, $\epsilon \equiv \ln(\lambda)$, and $\dot{\epsilon} = \dot{L}_z/L_z$. Transverse directions are either allowed to vary at fixed (typically zero) pressure, or controlled as described in Table 13.1. For fundamental GSH studies, compressive ($\dot{\epsilon} < 0$) rather than tensile ($\dot{\epsilon} > 0$) deformation is often preferred since compression suppresses void formation and strain localization, that is, it promotes homogeneous deformation. This is also typical experimental practice [3]. Applied strain rates $|\dot{\epsilon}|$ are typically in the range $10^{-6}/\tau - 10^{-4}/\tau$. The LJ time unit τ maps to $10^{-10.5\pm1}$ s [42], so the lower $|\dot{\epsilon}|$ corresponds to strain rates $10^{-4.5\pm1}$ s^{-1}.

13.5 INSIGHTS FROM SIMULATION AND THEORY, 2005–2011

The theoretical polymer science community responded enthusiastically to Kramer's challenge [34]. Over the next several years, multiple workers, including Lyulin, Hoy, Robbins, Riggleman, Schweizer, Larson, and their collaborators, performed simulations [35,38–41,47,64–74] and developed microphysics-based analytic theories [50–57]. These coincided with a renewal of experimental interest in the field [7,8,33,75–86], and much fruitful experiment–theory interplay. Much of this work amounted to bringing new perspectives and tools to the old problem. Taken together, the resulting developments revolutionized our understanding not only of GSH, but of glassy polymer mechanics more generally.

For GSH, the key new insights were: (i) strain hardening is fundamentally nonentropic. Instead it is fundamentally *viscoelastoplastic*. Hardening is controlled by many of the same mechanisms that control plastic flow; indeed, G_R scales linearly with σ_{flow} when pressure, T, or strain rate is varied. (ii) The initial stages of strain hardening (before the "dramatic" regime) are not dominated by the entanglement network. Instead,

"Gaussian" hardening (linear in $g(\bar{\lambda})$) reflects the increased plasticity necessary to produce large-scale affine deformation of chains while maintaining their connectivity. Only in the dramatic regime where stress is supralinear in $g(\lambda)$ do entanglements play a dominant role, but then their contribution to the stress is associated with energetic terms rather than entropic ones. (iii) The magnitude of G_R is therefore set by local interactions at the Kuhn segment scale and below. Relevant physical quantities setting G_R include the Kuhn length l_K, the intermonomer (noncovalent) binding energy u_0, and monomer diameter a. A natural stress scale u_0/a^3 has the same order of magnitude as both G_R and σ_{flow}. (iv) Correct microscopic theories of GSH must incorporate both *glassy-state* physics (for example, activated hopping over energy barriers [51–55]) and *melt-like* relaxation mechanisms.

Results (i–ii) (and to some extent (iii) and (iv)) have been extensively supported by a wide variety of recent experiments, microscopic theories, and constitutive modeling efforts [51–55,77,80–82,87,88], which have provided strong support to the notion that entanglements play only a secondary role in glassy-polymeric strain hardening, at least for the majority of synthetic polymers and in the Gaussian hardening regime. The "secondary" role is indicated by the fact that these works analyze and/or predict Gaussian hardening without invoking entanglements directly. Other recent experiments have shown that melt-like relaxations *are* important in deformed glasses [7,8,77,89], but also (and critically) that segmental dynamics and larger-scale relaxation dynamics are nontrivially coupled below T_g [90], that is, that one *cannot* assume that polymer chains in deformed glasses explore their rheological tubes in the same way as they do in rubbers above T_g.

In the following sections, we greatly expand on each of these points, and also discuss several additional insights. While we focus on bead–spring simulations, at least qualitatively equivalent results are obtained with united-atom models [36,37,91].

13.5.1 NONENTROPIC, VISCOELASTOPLASTIC NATURE OF GSH

Hoy and Robbins showed [65] that the bead–spring model can accurately capture many aspects of GSH experiments, including both Gaussian and "Langevin" (supralinear in $|g(\bar{\lambda})|$) hardening. By examining temperatures ranging from zero to near T_g, they found $G_R(T) \simeq G_R^0(1 - T/T_g)$, where G_R^0 is the zero-T value. This extension of van Melick's [33] result to low T rather conclusively proved that GSH cannot be *primarily* entropic. Instead, they showed [65] that strain hardening is closely associated with plastic flow. Stress–strain curves obtained at different strain rates, or in systems with different LJ potential cutoff radii r_c, nearly collapse when scaled by σ_{flow}. The latter collapse shows that stress is largely set by the amount of material that must be mobilized to produce plastic deformation, that is, both increase with increasing r_c. That the functional form of stress increase, $\sigma \sim g(\lambda)$, is the same as for entropic hardening remained somewhat mysterious; a potential explanation is given in Section 13.5.5.

Follow-up work by the same authors [66,67] both placed these results on a firm microscopic foundation and greatly extended them. Simulated stress–strain curves for a wide range of chain stiffnesses (and hence N_e) are well fit by Equation 13.3, but as discussed above, the fit parameters such as ρ_e exhibit unphysical trends. A key observation of References 66 and 67 was that in the limit $T \to 0$, the dissipative component of stress (σ^Q from Equation 13.5) is directly proportional to the rate R_p (per unit ϵ) of local, monomer-scale plastic rearrangements; R_p is the percentage of atoms that have undergone changes in the neighbors in their first coordination shell over a strain increment $\delta\epsilon$. Figure 13.5 shows results for σ_Q and R_p in uniaxial-stress compression of two systems with very different hardening responses. The lower curves are for a loosely entangled system that exhibits nearly Gaussian hardening, while the upper curves are for a tightly entangled system that exhibits strongly Langevin hardening with significant energetic contributions. For both systems, the correlation of σ^Q with R_p is clear; large fluctuations in both quantities match nearly perfectly. Similar results were found in plane strain, and for a wider variety of systems (including even very short-chain systems that produce nearly perfect plastic flow at a constant σ_{flow}). Experiments [78] also suggest that the energetic cost of local rearrangements is a major contributor to the hardening stress.

Another key observation of these studies involved the energetic stresses σ^U. Figure 13.6a shows σ, σ^Q, and σ^U for a tightly entangled system deformed in plane strain. The thermal component of stress remains essentially linear (Gaussian) even when the overall stress is highly supralinear in $g(\bar{\lambda})$. This means that the

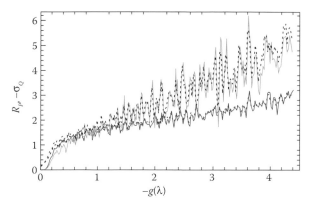

Figure 13.5 Correlation of dissipative stress with plasticity: Solid lines show the rate of plastic rearrangements R_p as a function of $g(\lambda)$ for well-entangled systems with $k_{bend}/u_0 = 1.5$ (upper curve) and $k_{bend} = 0$ (lower curve) at $T = 0$. Dashed lines show the corresponding dissipative stresses σ_Q. (Adapted from R. S. Hoy and M. O. Robbins, *Phys. Rev. Lett.*, 99, 117801, 2007.)

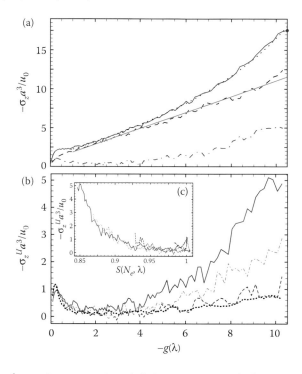

Figure 13.6 Breakdown of stress into energetic and dissipative terms, and relation to subaffine deformation on the scale of the entanglement mesh: (a) Results for σ_z (red solid line), σ_z^Q (dashed black line), and σ_z^U (blue dot-dashed line) for plane strain compression of a tightly entangled ($k_{bend}/u_0 = 2.0$) system at $T = 0$. The dotted black line shows a fit of σ_z to a modified eight-chain model [67], and the straight green line is a linear (Gaussian) fit to σ_z^Q. (b) Energetic stress σ_z^U for plane strain compression at $T = 0$ for systems ranging from tightly to loosely entangled: $k_{bend}/u_0 = 2.0$ (red solid line), 1.5 (green dot-dashed line), 0.75 (blue dashed line), and 0 (black dotted line). The inset (c) shows the same stresses and systems, but with σ_z^U plotted against $S(N_e, \lambda)$ (Equation 13.9); strain increases going from right to left in the plot. (Adapted from R. S. Hoy and M. O. Robbins, *Phys. Rev. E*, 77, 031801, 2008.)

nonlinear, "Langevin" contribution is (at least for this system) directly associated with energetic terms rather than entropic ones. Results for other (less tightly entangled) systems are consistent with these ideas, as shown in Figure 13.6b; as N_e/C_∞ decreases from ~50 to ~7, both σ^U and its nonlinear increase with strain grow dramatically. Note that both the fractions σ^U/σ and their trends with strain and entanglement density are consistent with experiments [26].

These trends can be explained by examining how chain configurations evolve under applied deformation. On the largest scales, that is, on chemical distances n comparable to the chain length N, deformation is affine, but for $n \sim N_e$, it can be significantly subaffine. For monomers separated by chemical distance n and mean-squared Euclidean distance $\langle R^2(n) \rangle$, the inset to Figure 13.6 shows a plot of stress versus the (sub)affinity measure

$$S(n, \bar{\lambda}) = \frac{\langle R^2(n; \bar{\lambda}) \rangle}{\langle \tilde{g}(\bar{\lambda}) R^2(n; \bar{I}) \rangle}, \tag{13.9}$$

where the identity matrix $\bar{I}$ in the denominator indicates values in the undeformed glass, and brackets indicate averaging over all chains. When deformation of chain configurations at scale n is affine to the macroscopic stretch $\bar{\lambda}$ (Equation 13.2), $S(n, \bar{\lambda}) = 1$. In general, of course, chain configurations relax via both thermally and stress-activated mechanisms during active deformation, and this relaxation reduces S; in turn, the condition $S(n, \bar{\lambda}) < 1$ requires both plastic deformation and energy transfer to internal terms (i.e., U_{LJ}, U_{bond}, and U_{bend}) to maintain local stress balance. As shown in the inset plot, the rate of this internal energy increase (σ^U) correlates well with subaffine stretching of chains at the entangled segment level. Note that $S(n, \bar{\lambda})$ always increases with increasing n and decreases with increasing strain; thus, at equal strains, it is smaller for more tightly entangled systems, explaining *why* they exhibit larger σ^U.

Other simulations [30,38] and investigation of related metrics show consistent trends. Hoy showed [35] that the subaffine stretching on scales $n \sim N_e$ is directly associated with correlated bond tension fluctuations over the same chemical scale. Figure 13.7a shows results for the correlations of the tensions $T = \partial U_{FENE}/\partial \ell$ along chain backbones in a tightly entangled system:

$$\mathcal{T}(n) = \langle T_i T_{i+n} - \langle T \rangle^2 \rangle, \tag{13.10}$$

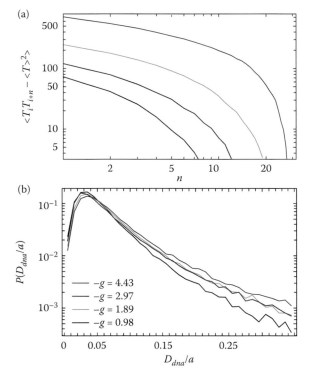

Figure 13.7 Effect of stretching of entanglement network during dramatic hardening: variations with strain of (a) bond tension correlation fluctuations along chain backbones (Equation 13.10), and (b) the probability distribution of nonaffine jump sizes (Equation 13.11), for a tightly entangled ($k_{bend}/u_0 = 2.0$) system uniaxially compressed to a true strain $\epsilon = -1.5$ at $T = 0$. The legend relating line colors to g-values applies to both panels. (Figure adapted from R. S. Hoy, *J. Polym. Sci., Part B: Polym. Phys.*, 49, 979, 2011.)

$T(n)$ grows both in magnitude and correlation length with increasing strain. For strains in the dramatic hardening regime, $T(n) \sim exp(-2n/N_e)$; the factor of 2 suggests a binary suppression of relaxation, wherein tension is concentrated at increasingly localized entanglement points (with two chains/entanglement) and decorrelates between entanglement points. In contrast, bond tensions in loosely entangled systems (not shown) are much smaller and less correlated. This is of interest since chain tension relief by covalent bond scission is a key mechanism leading to brittle fracture [3,23].

Such correlations alter the character of the microscopic, plastic events as measured by examining the distribution of plastic jump sizes and the energy dissipated per plastic event [35]. Plastic "jumps" (locally nonaffine motions) can be characterized using the differential nonaffine displacement

$$D_{dna} = \left| \vec{r}_{k+1} - \frac{\bar{\lambda}_{k+1}}{\bar{\lambda}_k} \vec{r}_k \right| \tag{13.11}$$

of monomers over small finite strain increments $\delta\epsilon$. Here, $\vec{r}_k$ is the position of a particle, and $\bar{\lambda}_k$ is the macroscopic stretch tensor, at strain $|\epsilon| = k\delta\epsilon$. Figure 13.7b shows the probability distribution $P(D_{dna})$ for such jumps to occur in a tightly entangled system deformed at zero temperature. The tails of $P(D_{dna})$ become longer with increasing stress and strain [92]. This effect, which is most pronounced for tightly entangled chains in the dramatic hardening regime, arises because correlated bond tension on the scale $n \sim N_e$ increases both the domain size and intensity (i.e., the typical monomer-jump size $\langle D_{dna} \rangle$) of local plastic rearrangements, which in turn increases the energy they dissipate [35]. In contrast, results for $P(D_{dna})$ in loosely entangled systems nearly collapse [92] because bond tension correlations remain small.

All these results are consistent with "crossovers" (Figure 13.1) from unary to binary relaxation as chains stretch between entanglements, and from noncooperativity to cooperativity between deformations occurring at the scale of the entanglement mesh (φ) and local plastic rearrangements at the monomer or Kuhn scale [35]. These dynamically defined crossovers correspond to the mechanically defined crossover from Gaussian to dramatic hardening. Unary relaxation can roughly be thought of as relaxation wherein the relaxing constituents evolve independently of one another, whereas binary relaxation is controlled by two-body correlations between these constituents [35]. However, since the crossovers remain *very* poorly understood, the rest of this chapter will focus on results for loosely and moderately entangled systems ($N_e > 10C_\infty$), which (in experiments [13]) are far more commonly encountered than tightly entangled ones.

13.5.2 CHAIN LENGTH DEPENDENCE

A third key observation of References 65–67 was that if fracture is suppressed, strain hardening can occur in polymers that are only weakly entangled ($N \sim N_e$) and cannot form the entangled network assumed present in rubber elasticity theories. Such behavior is not observable in experiments since samples with $N < N_e$ undergo brittle fracture at small strains, but in simulations, fracture under compressive deformation is suppressed by the periodic boundary conditions. As described below, this observation led to several useful insights.

Figure 13.8a illustrates strain hardening in systems of chains as short as $N_e/4$. As N increases, σ exhibits a continuous crossover from minimal strain hardening to the asymptotic, well-entangled behavior. No particular "discreteness" is apparent as N passes either N_e or the "critical" length $N_c \sim 2.3N_e$ [93]. Furthermore, as N increases, results follow the asymptotic behavior of well-entangled ($N/N_e > 4$) chains to progressively larger $|g|$. This result (and similar observations in Lyulin et al. [38] and Vorselaars et al. [41]) showed unambiguously that strain hardening does not require the presence of an entanglement network. Instead, it only requires that some order parameter associated with chain configurations continues to increase with increasing strain.

One such order parameter is the (in general subaffine) large-scale chain stretch λ_{eff}, defined in Figure 13.2. As shown in Figure 13.8b, strain hardening in weakly entangled systems can be mapped to that of well-entangled systems if the macroscopic deformation $\bar{\lambda}$ in Equation 13.1 is simply replaced by $\bar{\lambda}_{eff}$:

$$\sigma(\bar{\lambda}) = \sigma_{flow} + G_R^\infty g(\bar{\lambda}_{eff}), \tag{13.12}$$

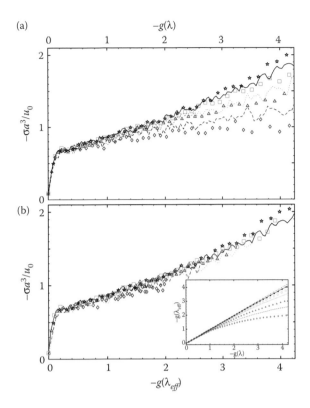

Figure 13.8 Scaling of stress with chain-scale rather than macroscopic deformation: (a) Stress during uni-axial compression at $k_B T = 0.2u_0$ and strain rate $\dot{\epsilon} = -10^{-5}/\tau_{LJ}$ for $k_{bend} = 0.75u_0$ systems of various chain lengths. Successive curves from bottom to top are for $N = 10$ (◊), 16 (– – –), 25 (Δ), 40 (···), 70 (squares), 175 (–), and 350 (★). (b) Stresses for different N collapse when plotted against $g(\lambda_{eff})$. The inset plots $g(\lambda_{eff})$ versus $g(\lambda)$ for the same systems; the light dot-dashed line corresponds to $\lambda_{eff} = \lambda$. (Adapted from R. S. Hoy and M. O. Robbins, *Phys. Rev. Lett.*, 99, 117801, 2007.)

where G_R^∞ is the value of G_R in the long-chain limit. Data for different chain lengths collapse onto a single curve even though N is as much as 4 times smaller than N_e. Similar results are found [66,67] for other T, N_e, and deformation modes. Deviations are only seen for $N \lesssim 5C_\infty$, that is, when chains are less than several persistence lengths long and hence can no longer be viewed as Gaussian random walks. These results show that entanglements play only an indirect role in Gaussian strain hardening; in the Gaussian regime, their main role appears to be in forcing λ_{eff} to follow the global stretch λ. They also help to explain the large magnitude of G_R and the fact that it grows with decreasing T, as is typical for the plastic flow stress σ_{flow}.

13.5.3 SCALING OF G_R WITH σ_{FLOW}

Motivated in part by these studies, Govaert et al. [79] examined the relationship between G_R and σ_{flow} for five different glassy polymers (polycarbonate, polystyrene, polymethyl methylacrylate [PMMA], polyethylene terephthalate [PETG], and polyphenyl ether [PPE]) over a broad range of strain rates and temperatures. In all cases, their results were well fit to the linear relationship

$$G_R = C_0 + C_1 \sigma_{flow}. \tag{13.13}$$

Selected results from Reference 79 are shown in Figure 13.9a. The values of C_0 and C_1, of course, depend strongly on chemistry; as expected, C_1 is larger for stiffer, more densely entangled polymers. They also (less obviously *a priori*) depend strongly on which test parameters are varied; for example, experiments that varied

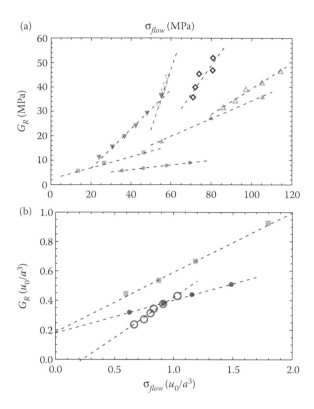

Figure 13.9 Scaling of G_R with σ_{flow}. (a) Experimental results from Reference 79. Symbols show data for PS (circles), PETG (squares), PC (downward triangles), PMMA (upward triangles), and PPE (diamonds). (b) Simulation results from Reference 68. Symbols show data for $k_{bend} = 0.75$ (circles) and $k_{bend} = 1.5$ (squares). Note that the Kremer–Grest bead–spring model's unit of stress [94] (u_0/a^3) is of the order 50 MPa, so simulation results for G_R and σ_{flow} are comparable to the experimental values. In both panels, solid symbols show results for constant strain rate (varying T) and empty symbols show results for constant T (varying $|\dot\epsilon|$); both G_R and σ_{flow} always increase with increasing $|\dot\epsilon|$ and decreasing T. Dashed lines in both panels show fits to Equation 13.13.

T at fixed $|\dot\epsilon|$ always yielded larger C_0 than experiments that varied $|\dot\epsilon|$ at fixed T. In fact, the latter protocol can even produce negative C_0; the only system that did not have a negative C_0 in this case was PETG, which also showed the least variation in G_R and σ_{flow} over the tested range.

Figure 13.9b shows that bead–spring simulation results [68] exhibit the same linear relation (Equation 13.13). Results for temperature dependence from systems with two different values of chain stiffness ($k_{bend}/u_0 = 0.75$ and 1.5) are shown by solid symbols. Stiffer chains produce a larger C_1 (as expected, since they are more densely entangled), but similar C_0. Rate dependence for $k_{bend} = 0.75u_0$ is shown with empty symbols; as in the experiments, C_1 is larger than in the corresponding T-dependence studies, and C_0 is negative.

The negative values of C_0 observed [68,79] when rate is varied are rather counterintuitive, since they imply G_R would vanish and then become negative as $T \to T_g$. However, these negative values of C_0 are not problematic since the weak logarithmic dependence of G_R on strain rate [65,76] makes attaining negative values impossible in practice. Negative σ_{flow} are of course physically meaningless. It is likely that Equation 13.13 simply breaks down in the $T \to T_g$ limit, but this has not yet been systematically explored in either experiments or simulations.

In contrast, positive C_0 implies that strain hardening can still be observed even when $\sigma_{flow} \to 0$, for example, as $T \to T_g$. This is understandable in terms of the picture suggested by References 66 and 67, wherein strain hardening is associated with the increasing number and size of plastic arrangements required to maintain chain connectivity. Flow stress drops to zero as $T \to T_g$ because thermal fluctuations ($k_B T$) become large enough to activate local, segment-scale rearrangements at the given strain rate. However, the same thermal fluctuations will not in general be large enough to activate the increasingly spatially correlated rearrangements needed to maintain chain connectivity at higher strains, leading to a nonzero G_R even when $\sigma_{flow} \to 0$.

Other simulations have shown that Equation 13.13 holds in even more general conditions. Vorselaars et al. [40] showed it is obeyed when the applied hydrostatic pressure P is varied (G_R and σ_{flow} both increase linearly with P). Ge and Robbins studied GSH in preoriented systems [71], and found that for varying chain-scale preorientation λ_{eff}^{pre}, G_R and σ_{flow} both increase by approximately a factor $g(\lambda_{eff}^{pre})$ as chains are increasingly preoriented along the strain direction. Finally, a recent experiment [95] has shown that Equation 13.13 is also obeyed in glassy nano- and micro-composite PMMA; both σ_{flow} and G_R increase in the same manner with increasing filler volume fraction v_p, in such a way that these systems also obey Equation 13.13.

One way of more easily understanding these developments is to note that combining Equation 13.13 with Equation 13.1 produces the relation

$$\sigma(\bar{\lambda}) = \sigma_{flow} + (C_0 + C_1 \sigma_{flow}) g(\bar{\lambda}). \tag{13.14}$$

Here, the parameters σ_{flow}, C_0, and C_1 are all functions of some set ($\{\Gamma\}$) of control variables. While ($\{\Gamma\}$) includes many factors (for example, $T, P, |\dot{\epsilon}|$, chain stiffness, and the details of microscopic interactions), and both the values of C_0 and C_1 and their ratio depend on which factors within the set are varied in the given experiment, the dependencies of all three parameters on any given factor are similar.

13.5.4 MICROSCOPICALLY CONTROLLED GSH: MEAN-FIELD BEHAVIOR

Despite the above-mentioned work, much controversy remained in the field, and claims that entanglements do not play the dominant role in GSH were often treated with suspicion. Experimentalists in particular cited multiple experiments showing $G_R \propto \rho_e$, noted that studies of variably cross-linked systems [33,96–98] suggest that physical entanglements and chemical cross-links affect G_R in the same manner, and claimed that even Gaussian hardening therefore must be dominated by entanglement structure. However, there are other explanations for these observations. For example, monodisperse melts obey the scaling relation $G_N^0 l_K^3 \propto (l_K/p)^3$, where p is the packing length [60,63]. The proportionality $G_R \propto G_N^0$ [34] implies $G_R \propto p^{-3}$. These "microscopic" lengths (l_K and p) are both considerably smaller than the "mesoscopic" length scales [22] corresponding to entanglement structure (the mesh size $\varphi \sim \rho_e^{-1/3}$ or the tube diameter $d_{tube} \sim (N_e \ell_0 l_K)^{1/2}$). Dominance of the mesoscale lengths would imply very different controlling mechanisms for GSH than dominance of the microscale lengths.

Studies of *bidisperse* systems provided a way to resolve this "meso–micro" ambiguity. For systems exhibiting Gaussian hardening, both simulations [69] and experimental studies [86] of bidisperse systems provided strong evidence that Kuhn-scale structure (and hence the corresponding microscopic interactions and energy scales) is the more fundamental controlling factor. Specifically, Hoy and Robbins compared [69] the mechanical responses of mixtures of two very different chain lengths N_{long} and N_{short}, and showed that when a weight fraction f of short chains is mixed with a weight fraction $(1-f)$ of long chains, the stress in mixtures is equal to the weighted average of the stresses in monodisperse systems. Figure 13.10 shows stress–strain curves for both pure systems and $f = 0.5$ mixtures of $k_{bend}/u_0 = 0.75$ bead–spring chains with $N_{long} = 350 \gg N_e$ and $N_{short} = 25 < N_e$. The response for the mixture (green line) lies directly between the results for pure systems with $f = 0$ and 1. The dashed line shows the weighted average

$$\sigma_{ave} = (1-f)\sigma_{long} + f\sigma_{short} \tag{13.15}$$

of the pure system results (σ_{long} and σ_{short}) for $f = 0.5$. Clearly, the actual stress in the mixture is equal to this weighted average, to within statistical uncertanties. This "stress superposition principle" also holds [69] for different N, T, f, and for all k_{bend} such that chains in pure systems are at most moderately entangled.

In all cases, when $N_{short} < N_e$, short chains contribute minimally to strain hardening, and G_R scales linearly with $(1-f)$. Since ρ_e in mixtures scales as a different power of f (i.e., $\rho_e = \rho_e^0 (1-f)^2$ [99]), this result

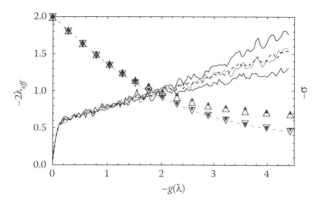

Figure 13.10 Mean-field behavior of chains: Blue, red, and green lines show stress–strain results for the following uniaxially compressed systems with $k_{bend}/u_0 = 0.75$, at $k_B T/u_0 = 0.2$: monodisperse long ($N_{long} = 350$: blue), monodisperse short ($N_{short} = 25$: red), and a bidisperse mixture of short and long chains with $f = 0.5$ (green). The black dashed line shows the weighted average $\sigma_{avg} = (\sigma_{long} + \sigma_{short})/2$ corresponding to this value of f (from Equation 13.15). Triangles show the variation of λ_{eff} with strain in the same simulations, for short chains (red) and long chains (blue); solid (open) triangles indicate results for pure systems (the bidisperse mixture). The gray dashed curve indicates affine deformation ($\lambda_{eff} = \lambda$) and is included to facilitate comparison of the data with the affine result expected in the large-N limit. (Figure adapted from R. S. Hoy and M. O. Robbins, *J. Chem. Phys.*, 131, 244901, 2009.)

is *fundamentally* inconsistent with entanglement-dominated GSH. Despite the different stress curvatures ($|\partial^2\sigma/\partial g^2|$) for the long and short chains in Figure 13.10a, their contributions to the stress in mixtures are simply additive. Indeed, the only discrepancies from stress superposition (Equation 13.15) were observed at in the dramatic hardening regime, and were associated with large σ^U.

Examining the changes in large-scale chain configurations during deformation explains these results. Reference 69 showed that the average large-scale orientations ($\bar{\lambda}_{eff}^{long}$, $\bar{\lambda}_{eff}^{short}$) of short and long chains are essentially the same in monodisperse systems and bidisperse mixtures, even when entangled and unentangled chains are mixed. Symbols in Figure 13.10 show results for $\lambda_{eff}(\lambda)$ for the same systems discussed above. $\lambda_{eff}^{short} > \lambda_{eff}^{long}$ because short chains relax back toward isotropic configurations ($\lambda_{eff} = 1$) during the compressive ($\lambda < 1$) deformation. Remarkably, λ_{eff}^{long} and λ_{eff}^{short} do not themselves depend on f. In other words, the orientation of the long (short) chains does not depend on whether they are in a pure system or mixed with a significant fraction of short (long) chains. This f-independence of λ_{eff} holds over the same broad range of conditions as the stress superposition principle, and implies that over this range, chains relax independently of one another on scales comparable to their radius of gyration R_g. Independent relaxation has the critical implication (for theory) that in this range, large-scale chain relaxation can be treated at the *mean-field* level.

In other words, the natural tendency of chains to deform affinely with the macroscopic stretch ($\bar{\lambda}_{eff} = \bar{\lambda}$) is opposed by relaxation mechanisms that favor subaffine deformation ($|\ln(\lambda_{eff})| < |\ln(\lambda)|$), and the results of Reference 69 suggested that this competition can be treated (theoretically) by assuming that individual chains relax independently within an effective, glassy "medium." Note that entanglements (if viewed as localized constraints similar to chemical cross-links) do not and indeed *cannot* enter directly in such picture. Since weakly entangled and well-entangled chains produce the same responses at small $|g|$ (Figures 13.8 and 13.10), constraints on their rearrangements in this regime must be determined primarily by microscale (for example, frictional) interactions with the surrounding matrix, except in the dramatic regime, when entanglements become more effective than local matrix friction in forcing affine chain deformation.

It is important to point out that none of these results conflict with experiments that found $G_R \propto \rho_e$. In the mixtures employed in experiments on bicomponent systems (for example, [33,96–98]), ρ_e is a linear average of its values in the corresponding pure systems, so these systems also obey stress superposition. Analogous experiments on polymer mixtures where ρ_e is a nonlinear function of f could test the superposition principle further, but none yet seem to have been performed.

13.5.5 A MEAN-FIELD THEORY OF GSH

Hoy and O'Hern incorporated all of the observations discussed above (except those for σ^U) into a simple mean-field theory for GSH [70]. Its key novel features were (i) associating both σ_{flow} and the asymptotic ($N \gg N_e$) hardening modulus (G_R^∞) with microscopic parameters such as u_0, a, ℓ_0, and l_K rather than with ρ_e; (ii) predictively associating Gaussian hardening with the increasing size of correlated plastic events; and (iii) predicting the chain length dependence of λ_{eff} (and thus, through Equation 13.12, σ). Polymeric strain hardening is thus cast as plastic flow in an anisotropic medium wherein flow stress increases with large-scale chain orientation. Although the theory is simple, its combination of features (i–iii) correctly predicted several results from new bead–spring simulations [70], including quantitatively accurate predictions of (iv) the continuous crossover from perfect plasticity to "Gaussian" (neo-Hookean) [13] strain hardening with increasing N (similar to that shown in Figure 13.8) and (v) the subaffine deformation $\lambda_{eff}(\lambda)$ in actively deforming systems.

Prediction (ii) was implemented as follows. Following common practice [9,16], mechanical work W and stress σ were separated into "segmental" and "polymeric" terms: $W = W^s + W^p$, where W^s accounts for the plastic flow stress in the absence of hardening, and W^p accounts for the viscoplastic component of strain hardening. For convenience, W^s was assigned a standard viscous-yield form

$$W^s = \frac{u_0}{a^3}\left(\epsilon + \epsilon_y \exp\left(\frac{-\epsilon}{\epsilon_y}\right)\right), \tag{13.16}$$

where ϵ_y is the yield strain and u_0/a^3 is the characteristic stress scale for athermal plasticity.

Following Haward [13], W^p was assumed to scale with the density of coherently relaxing contours $\rho_{cr} = \sqrt{6}\rho\ell_0/(NR_c)$, where ρ is monomer number density and Gaussian polymers have chain statistics defined by $R_c^2 \equiv \ell_0 l_K N$. The natural correlation length scale for the associated, locally *coherent* [70] plastic rearrangements is R_c, and the associated volume is $V = R_c^3/6^{3/2}$. Thus, W^p and σ_p are given by

$$\sigma_p = \frac{\Delta W^p}{\Delta \ln(\lambda)}, \quad \text{where} \quad \Delta W^p \simeq (u_0/a^3)\Delta(\rho_{cr}V). \tag{13.17}$$

In deformed systems, $R_c^2(\bar{\lambda}) = 3l_0 l_K N\tilde{g}(\lambda_{eff})$, and V varies with strain accordingly, giving

$$\Delta W^p = \frac{(u_0/a^3)\rho l_0^2 l_K \Delta(\tilde{g}(\lambda_{eff}))}{2}. \tag{13.18}$$

This result combined with Equation 13.16 gives a prediction for the athermal (zero-T) stress:

$$\sigma(\epsilon, \epsilon_{eff}) = \frac{u_0}{a^3}\left(\frac{1 - \exp(-\epsilon/\epsilon_y) + \rho l_0^2 l_K |g(\epsilon_{eff})|}{2}\right), \tag{13.19}$$

where $g(\epsilon) = (3/2)\partial\tilde{g}/\partial\epsilon$. Note that Equation 13.19 assumes that all contributions to σ scale with a single energy density (i.e., stress) u_0/a^3. While this approximation neglects some of the effects discussed above (for example, in Section 13.5.3), the "intercept" term C_0 in the linear relationship between σ_{flow} and G_R (Equation 13.13) is often fairly small compared to the linear term, especially for T well below T_g [69,79], so the approximation is reasonable. Reference 70 showed that Equation 13.19 in its current (minimalistic) form adequately predicts the strain hardening response for flexible ($k_{bend} = 0$) bead–spring chains, and in particular its variation with N.

Equation 13.19 has two particularly interesting features. First, it predicts that the asymptotic (large-N) strain hardening modulus is $G_R^\infty = (u_0/a^3)\rho l_0^2 l_K$. This value reflects the microscopic picture of GSH discussed above, and furthermore is much closer to the effective "constraint" density measured in nuclear magnetic resonance (NMR) experiments on deformed glassy samples [100] than to the entropic prediction $G_R = \rho_e k_B T$. Second, its prediction $G_R \propto l_K$ is consistent with the well-established result that straighter chains are harder to plastically deform [3,13], and explains this by relating the "activation" volume V to Kuhn length, that is, $V \propto l_K^{3/2}$. Note however that the power of l_K with which σ^p scales is sensitive to the theoretical assumptions, specifically the one implicitly present in the above definition of ρ_{cr}. Other definitions can produce $G_R \propto l_K^{3/2}$ or l_K^3, and determining which one is "best" will require further work.

For finite-N chains, large-scale relaxation during deformation (prediction (iii)) was calculated using a simple Maxwell-like model to predict $\lambda_{eff}(\lambda)$. In terms of $\epsilon_{eff} \equiv \ln(\lambda_{eff})$, the governing equation is

$$\dot{\epsilon}_{eff} = \dot{\epsilon} - \frac{\epsilon_{eff}}{\tau} .$$
(13.20)

Equation 13.20 is a standard "fading memory" form implying chains "forget" their large-scale orientation at a rate τ^{-1}. In other words, τ is the time scale over which ϵ_{eff} will relax toward its "equilibrium" value $\epsilon_{eff} = 0$. Nonetheless, since it employs λ_{eff} rather than λ as its key mesovariable, it differs critically from the many (for example, [9,16,17,101]) previous Maxwell-like models of glassy polymer mechanics.

Reference 70 made prediction (iii) as follows. Assuming that large-scale chain relaxation is tightly coupled to segment-scale relaxation (specifically, that the chain-scale relaxation time τ is proportional to the segmental [alpha] relaxation time τ_α), and that the chain length dependence of τ is given by a power law as is common in melt rheology [22], yields

$$\tau \sim N^\gamma \tau_\alpha \text{ (incoherent)}$$
(13.21)

Here, "incoherent" indicates that events that alter λ_{eff} have no favored "direction," that is, increases and decreases in λ_{eff} are equally likely. The independent relaxation of chains described above (Figure 13.10) suggests that their large-scale dynamics are essentially "Rouse-like" [22], implying $\gamma = 2$. Although a prediction that relaxation of $N > N_e$ chains in glasses obeys the same power law as relaxation of $N < N_e$ chains in *melts* is rather bold, it is supported by recent experiments [7,8] showing glasses exhibit melt-like relaxation mechanisms; see Chapter 10. It is also consistent with recent dielectric spectroscopy experiments [90] indicating that glasses possess strongly coupled segment-scale and chain-scale relaxations.

In *actively deforming systems*, plastic events that reduce $|\epsilon_{eff}| \equiv |\ln(\lambda_{eff})|$ are greatly favored over those that increase these terms. Assuming that the $|\epsilon_{eff}|$-reducing events dominate, and that relaxation therefore becomes "coherent," yields

$$\tau \sim N^{\gamma-1} \tau_\alpha \text{ (coherent)}.$$
(13.22)

The factor-of-N reduction in τ occurs because distant parts of chains move cooperatively, unlike the incoherent case. In practice, coherent relaxation is forced by the stiffness of the covalent backbone bonds, which have (nearly) constant length ℓ_0 and so maintain (nearly) constant chain contour length $L = (N-1)\ell_0$. Within the framework of Equations 13.21 and 13.22, when relaxation is strain activated, both τ_α and τ are reduced by a factor $\sim \dot{\epsilon}$ (i.e., $\tau_\alpha \propto \dot{\epsilon} \tau_\alpha^0$ for $|\dot{\epsilon}| < 1$), so the overall reduction in τ during active deformation scales as $N\dot{\epsilon}$. This reduction is consistent with the $O(\dot{\epsilon})$ reduction in τ_α during active deformation observed in both experiments [6–8] and simulations [102,103] and discussed in Chapters 10 and 11. Note that this reasoning assumes ϵ_{eff} and $\dot{\epsilon}$ have the same sign—a limitation of the theory that needs to be addressed before it can treat Bauschinger-type effects [4,71,80].

Constant strain rate deformation and constant-strain relaxation are two of the most commonly performed mechanical experiments. In a constant strain rate experiment, assuming τ is independent of ϵ, that is, assuming polymer glasses are *linearly* viscoplastic, the solution to Equation 13.20 is

$$\epsilon_{\mathit{eff}}(\epsilon) = \dot{\epsilon}\tau\left(1 - \exp\left(\frac{-\epsilon}{\dot{\epsilon}\tau}\right)\right) \equiv \dot{\epsilon}\tau\left(1 - \exp\left(\frac{-t}{\tau}\right)\right). \tag{13.23}$$

Figure 13.11a shows the predictions of Equation 13.23 for the evolution of $\epsilon_{\mathit{eff}}(\epsilon)$ in systems (with a wide range of N) compressively strained to $\epsilon = -1.0$ at constant rate. Lines assume coherent chain relaxation with $\gamma = 2$ (i.e., $\tau \propto \dot{\epsilon}^{-1}N^{\gamma-1}$) during active deformation, consistent with Equation 13.22. Symbols show bead–spring simulation results for the same values of N, for $k_{bend} = 0$ systems deformed at $(T = 0.2, \dot{\epsilon} = 10^{-5}\tau)$. $N = 500$ chains orient nearly affinely during strain $(\epsilon_{\mathit{eff}} \approx \epsilon)$, while short chains orient much less. Clearly, the agreement is excellent; data are quantitatively consistent with $\tau \propto N^{\gamma-1}$ and $\gamma = 2$. Given the large variation in stress as strain increases, it is remarkable that a single relaxation rate theory fits ϵ_{eff} so well. However, this is consistent with the enhanced dynamical homogeneity and narrowed relaxation spectrum observed in similar models [102,103] in the post-yield regime; cf. Section 13.5.6.

To predict chain-length-dependent stress response, one simply plugs τ, N, and the form of $\tilde{g}$ for the given deformation protocol (Table 13.1) into Equation 13.19. Figure 13.11b shows predictions for $\sigma(\epsilon)$ in uniaxial compression at various N. Solid lines assume $\tau \propto N^{\gamma-1}$ with $\gamma = 2$ as discussed above. The predicted large-strain ($\epsilon \gg \epsilon_y$) mechanical response [70] varies continuously from perfect-plastic flow ($\sigma^*(\epsilon) \to$ a constant σ_{flow}) to network-like polymeric response ($\sigma^*(\epsilon) \to 1 + \rho l_0^2 l_K g(\epsilon)$; cf. Equation 13.24) as $\dot{\epsilon}\tau$ varies from zero to ∞ (as N increases). In betweeen these limits, the response is "polymeric viscoplasticity" (or more accurately, "viscoelastoplasticity," since ϵ_{eff} is treated as microreversible). Since the model does not treat stress relaxation after cessation of deformation, $\sigma(\epsilon)$ may be regarded as an orientation-dependent plastic flow stress. Results for tension are not presented here because the model of Reference 70 includes no asymmetry between tension and compression. This asymmetry has been treated in other work (see, for example, References 17 and 104) and could in principle be treated by modifying W^s appropriately.

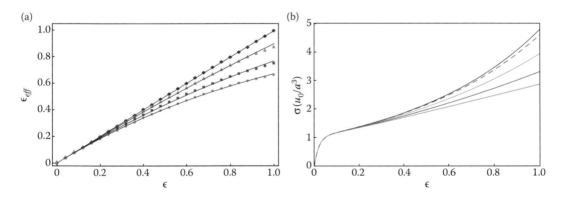

Figure 13.11 Theoretical predictions for chain length dependence of λ_{eff} and stress: (a) ϵ_{eff} versus ϵ for uniaxial compression: comparison of theory and simulation results. Symbols from top to bottom are bead–spring simulation results for flexible chains (at $T = 0.2$ and $|\dot{\epsilon}| = 10^{-5}/\tau$) of lengths $N = 500$ (blue), $N = 36$ (green), $N = 18$ (brown), and $N = 12$ (gray), while lines show predictions of Equation 13.23 for these N; predictions assume $\dot{\epsilon}\tau = BN^{\gamma-1}$ with the prefactor $B = 0.1$ suggested by Reference 50. (b) Stress–strain curves predicted by Equation 13.19 for the same N (from top to bottom), made using values of τ obtained from the left panel. Solid curves assume chain-scale relaxation is coherent ($\tau \propto N^{\gamma-1}$) during active deformation. The dashed curve, presented only for comparison, presents the prediction for $N = 12$ where this relaxation is *incoherent* ($\tau \propto N^\gamma$); corresponding incoherent-relaxation predictions for $N \gtrsim 50$ are not distinguishable from their large-N limit on the scale of this plot. In both panels, $\gamma = 2$. (Figure adapted from R. S. Hoy and C. S. O'Hern, *Phys. Rev. E*, 82, 041803, 2010.)

For the purpose of contrast, the dashed line in Figure 13.11b shows a prediction (Equation 13.19) assuming that relaxation remains incoherent (i.e., $\tau \propto \dot{\epsilon}^{-1} N^{\gamma}$). The solid lines are qualitatively consistent with simulation results [65,66,70] while the dashed lines (including other lines not shown) are inconsistent. Both show increasing strain hardening with increasing N, but in the latter case hardening increases much faster and saturates at a much lower value of N than is realistic. For example, the $\tau \propto N^{\gamma}$ predictions for $N = 40$ and $N = 500$ are indistinguishable on the scale of the plot.

The stress predicted by Equation 13.19 is not temperature-dependent. Imposing a $(1 - T/T_g)$ scaling of stress and scaling by u_0/a^3 yields

$$\sigma^*(\epsilon_{eff}) = A \frac{\sigma(\epsilon_{eff})}{u_0/a^3}, \tag{13.24}$$

Here, A is a prefactor arising from the neglect of prefactors in the above analysis; all "thermal" aspects of the theory [70] are implicitly wrapped into A and τ_α. Such a scaling analysis does not account for additional changes in τ_α arising from other causes, for example, increased mobility associated with yield, but the simple mathematical form of Equation 13.24 allows one to insert predictions for A and τ_α from more sophisticated theoretical treatments (for example, [54]) directly into Equation 13.19. The agreement of the solid lines with simulation results reported in Hoy and O'Hern [70] was only adequate at small strains, largely because of the ad hoc usage of Equation 13.16 to model the small-strain response; inserting more sophisticated predictions could improve it.

It is important to test this and other [50–57] microscopic theories of GSH. Fortuitously, λ_{eff} can now be accurately measured in scanning near-field optical microscopy (SNFOM) experiments [105–107], which can discriminate values for different chain lengths in a bidisperse system. Reference 105 showed that λ_{eff} falls behind λ for entangled chains deformed slightly above T_g. Reference 107 extended the power of this method; careful SNFOM observations showed that relaxation of the long chains in bidisperse films (that were rapidly deformed, then allowed to relax with λ held fixed) was inconsistent with the predictions of tube theory and common sliplink models. Analogous studies, well below T_g, could be performed to test the theory developed above, but unfortunately (as of 2015), none have yet been published. Modern neutron scattering techniques might be employed for the same purpose.

13.5.6 BEYOND THE MEAN FIELD: EFFECTS OF ELASTIC HETEROGENEITY

The above-described f-independence, stress superposition, and more generally the validity of mean-field treatments are expected to break down in the dramatic regime where strain hardening is supralinear in $g(\lambda)$. This is so because in the dramatic regime, stretching of chains between entanglements (and thus "binary" effects, for example, correlations between the configurations of 2 or more chain segments) become important. They should also break down as $T \rightarrow T_g$, since in this regime, cooperative relaxation effects are known to become important. For example, large-scale relaxation times of short chains in bidisperse melts are known to increase in the presence of long chains [108,109]. Finally, mean-field treatments should also fail when $N_{short} \ll N_e$ and f is large enough that the high density of chain ends significantly changes the density and internal friction of the glass, for example, in glasses wherein oligomers are added as (anti)plasticizers [74].

Other failings of mean-field treatments are more subtle and involve *heterogeneities*. Kuhn-segment-scale packing in polymer glasses is very sensitive to preparation history. For example, rapidly quenched systems are "less ordered" in the sense that they pack less efficiently and occupy higher regions of their free-energy landscapes than their slowly annealed counterparts [5,110]. Their greater structural heterogeneity necessarily produces correspondingly greater elastic heterogeneity (spatially localized fluctuations of the local bulk and/or shear modulus). These fluctuations turn out to be very significant for mechanical response. In glasses with spatially heterogeneous elastic properties, local "yield" (i.e., onset of plastic flow) occurs at different stresses/strains at different points within the sample. While it is clear upon a moment's reflection that this process is associated with the softer mechanical response of quenched systems [110], quantitative prediction of yield behavior based on these ideas (even in simulations) became possible only very recently.

Local yield can be predicted using theoretical developments such as "soft spots" (regions with low elastic moduli). Identification of soft spots was shown by Rottler and collaborators [111–113] to allow suprisingly accurate prediction of the highly localized plastic deformations (i.e., local yield) that are a precursor to macroscopic yield. An example of how these factors can affect glassy polymer mechanics is given in Figure 13.12, which shows the spatially heterogeneous elastic moduli in two different (undeformed) systems studied by

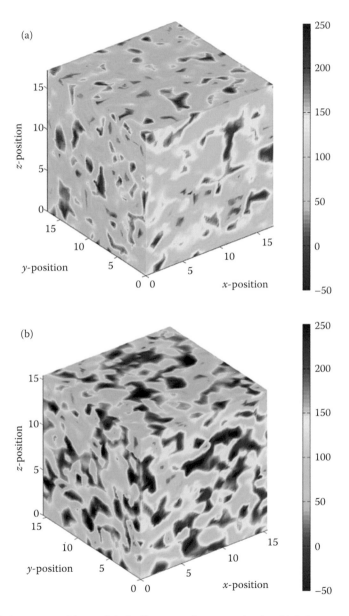

Figure 13.12 Elastic heterogeneities and their effect on polymer mechanics: (a) Color maps of the local elastic moduli of model polymer glasses [73]. Dark blue domains correspond to elastically unstable regions, while dark red domains correspond to the stiffest regions. (b) The same moduli in a system with added antiplasticizers that increase the material density while reducing the fragility of glass formation and T_g. Rapidly quenched systems should be both "softer" and more heterogeneous than slowly cooled systems, while stiffer chains should produce more fragile glasses [114] and correspondingly enhanced heterogeneity. (Figure adapted from R. A. Riggleman, J. F. Douglas, and J. J. de Pablo, *Soft Matter*, 6, 292, 2010.)

Riggleman et al. [74]. Red coloring indicates stiff regions, while blue coloring indicates soft regions possessing small or even negative local bulk modulus. References 73, 74, and 111–113 identified a clear relationship between fluctuations in local moduli and the propensity for plastic events (irreversible displacements) during applied deformation. Local plastic flow begins in low-modulus regions, well before the macroscopic yield strain is reached. Near yield, plastic event frequencies increase rapidly, and segmental relaxation times both decrease on average and broaden in distribution. Then, in the Gaussian hardening regime, the fluctuations diminish and relaxation becomes more homogeneous. Experimental consequences of these effects include, for example, the well-known result that the more efficient (more homogeneous) segment-scale packing of systems that have been slowly cooled from the melt produces not only a stiffer elastic response and larger yield stress, but also *qualitatively* different strain localization, for example, shear bands that are narrow and "sharp" rather than broad and diffuse [23]. The results of References 73,74, and 111–113 have also proven important in understanding the speedup of relaxation upon yield, and its subsequent slowdown upon initial hardening, observed in recent experiments [6–8,89].

13.6 OUTLOOK

It is clear that the same complexities that affect yield (as discussed in the above section) will also affect larger-strain deformation, and additional ones related to chain connectivity will also be increasingly important. Therefore, continuing to improve our understanding of the effects of (dynamically evolving) elastic heterogeneities in deformed polymer glasses will likely be quite important to explaining the variety of behavior depicted in Figure 13.1, including strain hardening. One particularly intriguing possibility is that the "hard spots" (i.e., high-modulus regions) *are in fact more important than rheological entanglements* in controlling strain hardening. However, whether this is true remains unknown because the ideas discussed in Riggleman et al. [73,74], Rottler et al. [111], Schoenholz et al. [112], Smessaert and Rottler [113], Kumar et al. [114] and Manning and Liu [115] have not yet been tested across a wide range of chain stiffnesses, or in the dramatic hardening regime. Quantitatively determining how differences in "average" structure (for example, between different polymers) combine with local heterogeneities to produce different macroscopic mechanical response, and then using this knowledge to accurately predict failure of amorphous materials, is currently a "grand challenge" for soft matter physicists [112], so it is to be hoped such tests will be performed in the near future.

Energy landscape analyses [5] should prove highly useful in this effort, and should also yield additional insights. These are required for the soft-spot analyses, but equally (if not more) important results will be obtained by examining how polymer glasses traverse their strain-dependent energy landscapes. Chung and Lacks [116–119] have shown that the nature of this traversal is tremendously important in controlling temperature- and rate-dependent mechanical properties. For example, "fold catastrophes" occur when the configuration (i.e., the set of all monomers' positions) corresponds to an energy minimum that then disappears under further deformation. These catastrophes produce local plastic deformation, as locally stored elastic energy is released and dissipated as heat. However, References 116–119 examined only short, weakly entangled chains that did not produce strong strain hardening, and moreover focused on shear. Studies that examine the role of fold catastrophes in the dramatic hardening and incipient-fracture regimes will shed much new light on GSH. One expects that the frequency of fold events will increase sharply at the prefracture stress peak, or equivalently the fracture strain (Figure 13.1) is approached, and that individual fold events will become larger (involve more monomers, over a more spatially extended region). Such effects may be quite important in understanding fracture, and ultimately developing the ablity to accurately predict its occurrence.

Other key GSH-related questions to be answered, together with potential strategies for investigating them, include: (i) How, precisely, does the "entanglement tube" formed by the uncrossability constraints of surrounding chains [22] affect plastic deformation? Sufficiently long chains can deorient on large scales only by moving along their tubes (whether via reptation, chain retraction, or other mechanisms of backsliding against the principal stretch). Such processes are known to dominate polymer-melt rheology [22,120,121], and results for λ_{eff} [68,70] have indicated that something similar happens in loosely and moderately entangled glasses. However, it is clear that the answer to this question is not a simple one. (ii) How does polymeric structure on scales from l_K to R_g affect spatial correlations of localized plastic events? Spatial correlations of nonaffine deformation (magnitude and direction) are known to relate to plastic event size in atomic glasses [122–125], but for polymers, the character (let alone the quantitative features) of these relations are unknown. Almost certainly they also depend on the polymeric nature of the system through the degree of polymerization N [66,67] and the chain stiffness (which sets the "tightness" of the entanglement mesh [63]). However, there have not yet been any comprehensive investigations of the relation of plastic-event correlations to *interchain* Kuhn-scale structure. (iii) How does plasticity vary with deformation mode? While this is certainly known to some extent from experiments [23] and simulations [91] (for example, crazing promotes chain scission while uniaxial stress compression promotes homogeneous deformation [13]), follow-up studies that quantitatively compare multiple metrics of plasticity for the same systems, across a far broader range of deformation protocols and principal stresses than has been considered previously, should be very helpful.

Finally, it must be noted that some workers still hold markedly dissenting views of the nature of GSH. Argon's recent book [4] argues for the entropic picture using the increased mobility under deformation [6] as evidence, and also argues that the simulation studies mentioned herein are consistent with this view. Needless to say, the present author does not agree with these viewpoints.

In conclusion, it is clear that much more work is necessary before we can quantitatively predict stress–strain and stress–relaxation curves for (arbitrarily composed and prepared) polymer glasses over the entire response range depicted in Figure 13.1. In particular, improved understanding of the effects of chain stiffness is required. It seems certain that microscopic structural detail at the Kuhn scale (for example, chemistry-dependent effects) exerts significant influence on segmental relaxation processes, and it is probable that these effects couple to chain-scale relaxation (see, for example, [41,79,90,126,127]). Studies using more chemically realistic models, or real polymers, would be welcome.

REFERENCES

1. P. I. Vincent, *Polymer*, 1, 425, 1960.
2. I. M. Ward and J. Sweeney, *Mechanical Properties of Solid Polymers* (John Wiley & Sons, Sussex), 2012.
3. R. N. Haward and R. J. Young, eds., *The Physics of Glassy Polymers* (Chapman & Hall, London), 2nd edn., 1997.
4. A. S. Argon, *The Physics of Deformation and Fracture of Polymers* (Cambridge Solid State Science, Cambridge), 2013.
5. F. H. Stillinger, *Science*, 267, 1935, 1995.
6. L. S. Loo, R. E. Cohen, and K. K. Gleason, *Science*, 288, 5463, 2000.
7. H. N. Lee, R. A. Riggleman, J. J. de Pablo, and M. D. Ediger, *Macromolecules*, 42, 4238, 2009.
8. H. N. Lee, K. Paeng, S. F. Swallen, and M. D. Ediger, *Science*, 323, 231, 2009.
9. R. N. Haward and G. Thackray, *Proc. Roy. Soc. Lond.*, 302, 453, 1968.
10. L. R. G. Treloar, *The Physics of Rubber Elasticity* (Clarendon Press, Oxford), 1975.
11. J. D. Ferry, *Viscoelastic Properties of Polymers* (Wiley, New York), 1980.
12. A. Cross and R. N. Haward, *Polymer*, 19, 677, 1978.
13. R. N. Haward, *Macromolecules*, 26, 5860, 1993.
14. M. C. Boyce, E. M. Arruda, and R. Jayachandran, *Polym. Eng. Sci.*, 34, 716, 1994.
15. E. M. Arruda and M. C. Boyce, *J. Mech. Phys. Solids*, 41, 389, 1993.
16. E. M. Arruda and M. C. Boyce, *Int. J. Plast.*, 9, 697, 1993.
17. M. C. Boyce, D. M. Parks, and A. S. Argon, *Mech. Mat.*, 7, 15, 1988.
18. A. S. Argon, *Philos. Mag.*, 28, 39, 1973.
19. A. S. Argon and M. I. Bessonov, *Philos. Mag.*, 35, 917, 1977.
20. M. C. Boyce and E. M. Arruda, *Rubber Chem. Tech.*, 73, 504, 2000.
21. O. A. Hasan and M. C. Boyce, *Polym. Eng. Sci.*, 35, 331, 1995.
22. M. Doi and S. F. Edwards, *The Theory of Polymer Dynamics* (Clarendon Press, Oxford), 1986.
23. E. J. Kramer, *Adv. Polym. Sci.*, 52/53, 1, 1983.
24. E. J. Kramer and L. L. Berger, *Adv. Polymer Sci.*, 91, 1, 1990.

25. J. Richeton, S. Ahzi, K. S. Vecchio, F. C. Jiang, and A. Makradi, *Int. J. Solids Stuct.*, 44, 7938, 2007.
26. O. A. Hasan and M. C. Boyce, *Polymer*, 34, 5085, 1993.
27. D. Rittel, *Mech. Mat.*, 31, 131, 1999.
28. Z. Li and J. Lambros, *Int. J. Solids Struct.*, 38, 3549, 2001.
29. J. S. Bergström and M. C. Boyce, *J. Mech. Phys. Solids*, 46, 931, 1998.
30. C. Chui and M. C. Boyce, *Macromolecules*, 32, 3795, 1999.
31. R. N. Haward, *Polymer*, 28, 1485, 1987.
32. C. P. Buckley and D. C. Jones, *Polymer*, 36, 3301, 1995.
33. H. G. H. van Melick, L. E. Govaert, and H. E. H. Meijer, *Polymer*, 44, 2493, 2003.
34. E. J. Kramer, *J. Polym. Sci., Part B: Polym. Phys.*, 43, 3369, 2005.
35. R. S. Hoy, *J. Polym. Sci., Part B: Polym. Phys.*, 49, 979, 2011.
36. J. L. Bouvard, D. K. Ward, D. Hossain, S. Nouranian, E. B. Marin, and M. F. Horstmeyer, *J. Eng. Mater. Tech.* 131, 041206, 2009.
37. J.-L. Barrat, J. Baschnagel, and A. Lyulin, *Soft Matter*, 6, 3430, 2010.
38. A. V. Lyulin, N. K. Balabaev, M. A. Mazo, and M. A. J. Michels, *Macromolecules*, 37, 8785, 2004.
39. A. V. Lyulin, B. Vorselaars, M. A. Mazo, N. K. Balabaev, and M. A. J. Michels, *Europhys. Lett.*, 71, 618, 2005.
40. B. Vorselaars, A. V. Lyulin, and M. A. J. Michels, *J. Chem. Phys.*, 130, 074905, 2009.
41. B. Vorselaars, A. V. Lyulin, and M. A. J. Michels, *Macromolecules*, 42, 5829, 2009.
42. K. Kremer and G. S. Grest, *J. Chem. Phys.*, 92, 5057, 1990.
43. C. F. Abrams and K. Kremer, *J. Chem. Phys.*, 115, 2776, 2001.
44. G. S. Grest, M. Pütz, R. Everarers, and K. Kremer, *J. Non-Cryst. Solids*, 274, 139, 2000.
45. A. D. Mulliken and M. C. Boyce, *Int J. Solids Struct.*, 43, 1331, 2006.
46. S. Sarva, A. D. Mulliken, and M. C. Boyce, *Int. J. Solids Struct.*, 44, 2381, 2007.
47. R. A. Riggleman, G. Toepperwein, G. J. Papakonstantopoulos, J. L. Barrat, and J. J. de Pablo, *J. Chem. Phys.*, 130, 244903, 2009.
48. G. N. Toepperwein and J. J. de Pablo, *Macromolecules*, 44, 5498, 2011.
49. G. N. Toepperwein, N. C. Karayiannis, R. A. Riggleman, M. Kröger, and J. J. de Pablo, *Macromolecules*, 44, 1034, 2011.
50. K. Chen and K. S. Schweizer, *Europhys. Lett.* 79, 26006, 2007.
51. K. Chen and K. S. Schweizer, *Macromolecules*, 41, 5908, 2008.
52. K. Chen and K. S. Schweizer, *Phys. Rev. Lett.*, 102, 038301, 2009.
53. K. Chen and K. S. Schweizer, *Phys. Rev. E*, 82, 041804, 2010.
54. K. Chen, E. J. Saltzman, and K. S. Schweizer, *Annu. Rev. Condens. Matter Phys.*, 1, 277, 2010.
55. K. Chen and K. S. Schweizer, *Macromolecules*, 44, 3988, 2011.
56. S. M. Fielding, R. G. Larson, and M. E. Cates, *Phys. Rev. Lett.*, 108, 048301, 2012.
57. S. M. Fielding, R. L. Moorcroft, R. G. Larson, and M. E. Cates, *J. Chem. Phys.*, 138, 12A504, 2013.
58. C. Bennemann, W. Paul, K. Binder, and B. Dünweg, *Phys. Rev. E*, 57, 843, 1998.
59. M. E. Mackura and D. S. Simmons, *J. Polym. Sci., Part B: Polym. Phys.*, 52, 134, 2013.
60. R. Everaers, S. K. Sukumaran, G. S. Grest, C. Svaneborg, A. Sivasubramanian, and K. Kremer, *Science*, 303, 823, 2004.
61. R. S. Hoy and M. O. Robbins, *Phys. Rev. E*, 72, 061802, 2005.
62. R. S. Hoy, K. Foteinopoulou, and M. Kröger, *Phys. Rev. E*, 80, 031803, 2009.
63. L. J. Fetters, D. J. Lohse, S. T. Milner, and W. W. Graessley, *Macromolecules*, 32, 6847, 1999.
64. J. Li, T. Mulder, B. Vorselaars, A. V. Lyulin, and M. A. J. Michels, *Macromolecules*, 39, 7774, 2006.
65. R. S. Hoy and M. O. Robbins, *J. Polym. Sci., Part B: Polym. Phys.*, 44, 3487, 2006.
66. R. S. Hoy and M. O. Robbins, *Phys. Rev. Lett.*, 99, 117801, 2007.
67. R. S. Hoy and M. O. Robbins, *Phys. Rev. E*, 77, 031801, 2008.
68. M. O. Robbins and R. S. Hoy, *J. Polym. Sci., Part B: Polym. Phys.*, 47, 1406, 2009.
69. R. S. Hoy and M. O. Robbins, *J. Chem. Phys.*, 131, 244901, 2009.
70. R. S. Hoy and C. S. O'Hern, *Phys. Rev. E*, 82, 041803, 2010.

71. T. Ge and M. O. Robbins, *J. Polym. Sci., Part B: Polym. Phys.*, 48, 1473, 2010.

72. R. A. Riggleman, K. S. Schweizer, and J. J. de Pablo, *Phys. Rev. E*, 77, 041502, 2008.

73. R. A. Riggleman, H.-N. Lee, M. D. Ediger, and J.-J. de Pablo, *Soft Matter*, 6, 287, 2010.

74. R. A. Riggleman, J. F. Douglas, and J. J. de Pablo, *Soft Matter*, 6, 292, 2010.

75. R. B. Dupaix and M. C. Boyce, *Polymer*, 46, 4827, 2005.

76. M. Wendlandt, T. A. Tervoort, and U. W. Suter, *Polymer*, 46, 11786, 2005.

77. P. J. Hine, A. Duckett, and D. J. Read, *Macromolecules*, 40, 2782, 2007.

78. F. Casas, C. Alba-Simionesco, H. Montes, and F. Lequeux, *Macromolecules*, 41, 860, 2008.

79. L. E. Govaert, T. A. P. Engels, M. Wendlandt, T. A. Tervoort, and U. W. Suter, *J. Polym. Sci., Part B: Polym. Phys.*, 46, 2475, 2008.

80. D. J. A. Senden, J. A. W. van Dommelen, and L. E. Govaert, *J. Polym. Sci., Part B: Polym. Phys.*, 48, 1483, 2010.

81. M. Wendlandt, T. A. Tervoort, and U. W. Suter, *J. Polym. Sci., Part B: Polym. Phys.*, 48, 1464, 2010.

82. D. S. A. de Focatiis, J. Embery, and C. P. Buckley, *J. Polym. Sci., Part B: Polym. Phys.*, 48, 1449, 2010.

83. H. N. Lee and M. D. Ediger, *Macromolecules*, 43, 5863, 2010.

84. H. N. Lee and M. D. Ediger, *J. Chem. Phys.*, 144, 014901, 2010.

85. K. Paeng, H. N. Lee, S. F. Swallen, and M. D. Ediger, *J. Chem. Phys.*, 134, 024901, 2011.

86. J. D. McGraw and K. Dalnoki-Veress, *Phys. Rev. E*, 82, 021802, 2010.

87. D. J. A. Senden, J. A. W. van Dommelen, and L. E. Govaert, *J. Polym. Sci., Part B: Polym. Phys.*, 50, 1589, 2012.

88. D. J. A. Senden, S. Krop, J. A. W. van Dommelen, and L. E. Govaert, *J. Polym. Sci., Part B: Polym. Phys.*, 50, 1757, 2012.

89. B. Bending, K. Christison, J. Ricci, and M. D. Ediger, *Macromolecules*, 47, 800, 2014.

90. J. Hintermeyer, A. Herrmann, R. Kahlau, C. Goiceanu, and E. A. Rössler, *Macromolecules*, 41, 9335, 2008.

91. J. Rottler, *J. Phys. Condens. Matter*, 21, 463101, 2009.

92. M. Warren and J. Rottler, *J. Chem. Phys.*, 133, 164513, 2010.

93. L. J. Fetters, D. J. Lohse, D. Richter, T. A. Witten, and A. Zirkel, *Macromolecules*, 27, 4639, 1994.

94. J. Rottler, S. Barsky, and M. O. Robbins, *Phys. Rev. Lett.*, 89, 148304, 2002.

95. J. Jancar, R. S. Hoy, A. J. Lesser, E. Jancarova, and J. Zidek, *Macromolecules*, 46, 9409, 2013.

96. C. G'Sell and A. Souahi, *J. Eng. Mater. Tech.*, 119, 223, 1997.

97. Z. Bartczak, *Macromolecules*, 38, 7702, 2005.

98. Z. Bartczak, *J. Polym. Sci., Part B: Polym. Phys.*, 48, 276, 2010.

99. W. W. Graessley and S. F. Edwards, *Polymer*, 22, 1329, 1981.

100. M. Wendlandt, T. A. Tervoort, J. D. van Beek, and U. W. Suter, *J. Mech. Phys. Solids*, 54, 589, 2006.

101. B. Gross and R. M. Fuoss, *J. Polym. Sci.*, 19, 39, 1956.

102. G. J. Papakonstantopoulos, R. A. Riggleman, J.-L. Barrat, and J. J. de Pablo, *Phys. Rev. E*, 77, 041502, 2008.

103. M. Warren and J. Rottler, *Phys. Rev. Lett.*, 104, 205501, 2010.

104. J. Rottler and M. O. Robbins, *Phys. Rev. E*, 68, 011507, 2003.

105. T. Ube, H. Aoki, S. Ito, J. Horinaka, and T. Takigawa, *Polymer*, 48, 6221, 2007.

106. T. Ube, H. Aoki, S. Ito, J. Horinaka, T. Takigawa, and T. Masuda, *Polymer*, 50, 3016, 2009.

107. T. Ube, H. Aoki, S. Ito, J. Horinaka, and T. Takigawa, *Soft Matter*, 8, 5603, 2012.

108. A. R. C. Baljon, G. S. Grest, and T. A. Witten, *Macromolecules*, 28, 1835, 1995.

109. C. M. Ylitalo, J. A. Kornfield, G. G. Fuller, and D. S. Pearson, *Macromolecules*, 24, 749, 1991.

110. H. E. H. Meijer and L. E. Govaert, *Prog. Polym. Sci.*, 30, 915, 2005.

111. J. Rottler, S. S. Schoenholz, and A. J. Liu, *Phys. Rev. E*, 89, 042304, 2014.

112. S. S. Schoenholz, A. J. Liu, R. A. Riggleman, and J. Rottler, *Phys. Rev. X*, 4, 031014, 2014.

113. A. Smessaert and J. Rottler, *Soft Matter*, 10, 8533, 2014.

114. R. Kumar, M. Goswami, B. G. Sumpter, V. N. Novikov, and A. P. Sokolov, *Phys. Chem. Chem. Phys.*, 15, 4604, 2013.

115. M. L. Manning and A. J. Liu, *Phys. Rev. Lett.*, 107, 108302, 2011.

116. Y. G. Chung and D. J. Lacks, *Macromolecules*, 45, 4416, 2012.

117. Y. G. Chung and D. J. Lacks, *J. Phys. Chem. B*, 116, 14201, 2012.

118. Y. G. Chung and D. J. Lacks, *J. Chem. Phys.*, 136, 124907, 2012.

119. Y. G. Chung and D. J. Lacks, *J. Polym. Sci., Part B: Polym. Phys.*, 50, 1733, 2012.

120. T. C. B. McLeish, *Adv. Phys.*, 51, 1379, 2002.

121. Y. Y. Lu, L. J. An, S. Q. Wang, and Z. G. Wang, *ACS Macro Lett.*, 2, 561, 2013.

122. C. Maloney and M. O. Robbins, *Chaos*, 17, 041105, 2007.

123. C. E. Maloney and M. O. Robbins, *Phys. Rev. Lett.*, 102, 225502, 2009.

124. M. L. Falk and C. E. Maloney, *Eur. Phys. J. B*, 4, 405, 2010.

125. M. L. Falk and J. S. Langer, *Ann. Rev. Condens. Matter Phys.*, 2, 373, 2011.

126. F. M. Capaldi, M. C. Boyce, and G. C. Rutledge, *Phys. Rev. Lett.*, 89, 175505, 2002.

127. A. V. Lyulin and M. A. J. Michaels, *J. Non-Cryst. Solids*, 352, 5008, 2006.

A comparison of constitutive descriptions of the thermo-mechanical behavior of polymeric glasses

GRIGORI A. MEDVEDEV AND JAMES M. CARUTHERS

14.1 INTRODUCTION

Diverse, nonlinear thermo-mechanical behavior has been observed experimentally for polymeric glasses (see Chapter 4), where a number of constitutive models have been proposed to describe various aspects of these experiments. Although there have been numerous papers comparing the predictions of a specific constitutive model with experimental data, these studies have primarily focused on the prediction of just that constitutive model for a relatively limited set of data. Moreover, there has not been a definitive review where the predictions from different constitutive models have been critically compared to the same data set in order to determine which components of a constitutive model are needed to predict a given class of experiments. To the extent that a constitutive model is a precise statement of a hypothesis concerning the physical processes that give rise to the behavior, the paucity of comprehensive analyses of the predictions of various constitutive models is an indicator of the lack of a generally agreed-upon understanding of the processes that control the deformation of glasses. The objective of this chapter is to begin a critical and comprehensive analysis of the collection of constitutive models that have been proposed to describe aspects of the thermo-mechanical behavior of polymeric glasses. We anticipate that this type of analysis will provide a more solid foundation for future analyses, creating a better point of departure for an improved understanding of the glassy state.

The goal of this chapter is not to describe all the proposed models or to give a historical perspective of how various classes of models have been developed, although we will attempt to give credit to the pioneering contributions. Rather, the objective of this chapter is to identify the strengths and weaknesses of the various types of constitutive models. With this objective in mind, the following approach will be used:

1. Key features of the experimental data that, in our opinion, expose the underlying physics of glassy behavior will be identified, where many of these features are challenging to predict via a constitutive model. The proposed list of features is not final, where the results of future experiments should be added once their significance is recognized.
2. Models in the literature that can (at least in principle) predict several of the important features will be identified and then analyzed in more detail versus the key experimental features. This includes the models for which the prediction of some experiments may not have been reported by the authors of these models, but where a straightforward pathway exists to obtain such predictions. Because it is impossible to analyze in detail all constitutive models that have been proposed to describe polymeric glasses, there will be an attempt to determine the underlying reason for certain predictions so that one can make inferences about all the models in that class.

This approach has been designed from a "scientific" perspective such that if a constitutive model has properly captured the underlying physical mechanisms, then, it should be able to describe, at least qualitatively, a number of apparently disconnected phenomena without the continual introduction of new model functions/parameters. Thus, the qualitative description of a wide variety of nonlinear phenomena is viewed as a more important first step than a quantitative description of a limited set of data. An alternative "engineering" approach would be to focus on higher-fidelity descriptions for a limited range of phenomena, where introduction of more model functions/parameters would not be disparaged—an approach that may have significant practical value. Finally, one should note that this approach for model discrimination is not completely objective, because it depends upon the specific list of the experimental features on which the constitutive predictions will be evaluated.

In this chapter, we will consider the ductile behavior of glasses subjected to arbitrary thermo-deformational histories. Because the constitutive models are inherently nonlinear, the response for a given set of boundary conditions can be spatially heterogeneous; however, in this chapter, we will only be looking at the constitutive response, and not how the material's inherent thermo-mechanical behavior interacts with the conservation equations of continuum mechanics to produce spatially inhomogeneous deformations. We believe that one must first understand the spatially homogeneous, intrinsic, material response prior to describing situations with spatial inhomogeneities. Also, this chapter will focus exclusively on the ductile behavior and not on the ultimate failure via brittle fracture, craze formation, and so on. This is not to say that the ultimate failure of a polymeric glass is not important (see Chapter 12); however, since nonlinear, time/rate-dependent deformation precedes the ultimate failure event, an understanding of the ductile constitutive response is a natural prerequisite for understanding ultimate failure.

One more comment is in order before we present our list of the key experiments. Once its parameters are set, a constitutive model should be able to quantitatively describe any experiment in the evaluation data set as well as predict the response for new experiments. Naturally the model parameters will have different values for different materials; thus, the entire set of experiments for testing the model should be performed on the same single, well-characterized material. However, such a unified data set does not currently exist, where the various types of experiments reported in the literature have been performed on different glassy materials. Consequently, in the strictest sense, the task we set out to do in the previous paragraph—to test the existing constitutive models against a discriminating set of experiments—cannot be accomplished. However, the proposed task is not meaningless even acknowledging today's lack of a comprehensive experimental data set. Specifically, one should establish if a candidate model is capable of predicting the qualitative features of the data before attempting quantitative fitting; thus, if the set of experiments contains discriminating features (even if it employs multiple materials), then, evaluation of the models with respect to their ability to predict the qualitative features of the data is valuable.

14.2 KEY EXPERIMENTAL FEATURES OF THE THERMO-MECHANICAL BEHAVIOR OF GLASSES

A detailed discussion of the "thermo-mechanical signatures of the glass" is available in Chapter 4; thus, we will not provide here a full description of each phenomenon with references. To evaluate multiple constitutive equations with respect to all the signatures identified in Chapter 4 would be overwhelming; thus, we will only employ a subset of the signatures as key "features" which will be evaluated for the various constitutive models. A summary of the key features to be employed in this chapter will now be presented.

14.2.1 PHYSICAL AGING

Physical aging, or equivalently structural relaxation, is at the heart of glassy behavior as it highlights the non-equilibrium nature of the glassy state [1]. A constitutive description must include physical aging, where some proposed constitutive models (as we will discuss later) deal with aging only operationally, that is, the effect of aging on the nonlinear response has been parameterized as a function of sub-Tg annealing time as opposed to being predicted for an arbitrary thermal history.

Feature 1.1—The thermo-mechanical behavior of a polymeric glass depends upon the thermal history including sub-Tg physical aging.

14.2.2 LINEAR VISCOELASTIC BEHAVIOR

For sufficiently small deformations, glassy materials exhibit linear viscoelastic creep, stress relaxation, and dynamic behavior [2]. Since many features of the nonlinear mechanical behavior are known to be strongly dependent on the strain rate, there is a temptation to postulate constitutive equations with evolution equations for the internal variables that explicitly depend upon the strain rate (see Section 14.5 of this chapter). However, objectivity requires that a proper tensorial form of the strain rate tensor be used, for example, the second invariant of the rate-of-deformation tensor; and, when an infinitesimal sinusoidal deformation is applied at high frequencies or for sufficiently long times, this type of model may predict a nonlinear response, which is contrary to the requirement of a linear viscoelastic limit.

Feature 2.1—Glassy polymers exhibit linear viscoelastic behavior for infinitesimal deformations.

The shear relaxation spectrum is significantly wider than the bulk relaxation spectrum at least for the case of polymeric glasses. The reason for this qualitative difference between the linear viscoelastic shear and bulk spectra is not obvious; specifically, both the Rouse model and the reptation model predict that the spectrum is the same in shear and bulk [3].

Feature 2.2—The shear relaxation spectrum for a polymeric glass is significantly wider than the bulk spectrum.

14.2.3 ISOTROPIC NONLINEAR MECHANICAL BEHAVIOR; VOLUME RELAXATION

Glasses exhibit nonlinear, time-dependent volume relaxation when the material experiences changes in temperature in the Tg region, where the gold standard is the set of experiments of Kovacs on poly(vinyl acetate) [4]. These experiments established several important features of glassy relaxation behavior, including nonlinearity, asymmetry of approach (following up- and down-jumps to the same temperature), memory, and the tau-effective expansion gap. There have been a number of attempts to fit the Kovacs data, where some of these features are easier to predict than others [5–12]. For example, nonlinearity follows naturally from the assumption that the instantaneous response depends on the amount of structural relaxation that had occurred in the material prior to the experiment. Similarly, the asymmetry of approach is achieved by postulating that the relaxation time is a function of the structural variable. The memory effect does not even require a structure-dependent relaxation time, where all that is needed is that there is a spectrum of relaxation times. This does not mean that quantitative fitting of the Kovacs data is trivial—only that the qualitative features of nonlinearity, asymmetry, and memory are straightforward to obtain.

The hard-to-predict features of the Kovacs volume relaxation data include the short time anneal experiment and the tau-effective paradox. The short time anneal experiment is when a glass at equilibrium is (i) rapidly quenched below Tg, (ii) allowed to relax for a short period of time, (iii) the temperature is increased back to the initial temperature, and (iv) the volume relaxation is monitored. One would anticipate that this would be a relatively easy experiment to predict, since the material has only spent a small amount of time relaxing in the glassy state; however, this is not the case, where most models of structural relaxation have difficulty in predicting this feature. The tau-effective paradox refers to observation that the effective rate of relaxation as equilibrium is approached after up-temperature jumps from different initial temperatures to the same final temperature is different. One would expect that since the final temperature is the same, as the material approaches the same equilibrium volume, the rate of relaxation would be the same; however, there is half an order-of-magnitude difference in the effective rate of relaxation even when the deviation from the equilibrium volume is exceedingly small. These data present a significant challenge for any model that assumes that there is an underlying spectrum of the relaxation times, because at the end of the relaxation, only the longest relaxation time is active and hence the rate of the final approach to equilibrium is the same for all thermal histories. The fact that this is not the case requires reevaluation of the basic constitutive model assumptions such as thermorheological simplicity.

Feature 3.1—The basic features of volume relaxation, including nonlinearity, asymmetry, and memory
Feature 3.2—"Short time anneal" volume relaxation
Feature 3.3—"Tau-effective paradox"

14.2.4 ANISOTROPIC NONLINEAR MECHANICAL BEHAVIOR

The time-dependent nonlinear mechanical response has been studied in uniaxial extension, uniaxial compression, and shear for a number of polymeric glasses (see Chapter 4 for details). As shown schematically in Table 14.1, the well-known nonlinear stress–strain curve is a common feature in polymeric glasses with its dependence upon strain rate and sub-Tg aging. Creep also exhibits highly nonlinear behavior as illustrated in Table 14.1 that also depends upon sub-Tg physical aging. It has been observed [13,14] that the log(strain rate) versus strain curve in nonlinear creep is qualitatively the mirror image of the stress–strain curve during a constant strain rate experiment (see Chapter 4 for details). This is an important observation, because it connects the nonlinear creep behavior to the nonlinear stress–strain behavior.

When a material is deformed isothermally, there can be heat exchanged between the material and the surroundings. In a very limited number of experiments, this heat flow has been measured using deformation calorimetry [15]. Consider the results of deformation calorimetry performed during constant strain rate uniaxial compression, where the specimen is taken through the yield and post-yield flow. Of particular significance, neither the heat flow nor the internal energy exhibits an abrupt change in behavior at yield. Also, for glassy materials, the internal energy only reaches saturation well into the post-yield flow regime, where it constitutes up to 50% of the total work of deformation, which should be compared to less than 3% for the plastic flow of crystalline materials [16]. This observation appears to put in doubt the description of the deformation of polymeric glasses as a plasticity phenomenon; specifically, from a plasticity perspective, the internal energy is expected to reach maximum at yield, after which the work of deformation is almost entirely converted into heat. Deformation-induced heat flow is obviously important to the overall energy balance associated with deformation, although most constitutive models are silent with respect to the enthalpic component of the deformation.

Although most experimental studies involve a constant strain rate or single-step creep, multistep deformations are a much more stringent test of a constitutive model, where a number of discriminating multistep experiments are shown schematically in Table 14.2. Specific multistep experiments include:

1. *Strain rate switching*: In this experiment, the sample is loaded through yield at a strain rate $\dot{\varepsilon}_1$ and then deformation continues at a strain rate $\dot{\varepsilon}_2$. After the change in strain rate, there is a stress overshoot if $\dot{\varepsilon}_2 \gg \dot{\varepsilon}_1$ and stress undershoot if $\dot{\varepsilon}_2 \ll \dot{\varepsilon}_1$ [17]. A constitutive model should be able to describe the response to the new strain rate; however, as will be shown subsequently, there are several constitutive models that can only describe the first stress overshoot, but not the second.

2. *Loading–unloading–reloading*: When the sample is loaded at a constant strain rate, unloaded at a constant strain rate to zero stress, and then reloaded at a constant strain rate, the material exhibits (i) a large permanent set when initially loaded past yield that only marginally relaxes with time, but which can be completely erased when the material is heated above Tg, (ii) smooth unloading stress response with no shoulder or kink, i.e., the so-called "reverse" or "second" yield is not experimentally observed,

Table 14.1 Features of single-step nonlinear stress–strain and creep experiments

Constant strain rate loading

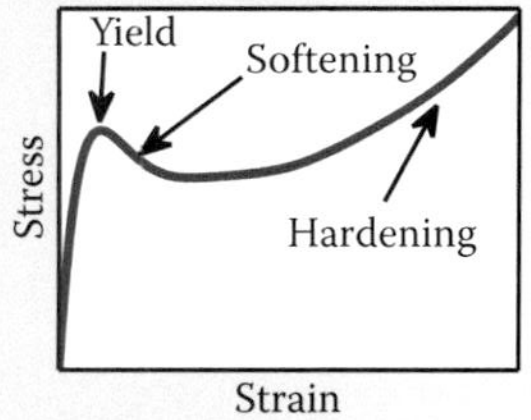

Figure 1.1 The stress–strain curve consists of the following distinct stages characterized by the sign of the tangent modulus (i.e., the derivative $d\sigma/d\varepsilon$): (I) stress buildup that is initially linear and then nonlinear; (II) yield point, $d\sigma/d\varepsilon = 0$; (III) post-yield softening, $d\sigma/d\varepsilon < 0$; (IV) flow, $d\sigma/d\varepsilon = 0$; and (V) hardening.

Comments:

- Qualitatively the same in extension, compression, and shear

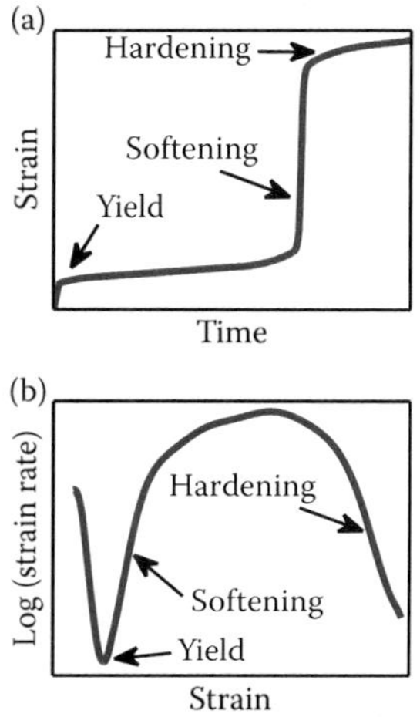

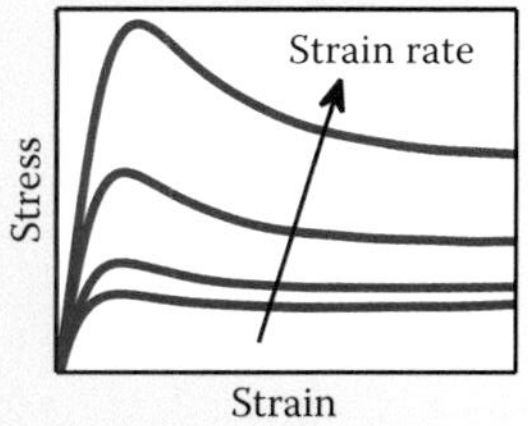

Figure 1.3 The magnitude of stress overshoot with subsequent softening increases with strain rate.

Comments:

- Readily observed close to Tg
- At high strain rates adiabatic heating is possible; however, at temperatures close to Tg adiabatic heating is negligible [26]

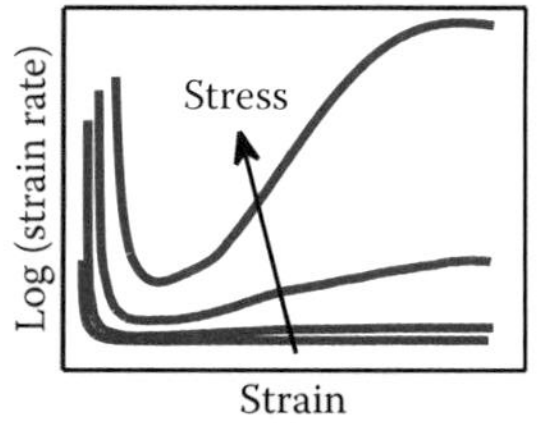

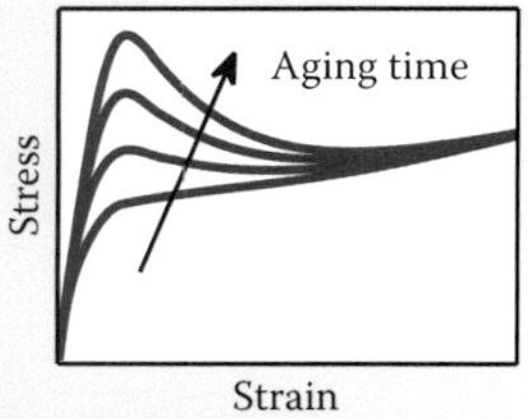

Figure 1.5 Magnitude of stress overshoot with subsequent softening increases with aging time.

Creep

Figure 1.2 The log strain rate versus strain curve (i.e., b) consists of the following distinct stages characterized by the strain rate: (I) primary, $d\dot\varepsilon/dt$ decreases; (II) secondary, $d\dot\varepsilon/dt = $ Const; (III) tertiary, $d\dot\varepsilon/dt$ increases; (IV) $d\dot\varepsilon/dt = $ Const but higher than during Stage II; and (V) $d\dot\varepsilon/dt$ decreases [14].

Comments:

- The third and fourth stages of creep have not been experimentally observed in shear

Figure 1.4 Magnitude of tertiary creep increases with stress.

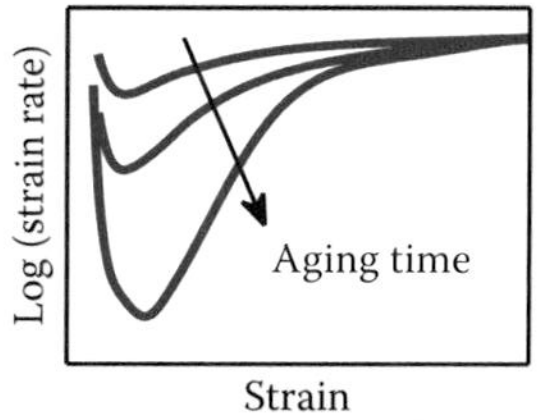

Figure 1.6 Magnitude of tertiary creep increases with aging time [27].

Table 14.2 Schematics of multistep deformation experiments

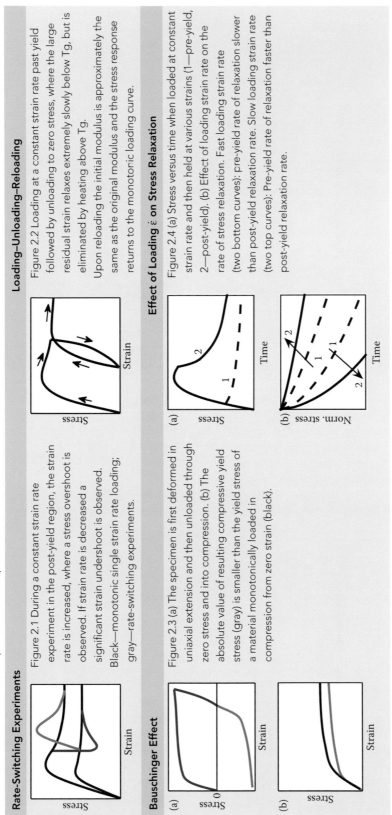

Rate-Switching Experiments

Figure 2.1 During a constant strain rate experiment in the post-yield region, the strain rate is increased, where a stress overshoot is observed. If strain rate is decreased a significant strain undershoot is observed. Black—monotonic single strain rate loading; gray—rate-switching experiments.

Bauschinger Effect

Figure 2.3 (a) The specimen is first deformed in uniaxial extension and then unloaded through zero stress and into compression. (b) The absolute value of resulting compressive yield stress (gray) is smaller than the yield stress of a material monotonically loaded in compression from zero strain (black).

Loading–Unloading–Reloading

Figure 2.2 Loading at a constant strain rate past yield followed by unloading to zero stress, where the large residual strain relaxes extremely slowly below Tg, but is eliminated by heating above Tg. Upon reloading the initial modulus is approximately the same as the original modulus and the stress response returns to the monotonic loading curve.

Effect of Loading $\dot{\varepsilon}$ on Stress Relaxation

Figure 2.4 (a) Stress versus time when loaded at constant strain rate and then held at various strains (1—pre-yield, 2—post-yield). (b) Effect of loading strain rate on the rate of stress relaxation. Fast loading strain rate (two bottom curves): pre-yield rate of relaxation slower than post-yield relaxation rate. Slow loading strain rate (two top curves): Pre-yield rate of relaxation faster than post-yield relaxation rate.

(Continued)

Table 14.2 (Continued) Schematics of multistep deformation experiments

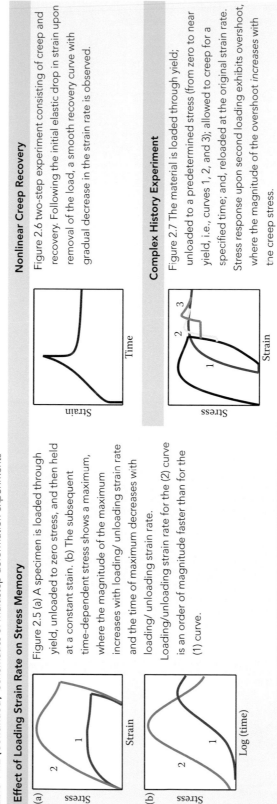

Effect of Loading Strain Rate on Stress Memory

Figure 2.5 (a) A specimen is loaded through yield, unloaded to zero stress, and then held at a constant stain. (b) The subsequent time-dependent stress shows a maximum, where the magnitude of the maximum increases with loading/ unloading strain rate and the time of maximum decreases with loading/ unloading strain rate.

Loading/unloading strain rate for the (2) curve is an order of magnitude faster than for the (1) curve.

Nonlinear Creep Recovery

Figure 2.6 two-step experiment consisting of creep and recovery. Following the initial elastic drop in strain upon removal of the load, a smooth recovery curve with gradual decrease in the strain rate is observed.

Complex History Experiment

Figure 2.7 The material is loaded through yield; unloaded to a predetermined stress (from zero to near yield, i.e., curves 1, 2, and 3); allowed to creep for a specified time; and, reloaded at the original strain rate. Stress response upon second loading exhibits overshoot, where the magnitude of the overshoot increases with the creep stress.

(iii) the initial modulus upon reloading that is nearly the same as the modulus observed on the initial loading, and (iv) during reloading, the stress response returns to the single-loading stress–strain curve with no overshoot [18]. The fact that the sample completely recovers when heated above Tg clearly shows that there is no permanent damage to the material.

3. *Bauschinger effect*: In a two-step experiment where the initial deformation is uniaxial extension followed by uniaxial compression, the compressive yield stress is lower than the yield stress observed in a single-step compression from zero strain. This phenomenon is called the Bauschinger effect [19]. The switch from extension to compression in a single experiment presents an experimental challenge, and predicting the Bauschinger effect provides a significant challenge for many constitutive models.

4. *Nonlinear stress relaxation*: In this experiment, the polymer glass is deformed at a constant strain rate to a predetermined strain, which is then held constant. The initial rate of stress relaxation decreases with strain if the loading strain rate is slow, but increases with strain if the loading strain rate is fast [20]. The increase in the instantaneous rate of relaxation with strain is consistent with constitutive models that assume that deformation increases mobility, but the data where the mobility decreases with increasing strain are problematic for many constitutive models.

5. *Stress memory*: In the stress memory experiment, the glass is loaded at a constant strain rate through yield, unloaded at a constant strain rate to zero stress, and then the strain is held constant. The subsequent stress response, which is called stress memory, exhibits an overshoot followed by relaxation back to zero. Both the location of the overshoot along the time axis and its magnitude depend on the strain rate during loading and unloading [21]. Consider a material that is deformed (via loading–unloading) to the same post-yield strain at two different strain rates. At the point where the stress memory observation begins, the strain rate is zero, the stress is zero, and the residual strain is the same; however, the time of the maximum stress overshoot is significantly different. This significant dependence of the stress response on loading/unloading strain rate even though the stress, strain rate, and strain are the same is a challenge for constitutive models.

6. *Two-step creep and recovery*: In this experiment, the load is removed during nonlinear creep, where the resulting strain response during recovery is smooth with the rate of recovery decreasing gradually. If the applied stress is a significant fraction of the yield stress in a constant strain rate experiment, there is a large residual strain that persists for extremely long times, although complete recovery is observed when the material is heated above Tg. These data challenge a number of constitutive models, where the deformation-induced mobility mechanism needed to describe nonlinear creep then makes the recovery abruptly cease upon unloading versus the experimental data, where there is a smooth recovery response.

7. *Loading–creep–reloading*: In this experiment, the glassy polymer is loaded at a constant strain rate, allowed to creep at a nonzero stress, and then reloaded at a constant strain rate. If the creep stress is small, there is no stress overshoot upon reloading; however, if the creep stress is slightly below the post-yield flow stress, there is a significant overshoot upon reloading [18]. This behavior has not been predicted as yet by any existing constitutive model.

In all of the experiments above, there is a dependence on temperature, thermal history (i.e., annealing time, etc.), deformation geometry (i.e., extension, compression, or shear) and the strain rate or the magnitude of the applied stress; however, the qualitative features of the polymeric glasses' response are to first order invariant to these details of thermo-deformational history.

The rotational mobility of a molecular dye has been monitored during both nonlinear creep and constant strain rate loading, where it was observed that (i) the average relaxation time decreases by several orders-of-magnitude during deformation, (ii) the width of the relaxation time spectrum (as measured via Kohlrausch [22]–Williams–Watts [23] [KWW] parameter β) decreases dramatically during deformation, and (iii) both the increase in mobility and the spectral narrowing are reversed when the load is removed [24,25]. This is a very important result because the change in the width of the relaxation spectrum upon deformation appears to eliminate thermorheologically simple constitutive models, where it is assumed that the shape of the relaxation spectrum is invariant to changes in temperature and deformation.

Feature 4.1—The yield stress is

a. A linear, increasing function of log (strain rate) for over 4–5 orders-of-magnitude; however, over a larger range of strain rates then it shows some curvature.
b. A nearly linear, decreasing function of temperature, where it approaches zero at Tg.
c. An increasing function of aging time, where once equilibrium is approached it levels off.
d. Well described by the pressure-modified von Mises criterion. It is possible to achieve yield in dilatation using a longitudinal deformation [28].

Feature 4.2—Stress–strain behavior:

a. The nonlinear σ versus ε curve in a constant strain rate deformation is the mirror image of the $\log \dot{\varepsilon}$ versus ε curve in a nonlinear creep experiment.
b. The magnitude of the stress overshoot in a constant strain rate deformation increases with sub-Tg aging time. The minimum in the $\log \dot{\varepsilon}$ versus ε curve in a nonlinear creep experiment decreases with sub-Tg aging time.
c. The magnitude of the stress overshoot in a constant strain rate deformation increases with strain rate. The magnitude of tertiary creep increases with the applied stress in nonlinear creep.
d. Strain hardening is observed at large strains for a constant strain rate deformation. Stage V creep is observed at large strains.

Feature 4.3—The internal energy stored during deformation does not exhibit an abrupt change at yield, reaching a maximum in the "plastic flow" regime where it accounts for 30%–50% of the work. The heat generated during deformation does not exhibit an abrupt change at yield.

Feature 4.4—In rate-switching experiments, there is a stress overshoot if the strain rate increases and a stress undershoot if the strain rate decreases.

Feature 4.5—In loading–unloading (to zero stress)–reloading experiments:

a. Unloading after the material that has been deformed past yield results in large residual strain that relaxes extremely slowly below Tg; however, the residual strain is eliminated upon heating above Tg.
b. There is no "second" or "reverse" yield in the unloading stress–strain curve.
c. The initial modulus upon reloading is nearly the same as for original loading from zero strain.
d. Upon reloading, the stress–strain curve returns to the single-loading stress–strain curve with no stress overshoot.

Feature 4.6—The Bauschinger effect: the yield stress in compression following an initial extension deformation is lower than the yield stress in a single-step compression from zero strain.

Feature 4.7—In a stress relaxation experiment comprised of constant strain rate loading to a specified strain and then holding that strain constant, the initial rate of stress relaxation (i) decreases with strain if the loading strain rate is slow, but (ii) increases with strain if the loading strain rate is fast.

Feature 4.8—In a stress memory experiment, the stress exhibits an overshoot before relaxing back to zero, where the location along the time axis of the overshoot and its magnitude depend on the strain rate during the previous loading and unloading.

Feature 4.9—In a nonlinear creep-recovery experiment, the recovery response is smooth and the residual strain persists for exceedingly long times, when the applied stress is sufficiently large and the temperature is below Tg.

Feature 4.10—In a constant strain rate loading–creep–reloading experiment, there is no stress overshoot upon the second constant strain rate loading if the creep stress is small versus a significant overshoot if the creep stress is close to the post-yield flow stress.

Feature 4.11—Using a molecular dye as a reporter of the mobility, in nonlinear creep and nonlinear constant strain rate deformation (both in uniaxial extension), the average relaxation time and the width of the relaxation spectrum decrease with deformation. Both the average relaxation time and spectral width recover upon removal of the load.

14.2.5 ENTHALPY RELAXATION

The enthalpy relaxation in polymeric glasses has been a subject of numerous studies (see Section 4.4 in Chapter 4). The key findings include (i) the transition from the equilibrium asymptotic heat capacity to the glassy asymptotic heat capacity upon cooling and (ii) the appearance of a heat capacity overshoot peak upon heating through the glass transition. The height and position of the overshoot depend on the thermal history, including the sub-Tg annealing time and temperature. It is obvious that enthalpy relaxation occurs simultaneously with isotropic volume relaxation; however, most modeling activities treat these as disconnected phenomena. By including the enthalpy relaxation into the list of critical features, we hope to encourage future modeling efforts to combine both phenomena in a unified description.

Feature 5.1—Upon constant rate heating from the glassy into equilibrium state, the heat capacity exhibits an overshoot, where the height and position of the overshoot depend on thermal history, i.e., the annealing time and temperature.

14.2.6 SUMMARY

The set of experiments detailed above will significantly challenge any prospective constitutive model for polymeric glasses. There are of course certain "signatures" of glassy behavior that have not been included in this set, for example, the observation of the broad exothermic heat capacity peak when heating a glassy material that was previously subjected to a large mechanical deformation (see Section 4.9.3 in Chapter 4). Nevertheless, this set of experiments is significantly more challenging than any data set that has been used to-date to evaluate constitutive models for polymeric glasses. Moreover, these experiments expose interactions between extension and compression (i.e., Feature 4.6), reversal of loading direction (i.e., Feature 4.5), differences/similarities between creep and constant strain rate experiments (i.e., Feature 4.2a), isotropic and multiaxial deformation (i.e., Features 3.1 through 3.3, and 4.1d), mechanical versus thermal contributions to nonlinear deformation (Feature 4.3), and changes in the shape of the relaxation spectrum with deformation (Feature 4.11). We make the bold postulate that for a constitutive model to naturally capture this diversity of thermo-mechanical behaviors, the constitutive model cannot be just a mathematical construct, but rather must have a structure that aligns with the underlying physical processes in polymeric glasses.

14.3 GLASSY CONSTITUTIVE MODELS

Prior to the discussion of particular constitutive equations for polymeric glasses, there are standard requirements imposed by continuum mechanics that must be recognized [29]; specifically,

- Stress and strain are tensorial quantities and thus a constitutive description has to be in a proper three-dimensional tensorial form that is objective (i.e., frame indifferent). One-dimensional versions of three-dimensional models maybe useful for elucidating particular mechanisms and for making sure that limiting behaviors are meaningful; however, transformation of an initially one-dimensional model into three-dimensional form is nontrivial. A number of one- and two-dimensional models have been proposed for nonlinear mechanical behavior of glasses, e.g., the Chen–Schweizer model [30] and the Bulatov–Argon model [31], but they will not be considered further in this chapter.
- A constitutive model must make predictions for arbitrary thermal and deformational histories. For example, there are "constitutive models" for glasses tailored to predict the nonlinear stress–strain curve observed in a constant strain rate uniaxial deformation, where the relevant equations contain the axial strain rate $\dot{\varepsilon}$ in the denominator. Clearly this precludes their use for experiments like stress relaxation where strain is constant and hence $\dot{\varepsilon} = 0$. These types of models will not be considered in this chapter.
- A constitutive model has to be thermodynamically consistent, i.e., it cannot violate the Second Law of thermodynamics in the form of the Clausius–Duhem inequality [32]. A standard way of ensuring thermodynamic consistency is by deriving the equations of interest, i.e., the stress equation and the

entropy equation, from an appropriate Helmholtz free energy. Unfortunately, this is not always done, e.g., the ad hoc insertion of various additional dependencies such as dependence on the strain rate into the coefficients of a properly derived stress versus strain equation is problematic. Specifically, without going back to the Helmholtz free energy and re-deriving the stress equation for the new extended set of variables, there is no assurance that the Second Law is not violated. In this chapter, we point out if thermodynamic consistency has (or has not) been established for the model under consideration.

Finally, one should remember that a constitutive model is not an *ab initio* model, where there are certain features of the glassy state that are postulated as opposed to being predicted, for example, the temperature dependence of the linear viscoelastic relaxation behavior is super-Arrhenian.

Although acknowledgment of a potential underlying molecular mechanism might provide inspiration for a particular functional form of a constitutive equation, continuum mechanics does not require explicit use of any molecular mechanism. Nevertheless, there are molecular insights that can provide value; specifically,

1. Dynamic heterogeneity is an important feature of the glassy state, where a glass exhibits order-of-magnitude mobility differences between neighboring domains that persists for extremely long times [33,34]. These extremely long-lived spatial and temporal fluctuations on the nanoscale seem to be an inherent feature of the glassy state, although it is a matter of debate if they represent a critical organizing principle of the glass or if they are just the by-product of some more fundamental process.
2. Bulk metallic glasses exhibit similar nonlinear ductile mechanical behavior to that of polymeric glasses, except perhaps strain hardening [35]. This seems to indicate that a description of ductile glassy behavior that critically depends upon long polymeric chains is inappropriate, even if the objective is to describe the deformation behavior of polymeric glasses.

At this point, one should acknowledge that a fundamental model of the glassy state remains one of the outstanding problems in condensed matter physics; thus, construction of a molecular-based model for the nonlinear thermo-mechanics of glasses is not currently possible. Finally, terms like "free volume" and "energy landscape" are frequently invoked to rationalize various aspects of the glassy behavior; however, we are reluctant to use these terms, because they have become so malleable that they can seemingly accommodate almost any behavior observed experimentally.

There are two distinct classes of constitutive models that will be considered in this chapter: *continuum models* where all field variables are in the macroscopic limit and hence smooth; and, a *stochastic model* where the field variables mesoscopically fluctuate and thus acknowledges dynamic heterogeneity. The vast majority of constitutive models in the literature to-date have been continuum models, where the relatively recent development of a stochastic description grew out of (i) dissatisfaction with the performance of the traditional continuum models and (ii) a desire to explicitly acknowledge dynamic heterogeneity in the constitutive model. The stochastic model is in its infancy, where many issues remain unresolved; however, preliminary results are encouraging and thus we deemed it worthy of being included in this chapter.

With respect to continuum models, there have been historically two independent constitutive modeling tracks for polymeric glasses: plasticity and viscoelasticity. We will now review general features of these two approaches:

1. *Plasticity models*: A number of plasticity models have been developed to describe the nonlinear mechanical behavior of polymeric glasses [36–55]. Plasticity-based models were initially developed to describe just the stress–strain curve in a constant strain rate deformation with primary focus on the yield phenomenon. The main assumption of the plasticity framework is that the observed strain is a combination of elastic and plastic components, each with its own kinematic and constitutive relationships, where the plastic component becomes dominant at yield. There are conceptual difficulties with employing a plasticity-based description for glassy materials, including (i) the reversibility of the effects of deformation upon heating above Tg for slightly cross-linked polymers [25] and (ii) the observation from deformation calorimetry that the internal energy and dissipated heat during deformation show no qualitative change in behavior at yield [15]. Another problem is that the glass transition and other thermal history effects such as physical aging are not inherent in the plasticity framework and hence must be added as separate

mechanisms. Even recovering the linear viscoelastic behavior that is experimentally observed at small deformation represents a challenge for plasticity- based models, but a challenge that can be overcome. These difficulties notwithstanding, the mathematical formalism of the plasticity models has proven sufficiently flexible to describe many facets of the nonlinear deformation behavior of polymeric glasses, at least operationally, via postulating additional dependencies and/or introducing multiple internal variables.

2. *Viscoelastic models*: A number of nonlinear viscoelastic constitutive models have been proposed for describing glassy thermo-mechanics [6,56–65]. The strength of the viscoelastic (or, more accurately, thermoviscoelastic) approach is that all responses, including the nonlinear mechanical response, aging, enthalpy relaxation, etc. (and even the glass transition itself) are treated as various manifestations of the same underlying glassy behavior. The key assumption is that all the relevant physics is captured by the "material clock," which describes how the rate of relaxation depends on the state of the material [66,67]. From this perspective, when a material undergoes vitrification, its material clock slows down as compared to the laboratory time. Conversely, application of a load to a glassy material causes the material clock to accelerate, which then manifests as yielding and other nonlinear relaxation phenomena. Consequently, if the all-important material clock function (or functional) can be constructed, then, the complete description of all the glassy phenomena is achieved. However, the material clock approach may be inadequate, because (i) the correct formula relating the evolving state variables to the acceleration/deceleration of the relaxation processes in the material clock has not been used or (ii) a single material clock does not capture the underlying physics of the glassy state.

A comment is in order regarding the above classification of constitutive models. Many plasticity models have evolved so that there is a certain convergence of the plasticity and viscoelasticity approaches, where even the term "plastic" has been replaced by "viscoplastic" or "viscoelastic–viscoplastic" in the literature. This evolution is exemplified by the glass–rubber constitutive model (GRCM) of Buckley and coworkers. In the original version of the model [64,65], the standard plasticity assumption was made where the total strain is a combination of elastic and plastic strains. However, it was possible to eliminate both elastic and plastic strains from the final constitutive equation, that is, only the total strain was present. Thus, the original GRCM is a nonlinear viscoelastic model according to our classification, where plasticity models need independent constitutive equations for the elastic/viscoelastic and plastic components of the strain. In a subsequent modification to improve the prediction of post-yield softening, the plastic strain was explicitly included in the defining equations of the GRCM in addition to the total strain [51–53]. This modification formally places the new GRCM into the viscoplastic domain. Nevertheless, since the GRCM invokes the fictive temperature idea and the other models using fictive temperature are discussed within the viscoelastic framework, in this chapter, we include the GRCM in the viscoelastic section.

A stochastic constitutive model (SCM) has recently been developed where nanoscale fluctuations were included in the stress and entropy constitutive equations. A key feature of the SCM is that there is only one inherent relaxation process, where it is the fluctuations in the state variables that give rise to the distribution of relaxation times observed experimentally. There is no constraint on the shape of distribution of relaxation times other than each realization of the stress/entropy response to a particular thermo-deformational history is given by the same stochastic constitutive equations; thus, the shape of the relaxation time distribution can change with temperature (i.e., thermorheological complexity) and with deformation (i.e., as shown in Ediger's molecular dye rotation studies [24,25]). Because the shape of the relaxation time distribution evolves during the thermo-deformational history, the behavior of the SCM is quite different from the traditional continuum plastic, viscoelastic, and viscoplastic models for describing polymeric glasses.

We are now ready to address the main objective of this chapter—to critically test constitutive models that have been proposed to describe the thermo-mechanical behavior of polymeric glasses against a set of discriminating experiments. The remainder of the chapter will be organized as follows: in Section 14.4, both plastic and viscoplastic models will be analyzed; in Section 14.5, nonlinear viscoelastic models will be discussed; in Section 14.6, the structure and predictions of the stochastic constitutive model will be reviewed; and, finally in Section 14.7, there will be a discussion of the state of the art in constitutive modeling of polymeric glasses and some future challenges and opportunities.

14.4 ANALYSIS OF PLASTICITY-BASED MODELS

The main idea behind plasticity-based models is that the deviatoric strain is a combination of elastic and plastic components, where the plastic strain component is responsible for the flow-like behavior. The plastic flow cannot be initiated by isotropic deformation. Plastic flow requires that the deviatoric stress is greater than some threshold value (for example, the yield stress); thus, for infinitesimal deformations, the response is purely elastic with no time dependence. Consequently, the basic plasticity model (BPM) is incapable of predicting

- Physical aging (Feature 1.1)
- Linear viscoelasticity (Feature 2.1)
- Volume relaxation (Features 3.1 through 3.3)
- Enthalpy relaxation (Feature 5.1)

It is not surprising that plasticity-based models are unable to describe these features, since they were originally developed to just describe large anisotropic deformations. With the exception of volume and enthalpy relaxation, these shortcomings can be addressed by postulating that (i) parameters in a plasticity model change with the aging time, (ii) the infinitesimal response is viscoelastic rather than elastic, and (iii) the model parameters are given temperature dependencies so that the relaxation response dramatically changes at the glass transition. Although not elegant and involving the introduction of a large number of phenomenological parameters, these generalizations of the basic plasticity model make it a viable candidate to describe the mechanical behavior of polymeric glasses. The question now to be addressed is: can plasticity models describe Features 4.1 through 4.11?

The plasticity literature is enormous. In our opinion, there are three main branches of plasticity models for polymer glasses that are well illustrated by three specific constitutive models:

1. *Basic plasticity model*: Originally developed by Haward and Thackray [68] as a one-dimensional model and later generalized to three dimensions, this model has the essential features of plasticity.
2. *Eindhoven glassy polymer (EGP) model*: The EGP model [39–43,45,46,69] is fully tensorial, includes a distribution of relaxation processes, and includes the effects of temperature and thermal history.
3. *Anand models*: Anand and coworkers [47–50] have developed plasticity-based models that include internal variables in addition to the traditional continuum variables of temperature, strain, stress, etc., where the internal variables obey their own evolution equations.

The capabilities of each of these three models to describe Features 4.1 through 4.11 will be critically analyzed in the rest of this section. Other plasticity-based models for polymeric glasses will be discussed as part of the three classes given above, where we acknowledge that vastness of the plasticity literature precludes considering in detail all plasticity models that have been proposed.

14.4.1 BASIC PLASTICITY MODEL

The standard plasticity model was originally developed to describe ductile behavior of crystalline materials such as metals, where the condition for triggering the plastic flow is an "if-statement" that activates on a specific criterion that usually involves the stress tensor [70]. However, the "if-statement" is not essential, where a smooth but steep function of strain rate versus stress can effectively produce the same response. A nonlinear viscoelastic model can also employ a steep function that results in "if-statement" like behavior, for example, the viscosity in a generalization of the standard Maxwell model can be made a steep function of stress; thus, an abrupt change in deformation behavior does not differentiate between plasticity and viscoelasticity. The criterion differentiating plastic and viscoelastic models adopted in this chapter is that in a plastic (or viscoplastic) model, the plastic strain is an independent variable, whose evolution is governed by an independent equation; in contrast, in viscoelastic models, the total strain is the only independent deformation variable controlling the constitutive behavior.

14.4.1.1 HAWARD AND THACKRAY MODEL

The origin of the modern plasticity models for glassy polymers is found in the pioneering work of Haward and Thackray [68], where earlier contribution of Tobolsky and Eyring [71] should be acknowledged. Although the Haward–Thackray model is (i) one dimensional (and hence not truly constitutive) and (ii) really viscoelastic rather than plastic based upon the criterion given above, it is the origin of much of the work in plastic constitutive modeling of polymeric glasses. Thus, the Haward–Thackray model has considerable didactic importance, where the essential components of a nonlinear viscosity-based plasticity models were introduced. The Haward–Thackray model has a spring and a dashpot connected sequentially, where the total axial strain is the sum of the spring and dashpot strains which may be called "elastic" and "plastic." Specifically,

$$\varepsilon = \varepsilon_e + \varepsilon_p \tag{14.1}$$

Equation 14.1 is only applicable for the case of infinitesimal deformations. There is a second spring that is parallel to the spring–dashpot element; thus, the axial stress is a sum of two terms—the driving stress (for consistency with the majority of the literature it is given the subscript "s") acting on the spring–dashpot component and the hardening stress (since it was assumed to have stemmed from rubbery elasticity it was given the subscript "r") acting on the second spring

$$\sigma = \sigma_s + \sigma_r \tag{14.2}$$

In the Haward–Thackray model, the driving stress term, which contains the dashpot viscosity η, is responsible for yielding and the σ_r term is responsible for the post-yield hardening. The key to the model is that the viscosity is a strong function of the driving stress, where Haward and Thackray used the modified Eyring [72] expression

$$\eta \sim \left[\sinh\left(\frac{\sigma_s(1+\varepsilon_p)V^\star}{k_B\theta} \right) \right]^{-1} \tag{14.3}$$

where θ is the temperature, k_B is the Boltzmann constant, and $V^\star$ is the activation volume. The original Eyring model does not have the $1 + \varepsilon_p$ factor, where it was added in the Haward–Thackray model in order to describe post-yield strain softening, although the predicted decrease is too gradual when compared to experimental data.

14.4.1.2 FORMULATION OF THE BASIC PLASTICITY MODEL

The Haward–Thackray model is the origin of plasticity models used for describing polymeric glasses, but as a one-dimensional model it is not constitutive. However, many plasticity models that followed have used the Haward–Thackray ideas contained in Equations 14.1 through 14.3 as a point of departure. Thus, in order to expose the basic structure and predictive capabilities of the plasticity class of constitutive models, we will develop the tensorial form of the Haward–Thackray model which we designate as the basic plasticity model (BPM). The one-dimensional Equation 14.1 is replaced with the elastic–plastic decomposition [73]

$$\mathbf{F} = \mathbf{F}_e \cdot \mathbf{F}_p \tag{14.4}$$

where $\mathbf{F}$ is the deformation gradient tensor. In the BPM, the plastic deformation will be considered incompressible and irrotational; thus, there is a particularly simple form of the kinematics given by

$$\dot{\mathbf{F}}_p = \mathbf{D}_p \cdot \mathbf{F}_p \tag{14.5}$$

where $\mathbf{D}_p$ is the symmetric plastic rate-of-deformation tensor. To complete the description of the evolution of the plastic strain, "a flow rule" is required, which is postulated to have the form

$$\mathbf{D}_p = \frac{\boldsymbol{\sigma}_s^d}{2\eta\left(\theta, \boldsymbol{\sigma}_s^d\right)} \tag{14.6}$$

where $\boldsymbol{\sigma}_s^d$ is the deviatoric (i.e., $\boldsymbol{\sigma}^d = \boldsymbol{\sigma} - 1/3\ \mathrm{tr}(\boldsymbol{\sigma})\mathbf{I}$) part of the driving stress tensor and η is the viscosity. The key postulate is that the viscosity is a function of temperature and stress consistent with the Eyring expression [72] given by

$$\eta\left(\theta, \boldsymbol{\sigma}_s^d\right) = \eta_0 \exp\left(\frac{\Delta U}{k_B\theta}\right)\frac{\bar{\sigma}_s V^*}{k_B\theta}\left[\sinh\left(\frac{\bar{\sigma}_s V^*}{k_B\theta}\right)\right]^{-1} \qquad \bar{\sigma}_s = \sqrt{\frac{\left(\boldsymbol{\sigma}_s^d : \boldsymbol{\sigma}_s^d\right)}{2}} \tag{14.7}$$

where $\bar{\sigma}_s$ is the equivalent driving stress and ΔU is the activation energy. Just as in the one-dimensional case, the total Cauchy stress is a sum of the driving stress and the hardening stress

$$\boldsymbol{\sigma} = \boldsymbol{\sigma}_s + \boldsymbol{\sigma}_r = \boldsymbol{\sigma}_s(\mathbf{B}_e) + \boldsymbol{\sigma}_r(\mathbf{B}) \tag{14.8}$$

where $\mathbf{B} = \mathbf{F} \cdot \mathbf{F}^T$ is the left Cauchy–Green strain tensor. The driving stress is a function of the elastic strain and the hardening stress is a function of total strain.

14.4.1.3 PREDICTIONS OF THE BPM

Let's now examine the predictions of BPM in a constant strain rate uniaxial deformation. Assuming that the contribution of the hardening stress at yield is negligible, the total strain rate becomes equal to the plastic strain rate; thus, the uniaxial strain rate $\dot{\varepsilon}$ is given by

$$\dot{\varepsilon} = \frac{1}{2\eta_0}\exp\left(-\frac{\Delta U}{k_B\theta}\right)\frac{k_B\theta}{V^*}\sinh\left(\frac{\sigma_Y V^*}{k_B\theta}\right) \tag{14.9}$$

where σ_Y is the yield stress. When the yield stress is sufficiently large, that is, $\sigma_Y V^*/k_B\theta \gg 1$, Equation 14.9 can be converted into

$$\sigma_Y = \frac{k_B\theta}{V^*}\ln\left(\frac{\dot{\varepsilon}}{\dot{\varepsilon}_0}\right) + \frac{\Delta U}{V^*} \tag{14.10}$$

with the characteristic strain rate $\dot{\varepsilon}_0 = k_B\theta/4V^*\eta_0$ being a weak function of temperature.

The yield stress predictions in a constant strain rate experiment given in Equation 14.10 compare well with experiments and this is the reason for enduring popularity of the plasticity models with the Eyring viscosity. Specifically, Equation 14.10:

1. Captures the experimentally observed linear dependence of yield stress on the logarithm of strain rate (Feature 4.1a).
2. Predicts linear decrease in yield stress with temperature, since for $\dot{\varepsilon} < \dot{\varepsilon}_0$ the coefficient of the linear temperature dependence in Equation 14.10 is negative (Feature 4.1b). By an appropriate choice of the parameters V^* and $\dot{\varepsilon}_0$, the yield stress becomes zero at the glass transition temperature.
3. Allows for straightforward incorporation of the effect of the hydrostatic pressure on yield. The modification of Equation 14.7 to include hydrostatic pressure, p, is given by

$$\eta\left(\theta,\sigma_s^d\right) = \eta_0 \exp\left(\frac{\Delta U}{k_B \theta}\right) \frac{\bar{\sigma}_s V^*}{k_B \theta} \left[\sinh\left(\frac{\bar{\sigma}_s V^*}{k_B \theta}\right)\right]^{-1} \exp\left(\mu \frac{pV^*}{k_B \theta}\right) \tag{14.11}$$

Since viscosity controls the onset of plastic flow, the yield surface is the constant viscosity surface. Noticing that the equivalent driving stress as defined in Equation 14.7 is (within a numerical factor) the octahedral stress and using the assumption that at yield $\bar{\sigma}_s V^*/k_B \theta \gg 1$ to replace the hyperbolic sine with an exponential function, Equation 14.11 results in the pressure-modified von Mises yield criterion that is in agreement with the experimental data (Feature 4.1d).

Although the predictions of the BPM with respect to yield are impressive, there are several conceptual deficiencies; specifically,

1. Plasticity-based models contain no physical mechanism pertaining to glass transition and hence no means to account for the effects of thermal history and aging on the nonlinear mechanical response.

2. The values of parameters V^* and $\dot{\varepsilon}_0$ in Equation 14.10 needed to fit the data for polymeric glasses are troubling, e.g., $V^* = 3$ nm^3 and $\dot{\varepsilon}_0 = 10^{30}$ s^{-1} for polycarbonate [74]. The fact that everything "works out in the end" (due to the presence of the logarithm) does not change the observation that such an enormous characteristic strain rate is aphysical. Also, in the Eyring model, the activation volume is related to the distance between adjacent molecular layers (i.e., angstroms not nanometers); thus, V^* for polycarbonate is too large by a factor of 10^3 for glasses as was pointed out by Eyring in the original 1936 paper [72]. If the "active" unit is not an atom but rather a segment of a polymer chain, then, the activation volume is still too large by "only" a factor of 10 as was pointed out by Haward and Thackray [68].

3. The deformation calorimetry experiments of Oleinik and coworkers (see in Section 4.9.2 in Chapter 4) indicate a smooth evolution of both internal energy and heat as the polymer glass is deformed through the yield point. However, the key idea of a plastic model is that there is a change from elastic behavior (where energy is stored) to plastic behavior (where energy is dissipated) at the yield point, which is not supported by the deformation calorimetry experiments. Also, in polymeric glasses, the internal energy constitutes a significant portion of the work of deformation, which is dramatically different from crystalline materials like metals where after yielding the majority of the mechanical work goes into heat. Thus, the thermal response of polymeric glasses is qualitatively different than that of crystalline materials, which are generally believed to be aligned with the plastic flow model [75].

Because of these concerns, in our opinion, there should be real questions about the applicability of plasticity-based models for describing the deformation behavior of glassy polymers, irrespective of any success in predicting the nonlinear mechanical behavior. However, there has been considerable research effort in using plasticity models that have extended the ideas of Haward and Thackray, where it is only appropriate that in this chapter, there is an examination of the ability of these models to describe the thermo-mechanical behavior of polymeric glasses.

The objective now is to determine the mechanical response for experimental Features 4.2 through 4.10. For simplicity, it will be assumed that both the driving and hardening stresses are neo-Hookean and incompressible; specifically,

$$\sigma_s = G\mathbf{B}_e^d \qquad \sigma_r = G_r \mathbf{B}^d \tag{14.12}$$

where G is the shear modulus and G_r is the hardening modulus. Haward and Thackray assumed that the physical mechanism behind the hardening stress is rubber elasticity, where the material is assumed to transition from the glassy state into rubbery state at yield. Since the Haward and Thackray paper, many authors had taken this assumption for granted; as a result, much of the early modeling effort was focused on improving the description of the elastic response. However, more recently a convincing case has been made that the rubbery elasticity is not the correct mechanism (see Senden et al. [46] and also Chapter 13 of this book). The key points are: (i) the hardening modulus required to fit the data is at least an order-of-magnitude larger

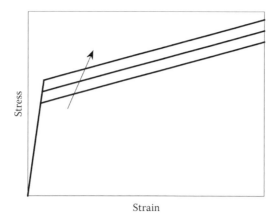

Figure 14.1 Stress–strain curves predicted by the BPM for constant strain rate uniaxial deformation. Arrow indicates increase in strain rate.

than the rubbery modulus at a reasonable cross-link density [76]; (ii) the rubbery elastic mechanism is entropic implying that its modulus has to increase with temperature, however, experimentally the hardening modulus decreases with temperature [77]; (iii) rubber is nearly incompressible, hence there should not be a dependence of the hardening modulus on pressure, although experimentally there is a pressure dependence [78]; and (iv) experimentally there is a dependence of strain hardening on strain rate which is inconsistent with a purely elastic contribution [44]. Recently, Senden et al. [46] have included a viscous-hardening contribution which has addressed some of these shortcomings.

The stress–strain curve produced by the BPM for a uniaxial constant strain rate deformation is shown in Figure 14.1. The pre-yield stress–strain curve is linear elastic; there is a strain rate-dependent yield stress as discussed previously; and, the post-yield stress–strain is again linear elastic with a tangent modulus given by G_r. The BPM does not exhibit post-yield softening or tertiary creep, although both are observed experimentally. Post-yield softening was added by Boyce, Park, and Argon (BPA) [36] by appending the factor $\exp(-D)$ to the Eyring viscosity given in Equation 4.7; specifically,

$$\eta\left(\theta,\sigma_s^d,D\right)=\eta\left(\theta,\sigma_s^d\right)\exp(-D) \tag{14.13}$$

where it was postulated that the evolution of the internal variable D is governed by

$$\dot{D}=h\dot{\bar{\varepsilon}}_p(D_\infty-D) \qquad D(t=0)=D_0 \tag{14.14}$$

$\dot{\bar{\varepsilon}}_p$ is the equivalent plastic strain rate. If $D_\infty > D_0$, the BPA model with Equations 14.13 and 14.14 exhibits post-yield softening as shown in Figure 14.2, although there is no dependence of the post-yield softening on strain rate (compare with Figure 4.33a in Chapter 4). This is because the range of decrease in the internal variable D, once triggered by the plastic flow, is always from D_0 to D_∞ regardless of the strain rate. An equally important observation is that in a given experiment, the "softening" described by Equation 14.14 can only happen once; as a result, the additional stress overshoot peak observed in the strain rate-switching experiment (Feature 4.4) is not predicted by the BPA model.

The overall performance of the BPA model in case of uniaxial compression was critically assessed by Dreistadt et al. [18], where representative results are shown in Figure 14.3. Examining the fit to the experimental data in Figure 14.3, two obvious deficiencies are (i) the incorrect shape of the predicted stress–strain curve in the pre-yield region and the abrupt change in the tangent modulus during unloading (this is the so-called "second" or "reverse" yielding [47]) which is not present in the data (i.e., Feature 4.5b). The origin of the spurious "second yield" is in the total stress decomposition as given in Equation 14.8 and the definition of the

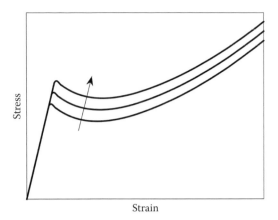

Figure 14.2 Stress–strain curves predicted by the BPA model for constant strain rate uniaxial deformation. Arrow indicates increase in strain rate.

tensorially correct equivalent stress given in Equation 14.7. Specifically, on unloading, the initial decrease in total strain is due entirely to the elastic strain component, since the plastic strain rate cannot change sign until the driving stress changes sign according to the flow rule given in Equation 14.6. As a result of this decrease in the elastic strain, the driving stress given by Equation 14.12 rapidly decreases, where the plastic strain is arrested because the equivalent driving stress is below the yield stress value. As the unloading continues, the elastic strain and hence driving stress become negative; however, the total stress as plotted in Figure 14.3 remains positive due to the hardening stress contribution. Eventually the negative driving stress reaches the point where its absolute value, that is, the equivalent driving stress as given in Equation 14.7, exceeds the yield stress in the Eyring viscosity expression in Equation 14.7. At that moment, the "reverse yield" occurs, where the plastic flow is activated, and from that point on, the decrease in total strain is due to the plastic strain. During this plastic flow stage, the tangent modulus is controlled by the hardening modulus as seen in Figure 14.3. Considering the response shown in Figure 14.3, the BPA cannot predict the Bauschinger effect (Feature 4.6). The BPA model encounters even more difficulties when dealing with multiple-step loading and reloading mechanical histories as documented by Dreistadt et al. [18] (i.e., Features 4.5 and 4.10).

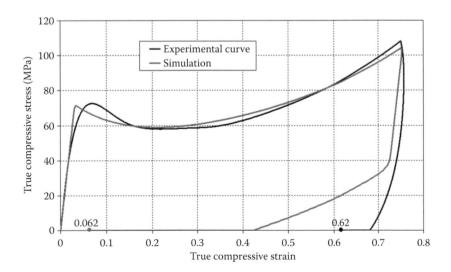

Figure 14.3 Stress–strain curve predicted by the BPA model for constant strain rate uniaxial loading–unloading deformation. (Reprinted from *Mater. Des.*, 30, C. Dreistadt et al., 3126, Copyright 2009, with permission from Elsevier.)

In an attempt to alleviate some of the aforementioned problems, Nguyen et al. [55] modified the BPA model by introducing a fictive temperature-dependent factor into the viscosity expression, which allowed for describing the glass transition and structural relaxation. However, the reliance on the phenomenological Equation 14.14 by Nguyen et al. means that all the shortcomings experienced by the original BPA model vis-à-vis nonlinear deformation experiments and in particular multistep experiments remain.

Recognition of the shortcomings of the basic plasticity model of Haward and Thackray as embodied in the BPM and BPA models has provided a major impetus for further development of the plasticity models. Two main approaches have been pursued: (i) replacing the single relaxation time with a spectrum of relaxation times in order to incorporate linear viscoelastic behavior and (ii) preserving a single relaxation time structure, but making nonlinearity much stronger and more intricate so that it manifests even at relatively small deformations. Representative examples of these approaches are (i) the multi-relaxation time version of the Eindhoven glassy polymer (EGP) model which will be analyzed in Section 14.4.2 and (ii) a series of plasticity models due to Anand and coworkers which will be analyzed in Section 14.4.3.

14.4.2 EINDHOVEN MODEL FOR GLASSY POLYMERS: EGP MODEL

Buckley and coworkers [65] developed a plasticity model with multiple relaxation times (or equivalently a multimodal model), where in the linear viscoelastic limit, the response reduces to that of a generalized Maxwell model. In an independent effort, the Eindhoven group has developed (i) an early multimodal version of plasticity model [39], where the lack of the correct linear viscoelastic behavior in the BPM has been overcome; (ii) a more sophisticated single-mode version of plasticity model, where the effects of post-yield softening and thermal history have been addressed [43]; and (iii) the recent multimodal version of plasticity model which combines the features present in (i) and (ii) that has been designated as the Eindhoven model for glassy polymers or EGP [45]. The EGP model is one of the most thoroughly studied constitutive models for polymeric glasses; thus, we will examine it in detail as the example for this class of plasticity models.

The key points of the EGP model will now be discussed. The stress decomposition follows the Haward–Thackray approach in Equation 14.8. The driving stress tensor is a sum of the hydrostatic (superscript "h") and deviatoric (superscript "d") contributions as given by

$$\sigma = \sigma_s^h + \sigma_s^d + \sigma_r = K(J-1)\mathbf{I} + \sum_{i=1}^{N} \sigma_{s,i}^d + \sigma_r \tag{14.15}$$

where K is the bulk modulus and $J = \det(\mathbf{F})$ is the volume change with respect to reference state. The deviatoric driving stress consists of N modes. The plastic deformation is assumed to be isochoric, that is, $\det(\mathbf{F}_p) = 1$; the hardening stress σ_r is a function of the total strain; and, the driving stress is a function of the elastic strain. In the simplest version of the EGP model [45], the functional forms of the driving and hardening stress are given by

$$\sigma_{s,i}^d = G_i J^{-(2/3)} \mathbf{B}_{e,i}^d \qquad \sigma_r = G_r J^{-(2/3)} \mathbf{B}^d \tag{14.16}$$

where each of the deviatoric driving stress modes has its own spectral strength G_i and G_r is the hardening modulus. In a recent version of the EGP [46], the Edwards–Vilgis model [79] was used to describe the hardening stress contribution instead of the neo-Hookean expression given in Equation 14.16. Edwards–Vilgis model for σ_r was also used by Buckley and coworkers [51,52,64]. The advantage of the Edwards–Vilgis model over the neo-Hookean model is seen at large strains greater than 60% [46], which are not the focus of this chapter.

The flow rule for the rate-of-deformation tensor for each plastic mode is given by

$$\mathbf{D}_{p,i} = \frac{\sigma_{s,i}^d}{2\eta_{0i}} \frac{1}{a\left(\theta, \sigma_s^d, \overline{\gamma}_p\right)} \tag{14.17}$$

Each mode has its own reference viscosity, η_{0i}, but all modes are affected equally by the $a\left(\theta, \boldsymbol{\sigma}_s^d, \overline{\gamma}_p\right)$ shift function that contains the effects of temperature, total driving stress, and the equivalent plastic strain $\overline{\gamma}_p$. The equivalent plastic strain is controlled by the $i = 1$ mode; specifically, for uniaxial deformation

$$\overline{\gamma}_p = \sqrt{3}\,|\ln(\lambda_{p,1})| \tag{14.18}$$

where $\lambda_{p,1}$ is the axial stretch ratio for the plastic mode number 1. The key to the EGP model is the form of the shift factor in Equation 14.17, which is given by

$$a\left(\theta, \boldsymbol{\sigma}_s^d, \overline{\gamma}_p\right) = \exp\left(\frac{\Delta U}{k_B\theta}\right)\exp\left(\mu\,\frac{pV^*}{k_B\theta}\right)\frac{\overline{\sigma}_s V^*}{k_B\theta}\left[\sinh\left(\frac{\sigma_s V^*}{k_B\theta}\right)\right]^{-1}\exp(S_{age}R(\overline{\gamma}_p)) \tag{14.19}$$

As compared to Equation 14.11, a new exponential factor appears in the right-hand side that contains the product of the parameter S_{age} and a function of the equivalent plastic strain $R(\overline{\gamma}_p)$. The function $R(\overline{\gamma}_p)$ is unity in the absence of plastic strain, that is, $R(0) = 1$ and decreases toward zero with increase in the equivalent plastic strain, for example, $R(0.4) = 0.1$. The reason for including $R(\overline{\gamma}_p)$ is to describe post-yield softening in constant strain rate deformations as well as the Stage III (tertiary) creep in constant stress deformations. The role of the parameter S_{age} is to capture the effects of sub-Tg aging. The spectrum employed in the multi-relaxation time version of the EGP model is shown in Figure 14.4, where the relaxation time is $\tau_i = \eta_{0,i}/G_i$ for each mode. The spectrum is a wedge with the exception of the first mode (i.e., the rightmost G_i in Figure 14.4 with the relaxation time $\tau_1 = 6 \times 10^8$ s), where in the single relaxation time version of the EGP model only the first mode is present. The values of all EGP model parameters that will be used in the subsequent predictions are for poly(carbonate) (PC) from Reference 45.

STRESS–STRAIN CURVES

The stress–strain curves predicted by the EGP model for uniaxial compression at a constant strain rate are shown in Figure 14.5. All the characteristic regions are predicted, including a nearly elastic pre-yield buildup of stress, a non-abrupt yield, post-yield softening, flow, and finally hardening. The prediction of the post-yield strain softening is a result of (i) the plastic strain dependence of $a\left(\theta, \boldsymbol{\sigma}_s^d, \overline{\gamma}_p\right)$ via the function $R(\overline{\gamma}_p)$ and (ii) the fact that the parameter S_{age} is not zero (i.e., the value of $S_{age} = 27$ from Reference 43 was used). The case of $S_{age} = 0$ corresponds to a rapidly quenched sample, where no post-yield strain softening is predicted in agreement with experimental observations (Feature 4.2b). The EGP model predicts that both the upper yield (i.e., the yield proper) and the lower yield (i.e., the flow stress) stresses are equally affected by the strain rate;

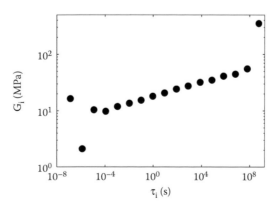

Figure 14.4 Multimodal EGP model spectrum for PC. Modes are numbered from right to left. Values are from Reference 45.

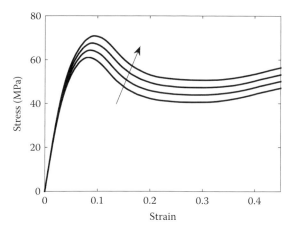

Figure 14.5 EGP model predictions for constant strain rate deformation in uniaxial extension of PC at 23°C (Tg-124°C). Arrow indicates increase in strain rate; the curves shown are for strain rates 10^{-5}, 10^{-4}, 10^{-3}, and 10^{-2} s^{-1}. EGP model parameter values are from Reference 45.

consequently, the stress–strain curves in Figure 14.5 corresponding to different strain rates are parallel to each other and the amount of post-yield strain softening is strain rate independent. This prediction is in a good agreement with experiments for PC at room temperature (i.e., at Tg-122°C); however, the experimental data at 125°C (i.e., at Tg-22°C) look very differently as illustrated in Figure 14.6. A strong dependence of the stress overshoot on strain rate at temperatures close to Tg which is similar to that in Figure 14.6 was also observed for PMMA as discussed in Section 4.6.3 of Chapter 4. The EGP model does not predict the strain rate dependence of the post-yield softening at any temperature; specifically, for the same thermal history (and thus the same S_{age}), the predicted stress–strain curves for different strain rates will remain parallel to each other as shown in Figure 14.5. This is a serious problem as it is embedded in the structure of the viscosity expression in Equation 14.19, where there does not seem to be an obvious way to correct this deficiency.

STRAIN RATE-SWITCHING EXPERIMENT

The EGP model predictions for the effect of changing strain rate during the course of the uniaxial compressive deformation are shown in Figure 14.7. When there is an increase/decrease in the strain rate, the stress response rapidly changes to the same stress response as if the material were always deformed at the final

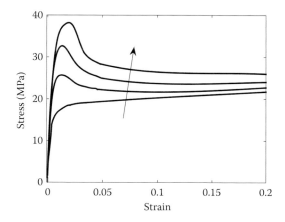

Figure 14.6 Experimental stress–strain curves for PC in uniaxial extension at 125°C; strain rates from bottom-to-top (as indicated by the arrow) are: 2×10^{-5}, 2×10^{-4}, 2×10^{-3}, and 2×10^{-2} s^{-1}. (With kind permission from Springer Science+Business Media: *J. Mater. Sci.*, 27, 1992, 5031, C. G'Sell et al.)

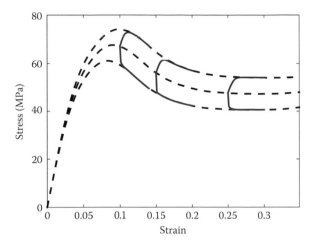

Figure 14.7 EGP model predictions of two-stage loading experiment. Dashed lines bottom to top: single-stage loading at 10^{-5}, 10^{-3}, and 10^{-1} s^{-1}; red/blue lines—switch from the strain rate of 10^{-3} s^{-1} to faster/slower strain rate of 10^{-1} /10^{-5} s^{-1} at strains of 0.1, 0.15, and 0.25. EGP model parameters for PC [45].

strain rate, but there is no undershoot or overshoot in the stress response as the strain rate is switched. The EGP predictions are substantially different from the experimental observations shown in Table 14.2, where a significant overshoot is observed when switching from slow to fast strain rate as well as a large undershoot when switching from fast to slow (Feature 4.4). The reason for this failure in the EGP model is straightforward. As stated earlier, the strain-softening behavior in the EGP is due to existence of the function $R(\overline{\gamma}_p)$ in the viscosity given in Equation 14.19. Once the plastic flow has been triggered, which happens near the yield point, the softening is controlled by the value of effective plastic strain $\overline{\gamma}_p$ but is unaffected by the strain rate. This indicates that the EGP prediction of the strain softening is probably a parameterization for one deformation history rather than an expression of the underlying physical mechanism.

UNLOADING AND THE BAUSCHINGER EFFECT

As pointed out by the authors of the EGP model, the EGP model predicts that when a glassy polymer is first loaded to a large strain and then unloaded at the same absolute value of the strain rate, the unloading stress–strain curve exhibits a sharp change in the tangent modulus as shown in Figure 14.8 (solid line) [81]. As discussed earlier,

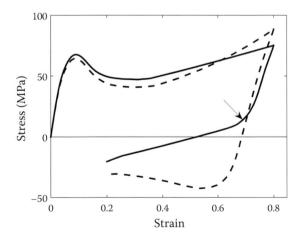

Figure 14.8 EGP model prediction of loading–unloading experiment. Solid—EGP using Equation 14.19; "second yield" is indicated by arrow. Dashed—modified EGP Equation 14.20. EGP model parameters for PC [45].

this aphysical behavior upon unloading is exhibited by not just the EGP model, but all viscoplastic models where the nonlinear viscosity is controlled by the equivalent driving stress. If the unloading deformation continues at negative stresses, that is, extension becomes compression, the predicted compressive Young's modulus is dramatically lower than the one experimentally observed, that is, these models predict too strong of a Bauschinger effect. To overcome this deficiency, Senden et al. [46] argued that strain hardening in polymeric glasses is caused by a combination of two effects: the elastic hardening described by the term σ_r with the modulus G_r in Equation 14.15 and a viscous-hardening contribution due to a newly postulated dependence of the Eyring viscosity on the total strain. Senden et al. [46] modified the Eyring viscosity in Equation 14.19 as given below

$$a\left(\theta,\boldsymbol{\sigma}_s^d,\overline{\gamma}_p,\mathbf{B}\right)=a\left(\theta,\boldsymbol{\sigma}_s^d,\overline{\gamma}_p\right)\exp\left(\left(\frac{A_1}{k_B\theta}-A_2\right)I_r(\mathbf{B})\right) \qquad I_r(\mathbf{B})=\frac{1}{2}J^{-(4/3)}(\mathbf{B}^d:\mathbf{B}^d) \qquad (14.20)$$

If temperature θ and the parameters A_1 and A_2 are such that the coefficient in front of $I_r(\mathbf{B})$ in Equation 14.20 is positive, then, an increase in strain results in an increase in viscosity and, hence, effective hardening. To summarize all the functions performed by the $a\left(\theta,\boldsymbol{\sigma}_s^d,\overline{\gamma}_p,\mathbf{B}\right)$ shift factor in the EGP model, it (i) decreases with the driving stress σ_s causing yield, (ii) decreases with the plastic strain $\overline{\gamma}_p$ causing post-yield softening, and (iii) increases with total strain $\mathbf{B}$ causing hardening. The EGP model prediction of loading–unloading experiment using modified Equation 14.20 is shown in Figure 14.8 as the dashed line, where the aphysical second yield has disappeared and the Bauschinger effect has been qualitatively predicted.

STRESS RELAXATION

The single relaxation time BPM as well as the earlier version of the EGP model [43] does not predict any stress relaxation prior to yield contrary to experimental data (see Figure 4.41 in Chapter 4). In the multimodal plasticity models, this problem has been overcome, where the question now is to determine how the rate of stress relaxation changes with strain. Experiments show that at slower strain rates, the initial rate of stress relaxation (estimated from the normalized stress vs. time plot) decreases monotonically from pre-yield to post-yield, while at faster strain rates, the initial rate of stress relaxation increases monotonically from pre-yield to post-yield (Feature 4.7). As illustrated in Figure 14.9, this reversal of the ordering of the stress relaxation curves is not predicted by the EGP model. The initial rate of stress relaxation is predicted to decrease with the loading strain rate; however, the pre-yield stress relaxation is always slower than post-yield stress relaxation irrespective of the loading strain rate.

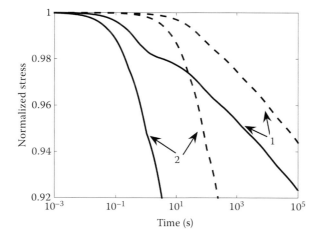

Figure 14.9 EGP model prediction of stress relaxation following uniaxial loading at strain rates: 1–10⁻⁴ s⁻¹, 2–10⁻² s⁻¹. The material was deformed to pre-yield strain (50% of the yield stress for each strain rate)—dashed line; post-yield strain (end of the strain softening)—solid line. "Shoulders" appearing on the curves are the result of the discreteness of the relaxation spectrum. EGP model parameters for PC [45].

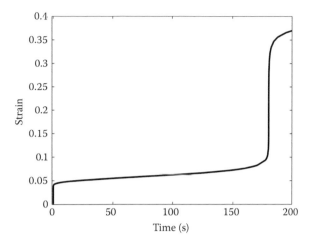

Figure 14.10 EGP model prediction of single-step creep in uniaxial extension for an applied stress of 40 MPa. EGP model parameters for PC [45].

CREEP AND RECOVERY

For single-step creep, the nonlinear strain versus time curve predicted by the EGP model is shown in Figure 14.10. The multiple relaxation time version of EGP model qualitatively predicts Stages I through V of creep. The reason that the EGP qualitatively describes nonlinear creep is that the nonlinear creep response is a mirror image of the stress–strain curve at constant strain rate (Feature 4.2) and the EGP model does a good job of describing the nonlinear stress–strain behavior as shown in Figure 14.5. However, the EGP nonlinear creep predictions will have similar limitations as already pointed out with respect to the constant strain rate experiments—i.e., the effect of thermal history on the response is not predicted, but rather parameterized via a different value of S_{age} for each thermal history.

With respect to creep recovery, after the removal of the axial load, the EGP model prediction shown in Figure 14.11 is very abrupt, whereas the experimentally measured response is smooth (Feature 4.9). The reason for the predicted behavior is straightforward: when the load is removed, the Eyring viscosity instantly reverts all relaxation times back to their large pre-yield values, with the result that no time-dependent plastic deformation takes place. A similar problem was reported by Dooling et al. [65] using their multimodal

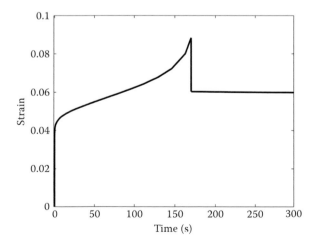

Figure 14.11 EGP model prediction of creep and recovery from creep for an applied stress of 40 MPa. EGP model parameters for PC [45].

viscoplastic model, where Dooling et al. state that "the model fails to exhibit the enhanced rate of initial recovery" observed experimentally.

In summary, the EGP [45,46] is able to successfully describe

1. The basic linear viscoelastic behavior in shear (Feature 2.1)
2. The yield stress dependence on temperature, strain rate, and pressure (Features 4.1a, b, d)
3. The stress–strain curve for a constant strain rate loading and single-step creep (Features 4.2a, d)
4. Some of the multistep experiments, including the Bauschinger effect, and stress memory (Features 4.5, 4.6, and 4.8)

However, the EGP model has the following deficiencies:

5. The volume relaxation is not predicted (Features 3.1 through 3.3)
6. Aging effects are accounted for via the parameter S_{age} in the viscosity expression Equation 14.19, which must be manually set for each thermal history (Features 1.1, 4.1c, and 4.2b)
7. The dependence of the magnitude of the post-yield softening on strain rate is not predicted (Feature 4.2c)
8. Overshoot and undershoot peaks on the stress–strain curve for two-stage loading experiments are not predicted (Feature 4.4)
9. The initial rate of stress relaxation increases from pre-yield strains to post-yield strains irrespective of the loading strain rate; whereas, experimentally the order is reversed for slow strain rates (Feature 4.7)
10. Recovery from large creep ceases much too abruptly compared to experiment (Feature 4.9)
11. The enthalpy relaxation is not addressed (Feature 5.1)

14.4.3 Viscoplastic models with internal variables; Anand models

In the Buckley and EGP models described in the previous section, only the conventional set of continuum thermodynamic variables, that is, strain, plastic strain, and stress, are employed. There are plasticity models that employ additional internal variables—variables that are governed by their own, independent evolution equations. From this perspective, the fictive temperature is not an internal variable if it is simply a function of volume (which is an invariant of the strain tensor); on the other hand, fictive temperature is an internal variable if it is defined via an independent evolution equation. Internal variables are formally permissible in continuum mechanics and afford more flexibility when developing constitutive equations; however, they must be used with care. First, in order to ensure thermodynamic consistency (i.e., that the Second Law of thermodynamics is not violated), model parameters cannot just be arbitrarily allowed to evolve with time; rather, if internal variables are postulated, they need to be first included in a tensorially appropriate manner in the Helmholtz free energy, where the thermodynamically consistent constitutive equations for stress, entropy, and the internal variable are then derived from the Helmhotz free energy [29]. Second, since the evolution equations for the internal variables are initial value differential equations, the initial values of these internal variables can be used as adjustable model parameters only once—at the beginning of the first stage of a multistage thermo-deformational history. It is inappropriate to arbitrarily change the initial values of the internal variables during subsequent stages of the deformation; specifically, the initial values of the internal variables for a second stage of deformation (for example, the recovery part of a creep-recovery experiment) are not new initial conditions, but a continuation of the variables from the end of the previous stage of the deformation. Finally, the internal variables may make the problem of ensuring that the model possesses the correct linear viscoelastic limit worse by creating spurious effects not observed experimentally.

In an extension to the standard viscoplastic approach, Hasan and Boyce used three internal variables, each with its own evolution equation, in the expression for the Erying viscosity [38]. The Hasan–Boyce model was able to describe well both (i) single-step constant strain rate loading and (ii) single-step nonlinear creep of PMMA in uniaxial compression. No attempt was made to determine if the Hasan–Boyce model is, or is not, thermodynamically consistent. The form of the evolution equation for two of the internal variables is such that these variables can only monotonically approach their equilibrium values, which is what happens during the first loading step; however, when the two-step loading–unloading experiments were modeled, these internal variables are reported to exhibit non-monotonic behavior as shown in Figure 14 in Hasan and Boyce [38]. Thus, there is an inconsistency either in model formulation or implementation.

Anand and coworkers have developed internal variables- based plasticity models that have been derived from the appropriate Helmholtz free energy and thus are thermodynamically consistent. There are two distinct versions of these models, which we designate as Anand06 [48] and Anand09 [49,50]. In the rest of this section, we will focus on the two Anand plasticity models, where we will only consider the irrotational case needed for uniaxial deformations. For the general case readers are referred to the original papers. Note that an approach closely related to that of Anand has been pursued by Bouvard et al. [54].

14.4.3.1 ANAND06 MODEL

The total Cauchy stress is assumed to be the sum of the driving and hardening components; specifically,

$$\sigma = \sigma_s + \sigma_r = \left[2GH_e^d + Ktr(H_e)I\right] + \mu_r \frac{\lambda_L}{3\bar{\lambda}} \mathcal{L}^{-1}\left(\frac{\bar{\lambda}}{\lambda_L}\right) J^{-2/3} B^d \tag{14.21}$$

where μ_r is the hardening modulus, $\mathcal{L}^{-1}$ is the inverse Langevin function, λ_L is a finite extensibility parameter, and the effective stretch is defined as

$$\bar{\lambda} = J^{-1/3}\sqrt{\frac{tr(C)}{3}} \tag{14.22}$$

Examining Equations 14.21 and 14.22, the driving stress is determined by the elastic strain (via the elastic Hencky strain tensor H_e) and the hardening stress is determined by the total strain (via the left and right Cauchy–Green tensors B and C, respectively). The total strain is decomposed into the elastic and plastic components according to $F = F_e \cdot F_p$. The flow rule is formulated as $\dot{F}_p = D_p \cdot F_p$ where the plastic flow rate is a sum of $N+1$ processes

$$D_p = \sum_{i=0}^{N} D_{p,i} = v_0 \sum_{i=0}^{N} \left[\frac{\bar{\sigma}_{eff,i}}{s_i - 1/3\alpha_p\, tr(\sigma_s)}\right]^{1/m_i} N_{p,i}^d \tag{14.23}$$

$N_{p,i}^d$ is the tensor of the direction of the plastic flow, which is deviatoric and has a unit norm. α_p is the pressure coefficient of yield stress and v_0 is a parameter. The important features of Equation 14.23 distinguishing it from the corresponding expression of the EGP model (given in Equation 14.17) are:

  i. The stress component affecting the plastic flow is not the entire driving stress, but rather the driving stress minus the so-called back stress

$$\sigma_{eff,i}^d = \sigma_s^d - \sigma_{back,i}^d \qquad \bar{\sigma}_{eff,i} = \sqrt{\frac{1}{2}\left(\sigma_{eff,i}^d : \sigma_{eff,i}^d\right)} \tag{14.24}$$

 ii. The dependence of the flow rate on stress is not of the Eyring form, rather it is a power law with the exponent $1/m_i$ (which is potentially different for each of the $N+1$ processes)
iii. The "yield stress" for each process is not a constant (at a given temperature), but an internal variable s_i that can evolve with time

The back stresses are controlled by (tensorial) internal variables

$$\sigma_{back,i} = \mu_i A_i \tag{14.25}$$

where the back stress moduli μ_i are themselves internal variables. The complete set of the evolution equations with initial conditions for all the internal variables in the Anand06 model is given by

$$\dot{\mathbf{A}}_i = \mathbf{D}_{p,i}\mathbf{A}_i \quad \mathbf{A}_i(0) = \mathbf{I} \quad i = 0,\ldots,N \tag{14.26}$$

$$\dot{\mu}_i = c_i\left(1 - \frac{\mu_i}{\mu_{cv,i}}\right)\dot{\varphi} \quad \mu_i(0) = \mu_{i0} \quad i = 0,\ldots,N \tag{14.27}$$

$$\dot{s}_0 = h_0\left(1 - \frac{s_0}{s_{cv}}\frac{1}{1+b(\varphi_{cv}-\varphi)}\right)v_0\left[\frac{\bar{\sigma}_{eff,0}}{s_0 - 1/3\alpha_p tr(\sigma_s)}\right]^{1/m_0} \quad s_0(0) = s_{00} \tag{14.28}$$

$$\dot{\varphi} = g_0\left(\frac{s_0}{s_{cv}} - 1\right)v_0\left[\frac{\bar{\sigma}_{eff,0}}{s_0 - 1/3\alpha_p tr(\sigma_s)}\right]^{1/m_0} \quad \varphi(0) = \varphi_0 \tag{14.29}$$

The remaining internal variables s_i are assumed to be constant

$$s_i(t) = s_{i0} \quad i = 1,\ldots,N \tag{14.30}$$

Here $\{c_i, \mu_{cv,i}, \mu_{i0}, h_0, s_{cv}, b, \varphi_{cv}, s_{00}, g_0, \varphi_0, s_{i0}\}$ are adjustable model parameters. Those parameters with the index i have to be set separately for each mode, resulting in an extremely large number of parameters. Since the power exponent parameters m_i are small as compared to unity, the plastic strain rate expressions, that is, Equations 14.23 and also 14.28 and 14.29 are steep functions of the ratio of the effective stress to the "yield stress" represented by the internal variable s_i. As a result, the ith plastic flow is turned on and off depending on this ratio. Similar to the EGP model, the main mode in the Anand06 model (i.e., the one characterized by the internal "yield stress" variable s_0) is responsible for the yield proper and the post-yield softening behavior. The role of the remaining N modes is to improve prediction of (i) the shape of the pre-yield stress–strain curve and (ii) the stress–strain curve on unloading.

Predictions of the Anand06 model will be analyzed, where the model parameters for $N = 3$ modes are from Reference 48 for PMMA. Two initial points are in order. First, the viscoplastic response for an infinitesimal deformation is described by the plastic strain rates $\mathbf{D}_{p,i}$ where the strain rate is

$$v_0\left[\frac{\sigma}{s_{10}}\right]^{1/m_1} \tag{14.31}$$

In arriving at Equation 14.31, it is assumed that the internal variables s_{i0} are ordered as $s_{10} < \cdots < s_{i0} < \cdots < s_{N0}$ as was done in Anand and Ames [48]. The expression in Equation 14.31 is a nonlinear function of the applied stress σ and thus the Anand06 model does not have the correct linear viscoelastic limit when the deformation is infinitesimal. Second, the Anand06 model does not contain any dependence on (i) temperature and (ii) thermal history and, hence, the sub-Tg aging time. Presumably the model will have to be re-parameterized for each new experimental condition, which is a daunting task considering the large number of parameters involved. Setting this concern aside, in what follows, the Anand06 model is assessed on its own terms—on the ability to predict the various mechanical behaviors of a polymeric material at a given temperature in the glassy state.

Constant strain rate experiments

The combination of the mathematical structure of the Anand06 model with its significant number of adjustable parameters enables excellent predictions of the stress–strain curve observed in constant strain rate uniaxial extension and compression as well as the unloading and reloading (see the figures in Anand and Ames [48] and Ames [82]). As shown in Figure 14.12, the model predicts that magnitude of the stress overshoot upon

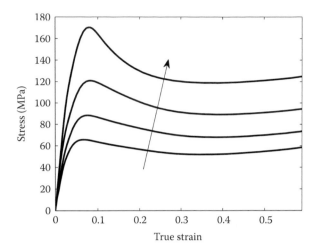

Figure 14.12 Anand06 model predictions of the strain rate effect on the stress–strain curves in uniaxial compression at true strain rates of from bottom-to-top 10^{-7}, 10^{-5}, 10^{-3}, and 10^{-1} s^{-1}. Anand06 model parameters for PMMA (Tg = 105°C) at 23°C [48].

loading depends upon the strain rate as is observed experimentally (see Figure 4.33a in Chapter 4). This is a major advantage over the EGP model, where the predicted stress overshoot is strain rate independent at all temperatures. However, the non-Eyring form of the plastic flow rule that enables the strain rate dependence of the stress overshoot in the Anand06 model, that is, Equation 14.23 comes at a price. As shown in Figure 14.13, the dependence of the lower yield stress on the log (strain rate) is never linear, which is contrary to the experimental data (see Section 4.6.2 in Chapter 4). This failure does not appear to be critical, as it may be possible to use the Eyring-based flow rule in place of the power law-based flow rule as will be shown later in this section with regard to the Anand09 model.

Strain rate-switching experiment

The predictions of the Anand06 model when the strain rate is switched during the loading are shown in Figure 14.14. There is a small overshoot when the strain rate is increased as compared to the EGP model where there is no overshoot; however, the magnitude of the overshoot is significantly less than that observed experimentally (Feature 4.4). There is no undershoot when the strain rate is decreased in contrast to the experimental data.

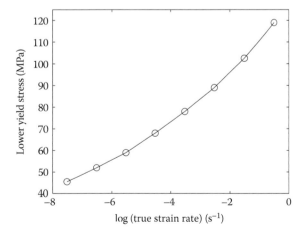

Figure 14.13 Anand06 model prediction of the (lower) yield stress in constant strain rate uniaxial compression. Anand06 model parameters for PMMA (Tg = 105°C) at 23°C [48].

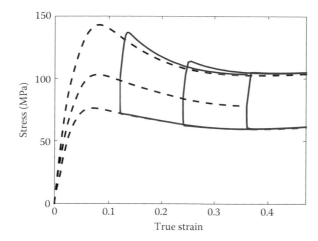

Figure 14.14 Anand06 model prediction of the strain rate switch experiment in uniaxial compression; rate switch from 10^{-3} to 10^{-1} s^{-1}—red, rate switch from 10^{-3} to 10^{-5} s^{-1}—blue; dashed curves—single rate loading at the true strain rates from bottom-to-top 10^{-5}, 10^{-3}, and 10^{-1} s^{-1}. Anand06 model parameters for PMMA (Tg = 105°C) at 23°C [48].

Stress relaxation and stress memory

The Anand06 model does not predict the correct nonlinear stress relaxation response (Feature 4.7); specifically, when the material is first loaded at a constant rate to a pre-yield, yield, or post-yield strain, the subsequent rate of stress relaxation does not change monotonically with the initial strain as is observed experimentally. This is because depending on the strain, the individual plastic modes $s_{10} < \cdots < s_{i0} < \cdots < s_{N0}$ in Anand06 become active in a discrete manner, resulting in the rate of stress relaxation being faster and then slower and then faster again. Similar behavior also manifests in the incorrect prediction of the stress memory experiment (Feature 4.8). The stress memory experiment [21] is a three-step history—(i) constant strain rate loading through yield, (ii) strain rate unloading at the same rate to zero stress, and (iii) holding the strain constant while monitoring the subsequent evolution of stress. This deformation protocol produces a stress overshoot, where the location and height of the peak depends strongly on the loading/unloading rate during steps (i) and (ii). The Anand06 model predictions are shown in Figures 14.15 and 14.16, where the loading and unloading response is shown in Figure 14.15 and the subsequent stress memory predictions

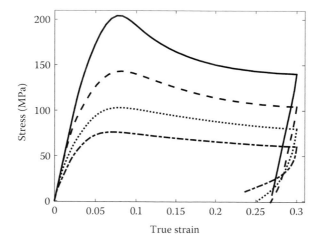

Figure 14.15 Anand06 model prediction of loading–unloading to zero stress experiment in uniaxial compression at true strain rates of 10^{-6} s^{-1}—dash-dotted, 10^{-4} s^{-1}—dotted, 10^{-2} s^{-1}—dashed, and 10^{-0} s^{-1}—solid. Anand06 model parameters for PMMA (Tg = 105°C) at 23°C [48].

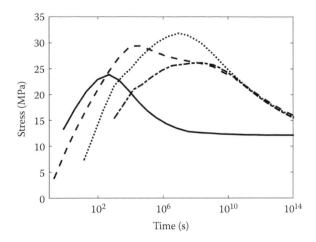

Figure 14.16 Anand06 model prediction of stress memory response in uniaxial compression following the loading/unloading deformations shown in Figure 4.15 (see details in text). Anand06 model parameters for PMMA (Tg = 105°C) at 23°C [48].

are shown in Figure 14.16. The first observation is the significant change in shape of the unloading curve depending upon the strain rate, where for slower rates, there is an abrupt change in the tangent modulus during unloading, which is not consistent with the experiment [50]. As shown in Figure 14.16, the time of the maximum in stress memory response increases significantly as the loading/unloading strain rate decreases, which is qualitatively consistent with the experimental data; however, the height of the stress memory peak is experimentally observed to decrease monotonically with decrease in the loading/unloading rate, which is not what is shown in Figure 14.16. However, the epoxy material used in the experimental studies of stress memory is different from the PMMA used to determine the parameters in the Anand06 model, where perhaps, there would be no qualitative difference between stress overshoot height predictions versus experiment if the model and data were for the same material. Finally, the Anand06 model predicts incomplete recovery of stress as seen by the long time plateau in Figure 14.16—this is due to the elastic hardening stress at the final strain. However, the longtime plateau is not seen in the experimental data of Kim et al. [21]. Perhaps, this is a deficiency in the Anand06 model or perhaps because the data in Kim et al. [21] were for strains that were significantly lower than those used to generate the predictions in Figures 14.15 and 14.16 so that the hardening had not yet begun.

Creep and recovery

The predictions of nonlinear creep and recovery from creep by the Anand06 model are shown in Figure 14.17. All five stages of creep are predicted and the recovery behavior qualitatively agrees with experimental observation. This is in contrast to the EGP model, where the rate of recovery abruptly stopped just after unloading. The reason for the higher rate of creep recovery in the Anand06 model is the presence of significant back stresses in the plastic flow rates expressions $\mathbf{D}_{p,i}$ in Equation 14.23. In the EGP model, the back stress is the hardening stress, which is small compared to the yield stress for typical strain values reached during primary and secondary creep; consequently, upon load removal, this back stress is insufficient to induce plastic flow and the creep recovery shuts down. In contrast, in the Anand06 model, the back stresses are independent internal variables whose values are not controlled by the strain reached during creep; thus, upon load removal, the recovery occurs if the back stress $\sigma_{back,1}$ is not negligible as compared to its "yield stress" s_{10}—a condition that is satisfied under most experimental conditions.

Loading–creep–reloading experiment

Predictions of the Anand06 model for the complex deformation history of (i) constant strain rate uniaxial compression loading through yield, (ii) creep at a stress σ for a time period t_{creep}, and (iii) reloading at the

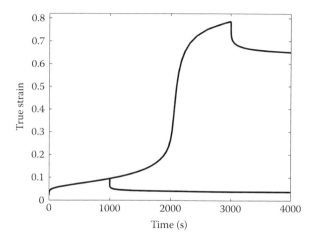

Figure 14.17 Anand06 model predictions of creep and recovery in uniaxial compression for creep recovery in both Stage II and Stage V creep for an applied stress of 90 MPa. Anand06 model parameters for PMMA (Tg = 105°C) at 23°C [48].

original strain rate are shown in Figure 14.18 for several values of σ with a fixed $t_{creep} = 2 \times 10^3$ s. Experimental data were shown in Figure 4.12 in Chapter 4 for PC. As compared to the experiment, the magnitude of the overshoot is too small and there is no significant increase in its magnitude with σ. However, there is a hint of an increase when comparing the curves for the lowest and the second lowest stress; thus, there may be an opportunity for improving the predictions by changing the values of the model parameters.

In summary, the Anand06 constitutive model is able to describe

1. The stress–strain curve in constant strain rate deformation and all the stages of nonlinear creep. (Feature 4.2a)
2. The dependence of the stress overshoot on strain rate. (Feature 4.2c)
3. The smooth recovery from creep. (Feature 4.9)

Difficulties with the Anand06 model are:

4. Temperature and aging effects are not addressed. (Features 1.1 and 4.1b,c)

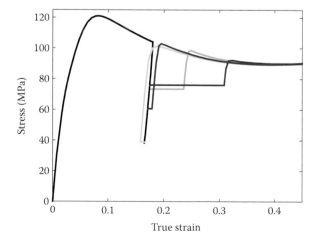

Figure 14.18 Anand06 model prediction of constant strain rate loading, creep, and reloading (at the same strain rate) in uniaxial compression. The true strain rate was 2×10^{-3} s^{-1}. The creep stress of 39 MPa (cyan), 59 MPa (blue), 72 MPa (green), and 75 MPa (red) was applied for 2000 s before reloading at 2×10^{-3} s^{-1}. Anand06 model parameters for PMMA (Tg = 105°C) at 23°C [48].

5. It does not have the correct linear viscoelastic limit. (Feature 2.1)
6. It does not describe volume and enthalpy relaxation. (Features 3.1 and 5.1)
7. Unloading stress–strain response after post-yield strains exhibits a strong strain rate dependence contrary to experiments. (Feature 4.5)
8. Overshoot and especially the undershoot peaks on the stress–strain curve for two-stage loading experiments are not predicted. (Feature 4.4)
9. The initial rate of stress relaxation in the pre-yield strain region is non-monotonic with strain. (Feature 4.7)
10. In the stress memory experiments, the predicted dependence of the stress overshoot peak on strain rate is incorrect. (Feature 4.8)
11. The stress response for the constant strain rate loading–creep–reloading deformation history is not predicted, at least using the current set of model parameters. (Feature 4.10)

14.4.3.2 ANAND09 MODEL

Unlike the earlier Anand06 model [48], the Anand09 model [49,50] is not a multimodal model; consequently, it is not concerned with having the correct linear viscoelastic limit, focusing instead on the nonlinear mechanical response. The challenge of predicting the observed shape of the stress–strain curve is addressed via a set of intricately coupled nonlinear equations for internal variables, where this is achieved without employing a spectrum of plastic processes. The Anand09 model, like the Anand06 model, does not include the effect of thermal history on the mechanical behavior, which is a serious shortcoming for any constitutive model for polymeric glasses. However, the Anand09 approach enables incorporation of temperature without the daunting task of determining the temperature dependence of the parameters for all the modes as would be required in the Anand06 model.

Just like the Anand06 model and the EGP model (as well as other models following in the Haward and Thackray footsteps), the Anand09 model postulates that the total stress is decomposed into the driving and hardening terms given by

$$\sigma = \sigma_s + \sigma_r = J^{-1}\left[2G\mathbf{H}_e^d + K\,tr(\mathbf{H}_e)\mathbf{I} - 3K\alpha(\theta - \theta_0)\mathbf{I}\right] + \mu_r J^{-5/3}\left(1 - \frac{I_1 - 3}{\lambda\sqrt{3}}\right)^{-1}\mathbf{B}^d \tag{14.32}$$

Comparing Equation 14.32 with Equation 14.21, the temperature dependence has been introduced into the driving stress term and the Gent [83] function for describing rubber elasticity has been used in place of the inverse Langevin function for describing the strain-hardening contribution. The effective stress that controls the plastic flow is obtained by subtracting the back stress from the driving stress in a similar fashion to Equation 14.24; specifically,

$$\boldsymbol{\sigma}_{eff}^d = \boldsymbol{\sigma}_s^d - \boldsymbol{\sigma}_{back}^d \qquad \overline{\sigma}_{eff} = \sqrt{\frac{1}{2}\left(\boldsymbol{\sigma}_{eff}^d : \boldsymbol{\sigma}_{eff}^d\right)} \tag{14.33}$$

Unlike the EGP and BPA models where the "back stress" is the same as the hardening stress, the back stress in the Anand09 is an internal variable whose evolution is governed by a separate equation.

The most important distinction of the Anand09 model is the existence of the yield criterion invariant given by

$$Y = \overline{\sigma}_{eff} - \left(S_1 + S_2 - \alpha_p \frac{1}{3}tr(\boldsymbol{\sigma}_s)\right) \tag{14.34}$$

where S_1 and S_2 are internal scalar variables. The plastic flow is either "on" or "off" depending on the criterion

$$\mathbf{D}_p = \mathbf{v}_p \frac{\boldsymbol{\sigma}_{eff}^d}{2\overline{\sigma}_{eff}} \qquad \mathbf{v}_p = \begin{cases} 0, & Y \leq 0 \\ \left[\mathbf{v}_0 \exp\left(-\frac{\Delta U}{k_B \theta} \right) \left[\sinh\left(\frac{YV^*}{2k_B \theta} \right) \right]^{1/m} \right], & Y > 0 \end{cases} \tag{14.35}$$

where α_p, $\mathbf{v}_0$, and m are model parameters. Comparing Equation 14.35 to Equation 14.23, the plastic strain rate expression in Equation 14.35 is of the Eyring form, since in the vicinity of the yield sinh ~ exp where the factor $1/m$ just rescales the yield stress. The "if" statement in Equation 14.35 shows that the Anand09 model is more in the spirit of plasticity framework than the Anand06 model or the EGP model, where the plastic strain rate is never formally zero although it may be small depending on the relevant variables.

The yield criterion Equation 14.34 is more complicated than just a stress surface (for example, the pressure- modified von Mises criterion) as it depends on the current values of several internal variables. This means that the same value of applied stress (or just the driving stress) may or may not produce plastic flow depending on how the current state has been reached. The back stress is a function of the internal tensorial variable $\mathbf{A}$ given by

$$\boldsymbol{\sigma}_{back} = B \ln \mathbf{A} \tag{14.36}$$

where B is a model parameter. The equations governing evolution of the internal variables with the associated initial conditions are

$$\dot{\mathbf{A}} = \mathbf{D}^p \mathbf{A} + \mathbf{A} \mathbf{D}^p - \gamma \mathbf{A} \ln \mathbf{A} \mathbf{v}_p \quad \mathbf{A}(0) = \mathbf{I} \tag{14.37}$$

$$\dot{S}_1 = h_1 (b(\varphi^* - \varphi) - S_1) \mathbf{v}_p \quad S_1(0) = S_{1i} \tag{14.38}$$

$$\dot{S}_2 = h_2 (\overline{\lambda}_p - 1) \left(S_2^* - S_2 \right) \mathbf{v}_p \quad S_2(0) = S_{2i} \tag{14.39}$$

φ in Equation 14.38 is another internal variable with the evolution equation

$$\dot{\varphi} = g(\varphi^* - \varphi) \mathbf{v}_p \quad \varphi(0) = \varphi_i \tag{14.40}$$

where

$$\varphi^* = \begin{cases} \varphi_r \left[1 + \left(\frac{\theta_c - \theta}{k} \right)^r \right] \left(\frac{\mathbf{v}_p}{\mathbf{v}_r} \right)^s, & \theta \leq \theta_c \\ 0, & \theta > \theta_c \end{cases} \tag{14.41}$$

$$\theta_c = \begin{cases} \theta_g + n \ln\left(\frac{\mathbf{v}_p}{\mathbf{v}_r} \right), & \mathbf{v}_p > \mathbf{v}_r \\ \theta_g, & \mathbf{v}_p \leq \mathbf{v}_r \end{cases} \tag{14.42}$$

$\{h_1, b, S_{1i}, g, \varphi_i, \varphi_r, k, r, s, \mathbf{v}_r, n, h_2, S_2^*\}$ are model parameters. The total number of parameters is large, where most of parameters values cannot be obtained from independent experiments but must be determined from the fitting of nonlinear thermo-mechanical data. Moreover, since physical aging is not explicitly predicted, at least some of the parameters will have to be re-optimized for different thermal histories.

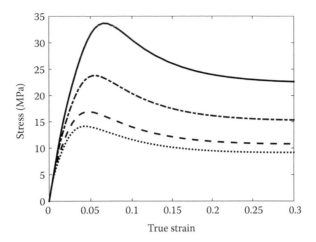

Figure 14.19 Anand09 model predictions of constant strain rate compression data for PMMA at 110°C that is shown in Figure 4.20 (only response for the true strains of up to 0.3 is shown). The true strain rates from top to bottom are 10^{-1}, 10^{-2}, 10^{-3}, and 3×10^{-4} s^{-1}. Anand09 model parameters for PMMA (Tg = 115°C) [50].

As already mentioned, the Anand09 model is designed to describe the anisotropic nonlinear thermo-mechanical behavior. It does not predict a correct linear viscoelastic response in the limit of small deformations. Neither does it predict the thermo-mechanical response for isotropic deformations, that is, volume relaxation. However, the objectives of Anand09 were to describe the large anisotropic response of glasses, where we will now examine its predictions for these types of deformations.

Constant strain rate loading

Anand09 model predictions for uniaxial compression of PMMA at Tg-5°C are shown in Figure 14.19, where the corresponding experimental data are shown in Figure 14.20. An important feature of the response is the progressive increase in the post-yield softening with the increase in strain rate, where at the slowest strain rate studied (i.e., 3×10^{-4} s^{-1}) the overshoot peak is significantly reduced. Thus, the Anand09 model is able to capture the strain rate dependence of the stress overshoot for PMMA, where it remains to be seen if the strain rate dependence could be made even more pronounced by adjusting the model parameters as would

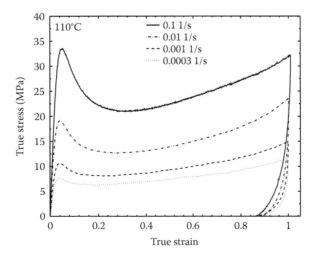

Figure 14.20 Constant true strain rate uniaxial compression of PMMA at 110°C (Tg-5°C) at the true strain rates indicated in the figure. (Reprinted from *Int. J. Plast.*, 25, N. M. Ames et al., 1495, Copyright 2009, with permission from Elsevier.)

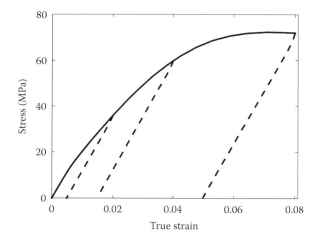

Figure 14.21 Anand09 model predictions of uniaxial compression and unloading of PMMA at 60°C (Tg-55°C) at the true strain rate of 10^{-3} s^{-1}. Loading—solid, unloading—dashed. Anand09 model parameters for PMMA (Tg = 115°C). (Adapted from N. M. Ames et al., *Int. J. Plast.*, 25, 1495, 2009.)

be required for PC (see Figure 14.6). This is an improvement over the EGP model, where the magnitude of the overshoot does not depend on the strain rate. The shapes of the stress–strain curves as predicted by Anand09 model in Figure 14.19 look similar to the ones observed experimentally. This is a distinct improvement as compared to the single modal basic plasticity model or the BPA model for which the pre-yield shape of the curve is not captured and the change in the tangent modulus at yield is too abrupt (i.e., Figures 4.1 through 4.3). It is interesting that the EGP model overcomes this difficulty by introducing the spectrum of the relaxation times such that the response of all modes except the first mode is viscoelastic; in contrast, in the Anand09 model, the smooth yield response is a result of the evolution of the various internal variables, where nontrivial amounts of plastic response occur early in the deformation.

The strongly nonlinear structure of the Anand09 model is clearly evident when the unloading response from a pre-yield strain is considered. The Anand09 model predictions using PMMA parameters are shown in Figure 14.21, where the typical experimental response is shown in Figure 14.22. It is clear from comparing Figures 4.21 and 4.22 that the Anand09 model's pre-yield unloading predictions are qualitatively incorrect.

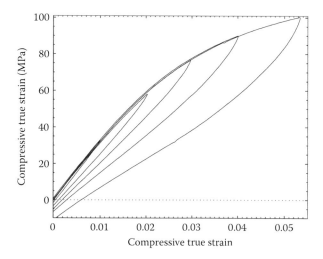

Figure 14.22 Uniaxial compression and unloading of PMMA at 25°C at the true strain rate of 3 × 10^{-4} s^{-1}. (Adapted from N. M. Ames, Massachusetts Institute of Technology, 2003.)

Specifically, the Anand09 model predicts that the unloading from the pre-yield strains exhibits a constant tangent modulus that is large as compared to the loading tangent modulus just prior to the beginning of the unloading; consequently, both the residual strain when the stress becomes zero and the amount of hysteresis are over-predicted by the Anand09 model. This is a common theme with the models relying on delicately balanced nonlinear contributions to describe what is essentially an effect of a viscoelastic spectrum; specifically, it is possible to set up the parameters of the nonlinear model to predict a single step loading well, but upon application of the next step (in this case unloading) the delicate balance often breaks.

Strain rate-switching experiment

The predictions of the strain rate-switching experiments by the Anand09 model are shown in Figure 14.23. The Anand09 model predicts both the over- and under-shoot when the strain rate is respectively increased or decreased; although the magnitudes of the peaks are less than that observed experimentally; nevertheless, the predictions of Anand09 are an improvement as compared to the EGP model and the Anand06 models. However, there are two issues to note: (i) the peaks originating from just prior to yield and at yield are not predicted by the model and (ii) the return to the stress–strain curve corresponding to the new strain rate occurs over a significantly longer strain range in the prediction as compared to the experiments shown in Figure 4.10 in Chapter 4.

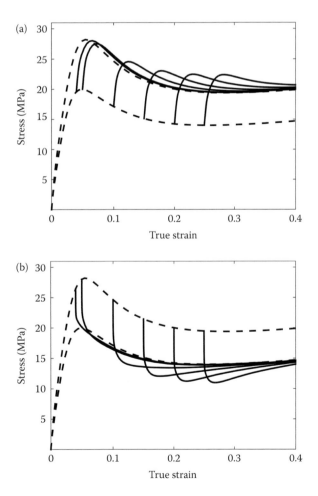

Figure 14.23 Two-stage loading experiments for PMMA predicted by the Anand09 model in uniaxial compression at 100°C (Tg-15°C). Dashed lines—monotonic loading at true strain rates of 10^{-3} s^{-1}—upper lines, 10^{-5} s^{-1}—lower lines. Solid lines: (a)—strain rate switch from 10^{-5} to 10^{-3} s^{-1}; (b)—strain rate switch from 10^{-3} to 10^{-5} s^{-1}. Anand09 model parameters for PMMA (Tg = 115°C) [50].

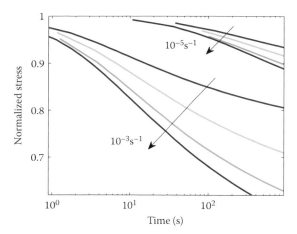

Figure 14.24 Anand09 model prediction of stress relaxation following uniaxial compressive loading at 100°C (Tg-15°C) at true strain rates indicated. Stress relaxation starts at pre-yield ($\sigma_Y/3$)—blue, yield—cyan, half-way through post-yield softening—green, end of strain softening—red. Arrows indicate increase in strain. Anand09 model parameters for PMMA (Tg = 115°C) [50].

Stress relaxation and stress memory

A traditional single relaxation time plasticity model does not predict any stress relaxation prior to yield; however, in the Anand09 model, stress relaxation does occur even at relatively small strains due to the early onset of nonlinear behavior. The corresponding rate of stress relaxation at the small strains (i.e., strains much less than 1%) behaves non-monotonically with strain, a feature that is not observed experimentally. At higher strains, beginning from 1% (i.e., approximately a sixth of the yield strain in Figure 14.21) and through yield and post-yield softening region, the rate of stress relaxation increases with strain as illustrated in Figure 14.24. The increase in the rate of relaxation with increasing strain is observed at all strain rates and thus the Anand09 model does not account for the strain rate dependence of the nonlinear stress relaxation seen experimentally, that is, Feature 4.7.

The stress memory behavior (i.e., the existence of the stress overshoot after the material is loaded through yield, unloaded to zero stress, and held at the resulting strain value) is not predicted by Anand09 model (Feature 4.8). Since the stress overshoot is not predicted, it is not possible to assess its dependence on the loading/unloading strain.

Bauschinger effect

The Anand09 model prediction of the Bauschinger effect using PMMA parameters is shown in Figure 14.25. The yield stress in extension during the first loading is 50 MPa whereas the yield stress in compression (not shown) is −65 MPa. If extension is followed by unloading and compression, the Anand09 model predicts that yield will occur near −20 MPa; thus, the Anand09 model predicts the Bauschinger effect, although it is too strong as compared to the experimental data (see Figure 4.14 in Chapter 4). The compression curve has a tangent modulus that equals the post-yield hardening modulus during extension, whereas the experimental stress–strain curve during the compression is flat. The value of the compressive yield stress for a sample that has first been deformed in uniaxial extension as well as the slope of the compression curve can perhaps be improved by adjusting Anand09 model parameters.

Creep and recovery

The true (i.e., Hencky) strain versus time curve predicted by the Anand09 model is shown in Figure 14.26, where the load was applied instantaneously at zero time. In agreement with the experimental data, the model qualitatively predicts all five stages of nonlinear creep. The Anand09 model predicts that at the lowest stress shown, there is no tertiary creep, which is in qualitative agreement with the observations in both compression and extension (see Reference 14 and references therein).

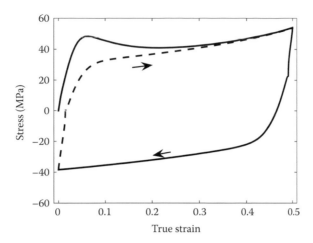

Figure 14.25 Anand09 model prediction of Bauschinger effect and cyclic deformation at 60°C at a true strain rate of 10^{-3} s^{-1}; initial loading (extension) and unloading—black, second loading—dashed. Anand09 model parameters for PMMA (Tg = 115°C) [50].

Creep-recovery predictions of the Anand09 model are shown in Figure 14.27, where there are two distinct regimes: (1) below a critical value of the strain, the recovery after initial strain drop is nonexistent on a time scale commensurate with the loading time and (2) above this critical strain, the recovery is smooth with a gradually decelerating rate of recovery. The predicted recovery behavior is contrary to the experiments, where there is no critical strain at which the character of the recovery curve changes. To understand the source of this incorrect Anand09 model prediction, it is instructive to consider the behavior of the plastic strain rate ν_p described by Equation 14.35 during creep and recovery. The plastic strain rate versus time plot is shown in Figure 14.28 corresponding to the creep-recovery predictions in Figure 14.27, where two cases are presented: recovery after 250 s of creep and recovery after 300 s of creep. Following 250 s of creep, the removal of the load of 24 MPa causes the plastic strain rate to drop by more than two orders-of-magnitude; alternatively, after 300 s of creep, the removal of the same load causes only a factor of 2 decrease in the plastic strain rate. This difference is due to the fact that in the course of creep from 250 to 300 s the internal variable S_1 has decreased. Considering the sharpness of the function in Equation 14.35, this extra relaxation of S_1 has a dramatic impact. This effect is embedded in the structure of Anand09 model and it is unclear how this aphysical prediction can be avoided, where just a change in the values of the model parameters will not be sufficient.

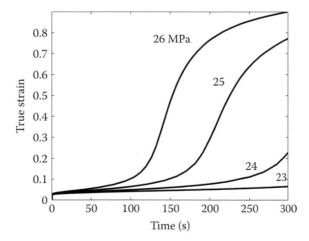

Figure 14.26 Anand09 model prediction of single-step creep in uniaxial compression at 100°C (Tg-15°C) at the applied stresses indicated in the figure. Anand09 model parameters for PMMA (Tg = 115°C) [50].

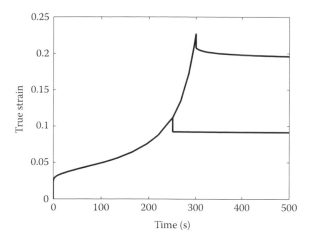

Figure 14.27 Anand09 model prediction of creep and recovery from creep in uniaxial compression at 100°C (Tg-15°C) at an applied stress of 24 MPa. Anand09 model parameters for PMMA (Tg = 115°C) [50].

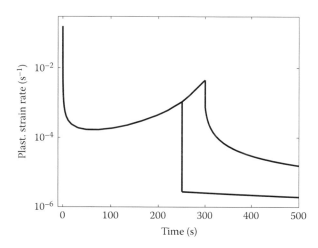

Figure 14.28 Anand09 model prediction of plastic strain rate during creep and recovery for the same strain vs. time curves shown in Figure 4.27 (see text).

Loading–creep–reloading experiment

The prediction of the Anand09 model for the deformation history that starts with constant strain rate loading in compression followed by creep with an applied stress σ for time t_{creep} followed by reloading at the original strain rate is shown in Figure 14.29. The Anand09 model captures some features of this experiment; specifically, there is a noticeable overshoot upon reloading. However, the magnitude of the overshoot is too small and most importantly does not increase with σ in contrast to the experimental data (see Figure 4.37 in Chapter 4). It is possible that these predictions can be improved by changing the Anand09 model parameters.

Summary of Anand09 model predictions

The Anand09 model is able to describe a number of the important features of the mechancial behavior of glassy polymers including:

1. The stress–strain curve in constant strain rate deformation and the strain versus time curve in nonlinear creep (Features 4.2a,c,d)
2. The yield stress dependence on the temperature and strain rate (Features 4.1a,b)

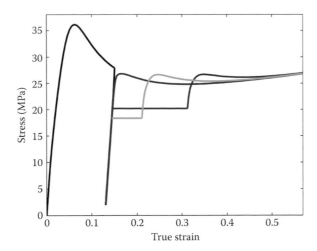

Figure 14.29 Anand09 model prediction of constant strain rate loading, creep, and reloading at the same strain rate in uniaxial compression at 100°C (Tg-15°C) using the true strain rate of 10^{-3} s^{-1}. The creep time was 100 s and the applied stress during creep was 2 MPa—blue; 18 MPa—green; and 20 MPa—red. Anand09 model parameters for PMMA (Tg = 115°C) [50].

3. The Bauschinger effect (Feature 4.6)
4. The strain rate-switching experiment (Feature 4.4)

   Problems with the Anand09 model include:

5. The specific volume response to a temperature change is elastic versus the time-dependent response observed experimentally (Features 3.1–3.3)
6. Aging effects are not addressed (Features 1.1, 4.1c, and 4.2b)
7. The stress response on unloading from pre-yield strains is not predicted correctly (Feature 4.5)
8. The initial rate of stress relaxation increases from pre-yield strains to post-yield strains irrespective of the loading strain rate (Feature 4.7)
9. Stress memory is not predicted (Feature 4.8)
10. Recovery from creep stops too abruptly after removal of the load for smaller strains (Feature 4.9)
11. The loading–creep–loading experiment is not predicted (Feature 4.10)

   Since Anand09 is derived in a thermodynamically consistent fashion from the Helmholtz free energy, the Anand09 model is not completely silent with respect to enthalpy response; unfortunately, no enthalpy relaxation predictions have been made to-date.

14.5 VISCOELASTICITY-BASED MODELS

The point of departure for viscoelastic models of the glass is linear viscoelasticity. The non-isothermal, three-dimensional constitutive equation for a linear viscoelastic material is given by Christensen and Naghdi [84] and Christensen [85]

$$\sigma(t) = \int_{-\infty}^{t} K(t-s)\mathbf{I}\frac{d}{ds}tr(\varepsilon(s))ds + \int_{-\infty}^{t} 2G(t-s)\frac{d}{ds}\varepsilon^d(s)ds + \int_{-\infty}^{t} A(t-s)\mathbf{I}\frac{d}{ds}\theta(s)ds \qquad (14.43)$$

where $\varepsilon(t)$ is the infinitesimal strain tensor, $\varepsilon^d(t)$ is the deviatoric part of $\varepsilon(t)$, and $\theta(t)$ is the temperature history. The first integral in Equation 14.43 describes the stress response to a volumetric deformation, the second term describes the effect of shear, that is, the deviatoric contribution to the overall stress, and the third term is the thermal stress that accrues from the isochoric response to a temperature perturbation. The linear viscoelastic bulk modulus $K(t)$, shear modulus $G(t)$, and thermal stress $A(t)$ are all required by thermodynamics to be nonincreasing functions of time [86], and are often expressed in terms of a Prony series given by

$$K(t) = \sum_{i=1}^{n} K_i \exp\left(-\frac{t}{\tau_i}\right) \qquad G(t) = \sum_{i=1}^{n} G_i \exp\left(-\frac{t}{\tau_i}\right) \qquad A(t) = \sum_{i=1}^{n} A_i \exp\left(-\frac{t}{\tau_i}\right) \tag{14.44}$$

where $\{K_i\}$, $\{G_i\}$, and $\{A_i\}$ are respectively the bulk, shear, and thermal stress spectral strengths associated with the discrete relaxation time spectrum $\{\tau_i\}$. The effect of temperature on the linear viscoelastic relaxation is described in terms of a temperature-reduced time t/a_T [2], where a_T is the well-known time–temperature shift factor that is typically parameterized via the empirical Williams–Landel–Ferry (WLF) [87] or Vogel–Tammann–Fulcher (VTF) [88–90] equations. For the case of a nonisothermal history, the temperature-reduced time is [66,67]

$$t_T^* = \int_0^t \frac{d\xi}{a_T[\theta(\xi)]} \tag{14.45}$$

Thus, the nonisothermal linear viscoelastic constitutive equation is given by

$$\sigma(t) = \int_{-\infty}^{t} K\left(t_T^* - s_T^*\right)\mathbf{I}\frac{d}{ds}\operatorname{tr}(\varepsilon(s))ds + \int_{-\infty}^{t} 2G\left(t_T^* - s_T^*\right)\frac{d}{ds}\varepsilon^d(s)ds + \int_{-\infty}^{t} A\left(t_T^* - s_T^*\right)\mathbf{I}\frac{d}{ds}\theta(s)ds \tag{14.46}$$

The key assumption in Equation 14.46 is that the temperature history can compress or expand the time scale of the relaxation, but from the perspective of the temperature- reduced time t_T^* all processes are similar.

The remainder of this section is organized as follows. First, we will discuss viscoelastic models that have been developed to describe isotropic volume or enthalpy relaxation. Next, quasi-linear viscoelastic constitutive models for polymeric glasses will be analyzed; finally, the nonlinear thermoviscoelastic constitutive model developed by Caruthers and coworkers [61,62] will be discussed.

14.5.1 ISOTROPIC VISCOELASTIC MODELS

Both the specific volume and enthalpy exhibit distinctive relaxation processes in the glass transition region that should be captured by any constitutive model for polymeric glasses. Moreover, since the nonlinear, anisotropic relaxation behavior of polymeric glasses is quite sensitive to the thermal history used to form the glass, it is difficult to see how any glassy constitutive model can be effective if it does not capture the isotropic processes used to form the glass. Although not formally constitutive models, nonlinear viscoelastic models have been developed to describe isotropic volume relaxation. It has been proposed that isobaric volume relaxation $V(t)$ to an arbitrary thermal history $\theta(t)$ is given by Moynihan et al. [91]

$$\frac{V(t)}{V_r} - 1 = \int_{-\infty}^{t} \alpha(t^* - s^*)\frac{d\theta}{ds}ds \qquad \alpha(t) = \alpha_g + (\alpha_\infty - \alpha_g)\sum_{i=1}^{N} g_i\left(1 - \exp\left(-\frac{t}{\tau_i}\right)\right) \tag{14.47}$$

where $\alpha_g = \alpha(t = 0)$ is the asymptotic glassy coefficient of thermal expansion, $\alpha_\infty = \alpha(t \to \infty)$ is the coefficient of thermal expansion for a fully equilibrated material, the spectral weights $g_i > 0$ are normalized such that

$\sum g_i = 1$, and V_r is the volume in a reference state, which is typically chosen at the conventional glass transition point. The material time in Equation 14.47 is the generalization of Equation 14.45 given by

$$t^* = \int_0^t \frac{d\xi}{a[\theta(\xi), V(\xi)]} \tag{14.48}$$

The meaning of Equation 14.48 is that the history dependence of the material time results not only from the temperature history but also from other time-dependent thermodynamic variables, in this case the specific volume. Equation 14.47 can be rewritten in terms of the volume departure from equilibrium

$$\delta = \frac{V - V_\infty(\theta)}{V_\infty(\theta)} \tag{14.49}$$

giving rise (for temperatures not too far from Tg) to the defining equation for volume relaxation

$$\delta = -\Delta\alpha \sum_{i=1}^N g_i \int_{-\infty}^t \exp\left(-\frac{t^* - s^*}{\tau_i}\right) \frac{d\theta}{ds} ds \tag{14.50}$$

where $\Delta\alpha = \alpha_\infty - \alpha_g$. Equation 14.50 is the integral form of the Kovacs, Aklonis, Hutchinson, and Ramos (KAHR) model [57]. The set of $\{\tau_i\}_{i=1}^N$ with the corresponding $\{g_i\}_{i=1}^N$ constitutes the spectrum of the relaxation times, where the spectral shape is arbitrary. In Equation 14.50, it is assumed that the spectrum shifts as a whole with the shift factor a, which enters via Equation 14.48, that is, the KAHR model is thermorheologically simple. The expression for the shift factor used in KAHR was proposed by Narayanaswamy and Tool [92,93], where the dependence on volume is via the fictive temperature θ_f given by

$$\log a = \frac{x\Delta h}{R\theta} + \frac{(1-x)\Delta h}{R\theta_f} - \frac{\Delta h}{R\theta_r} \tag{14.51}$$

where x and Δh are model parameters. The fictive temperature dependence on temperature and volume in the glassy state (i.e., $\theta < \theta_g$) is given by Moynihan et al. [91] as

$$\int_{\theta_f}^{\theta_\infty} (\alpha_\infty - \alpha_g) d\theta' = \int_\theta^{\theta_\infty} (\alpha - \alpha_g) d\theta' \tag{14.52}$$

where θ_∞ is a temperature far enough above θ_g so that the equilibrium response occurs and α is the coefficient of thermal expansion at the current temperature (the reader is reminded that we use θ_g and Tg interchangeably). For a typical polymeric glass, the quantities α_∞ and α_g are such that Equation 14.52 results in

$$\theta_f(\theta, \delta) = \theta + \frac{\delta}{\Delta\alpha} \tag{14.53}$$

Upon substitution into Equation 14.51 and linearization, the dependence of log a on temperature and volume is given by

$$\log a(\theta, \delta) = -\frac{\Delta h}{R\theta_r^2}\left[(\theta - \theta_r) + (1-x)\frac{\delta}{\Delta\alpha}\right] \tag{14.54}$$

Equations 14.48, 14.50, and 14.54 form a complete set that predicts the volume response to arbitrary thermal histories, where the key assumptions are (i) relaxation occurs on a material time t^* that includes the nonlinear effects of both temperature and volume and (ii) the material is thermorheologically simple such that the shape of the spectrum $\{g_i\}$ in Equation 14.50 does not change during the nonlinear volume relaxation process.

The KAHR model [57] embodied in Equations 14.48, 14.50, and 14.54 can describe many features of the Kovacs' PVAc volume relaxation data [7]; however, as discussed by Ng and Aklonis [5] and Medvedev et al. [11] the KAHR model is incapable of predicting correctly the "short anneal" and the "expansion gap" data (i.e., Features 3.2 and 3.3). Note that the predictions of the KAHR model do not change if Equation 14.51 for the shift factor is used instead of Equation 14.54 as was done by Knauss and Emri [6]. The failure to predict these features of the specific volume relaxation is somewhat surprising, since the KAHR model puts no constraints on the $\{g_i\}$ spectrum of relaxation times. Medvedev et al. [11] argued that the cause of this failure in the KAHR model is the assumption of thermorheological simplicity.

If volume in Equation 14.50 is replaced with the fictive temperature θ_f as the variable describing structural relaxation then

$$\theta_f - \theta_\infty = \int_{-\infty}^{t} \sum_{i=1}^{N} g_i \left(1 - \exp\left(-\frac{t^* - s^*}{\tau_i} \right) \right) \frac{d\theta}{ds} ds \tag{14.55}$$

The fictive temperature can be defined not just in terms of the specific volume, but also for other thermodynamic variables such as enthalpy. For the case of enthalpy, the coefficient of thermal expansion in the definition of the fictive temperature Equation 14.52 is replaced with the isobaric heat capacity. It should be noted that fictive temperatures for different properties do not coincide [94]. The heat capacity analog of Equation 14.52 results in

$$\frac{d\theta_f}{d\theta} = \frac{C_p - C_{pg}(\theta)}{C_{p\infty}(\theta_f) - C_{pg}(\theta_f)} \tag{14.56}$$

where $C_{p\infty}$ and C_{pg} are the asymptotic isobaric heat capacities in the equilibrium and glassy states, respectively. Equations 14.51, 14.55, and 14.56 form a complete set, known as the Tool [93]–Narayanaswamy [95]–Moynihan [91] (TNM) model that has been used by many authors to fit the enthalpy relaxation data [96,97]. However, just like the KAHR model for volume relaxation, the TNM enthalpy relaxation model has several known problems beginning with the observation that the model parameters present in Equation 14.51, that is, $\Delta h/R$ and x, as well as the width of the relaxation spectrum turn out to be different for volume versus enthalpy relaxation [98]. Moreover, even if one just considers the enthalpy relaxation, different TNM model parameters are required to fit the experimental data for different thermal histories [97,99], although model parameters must be constants independent of thermal history.

14.5.2 QUASI-LINEAR VISCOELASTIC MODELS

The first step toward extending the linear viscoelastic and nonlinear isotropic models presented in the previous two sections are the quasi-linear models, where the material time t^* that previously depended on just the isotropic variables such as temperature, volume, or fictive temperature is replaced by a more general function that in addition depends upon other continuum variables such as strain, stress, and so on. The material time t^* (instead of t_T^*) is used with the linear viscoelastic constitutive form given in Equation 14.46. If the deformation contributions to the "a" shift function are strong enough, the constitutive response can be highly nonlinear even though the underlying viscoelastic continuum model is linear—hence the term quasi-linear. Although it is not formally correct to employ the infinitesimal strain tensor ε for finite deformations, the typical yield strain in polymeric glasses is of order 5%, where the infinitesimal strain tensor may be an acceptable approximation. An example of quasi-linear viscoelastic model is the proposal by Knauss and Emri [6] given by

$$\sigma(t) = \int_{-\infty}^{t} K(t^* - s^*) \mathbf{I} \frac{d}{ds} \mathrm{tr}(\varepsilon(s)) ds + \int_{-\infty}^{t} 2G(t^* - s^*) \frac{d}{ds} \varepsilon^d(s) ds \tag{14.57}$$

The key nonlinearity is the "a" shift function, where Knauss and Emri used a generalization of the Doolittle fractional free volume formula [100] given by

$$\log a(t) = \frac{b}{2.303} \left[\frac{1}{f} - \frac{1}{f_0} \right] = \frac{b}{2.303} \left[\frac{1}{f_0 + \Delta f_\theta + \Delta f_\sigma} - \frac{1}{f_0} \right] \tag{14.58}$$

where changes to the fractional free volume can result from both changes in temperature

$$\Delta f_\theta = A \int_{-\infty}^{t} \alpha(t^* - s^*) \frac{d\theta}{ds} ds \tag{14.59}$$

and via deformation induced dilatation given by

$$\Delta f_\sigma = B \int_{-\infty}^{t} \beta(t^* - s^*) \frac{d}{ds} \mathrm{tr}(\sigma(s)) ds \tag{14.60}$$

where b, f_0, A, B are model parameters and $\alpha(t)$ and $\beta(t)$ are the viscoelastic coefficient of thermal expansion and bulk compliance, respectively. The form of Equation 14.58 assumes that the log a shift factor (i) is due to the time-dependent temperature and dilatation, but (ii) does not depend upon the deviatoric component of the strain. Critically, the shift factor affects equally every relaxation time present in the viscoelastic bulk and shear moduli in Equation 14.57, that is, the Knauss and Emri model is thermorheologically simple (or more appropriately "time shift invariant," which means that the shape of the relaxation spectrum is invariant to changes in both temperature and deformation). A conceptually similar model was proposed by Shay and Caruthers [59], where instead of the fractional free volume f the shift factor was a function of the "hole fraction" obtained from solving the Simha–Somcynsky equation of state [101]. An important feature of Equation 14.60 is that the dependence on the hydrostatic component of stress is linear; consequently, it predicts acceleration of the rate of relaxation in uniaxial extension (via Poisson's ratio that effects a dilation) and deceleration in uniaxial compression such that yield in compression is not predicted.

To address the lack of compressive yield, Popelar and Liechti [58] proposed an empirical modification of the free volume expression in Equation 14.60 to include the deviatoric strain

$$\log a = \frac{b}{2.303} \left[\frac{1}{f} - \frac{1}{f_0} \right] + \frac{b_s}{2.303} \left[\frac{1}{f_s + \bar{\varepsilon}_{eff}} - \frac{1}{f_s} \right] \qquad \bar{\varepsilon}_{eff} = \sqrt{\frac{2}{3} (\boldsymbol{\varepsilon}^d : \boldsymbol{\varepsilon}^d)} \tag{14.61}$$

As an improvement over the Knauss–Emri model, the Popelar–Liechti model does qualitatively predict yield not only in uniaxial extension but also in compression and shear (i.e., Feature 4.1d); however, it also predicts post-yield softening that continues to increase with the strain. To recover the experimentally observed constant flow stress, Popelar and Liechti required that the model "parameter" b_s decrease with $\bar{\varepsilon}_{eff}$. In addition to this problem, since Popelar–Liechti is a "strain clock" model, it does not predict the effect of strain rate on the shape of the stress–strain curve (i.e., Feature 4.2c). More significantly, when multiple-step experimental protocols are considered (for example, loading, unloading) and reloading, the shape of the reloading curve is different from that of the original loading curve, which is contrary to the experiments.

Schapery [56] proposed that the log a shift function depended upon either stress or strain. No specific functional form of the stress/strain dependence was postulated, but the dependence was obtained from shifting (both vertically and horizontally) the nonlinear creep compliance curves or stress relaxation curves, respectively. The "stress clock" version obtained following Schapery procedure was able to describe creep recovery using the preceding creep as model input in case of primary and secondary creep. However, the model fails to predict tertiary creep and post-yield softening in constant strain rate deformation.

Buckley and Jones [64] developed a nonlinear viscoelastic model (which we designate Buckley95), where the log a shift factor was the product of (i) a fictive temperature term like Equation 14.51 in order to describe the effects of formation of the glass and (ii) the Eyring viscosity like Equation 14.7 to describe the effects of deformation. The Buckley95 model is a "stress clock" model as the variable controlling Eyring viscosity is stress in contrast to the "strain clock" model of Popelar and Liechti (14.61). Nevertheless, both models have a similar underlying structure, where the overall shift factor is the combination of a fictive temperature term to address the effects of thermal history on the glass formation and a second term that includes the effects of deformation on the shift factor. As a purely "stress clock" model, the Buckley95 model does not predict (i) the post-yield softening in constant strain rate deformation and (ii) tertiary creep. To rectify this difficulty, in a subsequent model that we designate as Buckley04 [51,53], Buckley and coworkers revert to a plasticity model that employs a log a shift function that now employs a new fictive temperature that depends upon the plastic strain given by

$$\theta_f = \theta_{f0} + \Delta\theta_f \left[1 - \exp\left(-\frac{\bar{\varepsilon}_p}{\varepsilon_{p0}} \right) \right] \qquad \bar{\varepsilon}_p = \int_0^t \sqrt{\frac{2}{3}(\mathbf{D}_p(s) : \mathbf{D}_p(s))} \, ds \qquad (14.62)$$

where θ_{f0} is the fictive temperature in the absence of deformation and $\Delta\theta_f$ and ε_{p0} are model parameters. The resulting Buckley04 model includes not only a "stress clock" but also a "plastic strain clock" rendering it similar to the EGP model with qualitatively similar strengths and weaknesses as already discussed in Section 14.4.2.

Fielding et al. [102] proposed a viscoelastic constitutive model for glassy polymers based on modification of the upper-convected Maxwell model [103], where the relaxation time is an independent variable obeying the following evolution equation for the case of uniaxial deformation:

$$\dot{\tau} = 1 - (\tau - \tau_0)\mu \, |\dot{\varepsilon}| \qquad (14.63)$$

where τ_0 and μ are model parameters and $\dot{\varepsilon}$ is the axial Hencky strain rate. This quasi-linear (in the terminology of this chapter) model produces good predictions for (i) a single-step nonlinear creep, that is, all five stages of creep were predicted (see Feature 4.2) and (ii) a single-step constant strain loading including post-yield softening and hardening. However, the Fielding et al. model is a "strain rate clock" model, which suffers from the problems shared by all the strain rate-based models; specifically,

1. In the absence of deformation (i.e., $\dot{\varepsilon} = 0$) Equation 14.63 describes physical aging. It follows from Equation 14.63 that an infinitesimally small oscillatory deformation will arrest physical aging, where the relaxation time reaches a steady state that depends on the frequency of the applied deformation. This is not observed experimentally.
2. Multistep nonlinear deformations, where the relaxation time at the beginning of the current step has the value reached at the end of the previous step, pose a severe problem. To give but two examples: (i) At the beginning of the first loading (at constant strain rate) of an aged sample, the predicted relaxation time is large; in contrast, after passing through the yield and the strain-softening region

followed by the unloading, the Fielding et al. model predicts that the relaxation time at the beginning of the reloading will be several orders-of-magnitude smaller. Thus, the predicted response on reloading will differ dramatically from the one on first loading contrary to experiments. (ii) In the creep and recovery experiment, there was an *ad hoc* change in the Fielding et al. model elastic modulus by an order-of-magnitude during recovery as compared to the modulus used to describe the initial creep, where without this drastic change, the model predictions qualitatively disagree with the experimental data.

In summary, the isotropic, nonlinear viscoelastic models are able to describe a number of features of volume and enthalpy relaxation in the glass transition region, but are not able to describe the short time annealing experiments, the tau-effective paradox, and obviously anisotropic deformations. A number of quasi-linear viscoelastic models have been developed, where they have had some success is describing one class of experiments, for example, constant strain rate uniaxial extension; but, they have not been successful in unifying the diversity of thermo-mechanical behaviors exhibited by polymeric glasses. Examining the history of this approach, it appears that every time a new experiment is added to the evaluation data set, a new functional dependence appears in the log a shift function, and many of the features discussed in Section 14.2 have not even yet been considered. An even more serious issue with the quasi-linear approach is that as various terms are added to the proposed constitutive models, there is no assurance that the resulting models are thermodynamically consistent.

14.5.3 TVEM OF CARUTHERS AND COWORKERS

To address concerns about the thermodynamic consistency of the evolving class of quasi-linear viscoelastic models, Caruthers and coworkers pursued a systematic approach to constructing a general framework for the viscoelastic description of polymeric glasses using the principles of Rational Mechanics as put forward by Coleman and Noll [86,104]. The series of papers by Caruthers and coworkers comprise in chronological order and in terms of growing generality (1) a purely viscoelastic model with no thermal effects, where the shift factor is a function of free volume [59]; (2) the thermoviscoelastic model (TVEM) of Shay and Caruthers [60] that used tensorially correct finite strain measures and where the shift factor is controlled by the Adam–Gibbs [105] configurational entropy which was a functional of thermal and deformational history; (3) the general derivation by Lustig et al. [61] where the rigorous machinery of Rational Mechanics was deployed for a material that relaxes on a t^* material time scale that was controlled by the configurational entropy that is a functional of the thermo-deformational history; and finally (4) the thermoviscoelastic model (TVEM) of Caruthers et al. [62], where the shift factor is controlled by the configurational internal energy that is also a functional of the thermo-deformational history. The predictions of the earlier constitutive models in this sequence had various shortcomings similar to the type of difficulties described earlier in this chapter for the various plasticity-based and quasi-linear viscoelastic models, and consequently will not be discussed in detail. However, the TVEM has shown some promise for semiquantitative predictions using a single set of model parameters for a diverse set of nonlinear time-dependent thermo-mechanical responses, including enthalpy relaxation, volume relaxation, and yield in extension and compression [106]. A more extensive discussion of the model structure and predictions of TVEM will be presented following a brief derivation of the TVEM model.

14.5.3.1 DERIVATION OF THE TVEM

The TVEM for glassy polymers follows the general framework of the Rational Mechanics developed by Coleman and Noll [86,104]. In contrast to the original Coleman and Noll framework, in the derivation of the TVEM model [62], it was postulated that the relaxation response of glassy materials should be described using the "material time," t^*, rather than the laboratory time in order to capture how the evolving state of the glass affects the viscoelastic relaxation. This is a significant change to the original Coleman and Noll framework that allowed for nonlinear strain history contributions to the constitutive functional, but where relaxation was on the normal laboratory time t; in contrast, the relaxation time scale in all the constitutive functionals in the TVEM also have nonlinear contributions from the deformation via t^*. It was not clear *a priori* that a constitutive model with nonlinear contributions in the time scale for relaxation would be thermodynamically consistent. Also, implicit in the TVEM is the assumption of time shift invariance of the relaxation spectra. Lustig et al. [61] showed that all thermodynamic quantities are derived from a single Helmholtz free energy functional, which ensures thermodynamic consistency, and where the relaxation was on the material time scale t^*. The free energy functional was expanded in a Frechet series, where the TVEM makes an additional assumption that third- and higher-order Frechet terms can be neglected. The TVEM Helmholtz free energy through second-order terms is

$$
\begin{aligned}
\psi = \psi_\infty &+ \frac{1}{2}\Delta K V_r \int_{-\infty}^{t}\int_{-\infty}^{t} ds du\, f_1(t^*-s^*,t^*-u^*)\frac{dI_1}{ds}\frac{dI_1}{du} + 2\Delta A \int_{-\infty}^{t}\int_{-\infty}^{t} ds du\, f_3(t^*-s^*,t^*-u^*)\frac{dI_1}{ds}\frac{d\theta}{du} \\
&+ \Delta C \int_{-\infty}^{t}\int_{-\infty}^{t} ds du\, f_4(t^*-s^*,t^*-u^*)\frac{d\theta}{ds}\frac{d\theta}{du} + \Delta G V_r \int_{-\infty}^{t}\int_{-\infty}^{t} ds du\, f_2(t^*-s^*,t^*-u^*)\frac{d\mathbf{H}^d}{ds}:\frac{d\mathbf{H}^d}{du}
\end{aligned}
\tag{14.64}
$$

Here $\psi_\infty = \psi_\infty(\theta, \mathbf{H})$ is the equilibrium Helmholtz free energy, which is a function of temperature and the Hencky strain $\mathbf{H}$; $I_1 = \mathrm{tr}\,(\mathbf{H}) = \ln\,(V/V_r)$ is the first invariant of the Hencky strain tensor; the superscript "d" denotes the deviator of $\mathbf{H}$; V is the specific volume, and V_r is specific volume in a reference state. The coefficients in Equation 14.64 are expressed in terms of material parameters in the rubbery and glassy state (see discussion below); the functions $\{f_i(s,u)\}_{i=1,2,3,4}$ are the memory kernels, which are symmetric with respect to their two arguments. In general, the coefficients in the free energy expansion given in Equation 14.64 can be functions of the current temperature and the current strain [62]; however, this "higher fidelity" formulation, although much more involved mathematically, results in only quantitative rather than qualitative improvement of model predictions and thus will not be considered here. The stress and the entropy are obtained by taking the Frechet derivatives of Equation 14.64 over the strain history and the temperature history, respectively. After some additional algebraic manipulations, the Cauchy stress tensor for irrotational deformations is given by

$$
\begin{aligned}
\sigma = \sigma_\infty(\theta,\mathbf{H}) &+ \frac{V_r}{V}\left[\Delta K \int_{-\infty}^{t} ds\, f_1(t^*-s^*,0)\frac{dI_1}{ds} + \frac{2}{V_r}\Delta A \int_{-\infty}^{t} ds\, f_3(t^*-s^*,0)\frac{d\theta}{ds}\right]\mathbf{I} \\
&+ \frac{V_r}{V}2\Delta G \int_{-\infty}^{t} ds\, f_2(t^*-s^*,0)\frac{d\mathbf{H}^d}{ds}
\end{aligned}
\tag{14.65}
$$

where $\sigma_\infty(\theta, \mathbf{H})$ is equilibrium stress. The entropy is given by

$$
\eta = \eta_\infty - 2\Delta A \int_{-\infty}^{t} ds\, f_3(t^*-s^*,0)\frac{dI_1}{ds} - 2\Delta C \int_{-\infty}^{t} ds\, f_4(t^*-s^*,0)\frac{d\theta}{ds}
\tag{14.66}
$$

The expressions for σ and η for arbitrary deformations, including shear, are found in Caruthers et al. [62]. In the original paper, the generalized shift factor is based on the configurational internal energy, which was obtained from the total internal energy $e(t)$ that was determined self-consistently from $e(t) = \psi(t) + \theta\eta(t)$. In order to slightly improve predictions, Adolf et al. [63] modified the coefficients in the "internal energy" $e(t)$ functional to now allow them to be arbitrary and hence be optimized from nonlinear relaxation data. In the Adolf et al. [63] version of TVEM, the log a functional is given by

$$
\log a = C_1\left(\frac{C_2''}{C_2''+N}-1\right)
\tag{14.67}
$$

where

$$N = \left[(\theta - \theta_r) - \int_{-\infty}^{t} ds\, f_4(t^* - s^*, 0) \frac{d\theta}{ds} \right] + C_3 \left[I_1 - \int_{-\infty}^{t} ds\, f_3(t^* - s^*, 0) \frac{dI_1}{ds} \right]$$
$$+ C_4 \int_{-\infty}^{t} \int_{-\infty}^{t} ds\, du\, f_2(t^* - s^*, t^* - u^*) \frac{d\mathbf{H}^d}{ds} : \frac{d\mathbf{H}^d}{du}$$

(14.68)

Equations 14.67 and 14.68 have a standard WLF form [87] for an equilibrium material, where C_1 is the first WLF coefficient. The parameter C_2'' is given by

$$C_2'' = C_2(1 + C_3 \alpha_\infty)$$

(14.69)

where C_2 is the second WLF coefficient. C_3 and C_4 are model parameters. In deriving Equation 14.69, it was assumed that in the equilibrium state $I_1 = \alpha_\infty(\theta - \theta_r)$, where α_∞ is the equilibrium coefficient of thermal expansion.

The material coefficients in the free energy expansion Equation 14.64 can be determined from limiting values in the rubber state and deep in the glass where there is no relaxation; specifically, the material coefficients for the bulk modulus K and shear modulus G are given by

$$\Delta K = K_g - K_\infty \quad \Delta G = G_g - G_\infty$$

(14.70)

where subscript "g" denotes the asymptotic glassy value and subscript "∞" denotes the equilibrium (i.e., rubbery) value. The coefficients ΔA and ΔC are determined by invoking the "ideal glass" approximation, which assumes that the transition into glass on cooling occurs at a point such that the material above this glass formation point is equilibrated (i.e., the memory kernels $f_i = 0$) and below that point, the material is completely frozen with no relaxation (i.e., the memory kernels $f_i = 1$). Using this ideal glass approach, one can show that

$$\frac{2\Delta A}{V_r} = K_\infty \alpha_\infty - K_\infty \alpha_g - \Delta K \alpha_g = -(K_g \alpha_g - K_\infty \alpha_\infty)$$

(14.71)

and

$$C_{pg}(\theta) = C_{p\infty}(\theta) + \theta \left[V_r K_g \alpha_g^2 - V K_\infty \alpha_\infty^2 + (V - V_r) K_\infty \alpha_\infty \alpha_g - 2\Delta C \right]$$

(14.72)

where volume V is the asymptotic glassy volume corresponding to a given θ and p. Equation 14.72 shows that the model parameter ΔC controls both the discontinuous decrease in heat capacity at the glass transition point Tg and the slope of the glassy heat capacity asymptote with temperature. Experimentally these quantities are generally independent from each other versus the TVEM where both are controlled by the single-model constant ΔC. If the decrease in heat capacity at Tg is used to determine the value of ΔC, then

$$\theta_r 2\Delta C = C_{p\infty}(\theta_r) - C_{pg}(\theta_r) + \theta_r V_r \left[K_g \alpha_g^2 - K_\infty \alpha_\infty^2 \right]$$

(14.73)

Using the expressions in Equations 14.70, 14.72, and 14.73, the materials parameters ΔG, ΔK, ΔA, and ΔC are determined from time-independent data well-above Tg and well-below Tg.

The memory kernels are postulated to be the stretched exponential (KWW) functions, which are in turn expanded into a Prony series for subsequent use in the TVEM. Specifically,

$$f_i(t,0) = \exp\left[-\left(\frac{t}{\tau_i^{KWW}}\right)^{\beta_i^{KWW}}\right] = \sum_k f_k^{(i)} \exp\left(-\frac{t}{\tau_k^{(i)}}\right) \tag{14.74}$$

Thus, each memory kernel has two parameters τ_{KWW} and β_{KWW}. An important issue is how to interpret the double-integral memory kernels in Equations 14.64 and 14.68. Two alternatives (there are other possibilities as well) that both result in Equation 14.74 are [62]

$$\text{Case I} \quad f_i(s,u) = \sum_k f_k^{(i)} \exp\left(-\frac{s+u}{\tau_k^{(i)}}\right) \tag{14.75}$$

$$\text{Case II} \quad f_i(s,u) = f_i(s,0)f_i(0,u) = \sum_k f_k^{(i)} \exp\left(-\frac{s}{\tau_k^{(i)}}\right) \sum_m f_m^{(i)} \exp\left(-\frac{u}{\tau_m^{(i)}}\right) \tag{14.76}$$

Using Equations 14.76 and 14.65, the term in N with the coefficient C_4 can be transformed as follows:

$$\begin{aligned}
&C_4 \int_{-\infty}^{t}\int_{-\infty}^{t} ds\,du\, f_2(t^*-s^*,t^*-u^*)\frac{d\mathbf{H}^d}{ds}:\frac{d\mathbf{H}^d}{du} \\
&= C_4 \int_{-\infty}^{t} ds\, f_2(t^*-s^*,0)\frac{d\mathbf{H}^d}{ds}:\int_{-\infty}^{t} du\, f_2(t^*-u^*,0)\frac{d\mathbf{H}^d}{du} \\
&= C_4\left\{\left(\frac{1}{2\Delta G}\frac{V}{V_r}\right)^2 [(\sigma-\sigma_\infty):(\sigma-\sigma_\infty)] + \frac{1}{3\Delta G}\frac{V}{V_r}\phi(t)(p-p_\infty) + [\phi(t)]^2\right\}
\end{aligned} \tag{14.77}$$

where

$$\phi(t) \equiv \frac{1}{2\Delta G}\left[\Delta K \int_{-\infty}^{t} ds\, f_1(t^*-s^*,0)\frac{dI_1}{ds} + \frac{2}{V_r}\Delta A\int_{-\infty}^{t} ds\, f_3(t^*-s^*,0)\frac{d\theta}{ds}\right] \tag{14.78}$$

Examining the last line in Equation 14.77, it is a *function* of current stress and a *functional* of the temperature and deviatoric strain histories via $\phi(t)$. Thus, Case II is a "stress clock" model. There are several experiments that the stress clock has no hope of predicting. Consider for example a creep experiment for a material equilibrated at a temperature slightly below Tg, where equilibrium can be achieved within a reasonable laboratory time. For an equilibrium material, all the history integral terms will have relaxed to zero; therefore, upon application of the load, the log a shift factor (14.68) will depend only on stress, which is constant. Only primary and secondary creep will be predicted with no acceleration in the rate of creep, that is, tertiary creep is not possible, no matter how large the applied stress, which is contrary to experiment. As another example, the reversal of the ordering of the rate of stress relaxation when switching from fast loading strain rate to slow (i.e., Feature 4.7) cannot be explained by a stress-clock model [20]. Thus, the Case II TVEM is inadequate and will not be discussed further. In the remainder of this section, the focus will be on the Case I version of the TVEM as given in Equation 14.75.

The values of the material coefficients ΔK, ΔG, ΔA, and ΔC are fully determined by the equilibrium and asymptotic glassy material properties using Equations 14.70, 14.71, and 14.73. The WLF parameters C_1

and C_2 are independently set by the linear viscoelastic relaxation time versus temperature data above Tg. The relaxation functions $\{f_1, f_2, f_3, f_4\}$ are parameterized via the discrete spectra and are well defined in the linear limit; thus, they could, in principle, be determined using linear viscoelastic experiments. For example, the bulk relaxation kernel f_1 can be determined from the volume response of an equilibrium material to an infinitesimal pressure jump and the shear relaxation kernel f_2 from the stress response to an infinitesimal shear deformation. The important point is that for these *linear* experiments the log a shift factor will be constant. The relaxation kernels f_3 and f_4 as well as the coefficient C_3 in the log a expression (i.e., Equation 14.68) could be determined from linear viscoelastic physical aging studies, for example, using the Struik protocol [107] for linear viscoelastic creep. This would leave only the value of parameter C_4 to be set from *nonlinear* mechanical experiments. The program outlined above is appealing in principle, but it is challenging in practice, because an appropriately diverse data for a single well- characterized material is not available. As a result, Adolf et al. [63,106] employed a different approach, where the memory kernels and the coefficients C_3 and C_4 were determined by optimizing the TVEM predictions of enthalpy relaxation and nonlinear mechanical data that included (i) constant strain rate stress–strain curves in uniaxial extension and compression [106] and (ii) uniaxial creep in extension [63]. This parameter optimization was done for a DGEBA/DEA epoxy, where it was determined that

1. The memory kernels related to isotropic processes were the same, i.e., $f_1 = f_3 = f_4$ with KWW parameters $\tau_V^{KWW} = 6\,\text{s}$ and $\beta_V^{KWW} = 0.24$ if only constant strain rate extension/compression data was used [106]. However, in a later paper, nonlinear creep data were included in the optimization data set, and the optimized parameter values for the isotropic processes were found to be $\tau_V^{KWW} = 6\,\text{s}$ and $\beta_V^{KWW} = 0.14$ [63].
2. For the shear memory kernel f_2 the KWW parameters are $\tau_s^{KWW} = 0.12\,\text{s}$ and $\beta_s^{KWW} = 0.22$ [106].

It is concerning that the β_V^{KWW} parameter for f_2 is so small, thus indicating a very broad spectrum, and changed significantly when nonlinear creep was added to the optimization data set. The predictions of the TVEM that will now be shown using the models parameters optimized from the data set that included nonlinear creep.

14.5.3.2 PREDICTIONS OF THE TVEM

The TVEM predictions considered here are based on the latest published parameter set for the DGEBA/DEA epoxy material [63]. The temperature dependence of both the PVT response and heat capacity (i.e., enthalpy) in the glass transition region is predicted by TVEM, since these data were included in the optimization data set. This is a significant improvement over many of the viscoplastic and quasi-linear viscoelastic constitutive models described earlier, where no predictions of the enthalpic response have been made (in fact, these constitutive models with the exception of Anand09 are by construction silent with respect to enthalpy).

Linear viscoelastic behavior

Bulk versus shear spectrum

As described above, the linear viscoelastic bulk and shear moduli should in principle be inputs to TVEM. However, the memory kernels were not determined from linear viscoelastic data, but rather set on the basis of fitting the nonlinear mechanical data as well as enthalpy relaxation: thus, the TVEM predictions of linear viscoelastic response can be used for model validation. It is troubling that the TVEM bulk spectrum is at least as broad (using the parameter set from Reference 106) or significantly broader (using the most recent parameter set [63]) than the shear spectrum. This is contrary to the experimental findings described in Chapter 4, Section 4.2.2 (Feature 2.2).

Physical aging

Physical aging for the DGEBA/DEA epoxy was probed via stress relaxation, where the specimen was cooled to 50°C (i.e., Tg-25°C) at 1°C/min and aged for various times t_{age} before a 0.5% shear strain was applied [106]. Superposition of the various stress relaxation curves was effected by horizontal shifting, where the values of the shift factor are plotted in Figure 14.30. The log a versus t_{age} dependence is described reasonably well by TVEM using the parameter set from Reference 106; however, if the new parameter set is used (i.e., the

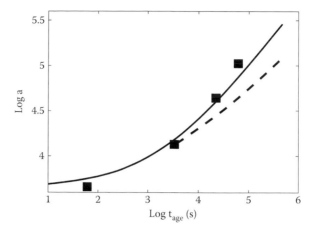

Figure 14.30 TVEM prediction of aging effect on log a for a DGEBA/DEA epoxy resin at 50°C (Tg-15°C). Experimental data [106]—markers; model parameters from Reference 106—solid; model parameters from Reference 63—dashed.

one proposed in Adolf et al. [63] to better describe the creep data), the prediction worsens, although it is not qualitatively incorrect.

Nonlinear volume relaxation

There is no data set on volume relaxation for DGEBA/DEA epoxy material that is comparable in richness with the Kovacs data set for PVAc [4]; therefore, assessment of the ability of TVEM to quantitatively predict volume relaxation data for a DGEBA/DEA material is subject for future study. Nevertheless, it is possible to qualitatively evaluate the TVEM performance with respect to the two critical signatures of volume relaxation: the "short anneal" experiment (Feature 3.2) and the tau-effective "expansion gap paradox" (Feature 3.3).

Short anneal experiment

The TVEM predictions of the annealing experiment are shown in Figure 14.31, where the material was subjected to the following thermal history: (i) the material was initially at equilibrium at 85°C (Tg + 10°C);

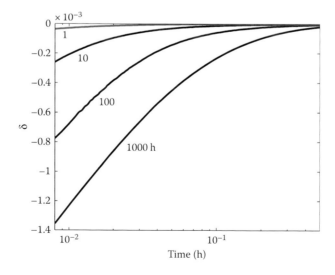

Figure 14.31 Short anneal prediction by TVEM using the parameter set from Reference 63 for DGEBA/DEA (Tg = 75°C). Thermal history: down-jump from 85°C to 70°C; annealing for time t_a indicated in the figure at 70°C; up-jump from 70°C to 85°C; and volume response at 85°C.

(ii) a temperature down-jump to 70°C where duration of the jump was 10^{-2} h; (iii) the material was annealed for a time t_a at 70°C; and a temperature up-jump back to 85°C was applied, where the specific volume response at 85°C is shown in Figure 14.31. These thermal histories are similar to the history employed by Kovacs (see Figure 9 in Kovacs [4]), where the temperatures have been corrected for the change in Tg which was 31°C for Kovacs' PVAc and 75°C for the DGEBA/DEA epoxy. The results for various t_a are shown in Figure 14.31. The TVEM predictions for the "short anneal" experiment with $t_a = 1$ h are similar to the prediction of the KAHR model discussed in Section 14.5.1. The TVEM predicts almost no volume relaxation, which is in disagreement with the experimental data of Kovacs shown in Figure 4.17.

Expansion Gap

The volume expansion following the temperature up-jump from various sub-Tg initial temperatures, θ_i, to the final temperature $\theta_f = 85$°C was determined for the TVEM model, where the material at the initial temperature was at equilibrium. If the normalized volume departure from the equilibrium volume at the final temperature is denoted as $\delta = (V - V_e(\theta_f))/V_e(\theta_f)$, then the effective relaxation time is obtained according to

$$\tau_{eff} = -\frac{1}{\delta}\frac{d\delta}{dt} \tag{14.79}$$

The plot of τ_{eff} versus δ for various initial temperatures as predicted by TVEM is shown in Figure 14.32. Again, similar to the prediction by the KAHR model and unlike the Kovacs PVAc data, TVEM predictions do not exhibit the "expansion gap," that is, Feature 3.3. Specifically, the curves corresponding to different initial temperatures clearly begin to merge at δ values as large as 10^{-3}, where the data show up to half a logarithmic decade difference in tau-effective as shown in Figure 4.19 in Chapter 4. Ng and Aklonis [5] argued based on extensive numerical experimentation with the spectra of various shapes that it was not possible to describe the "expansion gap" using the KAHR type of volume relaxation models. The inability of the TVEM to predict the "expansion gap" follows the pattern of other models that assume thermorheological simplicity.

Enthalpy relaxation

As already mentioned, the TVEM is the only constitutive model to-date that has made predictions for enthalpy relaxation. The protocol of the DSC experiment is as follows [106]: the material is cooled at 5°C/min from well above Tg to the indicated aging temperature; the material is aged for 2 h; then the material is further cooled to room temperature; and finally, the material is heated at 5°C/min to a temperature well above

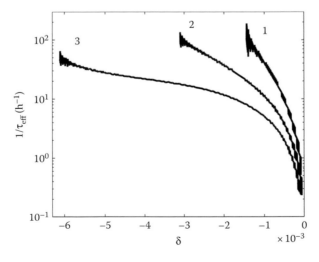

Figure 14.32 τ-effective expansion gap prediction by TVEM using the parameter set from Reference 63 for DGEBA/DEA (Tg = 75°C). The volume relaxation was after an up-jump from θ_i to $\theta_f = 85$°C. The initial temperatures were 1—$\theta_i = 80$°C; 2—$\theta_i = 75$°C; and, 3—$\theta_i = 65$°C; δ and τ_{eff} are defined in text.

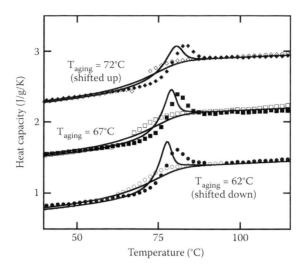

Figure 14.33 TVEM prediction of the effect of sub-θ_g aging on the heat capacity peak measured during cooling—open symbols and heating—filled symbols (see text for details concerning temperature history). TVEM predictions are solid lines. Experimental data are for DGEBA/DEA epoxy resin (Tg = 75°C). (Reprinted from *Polymer*, 45, D. B. Adolf, R. S. Chambers, and J. M. Caruthers, 4599, Copyright 2004, with permission from Elsevier.)

Tg, during which the heat capacity is measured. Representative results are shown in Figure 14.33, where the effect of varying the aging temperature on the location and height of the endothermic heat capacity peak is illustrated. The "older" DGEBA/DDS parameter set is used to obtain curves in Figure 14.33 [106]. The agreement between TVEM prediction and the data is semiquantitative, where the basic features of the heat capacity overshoot are observed although the height and location of the overshoot peak are not exact. When the "new" DGEBA/DEA parameter set in Adolf et al. [63] is used, the agreement with experiment worsens.

Nonlinear anisotropic deformations

Stress–strain curve with yield

For a constant strain rate uniaxial deformation, the TVEM predicts yield, a gradual post-yield softening, and no strain hardening. A similar stress response is predicted in both uniaxial compression and extension, where compressive yield stress is approximately 30% higher than the yield stress in extension, which agrees with experiments (Feature 4.1d). A representative stress–strain curve is shown in Figure 14.34, where the effect of aging time is shown (Feature 4.2b). The lack of the post-yield softening is especially apparent in case of an aged material, where experimentally the stress drop is quite pronounced. It should be noted, however, that the amount of post-yield softening predicted by TVEM, although too gradual as compared to the experimental stress–strain curves, does increase with the strain rate (i.e., compare the 10^{-5}, 10^{-3}, and 10^{-1} s^{-1} strain rate curves in Figure 14.36) in qualitative agreement with the experimental observations (Feature 4.2c). The dependence of yield stress on temperature is well captured by the TVEM (see data in Figure 38 in Adolf et al. [106]). The strain rate dependence of the yield stress is shown in Figure 14.35, where the nearly linear dependence of the yield stress on the logarithm of strain rate for the typically accessible range is predicted by TVEM (Features 4.1a). This is significant, because the TVEM expression for the relaxation time given by Equations 14.67 and 14.68, unlike the corresponding Eyring expression, does not explicitly depend on stress, but rather is quadratic in strain rate (i.e., the double-integral term in Equation 14.68). Thus, although the TVEM employs a different constitutive formalism with a different functional form for the relaxation time (i.e., Equations 14.67 and 14.68 vs. Equation 14.7), the TVEM is also capable of producing linear dependence of the yield stress versus log (strain rate). One of the major arguments for using the Eyring model (with all its known shortcomings as described in Section 14.4) was its ability to predict the strain rate dependence of the yield stress, where the TVEM provides an alternative approach that also results in the linear dependence of the yield stress on the logarithm of the strain rate.

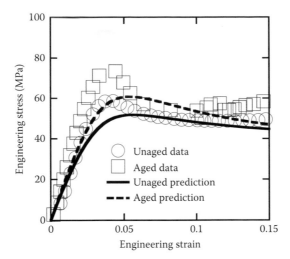

Figure 14.34 TVEM prediction of the stress–strain curve in uniaxial compression versus experimental data for DGEBA/DEA at 59°C (Tg-16°C). (Reprinted from *Polymer*, 45, D. B. Adolf, R. S. Chambers, and J. M. Caruthers, 4599, Copyright 2004, with permission from Elsevier.)

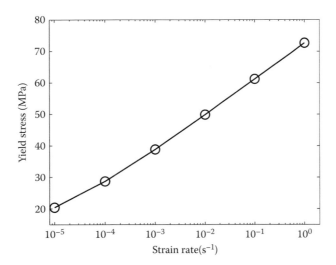

Figure 14.35 TVEM prediction of strain rate effect on yield in uniaxial compression at 50°C (Tg-25°C). TVEM simulation used the DGEBA/DEA parameter set from Reference 63.

Strain rate-switching experiment

TVEM predictions using the DGEBA/DEA parameter set are shown in Figure 14.36 for uniaxial compression experiments, where the strain rate that is initially at 10^{-3} s^{-1} is suddenly switched to either a 100 times faster or a 100 times slower rate. Upon switching of the strain rate, the stress response rapidly moves to the stress–strain curve at the final strain rate with a minimal overshoot/undershoot in cases where the switching has been exercised in the vicinity of yield. At larger switching strains, the overshoot becomes more pronounced. In contrast, the experimental data for PMMA at 100°C (i.e., Tg-15°C) exhibit a very pronounced overshoot/undershoot even when the strain rate was changed close to yield (see Chapter 4, Section 4.6.5). The inability of the TVEM to predict the qualitative features of the strain rate- switching experiment is a significant shortcoming (Feature 4.4).

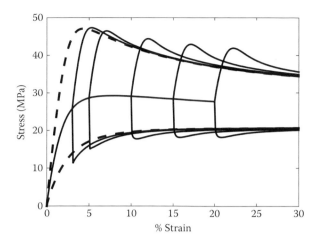

Figure 14.36 TVEM prediction of strain rate switch in uniaxial compression experiments at 50°C (Tg-25°C) using the DGEBA /DEA parameter set from Reference 63. Dashed—monotonic loading at strain rate: bottom—10^{-5} s^{-1} and top—10^{-1} s^{-1}; solid—switch from strain rate 10^{-3} s^{-1} to either 10^{-1} s^{-1} or 10^{-5} s^{-1}.

Unloading and reloading deformation; Bauschinger effect

The TVEM predictions of the stress response for a multistep loading–unloading and reloading deformation at a constant strain rate is shown in Figure 14.37 for the DGEBA/DEA epoxy glass at 50°C (i.e., Tg-25°C). Upon unloading to zero stress and reloading, the stress–strain curve is very similar to that for monotonic loading, and the initial modulus on reloading is similar to that for the original loading response from zero strain, where both predictions are in qualitative agreement with experimental data (Feature 4.5). If the second unloading that was initiated at approximately 20% axial strain is continued through zero axial stress into compression all the way back to zero strain, the TVEM predicts yield in compression. The predicted yield stress in compression following extension has an absolute value of 45 MPa, which should be compared to 40 MPa for a single-step compressive stress–strain curve (dashed curve in Figure 14.37). Thus, the Bauschinger effect is not predicted, where in the Bauschinger effect, the compressive yield stress for a single- step deformation from

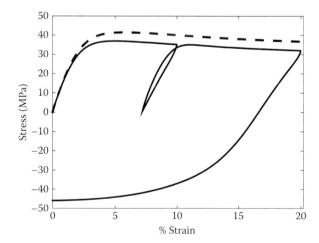

Figure 14.37 TVEM predictions of multistep uniaxial experiments at 50°C (Tg-25°C) with a strain rate of 10^{-3} s^{-1}. Simulation used the DGEBA/DEA parameter set from Reference 63. Solid—multistep deformation starts in extension at 0% strain and proceeds to 10% strain followed by unloading to zero stress and reloading to 20% strain; at this point unloading begins which passes through zero stress and is continued in compression to 0% strain. Dashed—single-step compression from 0% strain where the absolute value of stress is shown.

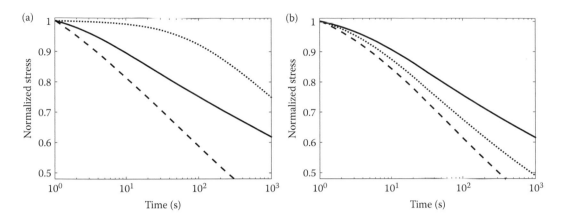

Figure 14.38 TVEM prediction of stress relaxation in uniaxial compression at 50°C (Tg-25°C). Simulation used the DGEBA/DEA parameter set from Reference 63. (a)—loading strain rate $10^{-2}\,s^{-1}$, (b)—loading strain rate $10^{-4}s^{-1}$; strain: pre-yield (1%) —dashed, yield (5–7%)—dotted, and post-yield (20%)—solid. The stress is normalized by the value at the beginning of relaxation.

zero strain is larger than the yield stress in compression following extension. For example, a 25% decrease in the compressive yield stress of the pre-extended sample versus a sample that was compressed from zero strain was observed experimentally, albeit for a different material (PC) and at Tg-124°C (see Section 4.6.6 in Chapter 4). In addition, the shape of the predicted stress–strain curve in compression is qualitatively different than that observed experimentally [81] (see Chapter 4, Figure 4.39); specifically, the "yield" in compression following extension is much more gradual and over a much broader strain range than the straight compressive yield shown as the dashed curve in Figure 14.37. Thus, the TVEM does not predict the Bauschinger effect (Feature 4.6), where the value of the compressive yield stress following pre-extension is too large and the shape of the compressive yield stress is different from that observed experimentally.

Stress relaxation

A particularly interesting feature of nonlinear stress relaxation of polymeric glasses is the reversal of the strain dependence of the rate of stress relaxation as the strain rate of loading is changed (Feature 4.7). The TVEM prediction of the nonlinear stress relaxation response is shown in Figure 14.38, where the stress relaxation begins at three different locations on the stress–strain curve: pre-yield, at yield, and post-yield. The stress relaxation is shown for a slow ($10^{-4}\,s^{-1}$) and a fast ($10^{-2}\,s^{-1}$) loading strain rate, where the stress is normalized by the initial stress value at the beginning of the relaxation. For slow loading, the TVEM predicts that the rate of stress relaxation monotonically decreases with strain from pre-yield to post-yield. For fast loading, the TVEM predicted behavior is non-monotonic, where the rate of stress relaxation is the fastest pre-yield, but slows down dramatically slightly past yield and then becomes faster well into the post-yield strain region, although it is still not as fast as pre-yield rate of relaxation. This is not what is observed experimentally as described in Chapter 4, Section 4.6.7, where the ordering of the stress relaxation curves with strain is either increasing for a slow loading rate or decreasing for a fast loading rate.

Stress memory

The stress memory experiment comprises a three-step history: constant strain rate loading through yield; unloading at the same strain rate to zero stress; and, holding the strain constant while monitoring the resultant stress. The TVEM predictions of stress response during the final step of the deformation history are shown in Figure 14.39 for two loading strain rates. The TVEM model predicts (i) the height of stress maximum increases with loading/unloading strain rate and (ii) the location of the maximum shift to shorter times with increasing strain (Feature 4.8), where both of these predictions are in qualitative agreement with the experimental data.

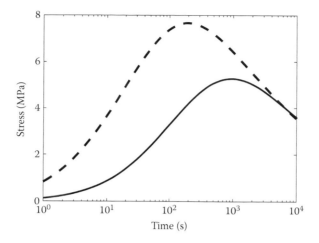

Figure 14.39 TVEM prediction of stress memory experiment in uniaxial extension at 55°C (Tg-20°C) using the DGEBA/DEA parameter set from Reference 63. Stress memory response for strain rate of 10^{-3} s^{-1}—solid; 10^{-2} s^{-1}—dashed.

Nonlinear creep

The nonlinear creep response for glassy polymers may include as many as five distinct stages (see Section 14.2 in this chapter and Chapter 4, Section 4.7.1). Stage V of creep is a manifestation of the same mechanism that is responsible for the post-yield hardening observed in constant strain rate deformation at large strain; consequently, the TVEM does not predict Stage V creep because it does not predict strain hardening in a constant strain rate deformation. Stages III and IV of creep are thought to be inextricably connected to the post-yield softening and the steady-state flow response in the constant strain rate experiment, respectively [14]. Since the TVEM does not exhibit steady-state behavior and only predicts a very gradual post-yield softening, one would expect that the TVEM will not predict Stages III and IV; however, the situation is somewhat more complicated. The TVEM predictions of a single-step creep in uniaxial extension and compression are shown in Figure 14.40. In extension, a tertiary creep-like response is predicted, where at some point, the rate of creep begins to rapidly increase and becomes unbounded (i.e., exhibits a singularity); in contrast, in compression, tertiary creep is not predicted by the TVEM, at least using the DGEBA/DEA parameter set given in Adolf et al. [63]. The tertiary

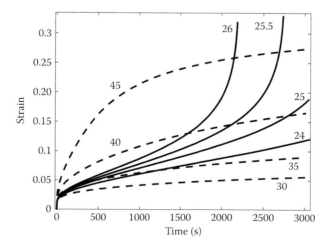

Figure 14.40 TVEM prediction of uniaxial creep using the DGEBA/DEA parameter set from Reference 63. The material was cooled at 1°C/min to 50°C (Tg-25°C). Solid—extension; dashed—compression, where the absolute values of the stress in MPa are indicated in the figure.

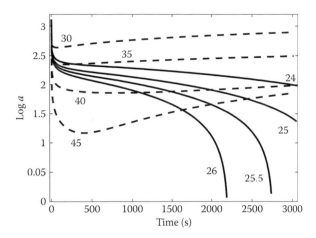

Figure 14.41 TVEM prediction of log a during uniaxial creep in extension (solid) and compression (dashed) corresponding to the same conditions as in Figure 14.40.

creep-like behavior in extension is a result of a positive feedback mechanism caused by the presence of the strain rate-dependent term, that is, the term with the coefficient C_4 in Equation 14.68, that works as follows: (i) an increase in the strain rate results in decrease in the relaxation time, (ii) faster relaxation means that the memory kernels in the stress integrals in Equation 14.65 decay faster, and (iii) to support the applied stress the strain rate has to increase further. The process accelerates in this positive feedback loop. The abrupt acceleration of the relaxation time at some point during extension creep is shown in Figure 14.41. The reason why this feedback mechanism does not work in compression is as follows: the volume term in log a in Equation 14.68, that is, the term with the coefficient C_3, has a different sign in compression than in extension; when the memory kernel decays faster, this term becomes more negative; which slows down the relaxation, where the log a is increasing during compressive creep as shown in Figure 14.41. An important caveat is that the result shown in Figure 14.40 (i.e., singularity observed in extension but not in compression) is when the material was not aged for a very long time. For a different thermal history, the TVEM predicts that tertiary creep with a singularity may occur in compression, where the singular behavior depends critically on the interplay of the positive and negative feedback terms in the log a function. In contrast to the predictions of the TVEM, the tertiary creep observed experimentally at sufficiently large applied stress occurs in both extension and compression for all thermal histories. Therefore, the TVEM prediction of tertiary creep is problematic for two reasons: (i) when the tertiary creep is predicted, it is too strong, that is, there is an infinite increase in the strain rate versus the finite increase observed experimentally, and (ii) in compression, the increase in strain rate associated with tertiary creep is extremely sensitive to details of the thermal history, where experimentally it is a robust phenomenon, occurring for a wide range of thermo-deformational histories.

Creep recovery

Recovery from nonlinear creep predicted by the TVEM is shown in Figures 14.42 and 14.43. In Figure 14.42, the deformation is uniaxial extension and the predicted response is for a quenched material, where the predictions are in qualitative agreement with the experiments; specifically, the rate of recovery upon removal of the axial stress is initially quite rapid and then becomes progressively slower (i.e., Feature 4.9). The TVEM predictions of the recovery response for a material equilibrated below Tg prior to uniaxial compression are shown in Figure 14.43, where the predictions are qualitatively incorrect; specifically, there is no smooth recovery following the instantaneous removal of the load with perhaps the exception of the largest strain. The reason that there is some long-term relaxation in the recovery response in the quenched material is due to the relaxation of the temperature and volume history terms in the stress equation, that is, Equation 14.65, and the log a equation, that is, Equation 14.68. When the deviatoric strain history terms abruptly decrease upon removal of the load, it is the temperature and volume history terms that are responsible for the smooth recovery behavior in the quenched material case. However, when the material is aged to equilibrium as is the case

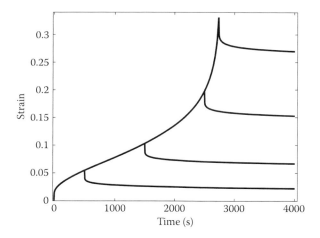

Figure 14.42 TVEM prediction of creep in uniaxial extension at 50°C (Tg-25°C) with an applied stress of 25.2 MPa using the DGEBA/DEA parameter set from Reference 63. The material was cooled at 1°C/min to 50°C prior to application of the creep stress.

for the predictions shown in Figure 14.43, the temperature and volume history terms have already relaxed to zero, leaving the deviatoric strain term as the only term that contributes to the response, and consequently there is no relaxation in the recovery response.

Loading–creep–reloading experiment

In Figure 14.44, the TVEM predictions of the stress response are shown for a complex deformation history comprising (i) loading in uniaxial extension at a constant strain rate through yield, (ii) loading at a slower strain rate for a time $t_{creep} = 160$ s, and (iii) reloading at the original strain rate. This deformation history mimics the experimental history where in step (ii) the sample was partially unloaded and the stress was kept constant. Although not exactly the same, we believe that the three-step strain rate deformation history used to generate the predictions in Figure 14.44 will have a similar effect to the one where constant stress is applied during step (ii), that is, the experiment shown in Figure 4.37 in Chapter 4. The key feature for this deformation history is the increase of the stress overshoot peak observed on second loading with the value of stress during step (ii) (Feature 4.10). As compared with the experimental data, the TVEM predicted overshoots

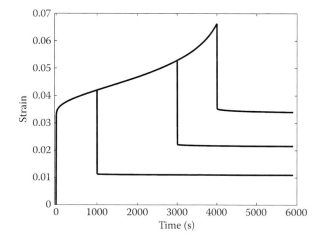

Figure 14.43 TVEM prediction of creep in uniaxial compression at 50°C (Tg-25°C) with an applied stress of 95.5 MPa using the DGEBA/DEA parameter set from Reference 63. The material was at equilibrium at 50°C prior to application of the creep stress.

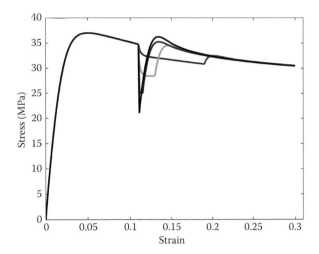

Figure 14.44 TVEM prediction of multistep deformation in uniaxial compression at 50°C (Tg-25°C) consisting of (i) constant strain rate loading at the strain rate of 10^{-3} s^{-1} to strain of 0.12, (ii) constant strain rate loading at a slower strain rate to imitate the creep deformation for 160 s, and (iii) reloading at the strain rate of 10^{-3} s^{-1}. The strain rate during step (ii) was 10^{-5} s^{-1}—black; 2×10^{-5} s^{-1}—blue; 10^{-4} s^{-1}—green; and 5×10^{-4} s^{-1}—red. In the simulation the DGEBA/DEA parameter set from Reference 63 was used.

are in the incorrect order, that is, the TVEM predicts that the magnitude of the overshoot during the second constant strain rate deformation decreases as the creep stress increases; consequently, this complex, three-step deformation history poses a real challenge to the TVEM.

Internal energy stored during deformation

As described in Chapter 4, Section 4.9.2, the deformation calorimetry experiments provide information about the evolution of internal energy during large constant strain rate deformation. Specifically, it is found that the internal energy increases smoothly through yield and post-yield softening, where it only levels off at large strains well into the flow regime (Feature 4.3). This picture is difficult to reconcile with the underlying premise of the plasticity-based models, where the internal energy is expected to stop increasing at yield. The TVEM does make predictions of the internal energy U during deformation, which is calculated as $U = \psi + \theta\eta$ using Equations 14.64 and 14.66. The result is shown in Figure 14.45 along with the stress–strain curve. It is

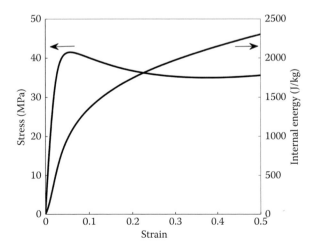

Figure 14.45 TVEM prediction of internal energy evolution during uniaxial compression at 50°C (Tg-25°C) at a constant strain rate of 10^{-3} s^{-1}. The DGEBA/DEA parameter set from Reference 63 was used in the simulation.

seen that the prediction is in general agreement with the experiment; specifically, there is no abrupt change in the internal energy versus deformation response at yield (Feature 4.3); however, as the deformation continues past yield, the TVEM predicts that the internal energy continues to rise albeit at a slower pace, where experimentally the internal energy becomes nearly independent of strain at large strains.

14.5.3.3 CONCLUSIONS

The TVEM is the only constitutive model in the literature to-date that has produced predictions for all types of experiments performed on polymeric glasses, including the PVT response, linear viscoelastic behavior, aging effects, volume relaxation, enthalpy relaxation, and the nonlinear mechanical response for both strain-controlled and stress-controlled deformations. The TVEM is able to qualitatively capture a number of features of the deformation of polymeric glasses including:

1. Physical aging. (Feature 1.1)
2. The basic features of nonlinear volume relaxation. (Feature 3.1)
3. Enthalpy relaxation. (Feature 5.1)
4. Internal energy as measured by the deformation calorimetry. (Feature 4.3)
5. The dependence of yield stress on temperature, strain rate, and aging time. (Feature 4.1)
6. The loading–unloading–reloading experiments. (Feature 4.5)
7. The stress memory experiment and the dependence of the stress overshoot on the strain rate of loading. (Feature 4.8)

Notwithstanding the predictive successes of the TVEM listed above, there are a number of serious difficulties encountered by the TVEM, including:

8. The linear viscoelastic shear relaxation spectrum (as determined by fitting the nonlinear relaxation data) is narrower than the bulk relaxation spectrum, which is contrary to experiments. (Feature 2.1)
9. In case of specific volume relaxation experiments, the TVEM does not qualitatively capture the "short anneal" and the expansion gap. (Features 3.2 and 3.3)
10. No hardening is predicted in constant strain rate experiments and there is no Stage V creep. (Feature 4.2d)
11. The predicted post-yield softening in a constant strain rate deformation is much more gradual than the one observed experimentally, especially for an aged material. (Features 4.2b, c)
12. In the strain rate-switching experiment, the magnitude of the predicted over- and under-shoot peaks is too small, especially when the strain rate switch is applied in the vicinity of yield. (Feature 4.4)
13. In the Bauschinger experiment, there is no decrease in the compressive yield stress following pre-extension and the shape of the compressive stress–strain curve does not show a sharp yield, but rather a slowly curving stress response. (Feature 4.6)
14. The TVEM incorrectly predicts the ordering of the nonlinear stress relaxation curves with strain for the case of fast loading rates. (Feature 4.7).
15. The strain rate increase during tertiary creep in extension is over-predicted with a singularity emerging, and tertiary creep in compression is not predicted (using the same thermal history and material parameters used in extension). (Features 4.2a, b)
16. For a material that has experienced considerable physical aging, the predicted creep recovery is abrupt with no long time relaxation. (Feature 4.9)
17. The constant strain rate loading–creep–reloading experiment is not predicted. (Feature 4.10)
18. The TVEM is thermorheologically simple, which is inconsistent with the dye reorientation results of Ediger. (Feature 4.11)

The TVEM has some compelling features; however, it also has some serious deficiencies when compared to a larger experimental data set as described in Section 14.2. The TVEM may be appropriately acknowledging a portion of the underlying physics that controls the thermo-mechanical behavior of polymeric glasses, but there appears to be something significant that is missing in the basic model framework that cannot be fixed by just changing values of some model parameters or by changing the functional form of log a.

14.6 STOCHASTIC CONSTITUTIVE MODEL

14.6.1 PERSPECTIVE

The constitutive models for glassy polymers discussed in the preceding sections have been continuum models, where all the relevant variables are fields that are smooth except at a finite number of interfaces. These models employ the well-known continuum postulate that a "material point" in the continuum contains a sufficient number of molecules so that all fluctuations are averaged out over the times of interest for typical experiments. The continuum postulate works exceedingly well in most cases, including liquids and crystalline solids, so it has been the default assumption in constitutive modeling of amorphous solids. However, there are at least two reasons to reconsider the basic continuum postulate: first, as was documented in the previous sections, even after decades of research, the existing constitutive models exhibit significant failures, suggesting that there perhaps is something structurally wrong with the current continuum approach; and second, the observation that glasses exhibit dynamic heterogeneity. Dynamic heterogeneity refers to spatial and temporal mobility heterogeneity on the 2 to 4 nm scale that persists for times that are at least as long as the times of interest in a macroscopic experiment, where dynamic heterogeneity has been firmly established experimentally (for references to the original experiments see References 11, 33, and 34). Clearly the existence of dynamic heterogeneity cannot be reconciled with the underlying assumption of continuum mechanics, where the fluctuations are averaged out.

Recently, a stochastic constitutive model (SCM) has been proposed that introduces dynamic heterogeneity into a constitutive description [108]. The primary assumption is that on the size scale of nanometers, the material still has a continuum-like description (vs. a particulate description used in molecular simulations), but where the meso-domains are subject to fluctuations. In the SCM (i) the temporal fluctuations experienced by an individual meso-domain are explicitly included, (ii) interactions between meso-domains are treated in a mean field sense, and (iii) the observed macroscopic quantities are obtained as ensemble averages.

The structure of the SCM approach is fundamentally different from traditional continuum mechanics. In traditional continuum mechanics, all field variables of interest are first temporally and spatially averaged, where the constitutive model then describes relationships between these averaged quantities. In contrast, in the SCM approach, the constitutive relationship is between the local, and hence fluctuating, variables, where the macroscopic stress and other thermodynamic variables are then determined as an average of the predictions of the local constitutive model. If the constitutive model was linear, there would be no difference between the two approaches; however, if there are strong nonlinearities in the local constitutive relationships, the SCM approach is quite different from traditional continuum mechanics. The SCM approach to date has shown promise in predicting several features of the polymeric glasses. We will first provide a brief introduction to the structure of the SCM and then describe its prediction for the suite of nonlinear thermo-mechanical properties of polymeric glasses.

14.6.2 STRUCTURE OF STOCHASTIC CONSTITUTIVE MODEL

The defining equations of the SCM are mathematically involved and will only be summarized, where a detailed derivation has been given elsewhere [108]. The key assumption of the SCM is that the mesoscopic, that is, local, mobility is described by $\log \hat{a}$ shift factor that depends upon (i) the macroscopic temperature and strain common to all meso-domains and (ii) the mesoscopic (i.e., local and instantaneous) stress and entropy. Since the size of a meso-domain is several nanometers, the fluctuations experienced by the local variables are large and have to be explicitly taken into account. The dimensionless, seven-component vector $\hat{\mathbf{x}}$ contains the fluctuating entropy and the six components of the fluctuating symmetric stress tensor, where the fluctuating quantities are denoted with a "hat." Consistent with equilibrium fluctuation thermodynamics [109], the magnitudes of the fluctuations in entropy, dilatational stress (i.e., pressure), and deviatoric stress are given respectively by

$$\sigma_S^2 = \frac{1}{L^3} V k_B C_{pg} \quad \sigma_K^2 = \frac{1}{L^3} 2 k_B \theta \frac{C_{pg}}{C_{Vg}} 3 K_g \quad \sigma_G^2 = \frac{1}{L^3} 2 k_B \theta \frac{C_{pg}}{C_{Vg}} 2 G_g \tag{14.80}$$

where L is the size of the meso-domain and C_v is the constant volume heat capacity. The components of $\hat{\mathbf{x}}$ are defined as

$$\hat{x}_0 = \sigma_S^{-1}(\hat{S} - S_\infty) \tag{14.81}$$

$$\begin{pmatrix} \hat{x}_1 \\ \hat{x}_2 \\ \hat{x}_3 \end{pmatrix} = \begin{pmatrix} 3^{-1/2}\sigma_K^{-1} & 3^{-1/2}\sigma_K^{-1} & 3^{-1/2}\sigma_K^{-1} \\ 0 & 2^{-1/2}\sigma_G^{-1} & -2^{-1/2}\sigma_G^{-1} \\ -2\cdot6^{-1/2}\sigma_G^{-1} & 6^{-1/2}\sigma_G^{-1} & 6^{-1/2}\sigma_G^{-1} \end{pmatrix} \begin{pmatrix} \hat{T}_{H11} - T_{H11}^\infty \\ \hat{T}_{H22} - T_{H22}^\infty \\ \hat{T}_{H33} - T_{H33}^\infty \end{pmatrix} \tag{14.82}$$

$$\hat{x}_4 = \sigma_G^{-1}\left(\hat{T}_{H12} - T_{H12}^\infty\right) \qquad \hat{x}_5 = \sigma_G^{-1}\left(\hat{T}_{H13} - T_{H13}^\infty\right) \qquad \hat{x}_6 = \sigma_G^{-1}\left(\hat{T}_{H23} - T_{H23}^\infty\right) \tag{14.83}$$

where S is the entropy and $\mathbf{T}_H$ is the stress tensor conjugate to the Hencky strain tensor $\mathbf{H}$. The equilibrium entropy, $S_\infty(\theta,\mathbf{H})$, and stress, $\mathbf{T}_H^\infty(\theta,\mathbf{H})$, are functions of the externally controlled temperature and strain. The dependence of the shift factor on the fluctuating variables contained in $\hat{\mathbf{x}}$ and the macroscopic variables is given by the expression

$$\log \hat{a}(\theta, V, \hat{\mathbf{x}}) = c_1 \left(\frac{e_{c,ref}^\infty}{e_c^\infty(\theta, V) + \hat{e}_c(\hat{\mathbf{x}})} - 1 \right) \tag{14.84}$$

Equation 14.84 assumes that the non-fluctuating, that is, $e_c^\infty(\theta,V)$, and the fluctuating contributions, that is, $\hat{e}_c(\hat{\mathbf{x}})$, in the denominator are additive. The non-fluctuating term in Equation 14.84 is constructed so that the WLF [87] form of $\log a$ is recovered above θ_g if (i) the fluctuations are formally set to zero, that is, $\hat{\mathbf{x}} = \mathbf{0}$, (ii) the specific volume is at equilibrium, and (iii) the material is at atmospheric pressure. Specifically,

$$e_c^\infty(\theta, V) = e_{c,ref}^\infty + 2\theta_{ref}\left[V_{ref}^{-1}\Delta A(V - V_{ref}) + \Delta C(\theta - \theta_{ref}) \right] \tag{14.85}$$

where

$$e_{c,ref}^\infty = 2\theta_{ref}c_2(\Delta A\alpha_\infty + \Delta C) \tag{14.86}$$

The various material constants present in Equations 14.85 and 14.86 have already been defined in Section 14.5 for the TVEM. The fluctuating term $\hat{e}_c(\hat{\mathbf{x}})$ is given by

$$\hat{e}_c(\hat{\mathbf{x}}) = n\theta_{ref}\sigma_S\hat{x}_0 - mV_{ref}\sigma_K\hat{x}_1 + \frac{1}{4}\frac{V_{ref}\sigma_G^2}{\Delta G}\left[\hat{x}_2^2 + \hat{x}_3^2 + 2\left(\hat{x}_4^2 + \hat{x}_5^2 + \hat{x}_6^2\right) \right] \tag{14.87}$$

where the coefficients n and m are model parameters. The expression in Equation 14.87 was inspired by the "configurational internal energy" clock model proposed by Caruthers et al. [62] in conjunction with the TVEM (see Section 14.5); however, examining the configurational internal energy, it does not have a pressure, that is, the $\hat{x}_1$, term. Thus, the effect of fluctuations on the local mobility as defined by Equations 14.84 and 14.87 is a model assumption. Finally, expanding upon a postulate by Robertson et al. [8], the mean field approximation is employed, where the fluctuating term $\hat{e}_c(\hat{\mathbf{x}})$ in Equation 14.84 is replaced with

$$\frac{1}{z}\hat{e}_c(\hat{\mathbf{x}}) + \frac{z-1}{z}\langle\hat{e}_c(\hat{\mathbf{x}})\rangle \tag{14.88}$$

where z is the coordination number and $<>$ indicates an ensemble average. The case of $z = 1$ corresponds to independent meso-domains and the case of $z \to \infty$ corresponds to the macroscopic average, where there are no fluctuations.

The dynamic equations for the fluctuating entropy and the six components of (symmetric) stress tensor are (i.e., Equations 52 and 53 in Medvedev and Caruthers [108])

$$d\hat{x}_0 = d\hat{r}_0 - \omega \left(\frac{2\Delta C}{\sigma_S} d\theta + \frac{2\Delta A}{\sigma_S} \frac{dV}{V_{ref}} \right) \tag{14.89}$$

$$\begin{pmatrix} d\hat{x}_1 \\ d\hat{x}_2 \\ d\hat{x}_3 \end{pmatrix} = \begin{pmatrix} d\hat{r}_1 \\ d\hat{r}_2 \\ d\hat{r}_3 \end{pmatrix} + \begin{pmatrix} 3^{1/2} k & 3^{1/2} k & 3^{1/2} k \\ 0 & 2^{-1/2} g & -2^{-1/2} g \\ -2 \cdot 6^{-1/2} g & 6^{-1/2} g & 6^{-1/2} g \end{pmatrix} \begin{pmatrix} dH_{11} \\ dH_{22} \\ dH_{33} \end{pmatrix} + 3^{1/2} \frac{2\Delta A}{\sigma_K V_{ref}} \begin{pmatrix} d\theta \\ 0 \\ 0 \end{pmatrix} \tag{14.90}$$

$$d\hat{x}_4 = d\hat{r}_4 + g \, dH_{12} \quad d\hat{x}_5 = d\hat{r}_5 + g \, dH_{13} \quad d\hat{x}_6 = d\hat{r}_6 + g \, dH_{23} \tag{14.91}$$

where the stochastic terms are given by

$$d\hat{r}_n = -\frac{dt}{\hat{a}\tau_0} \left[\hat{x}_n + \frac{\partial}{\partial \hat{x}_n} \ln(\hat{a}) \right] + \frac{1}{\sqrt{\hat{a}\tau_0}} \sqrt{2} d\hat{W}_n \quad n = 0, \ldots, 6 \tag{14.92}$$

and the dimensionless coefficients in Equations 14.90 and 14.91 are defined as

$$k = \frac{\Delta K}{\sigma_K} \quad g = \frac{2\Delta G}{\sigma_G} \tag{14.93}$$

Equations 14.84 through 14.93 are the defining stochastic differential equations (SDE) that include: (i) a single relaxation time Debye process (i.e., the first term in square brackets on the right-hand side of Equation 14.92); (ii) a vector of independent Wiener processes $d\hat{\mathbf{W}}$ that represents the fluctuations; and (iii) an additional drift term (i.e., the second term in square brackets on the right-hand side of Equation 14.92) needed to satisfy the requirement that the equilibrium distribution be Gaussian. Each of the dynamic equations (i.e., Equations 14.89 through 14.91) consists of a stochastic part in addition to a non-stochastic part that contains the effects of instantaneous changes in the macroscopic temperature and strain tensor. Note that the following property of the Hencky strain has been used in deriving Equation 14.89

$$\frac{dV}{V_{ref}} = dH_{11} + dH_{22} + dH_{33} \tag{14.94}$$

The reason for introducing the model parameter ω in Equation 14.89 is that the value of $\omega = 1$ corresponds to the case of S in the definition of $\hat{x}_0$ being the total entropy; however, following Adam and Gibbs [105], it is expected that the $\log \hat{a}$ function depends only on the configurational part of the total entropy. Thus, if $\hat{x}_0$ is configurational entropy only, then $0 < \omega < 1$. Since the configurational entropy is not experimentally measureable, ω is a parameter in the SCM.

The SCM describes the fluctuating stress and configurational entropy response to a given macroscopic strain and temperature history. Finally, the boundary conditions of the problem are an ensemble average of multiple realizations of Equations 14.89 through 14.92 given by

$$\mathbf{T_H} = \langle \hat{\mathbf{T}}_\mathbf{H} \rangle \tag{14.95}$$

Table 14.3 SCM parameters

References	Material	Basic			Conf. entropy	log a	
		τ_0 (s)	L (nm)	z	ω	n	m
[12]	PVAc	$5.0 \cdot 10^4$	2.7	7.0	0.75	0.3	0.185
[108]	PMMA	$1 \cdot 10^4$	9.0	1.0	1.0	1.0	0.0

The time evolution of the mesoscopic variables is obtained by solving the dynamic Equations 14.89 through 14.92 with the shift factor given by Equations 14.84 through 14.88 for each of the individual N domains, where the average stress for the entire ensemble is subject to the conditions Equation 14.95. The number of meso-domains (and hence of the simultaneously solved sets of SDEs) needed to obtain smooth ensemble averages is typically $N = 10^4$ or more, depending on the quantity of interest. The details of the numerical procedure are outlined in Medvedev and Caruthers [12,108].

14.6.2.1 PARAMETERS OF THE SCM

The SCM has six adjustable parameters, where the remaining model constants are determined by the independently measured asymptotic glassy and equilibrium material properties. Eventually optimization could be used to determine the six adjustable parameters. However, it is premature to attempt optimization until it is firmly established that the structure of the SCM can at least qualitatively describe a diverse set of nonlinear thermo-mechanical data like that given in Section 14.2 of this chapter. The SCM model parameters used for qualitative assessment of its predictive capabilities are listed in Table 14.3 for two cases—PVAc and PMMA, although it should be remembered that the parameters were not optimized over a large data set for PVAc and not at all for PMMA.

14.6.3 Predictions of the SCM

14.6.3.1 LINEAR VISCOELASTICITY

In the SCM, there is a single Debye relaxation time process (see Equation 14.92), where the relaxation time spectrum emerges due to fluctuations and is seen via the ensemble of meso-domains. The spectrum is determined by the probability density function (pdf) of the fluctuating $\log \hat{a}$ values of the domains. Examining the defining equations for $\log \hat{a}$, that is, Equations 14.84 through 14.87, the width of this distribution for an equilibrated material without deformation is controlled by the magnitudes of the fluctuations of entropy and stress given by Equations 14.80 and the mean field parameter z. The fluctuation magnitudes in turn depend on the domain size L.

Using the values for PVAc from Table 14.3, the distribution at 40°C is shown in Figure 6a in Medvedev and Caruthers [12], where its width is approximately five decades in qualitative agreement with the measurements of the bulk relaxation spectrum by McKinney and Belcher [110]. However, the predicted shear relaxation spectrum has the same width as the bulk relaxation spectrum, which does not agree with the experiments (Feature 2.1).

14.6.3.2 VOLUME RELAXATION

For an isotropic deformation, the SCM equations given in Equations 14.90 simplify; however, one should remember that even though the macroscopic deformation of interest is isotropic, there is no constraint that the mesoscopic fluctuating stress be isotropic. The SCM predictions of volume relaxation for the Kovacs PVAc data set are given in Figures 2 through 4 in Medvedev and Caruthers [12]. The SCM predictions are qualitatively correct for the full set of Kovacs thermal histories, where quantitative comparison between predications and data are quite good considering that only a few by-hand optimization steps were used. Of particular note is that the SCM adequately describes the "short anneal" experiment (Feature 3.2) as shown in Figure 14.46. This successful prediction is in contrast to the predictions by the KAHR model shown in Medvedev et al. [11] and the TVEM shown in Figure 14.31. Even more significantly, the SCM is the only model in the literature to date that predicts the "expansion gap" (Feature 3.3) as shown in Figure 14.47. Specifically, the effective relaxation time is

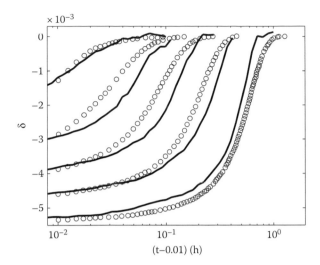

Figure 14.46 Annealing time effect on specific volume relaxation predicted by SCM. Down-jump from 40°C to 25°C, annealing for time t_a, and up-jump back to 40°C, where volume response at 40°C is shown. t_a from top to bottom: 0.3, 4, 28, 160, and 1500h; symbols—data, solid lines—predictions. (Reprinted with permission from G. A. Medvedev and J. M. Caruthers, *Macromolecules*, 48, 788. Copyright 2015, American Chemical Society.)

calculated from the current rate of volume relaxation according to Equation 14.79, where the "expansion gap" refers to observation that the curves for different initial temperatures do not merge up to the smallest measured specific volume departures from equilibrium. This is an important discriminating feature, since the material exhibits up to half an order of magnitude difference in τ_{eff} even though the temperature and volume are essential the same. The τ_{eff} predicted by the SCM exhibits expansion gaps that are close to those observed experimentally; in contrast, the various plasticity models discussed earlier do not even predict volume relaxation, while the KAHR model [5] and the TVEM (Figure 14.32) do not exhibit an expansion gap. A key difference in the SCM is that there is no constraint on the width of the relaxation time distribution (i.e., it is determined via the dynamics of the $\log \hat{a}$ pdf), where it is argued in Medvedev and Caruthers [12] that the success of the SCM in predicting the expansion gap is due to this inherent thermorheological complexity in the SCM.

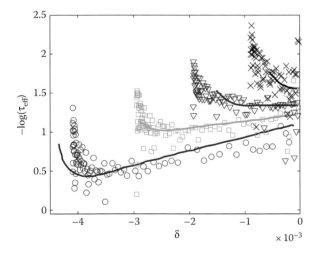

Figure 14.47 Expansion gap prediction by SCM for temperature up-jump from θ_i to $\theta_f = 40°C$. Circles—$\theta_i = 30°C$, squares—$\theta_i = 32.5°C$, triangles—$\theta_i = 35°C$; and crosses—$\theta_i = 7.5°C$. δ and τ_{eff} are defined in the text. Symbols—predictions, solid lines—data from Ref [4]. (Reprinted with permission from G. A. Medvedev and J. M. Caruthers, *Macromolecules*, 48, 788. Copyright 2015, American Chemical Society.)

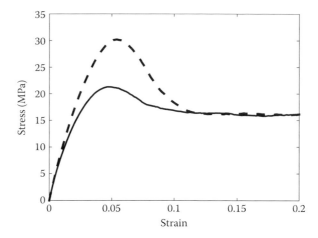

Figure 14.48 SCM prediction of aging time effect on the stress–strain curve in uniaxial extension for "PMMA" at 100°C at a strain rate of 10^{-3} s^{-1}. Aging times: solid—0.5 h, dashed—3 h. (Reprinted from *Polymer*, 74, G. A. Medvedev and J. M. Caruthers, 235, Copyright 2015, with permission from Elsevier.)

14.6.3.3 ANISOTROPIC DEFORMATION

Uniaxial loading

The SCM predictions for constant strain rate uniaxial experiments are shown in Figures 14.48 and 14.49. The stress–strain curves look quantitatively the same in extension and compression, where the compression yield stress is higher in qualitative agreement with experiments (Feature 4.1d). The SCM predicts yield, post-yield softening, and flow (Features 4.1 and 4.2b,c), but it does not predict hardening (Feature 4.2d). The increase in the stress overshoot peak with increasing aging time and strain rate is qualitatively predicted by SCM. The dependence of the yield stress and the flow stress on the strain rate in Figure 14.49 is stronger than that observed experimentally; however, no attempt was made to optimize parameters to fit the data. As discussed in Medvedev and Caruthers [108], the post-yield softening is not due to deformation induced changes in volume, because the material densifies at yield (see Figure 4.28 in Chapter 4) which would cause hardening instead of softening if volume were the key variable controlling mobility. In agreement with the experimental data, the SCM predicts a decrease in the specific volume after yield.

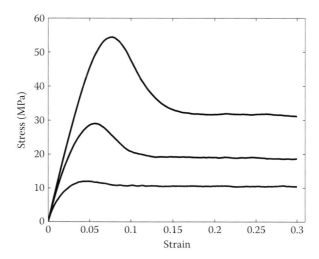

Figure 14.49 SCM prediction of strain rate effect on the stress–strain curve in uniaxial extension for "PMMA" at 100°C for 3 h aging time. Strain rates from bottom to top are 10^{-4}, 10^{-3}, and 10^{-2} s^{-1}.

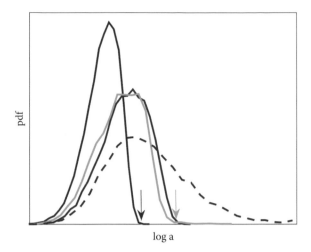

Figure 14.50 Schematic of the SCM prediction of the evolution of the relaxation time spectrum during deformation (see text for detailed explanation). (Reprinted from *Polymer*, 74, G. A. Medvedev and J. M. Caruthers, 235, Copyright 2015, with permission from Elsevier.)

The mechanism by which the SCM predicts the post-yield softening and its dependence on the strain rate and aging time is shown schematically in Figure 14.50. The blue curves is the distribution of the relaxation times of the meso-domains prior to deformation, where the solid blue curve is for a quenched material and the dashed blue curve is for an equilibrated material. Clearly the equilibrium spectrum contains a significant fraction of meso-domains with large relaxation times that result in the extended tail on the right hand side of the distribution. The red and green curves are $\log \hat{a}$ pdfs during the steady state stress response of a constant strain rate deformation (i.e., the flat portion of the stress–strain curve after yield), where the red curve describes the $\log \hat{a}$ pdf for a fast strain rate and the green curve for a slow strain rate. It was shown (see detailed explanation in Medvedev et al. [111]) that in a constant strain rate deformation the SCM predicts a cutoff on the $\log \hat{a}$ axis of the mobility pdf such that meso-domains cannot have relaxation times larger than this cutoff. The location of the cutoff depends upon the magnitude of the applied strain rate. Cutoffs in the mobility pdf for two strain rates are indicated in Figure 14.50 by arrows with the corresponding color. When a glass is subjected to a constant strain rate deformation the key physical process in the SCM is the evolution of the $\log \hat{a}$ pdf from the "blue" spectrum to either the "green" or "red" spectrum, depending on the applied strain rate. The following situations are possible:

1. If the material has been aged (i.e., the pre-deformation $\log \hat{a}$ pdf is the blue dashed line) there is a significant excess of the "rigid" meso-domains (i.e., domains with large relaxation times) as compared to the steady-state distribution shown by either the green or red lines. In the course of the deformation these "rigid" meso-domains will convert into "soft" meso-domains (i.e., domains with short relaxation times). Since "rigid" meso-domains support larger stress, there will be an initial stress overshoot, which will decay as the distribution shifts toward the steady-state distribution. The longer the aging time, the farther to the right the long time tail of the distribution will extend and, consequently, the stress overshoot will be larger in agreement with the experiment. The maximum possible overshoot is for a material that has been fully equilibrated.

2. If the material has been aged in the glassy state for only a short time, the pre-deformation $\log \hat{a}$ pdf is given by the solid blue line, where the appearance, or not, of the stress overshoot depends on the strain rate. For the case of a fast strain rate, the initial $\log \hat{a}$ pdf still has a significant excess of "rigid" meso-domains versus the steady-state $\log \hat{a}$ pdf given by the red curve, and thus a stress overshoot is predicted. On the other hand, in case of a slow strain rate with the green steady-state $\log \hat{a}$ pdf, the pre-deformation and the steady-state distributions are not significantly different; consequently, no stress overshoot is predicted in agreement with the experimental data.

The key to the above picture is that the meso-domains in the SCM are not permanently "rigid" (i.e., possess a long relaxation time) or "soft" (i.e., possess a short relaxation time). Specifically, an individual meso-domain travels stochastically along the $\log \hat{a}$ axis in Figure 14.50, where it is sometimes "rigid" and sometimes "soft." Deformation affects the probability of finding a meso-domain in a "rigid" or "soft" state thereby changing the distribution of the $\log \hat{a}$ value in the ensemble of meso-domains. Thus, according to the SCM the behavior of a glassy system in response to thermo-deformational history is controlled by the evolution of the meso-domains' mobility distribution, where the shape of the instantaneous distribution is not constant, that is, the material is thermorheologically complex or, more appropriately, the distribution is not time shift invariant. It is this change in the shape of the distribution that enabled the prediction of the post-yield stress softening and its dependence upon strain rate and annealing time— phenomena that could only be predicted via empirical parameterization by the traditional viscoelastic or viscoplastic constitutive models.

Strain rate-switching experiment

Figure 14.51 shows the predictions of the SCM for the stress response when the strain rate in a uniaxial extension is switched from slow to fast and vice versa. The SCM predictions are in qualitative agreement with the experiments shown in Chapter 4, Section 4.6.5, where the stress under- and overshoots after the rate switch have been observed (Feature 4.4). However, the predicted overshoot magnitude in case of a strain rate switch from slow to fast is lower than the stress overshoot magnitude observed during the monotonic loading at a fast strain rate. Since the model parameters were not optimized it is possible that the magnitude of the predicted overshoot can be improved. Although not minimizing the concern about the magnitude of the stress overshoots, the SCM is able to qualitatively predict the strain rate switching experiment—a feature of the thermo-mechanical behavior of polymeric glasses that has escaped, at least to-date, prediction by most of the traditional nonlinear viscoelastic and viscoplastic constitutive models.

Unloading–reloading deformation; Bauschinger effect

Figure 14.52 shows the SCM prediction of the stress response resulting from loading the glass in uniaxial extension and then unloading back to zero strain, where the stress becomes negative, that is, compression. The predicted compression portion of the loading–unloading curve exhibits a compressive yield stress (solid curve) that is only slightly lower than that for a single step compression from zero strain (dashed curve). In contrast, the data for PC [81] qualitatively shows (i) the compressive yield during the unloading to be approximately 75% of the single-step loading and (ii) a much sharper compressive yield during unloading. In a related experiment on PMMA, Anand and Ames [48] first deformed the material in compression and

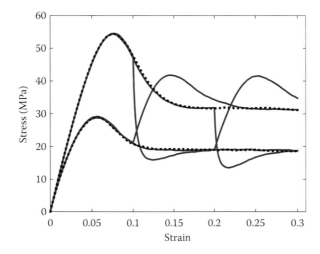

Figure 14.51 SCM prediction of strain rate switch experiment for "PMMA" at 100°C in uniaxial extension. Single step loading response—dotted lines with strain rates of 10^{-3} s^{-1} (bottom) and 10^{-2} s^{-1} (top). Strain rate down-jump—blue; strain rate up-jump—red.

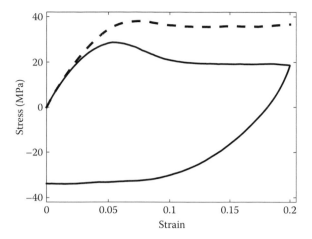

Figure 14.52 SCM prediction of Bauschinger effect for "PMMA" at 100°C at the strain rate of 10^{-3} s^{-1}. Solid line— extension–compression deformation protocol; dashed line— compression from zero strain where the absolute stress value is shown.

then applied uniaxial extension, where a non-abrupt yield transition upon unloading similar to that shown in Figure 14.52. The differences in the experimental data and the fact that the parameter space for the SCM has not been fully explored make it hard to assess at this time whether the SCM is able to qualitatively describe the Bauschinger effect, although currently the predicted magnitude of the compressive yield behavior after extension is too large.

Stress Relaxation

The SCM predictions of the nonlinear stress relaxation response for a material loaded in uniaxial extension to various strains are shown in Figure 14.53, where the qualitative behavior of interest is the effect of the strain rate during loading on the subsequent strain dependence of the rate of stress relaxation. The SCM predicts that for a fast strain rate the normalized stress relaxation is more rapid for a material deformed past yield than for a material which was only deformed to a pre-yield strain; in constrast, for a slow loading strain rate the normalized stress relaxation for a material deformed past yield is predicted to be slower that for a material where the stress relaxation starts pre-yield. This somewhat unexpected dependence of the ordering of the

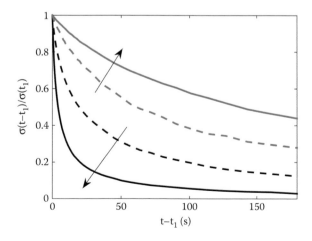

Figure 14.53 SCM prediction of the effect of loading strain rate on normalized stress relaxation for "PMMA" at 100°C at the strain rates of 10^{-4} s^{-1}—gray and 10^{-2} s^{-1}—black. Strain: pre-yield (at $\sigma_y/3$)—dashed, post-yield (at the end of softening)—solid. Arrows indicate the increase in strain.

strain dependence of the rate of stress relaxation on the loading strain rate was observed experimentally by Kim et al. [20] and is a significant challenge to traditional viscoelastic and viscoplastic constitutive models. The mechanism for this reversal of the rate of stress relaxation is naturally predicted via the evolution of $\log \hat{a}$ pdf shown schematically in Figure 14.50. Consider the situation where the pre-deformation distribution (i.e., the blue curve in Figure 14.50) is located between the steady-state distribution corresponding to the fast strain rate (i.e., the red curve) and the steady-state distribution corresponding to the slow strain rate (i.e., the green curve). The initial rate of stress relaxation can be thought of as a reporter of the instantaneous relaxation time spectrum that exists at the beginning of relaxation. In that case the stress relaxation at pre-yield strain is determined by the pre-deformation spectrum and the stress relaxation at a strain loacated in the steady-state/flow regime is determined by the corresponding steady-state spectrum. It follows that if the blue distribution is between the red distribution and the green distribution than the stress relaxation becomes faster when going from blue to red (i.e., the faster loading strain rate) and becomes slower when going from blue to green (i.e., the slower loading strain rate).

Stress memory

As shown in Figure 14.54 the SCM correctly predicts the dependence of the stress memory overshoot on the strain rate of loading/unloading (i.e., Feature 4.8) [21]. The stress memory experiment consists of constant strain rate loading to a post-yield condition, unloading at the same strain rate to zero stress, holding the strain constant, and observing the subsequent stress response. It was found that the stress first passes through a maximum before finally relaxing to an equilibrium value for that strain. The existence of the stress overshoot for this deformation history is predicted by even a linear viscoelastic model as long as there is more than a single relaxation time. The linear viscoelastic model predicts that the effect of the loading/unloading strain rate is as follows: the stress overshoot height is proportional to the strain rate and the stress overshoot location along the time axis is independent of the strain rate. Experimentally almost exact opposite is observed—the overshoot height is only weakly dependent on the strain rate of loading/unloading and the overshoot location is highly sensitive to the strain rate [21]. Figure 14.54 shows the SCM predictions that are qualitatively consistent with the experimental data.

Nonlinear creep

The SCM predictions of nonlinear creep are shown in Figures 14.55 through 14.57, where the model predicts the first four stages of creep (Feature 4.2a), but not the Stage V [14]. In agreement with experiment (see Chapter 4, Section 4.7.1) the SCM qualitatively describes the effects on the nonlinear creep of (i) the magnitude of the applied stress for a given thermal history and (ii) aging time at given applied stress. Specifically, the range and magnitude

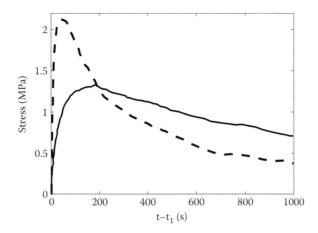

Figure 14.54 SCM predictions of stress memory effect in uniaxial extension at 100°C for "PMMA" at loading/unloading strain rates of 10^{-4} s^{-1} (solid) and 10^{-3} s^{-1} (dashed). (Reprinted from *Polymer*, 54, J. W. Kim, G. A. Medvedev, and J. M. Caruthers, 5993, Copyright 2013, with permission from Elsevier.)

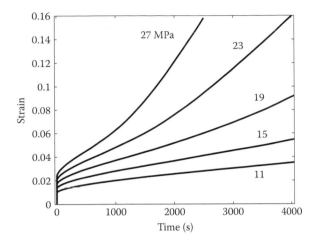

Figure 14.55 SCM prediction of creep in uniaxial compression for "PMMA" at 95°C after 3 h aging. The applied stresses are indicated in the figure. (Reprinted from *Polymer*, 74, G. A. Medvedev and J. M. Caruthers, 235, Copyright 2015, with permission from Elsevier.)

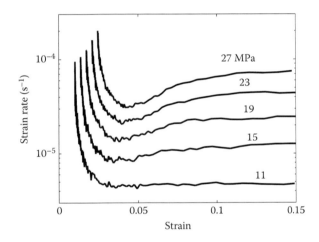

Figure 14.56 SCM prediction of strain rate versus strain curves corresponding to the creep curves shown in Figure 14.55. (Reprinted from *Polymer*, 74, G. A. Medvedev and J. M. Caruthers, 235, Copyright 2015, with permission from Elsevier.)

of the tertiary creep (as indicated by increase in the strain rate after the minimum in Figures 14.56 and 14.57) increases with both the applied stress and with the aging time; moreover, if the applied stress is small the SCM predicts that the resulting creep behavior, although nonlinear, does not result in Stage III tertiary creep in agreement with experimental data (see references in Medvedev and Caruthers [14]). The reasons for the appearance of the four different stages of creep are related to how the $\log \hat{a}$ pdf changes with strain rate as shown in Figure 14.50 in relation to the shape of the $\log \hat{a}$ pdf prior to deformation, but after the thermal history used to form the glass. A detailed discussion of the underlying nonlinear creep mechanism has been provided elsewhere [14], where the key feature is the evolution of the relaxation spectrum during the thermo-deformational history.

Recovery from creep

The predictions of the SCM for creep-recovery upon removal of the initial stress are shown in Figure 14.58, where the load has been removed at different creep times corresponding to the different stages of creep, that is, primary, secondary, and tertiary creep. In agreement with experiment [112] the SCM predicts that a significant residual strain may remain after removal of the creep load, where the amount of residual strain

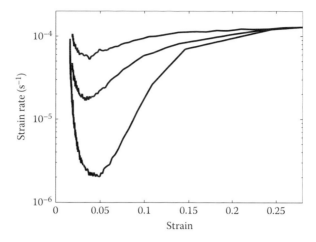

Figure 14.57 SCM predictions of aging time effect on the uniaxial extension creep of "PMMA" at 95°C for an applied stress of 17 MPa with aging times from top to bottom of 0.2 h, 3 h, and ∞ (i.e., an equilibrated material). (Reprinted from *Polymer*, 74, G. A. Medvedev and J. M. Caruthers, 235, Copyright 2015, with permission from Elsevier.)

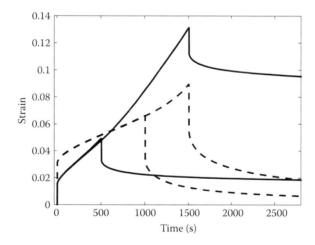

Figure 14.58 SCM predictions of creep and recovery for "PMMA" in uniaxial extension at 95°C. Solid: 18 MPa applied stress after 0.5 h aging; Dashed: 33 MPa applied stress on an equilibrated material.

dramatically increases as the creep prior to load removal changes from primary to secondary to tertiary creep. Also, in agreement with experiment, the SCM predicts recovery that is smooth (Feature 4.9) versus the abrupt transition after load removal predicted by most of the constitutive models considered in the previous sections.

Loading–creep–reloading deformation

The prediction of the SCM are shown in Figure 14.59 for the complex deformation history comprised of (i) loading in uniaxial extension at a constant strain rate through yield, (ii) loading at a slower strain rate for a time $t_{creep} = 160$ s, and (iii) reloading at the original strain rate. This deformation history mimics the experimental history where in step (ii) the sample was partially unloaded and the stress was kept constant. Although not exactly the same, we believe that the three-step strain rate deformation history used to generate the predictions in Figure 14.59 will have a similar effect to the one where constant stress is applied during step (ii), that is, the experiment shown in Figure 4.37 in Chapter 4 (i.e., Feature 4.10). The SCM predictions for this more complex deformation history are qualitatively incorrect, where the predicted σ dependence of the height of the second overshoot is in the wrong order. Experimentally the height of the second overshoot

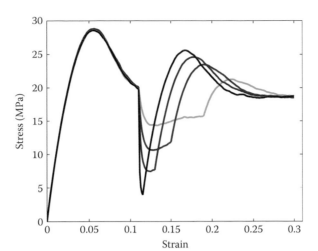

Figure 14.59 SCM prediction of multistep deformation in uniaxial extension for "PMMA" at 100°C consisting of (i) constant strain rate loading at the strain rate of 10^{-3} s^{-1} to strain of 0.11, (ii) constant strain rate loading at a slower strain rate to imitate the creep deformation for 160 s, and (iii) reloading at the strain rate of 10^{-3} s^{-1}. The strain rate during the step (ii) was 3×10^{-5} s^{-1}—black; 1.25×10^{-4} s^{-1}—red; 2.5×10^{-4} s^{-1}—blue; and 5×10^{-4} s^{-1}—green.

increases with σ whereas the SCM predicts the height of the overshoot *decreases* with σ. The loading–creep–reloading deformation history is a particularly challenging experiment, where the σ dependence of the height of the second overshoot is incorrectly predicted by all the constitutive models that have been examined to-date (with the possible exception of the Anand09 model, although the magnitudes of the overshoots predicted by Anand09 model are too small to draw a definitive conclusion).

Mobility increase during deformation

Using an optical, molecular probe rotation technique called photobleaching, Ediger and coworkers determined that during deformation (i) the average relaxation time decreases and (ii) the width of the relaxation times spectrum dramatically narrows as indicated by an increase in the KWW parameter β^{KWW} (i.e., Feature 4.11), although during the V Stage creep the average relaxation increases and the width of the spectrum broadens. The SCM correctly predicts (again with the exception on the hardening regime in constant strain rate deformation and Stage V creep) a decrease of the average relaxation time and narrowing of the spectrum that is correlated with the applied stress in creep experiments [14] and the strain rate in the constant strain rate experiment [108,111]. As shown in Figure 14.50 deformation causes the relaxation times spectrum to both shift toward shorter times and narrow as compared to the spectrum without the deformation. Also, both the shifting and the narrowing are more pronounced for faster strain rates (or equivalently larger creep loads) in agreement with the observations by Ediger and coworkers (see Chapter 4, Section 4.10.2).

14.6.3.4 ENTHALPY RELAXATION

Currently the SCM prediction of the enthalpy relaxation is not available, since the appropriate expression for the enthalpy is still being developed [113].

14.6.4 CONCLUSIONS

The SCM is able to qualitatively capture a number of important features of the thermo-mechanical behavior of polymeric glasses including:

1. Physical aging (Feature 1.1)
2. The specific volume relaxation associated with various step changes in temperature, including short time annealing and the tau-effective paradox. (Features 3.1 through 3.3)

3. The yield stress dependence on temperature, strain rate, aging time, pressure, and so on (Feature 4.1a–d)
4. The stress–strain curve in constant strain rate deformation and the nonlinear creep response; post-yield softening and its dependence on strain rate and aging time. (Features 4.2a–c)
5. The strain rate-switching experiment. (Feature 4.4)
6. The stress response upon loading–unloading–reloading. (Feature 4.5)
7. The effect of the strain rate of loading on the stress relaxation response. (Feature 4.7)
8. The stress memory experiment, including the effect of the loading/unloading strain rate on the over-shoot peak. (Feature 4.8)
9. The smooth recovery from nonlinear creep (Feature 4.9)
10. The decrease in the average relaxation time and the narrowing of the relaxation times spectrum under deformation. (Feature 4.11)

We believe the SCM is the only constitutive model to-date that has been able to capture points 2, 7, and 10 above. Notwithstanding the predictive success of the SCM listed above a number of challenges remain including:

11. The SCM predicts that the linear viscoelastic bulk and shear spectra are same; however, experimentally the two spectra are different with the shear spectrum extending to relaxation times that are orders-of-magnitude longer than the bulk spectrum. (Feature 2.1)
12. No post-yield strain hardening is predicted for constant strain rate experiments; and correspondingly, the SCM does not predict Stage V creep. (Feature 4.2d)
13. The SCM does not unambiguously predict the Bauschinger effect, although additional experimental data would provide value in solidifying this conclusion. (Feature 4.6)
14. With respect to the constant strain rate loading–creep–reloading deformation, the SCM predicts that the stress overshoot upon the second loading decrease with the creep stress, which is opposite to what is observed experimentally. (Feature 4.10)

14.7 DISCUSSION

In this chapter, the predictions of a number of constitutive models that have been proposed to describe the nonlinear relaxation behavior of polymeric glasses have been critically analyzed with respect to a suite of thermo-mechanical experiments. The objective has been to determine if the various constitutive models are able to at least qualitatively describe a number of discriminating features of the nonlinear thermo-mechanical behavior of polymeric glasses. In our opinion, it is important that the evaluation of a constitutive model is with respect to a wide diversity of data, where if the constitutive model effectively acknowledges the underlying deformation mechanisms it should be able to unify, perhaps imperfectly, seemingly disparate phenomena. We believe that this is the first review of constitutive models that have (i) employed a data set with a sufficiently wide diversity of thermo-mechanical phenomena and (ii) brought together in one place the various constitutive modeling approaches that have been proposed for polymeric glasses.

A summary of the detailed evaluation of the various constitutive models discussed in Sections 14.4 through 14.6 is provided in Table 14.4. In Section 14.2 there was a discussion of the various experimental Features, where a more complete discussion of the various experimental signatures of polymers in the glass transition region is provided in Chapter 4. The following constitutive models have been included in the summary evaluation in Table 14.4:

1. *Boyce, Parks, and Argon Plasticity Model (BPA)* [36] is a single relaxation time plasticity constitutive model that makes two improvements to the original Haward–Thackray [68] one-dimensional model. The BPA model (i) is properly tensorial and (ii) has added an "intrinsic" internal variable with its own evolution equation to describe the post-yield softening. The temperature effects come from standard Eyring model, where there are no thermal history effects.

Table 14.4 Summary of qualitative predictions of the various constitutive models for critical features in the thermo-mechanical behavior of polymeric glasses

Key to table:
Y = qualitative prediction
N = no qualitative fit to data
S = model is silent
? = predictions not known

Feature no.	Description of feature	BPA	EGP	Anand06	Anand09	QLVE-K/L	TVEM	Stochastic
Physical aging								
1.1	Effects of thermal history	S	Y	S	S	?	Y	Y
Linear viscoelasticity								
2.1	Linear viscoelastic limit	N	Y	N	N	Y	Y	Y
2.2	LVE shear spectrum wider than bulk spectrum	N	?	N	N	?	N	N
Volume relaxation								
3.1	Basic volume relaxation	S	S	S	S	Y	Y	Y
3.2	Short time anneal volume relaxation	S	S	S	S	N	N	Y
3.3	Tau-effective paradox for volume relaxation	S	S	S	S	N	N	Y
Anisotropic nonlinear mechanical behavior								
Constant strain rate yield								
4.1a	Yield stress linear function of log (strain rate)	Y	Y	N	Y	N	Y	Y
4.1b	Yield stress decreases linearly with temperature	Y	Y	S	Y	Y	Y	Y
4.1c	Yield stress increases with sub-Tg aging	S	Y	S	S	?	Y	Y
4.1d	Pressure-modified von Mises yield criterion	Y	Y	?	?	N/?	?	Y
Stress–strain curve								
4.2a	Nonlinear creep mirror image of stress–strain curve	N	Y	Y	Y	N/Y	N	Y
4.2b	Stress overshoot increases with sub-Tg aging	S	Y	S	S	?	N	Y
4.2c	Stress overshoot increases with strain rate	N	N	Y	Y	N	N	Y
4.2d	Tertiary creep increases with stress	N	?	Y	Y	N	N	Y
4.2e	Hardening/Vth stage of creep	Y	Y	Y	Y	N	N	N
4.3	Internal energy: 30%–50% of work; smooth function	S	S	S	?	S	Y	?

(Continued)

Table 14.4 (Continued) Summary of qualitative predictions of the various constitutive models for critical features in the thermo-mechanical behavior of polymeric glasses

Key to table:
Y = qualitative prediction
N = no qualitative fit to data
S = model is silent
? = predictions not known

Feature no.	Description of feature	BPA	EGP	Anand06	Anand09	QLVE-K/L	TVEM	Stochastic
4.4	Stress overshoot/undershoot with strain rate switching	N	N	N	Y	N	N	Y
	Loading–unloading–reloading deformation							
4.5a	Large residual strain when unloaded, but no flow	Y	Y	Y	Y	Y	Y	Y
4.5b	No "second yield"	N	Y	Y	Y	Y	Y	Y
4.5c	Initial modulus upon reloading same as first loading	Y	Y	Y	Y	Y	Y	Y
4.5d	No stress overshoot upon reloading	Y	Y	Y	Y	Y	Y	Y
4.6	Bauschinger effect	N	Y	N	Y	N	N	?
4.7	Nonlinear stress relaxation: relaxation rate changes order with strain rate	N	N	N	N	N	N	Y
4.8	Stress memory: location/magnitude of stress overshoot depends upon loading strain rate	N	Y	N	N	N	Y	Y
4.9	Nonlinear creep recovery is smooth	N	N	Y	N	Y	N	Y
4.10	Loading–creep–reloading deformation: stress overshoot on reloading increases with creep stress	N	N	?	?	N	N	N
4.11	Width of relaxation spectrum decreases during deformation	S	N	S	S	N	N	Y
	Enthalpy relaxation							
5.1	Heat capacity on heating through glass transition exhibits overshoot, which depends on thermal history	S	S	S	?	S	Y	?

2. *The Eindhoven Model for Glassy Polymers (EGP)* [45] is a multimodal plasticity model where the mobility depends upon the temperature, deviatoric stress (via the standard Eyring postulate), and plastic strain via a separate phenomenological function. The effect of thermal history is parameterized. The most recent version of the EGP model has an additional dependence on the total strain in order to describe the Bauschinger effect and the strain rate dependence of hardening.

3. *Anand06* is a multimodal plasticity model developed by Anand and Ames [48] that employs internal variables that satisfy their own evolution equations; however, unlike the EGP modes, the Anand06 "modes" are nonlinear processes. The model has no dependence on temperature or thermal history. In contrast to the BPA and EGP models, the back stress is not the hardening stress, but a separate function of the independent internal variables.

4. *Anand09* is a single relaxation time plasticity model developed by Anand and coworkers [50] that significantly reduced the number of model parameters in the Anand06 model by employing only one mode. This model is thermodynamically consistent [49]. Temperature effects are taken into account via (i) an Eyring-type viscosity and (ii) a phenomenological temperature dependence of the model parameters. The model does not account for thermal history effects.

5. *Quasi-Linear Viscoelastic Models (QLVE)* This refers to a collection of models that employ the linear viscoelastic stress constitutive equation, where relaxation occurs on a material timescale that depends in a nonlinear manner upon temperature and deformation. Various postulates have been made concerning what variable controls the deformation induced change in mobility, including free volume [6], stress [64], strain [58], and configurational entropy [60]. For purposes of the summary in Table 14.4, the QLVE-K refers to the model of Knauss and Emri [6], where the mobility depends upon free volume, and QLVE-L refers to the model of Popelar and Liechti [58], where the mobility depends on the deviatoric strain as well.

6. *Thermoviscoelastic Model (TVEM)* [62] This is a thermodynamically consistent, nonlinear viscoelastic model, where the stress, entropy and other state functions all emerge from the Helmholtz free energy. The mobility depends upon a functional of the configurational internal energy history that was inspired by the Adam–Gibbs [105] postulate that the mobility depends upon the configurational entropy.

7. *Stochastic Constitutive Model (SCM)* [12,108] The SCM explicitly acknowledges dynamic heterogeneity in the glassy state, where the constitutive relationship is local on the nanoscale and thus includes fluctuations. The local relaxation rate is assumed to have a functional form that is similar to the configurational internal energy used in the TVEM.

We believe that this collection of constitutive models exposes the strengths and weaknesses of the various modeling approaches used to describe polymeric glasses. The "features" and "models" in Table 14.4 should not be viewed as final, but rather as an initial attempt to capture the state of constitutive modeling of polymeric glasses where the future expansion of Table 14.4 is anticipated.

Table 14.4 is a summary of a considerable amount of information. We acknowledge that at this time there are several entries where the response is not currently known, where the "?" is meant to signify that the complexity of that particular constitutive model is such that exploration of the parameter space has not yet been carried out. It is anticipated that some of these deficiencies in Table 14.4 will be resolved in the future. Notwithstanding these valid concerns, some general conclusions can be drawn from the summary in Table 14.4:

1. The plasticity models are unable to describe the effects of the thermal history including sub-Tg physical aging, unless model "constants" are readjusted for different thermal histories.

2. The viscoplastic models are silent with respect to isotropic volume relaxation. The viscoelastic models predict some features of volume relaxation; however, only the SCM is able to predict the short annealing time experiments (Feature 3.1) and the tau-effective paradox (Feature 3.2).

3. Multimodal viscoplastic and viscoplastic models result in a proper linear viscoelastic limit for the shear response (Feature 2.1), although the viscoplastic models do not include viscoelastic bulk relaxation. The QLVE and TVEM models in principle can capture both the bulk and shear modulus, since the latters are input material functions to these models. However, it is troubling that when the

experimentally determined moduli are used in the TVEM versus the moduli determined via optimization of nonlinear deformation data, i.e., when the shear spectrum is wider than the bulk spectrum, the model predictions worsen. The SCM predicts the relaxation time distribution, where the bulk modulus has an appropriate width; however, the shear modulus is missing the long time contributions.

4. Most of the constitutive models do a reasonable job in describing the strain rate and temperature dependence of the yield stress (Features 4.1a, b), and perhaps some features of more complex deformation histories such as unloading (Feature 4.5a) and the initial modulus upon reloading (Feature 4.5c).

5. Traditional plastic and viscoelastic models begin to qualitatively fail for more complex loading histories, being unable to predict phenomena like the second overshoot/undershoot after strain rate switching (Feature 4.4) and effect of loading rate on the nonlinear stress relaxation (Feature 4.7).

6. Except for the TVEM, the traditional viscoelastic and viscoplastic constitutive models are silent with respect to the enthalpy response during changes in temperature (Feature 5.1) and deformation (Feature 4.3). This is a significant shortcoming, because the internal energy is between 30% and 50% of the work of deformation. There are no deformation enthalpy predictions as yet for the Anand09 or SCM models, although neither model is silent with respect to enthalpy.

7. With the exception of the SCM and the Anand06 models, nonlinear creep-recovery is a challenge for the rest of the constitutive models, where these models predict little if any relaxation upon removal of the load (Feature 4.9).

8. All the traditional viscoplastic and viscoplastic models implicitly assume that the shape of the relaxation time distribution is independent of temperature and deformation. In contrast, the SCM predicts the shape change of the relaxation time distribution during the thermo-deformational history which is in agreement with experimental data (Feature 4.11).

We fully acknowledge that in some cases a judgment was made concerning the capability of a particular constitutive model with respect to a given experiment as indicated in Table 14.4, where future analysis may change this judgment.

The assessment of constitutive modeling of the thermo-mechanical behavior of polymeric glasses summarized in Table 14.4 is sobering, especially considering the decades of modeling research. Currently there is not a single model that qualitatively captures all the key features of the thermo-mechanical behavior identified in Section 14.2. In addition, there are important features not even included in the Section 14.2 features list that constitutive modeling efforts have not even begun to address; for example, the occurrence of a broad exothermic peak during enthalpy relaxation of a glassy sample that had been previously subjected to a large deformation (see Chapter 4, Signature 9.4).

One must be careful not to overly disparage these substantial constitutive modeling efforts. As discussed in the Introduction there is a "polymer physics" approach and a "polymer engineering" approach to constitutive modeling. In the engineering approach the objective is to describe well a limited set of experiments, where the model predictions will only be used for very similar thermo-deformational histories—and, from this perspective the ability of a constitutive model to describe a wide range of features may not be relevant. In contrast, from a "polymer physics" perspective the constitutive model is supposed to at least acknowledge, if not fully capture, the underlying mechanism, where the lack of qualitative predictive success over a diverse set of experimental data indicates that the model is not properly capturing important components of the underlying physics.

Undoubtedly a major factor responsible for this state of affairs in the constitutive modeling of polymeric glasses is the general lack of a rigorous physical theory of the glass. Under these circumstances one possible course of action would be to wait until such a theory is developed, and instead focus on empirical parameterizations for specific engineering applications. We do not advocate this course of action. In our opinion, the flow of information is not only from the molecular level theories to the constitutive description—but also in reverse, where the intuition developed when attempting various macro- or meso-scopic relationships can inform fundamental understanding. In light of the above, one of the goals of this review is to make the information about the challenges encountered by constitutive models available to the broader glass research community that is currently focused almost exclusively on the physics of what happens when materials are isotropically cooled into the glassy state. Deformation is an equally important way to perturb the glassy state, where a fundamental understanding of the glassy state must encompass this important experimental axis.

We will now give our perspective on the state of constitutive modeling and potential pathways forward for the three major classes of models.

1. *Plasticity models*: The historical development path for these models started from the initial postulate of Erying, where empirical functions have been added to the models as the complexity of experimental data set was expanded. The number of adjustable parameter increases with each addition to the model. Because the foundations of the plasticity models are for equilibrium materials, incorporating the effects of thermal history on the mechanical behavior of the glass is difficult, since there is no inherent mechanism that accounts for the nonequilibrium behavior of the glass. To date the volumetric response of plasticity models for polymeric glasses is either ignored or modeled as perfectly elastic, which is orthogonal to much of polymer physics that believes that specific volume relaxation is a key feature of the glassy state. Moreover, in our experience the addition of new empirical functions to describe one feature of deformation (for example, the post-yield stress softening) often causes serious problem when considering additional deformation steps. The difficulty encountered with multiple deformation step experiments is quite problematic, because these large anisotropic deformations are exactly the type of experiments that plasticity models are supposed to be able to describe. In our opinion the domain of plasticity models is for describing the engineering response of polymeric glasses for a limited set of conditions, where their incorporation into large scale FEA code may provide real value to polymer engineers.

2. *Viscoelastic models*: The point of departure for the nonlinear viscoelastic models is linear viscoelasticity. The key assumptions are (i) that the dominant nonlinearity responsible for all observed glassy behaviors resides in the generalized shift factor, i.e., the log a function that describes how the material time scale is related to the laboratory time scale, (ii) the shift factor for all thermo-deformational histories is a function of the macroscopic averages of appropriate thermodynamic variables, and (iii) the entire spectrum of relaxation times moves with the log a shift function without any change in shape, i.e., thermorheological simplicity or, more appropriately, time shift invariance. The focus of constitutive modeling research has been on divining just the right form of the log a function (or functional) that can capture the entire glassy thermo-mechanical response, where log a models have included fractional free volume, stress, strain, configurational entropy, configurational internal energy, and various combinations of these quantities. Three-dimensional versions of these models naturally incorporate isotropic volume relaxation which is a key response in glasses, although these models cannot describe the short time anneal and tau-effective paradox. The nonlinear viscoelastic models have been able to describe some features of thermo-mechanical behavior of polymeric glasses; however, other features (for example, post-yield stress softening) are not predicted or only predicted by including additional relaxation functions. Assumption (iii) above is subject to question in light of the change in the width of the relaxation spectrum with deformation observed by Ediger and coworkers [24,25]. The nonlinear viscoelastic models naturally address physical aging without resorting to *ad hoc* parameterization like the ones employed in the plasticity models; thus, in our opinion, this class of constitutive models has captured some key features of the glass. However, the inability of these models to capture a wide range of nonlinear thermo-mechanical behaviors after decades of research indicates that continued searching for the magical log a function may be futile.

3. *Stochastic constitutive model*: In order to address the significant shortcoming of the plastic and viscoelastic classes of constitutive models, the recently developed SCM takes a different approach by explicitly incorporating nanoscale fluctuations, which is consistent with dynamic heterogeneity in the glassy state. The SCM is a constitutive model on the meso-scale, where the local stress is assumed to be a function of the local state of the system that includes fluctuations versus the traditional continuum approach that spatially and temporally averages all field variables prior to formulating a constitutive model. The SCM is a single relaxation time model, where the experimentally observed distribution of relaxation times is a consequence of the fluctuations. Although the SCM was not

explicitly built to fit any particular nonlinear phenomenon, it is able to naturally predict (without additional material functions and/or parameters) post-yield stress softening (Feature 4.2b), secondary and tertiary creep (Feature 4.2d), stress overshoot/undershoot for strain rate switching (Feature 4.4), the loading strain rate effect on nonlinear stress relaxation (Feature 4.7), stress memory (Feature 4.8), smooth nonlinear creep recovery (Feature 4.9), and the change in width of relaxation spectrum with deformation (Feature 4.11). These are features that are challenging or impossible to describe using the traditional viscoplastic or viscoelastic models, even with the addition of empirical functions. Notwithstanding the initial success of the SCM there are still some unresolved issues; specifically, (1) the determination of the underlying free energy of a nonlinear system undergoing large fluctuations is an outstanding problem in condensed matter physics, where the thermodynamic consistency of the SCM needs to be firmly established [113] and (2) there are a number of phenomena that cannot be predicted using the SCM, including most prominently the loading–creep–reloading response (Feature 4.10) and the observation that the shear relaxation spectrum is wider than the bulk relaxation spectrum (Feature 2.2).

With respect to constitutive modeling of polymeric glasses (from a "polymer physics" perspective), what is a viable path forward? We believe that the SCM offers the most reasonable path forward, because it naturally predicts, at least qualitatively, a significant fraction of the diverse thermo-mechanical behaviors observed experimentally and it acknowledges dynamic heterogeneity in the glassy state. There are however, several features that the SCM appears to be unable to capture. Of particular importance is the inability of the SCM to describe the long time response of the linear shear relaxation spectrum. A key assumption in the SCM was the use of the mean field approximation to reduce the spatial–temporal fluctuations of interacting mesodomains to just the temporal fluctuations at a material point. If local spatial interactions were included one would expect additional processes associated with the spatial relaxation. These spatial relaxation processes would potentially give rise to the longer time shear relaxation processes. It is not clear at this time how to incorporate interactions beyond the mean field in the SCM, but this could be a very profitable avenue for future research.

Although the focus of this chapter has been on the critical analysis of constitutive predictions, this analysis leads to some suggestions for future experimental research:

1. As described in Section 14.2, the experimental features were determined from a number of different polymeric glasses, where it would be invaluable to have a complete data set for a single, well-characterized material. This may not be critical at this stage of constitutive development, where just fitting the qualitative features of the thermo-mechanical behavior of different polymeric glasses is sufficiently discriminating to validate, or eliminate, a potential constitutive model. However, there will come a point at which quantitative fitting is important, and then a unified data set will be vital.

2. Enthalpy relaxation either during (Feature 4.3) or following deformation (Chapter 4, Signature 9.4) gives an important complementary perspective to the mechanical experiments. Although the limited data that are currently available for enthalpy relaxation during/after deformation are more than sufficient to cause considerable constitutive modeling headaches, additional enthalpy relaxation data for other thermo-deformational histories will be informative.

3. The difference between the linear viscoelastic bulk and shear relaxation spectra is an extremely discriminating feature for polymeric glasses with profound modeling consequences; however, currently there is only very limited data of high quality. Additional bulk modulus relaxation data are sorely needed.

4. There is a series of observations on the effect of preconditioning of polymeric glasses, e.g., via passing through a two-roll mill or applying large back-and-forth torsional deformation, prior to subsequent mechanical testing, where there is a dramatic change in response such as elimination of post-yield softening, and so on [41]. As far as we know, no attempts have been made to account for these effects by constitutive modeling. Perhaps this omission can be rectified if the study of the preconditioning is made more systematic, where the primary goal is to study the effect of the type/magnitude of the pre-deformation rather than just to have an empirical procedure on how to make materials more ductile.

5. It is well established that the thermo-mechanical properties of glasses critically depend upon the thermal history used to form the glass; however, many interesting mechanical studies are performed on commercial glassy materials where the glass transition history is unknown. For experiments to have value, the thermo-deformational history needs to start above Tg and then be fully reported for the cooling/annealing/deformation process.

The development of a fundamental understanding of deformation of polymeric glasses will depend upon the continued generation of high quality, discriminating data that will inform, inspire and eventually validate the physical ideas that are quantified in constitutive models.

This chapter should be viewed as an interim report on constitutive modeling of polymeric glasses, where we hope that this review will spark others to work toward a unified description of the thermo-mechanical behavior of polymeric glasses. Our expectation is that there will continue to be vigorous research activities at the intersection of discriminating experiments with insightful constitutive modeling—with the anticipation that fundamental understanding will eventually emerge.

ACKNOWLEDGMENT

This chapter was supported by National Science Foundation by Grant Number 1363326-CMMI.

LIST OF SYMBOLS

Symbols included in this list are the ones used in multiple sections throughout the chapter. Some of the plasticity models, for example, Anand06 and Anand09 have a large number of additional symbols that are used only once in conjunction with a specific model, so no ambiguity should arise as to their meaning. On the other hand, their inclusion here would only make this list of symbols difficult to use.

$\alpha(t)$, α_∞, α_g: viscoelastic coefficient of thermal expansion (CTE), asymptotic rubbery CTE, and asymptotic glassy CTE

a, $\log a$, $\log \hat{a}$: shift factor, logarithm of the shift factor, and logarithm of the mesoscopic shift factor (that fluctuates)

$\beta(t)$: viscoelastic compressibility

β_{KWW}: parameter of the KWW function

$\mathbf{B}$, $\mathbf{B}_e$: left Cauchy–Green strain tensor, elastic left Cauchy–Green strain tensor

$\mathbf{C}$: right Cauchy–Green strain tensor

C_1, C_2: WLF coefficients

$\mathbf{D}_p$: plastic symmetric rate-of-deformation tensor

ε: infinitesimal strain tensor

ε, $\dot{\varepsilon}$: axial strain, axial strain rate

ε_e, ε_p: axial elastic strain, axial plastic strain

$\mathbf{F}$, $\mathbf{F}_e$, $\mathbf{F}_p$: deformation gradient tensor, elastic deformation gradient tensor, and plastic deformation gradient tensor

G, G_r: shear modulus, hardening modulus

$\mathbf{H}$: Hencky strain tensor

K, K_∞, K_g: bulk modulus, asymptotic rubbery bulk modulus, and asymptotic glassy bulk modulus

σ, σ_s, σ_r: Cauchy stress tensor, driving stress, and hardening stress

$\bar{\sigma}_s$: equivalent driving stress

$\mathbf{T}$: stress tensor conjugate to the Hencky strain tensor

θ, θ_r: temperature, reference temperature

t^*: material time

τ: relaxation time

V, V_r: specific volume, reference specific volume

$\hat{\mathbf{x}}$: seven-component dimensionless vector describing the state of a meso-domain in the stochastic constitutive model; $\hat{x}_0$: configurational entropy, $\hat{x}_{1,...,6}$: components of the symmetric stress tensor

REFERENCES

1. G. B. McKenna, in *Long-Term Durability of Polymeric Matrix Composites*, edited by K. V. Pochiraju, G. Tandon, and G. A. Schoeppner (Springer, New York), 237, 2012.
2. J. D. Ferry, *Viscoelastic Properties of Polymers* (John Wiley & Sons, New York), 3rd edn., 1980.
3. D. M. and S. F. Edwards, *The Theory of Polymer Dynamics* (Oxford University Press, Oxford), 1986.
4. A. J. Kovacs, *Fortschritte Der Hochpolymeren-Forschung*, 3, 394, 1963.
5. D. Ng and J. J. Aklonis, in *Relaxation in Complex Systems*, edited by K. L. Ngai, and G. B. Wright (Naval Research Laboratory, Springfield, VA), 53, 1985.
6. W. G. Knauss and I. Emri, *Polym. Eng. Sci.*, 27, 86, 1987.
7. J. Greener, J. M. O'Reilly, and K. C. Ng, in *Structure, Relaxation, and Physical Aging of Glassy Polymers*, edited by R. J. Roe, J. M. O'Reilly, and J. Torkelson (Materials Research Society, Pittsburgh), 99, 1991.
8. R. E. Robertson, R. Simha, and J. G. Curro, *Macromolecules*, 17, 911, 1984.
9. R. E. Robertson, R. Simha, and J. G. Curro, *Macromolecules*, 21, 3216, 1988.
10. S. Vleeshouwers and E. Nies, *Macromolecules*, 25, 6921, 1992.
11. G. A. Medvedev, A. B. Starry, D. Ramkrishna, and J. M. Caruthers, *Macromolecules*, 45, 7237, 2012.
12. G. A. Medvedev and J. M. Caruthers, *Macromolecules*, 48, 788, 2015.
13. Y. Nanzai, *JSME Int. J., Ser. A*, 37, 149, 1994.
14. G. A. Medvedev and J. M. Caruthers, *Polymer*, 74, 235, 2015.
15. O. B. Salamatina, G. W. H. Hohne, S. N. Rudnev, and E. F. Oleinik, *Thermochimica Acta*, 247, 1, 1994.
16. M. B. Bever, D. L. Holt, and A. L. Titchener, *Progr. Mater. Sci.*, 17, 5, 1973.
17. Y. Nanzai, *Polym. Eng. Sci.*, 30, 96, 1990.
18. C. Dreistadt, A.-S. Bonnet, P. Chevrier, and P. Lipinski, *Mater. Des.*, 30, 3126, 2009.
19. G. E. Dieter, *Mechanical Metallurgy* (McGraw-Hill, London), 3rd edn., 1988.
20. J. W. Kim, G. A. Medvedev, and J. M. Caruthers, *Polymer*, 54, 3949, 2013.
21. J. W. Kim, G. A. Medvedev, and J. M. Caruthers, *Polymer*, 54, 5993, 2013.
22. R. Kohlrausch, *Annalen der Physik und Chemie*, 91, 179, 1854.
23. G. Williams and D. C. Watts, *Trans. Faraday Soc.*, 66, 80, 1970.
24. H.-N. Lee, K. Paeng, S. F. Swallen, and M. D. Ediger, *Science*, 323, 231, 2009.
25. H.-N. Lee, K. Paeng, S. F. Swallen, M. D. Ediger, R. A. Stamm, G. A. Medvedev, and J. M. Caruthers, *J. Polym. Sci. Pol. Phys.*, 47, 1713, 2009.
26. G. Binder and F. H. Muller, *Colloid Polym. Sci.*, 177, 129, 1961.
27. D. H. Ender, *J. Macromol. Sci., Part B: Physics*, 4, 635, 1970.
28. J. W. Kim, G. A. Medvedev, and J. M. Caruthers, *Polymer*, 54, 2821, 2013.
29. C. Truesdell and W. Noll, *The Non-Linear Field Theories of Mechanics* (Springer, Berlin), 3rd edn., 2004.
30. K. Chen and K. S. Schweizer, *Phys. Rev. E*, 82, 041804, 2010.
31. V. V. Bulatov and A. S. Argon, *Model. Simul. Mater. Sci. Eng.*, 2, 203, 1994.
32. C. Truesdell, *J. Arch. Ration. Mech. An.*, 1, 125, 1952.
33. H. Sillescu, *J. Non-Cryst. Solids*, 243, 81, 1999.
34. M. D. Ediger, *Annu. Rev. Phys. Chem.*, 51, 99, 2000.
35. J. Lu, G. Ravichandran, and W. L. Johnson, *Acta Materialia*, 51, 3429, 2003.
36. M. C. Boyce, D. M. Parks, and A. S. Argon, *Mech. Mater.*, 7, 15, 1988.
37. M. C. Boyce and E. M. Arruda, *Polym. Eng. Sci.*, 30, 1288, 1990.
38. O. A. Hasan and M. C. Boyce, *Polym. Eng. Sci.*, 35, 331, 1995.
39. T. A. Tervoort, E. T. J. Klompen, and L. E. Govaert, *J. Rheol.*, 40, 779, 1996.
40. T. A. Tervoort, R. J. M. Smit, W. A. M. Brekelmans, and L. E. Govaert, *Mech. Time-Depend. Mat.*, 1, 269, 1998.

41. L. E. Govaert, P. H. M. Timmermans, and W. A. M. Brekelmans, *J. Eng. Mater. Technol.*, 122, 177, 2000.
42. H. G. H. van Melick, L. E. Govaert, and H. E. H. Meijer, *Polymer*, 44, 3579, 2003.
43. E. T. J. Klompen, T. A. P. Engels, L. E. Govaert, and H. E. H. Mejer, *Macromolecules*, 38, 6997, 2005.
44. M. Wendlandt, T. A. Tervoort, and U. W. Suter, *Polymer*, 46, 11786, 2005.
45. L. C. A. van Breemen, E. T. J. Klompen, L. E. Govaert, and H. E. H. Meijer, *J. Mech. Phys. Solids*, 59, 2191, 2011.
46. D. J. A. Senden, S. Krop, J. A. W. van Dommelen, and L. E. Govaert, *J. Polym. Sci. Pol. Phys.*, 50, 1680, 2012.
47. L. Anand and M. E. Gurtin, *Int. J. Solids Struct.*, 40, 1465, 2003.
48. L. Anand and N. M. Ames, *Int. J. Plast.*, 22, 1123, 2006.
49. L. Anand, N. M. Ames, V. Srivastava, and S. A. Chester, *Int. J. Plast.*, 25, 1474, 2009.
50. N. M. Ames, V. Srivastava, S. A. Chester, and L. Anand, *Int. J. Plast.*, 25, 1495, 2009.
51. J. J. Wu and C. P. Buckley, *J. Polym. Sci. Pol. Phys.*, 42, 2027, 2004.
52. C. P. Buckley, P. J. Dooling, J. Harding, and C. Ruiz, *J. Mech. Phys. Solids*, 52, 2355, 2004.
53. D. S. A. de Focatis, J. Emberly, and C. P. Buckley, *J. Polym. Sci. Pol. Phys.*, 48, 1449, 2010.
54. J. L. Bouvard, D. K. Ward, D. Hossain, E. B. Marin, D. J. Bammann, and M. F. Horstemeyer, *Acta Mech.*, 213, 71, 2010.
55. T. D. Nguyen, H. J. Qi, F. Castro, and K. N. Long, *J. Mech. Phys. Solids*, 56, 2792, 2008.
56. R. A. Schapery, *Polym. Eng. Sci.*, 9, 295, 1969.
57. A. J. Kovacs, J. J. Aklonis, J. M. Hutchinson, and A. R. Ramos, *J. Polym. Sci. Pol. Phys.*, 17, 1097, 1979.
58. C. F. Popelar and K. M. Liechti, *Mech. Time-Depend. Mat.*, 7, 89, 2003.
59. R. M. J. Shay and J. M. Caruthers, *J. Rheol.*, 30, 781, 1986.
60. R. M. J. Shay and J. M. Caruthers, *Polym. Eng. Sci.*, 30, 1266, 1990.
61. S. R. Lustig, R. M. J. Shay, and J. M. Caruthers, *J. Rheol.*, 40, 69, 1996.
62. J. M. Caruthers, D. B. Adolf, R. S. Chambers, and P. Shrikhande, *Polymer*, 45, 4577, 2004.
63. D. B. Adolf, R. S. Chambers, and M. A. Neidigk, *Polymer*, 50, 4257, 2009.
64. C. P. Buckley and D. C. Jones, *Polymer*, 36, 3301, 1995.
65. P. J. Dooling, C. P. Buckley, and S. Hinduja, *Polym. Eng. Sci.*, 38, 892, 1998.
66. I. L. Hopkins, *J. Polym. Sci.*, 28, 631, 1958.
67. L. W. Moreland and E. H. Lee, *Trans. Soc. Rheol.*, 4, 233, 1960.
68. R. N. Haward and G. Thackray, *Proc. R. Soc. A*, 302, 453, 1968.
69. E. T. J. Klompen, T. A. P. Engels, L. C. A. van Breemen, P. J. G. Schreurs, L. E. Govaert, and H. E. H. Meijer, *Macromolecules*, 38, 7009, 2005.
70. M. Negahban, *The Mechanical and Thermodynamical Theory of Plasticity* (CRC Press, Boca Raton), 2012.
71. A. V. Tobolsky and H. Eyring, *J. Chem. Phys.*, 11, 125, 1943.
72. H. Eyring, *J. Chem. Phys.*, 4, 283, 1936.
73. E. H. Lee, *J. Appl. Mech.*, 36, 1, 1969.
74. D. J. A. Senden, J. A. W. van Dommelen, and L. E. Govaert, *J. Polym. Sci. Pol. Phys.*, 50, 1589, 2012.
75. L. Priester, *Grain Boundaries and Crystalline Plasticity* (ISTE Ltd, London), 2011.
76. E. J. Kramer, *J. Polym. Sci. Pol. Phys.*, 43, 3369, 2005.
77. T. A. P. Engels, L. E. Govaert, and H. E. H. Meijer, *Macromol. Mater. Eng.*, 294, 821, 2009.
78. W. A. Spitzig and O. Richmond, *Polym. Eng. Sci.*, 19, 1129, 1979.
79. S. F. Edwards and T. H. Vilgis, *Polymer*, 27, 483, 1986.
80. C. G'Sell, J. M. Hiver, A. Dahoun, and A. Souahi, *J. Mater. Sci.*, 27, 5031, 1992.
81. D. J. A. Senden, J. A. W. van Dommelen, and L. E. Govaert, *J. Polym. Sci. Pol. Phys.*, 48, 1483, 2010.
82. N. M. Ames, M.S. Thesis, Massachusetts Institute of Technology, 2003.
83. A. N. Gent, *Rubber Chem. Technol.*, 69, 59, 1996.
84. R. M. Christensen and P. M. Naghdi, *Acta Mech.*, 3, 1, 1967.
85. R. M. Christensen, *Theory of Viscoelasticity: An Introduction* (Academic Press, New York), 2nd edn., 1982.

86. B. D. Coleman, *Arch. Rational Mech. Anal.*, 17, 1, 1964.

87. M. L. Williams, R. F. Landel, and J. D. Ferry, *J. Am. Chem. Soc.*, 77, 3701, 1955.

88. H. Vogel, *Phys. Z.*, 22, 645, 1921.

89. G. Tammann, *J. Soc. Glass Technol.*, 9, 166, 1925.

90. G. S. Fulcher, *J. Am. Ceram. Soc.*, 8, 339, 1925.

91. C. T. Moynihan et al., *Ann. NY Acad. Sci.*, 279, 15, 1976.

92. O. S. Narayanaswamy, *J. Am. Ceram. Soc.*, 54, 491, 1971.

93. A. Q. Tool, *J. Am. Ceram. Soc.*, 29, 240, 1946.

94. G. B. McKenna and S. L. Simon, in *Handbook of Thermal Analysis and Calorimetry*, edited by S. Z. D. Cheng (Elsevier, Amsterdam), 49, 2002.

95. O. S. Narayanaswamy, *J. Am. Ceram. Soc.*, 54, 491, 1971.

96. I. M. Hodge, *J. Non-Cryst. Solids*, 169, 211, 1994.

97. J. J. Tribone, J. M. O'Reilly, and J. Greener, *Macromolecules*, 19, 1732, 1986.

98. P. Badrinarayanan and S. L. Simon, *Polymer*, 48, 1464, 2007.

99. G. B. McKenna and C. A. Angell, *J. Non-Cryst. Solids*, 131–133, 528, 1991.

100. A. K. Doolittle, *J. Appl. Phys.*, 22, 1471, 1951.

101. R. Simha and T. Somcynsky, *Macromolecules*, 2, 342, 1969.

102. S. M. Fielding, R. G. Larson, and M. E. Cates, *Phys. Rev. Lett.*, 108, 048301, 2012.

103. R. G. Larson, *Constitutive Equations for Polymer Melts and Solutions* (Butterworths, Boston), 1988, Butterworths Series in Chemical Engineering.

104. B. D. Coleman and W. Noll, *Rev. Mod. Phys.*, 33, 239, 1961.

105. G. Adam and J. H. Gibbs, *J. Chem. Phys.*, 43, 139, 1965.

106. D. B. Adolf, R. S. Chambers, and J. M. Caruthers, *Polymer*, 45, 4599, 2004.

107. L. C. E. Struik, *Physical Aging in Amorphous Polymers and Other Materials* (Elsevier, Amsterdam), 1978.

108. G. A. Medvedev and J. M. Caruthers, *J. Rheol.*, 57, 949, 2013.

109. L. D. Landau and E. M. Lifshitz, *Statistical Physics, Part 1* (Butterworth-Heinemann, Oxford), Third edn., 1980, Course of Theoretical Physics.

110. J. E. McKinney and H. V. Belcher, *J. Res. Natl. Bur. Stand. Sec. A*, 67, 43, 1963.

111. G. A. Medvedev, J. W. Kim, and J. M. Caruthers, *Polymer*, 54, 6599, 2013.

112. K. Chen, K. S. Schweizer, R. A. Stamm, E. Lee, and J. M. Caruthers, *J. Chem. Phys.*, 129, 184904, 2008.

113. G. A. Medvedev and J. M. Caruthers, *Ind. Eng. Chem. Res.*, 54, 10472, 2015.

Index

A

Accelerated aging, 197
Acceleration ratio, 387;
　　see also Active deformation
Activation volume, 303
Activation zones (AZs), 411
Active deformation, 383
　　accelerated dynamics, 385–387
　　acceleration ratio, 387
　　broad power law decay, 386
　　Debye–Waller factor, 384
　　elastic heterogeneity and location of failure,
　　　　383–385
　　inherent structure energies, 385
　　Kohlrausch–Williams–Watts law, 385
　　local shear moduli map, 384
　　segmental relaxation dynamics measurement,
　　　　386
Adam–Gibbs-inspired string model, 285;
　　see also Cooperative motion
　　of relaxation, 280–283
Adam–Gibbs theory (AG theory), 267;
　　see also Cooperative motion
Adhesion, work of, 224
AFM, *see* Atomic force microscope
Aging
　　in composite materials, 41–43
　　confinement effect of, 256
　　in dielectric permittivity and volume,
　　　　255–256
　　in polymer glass, 254
　　after quench, 322–323
　　rate, *see* Volume—relaxation
　　in semicrystalline polymers, 44
　　time shift factors, 43
Aging and rejuvenating, 304, 319;
　　see also Glass transition model
　　aging after quench, 322–323
　　dominant relaxation time, 321
　　equations for, 320–321
　　equilibrium and out-of-equilibrium situations,
　　　　321–322

Fokker–Planck equation, 321, 323–325
Kovacs memory effect, 320
reheating and temporal asymmetry, 323–325
relaxation time distributions, 321
temporal asymmetry, 320
Aging dynamics, 250, 262; *see also* Glass transition;
　　Physical aging
　　confinement effect of aging, 256
　　densification of glassy polymer, 256
　　dielectric permittivity and volume, 255–256
　　dynamical techniques, 244
　　investigated by DRS, 244–245
　　Kovacs effects, 250–252
　　memory and rejuvenation effects, 252–254
　　in polymer glass, 254
　　temperature dependence of complex dielectric
　　　　permittivity, 253
AG theory, *see* Adam–Gibbs theory
α process, 80, 243
α-relaxation, 163, *see* Cooperative motion
　　determination, 308
α relaxation time, 58, 79, 409;
　　see also Structural relaxation
　　Arrhenius plot of, 90
　　BSM with variable stiffness, 88
　　chain stiffness, 88
　　MCT and VFT parameters, 89
　　temperature dependence of, 88–93
　　Vogel–Fulcher–Tammann equation, 91
Amorphous polymeric materials, 23
Anand06 model, 477–483
Anand09 model, 483–491
Angell plot, 15
Anisotropy
　　decay data, 362
　　deformation, 518
　　nonlinear mechanical behavior, 455–460
Annealing
　　effect on volume relaxation, 517
　　experiments for PVAc, 123
　　time, 261
Areal density, 410, 411
Arrhenian dependence, 115